T0205295

Lecture Notes in Computer Science 13841

The series Lecture Notes in Computer Science (LNCS), including its subseries Lecture Notes in Artificial Intelligence (LNAI) and Lecture Notes in Bioinformatics (LNBI), has established itself as a medium for the publication of new developments in computer science and information technology research, teaching, and education.

LNCS enjoys close cooperation with the computer science R & D community, the series counts many renowned academics among its volume editors and paper authors, and collaborates with prestigious societies. Its mission is to serve this international community by providing an invaluable service, mainly focused on the publication of conference and workshop proceedings and postproceedings. LNCS commenced publication in 1973.

Lei Wang · Juergen Gall · Tat-Jun Chin ·
Imari Sato · Rama Chellappa
Editors

Computer Vision – ACCV 2022

16th Asian Conference on Computer Vision
Macao, China, December 4–8, 2022
Proceedings, Part I

Editors
Lei Wang
University of Wollongong
Wollongong, NSW, Australia

Tat-Jun Chin
University of Adelaide
Adelaide, SA, Australia

Rama Chellappa
Johns Hopkins University
Baltimore, MD, USA

Juergen Gall
University of Bonn
Bonn, Germany

Imari Sato
National Institute of Informatics
Tokyo, Japan

ISSN 0302-9743 ISSN 1611-3349 (electronic)
Lecture Notes in Computer Science
ISBN 978-3-031-26318-7 ISBN 978-3-031-26319-4 (eBook)
https://doi.org/10.1007/978-3-031-26319-4

This Springer imprint is published by the registered company Springer Nature Switzerland AG
The registered company address is: Gewerbestrasse 11, 6330 Cham, Switzerland

Preface

The 16th Asian Conference on Computer Vision (ACCV) 2022 was held in a hybrid mode in Macau SAR, China during December 4–8, 2022. The conference featured novel research contributions from almost all sub-areas of computer vision.

For the main conference, 836 valid submissions entered the review stage after desk rejection. Sixty-three area chairs and 959 reviewers made great efforts to ensure that every submission received thorough and high-quality reviews. As in previous editions of ACCV, this conference adopted a double-blind review process. The identities of authors were not visible to the reviewers or area chairs; nor were the identities of the assigned reviewers and area chairs known to the authors. The program chairs did not submit papers to the conference.

After receiving the reviews, the authors had the option of submitting a rebuttal. Following that, the area chairs led the discussions and final recommendations were then made by the reviewers. Taking conflicts of interest into account, the area chairs formed 21 AC triplets to finalize the paper recommendations. With the confirmation of three area chairs for each paper, 277 papers were accepted. ACCV 2022 also included eight workshops, eight tutorials, and one grand challenge, covering various cutting-edge research topics related to computer vision. The proceedings of ACCV 2022 are open access at the Computer Vision Foundation website, by courtesy of Springer. The quality of the papers presented at ACCV 2022 demonstrates the research excellence of the international computer vision communities.

This conference is fortunate to receive support from many organizations and individuals. We would like to express our gratitude for the continued support of the Asian Federation of Computer Vision and our sponsors, the University of Macau, Springer, the Artificial Intelligence Journal, and OPPO. ACCV 2022 used the Conference Management Toolkit sponsored by Microsoft Research and received much help from its support team.

All the organizers, area chairs, reviewers, and authors made great contributions to ensure a successful ACCV 2022. For this, we owe them deep gratitude. Last but not least, we would like to thank the online and in-person attendees of ACCV 2022. Their presence showed strong commitment and appreciation towards this conference.

December 2022

Lei Wang
Juergen Gall
Tat-Jun Chin
Imari Sato
Rama Chellappa

Preface

The 16th Asian Conference on Computer Vision (ACCV) 2022 was held in a hybrid mode in Macau SAR, China during December 4–8, 2022. The conference featured novel research contributions from almost all sub-areas of computer vision.

For the main conference, 836 valid submissions entered the review stage after desk rejection. Sixty-three area chairs and 959 reviewers made great efforts to ensure that every submission received thorough and high-quality reviews. As in previous editions of ACCV, this conference adopted a double-blind review process. The identities of authors were not visible to the reviewers or area chairs; nor were the identities of the assigned reviewers and area chairs known to the authors. The program chairs did not submit papers to the conference.

After receiving the reviews, the authors had the option of submitting a rebuttal. Following that, the area chairs led the discussions and final recommendations were then made by the reviewers. Taking conflicts of interest into account, the area chairs formed 21 AC triplets to finalize the paper recommendations. With the confirmation of three area chairs for each paper, 277 papers were accepted. ACCV 2022 also included eight workshops, eight tutorials, and one grand challenge, covering various cutting-edge research topics related to computer vision. The proceedings of ACCV 2022 are open access at the Computer Vision Foundation website, by courtesy of Springer. The quality of the papers presented at ACCV 2022 demonstrates the research excellence of the international computer vision communities.

This conference is fortunate to receive support from many organizations and individuals. We would like to express our gratitude for the continued support of the Asian Federation of Computer Vision and our sponsors, the University of Macau, Springer, the Artificial Intelligence Journal, and OPPO. ACCV 2022 used the Conference Management Toolkit sponsored by Microsoft Research and received much help from its support team.

All the organizers, area chairs, reviewers, and authors made great contributions to ensure a successful ACCV 2022. For this, we owe them deep gratitude. Last but not least, we would like to thank the online and in-person attendees of ACCV 2022. Their presence showed strong commitment and appreciation towards this conference.

December 2022

Lei Wang
Juergen Gall
Tat-Jun Chin
Imari Sato
Rama Chellappa

Organization

General Chairs

Gérard Medioni	University of Southern California, USA
Shiguang Shan	Chinese Academy of Sciences, China
Bohyung Han	Seoul National University, South Korea
Hongdong Li	Australian National University, Australia

Program Chairs

Rama Chellappa	Johns Hopkins University, USA
Juergen Gall	University of Bonn, Germany
Imari Sato	National Institute of Informatics, Japan
Tat-Jun Chin	University of Adelaide, Australia
Lei Wang	University of Wollongong, Australia

Publication Chairs

Wenbin Li	Nanjing University, China
Wanqi Yang	Nanjing Normal University, China

Local Arrangements Chairs

Liming Zhang	University of Macau, China
Jianjia Zhang	Sun Yat-sen University, China

Web Chairs

Zongyuan Ge	Monash University, Australia
Deval Mehta	Monash University, Australia
Zhongyan Zhang	University of Wollongong, Australia

AC Meeting Chair

Chee Seng Chan	University of Malaya, Malaysia

Area Chairs

Aljosa Osep	Technical University of Munich, Germany
Angela Yao	National University of Singapore, Singapore
Anh T. Tran	VinAI Research, Vietnam
Anurag Mittal	Indian Institute of Technology Madras, India
Binh-Son Hua	VinAI Research, Vietnam
C. V. Jawahar	International Institute of Information Technology, Hyderabad, India
Dan Xu	The Hong Kong University of Science and Technology, China
Du Tran	Meta AI, USA
Frederic Jurie	University of Caen and Safran, France
Guangcan Liu	Southeast University, China
Guorong Li	University of Chinese Academy of Sciences, China
Guosheng Lin	Nanyang Technological University, Singapore
Gustavo Carneiro	University of Surrey, UK
Hyun Soo Park	University of Minnesota, USA
Hyunjung Shim	Korea Advanced Institute of Science and Technology, South Korea
Jiaying Liu	Peking University, China
Jun Zhou	Griffith University, Australia
Junseok Kwon	Chung-Ang University, South Korea
Kota Yamaguchi	CyberAgent, Japan
Li Liu	National University of Defense Technology, China
Liang Zheng	Australian National University, Australia
Mathieu Aubry	Ecole des Ponts ParisTech, France
Mehrtash Harandi	Monash University, Australia
Miaomiao Liu	Australian National University, Australia
Ming-Hsuan Yang	University of California at Merced, USA
Palaiahnakote Shivakumara	University of Malaya, Malaysia
Pau-Choo Chung	National Cheng Kung University, Taiwan

Qianqian Xu	Key Laboratory of Intelligent Information Processing, Institute of Computing Technology, Chinese Academy of Sciences, China
Qiuhong Ke	Monash University, Australia
Radu Timofte	University of Würzburg, Germany and ETH Zurich, Switzerland
Rajagopalan N. Ambasamudram	Indian Institute of Technology Madras, India
Risheng Liu	Dalian University of Technology, China
Ruiping Wang	Institute of Computing Technology, Chinese Academy of Sciences, China
Sajid Javed	Khalifa University of Science and Technology, Abu Dhabi, UAE
Seunghoon Hong	Korea Advanced Institute of Science and Technology, South Korea
Shang-Hong Lai	National Tsing Hua University, Taiwan
Shanshan Zhang	Nanjing University of Science and Technology, China
Sharon Xiaolei Huang	Pennsylvania State University, USA
Shin'ichi Satoh	National Institute of Informatics, Japan
Si Liu	Beihang University, China
Suha Kwak	Pohang University of Science and Technology, South Korea
Tae Hyun Kim	Hanyang Univeristy, South Korea
Takayuki Okatani	Tohoku University, Japan/RIKEN Center for Advanced Intelligence Project, Japan
Tatsuya Harada	University of Tokyo/RIKEN, Japan
Vicky Kalogeiton	Ecole Polytechnique, France
Vincent Lepetit	Ecole des Ponts ParisTech, France
Vineeth N. Balasubramanian	Indian Institute of Technology, Hyderabad, India
Wei Shen	Shanghai Jiao Tong University, China
Wei-Shi Zheng	Sun Yat-sen University, China
Xiang Bai	Huazhong University of Science and Technology, China
Xiaowei Zhou	Zhejiang University, China
Xin Yu	University of Technology Sydney, Australia
Yasutaka Furukawa	Simon Fraser University, Canada
Yasuyuki Matsushita	Osaka University, Japan
Yedid Hoshen	Hebrew University of Jerusalem, Israel
Ying Fu	Beijing Institute of Technology, China
Yong Jae Lee	University of Wisconsin-Madison, USA
Yu-Chiang Frank Wang	National Taiwan University, Taiwan
Yumin Suh	NEC Laboratories America, USA

Yung-Yu Chuang National Taiwan University, Taiwan
Zhaoxiang Zhang Chinese Academy of Sciences, China
Ziad Al-Halah University of Texas at Austin, USA
Zuzana Kukelova Czech Technical University, Czech Republic

Additional Reviewers

Abanob E. N. Soliman
Abdelbadie Belmouhcine
Adrian Barbu
Agnibh Dasgupta
Akihiro Sugimoto
Akkarit Sangpetch
Akrem Sellami
Aleksandr Kim
Alexander Andreopoulos
Alexander Fix
Alexander Kugele
Alexandre Morgand
Alexis Lechervy
Alina E. Marcu
Alper Yilmaz
Alvaro Parra
Amogh Subbakrishna Adishesha
Andrea Giachetti
Andrea Lagorio
Andreu Girbau Xalabarder
Andrey Kuehlkamp
Anh Nguyen
Anh T. Tran
Ankush Gupta
Anoop Cherian
Anton Mitrokhin
Antonio Agudo
Antonio Robles-Kelly
Ara Abigail Ambita
Ardhendu Behera
Arjan Kuijper
Arren Matthew C. Antioquia
Arjun Ashok
Atsushi Hashimoto
Atsushi Shimada
Attila Szabo
Aurelie Bugeau
Avatharam Ganivada
Ayan Kumar Bhunia
Azade Farshad
B. V. K. Vijaya Kumar
Bach Tran
Bailin Yang
Baojiang Zhong
Baoquan Zhang
Baoyao Yang
Basit O. Alawode
Beibei Lin
Benoit Guillard
Beomgu Kang
Bin He
Bin Li
Bin Liu
Bin Ren
Bin Yang
Bin-Cheng Yang
BingLiang Jiao
Bo Liu
Bohan Li
Boyao Zhou
Boyu Wang
Caoyun Fan
Carlo Tomasi
Carlos Torres
Carvalho Micael
Cees Snoek
Chang Kong
Changick Kim
Changkun Ye
Changsheng Lu
Chao Liu
Chao Shi
Chaowei Tan
Chaoyi Li
Chaoyu Dong
Chaoyu Zhao
Chen He
Chen Liu
Chen Yang
Chen Zhang
Cheng Deng
Cheng Guo
Cheng Yu
Cheng-Kun Yang
Chenglong Li
Chengmei Yang
Chengxin Liu
Chengyao Qian
Chen-Kuo Chiang
Chenxu Luo
Che-Rung Lee
Che-Tsung Lin
Chi Xu
Chi Nhan Duong
Chia-Ching Lin
Chien-Cheng Lee
Chien-Yi Wang
Chih-Chung Hsu
Chih-Wei Lin
Ching-Chun Huang
Chiou-Ting Hsu
Chippy M. Manu
Chong Wang
Chongyang Wang
Christian Siagian
Christine Allen-Blanchette

Christoph Schorn
Christos Matsoukas
Chuan Guo
Chuang Yang
Chuanyi Zhang
Chunfeng Song
Chunhui Zhang
Chun-Rong Huang
Ci Lin
Ci-Siang Lin
Cong Fang
Cui Wang
Cui Yuan
Cyrill Stachniss
Dahai Yu
Daiki Ikami
Daisuke Miyazaki
Dandan Zhu
Daniel Barath
Daniel Lichy
Daniel Reich
Danyang Tu
David Picard
Davide Silvestri
Defang Chen
Dehuan Zhang
Deunsol Jung
Difei Gao
Dim P. Papadopoulos
Ding-Jie Chen
Dong Gong
Dong Hao
Dong Wook Shu
Dongdong Chen
Donghun Lee
Donghyeon Kwon
Donghyun Yoo
Dongkeun Kim
Dongliang Luo
Dongseob Kim
Dongsuk Kim
Dongwan Kim
Dongwon Kim
DongWook Yang
Dongze Lian
Dubing Chen
Edoardo Remelli
Emanuele Trucco
Erhan Gundogdu
Erh-Chung Chen
Rickson R. Nascimento
Erkang Chen
Eunbyung Park
Eunpil Park
Eun-Sol Kim
Fabio Cuzzolin
Fan Yang
Fan Zhang
Fangyu Zhou
Fani Deligianni
Fatemeh Karimi Nejadasl
Fei Liu
Feiyue Ni
Feng Su
Feng Xue
Fengchao Xiong
Fengji Ma
Fernando Díaz-del-Rio
Florian Bernard
Florian Kleber
Florin-Alexandru Vasluianu
Fok Hing Chi Tivive
Frank Neumann
Fu-En Yang
Fumio Okura
Gang Chen
Gang Liu
Gao Haoyuan
Gaoshuai Wang
Gaoyun An
Gen Li
Georgy Ponimatkin
Gianfranco Doretto
Gil Levi
Guang Yang
Guangfa Wang
Guangfeng Lin
Guillaume Jeanneret
Guisik Kim
Gunhee Kim
Guodong Wang
Ha Young Kim
Hadi Mohaghegh Dolatabadi
Haibo Ye
Haili Ye
Haithem Boussaid
Haixia Wang
Han Chen
Han Zou
Hang Cheng
Hang Du
Hang Guo
Hanlin Gu
Hannah H. Kim
Hao He
Hao Huang
Hao Quan
Hao Ren
Hao Tang
Hao Zeng
Hao Zhao
Haoji Hu
Haopeng Li
Haoqing Wang
Haoran Wen
Haoshuo Huang
Haotian Liu
Haozhao Ma
Hari Chandana K.
Haripriya Harikumar
Hehe Fan
Helder Araujo
Henok Ghebrechristos
Heunseung Lim
Hezhi Cao
Hideo Saito
Hieu Le
Hiroaki Santo
Hirokatsu Kataoka
Hiroshi Omori
Hitika Tiwari
Hojung Lee
Hong Cheng

Hong Liu
Hu Zhang
Huadong Tang
Huajie Jiang
Huang Ziqi
Huangying Zhan
Hui Kong
Hui Nie
Huiyu Duan
Huyen Thi Thanh Tran
Hyung-Jeong Yang
Hyunjin Park
Hyunsoo Kim
HyunWook Park
I-Chao Shen
Idil Esen Zulfikar
Ikuhisa Mitsugami
Inseop Chung
Ioannis Pavlidis
Isinsu Katircioglu
Jaeil Kim
Jaeyoon Park
Jae-Young Sim
James Clark
James Elder
James Pritts
Jan Zdenek
Janghoon Choi
Jeany Son
Jenny Seidenschwarz
Jesse Scott
Jia Wan
Jiadai Sun
JiaHuan Ji
Jiajiong Cao
Jian Zhang
Jianbo Jiao
Jianhui Wu
Jianjia Wang
Jianjia Zhang
Jianqiao Wangni
JiaQi Wang
Jiaqin Lin
Jiarui Liu
Jiawei Wang
Jiaxin Gu
Jiaxin Wei
Jiaxin Zhang
Jiaying Zhang
Jiayu Yang
Jidong Tian
Jie Hong
Jie Lin
Jie Liu
Jie Song
Jie Yang
Jiebo Luo
Jiejie Xu
Jin Fang
Jin Gao
Jin Tian
Jinbin Bai
Jing Bai
Jing Huo
Jing Tian
Jing Wu
Jing Zhang
Jingchen Xu
Jingchun Cheng
Jingjing Fu
Jingshuai Liu
JingWei Huang
Jingzhou Chen
JinHan Cui
Jinjie Song
Jinqiao Wang
Jinsun Park
Jinwoo Kim
Jinyu Chen
Jipeng Qiang
Jiri Sedlar
Jiseob Kim
Jiuxiang Gu
Jiwei Xiao
Jiyang Zheng
Jiyoung Lee
John Paisley
Joonki Paik
Joonseok Lee
Julien Mille
Julio C. Zamora
Jun Sato
Jun Tan
Jun Tang
Jun Xiao
Jun Xu
Junbao Zhuo
Jun-Cheng Chen
Junfen Chen
Jungeun Kim
Junhwa Hur
Junli Tao
Junlin Han
Junsik Kim
Junting Dong
Junwei Zhou
Junyu Gao
Kai Han
Kai Huang
Kai Katsumata
Kai Zhao
Kailun Yang
Kai-Po Chang
Kaixiang Wang
Kamal Nasrollahi
Kamil Kowol
Kan Chang
Kang-Jun Liu
Kanchana Vaishnavi Gandikota
Kanoksak Wattanachote
Karan Sikka
Kaushik Roy
Ke Xian
Keiji Yanai
Kha Gia Quach
Kibok Lee
Kira Maag
Kirill Gavrilyuk
Kohei Suenaga
Koichi Ito
Komei Sugiura
Kong Dehui
Konstantinos Batsos
Kotaro Kikuchi

Kouzou Ohara
Kuan-Wen Chen
Kun He
Kun Hu
Kun Zhan
Kunhee Kim
Kwan-Yee K. Wong
Kyong Hwan Jin
Kyuhong Shim
Kyung Ho Park
Kyungmin Kim
Kyungsu Lee
Lam Phan
Lanlan Liu
Le Hui
Lei Ke
Lei Qi
Lei Yang
Lei Yu
Lei Zhu
Leila Mahmoodi
Li Jiao
Li Su
Lianyu Hu
Licheng Jiao
Lichi Zhang
Lihong Zheng
Lijun Zhao
Like Xin
Lin Gu
Lin Xuhong
Lincheng Li
Linghua Tang
Lingzhi Kong
Linlin Yang
Linsen Li
Litao Yu
Liu Liu
Liujie Hua
Li-Yun Wang
Loren Schwiebert
Lujia Jin
Lujun Li
Luping Zhou
Luting Wang
Mansi Sharma
Mantini Pranav
Mahmoud Zidan Khairallah
Manuel Günther
Marcella Astrid
Marco Piccirilli
Martin Kampel
Marwan Torki
Masaaki Iiyama
Masanori Suganuma
Masayuki Tanaka
Matan Jacoby
Md Alimoor Reza
Md. Zasim Uddin
Meghshyam Prasad
Mei-Chen Yeh
Meng Tang
Mengde Xu
Mengyang Pu
Mevan B. Ekanayake
Michael Bi Mi
Michael Wray
Michaël Clément
Michel Antunes
Michele Sasdelli
Mikhail Sizintsev
Min Peng
Min Zhang
Minchul Shin
Minesh Mathew
Ming Li
Ming Meng
Ming Yin
Ming-Ching Chang
Mingfei Cheng
Minghui Wang
Mingjun Hu
MingKun Yang
Mingxing Tan
Mingzhi Yuan
Min-Hung Chen
Minhyun Lee
Minjung Kim
Min-Kook Suh
Minkyo Seo
Minyi Zhao
Mo Zhou
Mohammad Amin A. Shabani
Moein Sorkhei
Mohit Agarwal
Monish K. Keswani
Muhammad Sarmad
Muhammad Kashif Ali
Myung-Woo Woo
Naeemullah Khan
Naman Solanki
Namyup Kim
Nan Gao
Nan Xue
Naoki Chiba
Naoto Inoue
Naresh P. Cuntoor
Nati Daniel
Neelanjan Bhowmik
Niaz Ahmad
Nicholas I. Kuo
Nicholas E. Rosa
Nicola Fioraio
Nicolas Dufour
Nicolas Papadakis
Ning Liu
Nishan Khatri
Ole Johannsen
P. Real Jurado
Parikshit V. Sakurikar
Patrick Peursum
Pavan Turaga
Peijie Chen
Peizhi Yan
Peng Wang
Pengfei Fang
Penghui Du
Pengpeng Liu
Phi Le Nguyen
Philippe Chiberre
Pierre Gleize
Pinaki Nath Chowdhury
Ping Hu

Ping Li
Ping Zhao
Pingping Zhang
Pradyumna Narayana
Pritish Sahu
Qi Li
Qi Wang
Qi Zhang
Qian Li
Qian Wang
Qiang Fu
Qiang Wu
Qiangxi Zhu
Qianying Liu
Qiaosi Yi
Qier Meng
Qin Liu
Qing Liu
Qing Wang
Qingheng Zhang
Qingjie Liu
Qinglin Liu
Qingsen Yan
Qingwei Tang
Qingyao Wu
Qingzheng Wang
Qizao Wang
Quang Hieu Pham
Rabab Abdelfattah
Rabab Ward
Radu Tudor Ionescu
Rahul Mitra
Raül Pérez i Gonzalo
Raymond A. Yeh
Ren Li
Renán Rojas-Gómez
Renjie Wan
Renuka Sharma
Reyer Zwiggelaar
Robin Chan
Robin Courant
Rohit Saluja
Rongkai Ma
Ronny Hänsch
Rui Liu
Rui Wang
Rui Zhu
Ruibing Hou
Ruikui Wang
Ruiqi Zhao
Ruixing Wang
Ryo Furukawa
Ryusuke Sagawa
Saimunur Rahman
Samet Akcay
Samitha Herath
Sanath Narayan
Sandesh Kamath
Sanghoon Jeon
Sanghyun Son
Satoshi Suzuki
Saumik Bhattacharya
Sauradip Nag
Scott Wehrwein
Sebastien Lefevre
Sehyun Hwang
Seiya Ito
Selen Pehlivan
Sena Kiciroglu
Seok Bong Yoo
Seokjun Park
Seongwoong Cho
Seoungyoon Kang
Seth Nixon
Seunghwan Lee
Seung-Ik Lee
Seungyong Lee
Shaifali Parashar
Shan Cao
Shan Zhang
Shangfei Wang
Shaojian Qiu
Shaoru Wang
Shao-Yuan Lo
Shengjin Wang
Shengqi Huang
Shenjian Gong
Shi Qiu
Shiguang Liu
Shih-Yao Lin
Shin-Jye Lee
Shishi Qiao
Shivam Chandhok
Shohei Nobuhara
Shreya Ghosh
Shuai Yuan
Shuang Yang
Shuangping Huang
Shuigeng Zhou
Shuiwang Li
Shunli Zhang
Shuo Gu
Shuoxin Lin
Shuzhi Yu
Sida Peng
Siddhartha Chandra
Simon S. Woo
Siwei Wang
Sixiang Chen
Siyu Xia
Sohyun Lee
Song Guo
Soochahn Lee
Soumava Kumar Roy
Srinjay Soumitra Sarkar
Stanislav Pidhorskyi
Stefan Gumhold
Stefan Matcovici
Stefano Berretti
Stylianos Moschoglou
Sudhir Yarram
Sudong Cai
Suho Yang
Sumitra S. Malagi
Sungeun Hong
Sunggu Lee
Sunghyun Cho
Sunghyun Myung
Sungmin Cho
Sungyeon Kim
Suzhen Wang
Sven Sickert
Syed Zulqarnain Gilani
Tackgeun You
Taehun Kim

Takao Yamanaka
Takashi Shibata
Takayoshi Yamashita
Takeshi Endo
Takeshi Ikenaga
Tanvir Alam
Tao Hong
Tarun Kalluri
Tat-Jen Cham
Tatsuya Yatagawa
Teck Yian Lim
Tejas Indulal Dhamecha
Tengfei Shi
Thanh-Dat Truong
Thomas Probst
Thuan Hoang Nguyen
Tian Ye
Tianlei Jin
Tianwei Cao
Tianyi Shi
Tianyu Song
Tianyu Wang
Tien-Ju Yang
Tingting Fang
Tobias Baumgartner
Toby P. Breckon
Torsten Sattler
Trung Tuan Dao
Trung Le
Tsung-Hsuan Wu
Tuan-Anh Vu
Utkarsh Ojha
Utku Ozbulak
Vaasudev Narayanan
Venkata Siva Kumar Margapuri
Vandit J. Gajjar
Vi Thi Tuong Vo
Victor Fragoso
Vikas Desai
Vincent Lepetit
Vinh Tran
Viresh Ranjan
Wai-Kin Adams Kong
Wallace Michel Pinto Lira
Walter Liao
Wang Yan
Wang Yong
Wataru Shimoda
Wei Feng
Wei Mao
Wei Xu
Weibo Liu
Weichen Xu
Weide Liu
Weidong Chen
Weihong Deng
Wei-Jong Yang
Weikai Chen
Weishi Zhang
Weiwei Fang
Weixin Lu
Weixin Luo
Weiyao Wang
Wenbin Wang
Wenguan Wang
Wenhan Luo
Wenju Wang
Wenlei Liu
Wenqing Chen
Wenwen Yu
Wenxing Bao
Wenyu Liu
Wenzhao Zheng
Whie Jung
Williem Williem
Won Hwa Kim
Woohwan Jung
Wu Yirui
Wu Yufeng
Wu Yunjie
Wugen Zhou
Wujie Sun
Wuman Luo
Xi Wang
Xianfang Sun
Xiang Chen
Xiang Li
Xiangbo Shu
Xiangcheng Liu
Xiangyu Wang
Xiao Wang
Xiao Yan
Xiaobing Wang
Xiaodong Wang
Xiaofeng Wang
Xiaofeng Yang
Xiaogang Xu
Xiaogen Zhou
Xiaohan Yu
Xiaoheng Jiang
Xiaohua Huang
Xiaoke Shen
Xiaolong Liu
Xiaoqin Zhang
Xiaoqing Liu
Xiaosong Wang
Xiaowen Ma
Xiaoyi Zhang
Xiaoyu Wu
Xieyuanli Chen
Xin Chen
Xin Jin
Xin Wang
Xin Zhao
Xindong Zhang
Xingjian He
Xingqun Qi
Xinjie Li
Xinqi Fan
Xinwei He
Xinyan Liu
Xinyu He
Xinyue Zhang
Xiyuan Hu
Xu Cao
Xu Jia
Xu Yang
Xuan Luo
Xubo Yang
Xudong Lin
Xudong Xie
Xuefeng Liang
Xuehui Wang
Xuequan Lu

Xuesong Yang
Xueyan Zou
XuHu Lin
Xun Zhou
Xupeng Wang
Yali Zhang
Ya-Li Li
Yalin Zheng
Yan Di
Yan Luo
Yan Xu
Yang Cao
Yang Hu
Yang Song
Yang Zhang
Yang Zhao
Yangyang Shu
Yani A. Ioannou
Yaniv Nemcovsky
Yanjun Zhu
Yanling Hao
Yanling Tian
Yao Guo
Yao Lu
Yao Zhou
Yaping Zhao
Yasser Benigmim
Yasunori Ishii
Yasushi Yagi
Yawei Li
Ye Ding
Ye Zhu
Yeongnam Chae
Yeying Jin
Yi Cao
Yi Liu
Yi Rong
Yi Tang
Yi Wei
Yi Xu
Yichun Shi
Yifan Zhang
Yikai Wang
Yikang Ding
Yiming Liu
Yiming Qian
Yin Li
Yinghuan Shi
Yingjian Li
Yingkun Xu
Yingshu Chen
Yingwei Pan
Yiping Tang
Yiqing Shen
Yisheng Zhu
Yitian Li
Yizhou Yu
Yoichi Sato
Yong A.
Yongcai Wang
Yongheng Ren
Yonghuai Liu
Yongjun Zhang
Yongkang Luo
Yongkang Wong
Yongpei Zhu
Yongqiang Zhang
Yongrui Ma
Yoshimitsu Aoki
Yoshinori Konishi
Young Jun Heo
Young Min Shin
Youngmoon Lee
Youpeng Zhao
Yu Ding
Yu Feng
Yu Zhang
Yuanbin Wang
Yuang Wang
Yuanhong Chen
Yuanyuan Qiao
Yucong Shen
Yuda Song
Yue Huang
Yufan Liu
Yuguang Yan
Yuhan Xie
Yu-Hsuan Chen
Yu-Hui Wen
Yujiao Shi
Yujin Ren
Yuki Tatsunami
Yukuan Jia
Yukun Su
Yu-Lun Liu
Yun Liu
Yunan Liu
Yunce Zhao
Yun-Chun Chen
Yunhao Li
Yunlong Liu
Yunlong Meng
Yunlu Chen
Yunqian He
Yunzhong Hou
Yuqiu Kong
Yusuke Hosoya
Yusuke Matsui
Yusuke Morishita
Yusuke Sugano
Yuta Kudo
Yu-Ting Wu
Yutong Dai
Yuxi Hu
Yuxi Yang
Yuxuan Li
Yuxuan Zhang
Yuzhen Lin
Yuzhi Zhao
Yvain Queau
Zanwei Zhou
Zebin Guo
Ze-Feng Gao
Zejia Fan
Zekun Yang
Zelin Peng
Zelong Zeng
Zenglin Xu
Zewei Wu
Zhan Li
Zhan Shi
Zhe Li
Zhe Liu
Zhe Zhang
Zhedong Zheng

Contents – Part I

3D Computer Vision

Optimization Methods

3D Computer Vision

EAI-Stereo: Error Aware Iterative Network for Stereo Matching

Haoliang Zhao[1,4], Huizhou Zhou[2,4], Yongjun Zhang[1(✉)], Yong Zhao[1,3,4], Yitong Yang[1], and Ting Ouyang[1]

[1] State Key Laboratory of Public Big Data, Institute for Artificial Intelligence, College of Computer Science and Technology, Guizhou University, Guiyang 550025, Guizhou, China
zyj6667@126.com

[2] School of Physics and Optoelectronic Engineering, Guangdong University of Technology, Guangzhou 510006, China

[3] The Key Laboratory of Integrated Microsystems, Shenzhen Graduate School, Peking University,Beijing, China

[4] Ghost-Valley AI Technology, Shenzhen, Guangdong, China

Abstract. Current state-of-the-art stereo algorithms use a 2D CNN to extract features and then form a cost volume, which is fed into the following cost aggregation and regularization module composed of 2D or 3D CNNs. However, a large amount of high-frequency information like texture, color variation, sharp edge etc. is not well exploited during this process, which leads to relatively blurry and lacking detailed disparity maps. In this paper, we aim at making full use of the high-frequency information from the original image. Towards this end, we propose an error-aware refinement module that incorporates high-frequency information from the original left image and allows the network to learn error correction capabilities that can produce excellent subtle details and sharp edges. In order to improve the data transfer efficiency between our iterations, we propose the Iterative Multiscale Wide-LSTM Network which could carry more semantic information across iterations. We demonstrate the efficiency and effectiveness of our method on KITTI 2015, Middlebury, and ETH3D. At the time of writing this paper, EAI-Stereo ranks 1st on the Middlebury leaderboard and 1st on the ETH3D Stereo benchmark for 50% quantile metric and second for 0.5px error rate among all published methods. Our model performs well in cross-domain scenarios and outperforms current methods specifically designed for generalization. Code is available at https://github.com/David-Zhao-1997/EAI-Stereo.

1 Introduction

Stereo Matching is a fundamental vision problem in computer vision with direct real-world applications in robotics, 3D reconstruction, augmented reality, and autonomous driving. The task is to estimate pixel-wise correspondences of an

H. Zhao and H. Zhou—These authors contributed equally.

L. Wang et al. (Eds.): ACCV 2022, LNCS 13841, pp. 3–19, 2023.
https://doi.org/10.1007/978-3-031-26319-4_1

image pair and generate a displacement map termed disparity which can be converted to depth using the parameters of the stereo camera system.

Generally, traditional stereo matching algorithms [9,12,13] perform the following four steps: matching cost computation, cost aggregation, disparity computation and refinement [29]. These algorithms can be classified into global methods and local methods. Global methods [8,19,20,33] take advantage of solving the problem by minimizing a global energy function [5], which consumes time in exchange for accuracy. While local methods [1,15] only make use of neighboring pixels, which usually runs faster [44]. However, in open-world environments, it is difficult for traditional methods to achieve satisfactory results in textureless regions and regions with repetitive patterns, while traditional high-precision algorithms are often limited in terms of computational speed.

Recently, with the continuous research in convolutional neural networks, learning-based methods are widely used in the field of binocular stereo matching. Compared with traditional methods, learning-based methods tend to produce more accurate and smooth [23,35] disparity maps, and some of them also have advantages in computational speed [35,41]. However, to apply algorithms in real scenarios, there are still some challenges to be solved.

One challenge is that current methods do not perform well in recovering thin objects and sharp edges. Most current algorithms use a 2D CNN to extract features and then form a cost volume, which is fed into the following cost aggregation and regularization module composed of 2D or 3D CNNs. During this process, a large amount of high-frequency information is ignored, which leads to relatively blurry and lacking detailed disparity maps. However, stereo vision is often used in areas such as navigation, where it is important to recognize thin objects such as wires and highly reflective surfaces such as glass.

Another challenge is that current state-of-the-art stereo methods [23,36,38] use an iterative structure based on stock GRU which we found to be a bottleneck for the iterative model designed for stereo matching. A more efficient iterative structure is needed for performance improvements.

The other key issue is that learning-based algorithms are often not as effective as on specific datasets when applied to real-world scenarios due to their limited generalization capabilities [39].

In this work, we propose EAI-Stereo (Error Aware Iterative Stereo), a new end-to-end data-driven method for stereo matching.

The major contributions of this paper can be summarized as follows:

1. We propose an error-aware refinement module that combines left-right warping with learning-based upsampling. By incorporating the original left image that contains more high-frequency information and explicit calculating error maps, our refinement module enables the network to better cope with overexposure, underexposure as well as weak textures and allows the network to learn error correction capabilities which allows EAI-Stereo to produce extreme details and sharp edges. The learning-based upsampling method in the module can provide more refined upsampling results compared to bilinear

interpolation. We have carefully studied the impact of the module's microstructure on performance. From our experiments, we find that the structure improves generalization ability while improving performance. This approach is highly general and can be applied to all models that produce disparity or depth maps.
2. We propose an efficient iterative update module, called Multiscale Wide-LSTM, which can efficiently combine multi-scale information from feature extraction, cost volume, and current state, thus enhancing the information transfer between each iteration.
3. We propose a flexible overall structure that can balance inference speed and accuracy. The tradeoff could be done without retraining the network or even at run time. The number of iterations can also be determined dynamically based on the minimum frame rate.

2 Related Works

2.1 Data-driven Stereo Matching

Recently, data-driven methods dominate the field of stereo matching. Zbontar and LeCun proposed the first deep learning stereo matching method [46]. Mayer et al. proposed DispNetC [24], the first end-to-end stereo matching network.

In order to improve accuracy, 3D convolution was adopted by various of works [2,10,18,42,49]. Chang et al. propose PSMNet [2], a pyramid stereo matching network consisting of spatial pyramid pooling and several 3D convolutional layers. Taking advantage of the strong regularization effect of 3D convolution, PSMNet outperformed other methods at that time while 3D convolutions are very computationally expensive. To further increase accuracy, Zhang et al. proposes GANet [47] which approximates semi-global matching (SGM) [13] by introducing a semi-global guided aggregation (SGA) layer [47]. The accuracy is improved by cost aggregation from different directions, which improves the performance in occluded and textureless regions. However, these networks have limited ability to generalize across datasets. After training on simulated datasets, these feature maps tend to become noisy and discrete when the network is used to predict real-world scenes, and therefore output inaccurate disparity maps. To address this problem, DSMNet [48] improves the generalization ability of the network by adding domain normalization and a non-local graph-based filter, which also improves the accuracy. Shen et al. believe that the large domain differences and unbalanced disparity distribution across a variety of datasets limit the real-world performance of the model and propose CFNet [31], which introduces Cascade and Fused Cost Volume to improve the robustness of the network.

Due to the high computational cost, some methods come up with new ways to avoid the use of 3D convolutions. Xu et al. proposed AANet [41], which replaces the computationally intensive 3D convolution and improves the accuracy by using the ISA module and CSA module. While some researchers proposed a coarse-to-fine routine [32,37,43] to replace 3D convolution in order to further speed up inference. Tankovich et al. proposed HITNet [35], which introduced slanted plane hypotheses that allow performing geometric warping and upsampling operations more accurately which achieve a higher level of accuracy [35].

2.2 Iterative Network

With the development of deep learning, there is a tendency to add more layers to convolutional neural networks to achieve better accuracy. However, as the network gets deeper and deeper, the computational cost and the number of parameters have greatly increased. To address the problem, Neshatpour et al. proposed ICNN [27] (Iterative Implementation of Convolutional Neural Networks) which replaces the single heavy feedforward network with a series of smaller networks executed sequentially. Since many images are detected in early iterations, this method draws much less computational complexity.

IRR [16] first introduce a recurrent network for Optic Flow, which uses FlowNetS [6] or PWC-Net [32] as its recurrent module. This enables IRR to be able to achieve better performance by increasing its iterations. However, both FlowNetS and PWC-Net are relatively heavy when used as iterative modules, which limits the number of iterations. To address this problem, RAFT [36], proposed by Teed et al. uses GRU [3] as its iterative module to update the flow predictions and achieve state-of-the-art performance in optic flow. It is proved that RAFT has strong cross-dataset generalization ability while keeping a high efficiency in inference time [36].

In our work, we found that the stock GRU is becoming a bottleneck for the iterative model designed for stereo matching. To alleviate this problem, we proposed an improved iterative module to achieve better performance.

3 Approach

Our network takes a pair of rectified images I_l and I_r as input. Then the features are extracted and injected into the cost volume. The Multiscale Iterative Module retrieves data from the cost volume and iterates to update the disparity map. Finally, the iterated 1/4 resolution disparity map is fed into the Error Aware Refinement module, which can perform learned upsampling and error-aware correction to obtain the final disparity map (Fig. 1).

3.1 Multi-scale Feature Extractor

We use a ResNet-like network [11] as our feature extractor, feature maps of a pair of images I_l and I_r are extracted using two shared-weight extractors which are used to construct a 3D correlation volume following RAFT-Stereo [23]. The network consists of a sequence of residual blocks and then followed by two downsampling layers which are used to provide multi-scale information F_h, F_m and F_l for the following iterative Wide-LSTM modules. The spatial sizes of the features maps F_h, F_m and F_l are 1/4, 1/8 and 1/16 of the original input image size.

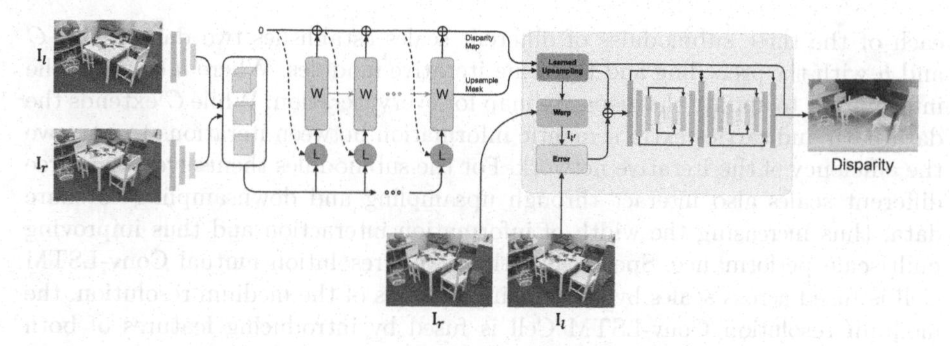

Fig. 1. The overall structure of EAI-Stereo. The left and right images are extracted by a weight-sharing feature extractor and the features are injected into the multiscale correlation volume. The following Wide-LSTM Module combines information from correlation volume, the previous iteration and the current disparity map to produce a disparity map and an unsampling mask which are used for our Error Aware Refinement. The refinement module upsamples the disparity map and then uses it to warp the right image to the left and calculate the error map. Then the combined information is fed to the hourglass models to output the final refined disparity map.

3.2 Iterative Multiscale Wide-LSTM Network

For iterative networks, the design of the iteration module has a significant impact on the network performance. For image tasks, most of the models take the stock GRU [3] as their iterative module. However, in our research, we found that the performance of the network could be increased by improving the iterative module. To address this problem, we propose an efficient iterative update module named Multiscale Wide-LSTM that can efficiently combine the information from feature extraction, cost volume, and current state, which also enhances the information transfer between each iteration. Experiments show that our model increases performance with a minor computational cost increase.

The network predicts a sequence of disparity maps $\{d_1, \ldots, d_n\}$ with an initial disparity map $d_0 = 0$. And the first hidden state h_0 is initialized using the information extracted by feature extractors. For each iteration, the LSTM module takes the previous hidden state h_{i-1}, the previous state of the disparity map d_{i-1}, and the information from the feature extraction F as inputs and then outputs Δd which adds to the current disparity: $d_i = d_{i-1} + \Delta d$. After n iterations, the iteration result d_n is fed into the refinement module for the final disparity map $d_{refined}$. We supervised our network by the following equation:

$$L_{regress} = \sum_{i=1}^{n-1} \gamma^{n-i} ||d_{gt} - d_i||_1 + ||d_{gt} - d_{refined}||_1, \ where\, \gamma = 0.9. \quad (1)$$

Multiscale Iterative Module. In our study, we found that for image tasks, the width of information transfer between iterative modules affects the model performance. Therefore, we widen the iterative modules in our design. In our module, each of the three submodules of different scales establishes two data paths, C and h with the preceding and following iterative modules. Where h contains the information to update the disparity map for every iteration. While C extends the data path and carries extra semantic information between iterations to improve the efficiency of the iterative network. For the submodules themselves, the three different scales also interact through upsampling and downsampling to share data, thus increasing the width of information interaction and thus improving multiscale performance. Specifically, the lowest resolution mutual Conv-LSTM Cell is fused across scales by introducing features of the medium resolution, the medium resolution Conv-LSTM Cell is fused by introducing features of both low and high resolution, and the highest resolution cell is fused by introducing features of medium resolution.

The multiscale fusion mechanism follows the following formulas:

$$C_l, h_l = MutualLSTMCell(C_{l_{prev}}, h_{l_{prev}}, ctx, pool(h_{m_{prev}})) \tag{2}$$

$$C_m, h_m = CLSTMCell(C_{m_{prev}}, h_{m_{prev}}, ctx, pool(h_{h_{prev}}), interp(h_{l_{prev}})) \tag{3}$$

$$C_h, h_h = CLSTMCell(C_{l_{prev}}, h_{l_{prev}}, ctx, disp, interp(h_{m_{prev}})) \tag{4}$$

where subscript l, m and h denote low, middle and high resolution respectively. $CLSTMCell$ is short for Conv-LSTM Cell. The low-resolution MutualLSTM-Cell not only takes $C_{l_{prev}}$ and $h_{l_{prev}}$ as input but also makes use of the features from downsampled middle resolution. The middle-resolution ConvLSTM-Cell takes advantage of using both downsampled high-resolution features and upsampled low-resolution features. For the highest resolution, the module not only makes use of upsampled middle resolution but also takes *disp* as input.

In general, low-resolution features have a larger perceptual field, which helps to improve the matching accuracy in textureless regions, while high-resolution features contain more high-frequency details, the combined use of this information can increase the perceptual field without adding much computational cost, thus improving the results. Another advantage of using multiscale is that we can use different iterative submodules at each scale, and the lower resolution feature maps have fewer pixels, which allows relatively time-consuming operations to be performed. In our model, the Mutual Conv-LSTM Cell is used only in the 1/16 resolution module. Experimental results show that using this module only at low resolution improves the performance with little change in computational cost (Fig. 2).

Mutual Conv-LSTM Cell. To further improve the performance of the iterative network, improvements have also been made to the cells that make up the iterative module. Currently, the use of iterative networks to process image data is gaining popularity. However, most of the networks simply use GRU [3] cell and replace the fully connected layers in them with convolutional layers. However, according to our observation, widening and increasing the hidden state of the network can improve the performance very well, so we use the LSTM [21] which has a performance improvement in this task with little difference in computational cost as our baseline.

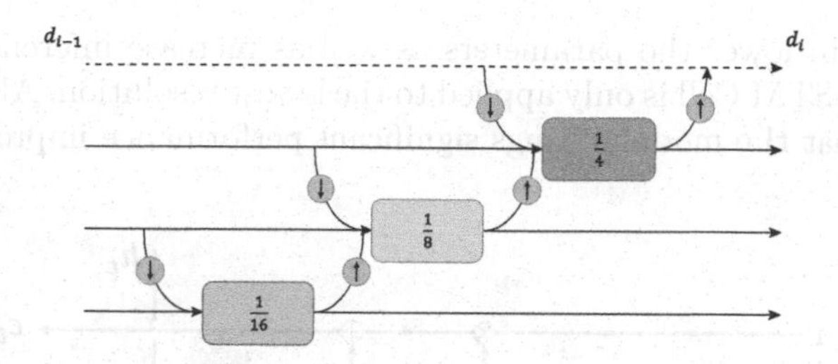

Fig. 2. Structure of Multiscale Iterative Module. Instead of fully connected fusion, our model transfer information between adjacent resolutions.

In ordinary LSTM [14], the input and hidden states do not interact much, but simply perform concatenation operations and then followed by various gate operations, which do not make good use of the input and hidden state information. In the field of natural language processing, there are several attempts to modify the LSTM cell to get better results. Inspired by the Multiplicative LSTM [21] and the MOGRIFIER LSTM [25], we propose the Mutual Conv-LSTM Cell.

In the Mutual Conv-LSTM Cell, the input x_t and the hidden state h_{t-1} interact following the formulas:

$$x_t^i = \delta Sigmoid(W_{conv_x} h_{t-1}^{i-1}) \odot x_t^{i-2}, for\ i \in \{x|x \leq n, x\%2 \neq 0\} \quad (5)$$

$$h_{t-1}^i = \delta Sigmoid(W_{conv_h} x_t^{i-1}) \odot x_{t-1}^{i-2}, for\ i \in \{x|x \leq n, x\%2 = 0\} \quad (6)$$

where n denotes the number of interactions between the input and the hidden state, W_{conv_x} and W_{conv_h} denote the weights of the two convolutional layers, δ denotes a constant to balance the distribution and is set to 2 in our experiments.

As depicted in Fig. 3, the convolution-processed feature maps of h_{t-1}^i are element-wise multiplied with x_t^i to generate a new hidden state. Similarly, the convolution-processed feature maps of x_t^i are element-wise multiplied with h_{t-1}^i to generate a new input. After interactions, the generated input x_t^i and hidden state h_{t-1}^n are processed with a procedure similar to a regular LSTM module following the equations:

$$f_t = Sigmoid(W_f \cdot [h_{t-1}, x_t] + b_f) \quad (7)$$

$$i_t = Sigmoid(W_i \cdot [h_{t-1}, x_t] + b_i) \quad (8)$$

$$\tilde{C}_t = tanh(W_i \cdot [h_{t-1}, x_t] + b_C) \quad (9)$$

$$C_t = f_t * C_{t-1} + i_t * \tilde{C}_t \quad (10)$$

$$o_t = Sigmoid(W_o \cdot [h_{t-1}, x_t] + b_o) \quad (11)$$

$$h_t = o_t * tanh(C_t) \quad (12)$$

The features from the input and hidden state are fully fused by multiple interactions, and the effective part of the features in both are retained and enhanced.

In our model, to lower the parameters as well as increase inference speed, the Mutual Conv-LSTM Cell is only applied to the lowest resolution. Ablation experiments show that the module brings significant performance improvement.

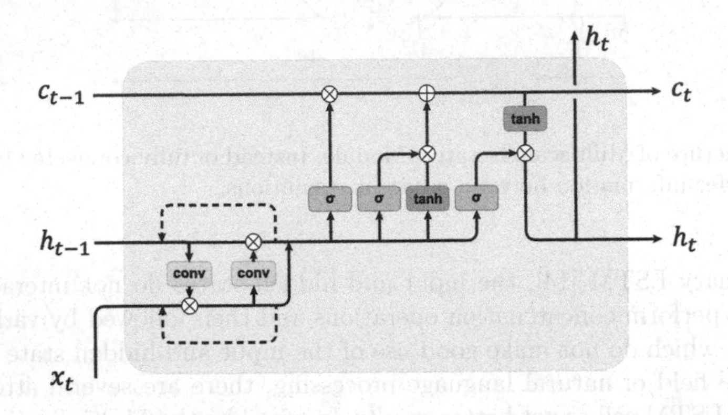

Fig. 3. Mutual Conv-LSTM Cell. The input and hidden state gate each other with convolution-processed features for n times. After interactions, the generated input and hidden state are processed with a procedure similar to a regular LSTM.

3.3 Error Aware Refinement

As we motioned before, a large amount of high-frequency information is ignored in the former structure of the model. In our refinement model, we aim to make full use of the former information and incorporate the high-frequency information from the original left image.

To make full use of the former information, we use learned upsampling to upsample the 1/4 resolution raw disparity map predicted by the LSTM network. Following RAFT [36], the highest resolution output is fed to a series of convolutional layers and generates an upsampling mask which is used to provide information to the convex upsampling. This method is proved to be much more efficient than bilinear upsampling. After the Learned Upsampling process, we get the disparity map of the same size as the original image. However, the disparity map is not error-aware processed at this point.

To incorporate the high-frequency information from the original left image and alleviate the problem of false matching, in Error Aware Module, we perform error perception by the following equations:

$$I_l^{'} = warp(I_r, disp) \tag{13}$$

$$e = I_l^{'} - I_l \tag{14}$$

$$I_{fuse} = Conv_{3\times3}([e, I_l]) \tag{15}$$

$$disp^{'} = hourglass([I_{fuse}, Conv_{3\times3}(disp)]) \tag{16}$$

where $I_l^{'}$ denotes the warped right image, e denotes the reprojection error, $disp^{'}$ denotes the refined disparity map.

The disparities are the correspondences between the left and right images. By using the warp method, the reconstructed left image can be calculated using the right image and the disparity map. Then subtraction is performed to get the error map which we called explicit error. As we motioned before, a large amount of high-frequency information is ignored during the former process. To alleviate this problem, we introduce the left image directly into the module, which composes of our implicit error. These two different forms of error improve the performance of our model in different aspects. We analyze them in the subsequent experiment section. By introducing more high-frequency information from the original left image, our model is therefore capable of recovering extreme details and sharp edges. Comparisons with the state-of-the-art methods are shown in Fig. 6.

In the Hourglass model, we reduced the number of its same-resolution convolution layers to streamline it. We tried deformable convolution [4] and dilation convolution [45], experimental results show that using deformable convolution is not as effective as dilation convolution. The reason behind it may be that deformable convolution is relatively weak for different scenes, while dilation convolution has a larger perceptual field and is capable of long-distance modeling (Fig. 4).

We have carefully studied the impact of the module's microstructure on performance. Details are shown in Table 5(a).

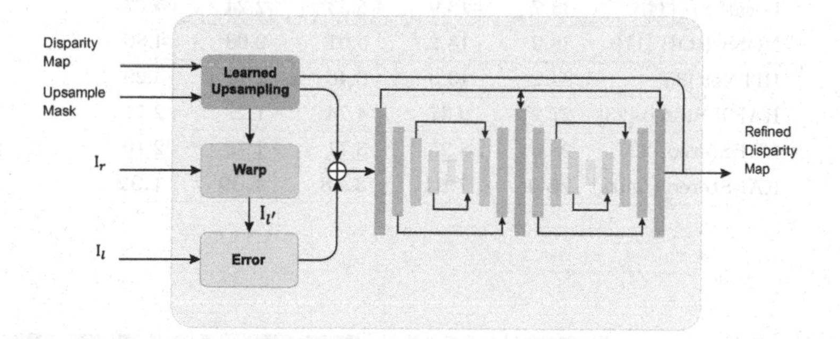

Fig. 4. Error Aware Refinement. The error map, the left image, and the original disparity map are passed into the hourglass to calculate the refined disparity map.

4 Experiments

EAI-Stereo is implemented in PyTorch and trained with two Tesla A100 GPUs. All models are trained using AdamW optimizer with a weight decay of $1e^{-5}$. Warm-up takes 1% of the whole training schedule. We used data augmentation in all experiments. The methods are saturation change, image perturbance, and random scales. For all the pretraining, we train our model on Scene Flow for 200k iterations with a learning rate of $2e^{-4}$.

We evaluate our EAI-Stereo with different settings using Scene Flow [24], KITTI-2015 [26], ETH3D [30] and Middlebury [28] datasets.

4.1 Middlebury

EAI-Stereo ranks 1st on the Middlebury test set, with an average error of 1.92% for all pixels, outperforming the next best method by 8.6%. Our method outperforms state-of-the-art methods on most of the metrics. See Table 1.

We fine-tune our model on the 23 Middlebury training images with a maximum learning rate of 2e-5 for 4000 iterations. Though Middlebury provides images with different color temperatures, we only use the standard images for training. Experiments show that EAI-Stereo is robust for various lighting conditions with simple data augmentation methods.

We also evaluate our EAI-Stereo on the Middlebury dataset without any fine-tuning, results are shown in Table 4, which prove the strong cross-domain performance of our model (Fig. 5).

Table 1. Results on the Middlebury stereo dataset V3 [28] leaderboard.

Method	bad 0.5 nonocc (%)	bad 1.0 nonocc (%)	bad 2.0 nonocc (%)	Avgerr nonocc (%)	Avgerr all (%)
LocalExp [34]	38.7	13.9	5.43	2.24	5.13
NOSS-ROB [17]	38.2	13.2	5.01	2.08	4.80
HITNet [35]	34.2	13.3	6.46	1.71	3.29
RAFT-Stereo [23]	27.2	9.37	4.74	1.27	2.71
CREStereo [22]	28.0	8.25	3.71	1.15	2.10
EAI-Stereo (Ours)	**25.1**	**7.81**	**3.68**	**1.09**	**1.92**

Fig. 5. Results on Middlebury dataset. Our EAI-Stereo recovers extreme details such as the spokes of the bicycle, toys on the table, and the subtle structures of plants.

4.2 ETH3D

For the ETH3D [30] dataset, we did not perform further fine-tune. Since the images are all in grayscale with many overexposed and underexposed areas. We use data augmentation to simulate the situation by setting saturation to 0, adjusting image gamma between 0.5 and 2.0, and adjusting image gain between 0.8 and 1.2.

At the time of writing this paper, EAI-Stereo ranks 1st on the ETH3D Stereo benchmark for 50% quantile metric and second for 0.5px error rate (see Table 2) among all published methods.

Table 2. Results on the ETH3D [30] leaderboard.

Method	bad 0.5 (%)	bad 1.0 (%)	50% quantile
AANet_RVC [41]	13.16	5.01	0.16
CFNet [31]	9.87	3.31	0.14
ACVNet [40]	10.36	2.58	0.15
HIT-Net [35]	7.83	2.79	0.10
RAFT-Stereo [23]	7.04	2.44	0.10
EAI-Stereo (Ours)	**5.21**	**2.31**	**0.09**

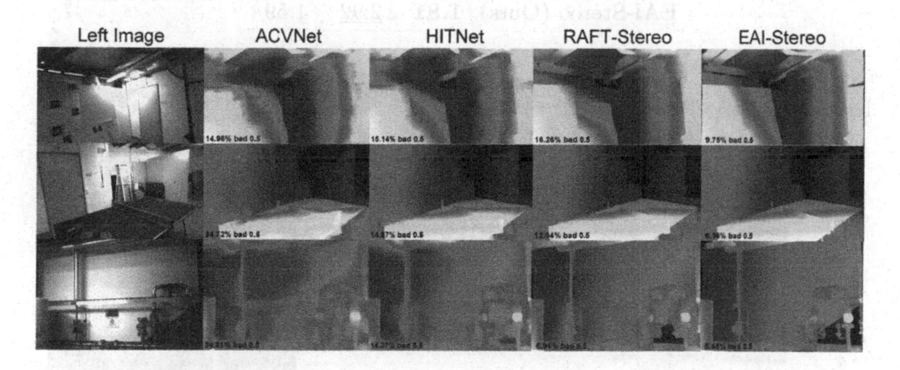

Fig. 6. Results on ETH3D compared to state-of-the-art methods. Bad 0.5 error is reported at the corners. EAI-Stereo shows advantages in recovering extreme details and sharp edges of the scenes such as the detailed structure of the pipes and valves. Our model is also capable of handling extreme overexposure and underexposure such as the reflective cardboard and pitch-black pipes on the roof of the top image.

4.3 KITTI-2015

We trained our model on the Scene Flow dataset and then fine-tuned our model on the KITTI training set for 6000 iterations with a batch size of 8 and a maximum learning rate of 2e-5. The images and disparity maps are randomly cropped with a resolution of 320×1024. We iterate the Multiscale Iterative Module 32 times.

Ground Truth values in the KITTI dataset are sparse and only cover the lower part of the image. We can observe from Fig. 7 that other methods fail to generalize in the upper part while our method recovers extreme details and sharp edges which proves the strong generalization performance of our model (Table 3).

Table 3. Results on the KITTI-2015 [26] leaderboard. Only published results are included. The best results for each metric are bolded, second best are underlined.

Method	D1-all	D1-fg	D1-bg
AcfNet [49]	1.89	3.80	1.51
AMNet [7]	<u>1.82</u>	3.43	1.53
OptStereo [39]	<u>1.82</u>	3.43	<u>1.50</u>
GANet-deep [47]	**1.81**	3.46	**1.48**
RAFT-Stereo [23]	1.96	**2.89**	1.75
HITNet [35]	1.98	3.20	1.74
CFNet [31]	1.88	3.56	1.54
EAI-Stereo (Ours)	**1.81**	<u>2.92</u>	1.59

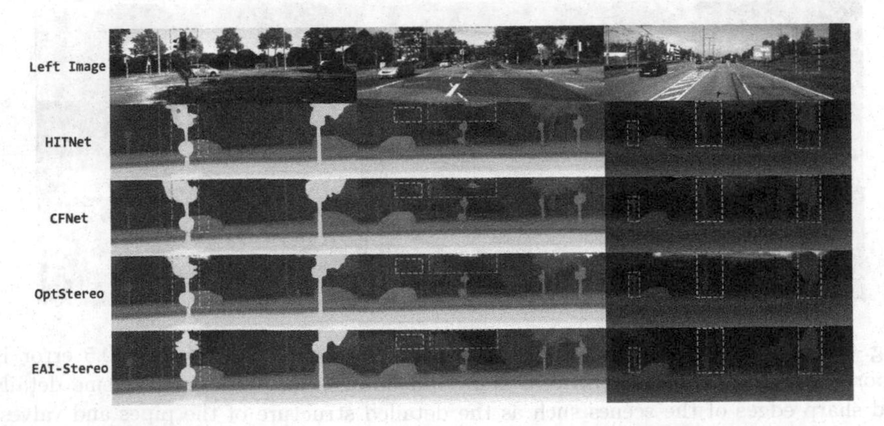

Fig. 7. Results on the KITTI-2015 test set compared to state-of-the-art methods. EAI-Stereo shows an advantage in recovering extreme details and sharp edges of the scenes. Zoom in for a better view. RAFT-Stereo [23] is not included in this comparison because it does not have an official submission to the benchmark.

4.4 Cross-domain Generalization

Generalization performance is crucial for real-world applications. Towards this end, we evaluated our model on three public datasets. We train our model on the Scene Flow dataset using the strategy exactly the same as pretrain and then use the weight for evaluation directly. In Table 4, we compare our model with some state-of-the-art methods and some classical methods. The comparison shows that our method outperforms DSMNet and CFNet, which are specifically designed for generalization performance, by a notable margin.

4.5 Ablations

We evaluate the performance of EAI-Stereo with different settings, including different architectures and different numbers of iterations.

Iterative Multiscale Wide-LSTM Network. We observe a significant performance leap (10.14% D1 error decrease on Scene Flow validation set and 4.80% EPE decrease on KITTI validation set) by using the wide LSTM module. Most iterative networks use GRUs as their iterative modules. However, we found that the performance of the network can be increased by refining the iterative module. A comparison between the GRU-based network and our iterative multiscale wide LSTM network is shown in Table 5(c). Using Mutual Conv-LSTM Cell at the lowest resolution can further improve the performance of the model. As shown in Table 5(c), this module led to 1.4% D1 error decrease on the Scene Flow validation set and 2.58% D1 error decrease on the KITTI validation set.

Error Aware Refinement. The Error Aware Refinement module is used to do the upsampling and refinement work. To verify and analyze the effects of our Error Aware Refinement module, we evaluate the different structures of the refinement module, and the results are shown in Table 5(c). Compared with the Wide LSTM baseline, Dilation Refinement decreases the D1-error by 2.81% on the Scene Flow validation set and 12.39% EPE (end-point-error) decrease on the KITTI validation set. Using deformable convolution is not as effective as dilation

Table 4. Cross-domain generalization experiments.

Method	KITTI2015 bad 3.0 (%)	Middlebury bad 2.0 (%)	ETH3D bad 1.0 (%)
PSMNet [2]	16.3	39.5	23.8
GANet [47]	11.7	32.2	14.1
DSMNet [48]	6.5	21.8	6.2
CFNet [31]	–	28.2	5.8
EAI-Stereo(Ours)	**6.1**	**14.5**	**3.3**

Table 5. Ablations Experiments.

(a) Ablations on refinement microstructures.

Hourglass	Error	Left image	Scene Flow D1	KITTI EPE	KITTI D1
			5.88	0.47	1.11
✓			5.85	0.47	0.89
✓	✓		5.84	0.40	0.85
✓		✓	5.76	0.41	0.89
✓	✓	✓	5.74	0.40	0.85

(b) Inference time.

Iters	Scene Flow EPE	Scene Flow D1	Time (ms)
5	0.596	7.326	92
7	0.539	6.527	100
10	0.510	6.046	132
16	0.495	5.821	154
32	0.491	5.661	236

(c) Ablations on different structures of the model.

Model	Conv GRU	Wide LSTM	Deform Refine	Dilation Refine	Mutual Conv LSTM	Scene Flow D1	KITTI EPE	KITTI D1
Baseline	✓					6.542	0.491	1.290
Wide LSTM		✓				5.879	0.468	1.108
EAI-Deform		✓	✓			5.840	0.410	0.850
EAI-Dilation		✓		✓		5.741	0.401	0.854
EAI-Mutual		✓		✓	✓	5.661	0.397	0.832

convolution, and we think the reason behind it may be that deformable convolution is relatively weak for different scenes, while dilated convolution has a larger perceptual field and is capable of long-distance modeling. Detailed comparisons are shown in Table 5(c).

Number of Iterations. Due to the structural improvements of our model, inference can be accelerated by reducing iterations while maintaining competitive performance. Since the model requires only a single training, the number of iterations can be adjusted after training, which increases the flexibility of the model. In practical applications, the number of iterations can also be inferred in the running state by giving a minimum frame rate, which is useful for scenarios with real-time requirements. Details are shown in Table 5(b).

5 Conclusion

We have proposed a novel error-aware iterative network for stereo matching. Several experiments were conducted to determine the structure of the module. Experiment results show that our model performs well on various datasets for both speed and accuracy while having a strong generalization performance.

Acknowledgements. This work is supported by Shenzhen Fundamental Research Program (JCYJ20180503182133411).

References

1. Bleyer, M., Gelautz, M.: Simple but effective tree structures for dynamic programming-based stereo matching. In: International Conference on Computer Vision Theory and Applications, vol. 2, pp. 415–422. SCITEPRESS (2008)
2. Chang, J.R., Chen, Y.S.: Pyramid stereo matching network. In: Proceedings of the IEEE Conference on Computer Vision and Pattern Recognition (CVPR), pp. 5410–5418 (2018)
3. Cho, K., van Merriënboer, B., Bahdanau, D., Bengio, Y.: On the properties of neural machine translation: Encoder-decoder approaches. In: Proceedings of SSST-8, Eighth Workshop on Syntax, Semantics and Structure in Statistical Translation, Doha, Qatar, pp. 103–111. Association for Computational Linguistics, October 2014. https://doi.org/10.3115/v1/W14-4012
4. Dai, J., et al.: Deformable convolutional networks. In: Proceedings of the IEEE International Conference on Computer Vision, pp. 764–773 (2017)
5. Dinh, V.Q., Munir, F., Sheri, A.M., Jeon, M.: Disparity estimation using stereo images with different focal lengths. IEEE Trans. Intell. Transp. Syst. **21**(12), 5258–5270 (2019)
6. Dosovitskiy, A., et al.: Flownet: learning optical flow with convolutional networks. In: Proceedings of the IEEE International Conference on Computer Vision, pp. 2758–2766 (2015)
7. Du, X., El-Khamy, M., Lee, J.: AmNet: deep atrous multiscale stereo disparity estimation networks. arXiv preprint arXiv:1904.09099 (2019)
8. Felzenszwalb, P.F., Huttenlocher, D.P.: Efficient belief propagation for early vision. Int. J. Comput. Vision **70**(1), 41–54 (2006)
9. Fife, W.S., Archibald, J.K.: Improved census transforms for resource-optimized stereo vision. IEEE Trans. Circuits Syst. Video Technol. **23**(1), 60–73 (2012)
10. Guo, X., Yang, K., Yang, W., Wang, X., Li, H.: Group-wise correlation stereo network. In: Proceedings of the IEEE Conference on Computer Vision and Pattern Recognition (CVPR), pp. 3273–3282 (2019)
11. He, K., Zhang, X., Ren, S., Sun, J.: Deep residual learning for image recognition. In: Proceedings of the IEEE Conference on Computer Vision and Pattern Recognition (CVPR), pp. 770–778 (2016). https://doi.org/10.1109/CVPR.2016.90
12. Heo, Y.S., Lee, K.M., Lee, S.U.: Robust stereo matching using adaptive normalized cross-correlation. IEEE Trans. Pattern Anal. Mach. Intell. **33**(4), 807–822 (2010)
13. Hirschmuller, H.: Stereo processing by semiglobal matching and mutual information. IEEE Trans. Pattern Anal. Mach. Intell. **30**(2), 328–341 (2007)
14. Hochreiter, S., Schmidhuber, J.: Long short-term memory. Neural Comput. **9**(8), 1735–1780 (1997)
15. Hosni, A., Rhemann, C., Bleyer, M., Rother, C., Gelautz, M.: Fast cost-volume filtering for visual correspondence and beyond. IEEE Trans. Pattern Anal. Mach. Intell. **35**(2), 504–511 (2012)
16. Hur, J., Roth, S.: Iterative residual refinement for joint optical flow and occlusion estimation. In: Proceedings of the IEEE Conference on Computer Vision and Pattern Recognition (CVPR), June 2019
17. Ji, P., Li, J., Li, H., Liu, X.: Superpixel alpha-expansion and normal adjustment for stereo matching. J. Vis. Commun. Image Represent. **79**, 103238 (2021)
18. Kendall, A., et al.: End-to-end learning of geometry and context for deep stereo regression. In: Proceedings of the IEEE International Conference on Computer Vision, pp. 66–75 (2017)

19. Klaus, A., Sormann, M., Karner, K.: Segment-based stereo matching using belief propagation and a self-adapting dissimilarity measure. In: 18th International Conference on Pattern Recognition (ICPR 2006), vol. 3, pp. 15–18. IEEE (2006)
20. Kolmogorov, V., Zabih, R.: Computing visual correspondence with occlusions using graph cuts. In: Proceedings Eighth IEEE International Conference on Computer Vision. ICCV 2001, vol. 2, pp. 508–515. IEEE (2001)
21. Krause, B., Lu, L., Murray, I., Renals, S.: Multiplicative LSTM for sequence modelling. arXiv preprint arXiv:1609.07959 (2016)
22. Li, J., et al.: Practical stereo matching via cascaded recurrent network with adaptive correlation. In: Proceedings of the IEEE Conference on Computer Vision and Pattern Recognition (CVPR), pp. 16263–16272 (2022)
23. Lipson, L., Teed, Z., Deng, J.: Raft-stereo: multilevel recurrent field transforms for stereo matching. In: 2021 International Conference on 3D Vision (3DV), pp. 218–227. IEEE (2021)
24. Mayer, N., et al.: A large dataset to train convolutional networks for disparity, optical flow, and scene flow estimation. In: Proceedings of the IEEE Conference on Computer Vision and Pattern Recognition (CVPR), pp. 4040–4048 (2016)
25. Melis, G., Kočiskỳ, T., Blunsom, P.: Mogrifier lstm. arXiv preprint arXiv:1909.01792 (2019)
26. Menze, M., Geiger, A.: Object scene flow for autonomous vehicles. In: Proceedings of the IEEE Conference on Computer Vision and Pattern Recognition (CVPR), pp. 3061–3070 (2015)
27. Neshatpour, K., Behnia, F., Homayoun, H., Sasan, A.: ICNN: an iterative implementation of convolutional neural networks to enable energy and computational complexity aware dynamic approximation. In: 2018 Design, Automation & Test in Europe Conference & Exhibition (DATE), pp. 551–556. IEEE (2018)
28. Scharstein, D., et al.: High-resolution stereo datasets with subpixel-accurate ground truth. In: Jiang, X., Hornegger, J., Koch, R. (eds.) GCPR 2014. LNCS, vol. 8753, pp. 31–42. Springer, Cham (2014). https://doi.org/10.1007/978-3-319-11752-2_3
29. Scharstein, D., Szeliski, R.: A taxonomy and evaluation of dense two-frame stereo correspondence algorithms. Int. J. Comput. Vision **47**(1), 7–42 (2002)
30. Schöps, T., et al.: A multi-view stereo benchmark with high-resolution images and multi-camera videos. In: Proceedings of the IEEE Conference on Computer Vision and Pattern Recognition (CVPR) (2017)
31. Shen, Z., Dai, Y., Rao, Z.: CFNet: cascade and fused cost volume for robust stereo matching. In: Proceedings of the IEEE Conference on Computer Vision and Pattern Recognition (CVPR), pp. 13906–13915 (2021)
32. Sun, D., Yang, X., Liu, M.Y., Kautz, J.: PWC-Net: CNNs for optical flow using pyramid, warping, and cost volume. In: Proceedings of the IEEE Conference on Computer Vision and Pattern Recognition (CVPR), pp. 8934–8943 (2018)
33. Sun, J., Zheng, N.N., Shum, H.Y.: Stereo matching using belief propagation. IEEE Trans. Pattern Anal. Mach. Intell. **25**(7), 787–800 (2003)
34. Taniai, T., Matsushita, Y., Sato, Y., Naemura, T.: Continuous 3d label stereo matching using local expansion moves. IEEE Trans. Pattern Anal. Mach. Intell. **40**(11), 2725–2739 (2017)
35. Tankovich, V., Hane, C., Zhang, Y., Kowdle, A., Fanello, S., Bouaziz, S.: HitNet: hierarchical iterative tile refinement network for real-time stereo matching. In: Proceedings of the IEEE Conference on Computer Vision and Pattern Recognition (CVPR), pp. 14362–14372 (2021)

36. Teed, Z., Deng, J.: RAFT: Recurrent all-pairs field transforms for optical flow. In: Vedaldi, A., Bischof, H., Brox, T., Frahm, J.-M. (eds.) ECCV 2020. LNCS, vol. 12347, pp. 402–419. Springer, Cham (2020). https://doi.org/10.1007/978-3-030-58536-5_24
37. Tonioni, A., Tosi, F., Poggi, M., Mattoccia, S., Stefano, L.D.: Real-time self-adaptive deep stereo. In: Proceedings of the IEEE Conference on Computer Vision and Pattern Recognition (CVPR), pp. 195–204 (2019)
38. Wang, F., Galliani, S., Vogel, C., Pollefeys, M.: IterMVS: iterative probability estimation for efficient multi-view stereo (2022)
39. Wang, H., Fan, R., Cai, P., Liu, M.: Pvstereo: pyramid voting module for end-to-end self-supervised stereo matching. IEEE Robot. Autom. Lett. **6**(3), 4353–4360 (2021)
40. Xu, G., Cheng, J., Guo, P., Yang, X.: ACVNet: attention concatenation volume for accurate and efficient stereo matching. In: Proceedings of the IEEE Conference on Computer Vision and Pattern Recognition (CVPR) (2022)
41. Xu, H., Zhang, J.: AANet: adaptive aggregation network for efficient stereo matching. In: Proceedings of the IEEE Conference on Computer Vision and Pattern Recognition (CVPR), pp. 1959–1968 (2020)
42. Yao, Y., Luo, Z., Li, S., Fang, T., Quan, L.: MVSNet: depth inference for unstructured multi-view stereo. In: Ferrari, V., Hebert, M., Sminchisescu, C., Weiss, Y. (eds.) ECCV 2018. LNCS, vol. 11212, pp. 785–801. Springer, Cham (2018). https://doi.org/10.1007/978-3-030-01237-3_47
43. Yin, Z., Darrell, T., Yu, F.: Hierarchical discrete distribution decomposition for match density estimation. In: Proceedings of the IEEE Conference on Computer Vision and Pattern Recognition (CVPR), pp. 6044–6053 (2019)
44. Yoon, K.J., Kweon, I.S.: Adaptive support-weight approach for correspondence search. IEEE Trans. Pattern Anal. Mach. Intell. **28**(4), 650–656 (2006)
45. Yu, F., Koltun, V.: Multi-scale context aggregation by dilated convolutions. arXiv preprint arXiv:1511.07122 (2015)
46. Zbontar, J., LeCun, Y.: Computing the stereo matching cost with a convolutional neural network. In: Proceedings of the IEEE Conference on Computer Vision and Pattern Recognition (CVPR), pp. 1592–1599 (2015)
47. Zhang, F., Prisacariu, V., Yang, R., Torr, P.S.: GA-Net: Guided aggregation net for end-to-end stereo matching. In: 2019 IEEE/CVF Conference on Computer Vision and Pattern Recognition (CVPR), Los Alamitos, CA, USA, pp. 185–194. IEEE Computer Society, June 2019. https://doi.org/10.1109/CVPR.2019.00027, https://doi.ieeecomputersociety.org/10.1109/CVPR.2019.00027
48. Zhang, F., Qi, X., Yang, R., Prisacariu, V., Wah, B., Torr, P.: Domain-invariant stereo matching networks. In: Vedaldi, A., Bischof, H., Brox, T., Frahm, J.-M. (eds.) Domain-invariant stereo matching networks. LNCS, vol. 12347, pp. 420–439. Springer, Cham (2020). https://doi.org/10.1007/978-3-030-58536-5_25
49. Zhang, Y., et al.: Adaptive unimodal cost volume filtering for deep stereo matching. In: Proceedings of the AAAI Conference on Artificial Intelligence, vol. 34, pp. 12926–12934 (2020)

Temporal-Aware Siamese Tracker: Integrate Temporal Context for 3D Object Tracking

Kaihao Lan, Haobo Jiang, and Jin Xie(✉)

Nanjing University of Science and Technology, Nanjing, China
{lkh,jiang.hao.bo,csjxie}@njust.edu.cn

Abstract. Learning discriminative target-specific feature representation for object localization is the core of the 3D Siamese object tracking algorithms. Current Siamese trackers focus on aggregating the target information from the latest template into the search area for target-specific feature construction, which presents the limited performance in the case of object occlusion or object missing. To this end, in this paper, we propose a novel temporal-aware Siamese tracking framework, where the rich target clue lying in a set of historical templates is integrated into the search area for reliable target-specific feature aggregation. Specifically, our method consists of three modules, including a template set sampling module, a temporal feature enhancement module and a temporal-aware feature aggregation module. In the template set sampling module, an effective scoring network is proposed to evaluate the tracking quality of the template so that the high-quality templates are collected to form the historical template set. Then, with the initial feature embeddings of the historical templates, the temporal feature enhancement module concatenates all template embeddings as a whole and then feeds them into a linear attention module for cross-template feature enhancement. Furthermore, the temporal-aware feature aggregation module aggregates the target clue lying in each template into the search area to construct multiple historical target-specific search-area features. Particularly, we follow the collection orders of the templates to fuse all generated target-specific features via an RNN-based module so that the fusion weight of the previous template information can be discounted to better fit the current tracking state. Finally, we feed the temporal fused target-specific feature into a modified CenterPoint detection head for target position regression. Extensive experiments on KITTI, NuScenes and waymo open datasets show the effectiveness of our proposed method. Source code is available at https://github.com/tqsdyy/TAT.

1 Introduction

Visual object tracking is a fundamental task in the computer vision field, and has achieved extensive applications such as autonomous driving [19] and robotics

---

Supplementary Information The online version contains supplementary material available at https://doi.org/10.1007/978-3-031-26319-4_2.

L. Wang et al. (Eds.): ACCV 2022, LNCS 13841, pp. 20–35, 2023.
https://doi.org/10.1007/978-3-031-26319-4_2

vision [4]. With the development of cheap LiDAR sensors, the point cloud-based 3D object tracking has obtained much more attention. Compared with the visual tracking using 2D images, point cloud data can effectively handle the challenges, such as the changes in the light condition and the object size. Generally, with the high-quality 3D bounding box in the first frame, 3D single object tracking aims to continuously evaluate the state of the object throughout the tracking video sequence. However, the sparsity of the point cloud and the noise interference still hinder its applications in the real world.

In recent years, inspired by the successful applications of Siamese network [1] in 2D object tracking, 3D tracker focuses on exploiting the Siamese network paradigm for object tracking. As a pioneer, SC3D [8] is proposed to perform the 3D Siamese object tracking by matching the embedded shape information in the template to a large number of candidate proposals in the search area. However, SC3D is time-consuming and not end-to-end. To this end, Qi *et al.* [25] proposed the Point-to-Box network (P2B) for object tracking. A PointNet++ [24] is employed to extract the features of the template and search area, which are then utilized to construct the target-specific feature for object position regression via the VoteNet [23]. Based on P2B, Zheng *et al.* [37] further proposed a box-aware feature embedding to capture the prior shape information of the object for robust object location. In addition, Hui *et al.* [10] proposed a voxel-to-BEV Siamese tracker to improve the tracking performance in the cases of sparse point clouds. In summary, current Siamese trackers mainly focus on exploiting a single template for target-specific feature generation while ignoring the rich temporal context information lying in the set of the historical templates.

In this paper, we propose a simple yet powerful temporal-aware Siamese tracking framework, where the high-quality historical templates are collected to learn the discriminative target-specific feature for object localization. Our key idea is to utilize a powerful linear attention mechanism for temporal context learning among the historical templates, which is then integrated into the search area for robust temporal-aware target information aggregation and object localization.

Specifically, our framework consists of three modules, including a template set sampling module, a temporal feature enhancement module and a temporal-aware feature aggregation module. In the template set sampling module, with the template as input, a lightweight scoring network is designed to evaluate the 3DIoU between the template and the ground-truth target so that the high-quality template set can be obtained by sampling the templates with high 3DIoUs. Then, taking as input the initial features of historical templates, the temporal feature enhancement concatenates the point features of all templates as the whole and feeds them into a linear attention module for efficient cross-template feature enhancement. Furthermore, the temporal-aware feature aggregation module constructs a feature matching matrix between each template feature and the search-area feature, which guides the target information transferring in each template into the search area for the target-specific search-area feature generation. In particular, a RNN-based (Recurrent Neural Network) module is employed to fuse multiple target-specific features in the collection order of the templates, based on the intuition that the target information in the latter templates tend to own a higher correlation with the current tracking state. Finally, with the learned target-specific feature, we utilize a modified CenterPoint [36] detection head for object position regression. Notably, benefitting from the target information aggregation from the historical templates, our method can still obtain effective target-specific feature representation in the case of occlusion or object missing. Extensive experiments verify the effectiveness of our proposed method.

The main contributions of our work are as follows:

- We propose a novel temporal-aware Siamese tracking framework, where the target information lying in the multiple historical templates is aggregated for the discriminative target-specific feature learning and the target localization.
- An effective 3DIoU-aware template selector is designed to collect high-quality templates as the historical template set.
- Our method can achieve state-of-the-art performance on multiple benchmarks and be robust in the cases of object occlusion or object missing.

2 Related Work

2.1 2D Object Tracking

In recent years, with the successful application of Siamese network [1] in the 2D object tracking, Siamese-based trackers [9,30,33,39] have become mainstream. The core idea of the Siamese tracker is to extract features from the template image and the current image by using the shared weight backbone network to ensure that the two images are mapped to the same feature space, and use the idea of similarity matching to locate the most similar part of the image to the template. However, the lack of depth information in RGB images makes it difficult for trackers to accurately estimate the depth of objects in the images. Therefore, scholars try to study the object tracking method based on RGB-D data, but it is still some efforts have been made on RGB-D object tracking. The subject approach used in RGB-D data-based trackers [2,22] is not much different from the 2D object tracking method, except that additional depth information enhances the tracker's ability to perceive depth information. Therefore, these methods still rely heavily on RGB information and still suffer from problems such as sensitivity to illumination changes and object size variations.

2.2 3D Single Object Tracking

Due to the characteristics of point cloud data, object tracking based on 3D point cloud can effectively avoid a series of problems in 2D image tracking field. Therefore, in recent years, much more work [6,27,32] focuses on 3D object tracking based on point clouds. As a pioneer, Giancola *et al.* [8] proposed a Siamese tracker named SC3D dedicated to 3D single object tracking (SOT). SC3D uses shape completion on the template to obtain the shape information of the target, generates a large number of candidate proposals in the search area and compares them with the template, and takes the most similar proposal as the current tracking result. Qi *et al.* [25] proposed a point-to-box network (P2B) for the problem that SC3D cannot be trained end-to-end, P2B uses feature augmentation to enhance the perception ability of the search area to a specific template, and then uses VoteNet [23] to locate the target in the search area. Zheng *et al.* [37] proposed a box-aware module based on P2B to enhance the network's mining of bounding box prior information. Furthermore, to enhance the tracking performance of the tracker for sparse point clouds, V2B proposed by Hui *et al.* [10] uses a shape-aware feature learning module and uses a voxel-to-BEV detector to regress the object center. Lately, Jiang *et al.* [12] proposed a two-stage Siamese tracker named RDT, which uses point cloud registration to achieve robust feature matching between the template and potential targets in the search area.

In the 3D SOT task, existing Siamese trackers use a single-template matching mechanism for object localization, which ignores the rich temporal templates information of historical tracking results. This makes the trackers suffer from low-quality template feature representation in the presence of noise interference or occlusion, leading to the tracking failure. Therefore, we consider designing a novel Temporal-aware Siamese Tracker to associate temporal templates and sufficiently mine temporal context to improve the tracking robustness.

2.3 3D Multi Object Tracking

Unlike SOT, most of the multi object tracking (MOT) algorithms follow the paradigm of "tracking-by-detection" [15,26,35], it including two stages: targets detection and targets association. Specifically, first, they use a detector to detect a large number of instance objects in each frame, and then use methods such as motion information to associate objects across frames. The difference between 3D-MOT and 2D-MOT is that 3D detection [17,23,28] and association [14] algorithms are used instead of 2D methods. As a pioneer, Weng *et al.* [34] uses PointRCNN [28] to detect instance targets per frame, and uses a 3D Kalman filter to predict object motion trajectories, finally using the Hungarian algorithm for detected objects matching. Recently, Luo *et al.* [18] attempted to unify detection and association into a unified framework, and achieved good results. In summary, 3D MOT pays more attention to the completeness of detection and the accuracy of association, and 3D SOT pays more attention to learning discriminative target-specific feature representation for object localization.

3 Approch

3.1 Problem Setting

In 3D single object tracking task, given the initial bounding box (BBox) of the object in the first frame, the tracker needs to continuously predict the BBoxes of the object throughout the tracking sequence. Specifically, an object BBox consists of nine parameters, including the object center coordinate (x, y, z), object size (l, w, h), and rotation angle (α, β, θ) (corresponding to three coordinate axes, respectively). Generally, in the 3D SOT field, we assume that the target size is fixed and the rotation direction is just around the z-axis. Therefore, the estimation of the object states will only contain the center coordinates (x, y, z) and rotation angle θ.

Our temporal-aware Siamese tracking framework mainly consists of three modules, including the template set sampling module (Sect. 3.2), the temporal feature enhancement module (Sect. 3.3) and the temporal-aware feature aggregation module (Sect. 3.4).

3.2 Template Set Sampling

Following the Siamese tracking paradigm, the traditional Siamese tracker takes as input a single template point cloud $\boldsymbol{P}^t = \{\boldsymbol{p}_i^t \in \mathbb{R}^3 | i = 1, 2, ..., N\}$ and a search area point cloud $\boldsymbol{P}^s = \{\boldsymbol{p}_j^s \in \mathbb{R}^3 | j = 1, 2, ..., M\}$, and matches the closest target object to the template in the search area. Instead, we focus on extracting the rich temporal context information from a set of collected templates $\boldsymbol{T} = \{\boldsymbol{P}_1^t, \boldsymbol{P}_2^t, ..., \boldsymbol{P}_k^t\}$ for robust object localization. We consider three sampling mechanisms for template set generation, including the random sampling, the closest sampling and the template score ranking sampling, and in Sect. 4.3, we will discuss the effects of different ways of generating template set:

Random Sampling. We randomly select k templates from the whole historical template buffer as the template set.
Closest Sampling. We select k templates closest to the current frame as the template set, which tends to be more related to the current frame.

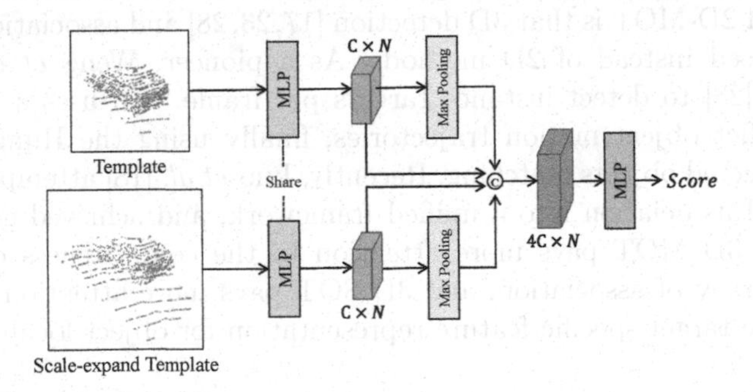

Fig. 1. The framework of the 3DIoU-aware template selection network.

Template Score Ranking Sampling. To collect high-quality templates, we construct a scoring network to predict the 3DIoU score between each historical template and the ground-truth target, and then select k templates with the highest scores as the template set. Specifically, as shown in Fig. 1, we exploit the original template $\boldsymbol{P}^t$ and the scale-expanded template $\boldsymbol{P}^{tg} = \{\boldsymbol{p}_i^{tg} \in \mathbb{R}^3 | i = 1, 2, ..., N\}$ as the network input, where the scale-expanded template is used to provide the necessary context information for reliable 3DIoU prediction. Then,

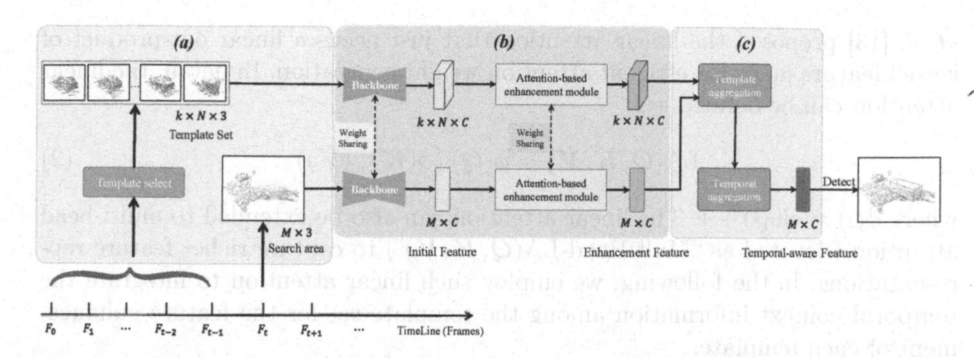

Fig. 2. The framework of our Temporal-aware Siamese Tracker. (a) **Template set sampling**: We first exploit the template selector to collect the high-quality template set. (b) **Siamese feature extraction**: Then, the weight-shared backbone network followed by a self-attention module is employed as the Siamese network to extract the point-level features of the template set and the search area. (c) **Target-specific feature fusion**: We integrate the target clue from k templates into the search area to learn k target-specific search area features, and an RNN-based fusion module is then employed to obtain the adaptively-fused target-specific feature. Finally, We pass the fused target-specific feature into a modified CenterPoint detector for object localization.

a weight-shared MLP is employed to extract their local point features $\boldsymbol{F}^t_{local} \in \mathbb{R}^{C\times N}$ and $\boldsymbol{F}^{tg}_{local} \in \mathbb{R}^{C\times N}$, and meanwhile a max-pooling function is performed to extract their global feature $\boldsymbol{F}^t_{global} \in \mathbb{R}^{C\times N}$ and $\boldsymbol{F}^{tg}_{global} \in \mathbb{R}^{C\times N}$, respectively. Finally, taking as input the concatenated features of the local and global features, an MLP is used to predict the score of the template:

$$Score = \text{MLP}(\text{Cat}(\boldsymbol{F}^t_{local}, \boldsymbol{F}^t_{global}, \boldsymbol{F}^{tg}_{local}, \boldsymbol{F}^{tg}_{global})). \tag{1}$$

We use PointNet++ [24] as the backbone network to extract the initial features of the template set and the search area. We denote the obtained initial search-area feature as $\boldsymbol{S} \in \mathbb{R}^{M\times C}$, and the initial feature of l-th template in the template set as $\boldsymbol{T}^l \in \mathbb{R}^{N\times C}$, where C indicates the feature dimension (Fig. 2).

3.3 Temporal Feature Enhancement

Temporal feature enhancement module aims to enhance the initial template features $\boldsymbol{T}^1, \boldsymbol{T}^2, ..., \boldsymbol{T}^k$ with the cross-template message passing based on the linear attention. In this section, we briefly introduce the linear attention mechanism, and then demonstrate how to exploit the temporal context information to enhance the feature representation of each template.

Linear-Attention Mechanism. The basic attention mechanism uses the dot-product attention [31] between the query $\boldsymbol{Q} \in \mathbb{R}^{N_q\times C}$ and key $\boldsymbol{K} \in \mathbb{R}^{N_k\times C}$ as the cross-attention weights for message passing. However, for the large-scale tasks, the dot-product attention is inefficient and usually needs high computational complexity for long-range relationship modeling. To relieve it, Katharopoulos *et al.* [13] proposed the linear attention that just needs a linear dot-product of kernel feature maps for efficient attention weight generation. In detail, the linear attention can be defined as:

$$\mathrm{LA}(\boldsymbol{Q}, \boldsymbol{K}, \boldsymbol{V}) = \phi(\boldsymbol{Q})\left(\phi(\boldsymbol{K})^{\top}(\boldsymbol{V})\right) \tag{2}$$

where $\phi(\cdot) = \mathrm{elu}(\cdot) + 1$. The linear attention can also be extended to multi-head attention (denoted as "MultiHead-LA($\boldsymbol{Q}, \boldsymbol{K}, \boldsymbol{V}$)") to capture richer feature representations. In the following, we employ such linear attention to integrate the temporal context information among the template set for the feature enhancement of each template.

Temporal Feature Enhancement. Given a set of initial template features, we first concatenate them to form an ensemble of the template features $\boldsymbol{T} = \mathrm{Concat}(\boldsymbol{T}^1, \boldsymbol{T}^2, ..., \boldsymbol{T}^k)$, where k is the template number of template set. Then, we treat them as a whole by reshaping $\boldsymbol{T} \in \mathbb{R}^{k \times N \times C}$ to $\boldsymbol{T} \in \mathbb{R}^{N_t \times C}$ $(N_t = k \times N)$ to satisfy the input shape of linear-attention module. The feature enhancement can be formulated as below:

$$\boldsymbol{T}^{'} = \text{MultiHead-LA}(\boldsymbol{T} + \boldsymbol{T}_p, \boldsymbol{T} + \boldsymbol{T}_p, \boldsymbol{T} + \boldsymbol{T}_p) \tag{3}$$

where $\boldsymbol{T}^{'} \in \mathbb{R}^{N_t \times C}$ is the enhanced template feature with the linear-attention module, and $\boldsymbol{T}_p \in \mathbb{R}^{N_t \times C}$ is the coordinate embedding of the template points via a MLP. Furthermore, the enhanced feature $\boldsymbol{T}^{'}$ is added as the residual item to the initial feature $\boldsymbol{T}$, followed by an instance normalization operation Ins. Norm$(\cdot)$:

$$\boldsymbol{T}^{''} = \mathrm{Ins.\,Norm}(\boldsymbol{T}^{'} + \boldsymbol{T}). \tag{4}$$

In addition, in order to improve the generalization ability of the network, we enhance $\boldsymbol{T}^{''}$ using a feed-forward neural network (FFN$(\cdot)$) followed by a instance normalization:

$$\widehat{\boldsymbol{T}} = \mathrm{Ins.\,Norm}(\mathrm{FFN}(\boldsymbol{T}^{''}) + \boldsymbol{T}^{''}). \tag{5}$$

Finally, for sufficient message passing among the template set, we iteratively perform the feature enhancement above in m times to achieve the deeper temporal context information aggregation. We will discuss the performance changes of different attention-iteration times m and different attention heads n in Sec. 4.3. Also, to ensure the feature-space consistency between the template set and the search area, we share the linear-attention module to the search area feature for feature transformation, i.e. $\boldsymbol{S} \rightarrow \widehat{\boldsymbol{S}} \in \mathbb{R}^{M \times C}$ (Fig. 3).

3.4 Temporal-Aware Feature Aggregation

The Siamese tracker focuses on constructing the feature similarity between the template and the search area, which guides the transferring of the target information from the template to the search area for the target-specific feature learning and the object localization. However, the current 3D Siamese trackers just exploit a single template for tracking while ignoring the rich template clue lying in the

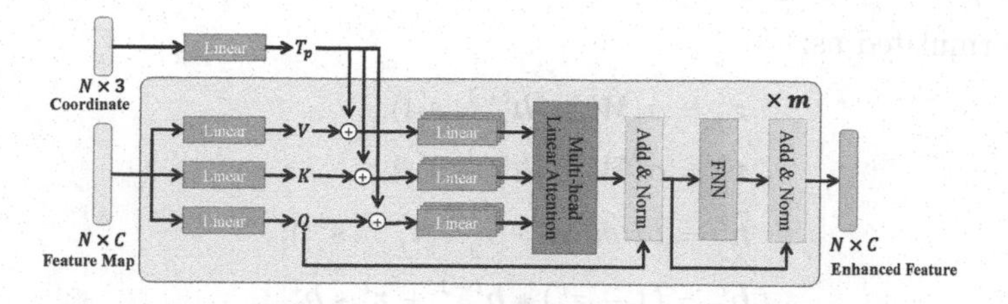

Fig. 3. Framework of the linear attention-based template enhancement module, which is iteratively performed m-times to achieve a richer temporal feature fusion.

historical templates, resulting in low tracking performance in some challenging cases (e.g., the object occlusion and the missing).

Target-specific Feature Learning. Instead of using a single template, we focus on exploiting the sampled high-quality template set (in Sect. 3.2) for target-specific feature learning. Specifically, taking as input the enhanced template-set features $\widehat{\boldsymbol{T}} \in \mathbb{R}^{k \times N \times C}$ and search area feature $\widehat{\boldsymbol{S}} \in \mathbb{R}^{M \times C}$, we construct the feature similarity map via the Cosine similarity for each template $\widehat{\boldsymbol{T}}^l$ $(1 \leq l \leq k)$ and the search area $\widehat{\boldsymbol{S}}$:

$$\mathrm{Sim}^l_{i,j} = \frac{\boldsymbol{t}_i{}^T \cdot \boldsymbol{s}_j}{\|\boldsymbol{t}_i\|_2 \cdot \|\boldsymbol{s}_j\|_2}, \forall \boldsymbol{t}_i \in \widehat{\boldsymbol{T}}^l, \boldsymbol{s}_j \in \widehat{\boldsymbol{S}}. \tag{6}$$

Then, based on the feature similarity above, we use it to guide the target-specific search-area feature learning. Specifically, for j-th search-area point $\boldsymbol{p}^s_j$, we gather the target information in its most related template point $\boldsymbol{p}^t_{x^*}$ (index $x^* = \mathrm{argmax}_i\ \mathrm{Sim}^l_{i,j}$) of the l-th template, and use it to guide the transferring of the information into the $\boldsymbol{p}^s_j$. The target information of template point $\boldsymbol{p}^t_{x^*}$ consists of the point coordinate $\boldsymbol{p}^t_{x^*}$, enhanced point feature $\boldsymbol{t}^l_{x^*} \in \widehat{\boldsymbol{T}}^l$ and the similarity score $\mathrm{Sim}^l_{x^*,j}$. These target information will be sent to a MLP together with the corresponding search area feature $\boldsymbol{s}_j \in \widehat{S}$ to build the fusion feature $\boldsymbol{s}^l_j = \mathrm{MLP}([\boldsymbol{s}_j, \boldsymbol{p}^t_{x^*}, \boldsymbol{t}^l_{x^*}, \mathrm{Sim}^l_{x^*,j}])$ of the j-th search-area point by the l-th template. Then, we can form the multiple target-specific features $\boldsymbol{S}^1_{fused}, \boldsymbol{S}^2_{fused}, \ldots, \boldsymbol{S}^k_{fused}$.

RNN-Based Temporal-Aware Feature Fusion. With the generated multiple target-specific features above, we exploit a GRU network (Gated Recurrent Unit, a popular RNN network) to fuse them, where the GRU network can adaptively assign higher fusion weights on the target-specific features from the latter templates while discounting the fusion weights of the features from the previous templates. It is mainly based on the intuition that compared to the previous template information, the latter ones tend to own a higher correlation with the current tracking state, thereby can benefit the current tracking performance. Among them, for the k features $\boldsymbol{s}^1_j, \boldsymbol{s}^2_j, \ldots, \boldsymbol{s}^k_j$ at the j-th point in the search area, the process of GRU network to fuse the historical template information can be formulated as:

$$\begin{aligned}
\boldsymbol{z}_j^t &= \sigma(\boldsymbol{W}^z \cdot [\boldsymbol{h}_j^{t-1}, \boldsymbol{s}_j^t]), \\
\boldsymbol{r}_j^t &= \sigma(\boldsymbol{W}^r \cdot [\boldsymbol{h}_j^{t-1}, \boldsymbol{s}_j^t]), \\
\widehat{\boldsymbol{h}}_j^t &= tanh(\boldsymbol{W} \cdot [\boldsymbol{r}_j^t * h_j^{t-1}, \boldsymbol{s}_j^t]), \\
\boldsymbol{h}_j^t &= (1 - \boldsymbol{z}_j^t) * \boldsymbol{h}_j^{t-1} + \boldsymbol{z}_j^t * \widehat{\boldsymbol{h}}_j^t,
\end{aligned} \tag{7}$$

where $\boldsymbol{W}, \boldsymbol{W}^z, \boldsymbol{W}^r$ are learnable parameter matrices, and $\boldsymbol{H} = \{\boldsymbol{h}_j^t \in \mathbb{R}^C | 1 \leq j \leq M, 1 \leq t \leq k\}$ is initialized to a zero matrix. We regard the output $\boldsymbol{H}^k \in \mathbb{R}^{M \times C}$ of the last layer as our final temporal-aware feature $\boldsymbol{S}_{final}$.

3.5 Loss Function

Templet Scoring Supervision. In the training phase, we first add the noise on the GT BBox of the target to generate the sample BBox and then generate the template by cropping the point cloud with the sample BBox. We use the IoU (Intersection over Union) between the GT BBox and the sample BBox as the GT score, and use the SmoothL1 Loss to supervise network training. Assuming that the GT score is S_{gt} and the predicted score of the network is S_{pred}, the loss of score supervision $\mathcal{L}_{score}$ can be written as:

$$\mathcal{L}_{score} = \text{SmoothL1}(S_{gt} - S_{pred}) \tag{8}$$

Detection Head Supervision. Based on the temporal-aware feature aggregation module, we obtain the temporal-aware fusion feature map, and we utilize the modified CenterPoint [36] detection network on this feature map to regress the target position. Following [10], we first voxelize the feature maps of each point by the averaging operation, and then use a stack of 3D convolutions to aggragte the features in the volumetric space. Next, we obtain the bird's eye view (BEV) feature map along the z-axis by max pooling operation. Finally, we aggregate the feature map using a stack of 2D convolution on the BEV feature map and use three different heads to regress the target position. Specifically, the three heads are 2D-center head, offset&rotation head, and z-axis head. We will use three losses to constrain them separately, and the details of the design can be found in [10], where we denote them as $\mathcal{L}_{detect}$.

The final loss function $\mathcal{L}$ is obtained by simply adding the two terms:

$$\mathcal{L} = \mathcal{L}_{score} + \mathcal{L}_{detect}. \tag{9}$$

4 Experiments

4.1 Experimental Settings

Implementation Details. Following [25], we randomly sample $N = 512$ for template point cloud P^t and $M = 1024$ for search area point cloud, and sample

$N = 512$ for scale-expand template point cloud in the template scoring module. We set the size of the template set $k = 8$. In our attention enhancement module, we employ the number of iterations $n = 2$ and the number of heads of multi-heads attention $m = 4$. And for the temporal-aware feature aggregation module, we use the GRU network to associate k-temporal features, which can be defined and used via *torch.nn.GRUCell* in PyTorch [21]. We use the Adam [16] optimizer to update the network's parameters, and set the learning rate from initial 0.001 decayed by 0.2 every 6 epochs. The network will converge by training for about 30 epochs. We implement our model with PyTorch [21] and deploy all experiments on a server with TITAN RTX GPU and Intel i5 2.2GHz CPU.

Datasets. We use the KITTI [7], nuScenes [3] and waymo open [29] datasets for our experiments. Among them, the KITTI dataset has 21 video sequences. Following the P2B [25], we split the sequences into three parts: sequences 0–16 for training, 17–18 for validation, and 19–20 for testing. In addition, the nuScenes dataset with 700 training video sequences and 150 validation video sequences. Since the nuScenes dataset only labels the ground truth in key frames, following V2B [10], we use the official toolkit to interpolate the corresponding labels for the unlabeled frames. For the waymo open dataset (WOD), it is currently one of the largest outdoor point cloud datasets. Pang *et al.* [20] established a 3D SOT benchmark based on the WOD, which we will use directly.

Evaluation Metrics. Following [25], we use *Success* and *Precision* criteria to measure the model performance. Specifically, *Success* is used to measure the IoU between the predicted BBox and the GT BBox, and *Precision* is used to measure the AUC (Area Under Curve) of the distance between the predicted BBox and the GT BBox centers from 0 to 2 m.

Data Pre-processing. The input to the core network consists of a template point cloud set, search area point cloud, and an additional scale-expand template point cloud is required for the template scoring module. For training, we enlarge the GT BBox of the previous frame by 2 m and plus random offset and crop the search area from the current frame point cloud. In order to build the template set, we will randomly select k frame from the first frame of the current tracking sequence to the current frame, a small random noise is added to the GT BBox corresponding to each frame, and it is concatenated with the point cloud of the first frame as the template set. In addition, we will enlarge the BBox by 1 m as the scale-expand template point cloud. For testing, we enlarge the predicted BBox of the previous frame by 2 m in the current frame and collect the points inside to generate the search area. The template set of the current frame will evaluate all the tracking results of the historical frame through the template scoring module, and select the k frames with the highest scores to be spliced with the first frame respectively. The scale-expand template corresponding to each template will be obtained by expanding the respective BBox by 1 m. For simplicity, in the subsequent discussion we name our **T**emporal-**A**ware Siamese **T**racker as **TAT**.

Table 1. The performance of different methods on the KITTI and nuScenes datasets. **Bold** and <u>underline</u> denote the best performance and the second-best performance of the compared methods, respectively. "Mean" denotes the average results of four categories.

	Metrics	*Success*					*Precision*				
	Category	Car	Pedestrian	Van	Cyclist	Mean	Car	Pedestrian	Van	Cyclist	Mean
	Frame Num	6424	6088	1248	308	14068	6424	6088	1248	308	14068
KITTI	SC3D [8]	41.3	18.2	40.4	41.5	31.2	57.9	37.8	47.0	70.4	48.5
	P2B [25]	56.2	28.7	40.8	32.1	42.4	72.8	49.6	48.4	44.7	60.0
	LTTR [5]	65.0	33.2	35.8	66.2	48.7	77.1	56.8	45.6	89.9	65.8
	BAT [37]	60.5	42.1	52.4	33.7	51.2	77.7	70.1	67.0	45.4	72.8
	PTT [27]	67.8	44.9	43.6	37.2	55.1	81.8	72.0	52.5	47.3	74.2
	PTTR [38]	65.2	<u>50.9</u>	52.5	65.1	58.4	77.4	<u>81.6</u>	61.8	90.5	77.8
	V2B [10]	70.5	48.3	50.1	40.8	58.4	81.3	73.5	58.0	49.7	75.2
	STNet [11]	<u>72.1</u>	49.9	<u>58.0</u>	<u>73.5</u>	<u>61.3</u>	**84.0**	77.2	**70.6**	<u>93.7</u>	<u>80.1</u>
	TAT (ours)	**72.2**	**57.4**	**58.9**	**74.2**	**64.7**	<u>83.3</u>	**84.4**	<u>69.2</u>	**93.9**	**82.8**
	Category	Car	Pedestrian	Truck	Bicycle	Mean	Car	Pedestrian	Truck	Bicycle	Mean
	Frame Num	15578	8019	3710	501	27808	15578	8019	3710	501	27808
nuScenes	SC3D [8]	23.9	13.6	28.9	16.1	21.5	26.2	15.1	26.4	18.5	22.9
	P2B [25]	32.7	18.1	28.1	18.5	27.6	35.5	25.0	25.8	**23.9**	30.9
	BAT [37]	32.9	<u>19.6</u>	29.4	17.8	28.3	35.3	**30.3**	26.2	21.9	32.4
	V2B [10]	<u>36.5</u>	19.4	<u>30.8</u>	<u>18.9</u>	<u>30.5</u>	<u>39.0</u>	26.9	<u>28.6</u>	22.0	<u>33.8</u>
	TAT (ours)	**36.8**	**20.7**	**32.2**	**19.0**	**31.2**	**39.6**	<u>29.5</u>	**28.9**	<u>22.6</u>	**34.9**

4.2 Result

Evaluation on KITTI Dataset. Following [25], we select the four most representative categories from all object categories in the KITTI dataset [7] for our experiments, including Car, Pedestrian, Van and Cyclist. We compare our method with the previous state-of-art approaches [5,8,10,11,25,27,37,38]. Each of these methods have published their results on KITTI, and we use them directly. As shown at the top of Table 1, our method outperforms other methods in most metrics. For the mean result of the four categories, our method can improve the *Success* from 61.3(STNet) to 64.7(TAT) and *Precision* from 80.1(STNet) to 82.8(TAT), which boosted by 5.5% and 3.4%, respectively.

Evaluation on nuScenes Dataset. We compare our method with the most typical four Siamese trackers [8,10,25,37] on the Car, Pedestrian, Truck and Bicycle categories of nuScenes dataset [3]. Following [10], since the nuScenes dataset is only labeled on key frames, we only report the performance evaluation on key frames. Compared with the KITTI dataset, the scenes of the nuScenes dataset are more complex and diverse, and the point cloud is more sparse, which greatly increases the challenge of the 3D SOT task. Nonetheless, as shown at the bottom of Table 1, our method can still achieve the best performance on the mean results of the four categories.

Evaluation on Generalization Ability. To evaluate the generalization ability of the model, we directly use the model trained on the corresponding classes of

Table 2. The performance of different methods on the waymo open dataset. Each category is divided into three levels of difficulty: Easy, Medium and Hard. "Mean" denotes the average results of three difficulty.

	Category	Vehicle				Pedestrian			
	Split	Easy	Medium	Hard	Mean	Easy	Medium	Hard	Mean
	Frame Num	67832	61252	56647	185731	85280	82253	74219	241752
Success	P2B [25]	57.1	52.0	47.9	52.6	18.1	17.8	17.7	17.9
	BAT [37]	61.0	53.3	48.9	54.7	19.3	17.8	17.2	18.2
	V2B [10]	64.5	55.1	52.0	57.6	27.9	22.5	20.1	23.7
	TAT (ours)	**66.0**	**56.6**	**52.9**	**58.9**	**32.1**	**25.6**	**21.8**	**26.7**
Precision	P2B [25]	65.4	60.7	58.5	61.7	30.8	30.0	29.3	30.1
	BAT [37]	68.3	60.9	57.8	62.7	32.6	29.8	28.3	30.3
	V2B [10]	71.5	63.2	62.0	65.9	43.9	36.2	33.1	37.9
	TAT (ours)	**72.6**	**64.2**	**62.5**	**66.7**	**49.5**	**40.3**	**35.9**	**42.2**

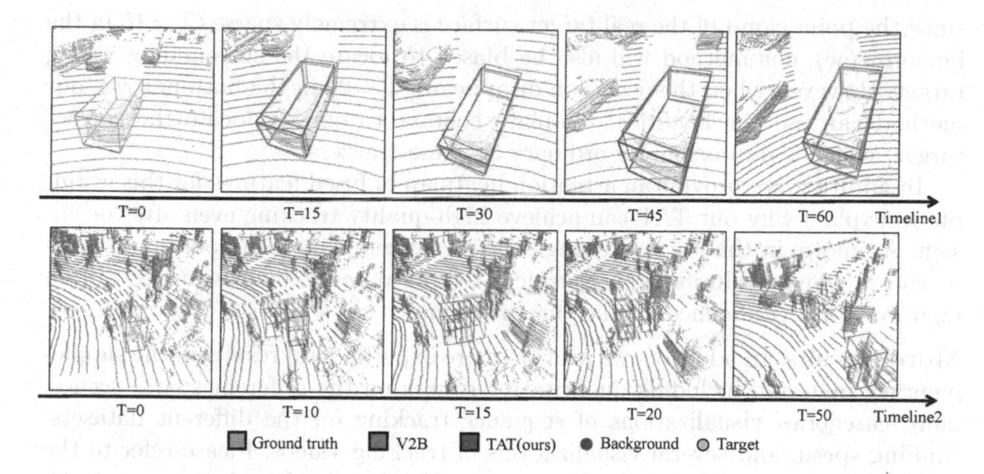

Fig. 4. The sequence tracking visualization of V2B and our TAT on the KITTI dataset. We color the GT BBoxes in green, while the BBoxes predicted by TAT and V2B are colored red and skyblue, respectively. In addition, we mark the points of target object in orange for better identification from the background. (Color figure online)

the KITTI dataset to evaluate its tracking performance on the WOD. Among them, the corresponding categories between WOD and KITTI are Vehicle → Car, Pedestrian → Pedestrian respectively. As shown in the Table 2, our method still shows excellent performance, which verifies that our model still has an advantage in generalization ability.

Visualization. As shown in Fig. 4, we show the visualization results of sequence tracking of car and pedestrian on the KITTI dataset. For the car, our method can achieve more precise localization. For the pedestrian, the existing Siamese trackers (such as V2B) are easy to match incorrectly when multiple pedestrians are adjacent or occluded. However, our method can accurately localize the target object among multiple candidates in complex scenes. In the case of occlusion,

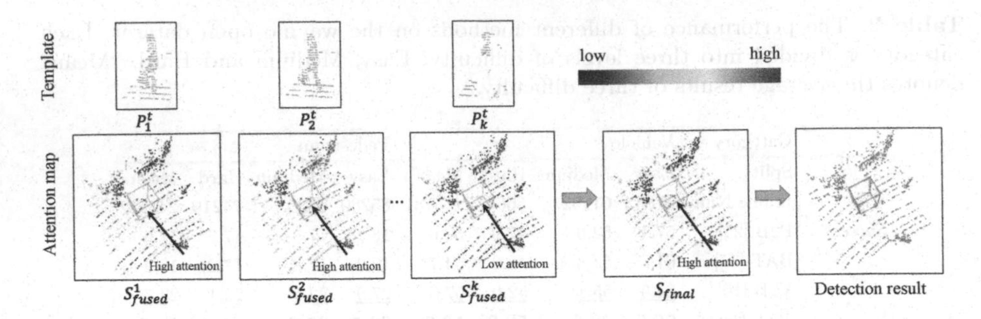

Fig. 5. The attention heatmap visualization of target-specific features, using historical templates to alleviate low-quality fusion features brought by recent low-quality templates, so that high-quality detection can still be achieved.

since the point cloud of the real target surface is extremely sparse (T = 15 in the bottom row), our method will also be biased to locate the surrounding wrong target. However, when the occlusion disappears (T = 20 in the bottom row), our method can use the historical template context to quickly locate the correct target, which is impossible for ordinary Siamese trackers.

In addition, we provide an attention heatmap of fused features for this example to explain why our TAT can achieve high-quality tracking even after occlusion. As shown in the Fig. 5, although the recent template is of poor quality due to the previous occlusion, the historical temporal context allows us to obtain high matching confidence for the current frame.

More Results. In addition, we provide more experimental results in the supplementary material, including quantitative results on the different sparse scenes, more categories visualizations of sequence tracking on the different datasets, running speed, and several visualizations of tracking videos. Please refer to the supplementary material for more experimental results and analyses.

4.3 Ablation Study

In this section, we design a rich ablation study to validate our proposed module and the effect of some hyperparameters on the results. We will simultaneously conduct experiments on the two main categories of car and pedestrian to provide comprehensive and reliable ablation study results.

Different Collection Strategies for Template Set. As we discussed earlier, the main purpose of introducing the template set is to use the historical successful tracking results to optimize the recent low-quality tracking. As shown at the bottom of Table 3, take the car category as an example, the performance of *Closest* sample method is significantly reduced by 2.6/3.2 (from 72.2/83.3 to 69.6/80.1), this further validates the effectiveness of our method for this situation. And compared with *Random* sample method, our proposed template score selector can bring performance gains of 0.7/0.6 (from 71.5/82.7 to 72.2/83.3) points.

Table 3. Ablation study on different template collection strategies and components. CSTS: Collection Strategies for Template Set. TFEM: Temporal Feature Enhancement Module. TFAM: Temporal-aware Feature Aggregation Module.

CSTS			Components		Car		Pedestrian	
Random	*Closest*	*Score*	TFEM	TFAM	*Success*	*Precision*	*Success*	*Precision*
		√			68.1 (-4.1)	79.3 (-4.0)	51.4 (-6.0)	78.1 (-6.3)
		√	√		69.0 (-3.2)	80.6 (-2.7)	52.1 (-5.3)	81.6 (-2.8)
		√		√	70.2 (-2.0)	81.5 (-1.8)	54.4 (-3.0)	83.0 (-1.4)
√			√	√	71.5 (-0.7)	82.7 (-0.6)	55.3 (-2.1)	83.3 (-1.1)
	√		√	√	69.6 (-2.6)	80.1 (-3.2)	52.7 (-4.7)	81.8 (-2.6)
		√	√	√	**72.2**	**83.3**	**57.4**	**84.4**

Temporal-Aware Components. The temporal-aware components consist of temporal feature enhancement module and temporal-aware feature aggregation module. For the temporal feature enhancement module, the dimensions of the input and output features are consistent, and we can directly remove it to verify the effectiveness of the module. For the temporal-aware feature aggregations module, we use the RNN-based network to associate the k fusion feature maps of different time series. In addition, we can also ignore the temporal relationship between templates, and use an MLP to obtain the final fusion feature map after concat k feature maps. As shown in the upper part of Table 3, take the car category as an example, these two modules can bring performance improvement of 2.0/1.8 (from 70.2/81.5 to 72.2/83.3) and 3.2/2.7 (from 69.0/80.6 to 72.2/83.3) respectively. These ablation experiments effectively verify that our proposed module can utilize the temporal context more effectively.

More Ablation Studies. We also investigate the effect of different numbers of templates and hyperparameters of attention modules on the results. Based on the experiments, the number of template sets is 8, the number of heads for multi-head attention is 4, and the number of iterations of the template enhancement module is 2 is a good experimental setting. For more experimental data and analysis, please refer to Supplementary Materials.

5 Conclusions

In this paper, we proposed a simple yet powerful Temporal-aware Siamese tracking framework, where we introduce the temporal feature enhancement module and the temporal-aware feature aggregation module into the architecture of the Siamese 3D tracking methods. Our method optimizes the current tracking process by correlating multi-frame template set and making full use of temporal context. In addition, we designed a simple yet effective 3DIoU-aware template selector to build a high-quality temporal template set. Our proposed method significantly improves the performance of 3D SOT on several benchmark datasets.

Acknowledgements. This work was supported by the National Science Fund of China (Grant No. 61876084).

References

1. Bertinetto, L., Valmadre, J., Henriques, J.F., Vedaldi, A., Torr, P.H.S.: Fully-convolutional Siamese networks for object tracking. In: Hua, G., Jégou, H. (eds.) ECCV 2016. LNCS, vol. 9914, pp. 850–865. Springer, Cham (2016). https://doi.org/10.1007/978-3-319-48881-3_56
2. Bibi, A., Zhang, T., Ghanem, B.: 3D part-based sparse tracker with automatic synchronization and registration. In: CVPR (2016)
3. Caesar, H., et al.: nuScenes: a multimodal dataset for autonomous driving. In: CVPR (2020)
4. Comport, A.I., Marchand, É., Chaumette, F.: Robust model-based tracking for robot vision. In: IROS (2004)
5. Cui, Y., Fang, Z., Shan, J., Gu, Z., Zhou, S.: 3D object tracking with transformer. arXiv preprint arXiv:2110.14921 (2021)
6. Fang, Z., Zhou, S., Cui, Y., Scherer, S.: 3D-SiamRPN: an end-to-end learning method for real-time 3d single object tracking using raw point cloud. IEEE Sens. J. (2020)
7. Geiger, A., Lenz, P., Urtasun, R.: Are we ready for autonomous driving? The Kitti vision benchmark suite. In: CVPR (2012)
8. Giancola, S., Zarzar, J., Ghanem, B.: Leveraging shape completion for 3D Siamese tracking. In: CVPR (2019)
9. Guo, Q., Feng, W., Zhou, C., Huang, R., Wan, L., Wang, S.: Learning dynamic Siamese network for visual object tracking. In: ICCV (2017)
10. Hui, L., Wang, L., Cheng, M., Xie, J., Yang, J.: 3D Siamese voxel-to-BEV tracker for sparse point clouds. In: NeurIPS (2021)
11. Hui, L., Wang, L., Tang, L., Lan, K., Xie, J., Yang, J.: 3D Siamese transformer network for single object tracking on point clouds. In: Avidan, S., Brostow, G., Cissé, M., Farinella, G.M., Hassner, T. (eds) Computer Vision – ECCV 2022. ECCV 2022. Lecture Notes in Computer Science, vol. 13662, pp. 293–310. Springer, Cham (2022). https://doi.org/10.1007/978-3-031-20086-1_17
12. Jiang, H., Lan, K., Hui, L., Li, G., Xie, J., Yang, J.: Point cloud registration-driven robust feature matching for 3d siamese object tracking. arXiv preprint arXiv:2209.06395 (2022)
13. Katharopoulos, A., Vyas, A., Pappas, N., Fleuret, F.: Transformers are RNNs: fast autoregressive transformers with linear attention. In: ICML (2020)
14. Kelly, A.: A 3d state space formulation of a navigation Kalman filter for autonomous vehicles. Carnegie-Mellon Univ Pittsburgh Pa Robotics Inst, Technical report (1994)
15. Kim, A., Ošep, A., Leal-Taixé, L.: EagerMOT: 3D multi-object tracking via sensor fusion. In: ICRA (2021)
16. Kingma, D.P., Ba, J.: Adam: a method for stochastic optimization. In: ICLR (2015)
17. Lang, A.H., Vora, S., Caesar, H., Zhou, L., Yang, J., Beijbom, O.: PointPillars: fast encoders for object detection from point clouds. In: CVPR (2019)
18. Luo, C., Yang, X., Yuille, A.: Exploring simple 3D multi-object tracking for autonomous driving. In: ICCV (2021)

19. Luo, W., Yang, B., Urtasun, R.: Fast and furious: real time end-to-end 3D detection, tracking and motion forecasting with a single convolutional net. In: CVPR (2018)
20. Pang, Z., Li, Z., Wang, N.: Model-free vehicle tracking and state estimation in point cloud sequences. In: IROS (2021)
21. Paszke, A., et al.: Pytorch: an imperative style, high-performance deep learning library. In: NeurIPS (2019)
22. Pieropan, A., Bergström, N., Ishikawa, M., Kjellström, H.: Robust 3D tracking of unknown objects. In: ICRA (2015)
23. Qi, C.R., Litany, O., He, K., Guibas, L.J.: Deep Hough voting for 3D object detection in point clouds. In: ICCV (2019)
24. Qi, C.R., Yi, L., Su, H., Guibas, L.J.: PointNet++: deep hierarchical feature learning on point sets in a metric space. In: NeurIPS (2017)
25. Qi, H., Feng, C., Cao, Z., Zhao, F., Xiao, Y.: P2b: Point-to-box network for 3D object tracking in point clouds. In: CVPR (2020)
26. Scheidegger, S., Benjaminsson, J., Rosenberg, E., Krishnan, A., Granström, K.: Mono-camera 3D multi-object tracking using deep learning detections and PMBM filtering. In: IV (2018)
27. Shan, J., Zhou, S., Fang, Z., Cui, Y.: PTT: Point-track-transformer module for 3D single object tracking in point clouds. In: IROS (2021)
28. Shi, S., Wang, X., Li, H.: PointrCNN: 3D object proposal generation and detection from point cloud. In: CVPR (2019)
29. Sun, P., et al.: Scalability in perception for autonomous driving: WAYMO open dataset. In: CVPR (2020)
30. Tao, R., Gavves, E., Smeulders, A.W.: Siamese instance search for tracking. In: CVPR (2016)
31. Vaswani, A., et al.: Attention is all you need. arXiv preprint arXiv:1706.03762 (2017)
32. Wang, L., Hui, L., Xie, J.: Facilitating 3D object tracking in point clouds with image semantics and geometry. In: PRCV (2021)
33. Wang, N., Zhou, W., Wang, J., Li, H.: Transformer meets tracker: exploiting temporal context for robust visual tracking. In: CVPR (2021)
34. Weng, X., Kitani, K.: A baseline for 3D multi-object tracking. arXiv preprint arXiv:1907.03961 (2019)
35. Wu, H., Han, W., Wen, C., Li, X., Wang, C.: 3D multi-object tracking in point clouds based on prediction confidence-guided data association. IEEE Trans. Intell. Transp. Syst. **23**, 5668–5677 (2021)
36. Yin, T., Zhou, X., Krahenbuhl, P.: Center-based 3d object detection and tracking. In: CVPR (2021)
37. Zheng, C., et al.: Box-aware feature enhancement for single object tracking on point clouds. In: ICCV (2021)
38. Zhou, C., et al.: PTTR: relational 3D point cloud object tracking with transformer. In: CVPR (2022)
39. Zhu, Z., Wang, Q., Li, B., Wu, W., Yan, J., Hu, W.: Distractor-aware Siamese networks for visual object tracking. In: Ferrari, V., Hebert, M., Sminchisescu, C., Weiss, Y. (eds.) ECCV 2018. LNCS, vol. 11213, pp. 103–119. Springer, Cham (2018). https://doi.org/10.1007/978-3-030-01240-3_7

Neural Plenoptic Sampling: Learning Light-Field from Thousands of Imaginary Eyes

Junxuan Li(✉), Yujiao Shi, and Hongdong Li

Australian National University, Canberra, Australia
{junxuan.li,yujiao.shi,hongdong.li}@anu.edu.au

Abstract. The Plenoptic function describes light rays observed from any given position in every viewing direction. It is often parameterized as a 5-D function $L(x, y, z, \theta, \phi)$ for a static scene. Capturing all the plenoptic functions in the space of interest is paramount for Image-Based Rendering (IBR) and Novel View Synthesis (NVS). It encodes a complete light-field (*i.e.*, lumigraph) therefore allows one to freely roam in the space and view the scene from any location in any direction. However, achieving this goal by conventional light-field capture technique is expensive, requiring densely sampling the ray space using arrays of cameras or lenses. This paper proposes a much simpler solution to address this challenge by using only a small number of sparsely configured camera views as input. Specifically, we adopt a simple Multi-Layer Perceptron (MLP) network as a universal function approximator to learn the plenoptic function at every position in the space of interest. By placing virtual viewpoints (dubbed 'imaginary eyes') at thousands of randomly sampled locations and leveraging multi-view geometric relationship, we train the MLP to regress the plenoptic function for the space. Our network is trained on a per-scene basis, and the training time is relatively short (in the order of tens of minutes). When the model is converged, we can freely render novel images. Extensive experiments demonstrate that our method well approximates the complete plenoptic function and generates high-quality results.

1 Introduction

Image-Based Rendering (IBR) for view synthesis is a long-standing problem in the field of computer vision and graphics. It has a wide range of important applications, *e.g.*, robot navigation, film industry, AR/VR applications. The plenoptic function, introduced by Adelson *et al.* [1], offers an ultimate solution to this novel view synthesis problem. The plenoptic function captures the visual appearances of a scene viewed from any viewing direction (θ, ϕ) and at any location (x, y, z). Once a complete plenoptic function (i.e. the light-field) for the entire space is available, one can roam around the space and synthesize free-viewpoint images simply by sub-sampling the plenoptic light-field.

Supplementary Information The online version contains supplementary material available at https://doi.org/10.1007/978-3-031-26319-4_3.

L. Wang et al. (Eds.): ACCV 2022, LNCS 13841, pp. 36–54, 2023.
https://doi.org/10.1007/978-3-031-26319-4_3

To model the plenoptic function, the best-known methods in the literature are the light field rendering and the lumigraph [20,25]. These approaches interpolate rays instead of scene points to synthesize novel views. However, they require the given camera positions to be densely or regularly sampled or restrict the target image to be a linear combination of source images. Unstructured light-field/lumigraph methods [5,13] were proposed to address this limitation; they do so by incorporating geometric reconstruction with light ray interpolation.

This paper introduces a novel solution for plenoptic field sampling from a few and often sparse and unstructured multi-view input images. Since a plenoptic function is often parameterized by a 5D function map, we use a simple Multi-Layer Perceptron (MLP) network to learn such functional map: the MLP takes a 5D vector as input and outputs an RGB color measurement, *i.e.*, $\mathbb{R}^5 \rightarrow \mathbb{R}^3$. However, capturing the complete plenoptic function for a scene remains a major challenge in practice. It requires densely placing many physical cameras or moving a camera (or even a commercial light-camera) to scan *every point and in every direction.*

To address this challenge, this paper uses an MLP to *approximate* the plenoptic function (*i.e.*, the entire light field), by placing thousands of virtual cameras (*i.e.*, imaginary eyes) during the network training. We use the available physical camera views, however a few and sparsely organized, to provide multi-view geometry constraints as the self-supervision signal to supervise the training of the MLP networks. We introduce *proxy-depth* as a bridge to ensure that the multi-view geometry relationship is well respected during the training process. Those "imaginary eyes" is sampled randomly throughout the space following a uniform distribution. We use proxy-depth to describe the estimated depth by the visual similarity among input images. Once the proxy-depth of a virtual ray is determined, we can retrieve candidate colors from input images and pass them to a color blending network to determine the real color.

2 Related Work

Conventional View Synthesis. Novel view synthesis is a long-standing problem in the field of computer vision and graphics [9,14,46]. Conventional methods use image colors or handcrafted features to construct correspondences between the views [16,42]. With the advance of deep networks, recent approaches employ neural networks to learn the transformation between input and target views implicitly [15,38,40,57,70]. In order to explicitly encode the geometry guidance, several specific scene representations are proposed, such as Multi-Plane Images (MPI) [17,35,55,60,69], and Layered Depth Images (LDI) [48,50,61]. Some Image-Based Rendering (IBR) techniques [7,11,21,42,44,45,49,59] warp input view images to a target viewpoint according to the estimated proxy geometry, and then blend the warped pixels to synthesize a novel view.

Panorama Synthesis. Zheng *et al.* [68] propose a layered depth panorama (LDP) to create a layered representation with a full field of view from a sparse set of images taken by a hand-held camera. Bertel *et al.* [4] investigate two blending methods for interpolating novel views from two nearby views, one is a linear blending, and the other is a view-dependent flow-based blending. Serrano *et al.* [47] propose to synthesize new views from a fixed viewpoint 360° video. Huang *et al.* [23] employ a typical depth-warp-refine procedure in synthesizing new views. They estimate the depth map for each input image and reconstruct the 3D point cloud by finding correspondences between input images using hand-crafted features. They then synthesize new views from the reconstructed point cloud. With panorama synthesis, scene roaming and lighting estimation [19,26] is possible for AR/VR applications.

Plenoptic Modeling. Early light-field/lumigraph methods [20,25] reduce the 5D representation (position (x, y, z) and direction (θ, ϕ)) of the plenoptic function to 4D ((u, v, s, t), intersection between two image planes). They do not require scene geometry information, but either require the camera grid is densely and regularly sampled, or the target viewing ray is a linear combination of the source views [6,30]. For unstructured settings, a proxy 3D scene geometry is required to be combined with light-field/lumigraph methods for view synthesis [5,13]. Recent methods [24,56,64,65] applied learning methods to improve light field rendering.

Neural Rendering. Deep networks have also demonstrated their capability of modeling specific scenes as implicit functions [31,32,36,39,41,52–54,58,66,67]. They encapsulate both the geometry and appearance of a scene as network parameters. They take input as sampled points along viewing rays and output the corresponding color and density values during the inference stage. The target image is then rendered from the sampled points by volume rendering techniques [33]. The denser the sampled points, the higher quality of rendered images. However, densely sampling points along viewing rays would significantly increase the rendering time, prohibiting interactive applications to real-world scenarios.

Neural Radiance Fields. Our idea of using MLP to learn light-field is similar Neural radiance fields [10,22,29,36], which estimate the radiance emitted at 3D location by a network; a recent work, DoNeRF [37] speed up the rendering process by only querying a few samples around the estimated depth; but they requires ground-truth depth map for each viewing ray during training; Fast-NeRF [18] is proposed to accelerate the rendering speed during inference by caching pre-sampled scene points. IBRNet [63] is another recent NeRF-based work that utilizes neighboring features and a transformer for image rendering. MVSNeRF [8] constructed a cost volume from nearby views and use these features to help regressing the neural networks. LFNs [51] proposed a neural implicit representation for 3D light field. However, their method is limited to simple objects and geometries.

There are key differences between ours and previous works. Previous methods focused on estimating the lights emitted at every location, in any direction, within a bounded volumetric region, often enclosing the 3D scene or 3D object of interest. In contrast, our method focuses on estimating all the light rays observed at any point in space, coming in any direction. In essence, our formulation is not only close to, but precisely is, the plenoptic function that Adelson and Bergen had

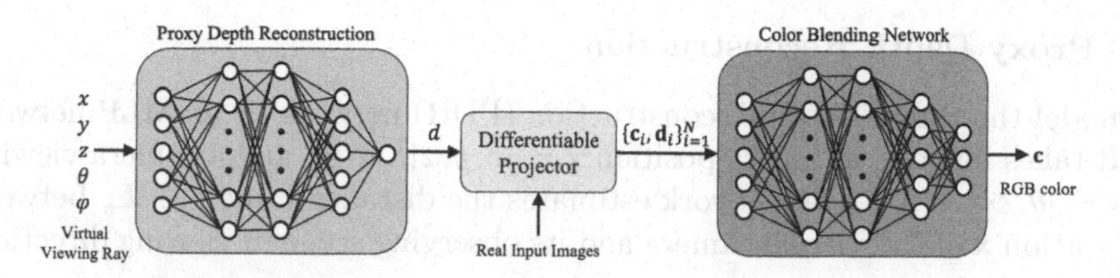

Fig. 1. The overall pipeline of the proposed framework. Our framework includes a proxy depth reconstruction (PDR) model to determine the depth of a virtual viewing ray, a differentiable ray tracer to retrieve corresponding colors from real input images, and a color blending network (CBNet) to recover the RGB color information.

contrived. In fact, in principle, our formulation can be extended to the original 7D plenoptic function by adding time and wavelength as new dimensions [3, 27,28]. Our method also offers a computational advantage over NeRF. Namely, when the model has been well-approximated, we can directly display the network output as rendered images without sampling points along viewing rays and then rendering them in a back-to-front order. Our model will significantly accelerate the rendering speed and facilitate interactive applications.

3 Neural Plenoptic Sampling

A complete plenoptic function corresponds to the holographic representation of the visual world. It is originally defined as a 7D function $L(x, y, z, \theta, \phi, \lambda, t)$ which allows reconstruction of every possible view (θ, ϕ) from any position (x, y, z), at any time t and every wavelength λ. McMillan and Bishop [34] reduce its the dimensionality from 7D to 5D by ignoring the time and wavelength for the purpose of static scene view synthesis. By restricting the viewpoints or the object inside a box, light field [25] and lumigraph [20] approaches reduce the dimensionality to four. Without loss of generality, this paper uses original 5D representations $L(x, y, z, \theta, \phi)$ for plenoptic function and focuses on the scene representation at a fixed time.

We model the plenoptic function by an MLP. However, a brute-force training of a network mapping from *viewing* position and direction to RGB colors is infeasible. The observed images only have a partial coverage of the input space. By using the above training method, the model may fit well on the observed viewpoints, but also generates highly-blurred images on the non-observed regions. Our experiments in Fig. 5 demonstrate this situation.

To address this problem, we introduce an Imaginary Eye Sampling (IES) method to fully sample the target domain. We evaluate a proxy depth to provide self-supervision by leveraging photo-consistency among input images. Our method firstly outputs a proxy depth for a virtual viewing ray from the imaginary eye we randomly placed in the scene. Then, we retrieve colors from input views by a differentiable projector using this depth. Lastly, the colors pass through a color blending network to generate the real color. Figure 1 depicts the overall pipeline of our framework.

3.1 Proxy Depth Reconstruction

We model the Proxy Depth Reconstruction (PDR) network by an MLP network F_Θ. It takes input as a camera position $\mathbf{x} = [x, y, z]^T \in \mathbb{R}^3$ and a camera viewing ray $\mathbf{v} = [\theta, \phi]^T \in \mathbb{R}^2$. The network estimates the distance value $d \in \mathbb{R}_+$ between the location $\mathbf{x}$ of the virtual camera and its observing scene in viewing direction,

$$d = F_\Theta(\mathbf{x}, \mathbf{v}), \tag{1}$$

where Θ represents the trainable network parameters.

We use a similar MLP structure from NeRF [36] to parameterize the neural plenoptic modeling. The difference is that NeRF approximates the emitting colors and transparency on the scene objects location, while our PDR model estimates the distance between the scene objects and observing cameras along the viewing direction.

3.2 Imaginary Eye Sampling

Since our purpose is to move around the scene and synthesize new views continuously, we need to sample the input space for the network training densely. However, in general, the camera locations of input images are sparsely sampled. The observed images only cover partial regions of such an input space.

To address this problem, we propose an Imaginary Eye Sampling (IES) strategy. We place thousands of imaginary eyes (virtual cameras) in the space of interest. Those imaginary eyes are randomly generated in the space to allow dense sampling of the plenoptic input space. By doing so, we are able to approximate a whole complete plenoptic function.

Here, note that we do not have ground-truth depths for supervision, even for real-observed images (viewing rays). In order to provide training signals, we propose a self-supervision method by leveraging photo-consistency among real input images.

3.3 Self-supervision via Photo-Consistency

Given a virtual camera at a random location $\mathbf{x}$ and a viewing direction $\mathbf{v}$, our network predicts a depth d between the observed scene point and the input camera location. The world coordinate $\mathbf{w}$ of the scene point is then computed as

$$\mathbf{w} = \mathbf{x} + d\mathbf{v}. \tag{2}$$

When the estimated depth d is at the correct value, the colors of its projected pixels on real observed images should be consistent with each other. Hence, we then use a differentiable projector $T(\cdot)$ to find the projected pixel of this scene point at real camera image planes. Denote $\mathbf{P}_i$ as the projection matrix of real camera i. The projected image coordinate of a 3D point $\mathbf{w}$ is computed as $[u_i, v_i, 1]^T = \mathbf{P}_i\mathbf{w}$. Our projector then uses bilinear interpolation to fetch information (*e.g.*, color) from the corresponding real images.

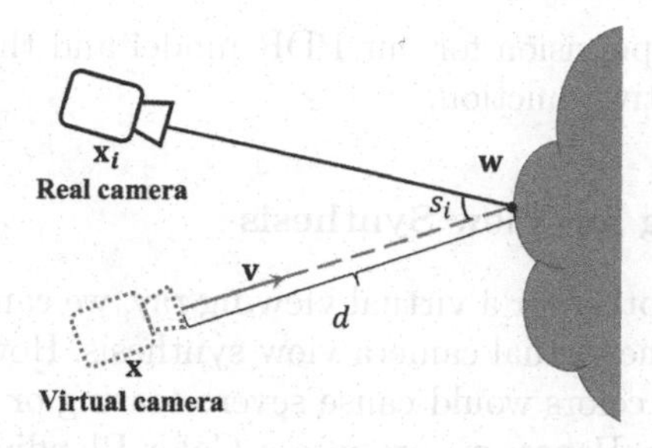

Fig. 2. For a virtual camera position $\mathbf{x}$ and viewing direction $\mathbf{v}$, we estimate a depth d between the scene point $\mathbf{w}$ and the camera location $\mathbf{x}$. By reprojecting the scene point to real cameras, we retrieve the color $\mathbf{c}_i$ and high-level feature $\mathbf{f}_i$ from the observed images. The cosine distance (angle) s_i between the virtual viewing direction and real viewing direction determine the influence of corresponding real cameras when calculating the photometric consistency.

By computing the photo-consistency (similarity) among the retrieved colors, we can measure the correctness of the estimated depth. In practice, we argue that only using the colors of the retrieved pixels is not accurate enough for this measurement because it cannot handle textureless and reflective regions. To increase the representative and discriminative ability, we propose to retrieve colors as well as high-level features from real input images for the photo-consistency measure.

Denote $\mathbf{f}_i$ and $\mathbf{c}_i$ as the retrieved features and colors from input camera i, respectively. The photo-consistency among all input cameras is defined as

$$\mathcal{L}_d = \sum_{i=1}^{N} s_i \left(\|\mathbf{c}^{\text{top}_k} - \mathbf{c}_i\|_1 + \lambda \|\mathbf{f}^{\text{top}_k} - \mathbf{f}_i\|_1 \right), \tag{3}$$

where $\|\cdot\|_1$ denotes the L_1 distance, N is the number of real input cameras, λ is the balance of the influence between color difference and the feature difference, s_i is the normalized weight assigned to each real camera i, and it is determined by the angle difference (cosine distance) between the virtual camera viewing ray $(\mathbf{w} - \mathbf{x})$ and the real camera viewing ray $(\mathbf{w} - \mathbf{x_i})$. Figure 2 illustrates the situation. Mathematically, it is expressed as

$$s_i = \frac{\cos(\mathbf{w} - \mathbf{x}, \mathbf{w} - \mathbf{x_i})}{\sum_{j=1}^{N} \cos(\mathbf{w} - \mathbf{x}, \mathbf{w} - \mathbf{x_j})}, \tag{4}$$

where $\cos(\cdot, \cdot)$ is the cosine of the angle spanned by the two vectors. The smaller of the angle between the virtual camera viewing ray and the real camera viewing ray, the larger s_i is. Given the weight for each input camera, the reference color $\mathbf{c}^{\text{top}_k}$ and feature $\mathbf{f}^{\text{top}_k}$ in Eq. 3 is computed as the average of top k retrieved colors and features

$$\mathbf{c}^{\text{top}_k} = \sum_{i \in \text{top}_k} \mathbf{c}_i / k, \qquad \mathbf{f}^{\text{top}_k} = \sum_{i \in \text{top}_k} \mathbf{f}_i / k. \tag{5}$$

We use Eq. 3 as the supervision for our PDR model and the network is trained to minimize this objective function.

3.4 Color Blending for View Synthesis

Given an estimated depth d for a virtual viewing ray, we can retrieve colors from real input images for the virtual camera view synthesis. However, a naive aggregation of the retrieved colors would cause severe tearing or ghosting artifacts in the synthesized images. Hence, we propose a Color Blending Network (CBNet) to blend the retrieved colors and tolerate the errors caused by inaccurate depths to synthesize realistic images.

In order to provide sufficient clues, we feed the direction differences between the reprojected real viewing rays (solid line in Fig. 2) and the virtual (target) viewing ray (dash line in Fig. 2) along with the retrieved colors to the color blending network. Formally, our CBNet is expressed as

$$\mathbf{c} = F_{\Phi}\left(\{\mathbf{c}_i, \mathbf{d}_i\}_{i=1}^{N}\right), \tag{6}$$

where Φ is the trainable parameter of the CBNet, $\mathbf{c}_i$ is the RGB color retrieved from the real camera i and $\mathbf{d}_i$ is the projection of the real viewing ray on the target virtual viewing ray, $\mathbf{c}$ is the estimated color of the virtual viewing ray. We employ a Pointnet network architecture for our CBNet. The supervision of our CBNet is the colors observed from real cameras, denoted as

$$\mathcal{L}_{\mathbf{c}} = \|\mathbf{c}^{*} - \mathbf{c}\|_1, \tag{7}$$

where $\mathbf{c}^{*}$ is the ground truth colors.

Unlike our PDR model, the CBNet is trained only on the observed images (viewing rays) since it needs the ground-truth color as supervision. Instead of remembering the color of each training ray, the CBNet is trained to learn a sensible rule for blending retrieved colors from real input views. Thus it is able to be generalized to unseen viewing rays. The PDR and the CBNet in our framework are trained separately. During the training of CBNet, we fix the model parameters of PDR to not destroy the learned patterns for the whole plenoptic space. For inference, a query viewing ray first passes through our PDR model to compute a depth value; its corresponding colors on real input views are then retrieved and fed into the CBNet to estimate the color information. Since it is a single feed-forward pass through the network, the rendering speed is rapid (less than one second when rendering a 1024 × 512 image).

4 Experiments

In this section, we conduct comprehensive experiments to demonstrate the effectiveness of the proposed algorithm. We use 360° panoramas captured by an omnidirectional camera for the plenoptic modeling, since its representation well

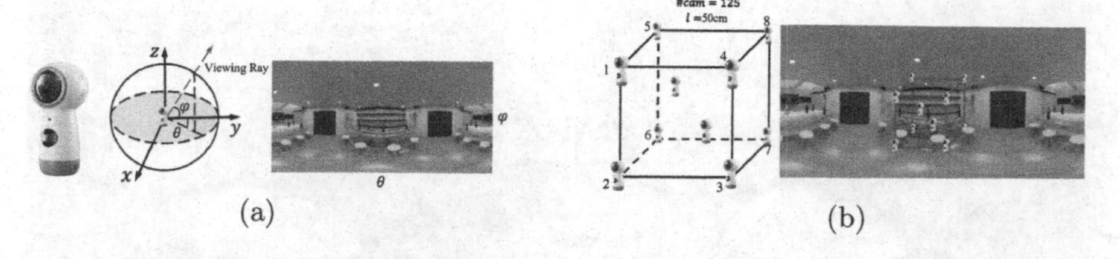

Fig. 3. (a) An illustration of an omni-directional camera and its captured light-field and a sample image. (b) An illustration of our camera arrangement for dataset generation. For each scene, we capture 125 omnidirectional images at different locations for evaluation. The cameras are positioned in a 50 × 50 × 50 centimeter volume (roughly) at the center of each scene.

Table 1. Quantitative comparison of our method and others given eight input views. Here, **bold** indicates the best results and underline denotes the second best results.

	Diningroom		Bar		Livingroom		Lounge		Average	
	PSNR↑	SSIM↑	PSNR↑	SSIM↑	PSNR↑	SSIM↑	PSNR↑	SSIM↑	PSNR↑	SSIM↑
FVS [44]	26.09	0.770	24.54	0.800	25.61	0.780	21.23	0.690	24.37	0.760
NeRF [36]	<u>37.54</u>	0.938	<u>33.95</u>	0.941	33.62	0.936	31.96	0.939	34.27	0.939
NeRF++ [67]	36.29	0.931	32.87	0.936	33.72	0.929	<u>34.05</u>	0.947	34.23	0.936
IBRNet [63]	**37.83**	0.953	**34.12**	<u>0.959</u>	33.39	0.941	32.35	0.953	<u>34.42</u>	0.952
Ours w/o imaginary eye	32.32	0.929	32.93	0.950	32.57	0.948	30.50	0.932	32.08	0.940
Ours w/o feature	36.03	<u>0.957</u>	33.47	0.954	<u>33.97</u>	<u>0.957</u>	32.17	<u>0.960</u>	33.91	<u>0.957</u>
Ours w/o weighting	32.69	0.931	29.57	0.903	30.81	0.919	29.18	0.925	30.56	0.920
Ours	36.62	**0.959**	33.86	**0.961**	**34.33**	**0.965**	**34.31**	**0.968**	**34.78**	**0.963**

aligns with the plenoptic function. We show an omnidirectional camera, its imaging geometry, and an example image in Fig. 3. The pixel coordinates of a 360° panorama correspond to the azimuth angle θ and the elevation angle ϕ of the viewing rays.

4.1 Datasets and Evaluations

Synthetic Dataset. When the plenoptic function has been correctly (approximately) modeled, we want to freely move across the space to synthesize new views. For the purpose of performance evaluation, we need to sample evaluation viewpoints densely in the space and their corresponding ground truth data. Hence, we propose to synthesize a dataset for our evaluation. Following recent novel view synthesis methods, we use SSIM and PSNR for the performance evaluation.

We use Blender [12] to synthesize images with freely moving camera viewpoints. Figure 3 shows the camera setting. Specifically, we randomly sample 125 points in a $50 \times 50 \times 50$ cm^3 volume within the space and synthesize corresponding omnidirectional images. The images are generated from four scenes, *i.e.*, "Bar", "Livingroom", "Lounge" and "Diningroom". We refer the readers to our qualitative comparisons for the visualization of sampled images from the four scenes. This evaluation set is adopted for all the experiments in this paper, although the input views and training methods might be changed.

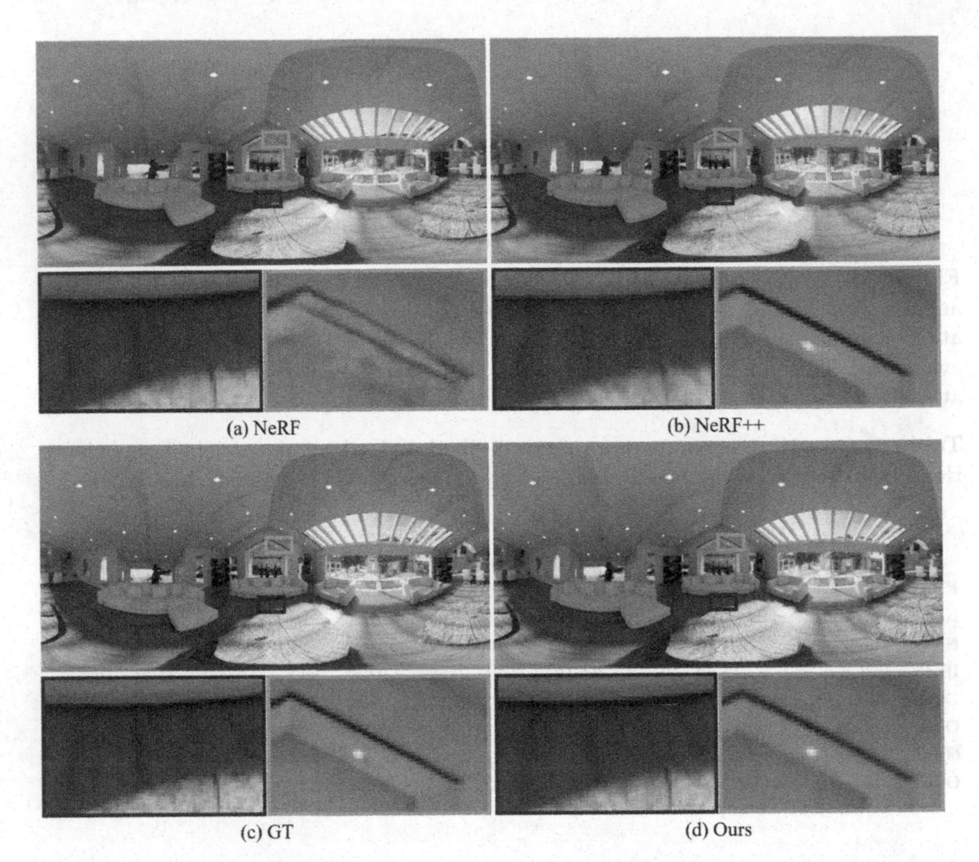

Fig. 4. Qualitative comparison with NeRF and NeRF++ on our generated scenes "Lounge". Our method generates sharper results than the comparison algorithms.

Real Dataset. To fully demonstrate the effectiveness of the proposed method, we also conduct experiments on real-world data. The real-world data we use are from [62], which sparsely captured two images per scene. We only provide qualitative results for visual evaluation, and interested readers are suggested to watch our supplementary video for more results.

4.2 Training Details

We train a separate plenoptic function for each scene. To approximate the sharp edge of real world objects and textures, our plenoptic function model usually has high frequency output in both the viewing rays and camera position. We encode the 5D input into Fourier features as the positional encoding [36] before feeding it into the proxy-depth reconstruction network. The PDR network F_Θ is designed following the structure of NeRF. It consists of 8 fully-connected (fc) layers with 256 hidden channels, and a ReLU activation layer follows each fc

Table 2. Quantitative comparison with volume-based method on two input views.

	Diningroom		Bar		Livingroom		Lounge	
	PSNR↑	SSIM↑	PSNR↑	SSIM↑	PSNR↑	SSIM↑	PSNR↑	SSIM↑
360SD-Net [62]	24.76	0.746	23.38	0.781	23.30	0.747	21.10	0.700
Ours (Vertical)	**27.54**	**0.910**	**27.29**	**0.918**	**28.20**	**0.907**	**26.13**	**0.888**
MatryODShka [2]	20.43	0.673	27.26	0.864	23.85	0.766	22.19	0.765
Ours (Horizontal)	**30.50**	**0.921**	**28.20**	**0.918**	**29.07**	**0.907**	**27.68**	**0.898**

Table 3. Training and testing time comparison with NeRF and NeRF++ given eight unstructured input views. The testing time is for rendering images with resolution of 512×1024.

	Training (hours)	Testing (seconds)
NeRF [36]	10	30
NeRF++ [67]	20	110
Ours	2	0.14

layer. We use the pretrained model from Shi *et al.* [49] as our feature extractor to compute photo consistency.

When training the MLP, we randomly sample a virtual camera at location $\mathbf{x}$ and draw an arbitrary viewing direction $\mathbf{v}$. Given this 5D input, the MLP estimates a proxy-depth d in the output, which is then self-supervised by the photometric consistency loss $\mathcal{L}_d$. The above network is end-to-end differentiable. Once we have sampled and trained the virtual camera domain thoroughly, the MLP for proxy-depth reconstruction is then frozen for the training of the color blending network later.

The CBNet takes a series color and direction $(\mathbf{c}_i, \mathbf{d}_i)$ to inference the output color of the plenoptic function. Its design follows the structure of the PointNet [43]. The observations from real cameras are firstly processed separately by three fc layers. Next, a max-pooling layer is applied to select the most salient features from them. We then employ a prediction layer to generate the color values $\mathbf{c}$.

In our experiments, we use 200k rays per iteration for the PDR network training, and 100k rays for the CBNet training. Our model is trained from scratch with an Adam optimizer. The learning rate is set to 5×10^{-4}. The PDR network takes around 30k iterations to converge, while the CBNet only takes 10k iterations. The complete model takes around one hour to converge in a NVIDIA RTX 3090 GPU.

4.3 Comparison with the State-of-the-Art

Comparison with NeRF Variants. We conduct experiments to compare with NeRF and its variants NeRF++ [67], IBRNet [63]. In this comparison, all of the methods take eight views as input. The quantitative evaluation results are

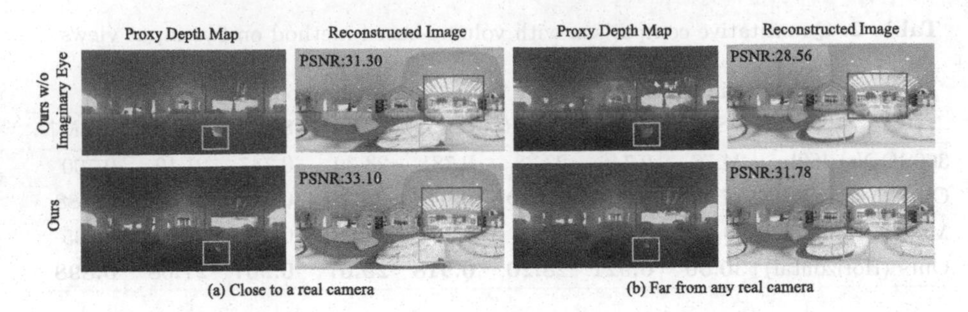

Fig. 5. Qualitative comparisons of our method w or w/o Imaginary Eye Sampling (IES). Without using IES, the image synthesized at a position far from any real camera (top right) suffers much lower quality compared to the one closer to a real camera (top left) (2.74 dB drop). When the IES is applied, the quality of both images (bottom left and bottom right) improves, and the PSNR gap decreases (1.32 dB).

presented in Table 1. Visual comparison is presented in Fig. 4. It can be seen that our method achieves better performance than NeRF and NeRF++ in most of the scenarios. Other NeRF variants aim to estimate the radiance emitted by scene points at any position and direction, while our method is designed to recover the irradiance perceived by an observer from any point and direction. Since NeRF and NeRF++ need to sample points along viewing rays and render them in a back-to-front order, they require hundreds of network calls when synthesizing an image. Thus their rendering time is very long. In contrast, our method directly outputs the color information given a viewing ray. Thus, our training and testing time are relatively shorter, as shown in Table 3.

Comparison with Color Blending Approaches. To demonstrate the effectiveness of our CBNet. We compare our CBNet with another image-based warping method FVS [44]. FVS is a neural network based color blending approach for view synthesis. We retrain their network on our datasets. The results are presented in the first row of Table 1. It is evident that our method achieves significantly better performance.

Comparison with Panorama Synthesis Approaches. We employ a deep-based method 360SD-Net [62] to estimate depth maps for input images. We then build a point cloud from the depth map and input images. The point cloud are warped and refined for novel view synthesis. Since 360SD-Net only takes two vertically aligned panoramas as input, we take the same vertical inputs in this comparison, denoted as "Our (Vertical)" in Table 2. We further compare with a multi-sphere-images-based method MatryODShka [2] on view synthesis. Note that MatryODShka only takes two horizontally aligned panoramas as input. For fair comparison, we take the same input and denoted as "Our (Horizontal)" in Table 2. The numerical evaluations in Table 2 demonstrate that our method significantly outperforms the conventional depth-warp-refine and multi-sphere-images procedure in synthesizing new views. Besides, the competing methods

Table 4. Quantitative comparison of different imaginary eye sampling (IES) regions (large or small). Larger imaginary eye sampling space contributes to higher image quality.

Scene	Lounge				Livingroom			
N	2		4		2		4	
IES Range	PSNR↑	SSIM↑	PSNR	SSIM↑	PSNR↑	SSIM↑	PSNR↑	SSIM↑
Small	24.82	0.8484	27.98	0.8995	26.77	0.8690	29.93	0.9144
Large	**26.13**	**0.8883**	**29.13**	**0.9193**	**28.20**	**0.9068**	**31.27**	**0.9383**

both requires a structured input (horizontally or vertically aligned). This limitation does not apply to our method.

4.4 Effectiveness of Imaginary Eye Sampling

We demonstrate the necessity and effectiveness of the imaginary eye sampling strategy. In doing so, we train our network only using real camera locations and directions, without any imaginary eye sampling, denoted as "Ours w/o Imaginary Eye". The quantitative results and qualitative evaluations are presented in Table 1 and Fig. 5 respectively. For better comparison, we select two images for visualization. One is close to a real camera, and the other is far from input cameras.

As illustrated by the results, the performance of "Ours w/o Imaginary Eye" is inferior to our whole pipeline. More importantly, the performance gap between images that are near and far from the real camera is significant. There is 2.75 dB difference in terms of PSNR metric. This demonstrates that the model learns better for training data while does not have the ability to interpolate similar-quality test data.

A network is usually good at learning a continuous representation from discrete but uniformly distributed samples in a general case. In our plenoptic modeling, the values of the input parameters (x, y, z, θ, ϕ) are continuous and always span in a large range, while the input images only cover small and sparsely sampled regions in the whole space. Hence, it is not surprising that the model can fit well in training data while interpolating low-quality images at camera locations far from real cameras. Using our imaginary eye sampling strategy, the performance gap between the two cases is reduced (1.32 dB in terms of PSNR). Furthermore, the quality of synthesized images on the location that is near to a real camera is further improved. This is owed to our photometric consistency self-supervision loss for the virtual eye training. It helps the learned model to encode the geometry constraints across different viewpoint images.

We also conduct comparison experiments on the imaginary eye sampling area (large or small). The results are shown in Table 4. We found that sampling on a larger region will allow more freedom on the moving space of rendering cameras, while the downside is that it requires longer training time.

Table 5. Quantitative evaluations on different input view numbers.

N	2		4		8		25	
	PSNR↑	SSIM↑	PSNR↑	SSIM↑	PSNR↑	SSIM↑	PSNR↑	SSIM↑
Lounge	26.13	0.8883	29.13	0.9193	34.31	0.9684	37.27	0.9775
Livingroom	28.20	0.9068	31.27	0.9383	34.33	0.9648	36.72	0.9746

4.5 Proxy-Depth from Colors and Features

In what follows, we conduct experiments to demonstrate why the features are required in our photometric consistency loss. In doing so, we remove the feature item in Eq. 3 and train our model again, denoted as "Ours w/o Feature". The quantitative results are presented in the third last row of Table 1.

Compared to pixel-wise RGB colors, features have a larger reception field that makes textureless regions discriminative, and the encoded higher level information is more robust to illumination changes and other noises. Thus, the reconstructed proxy depths from both RGB colors and features are more accurate than those purely from RGB colors. Consequently, the quality of synthesized images is facilitated. We present the qualitative illustrations in the supplementary material.

4.6 With or Without View-Direction Weighting

We also ablate the necessity of the view-direction based weighting in Eq. 3. In this experiment, we set the weighting term s_i to one, denoted as "Ours w/o weighting". The results are presented in the second last row of Table 1. Not surprisingly, the performance drops. This demonstrates the effectiveness of our weighting strategy.

4.7 Different Number of Input Views

Below, we conduct experiments on a different number of input views. For this experiment setting, we aim to investigate the performance difference when the input views are located on a line, a flat plane, and a cube, corresponding to 2, 4, 8, and 25 input views, respectively. Table 5 and Fig. 6 provide the quantitative and qualitative results, respectively. As shown by the results, when the input view number is reduced to 2, our method still generates acceptable quality novel view images. As the number of input views increases, the quality of the view synthesis improves rapidly. For 8 and 25 input views in this experiment, the input cameras are randomly sampled within the region (cube) of interest. This demonstrates that our method is not limited to structured settings and can synthesize free-viewpoint images from unstructured input images.

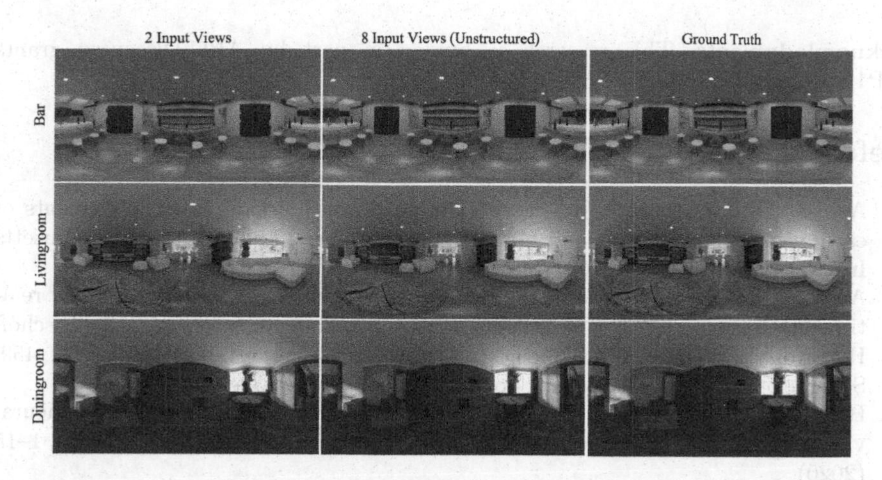

Fig. 6. Qualitative visualization of synthesized images by different view number and camera configurations. The three images are from our generated scenes "Bar", "Livingroom" and "Diningroom" respectively.

5 Conclusion

Capturing a complete and dense plenoptic function from every point and angle within a space has been the "holy grail" for IBR-based view synthesis applications. There is always a tension between how densely one samples the space using many real cameras and the total efforts and cost that one has to bear in doing this task. This paper proposes a simple yet effective solution to this challenge. By placing thousands of imaginary eyes (virtual cameras) at randomly sampled positions in the space of interest, this paper proposes a new neural-network-based method to learn (or to approximate) the underlying 5D plenoptic function. Real images captured by physical cameras are used as a teacher to train our neural network. Although those randomly placed imaginary eyes themselves do not provide new information, they are critical to the success of our method, as they provide a bridge to leverage the existing multi-view geometry relationship among all the views (of both real and virtual). Our experiments also validate this claim positively and convincingly. Our method produces accurate and high-quality novel views (on the validation set) and compelling visual results (on unseen testing images). We will release the code and models in this paper.

Limitations and Future Works: The training of the PDR network uses photo-consistency constraint. The proposed method may not work well when such constraint is violated, such as transparent and large areas of occlusion. While the followed CBNet can learn how to resolve these challenging scenarios to some extent, this is not a principled rule. We expect that more sophisticated solutions can be proposed in the future.

Acknowledgments. This research is funded in part by ARC-Discovery grants (DP190102261 and DP220100800).

References

1. Adelson, E.H., Bergen, J.R., et al.: The plenoptic function and the elements of early vision, vol. 2. Vision and Modeling Group, Media Laboratory, Massachusetts Institute of Technology (1991)
2. Attal, B., Ling, S., Gokaslan, A., Richardt, C., Tompkin, J.: MatryODShka: real-time 6DoF video view synthesis using multi-sphere images. In: Vedaldi, A., Bischof, H., Brox, T., Frahm, J.-M. (eds.) ECCV 2020. LNCS, vol. 12346, pp. 441–459. Springer, Cham (2020). https://doi.org/10.1007/978-3-030-58452-8_26
3. Bemana, M., Myszkowski, K., Seidel, H.P., Ritschel, T.: X-fields: implicit neural view-, light-and time-image interpolation. ACM Trans. Graph. (TOG) **39**(6), 1–15 (2020)
4. Bertel, T., Campbell, N.D., Richardt, C.: MegaParallax: casual 360° panoramas with motion parallax. IEEE Trans. Visual Comput. Graphics **25**(5), 1828–1835 (2019)
5. Buehler, C., Bosse, M., McMillan, L., Gortler, S., Cohen, M.: Unstructured lumigraph rendering. In: Proceedings of the 28th Annual Conference on Computer Graphics and Interactive Techniques, pp. 425–432 (2001)
6. Chai, J.X., Tong, X., Chan, S.C., Shum, H.Y.: Plenoptic sampling. In: Proceedings of the 27th Annual Conference on Computer Graphics and Interactive Techniques, pp. 307–318 (2000)
7. Chaurasia, G., Duchene, S., Sorkine-Hornung, O., Drettakis, G.: Depth synthesis and local warps for plausible image-based navigation. ACM Trans. Graph. (TOG) **32**(3), 1–12 (2013)
8. Chen, A., et al.: MVSNeRF: fast generalizable radiance field reconstruction from multi-view stereo. In: Proceedings of the IEEE/CVF International Conference on Computer Vision, pp. 14124–14133 (2021)
9. Chen, S.E., Williams, L.: View interpolation for image synthesis. In: Proceedings of the 20th Annual Conference on Computer Graphics and Interactive Techniques, pp. 279–288 (1993)
10. Chibane, J., Bansal, A., Lazova, V., Pons-Moll, G.: Stereo radiance fields (SRF): learning view synthesis for sparse views of novel scenes. In: Proceedings of the IEEE/CVF Conference on Computer Vision and Pattern Recognition, pp. 7911–7920 (2021)
11. Choi, I., Gallo, O., Troccoli, A., Kim, M.H., Kautz, J.: Extreme view synthesis. In: Proceedings of the IEEE International Conference on Computer Vision, pp. 7781–7790 (2019)
12. Community, B.O.: Blender - a 3D modelling and rendering package (2020). http://www.blender.org
13. Davis, A., Levoy, M., Durand, F.: Unstructured light fields. In: Computer Graphics Forum, vol. 31, pp. 305–314. Wiley Online Library (2012)
14. Debevec, P.E., Taylor, C.J., Malik, J.: Modeling and rendering architecture from photographs: a hybrid geometry-and image-based approach. In: Proceedings of the 23rd Annual Conference on Computer Graphics and Interactive Techniques, pp. 11–20 (1996)

15. Eslami, S.A., et al.: Neural scene representation and rendering. Science **360**(6394), 1204–1210 (2018)
16. Fitzgibbon, A., Wexler, Y., Zisserman, A.: Image-based rendering using image-based priors. Int. J. Comput. Vision **63**(2), 141–151 (2005)
17. Flynn, J., et al.: DeepView: view synthesis with learned gradient descent. In: Proceedings of the IEEE Conference on Computer Vision and Pattern Recognition, pp. 2367–2376 (2019)
18. Garbin, S.J., Kowalski, M., Johnson, M., Shotton, J., Valentin, J.: FASTNeRF: high-fidelity neural rendering at 200 fps. arXiv preprint arXiv:2103.10380 (2021)
19. Gera, P., Dastjerdi, M.R.K., Renaud, C., Narayanan, P., Lalonde, J.F.: Casual indoor HDR radiance capture from omnidirectional images. arXiv preprint arXiv:2208.07903 (2022)
20. Gortler, S.J., Grzeszczuk, R., Szeliski, R., Cohen, M.F.: The lumigraph. In: Proceedings of the 23rd Annual Conference on Computer Graphics and Interactive Techniques, pp. 43–54 (1996)
21. Hedman, P., Philip, J., Price, T., Frahm, J.M., Drettakis, G., Brostow, G.: Deep blending for free-viewpoint image-based rendering. ACM Trans. Graph. (TOG) **37**(6), 1–15 (2018)
22. Hedman, P., Srinivasan, P.P., Mildenhall, B., Barron, J.T., Debevec, P.: Baking neural radiance fields for real-time view synthesis. In: Proceedings of the IEEE/CVF International Conference on Computer Vision, pp. 5875–5884 (2021)
23. Huang, J., Chen, Z., Ceylan, D., Jin, H.: 6-DOF VR videos with a single 360-camera. In: 2017 IEEE Virtual Reality (VR), pp. 37–44. IEEE (2017)
24. Kalantari, N.K., Wang, T.C., Ramamoorthi, R.: Learning-based view synthesis for light field cameras. ACM Trans. Graph. (TOG) **35**(6), 1–10 (2016)
25. Levoy, M., Hanrahan, P.: Light field rendering. In: Proceedings of the 23rd Annual Conference on Computer Graphics and Interactive Techniques, pp. 31–42 (1996)
26. Li, J., Li, H., Matsushita, Y.: Lighting, reflectance and geometry estimation from 360 panoramic stereo. In: 2021 IEEE/CVF Conference on Computer Vision and Pattern Recognition (CVPR), pp. 10586–10595. IEEE (2021)
27. Li, T., et al.: Neural 3D video synthesis. arXiv preprint arXiv:2103.02597 (2021)
28. Li, Z., Xian, W., Davis, A., Snavely, N.: CrowdSampling the plenoptic function. In: Vedaldi, A., Bischof, H., Brox, T., Frahm, J.-M. (eds.) ECCV 2020. LNCS, vol. 12346, pp. 178–196. Springer, Cham (2020). https://doi.org/10.1007/978-3-030-58452-8_11
29. Lin, K.E., Yen-Chen, L., Lai, W.S., Lin, T.Y., Shih, Y.C., Ramamoorthi, R.: Vision transformer for NERF-based view synthesis from a single input image. arXiv preprint arXiv:2207.05736 (2022)
30. Lin, Z., Shum, H.Y.: On the number of samples needed in light field rendering with constant-depth assumption. In: Proceedings IEEE Conference on Computer Vision and Pattern Recognition, CVPR 2000 (Cat. No. PR00662), vol. 1, pp. 588–595. IEEE (2000)
31. Liu, L., Gu, J., Lin, K.Z., Chua, T.S., Theobalt, C.: Neural sparse voxel fields. arXiv preprint arXiv:2007.11571 (2020)
32. Martin-Brualla, R., Radwan, N., Sajjadi, M.S., Barron, J.T., Dosovitskiy, A., Duckworth, D.: NERF in the wild: neural radiance fields for unconstrained photo collections. arXiv preprint arXiv:2008.02268 (2020)
33. Max, N.: Optical models for direct volume rendering. IEEE Trans. Visual Comput. Graphics **1**(2), 99–108 (1995)

34. McMillan, L., Bishop, G.: Plenoptic modeling: an image-based rendering system. In: Proceedings of the 22nd Annual Conference on Computer Graphics and Interactive Techniques, pp. 39–46 (1995)
35. Mildenhall, B., et al.: Local light field fusion: practical view synthesis with prescriptive sampling guidelines. ACM Trans. Graph. (TOG) **38**(4), 1–14 (2019)
36. Mildenhall, B., Srinivasan, P.P., Tancik, M., Barron, J.T., Ramamoorthi, R., Ng, R.: NeRF: representing scenes as neural radiance fields for view synthesis. In: Vedaldi, A., Bischof, H., Brox, T., Frahm, J.-M. (eds.) ECCV 2020. LNCS, vol. 12346, pp. 405–421. Springer, Cham (2020). https://doi.org/10.1007/978-3-030-58452-8_24
37. Neff, T., et al.: DONeRF: towards real-time rendering of neural radiance fields using depth oracle networks. arXiv preprint arXiv:2103.03231 (2021)
38. Nguyen-Ha, P., Huynh, L., Rahtu, E., Heikkila, J.: Sequential neural rendering with transformer. arXiv preprint arXiv:2004.04548 (2020)
39. Niemeyer, M., Mescheder, L., Oechsle, M., Geiger, A.: Differentiable volumetric rendering: learning implicit 3D representations without 3D supervision. In: Proceedings of the IEEE/CVF Conference on Computer Vision and Pattern Recognition, pp. 3504–3515 (2020)
40. Park, E., Yang, J., Yumer, E., Ceylan, D., Berg, A.C.: Transformation-grounded image generation network for novel 3D view synthesis. In: Proceedings of the IEEE Conference on Computer Vision and Pattern Recognition, pp. 3500–3509 (2017)
41. Park, K., et al.: Deformable neural radiance fields. arXiv preprint arXiv:2011.12948 (2020)
42. Penner, E., Zhang, L.: Soft 3D reconstruction for view synthesis. ACM Trans. Graph. (TOG) **36**(6), 1–11 (2017)
43. Qi, C.R., Su, H., Mo, K., Guibas, L.J.: PointNet: deep learning on point sets for 3D classification and segmentation. In: Proceedings of the IEEE Conference on Computer Vision and Pattern Recognition, pp. 652–660 (2017)
44. Riegler, G., Koltun, V.: Free view synthesis. In: Vedaldi, A., Bischof, H., Brox, T., Frahm, J.-M. (eds.) ECCV 2020. LNCS, vol. 12364, pp. 623–640. Springer, Cham (2020). https://doi.org/10.1007/978-3-030-58529-7_37
45. Riegler, G., Koltun, V.: Stable view synthesis. arXiv preprint arXiv:2011.07233 (2020)
46. Seitz, S.M., Dyer, C.R.: View morphing. In: Proceedings of the 23rd Annual Conference on Computer Graphics and Interactive Techniques, pp. 21–30 (1996)
47. Serrano, A., et al.: Motion parallax for 360 RGBD video. IEEE Trans. Visual Comput. Graphics **25**(5), 1817–1827 (2019)
48. Shade, J., Gortler, S., He, L.w., Szeliski, R.: Layered depth images. In: Proceedings of the 25th Annual Conference on Computer Graphics and Interactive Techniques, pp. 231–242 (1998)
49. Shi, Y., Li, H., Yu, X.: Self-supervised visibility learning for novel view synthesis. In: Proceedings of the IEEE/CVF Conference on Computer Vision and Pattern Recognition, pp. 9675–9684 (2021)
50. Shih, M.L., Su, S.Y., Kopf, J., Huang, J.B.: 3D photography using context-aware layered depth inpainting. In: Proceedings of the IEEE/CVF Conference on Computer Vision and Pattern Recognition, pp. 8028–8038 (2020)
51. Sitzmann, V., Rezchikov, S., Freeman, W.T., Tenenbaum, J.B., Durand, F.: Light field networks: neural scene representations with single-evaluation rendering. arXiv preprint arXiv:2106.02634 (2021)

52. Sitzmann, V., Thies, J., Heide, F., Nießner, M., Wetzstein, G., Zollhofer, M.: DeepVoxels: learning persistent 3D feature embeddings. In: Proceedings of the IEEE Conference on Computer Vision and Pattern Recognition, pp. 2437–2446 (2019)
53. Sitzmann, V., Zollhöfer, M., Wetzstein, G.: Scene representation networks: continuous 3D-structure-aware neural scene representations. In: Advances in Neural Information Processing Systems, pp. 1121–1132 (2019)
54. Srinivasan, P.P., Deng, B., Zhang, X., Tancik, M., Mildenhall, B., Barron, J.T.: NERV: neural reflectance and visibility fields for relighting and view synthesis. arXiv preprint arXiv:2012.03927 (2020)
55. Srinivasan, P.P., Tucker, R., Barron, J.T., Ramamoorthi, R., Ng, R., Snavely, N.: Pushing the boundaries of view extrapolation with multiplane images. In: Proceedings of the IEEE Conference on Computer Vision and Pattern Recognition, pp. 175–184 (2019)
56. Srinivasan, P.P., Wang, T., Sreelal, A., Ramamoorthi, R., Ng, R.: Learning to synthesize a 4D RGBD light field from a single image. In: Proceedings of the IEEE International Conference on Computer Vision, pp. 2243–2251 (2017)
57. Sun, S.-H., Huh, M., Liao, Y.-H., Zhang, N., Lim, J.J.: Multi-view to novel view: synthesizing novel views with self-learned confidence. In: Ferrari, V., Hebert, M., Sminchisescu, C., Weiss, Y. (eds.) ECCV 2018. LNCS, vol. 11207, pp. 162–178. Springer, Cham (2018). https://doi.org/10.1007/978-3-030-01219-9_10
58. Thies, J., Zollhöfer, M., Nießner, M.: Deferred neural rendering: image synthesis using neural textures. ACM Trans. Graph. (TOG) **38**(4), 1–12 (2019)
59. Thies, J., Zollhöfer, M., Theobalt, C., Stamminger, M., Nießner, M.: Image-guided neural object rendering. In: International Conference on Learning Representations (2019)
60. Tucker, R., Snavely, N.: Single-view view synthesis with multiplane images. In: Proceedings of the IEEE/CVF Conference on Computer Vision and Pattern Recognition, pp. 551–560 (2020)
61. Tulsiani, S., Tucker, R., Snavely, N.: Layer-structured 3D scene inference via view synthesis. In: Ferrari, V., Hebert, M., Sminchisescu, C., Weiss, Y. (eds.) ECCV 2018. LNCS, vol. 11211, pp. 311–327. Springer, Cham (2018). https://doi.org/10.1007/978-3-030-01234-2_19
62. Wang, N.H., Solarte, B., Tsai, Y.H., Chiu, W.C., Sun, M.: 360SD-Net: 360° stereo depth estimation with learnable cost volume. In: 2020 IEEE International Conference on Robotics and Automation (ICRA), pp. 582–588. IEEE (2020)
63. Wang, Q., et al.: IBRNet: learning multi-view image-based rendering. In: Proceedings of the IEEE/CVF Conference on Computer Vision and Pattern Recognition, pp. 4690–4699 (2021)
64. Wu, G., Zhao, M., Wang, L., Dai, Q., Chai, T., Liu, Y.: Light field reconstruction using deep convolutional network on EPI. In: Proceedings of the IEEE Conference on Computer Vision and Pattern Recognition, pp. 6319–6327 (2017)
65. Yoon, Y., Jeon, H.G., Yoo, D., Lee, J.Y., So Kweon, I.: Learning a deep convolutional network for light-field image super-resolution. In: Proceedings of the IEEE International Conference on Computer Vision Workshops, pp. 24–32 (2015)
66. Yu, A., Ye, V., Tancik, M., Kanazawa, A.: PixelNeRF: neural radiance fields from one or few images. arXiv preprint arXiv:2012.02190 (2020)
67. Zhang, K., Riegler, G., Snavely, N., Koltun, V.: NeRF++: analyzing and improving neural radiance fields (2020)
68. Zheng, K.C., Kang, S.B., Cohen, M.F., Szeliski, R.: Layered depth panoramas. In: 2007 IEEE Conference on Computer Vision and Pattern Recognition, pp. 1–8. IEEE (2007)

69. Zhou, T., Tucker, R., Flynn, J., Fyffe, G., Snavely, N.: Stereo magnification: learning view synthesis using multiplane images. arXiv preprint arXiv:1805.09817 (2018)
70. Zhou, T., Tulsiani, S., Sun, W., Malik, J., Efros, A.A.: View synthesis by appearance flow. In: Leibe, B., Matas, J., Sebe, N., Welling, M. (eds.) ECCV 2016. LNCS, vol. 9908, pp. 286–301. Springer, Cham (2016). https://doi.org/10.1007/978-3-319-46493-0_18

3D-Yoga: A 3D Yoga Dataset for Visual-Based Hierarchical Sports Action Analysis

Jianwei Li[1(✉)], Haiqing Hu[1], Jinyang Li[1], and Xiaomei Zhao[2]

[1] Beijing Sports University, Beijing 100084, People's Republic of China
{jianwei,hhq,ljy}@bsu.edu.cn

[2] Shandong Jianzhu University, Jinan 250101, People's Republic of China
zhaoxiaomei20@sdjzu.edu.cn

Abstract. Visual-based human action analysis is an important research topic in the field of computer vision, and has great application prospect in sports performance analysis. Currently available 3D action analysis datasets have a number of limitations in sports application, including the lack of special sports actions, distinct class or score labels and variety of samples. Existing researches mainly use various special RGB videos for sports action analysis, but analysis with 2D features is less effective than 3D representation. In this paper, we introduce a new 3D yoga pose dataset (3D-Yoga) with more than 3,792 action samples and 16,668 RGB-D key frames, collected from 22 subjects performing 117 kinds of yoga poses with two RGB-D cameras. We have reconstructed 3D yoga poses with sparse multi-view data and carried out experiments with the proposed cascade two-stream adaptive graph convolutional neural network (Cascade 2S-AGCN) to recognize and assess these poses. Experimental results have shown the advantage of applying our 3D skeleton fusion and hierarchical analysis methods on 3D-Yoga, and the accuracy of Cascade 2S-AGCN outperforms the state-of-the-art methods. The introduction of 3D-Yoga will enable the community to apply, develop and adapt various methods for visual-based sports activity analysis.

Keywords: Yoga pose · Action analysis · 3D dataset · Human motion capture

1 Introduction

Human action analysis is an important and challenging problem in the field of computer vision, and in recent years it has been widely applied in intelligent sports. Intelligent sports action analysis can help to improve the athletes' competitive ability or promote public scientific fitness, and usually use human motion capture (Mocap) technology to obtain 3D human movements. The traditional inertia and optical motion capture systems can track and record human motion well, but these systems need to bind sensors or paste marker points on human body which may affect human motion and have not been popularized in public exercise. Therefore, visual-based markerless motion capture technologies have being increasingly researched and used in human sports analysis.

Supplementary Information The online version contains supplementary material available at https://doi.org/10.1007/978-3-031-26319-4_4.

L. Wang et al. (Eds.): ACCV 2022, LNCS 13841, pp. 55–71, 2023.
https://doi.org/10.1007/978-3-031-26319-4_4

In recent years, with the development of deep learning and large human action datasets, visual-based human action analysis has made remarkable progress. Human 3.6M [9] and NTU RGB+D [19] are two well known large scale 3D datasets, which are the potential resources for deep learning methods for human regular action recognition. Human 3.6M has 3.6 million human poses and corresponding RGB+D images, acquired by recording the performance of 11 subjects under 4 different viewpoints. NTU RGB+D 120 contains 114,480 video sequences and over 8 million frames, and has 120 action classes performed by 106 subjects captured from 155 views with Kinect cameras. To meet the increasing application requirements in intelligent sports, some human professional sports datasets (HPSDs) also have been presented in recent years. However, the data of most HPSDs are RGB images or videos collected from the Internet [16,23,29,34], and no 3D skeleton is provided. As the current mainstream method, deep learning needs large-scale human motion data to train a better model, which limits their effects on sports action analysis. The performance of algorithms of motion scoring or quality evaluation for sports training is still below the current application requirements. Especially for home fitness, such as yoga, the popularity rate is relatively high but lacking scientific guidance and feedback. Besides, a main challenge in sports action analysis is the accurate recovery of 3D human motion, and how to obtain accurate 3D human pose and analyze human action with limited information (such as data missing caused by occlusion) need to be further studied.

To address above issues, this paper proposes a hierarchical 3D yoga pose dataset called 3D-Yoga, which consists of 117 kinds of poses with 3,792 action samples, and each sample includes RGB-D image, human skeleton, pose label and quality score. Compared with existing sports action datasets, 3D-Yoga has several appealing properties: i) First, poses in 3D-Yoga are actually quite complex and challenging for 3D human pose estimation; ii) Second, data in 3D-Yoga is manually corrected and originally organized with hierarchical classification labels and pose quality scores; iii) Third, 3D-Yoga can be applied both to visual-based action recognition and action quality assessment tasks. To the best of our knowledge, 3D-Yoga is the first 3D sports dataset which covers actions task types including action recognition and quality assessment, and provides the corresponding RGB-D images and 3D skeleton. Based on 3D-Yoga, we propose a 3D skeleton fusion method to alleviate the occlusion problem and a hierarchical action analysis method for complex yoga poses. In summary the main **contributions** of this work are the followings:

- A new 3D sports action dataset with 117 categories of yoga poses performed by 22 subjects in various indoor environments;
- A sparse multi-view data alignment method to reconstruct 3D yoga poses to solve the severe self-occlusion problem;
- A hierarchical sports action analysis method through a cascade graph convolutional neural network for yoga pose classification and assessment.

2 Related Work

2.1 Skeleton-Based Action Recognition

Human action recognition (HAR) aims to identify what the action is, including action detection and action classification. Deep learning is currently the mainstream method for skeleton-based action recognition, where the most widely used models are RNNs and CNNs. RNN-based methods [5,6,19,20,31,38] usually model the skeleton data as a sequence of the coordinate vectors along both the spatial and temporal dimensions, where each vector represents a human body joint. CNN-based methods [12,14,18,33] generally model the skeleton data as a pseudo image based on the manually designed transformation rules. Both of the RNN-based and CNN-based methods fail to fully represent the structure of the skeleton data because the skeleton data are naturally embedded in the form of graphs rather than a vector sequence or a 2D grid. In recent years, graph convolutional networks (GCNs), which generalize convolution from image to graph, have been successfully adopted in many applications. ST-GCN [36] proposes a dynamic skeleton model which can automatically learn both the spatial and temporal patterns from images, and demonstrated to be effective in learning both spatial and temporal dependencies on skeleton graphs. Thus many improvements based on ST-GCN have emerged, such as ST-TR [28], 2S-AGCN [30], CTR-GCN [3], and so on. ST-TR models dependencies between joints using a spatial self-attention module to understand intra-frame interactions between different body parts and a temporal self-attention module to model inter-frame correlations. 2S-AGCN improves the ST-GCN method by splitting the adjacency matrix representing action features into three, already containing richer behavioral information. That includes both the first and second features of the skeleton data, which represent the joint coordinates and length and direction of the bone, respectively. CTR-GCN dynamically learns different topologies and effectively aggregates joint features in different channels for skeleton-based action recognition. For sports action recognition, Li et al. [15] introduce an efficient fitness action analysis based on 2D spatio-temporal feature encoding, which can be applied in artificial intelligence (AI) fitness system. Aifit [7] introduces an automatic 3D human-interpretable feedback models for fitness training. HDVR [8] proposes a hierarchical dance video recognition framework by estimating 3D human pose from the corresponding 2D human pose sequences, but the accuracy is limited because of the lack of ground-truth 3D annotations for training the proposed semi-supervised method.

2.2 Visual-Based Action Quality Assessment

Human action quality assessment (AQA) aims to automatically quantify the performance of the action or to score its performance. General methods deployed to compare action sequences are based on estimating the distance error or dynamic time warping. Deep learning methods for AQA can be divided into RGB video-based methods and skeleton-based methods. Algorithms based on RGB video generally extract features directly from images through deep learning models, such as C3D [26], I3D [1], and TSN [35], and then extract time domain features by LSTM, pooling, and so on. The final score prediction is performed by a fully connected neural network. ScoringNet [17] and SwingNet [23] are all based on such methods, and support fine-grained action classification and action scoring. These methods mainly focus on the visual activity information of the whole scene including the performer's body and background, but ignore the detailed joint interactions. Skeleton-based methods firstly detect the human skeleton in the image or video, and then model the correlation information between human joints, so as to realize human motion modeling and motion quality evaluation. Pan et al. [24] propose to learn

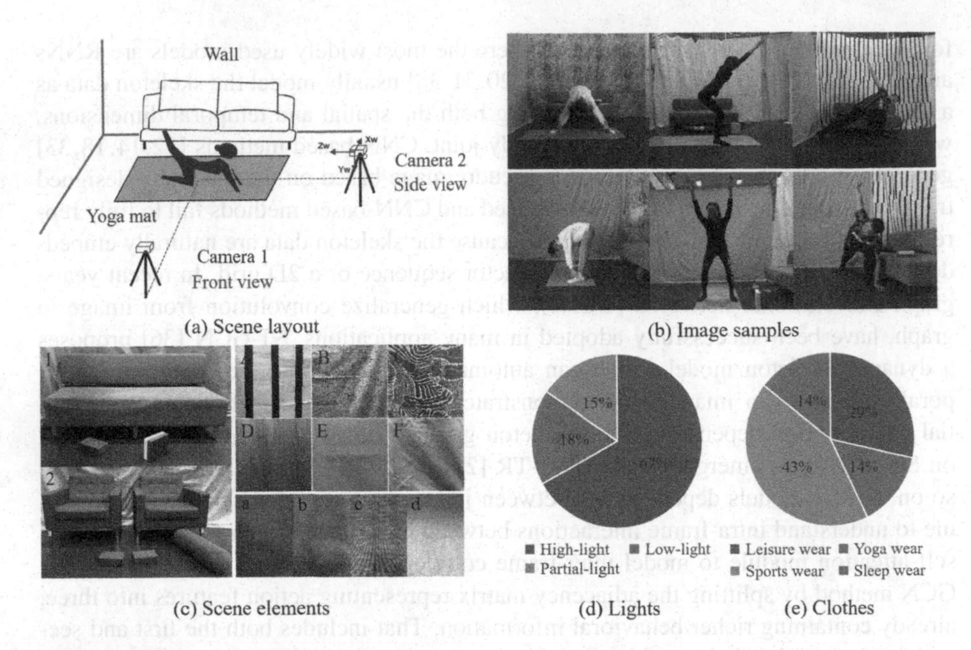

Fig. 1. The capturing of yoga poses in 3D-Yoga. (a) Scene layout. (b) Samples of yoga pose. (c) Scene elements: 1 and 2 are two indoor scenes; A, B, C, D, E, and F are the textures of the wall; a, b, c, and d are the textures of yoga mates. (d) Proportions of three light types. (e) Proportions of four cloth types.

the detailed joint motion based on the joint relation, which consists of a joint commonality module modeling the general motion for certain body parts and a joint difference module modeling the motion differences within body parts. These methods can better evaluate the motion information of human body when the skeleton joint is accurate and have good interpretability. SportsCap [2] introduces a multi-stream ST-GCN method to predict a fine-grained semantic action attributes, and adopts a semantic attribute mapping block to assemble various correlated action attributes into a high-level action label for the overall detailed understanding of the whole sequence, so as to enable various applications like sports action assessment or motion scoring. However, the athletes are often in very unusual postures (such as folding and bending), which leads to poor effect of human skeleton model on 2D sports data.

2.3 Sports Action Dataset

UCF-sport [32] is the first sports action dataset, and contains close to 200 action video sequences collected from various sports which are typically featured on broadcast television channels such as BBC and ESPN. Since then a number of sports motion datasets [16,23,29,34] used for action recognition have emerged. Diving48 [16] is a fine-grained 2D dataset for competitive diving, consisting of 18K trimmed video clips of 48 unambiguous dive sequences. Each dive sequence is defined by a combination of takeoff, movements in flight, and entry. GolfDB [23] is a 2D dataset created for general recognition applications in the sport of golf, and specifically for the task of golf swing sequencing. FineGym [29] provides coarse-to-fine annotations both temporally and semantically for gymnastics videos. There are three levels of categorical labels, and the temporal dimension is also divided into two levels, i.e., actions and sub-actions. Yoga-82 [34] is a hierarchical 2D dataset for large-scale yoga pose recognition with 82 classes. Each image contains one or more people doing the same yoga pose, and the picture information is complex, involving different backgrounds, different clothes and colors, and different camera view angles. Pose tutor [4] curates two other fitness datasets: Pilates-32, and Kungfu-7 datasets, and combines vision and 2D pose skeleton models in a coarse-to-fine framework to obtain pose class predictions. Some 2D datasets used for AQA task also have been proposed, such as MIT Olympic sports dataset [27], Nevada Olympic sports dataset [26], and AQA-7 [25], which predict the sports performance of diving, gymnastic vault, etc. FSD-10 [21] is a figure skating dataset for fine-grained sports content analysis, which collects 1,484 clips from the worldwide figure skating championships of 10 different actions in females and males programs. Existing sports action datasets used for AQA are mainly based on publicly available RGB images or videos, but have no 3D human skeleton pose. Action analysis methods with them may focus on the global texture features and tend to ignore the motion relationship within the human skeleton joints.

3 3D Yoga Dataset

In this section, we propose a new yoga pose dataset 3D-Yoga. We will introduce 3D-Yoga from data capturing, pose classification, pose assessment and data organization. The skeleton data will be made publicly available at https://3DYogabsu.github.io.

3.1 Yoga Pose Capturing

For yoga pose capturing, we deploy 22 subjects to perform yoga actions on the *daily yoga* (https://www.dailyyoga.com.cn) in two indoor scenes. The yoga poses are captured by two Microsoft Kinect Azure cameras from front view and side view simultaneously. As illustrated in Fig. 1 (a), two cameras are mutually orthogonal in each scene. The distances between two cameras and the center of yoga mat are 250 cm and 270 cm, respectively. Both cameras are 70 cm above the ground. Figure 1 (b) shows some image samples in 3D Yoga dataset. The top shows images captured from the front view, while the corresponding images collected from the side view are shown at the bottom.

To achieve a variety of data samples, environments of the scenes are changed during the capturing. In total, there are five backgrounds, three ambient lights, and four yoga mats. These scene elements are shown in Fig. 1 (c): 1–2 are two indoor scenes, A-F are the textures of the wall, and a-d are the textures of yoga mates. Figure 1 (d) shows proportions of high-light (67%), low-light (18%) and polarized light (15%) (light source on the side). 22 subjects (7 male and 15 female), ranging in age from 18 to 50, are asked to perform 158 consecutive actions in *beauty back build plan*, *relaxation sleep plan* and *menstrual period conditioning plan*. Each subject has different BMI and yoga skill, and wears various styles of clothes. Figure 1 (e) shows proportions of leisure wear (29%), yoga wear (14%), sports wear (43%) and sleep wear (14%). More specific information of the subjects is shown in the supplementary material.

Table 1. Design of the two-level hierarchical classification for yoga poses.

Classification I	Descriptions	# Labels	Examples
I. Dynamic pose	Consisted by two or more single forms of convergence.	1-4	
II. Sitting	The body pose with the pelvic bottom as the main support.	5-15	
III. Inversion	Inverted spine supported by the head, shoulder neck and upper limb.	16,17	
IV. Standing	Always based on foot as the main support type.	18-25	
V. Revolve	Spine move along its vertical axis.	26-35	
VI. Prone pose	Action supported by one of the complete surfaces of the body, or by multiple parts of one surface.	36-49	
VII. Support	Body is detached from the land, and the pose is mainly supported by the hand, elbow and foot.	50-57	
VIII. Balance	Maintain balance by regulating the limbs.	58-69	
IX. Bending	The body along a certain direction of folding or folding trend of action.	70-109	
X. Kneeling	Action supported by knee, shank and foot, belongs to kneeling type.	110-117	

3.2 Pose Classification

There are 158 yoga movements performed by each subject following four exercise sets in *daily yoga*. Since some yoga poses are repeated, such as *hunker pranayama* appearing three times, we merge the same movements. The final categories of yoga poses in 3D-Yoga are adjusted to 117, and each pose is different but covered all yoga formulas. As shown in Table 1, we design a two-level hierarchy to organize these 117 categories of yoga poses, in which the first level has 10 categories (listed in the first column) and the second level are the sub-categories (labeled in the third column) of yoga poses. For the first level classification, we provide their names, detailed definitions, corresponding labels of second-level, and posture examples. The second level classification is the detailed division of first level classification, which is described in the supplementary material with labels and specific pose names.

Table 2. Scoring examples for three subjects performing two different yoga poses. S denotes strength, B denotes balance, and P&T denotes pliable and tough.

Examples	Front view	Side view	Scores (S, B, P&T)
Split boat pose			(3, 3, 3)
Split boat pose			(2, 3, 1)
Easy warrior III pose			(3, 3, 1)

3.3 Pose Assessment

To provide a benchmark with domain knowledge for yoga pose quality evaluation, three yoga coaches with rich experiences (having coach cards) have completed the yoga scoring for each subject. Each sample is evaluated with two indicators, i.e., difficulty coefficient and completion score. The difficulty coefficient score is obtained by referring the standard of fitness yoga posture released on the national health yoga steering committee. The completion scoring standard is made by the coach and has four levels (0–3). The distribution of the completion score ranges from 0 to 3 in terms of strength (S), balance (B), pliable and tough (P&T). Table 2 gives scoring examples for three subjects performing two yoga poses, i.e., *split boat pose* and *easy warrior III pose*. The detail criteria of yoga pose quality assessment are described in the supplementary material. Since the scoring for the same sample from different coaches may be various, we compute an average value as the completion score. The final evaluation score $Score_{pose}^{level}$ for each sample is multiplied by the completion score and the difficulty coefficient:

$$Score_{pose}^{level} = P_{level} \times C_{pose}, \tag{1}$$

where P_{level} denotes the difficulty coefficient, and C_{pose} denotes the completion score.

3.4 Data Organization

There are 3,792 action samples and 16,668 key frames in 3D-Yoga, and the organization is shown in Fig. 2. Under the *Scene* directory, there are three folders: *Front*, *Side* and *Docs*. Under *Docs*, we provide file list, pose score and camera calibration information. There are 22 folders respectively for 22 subjects under the *Front* and *Side* folders. The folders of 7 male subjects are represented as M01, M02,..M07, and the folders of 15 female subjects are represented as F01, F02,..F15. The action label (A01, A02,...A10) represents the folders name of classification I, and the sub-action label (a01, a03,...a117) represents the folders name of classification II. In each sub-action folder corresponding to a motion segment, there are three sub-folders, i.e., *Color*, *Depth* and *Skeleton*.

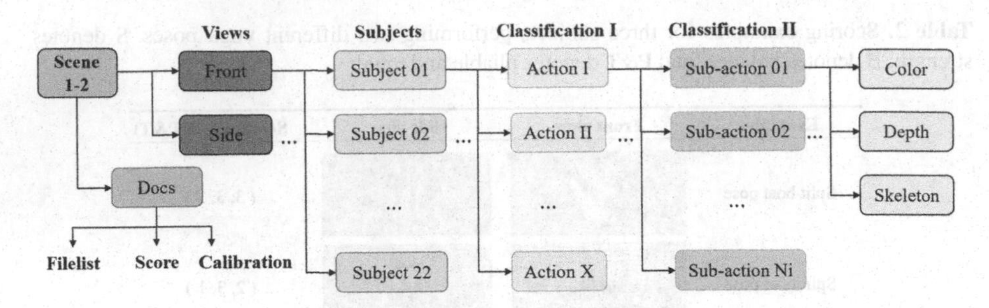

Fig. 2. Schematic representation of directory organization of the 3D-Yoga dataset.

The original unregistered full resolution depth (640 × 574) and color (1920 × 1080) images are captured by different sensors of the Kinect camera, while the skeletal joints are obtained through the bone tracking technology in Kinect SDK. In order to make the data easy applicable in many existing implementations, we provide both the color and the depth in a resolution of 1280 × 720 and register them both in the color camera coordinate system. Human body skeleton is composed of the three-dimensional position coordinates of the 32 main human body joints, and also is aligned in the color camera coordinate system. Skeleton data is stored in CSV format, and contains time stamps, personIDs, jointsIDs, 3D spatial coordinates and quaternions of the joints, confidences and 2D pixel coordinates of the joints.

We compare 3D-Yoga dataset with the state-of-the-art HPSDs, and Table 3 shows the comparison in terms of action classes, sample numbers, data types, sources and analysis tasks. It's obvious that 3D-Yoga is built through Mocap and provides more data type, which can be used for action recognition and AQA tasks.

4 Yoga Pose Analysis

The pipeline of the proposed yoga pose analysis method is shown in Fig. 3, which consists of data pre-processing and hierarchical pose analysis. The innovative points are elaborated in the following subsections.

Table 3. Comparison with the state-of-the-art professional sports datasets. R denotes action recognition task, and AQA denotes action quality assessment task.

Datasets	Classes	Samples	Data types	Sources	Tasks
UCF-sport [32]	10	150	RGB	BBC/ESPN	R
Sport-1M [11]	487	1,133,158	RGB	YouTube	R
Diving48 [16]	48	18,404	RGB	Internet	R
Yoga-82 [34]	82	28,478	RGB	Bing	R
FineGym [29]	99/288	~30K	RGB	Internet	R
FSD-10 [21]	10	1,484	RGB	YouTube	R&AQA
3D-Yoga	10/117	3,792	RGB-D+Skeleton	Mocap (Kinects)	R&AQA

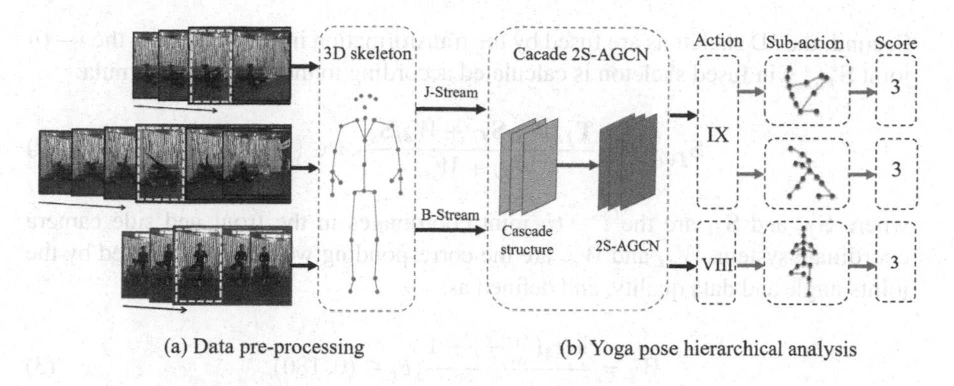

Fig. 3. The pipeline of our proposed hierarchical yoga pose analysis method.

4.1 Data Pre-processing

To get a complete and accurate 3D skeleton pose for yoga pose analysis, we process the action sequences mainly through the following three steps:

Camera Calibration: The calibration for Kinect cameras is performed before yoga pose capturing. More than 100 checkerboard images are selected in each calibration. The intrinsic matrix of each camera is obtained from Kinect azure SDK, while the geometric relationship of two cameras in front and side views is computed by the tool of stereo camera calibration with Matlab. The grids in the checkerboard are 10×15, and the actual side length of each grid is 5 cm. The average re-projection error of stereo camera calibration is 2.81 pixels.

Key Frame Extraction: Considering that most yoga movements are static processes, we use the key frame based method to analyze yoga poses. For each captured data sequence, we first carry out coarse extraction based on image similarity, and then extract frames of each pose based on the confidence value and depth value of the human skeleton. For each yoga action, we manually select the best time synchronization key frames captured by two Kinects, and retain key frame pairs with the most obvious features. For a few dynamic yoga poses with different sequence lengths, we select multiple key frames to describe the entire movement.

3D Skeleton Fusion: For each pair of key frame data, the alignment of 3D skeleton pose is executed as follows:

– First, we align the 3D points from the front and side views through ICP algorithm [37] with camera calibration result as an initial value. 3D points are calculated by corresponding depth images and camera intrinsic parameter. The transformation matrix $\mathbf{T}_{fs}$ between two views (from front to side) is obtained after ICP registration.

– Second, the 3D skeletons are fused by the transformation information, and the $i-th$ joint $\mathbf{S}^i_{fused}$ in fused skeleton is calculated according to the following formula:

$$\mathbf{S}^i_{fused} = \frac{\mathbf{T}_{fs} W_{fi} \mathbf{S}_{fi} + W_{si} \mathbf{S}_{si}}{W_{fi} + W_{si}}, i \in (0, 31) \tag{2}$$

where $\mathbf{S}_{fi}$ and $\mathbf{S}_{si}$ are the $i-th$ joint coordinates in the front and side camera coordinate system. W_{fi} and W_{si} are the corresponding weights, determined by the joints angle and data quality, and defined as:

$$W_i = \frac{\log_4(\frac{c_i+1}{z_i}) + 1}{1 + e^{\frac{\theta_i - 90}{10}}}, \theta_i \in (0, 180) \tag{3}$$

where z_i, θ_i and c_i are the depth value, joint angle and confidence coefficient respectively. The joint with high confidence and close to the camera has larger fusion weight. Equation 3 is a refinement of Jiang et al. [10].

– Finally, the 3D skeletons are further optimized by embedding a parametric human model [22] through minimizing the following energy function:

$$\mathbf{E}_{fused}(\theta, \beta) = w_{pro} \mathbf{E}_{pro} + w_{shape} \mathbf{E}_{shape} \tag{4}$$

where $\mathbf{E}_{pro}$ is the data term aligning 2D projections on each view to the 3D joints, $\mathbf{E}_{shape}$ penalizes human shape prior. w_{pro} and w_{shape} are balancing weights, and set to 1 in our experiments. θ and β are two optimized parameters, which control the bone length and the joint posture respectively.

4.2 Hierarchical Analysis

For yoga pose hierarchical analysis, we design the Cascade 2S-AGCN with a cascade structure and 2S-AGCN models to realize the coarse-grained to fine-grained yoga pose classification and specific yoga pose quality evaluation.

Cascade Structure: Cascade 2S-AGCN contains three stages and consists of a main network and three branches (two for pose classification and one for pose assessment). The main network is modelled after AlexNet [13] and each branch is a 2S-AGCN model. The details of Cascade 2S-AGCN are described in the supplementary material. After completing the former stage by 2S-AGCN, we choose a recall rate threshold to remove error samples that will not be used to train later stages. The advantage of this filtering mechanism is that since the main and branch networks are in a unified framework, the feature maps extracted at the beginning can be shared throughout the network, rather than features being collected at each layer of the network from the original data.

Graph Construction: Based on human skeleton model, we construct the skeleton information undirected graph with human skeleton link rule, in which the origin represents the key point, and the line segments represent the connection relationship of each joint point. The input of Cascade 2S-AGCN has two streams: B-stream and J-stream. For J-stream, we use a spatio-temporal graph to simulate the structured information between them along the spatio-temporal dimensions of these joints. The structure of the graphs contains not only joint point coordinate but also spatial constraints between adjacent key points. For B-stream, the input data are the length and direction of the skeleton. We set the middle point of the pelvis as the central point, the joint near the central point as the source joint, and the joint away from the central point as the target joint. Thus the joint is the key point, the bone is the vector from one point to another, the length of the vector is the length of the bone, and the direction of the vector is the direction of the bone. The main formula for the adaptive graph convolution is as follows:

$$\mathbf{f}_{out} = \sum_{k}^{\mathbf{K}_v} \mathbf{W}_k \mathbf{f}_{in} (\mathbf{A}_k + \mathbf{B}_k + \mathbf{C}_k), \tag{5}$$

where $\mathbf{f}$ denotes the feature map of the graph convolution, $\mathbf{W}_k$ is the weight vector of the 1×1 convolution operation, $\mathbf{K}_v$ denotes the number of subsets. $\mathbf{A}_k$ is an $N \times N$ adjacency matrix (N denotes the number of vertexes) and represents the physical structure of the human body. $\mathbf{B}_k$ also is an $N \times N$ adjacency matrix similar to $\mathbf{A}_k$, but the elements of $\mathbf{B}_k$ are parameterized and optimized together with the other parameters in the training process. $\mathbf{C}_k$ is a graph unique to each key frame that uses a classical Gaussian embedding function to capture the similarity between joints. The input data is in the form of $C_{in} \times T \times N$, encoded into $C_e \times T \times N$ with two embedding functions, i.e., θ and ϕ. Since the number of parameters for the convolution of 1×1 is quite large, the encoded $C_{in} \times T \times N$ is transformed into $N \times C_e T$ and $C_e T \times N$ through two conversion functions. These two matrices are multiplied to obtain the $N \times N$ similarity matrix of $\mathbf{C}_k$. Element $C_k^{i,j}$ in $\mathbf{C}_k$ represents the similarity between vertex v_i and v_j. Therefore, $\mathbf{C}_k$ is calculated by multiplying θ and ϕ:

$$\mathbf{C}_k = softmax({f_{in}}^T {W_{\theta k}}^T W_{\phi k} f_{in}), \tag{6}$$

where T denotes the temporal length, W_θ and W_ϕ are the parameters of θ and ϕ.

Pose Recognition and Assessment: The yoga pose recognition task is constructed with the first and second stages of the cascade network to obtain a coarse-to-fine pose classification. For example, the *sitting leg up* and *seated butt lift back-bending* both are recognized as *Bending* at the first stage, and then identified as their respective sub-actions (label 80 and label 96) at the second stage. The yoga pose quality for each sample is evaluated with the pose completion score and difficulty coefficient. We firstly use the third stage of the cascade network to predict a completion score for each pose performed by a subject, and then multiply the difficulty coefficient score based on the yoga level. For example, if the predicted completion score of *boat pose* ($P_{level} = 3$) is 2, the final evaluation score is 6.

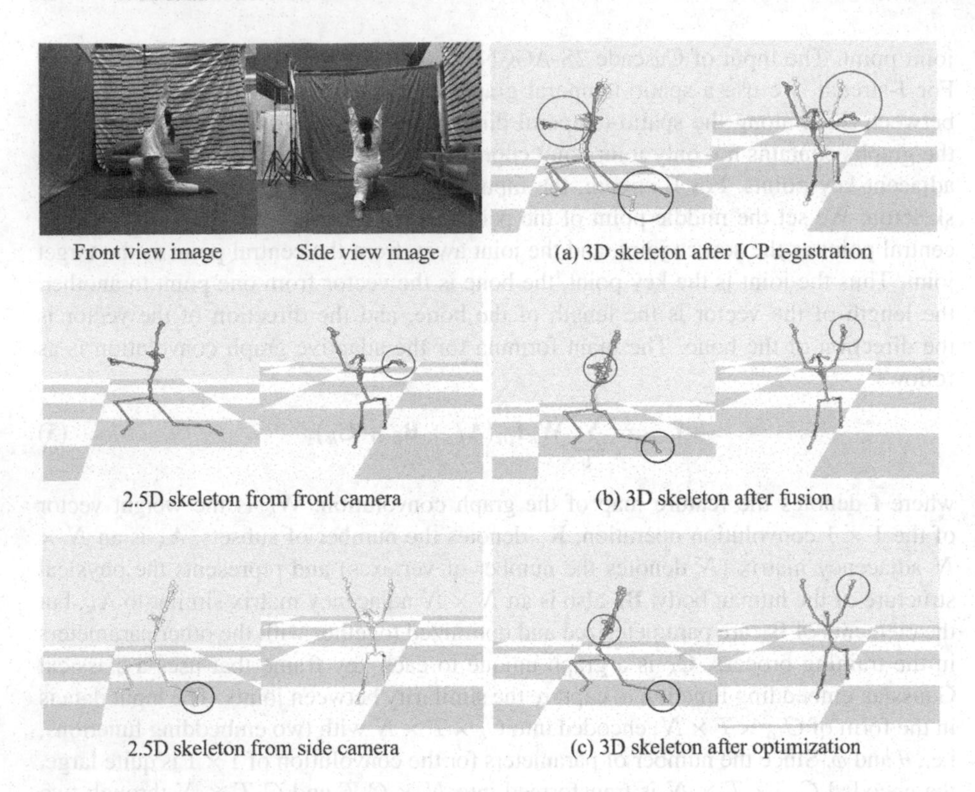

Fig. 4. Results of 3D skeleton fusion for a common yoga pose. The left column shows input data, and the right column shows output data.

5 Experiments and Analysis

In this section, we make experiments and analyses with yoga pose analysis methods on 3D-Yoga. For all experiments, we run our methods on a standard desktop PC with Intel Core i7-7700 3.6 GHz CPU.

5.1 Implementation Details

Training Details: As introduced in Sect. 4.2, there are three types of graphs in the adaptive graph convolutional block (**A**, **B**, **C**). Three-axis data is trained in the 2S-AGCN model, in which TCN and TGN are two predominant sub-models to extract time-series features and graph features from the raw data. The predicted result is the maximum classification probability of the *softmax* classifier through the `argmax` function. In our experiments, cross-entropy and AdamW are selected as the loss function and optimizer respectively, and Relu activation function is used to avoid gradient disappearance. During the training, we set the learning rate = 1e-3, seed = 42, weight_decay = 0.1, betas $\in$ (0.9–0.999), batch size = 32, and epoch = 80.

Table 4. Analysis accuracies (%) of yoga poses in 3D-Yoga. The best results are in bold.

Stages	Categories	Number	Front	Side	Combined	Fused
I	Coarse-gained	10	75.60	71.87	77.18	**80.42**
	Dynamic	4	77.6	68.86	**78.14**	73.73
	Sitting	11	74.65	69.91	85.64	**87.29**
	Inversion	2	**100**	93.48	**100**	92.31
	Standing	8	83.77	72.08	81.81	**90.40**
	Revolve	10	76.98	**79.86**	77.70	73.56
II	Prone	14	61.68	65.87	**76.05**	60.68
	Support	8	54.17	**62.50**	60.41	53.84
	Balance	12	82.96	82.22	80.00	**92.30**
	Bending	40	64.09	62.19	67.68	**75.52**
	Kneeling	8	60.43	69.06	66.91	**69.33**
	Fine-gained	117	76.10	74.13	74.92	**82.94**
III	Score	4	71.40	70.97	65.43	**72.57**

Experimental Data: To compare the performance of our methods, we have carried out experiments with four types data of 3D-Yoga: 1) 2.5D skeletons of front view; 2) 2.5D skeletons of side view; 3) combined 2.5D skeletons (both of front and side views); 4) fused 3D skeleton. We use 32 joints model provided by Kinect for experiments (1–3) and 23 joints provided by the SMPL model for the experiment (4). To ensure a better training effect, we split the dataset into a training set with 16 subjects and a validation set with 6 subjects, and use hierarchical sampling in the pose labels.

5.2 Ablation Studies

Yoga Pose Fusion: Due to human body self-occlusion, 2.5D skeleton data captured by the front or side camera always has some errors. In order to get an accurate 3D skeleton, we align 2.5D skeletons with the proposed method in Sect. 4.1. Figure 4 visualizes the fused results of 3D skeleton from the front view and side view for a common yoga pose. The left column shows color images and 2.5D skeletons (in camera coordinate system) captured by the front and side Kinects, and the right column shows 3D skeletons after ICP registration, fusion and optimization respectively. The main distinctions are circled in red and blue on the Figure. The contrasts show that we have effectively restored some 3D skeleton joints after the operations of ICP registration, fusion and optimization. The data quality analysis for 3D-Yoga dataset before and after 3D skeleton fusion is presented in supplementary material.

Yoga Pose Analysis: Table 4 shows the quantitative analysis accuracies of yoga poses with Cascade 2S-AGCN in terms of front view skeletons, side view skeletons, combined skeletons and fused skeletons. The first line is the recognition results of Classification I in the first stage, lines 2 through 13 are the recognition results of each sub-action categories and the average accuracies of Classification II in the second stage, and the bottom line is the prediction accuracy of completion score for each yoga pose in the third stage. Number is the kinds of yoga poses in each category. Except for individual

Table 5. Comparison of recognition accuracies (%) with the state-of-the-art methods on 3D-Yoga. The best results are in bold.

Methods	Front	Side	Combined	Fused
ST-GCN [36]	53.65	56.25	56.56	58.33
2S-AGCN [30]	58.90	55.82	56.86	66.84
CTR-GCN [3]	73.02	66.24	**76.40**	72.93
Ours	**76.10**	**74.13**	74.92	**82.94**

rare poses, most recognition accuracies by using combined and fused skeletons are generally higher than the accuracies of the front view or side view skeletons. With our fused 3D skeletons, the coarse-gained recognition (Classification I) accuracy is up to 80.42%, the fine-gained recognition (Classification II) accuracy is 82.94%, and the prediction accuracy for pose completion score is 72.57%. It verifies that the average performance of our yoga pose analysis method with 3D fused skeletons has been improved. More experiments results and analyses for yoga pose are provided in supplementary material.

5.3 Comparison with Other Methods

We also compare Cascade 2S-AGCN with the state-of-the-art methods on 3D-Yoga, and Table 5 shows comparison results (117 categories) of the recognition accuracies (%) with related methods: ST-GCN [36], 2S-AGCN [30], and CTR-GCN [3]. We run these methods on a NVIDIA RTX 1080Ti GPU, and use the same epoch equal to 80 in the comparison experiments. It can be seen that Cascade 2S-AGCN outperforms other methods and the accuracies of all methods are improved by using the fused 3D skeletons. A detailed version is provided in supplementary material.

6 Conclusions

In this work, we present a new 3D-Yoga dataset with RGB-D images, 3D skeletons, multi-level classification labels and pose scores. To evaluate 3D-Yoga, we perform a cascade graph-based convolution network to recognize coarse-grained to fine-grained yoga poses and assess the quality of each pose. Experiments have been carried out for hierarchical yoga pose analysis, and the results show that the proposed pose recognition and assessment methods have achieved a good performance. The introduction of 3D-Yoga will enable the community to apply, develop and adapt various deep learning techniques for visual-based sports activity analysis. We will further research more efficient and robust methods for sports action analysis with 3D-Yoga.

Acknowledgment. This work is assisted by Siqi Wang, Rui Shi, Ruihong Cheng, Yongxin Yan and Jie Liu, five students from the school of sport engineering of Beijing Sports University, who participated in data acquisition. This work is supported by the Open Projects Program of National Laboratory of Pattern Recognition under Grant No. 202100009, and the Fundamental Research Funds for Central Universities No. 2021TD006.

References

1. Carreira, J., Zisserman, A.: Quo vadis, action recognition? A new model and the kinetics dataset. In: Proceedings of the IEEE Conference on Computer Vision and Pattern Recognition, pp. 6299–6308 (2017)
2. Chen, X., Pang, A., Yang, W., Ma, Y., Xu, L., Yu, J.: SportScap: monocular 3D human motion capture and fine-grained understanding in challenging sports videos. Int. J. Comput. Vision **129**(10), 2846–2864 (2021)
3. Chen, Y., Zhang, Z., Yuan, C., Li, B., Deng, Y., Hu, W.: Channel-wise topology refinement graph convolution for skeleton-based action recognition. In: Proceedings of the IEEE/CVF International Conference on Computer Vision, pp. 13359–13368 (2021)
4. Dittakavi, B., et al.: Pose tutor: an explainable system for pose correction in the wild. In: Proceedings of the IEEE/CVF Conference on Computer Vision and Pattern Recognition, pp. 3540–3549 (2022)
5. Donahue, J., et al.: Long-term recurrent convolutional networks for visual recognition and description. In: Proceedings of the IEEE Conference on Computer Vision and Pattern Recognition, pp. 2625–2634 (2015)
6. Du, Y., Wang, W., Wang, L.: Hierarchical recurrent neural network for skeleton based action recognition. In: Proceedings of the IEEE Conference on Computer Vision and Pattern Recognition, pp. 1110–1118 (2015)
7. Fieraru, M., Zanfir, M., Pirlea, S.C., Olaru, V., Sminchisescu, C.: AiFit: automatic 3D human-interpretable feedback models for fitness training. In: Proceedings of the IEEE/CVF Conference on Computer Vision and Pattern Recognition, pp. 9919–9928 (2021)
8. Hu, X., Ahuja, N.: Unsupervised 3D pose estimation for hierarchical dance video recognition. In: Proceedings of the IEEE/CVF International Conference on Computer Vision, pp. 11015–11024 (2021)
9. Ionescu, C., Papava, D., Olaru, V., Sminchisescu, C.: Human3.6M: large scale datasets and predictive methods for 3D human sensing in natural environments. IEEE Trans. Pattern Anal. Mach. Intell. **36**(7), 1325–1339 (2014)
10. Jiang, Y., Song, K., Wang, J.: Action recognition based on fusion skeleton of two kinect sensors. In: 2020 International Conference on Culture-oriented Science & Technology (ICCST), pp. 240–244. IEEE (2020)
11. Karpathy, A., Toderici, G., Shetty, S., Leung, T., Sukthankar, R., Fei-Fei, L.: Large-scale video classification with convolutional neural networks. In: Proceedings of the IEEE conference on Computer Vision and Pattern Recognition, pp. 1725–1732 (2014)
12. Ke, Q., Bennamoun, M., An, S., Sohel, F., Boussaid, F.: A new representation of skeleton sequences for 3D action recognition. In: Proceedings of the IEEE Conference on Computer Vision and Pattern Recognition, pp. 3288–3297 (2017)
13. Krizhevsky, A., Sutskever, I., Hinton, G.E.: ImageNet classification with deep convolutional neural networks. In: Advances in Neural Information Processing Systems, vol. 25 (2012)
14. Li, B., Dai, Y., Cheng, X., Chen, H., Lin, Y., He, M.: Skeleton based action recognition using translation-scale invariant image mapping and multi-scale deep CNN. In: 2017 IEEE International Conference on Multimedia & Expo Workshops (ICMEW), pp. 601–604. IEEE (2017)
15. Li, J., Cui, H., Guo, T., Hu, Q., Shen, Y.: Efficient fitness action analysis based on spatio-temporal feature encoding. In: 2020 IEEE International Conference on Multimedia & Expo Workshops (ICMEW), pp. 1–6. IEEE (2020)
16. Li, Y., Li, Y., Vasconcelos, N.: RESOUND: towards action recognition without representation bias. In: Ferrari, V., Hebert, M., Sminchisescu, C., Weiss, Y. (eds.) ECCV 2018. LNCS, vol. 11210, pp. 520–535. Springer, Cham (2018). https://doi.org/10.1007/978-3-030-01231-1_32

17. Li, Y., Chai, X., Chen, X.: ScoringNet: learning key fragment for action quality assessment with ranking loss in skilled sports. In: Jawahar, C.V., Li, H., Mori, G., Schindler, K. (eds.) ACCV 2018. LNCS, vol. 11366, pp. 149–164. Springer, Cham (2019). https://doi.org/10.1007/978-3-030-20876-9_10
18. Liu, H., Tu, J., Liu, M.: Two-stream 3D convolutional neural network for skeleton-based action recognition. arXiv preprint arXiv:1705.08106 (2017)
19. Liu, J., Shahroudy, A., Perez, M., Wang, G., Duan, L.Y., Kot, A.C.: NTU RGB+ D 120: a large-scale benchmark for 3D human activity understanding. IEEE Trans. Pattern Anal. Mach. Intell. **42**(10), 2684–2701 (2019)
20. Liu, J., Shahroudy, A., Xu, D., Wang, G.: Spatio-temporal LSTM with trust gates for 3D human action recognition. In: Leibe, B., Matas, J., Sebe, N., Welling, M. (eds.) ECCV 2016. LNCS, vol. 9907, pp. 816–833. Springer, Cham (2016). https://doi.org/10.1007/978-3-319-46487-9_50
21. Liu, S., et al.: FSD-10: a fine-grained classification dataset for figure skating. Neurocomputing **413**, 360–367 (2020)
22. Loper, M., Mahmood, N., Romero, J., Pons-Moll, G., Black, M.J.: SMPL: a skinned multi-person linear model. ACM Trans. Graph. (TOG) **34**(6), 1–16 (2015)
23. McNally, W., Vats, K., Pinto, T., Dulhanty, C., McPhee, J., Wong, A.: GolfDB: a video database for golf swing sequencing. In: Proceedings of the IEEE/CVF Conference on Computer Vision and Pattern Recognition Workshops (2019)
24. Pan, J.H., Gao, J., Zheng, W.S.: Action assessment by joint relation graphs. In: Proceedings of the IEEE/CVF International Conference on Computer Vision, pp. 6331–6340 (2019)
25. Parmar, P., Morris, B.: Action quality assessment across multiple actions. In: IEEE Winter Conference on Applications of Computer Vision (WACV), pp. 1468–1476. IEEE (2019)
26. Parmar, P., Morris, B.T.: What and how well you performed? A multitask learning approach to action quality assessment. In: Proceedings of the IEEE/CVF Conference on Computer Vision and Pattern Recognition, pp. 304–313 (2019)
27. Pirsiavash, H., Vondrick, C., Torralba, A.: Assessing the quality of actions. In: Fleet, D., Pajdla, T., Schiele, B., Tuytelaars, T. (eds.) ECCV 2014. LNCS, vol. 8694, pp. 556–571. Springer, Cham (2014). https://doi.org/10.1007/978-3-319-10599-4_36
28. Plizzari, C., Cannici, M., Matteucci, M.: Skeleton-based action recognition via spatial and temporal transformer networks. Comput. Vis. Image Underst. **208**, 103219 (2021)
29. Shao, D., Zhao, Y., Dai, B., Lin, D.: FineGym: a hierarchical video dataset for fine-grained action understanding. In: Proceedings of the IEEE/CVF Conference on Computer Vision and Pattern Recognition, pp. 2616–2625 (2020)
30. Shi, L., Zhang, Y., Cheng, J., Lu, H.: Two-stream adaptive graph convolutional networks for skeleton-based action recognition. In: Proceedings of the IEEE/CVF Conference on Computer Vision and Pattern Recognition, pp. 12026–12035 (2019)
31. Song, S., Lan, C., Xing, J., Zeng, W., Liu, J.: An end-to-end spatio-temporal attention model for human action recognition from skeleton data. In: Proceedings of the AAAI Conference on Artificial Intelligence, vol. 31 (2017)
32. Soomro, K., Zamir, A.R.: Action recognition in realistic sports videos. In: Computer Vision in Sports, pp. 181–208 (2014)
33. Tran, D., Bourdev, L., Fergus, R., Torresani, L., Paluri, M.: Learning spatiotemporal features with 3D convolutional networks. In: Proceedings of the IEEE International Conference on Computer Vision, pp. 4489–4497 (2015)
34. Verma, M., Kumawat, S., Nakashima, Y., Raman, S.: Yoga-82: a new dataset for fine-grained classification of human poses. In: Proceedings of the IEEE/CVF Conference on Computer Vision and Pattern Recognition Workshops, pp. 1038–1039 (2020)

35. Xiang, X., Tian, Y., Reiter, A., Hager, G.D., Tran, T.D.: S3D: stacking segmental P3D for action quality assessment. In: 2018 25th IEEE International Conference on Image Processing (ICIP), pp. 928–932. IEEE (2018)
36. Yan, S., Xiong, Y., Lin, D.: Spatial temporal graph convolutional networks for skeleton-based action recognition. In: Thirty-Second AAAI Conference on Artificial Intelligence (2018)
37. Yang, C., Medioni, G.: Object modeling by registration of multiple range images. Image Vis. Comput. **10**(3), 145–155 (2002)
38. Zhang, P., Lan, C., Xing, J., Zeng, W., Xue, J., Zheng, N.: View adaptive neural networks for high performance skeleton-based human action recognition. IEEE Trans. Pattern Anal. Mach. Intell. **41**(8), 1963–1978 (2019)

NEO-3DF: Novel Editing-Oriented 3D Face Creation and Reconstruction

Peizhi Yan[1], James Gregson[2], Qiang Tang[2], Rabab Ward[1], Zhan Xu[2], and Shan Du[3(✉)]

[1] The University of British Columbia, Vancouver, BC, Canada
{yanpz,rababw}@ece.ubc.ca

[2] Huawei Technologies Canada, Burnaby, BC, Canada

[3] The University of British Columbia (Okanagan), Kelowna, BC, Canada
shan.du@ubc.ca

Abstract. Unlike 2D face images, obtaining a 3D face is not easy. Existing methods, therefore, create a 3D face from a 2D face image (3D face reconstruction). A user might wish to edit the reconstructed 3D face, but 3D face editing has seldom been studied. This paper presents such method and shows that reconstruction and editing can help each other. In the presented framework named NEO-3DF, the 3D face model we propose has independent sub-models corresponding to semantic face parts. It allows us to achieve both local intuitive editing and better 3D-to-2D alignment. Each face part in our model has a set of controllers designed to allow users to edit the corresponding features (e.g., nose height). In addition, we propose a differentiable module for blending the face parts and making it possible to automatically adjust the face parts (both the shapes and the locations) so that they are better aligned with the original 2D image. Experiments show that the results of NEO-3DF outperform existing methods in intuitive face editing and have better 3D-to-2D alignment accuracy (14% higher IoU) than global face model-based reconstruction. Code available at https://github.com/ubc-3d-vision-lab/NEO-3DF.

Keywords: 3D face editing · 3D-to-2D face alignment

1 Introduction

It has become popular for individuals recently to allow their faces to be used in 3D virtual reality (VR) applications and user-generated content games. For example, in a virtual conference or a virtual get-together meeting, the use of reconstructed

J. Gregson—This work was done when James Gregson was at Huawei Technologies Canada.

S. Du—This work was supported by the University of British Columbia (Okanagan) [GR017752].

Supplementary Information The online version contains supplementary material available at https://doi.org/10.1007/978-3-031-26319-4_5.

L. Wang et al. (Eds.): ACCV 2022, LNCS 13841, pp. 72–88, 2023.
https://doi.org/10.1007/978-3-031-26319-4_5

3D faces that resemble those of the real users helps the participants to identify the persons behind the avatars. Enabling a user to change some features, i.e. edit their reconstructed 3D face, can solve a main problem in existing 3D face reconstruction methods. As the reconstructed 3D face is usually not accurate enough, the user may want to correct their reconstructed 3D face. Also when the user is not satisfied with some parts of their actual face, then they can modify those parts. While 3D face reconstruction has been well-studied in the past two decades [3,8,9,17,35,38], editing of 3D reconstructed faces has not. We believe there is a great potential for deep learning-based 3D face editing in many real-world applications, including 3D gaming, film-making, VR, and plastic surgery. 3D face editing can also benefit 2D face editing because the 3D face can provide explicit geometry information to 2D generative models [7,27,37].

3D editing is a challenging task. Editing a 3D shape usually requires strong spatial thinking ability, and traditional 3D editing tools are designed for use by artists. Moreover, editing of 3D faces is even more difficult than editing many other 3D shapes. This is because people are very sensitive to the appearance of a human face. Minor modifications of a 3D face could cause very different feelings in people. Because facial parts such as eyebrows and nose are relatively independent, segmenting the 3D face into parts and independently editing each part makes 3D face editing easier to handle [13,14,34]. In this work, we want to take one step further to bridge single-image 3D face reconstruction (SIFR) and 3D face editing. Since ambiguities such as depth and lighting will cause the reconstructed 3D face to be different from the actual face, the SIFR is an ill-posed problem. Recent works on SIFR display limited improvement in reconstruction when compared to the 3D ground-truth [8,11], whereas 3D-to-2D alignment, i.e. aligning the 3D face with the image it was reconstructed from, is still an open challenge [25]. Therefore, we also explore improving the 3D-to-2D alignment by taking advantage of 3D face editing. Inspired by [17,38], we use 2D face semantic segmentation/parsing to provide ground truth in aligning the 3D face to the image. This requires us to segment our 3D face model based on the 2D face parsing, which is different from how a 3D face is segmented in [13,14,34].

In this paper, we propose a novel framework (which we named NEO-3DF) that allows the reconstructed 3D face to be edited locally. The features that can be edited are defined by the designer, but the extent of editing is determined by the user. Most SIFR methods use a global face model (i.e., the 3D face shape is modeled as a whole) which makes it extremely challenging for editing [8,11,21]. Our 3D face modeling however uses different sub-models for the different semantic segments (parts) of the face. This ensures that editing the shape of one part of a face does not cause changes in other parts, resulting in easier and more convenient editing. This approach also allows us to improve the 3D reconstruction accuracy using 2D supervision (see Sect. 3). We developed a way of using the As-Rigid-As-Possible (ARAP) method [33] to blend the face segments (parts), so the face appears natural at the segment boundaries. Importantly, we propose a differentiable version of ARAP that makes it possible to use gradient descent-based optimization to improve the alignment between the 3D reconstructed face segments and those of the original 2D image (i.e., 3D-to-2D alignment).

In summary, our main contributions are: (1) We proposed NEO-3DF, the first method that couples single-image 3D face reconstruction and intuitive 3D face editing. (2) The NEO-3DF is face-parts-based and follows 2D semantic face segmentation, which makes it possible for an automatic shape adjusting process. (3) We proposed a differentiable version of ARAP to allow us to use 2D face segmentation as guidance to adjust the reconstructed 3D face shape automatically. (4) We associated editing controllers with the latent space and justified the effectiveness of its use in an automatic 3D-to-2D alignment process.

2 Related Works

3D Face Modeling and Reconstruction. The 3D face is usually represented by the 3D mesh structure, where the vertices and facets define the surface of the 3D face. A straightforward way to create a 3D face is by using a weighted average of a linear combination of a set of example 3D faces [36]. An improved version of this approach, the 3D Morphable Face Model (3DMM), transforms the scanned and processed example 3D faces to a vector space representation, using Principal Component Analysis (PCA) [3]. The PCA captures the primary modes of variation in the pre-collected 3D face dataset. Although 3DMM was proposed two decades ago, it is still widely used in current human face-related research and applications, such as 3D face reconstruction [24,38], controllable face image generation [7,37], and 3D face animation [2]. In recent years, research on the use of deep learning to learn the general 3D face model shows the advantage of neural network-based 3D face models over traditional 3DMMs [4,29,35].

Single-image 3D face reconstruction (SIFR) is an important application of both 3DMMs and neural network-based 3D face models [9]. The task is to reconstruct a 3D face from a given single-view face image. In most cases, the camera intrinsic and lighting conditions are unknown. In addition, for images where the faces appear identical, the actual 3D shape of these faces can be different due to potential ambiguities. Thus, the SIFR is an ill-posed problem, and we can only approximately reconstruct a 3D face from a single image. However, compared with accurate 3D reconstruction methods, using expensive 3D scanning devices or multi-view images, SIFR is more accessible to users of consumer electronics. We can roughly group the existing SIFR methods as fitting-based methods [3,38] and learning-based methods [8,12,17,35]. Generally, both fitting-based and learning-based methods use a 3D face model such as linear 3DMM [8,11,17,38] or non-linear neural network-based model [12,35] to serve as a priori knowledge. The main difference is that the fitting-based methods will iteratively optimize the variables to make the generated 3D face look more like the face in the photo. The learning-based method uses another parametric model (usually a neural network) to regress those variables.

General-Purpose 3D Mesh Editing. Many 3D mesh editing methods allow the user to modify the location of some vertices (constraint vertices) to edit the 3D mesh. The cage-based deformation (CBD) method warps the space enclosed by "cages" to deform the mesh surface [19]. CBD is usually used for large-scale deformations such as human body postures and is not ideal for 3D face editing because the latter focuses more on local details. Building the appropriate "cages" is also a challenging task. The ARAP method is a detail-preserving mesh surface deformation method [33]. ARAP assumes that by making the local surface deformation close to rigid, one can preserve the local surface details. In editing 3D face shapes, neither CBD nor ARAP can prevent improper edits, this is because these methods do not have prior knowledge on 3D faces. In comparison, the blendshape methods solve the face shape editing problem by using a set of predefined shape offsets to deform a given face shape [22]. Although the concept of the blendshape is straightforward, developing intuitive meaningful blendshapes is labor-intensive. 3DMMs can be considered as a special type of blendshape method, where the blendshapes are derived from 3D scans and compressed by PCA. However, the eigenvectors do not have intuitive meaning.

3D Face Editing. In the video gaming industry, 3D face editing systems are usually designed by artists. For example, the artists can define meaningful controllers (e.g., nose height) with corresponding handcrafted blendshapes to allow users to easily change the shape of a 3D face [32]. Foti et al. proposed the Latent Disentanglement Variational Autoencoder (LD-VAE) and demonstrated that it can be used for constraint vertex-based face shape editing [13]. Because latent disentanglement in terms of the face parts is an objective during the training of LD-VAE, the local editing result is better than using a traditional 3DMM. Whereas this type of editing still requires the user to have some skills in arts. Ghafourzadeh et al. developed a face editing system that uses a face parts-based 3DMM (PB-3DMM) [14]. This method decomposes the face shape into five manually selected non-overlapping parts to gain local control. The models of these face parts are built using PCA on 135 3D face scans. They also use the method proposed by Allen et al. to find a linear mapping between anthropometric measurement space and the PCA coefficient spaces to achieve intuitive control/editing (such as changing the nose height) [1]. PB-3DMM is the first data-driven 3D face model that supports intuitive local editing. However, like traditional 3DMM [3], the PB-3DMM still relies on 3D scans, which are expensive to collect and process. Moreover, PB-3DMM uses a specially designed low-polygon mesh topology and is not optimized for single-image 3D face reconstruction and 3D-to-2D face alignment. We briefly summarize some important factors of PB-3DMM, LD-VAE and the proposed NEO-3DF in Table 1.

Table 1. Comparison of the NEO-3DF and other parts-based 3D face methods.

Method	Model	Mesh info. (vertices)	Face parsing	Intuitive editing	Parts-blending	Back-prop.	SIFR
PB-3DMM [14]	Multiple PCAs	Specially designed (6,014)	✗	✓	✓	✗	✗
LD-VAE [13]	Single VAE	FLAME [23] (71,928)	✗	✗	N/A	✓	✗
NEO-3DF (proposed)	Multiple VAEs	Modified BFM [8,26] (35,709)	✓	✓	✓	✓	✓

3 The Proposed Method

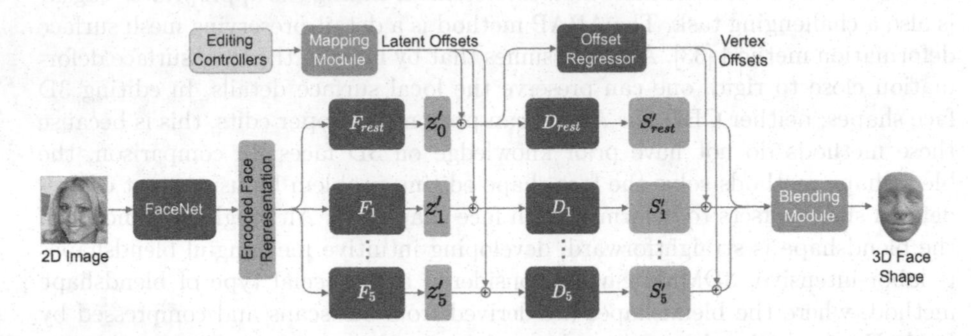

Fig. 1. Flowchart of the proposed NEO-3DF. FaceNet: The convolutional neural network encodes the 2D face image to a vector representation. Networks F_i estimate latent representations z'_i of corresponding face segments, and shape decoders D_i reconstruct the segment shape S'_i from z'_i. The Offset Regressor predicts offsets used to assemble the segment shapes to their correct location.

Overview. The proposed framework (see Fig. 1) supports single-view 3D face reconstruction and its further editing. In this work, we first reconstruct a neutral 3D face shape. The expression and texture of the 3D face can be added later through existing methods (such as [5,24]). We use a pre-trained FaceNet [31] as the face image encoder. In this work we segment our 3D face topology into six segments (parts). Each segment has its own editing controllers. These controllers determine the variables that can be changed in this segment (e.g., nose height). For each segment i, we have a regressor network F_i that estimates the latent encoding z'_i of the segment. Following each encoder F_i is the shape decoder D_i, which reconstructs the shape S'_i. The set of editing controllers enables intuitive face shape editing. To find the mapping between the editing controllers space and the neural network latent space, we use a similar way as used in [1,14]. The main idea is to construct a linear system representing the mapping function and solve the unknown parameters of the mapping function. The blending module uses ARAP optimization to derive an overall natural-looking face shape.

Semantic Face-Segment Based 3D Face Model. To develop a face model based on semantic face-segment, we use the Basel Face Model (BFM) [26] as the base model. We consider the face to have six segments: five semantic face segments (*eyebrows*, *eyes*, *nose*, *upper* and *lower lips*) and one segment for the *rest* of the face. Following [8], we remove the neck and ears from the original BFM mesh topology. The resulting 3D face mesh has 35,709 vertices. To automatically segment the 3D face, we use a similar approach as in [17]. The main idea is to render the randomly synthesized 3D faces to 2D, then run the 2D face parsing model to get 2D parsing labels. Finally, we can map the 2D pixel-level labels back to 3D vertices. This automatic 3D face segmentation process ensures the 3D vertex labels are consistent with the mainstream face image semantic parsing, which is essential in our following work on fine-tuning the model, using ground-truth 2D face parsing masks as supervision. Figure 2 (left) shows the face parts of our 3D face mesh topology.

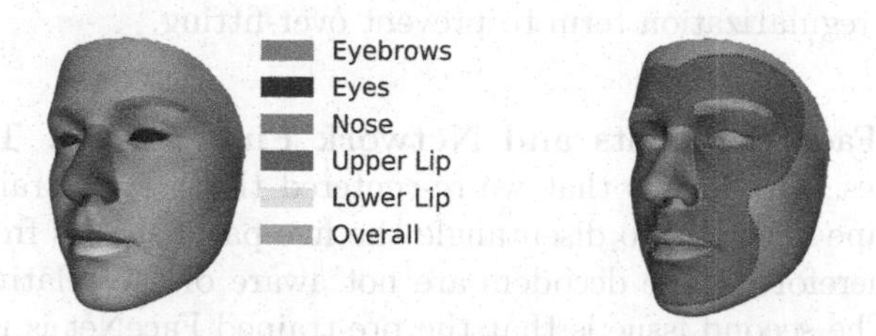

Fig. 2. Left: The six face segments considered. Right: Shape boundaries (semantic segment boundaries in red lines, outer boundary in blue line) and transition area (in green). (Color figure online)

We denote the 3D shape S as a set of vertices $S = \{v_i | v_i \in \mathbb{R}^3\}_{i \in V}$, where V is the set of vertex indices of the shape S, v_i represents the 3D location of the ith vertex in S. We use $S_k (k \in \{1..5\})$ to represent the five semantic face segments of eyebrows, eyes, nose, upper lip, and lower lip respectively. For convenience, we call S_ks the five-segments. The shape of the rest segment S_{rest} is a little different from S_ks, because it has a one-ring overlap with each of the S_ks. This one-ring overlap is important in our ARAP-based blending module which provides the curvature information at the segment boundaries (see Fig. 2).

We use Multi-Layer Perceptron (MLP) based Variational Autoencoders (VAEs) to learn the 3D shape representation. In total, we train six VAEs for each face segment i (E_i and D_i are encoder and decoder respectively). For S_{rest}, the latent representation is a vector $z_0 \in \mathbb{R}^{1\times 30}$, and for S_ks, $z_k \in \mathbb{R}^{1\times 10}$. We pre-process each segment shape in the dataset, except S_{rest}, to make the geometric center of the dataset's 3D bounding box as the origin of the 3D Cartesian coordinate system. The reason is that we expect the shape decoders of the five-segments to only care about the shape itself rather than the relative location of the segment on the face. The loss in training the VAEs is defined as:

$$\mathcal{L}_{VAE} = \lambda_{recon}\mathcal{L}_{recon} + \lambda_{KL}\mathcal{L}_{KL} + \lambda_{smooth}\mathcal{L}_{smooth}, \tag{1}$$

where $\mathcal{L}_{recon}$ is the shape reconstruction loss, $\mathcal{L}_{KL}$ is the Kullback-Leibler (KL) divergence loss, $\mathcal{L}_{smooth}$ is the cotangent-weight Laplacian loss, and the λs are loss term weights. Denote v and v' as the vertices of the ground-truth shape S and reconstructed shape S' respectively. We compute the reconstruction loss $\mathcal{L}_{recon}$ as $\sum_{i \in V} \|v_i - v'_i\|_2^2$. The $\mathcal{L}_{KL}$ is the KL divergence of the distribution of $z \in \mathbb{R}^{1\times d_z}$ from a multivariate normal distribution $\mathcal{N}(\mathbf{0}, \mathbf{I})$, where $\mathbf{I} \in \mathbb{R}^{d_z \times d_z}$ is the identity matrix and $\mathbf{0} \in \mathbb{R}^{1\times d_z}$ is a vector of zeros. The Laplacian loss $\mathcal{L}_{smooth}$ helps the decoder to generate smooth mesh surfaces.

Because the input to E_i is the 3D shape, we still need to couple the 2D face image encoding with the 3D shape decoders D_i. To do this, after VAE training we discard the VAE encoders E_i and add another set of regressor networks F_i between the 2D face image encoding and the 3D shape decoders D_i. To pre-train F_i, we freeze the parameters of the face image encoder and shape decoders. The loss function is the same as (1) except for discarding the $\mathcal{L}_{smooth}$ term. Here, $\mathcal{L}_{KL}$ acts as a regularization term to prevent over-fitting.

Assembling Face Segments and Network Fine-Tuning. There are two remaining issues. The first is that we re-centered the S_ks for training the corresponding shape decoders to disentangle the five-parts shapes from the global face shape. Therefore, these decoders are not aware of the relative position of S_ks to S_{rest}. The second issue is that the pre-trained FaceNet is not optimized for 3D face reconstruction.

To address the first issue, we train an offset regressor that takes S_{rest}'s latent representation z_0' as input to predict an offset $(0, \Delta \mathrm{y_k}, \Delta \mathrm{z_k})$ for each reconstructed segment shape S_k'. Because the human face is bilaterally symmetric, we do not add any offset along the x-axis. After adding the offsets, we can assemble the S_ks with S_{rest} to get a full 3D face $S'_{overall}$. Note that $S'_{overall}$ is likely to have visible segment boundaries without proper blending.

To address the second issue, we fine-tune all the networks except the shape decoders D_is to improve the reconstruction accuracy. Besides using the target shapes as supervision, we leverage the ground-truth 2D face parsing masks as additional supervision. The loss function for fine-tuning is defined as:

$$\mathcal{L} = \lambda_s \mathcal{L}_s + \lambda_p \mathcal{L}_p + \lambda_{KL} \mathcal{L}_{KL}. \tag{2}$$

$\mathcal{L}_s$ is the shape loss between $S_{overall}$ and $S'_{overall}$, the computation is the same as $\mathcal{L}_{recon}$. Define $\mathrm{P} \in \mathbb{R}^{\mathrm{H} \times \mathrm{W} \times 5}$ as the ground-truth parsing mask of S_ks, where H and W are the height and width of the mask. P_{ijk} represents the binary label that indicates whether a pixel at image location (i, j) belongs to the kth $(k \in \{1..5\})$ segment or not. We include the estimated expression and pose and use a differentiable renderer to render $S'_{overall}$ to 2D, denoted by P'. The renderer is modified to use the vertex one-hot semantic labels as five-channel colors for shading. To prevent the gradient from vanishing when the displacement between P and P' is too large, we follow the idea in [17] to use Gaussian filter (we set the standard deviation to be 1 and filter size to be 10×10) to soften both P and P'. The parsing loss is then defined as: $\mathcal{L}_p = \|\mathrm{P} - \mathrm{P}'\|_2$. Here, the $\mathcal{L}_{KL}$ acts as a regularization term.

Face Segments Blending. An evident challenge for the segment-based face model is that it is not straightforward to ensure the transition between segments is natural-looking. Simply smoothing the region between different face segments could destroy the high-frequency local structure. Our goal is to preserve the surface details as much as possible so that the final shape will still looks human and to make the transitions between face segments look natural without abrupt boundaries between segments. To achieve this, we use the As-Rigid-As-Possible (ARAP) [33] for mesh deformation optimization in our blending module.

Our blending module will only deform some areas on S'_{rest} that connects it to the five-segments. We call these areas the transition area (see Fig. 2 right). We define the transition area using the average BFM face shape. The transition area covers the vertices on S'_{rest} whose Euclidean distances are less than $30mm$ from their nearest vertices on the segment boundaries (see Fig. 2 right, the red lines). The outer contour of the transition area is the outer boundary (see Fig. 2 right, the blue lines). We use the ARAP method to deform the shape of the transition area. The constraints are formed by two groups of vertices – vertices from the five-segments lying on the segment boundaries and vertices from S'_{rest} lying on the outer boundary. In our experiments, we run the ARAP optimization for three iterations.

Intuitive Face Editing. We expect the editing to be intuitive. For example, one can increase/decrease the nose width by an adjustable strength. The overall idea is to find mappings between the pre-defined facial feature measurements space and the latent encoding space. Different from [14], our editing is not based on the exact measurement values (e.g., the exact width of the nose bridge). It is based on the direction (+ for increase, − for decrease) and strength (how many standard deviations from the mean shape). We refer to some studies on facial feature measurements [10,16,28,30] and derive a list of features for editing purposes (see supplementary material). We grouped the features by the six segments. For example, nose height is a feature of the nose segment.

We denote N as the number of face shapes in the dataset and m as the number of features of the shape S (S can be the shape of any of the six segments). For each S in the dataset, we record the measurements of its features (the measurement is done automatically using pre-defined measurement rules) as a row in the matrix $X \in \mathbb{R}^{N \times m}$, and also record its latent encoding vector z as a row in the latent matrix $Z \in \mathbb{R}^{N \times d_z}$. We compute the average face shape $\bar{S}$ over the dataset and record its latent encoding as $\bar{z}$. Next, we subtract the measurements of average shape $\bar{S}$ from each row in X and denote the resultant matrix as ΔX. Similarly, we subtract $\bar{z}$ from each row in Z and denote the resultant matrix as ΔZ. Assume there is a linear mapping $M \in \mathbb{R}^{m \times d_z}$ such that $\Delta X M = \Delta Z$, we can estimate M by solving the linear system $M = \Delta X^{\dagger} \Delta Z$, where $\Delta X^{\dagger}$ is the Moore-Penrose pseudo inverse of ΔX.

We define the editing controllers as a vector $c \in \mathbb{R}^{1 \times m}$. Each value in c controls a corresponding feature. We compute the standard deviation of each column in ΔX, and record it as $\Sigma \in \mathbb{R}^{1 \times m}$. Finally, we represent the editing as the modification of latent encoding space: $\tilde{z} = z + \Delta z = z + (c \circ \Sigma)M$, where z is the latent encoding of the original shape S, and $\circ$ is the Hadamard product operator. Using the shape decoder to decode the new latent encoding $\tilde{z}$, we get the edited shape. We need to compute an M and a Σ for each segment.

Differentiable Blending Module in Shape Adjusting. To further improve the 3D-to-2D alignment accuracy, we propose a novel differentiable version of ARAP in our blending module. Our differentiable ARAP is fully functional, transforming traditional ARAP optimization iterations into a sequence of differentiable operations through which loss function gradients can back-propagate. Our method is different from [38], which has proposed the use of the ARAP energy as a loss function but did not carry out the ARAP optimization itself. In contrast, our method back-propagates through the ARAP algorithm, including estimating neighborhood rotations and solving the linear system. In principle, our ARAP deformer module can be applied to many deep learning-based geometry processing tasks and used as a component in end-to-end training. However, here we use it to fine-tune the editing controllers due to the high vertex count of our facial model.

We define n_p as the number of vertices of the face mesh, and n_c as the number of constraint vertices. We represent the vertex positions of the 3D mesh as a matrix $P \in \mathbb{R}^{n_p \times 3}$, and the constraint vertex positions as a matrix $H \in \mathbb{R}^{n_c \times 3}$. We first build a sparse constraint matrix $C \in \mathbb{R}^{n_c \times n_p}$, such that $C_{ij} = 1$ only if the jth vertex of the mesh is the ith constraint vertex. Then we compute the combinatorial Laplacian of the 3D mesh denoted by $L \in \mathbb{R}^{n_p \times n_p}$. The reason for using the combinatorial Laplacian rather than the cotangent Laplacian is that the former only depends on the mesh topology so that we can reuse it on different 3D faces with the same mesh topology. We define $P' \in \mathbb{R}^{n_p \times 3}$ as the new vertex positions, R as the right-hand-side of linear system $AW = R$, where A is a sparse matrix (3) and $(W_{ij})_{i \in \{1..n_p\}, j \in \{1..3\}}$ is P'. The estimation of R needs both P and P', thus the optimization of P' is done by alternatively estimating R and solving the linear system $AW = R$ for multiple iterations. Because A only depends on the mesh topology, the system only needs to be formed and inverted once. Algorithm 1 is the pseudo-code of the proposed differentiable ARAP, where n_{iter} is the number of iterations we would like to run the ARAP optimization. In our experiment, we choose $n_{iter} = 3$. Please refer to the supplementary material for more details of our differentiable ARAP and the pseudo-code of *estimate_rhs*().

Unlike our fine-tuning of the network, we do not apply the Gaussian filter on both the ground-truth parsing P and the rendered parsing P′, instead we only apply distance-transform to P. We empirically find that this approach leads to a better result. Importantly, we directly adjust the editing controllers instead of adjusting the segment latent representations. In the optimization process, we only minimize $\|\mathrm{P} - \mathrm{P}'\|_2$. The optimization is performed on a single shape each time.

$$A = \begin{bmatrix} L^{\mathrm{T}}L & C^{\mathrm{T}} \\ C & \mathbf{0} \end{bmatrix} \in \mathbb{R}^{(n_p+n_c)\times(n_p+n_c)}. \tag{3}$$

Algorithm 1. Differentiable ARAP

Input: P, H, L, and A^{-1}
Output: P'
Require: $n_{iter} \geq 1$
$iter \leftarrow 0$
while $iter \leq n_{iter}$ **do**
  if $iter = 0$ **then**
    $R \leftarrow \begin{bmatrix} L^T LP \\ H \end{bmatrix}$ //Initialize the right-hand-side.
  else
    $R \leftarrow \begin{bmatrix} estimate\_rhs() \\ H \end{bmatrix}$ //Estimate new right-hand-side.
  end if
  $W \leftarrow A^{-1}R$ //Solve the linear system $AW = R$.
  $P' \leftarrow (W_{ij})_{i \in \{1..n_p\}, j \in \{1..3\}}$ //Update new vertex positions.
  $iter \leftarrow iter + 1$
end while

4 Experiments

Data and Training. The face images we use in our experiments are from CelebAMask-HQ dataset (30,000 images) [20] and FFHQ dataset (70,000 images) [15]. We run Deng et al.'s method (Deep 3DMM) [8] on both datasets to get the estimated BFM coefficients for each image. An alternative option is to use the synthesis-by-analysis method [3] to estimate the coefficients, but it would be slow. Then, we use the BFM coefficients to recover the 3D faces and get 100,000 3D faces. Because the CelebAMask-HQ dataset comes with ground-truth semantic face parsing masks, we use its first 5,000 images to form the validation set. The rest of the 95,000 images and their 3D reconstructions are used to train the models. We use the 25,000 training images from the CelebAMask-HQ dataset and their corresponding semantic parsing masks to fine-tune our network. We also use all the 95,000 3D reconstructions in the training set to compute the mappings for intuitive editing. The loss term weights are set as follows: $\lambda_{recon} = 1$, $\lambda_{KL} = 0.01$, $\lambda_{smooth} = 1$, $\lambda_s = 1$, $\lambda_p = 1$. We use Adam optimizer with a constant learning rate of 10^{-4}. The training batch size is 16. Our framework has an average per-vertex mean squared error of 9.0×10^{-3} on validation data (3D mesh data is measured in millimeters).

3D-to-2D Segments Alignment. In the single-image 3D face reconstruction task, the best that can be done is to make the rendered 3D face look as close as possible to the face in the image. It is challenging for existing global-based 3D face reconstruction methods to make each 3D semantic face segment better align with its corresponding 2D image semantic segment. We presume that the global-based 3D face model has strong constraints to ensure the entire 3D face looks as natural as possible. The drawback is that small face segments such as the eyes and the lips may not look realistic and are prone to be misaligned.

In addition, the human visual system is very sensitive to tiny differences in a face. Thus, the global-based methods are usually not able to provide satisfactory reconstructions. Some research works [6,18] try to mitigate this problem by using more realistic textures to cheat the human viewers. In this work, we take advantage of our segment-based face model to improve the shape accuracy by tuning each 3D semantic segment, so it better aligns with the image. Learning realistic texture will also benefit from better aligned 3D shapes [24]. We conjecture that improved registration of 3D geometry to semantic face segments will assist texture reconstruction models in learning realistic textures, although it is not a focus of the current work.

We use the average Intersection-over-Union (IoU) metric to evaluate the 3D-to-2D alignment accuracy. Table 2 compares the deep 3DMM (global method), PB-3DMM, LD-VAE, and the proposed NEO-3DF. Note that both PB-3DMM and LD-VAE are not designed for single-image 3D face reconstruction, and both methods originally use different mesh topologies other than BFM, which we use. To compare with PB-3DMM and LD-VAE, we use the BFM mesh topology and our parsing-based face segmentation. Because both PB-3DMM and LD-VAE are local models, we use the similar network architecture of NEO-3DF to give both methods the single-image 3D face reconstruction ability. For PB-3DMM, the shape decoders are from BFM. For LD-VAE, the shape decoder is pre-trained using the original LD-VAE code, but on our dataset. We fine-tuned both variants under the same setting as training NEO-3DF.

Table 2. 3D-to-2D alignment results on validation data. Metric: average IoU.

Part name	Deep 3DMM [8]	PB-3DMM [14]	LD-VAE [13]	NEO-3DF
Eyebrows	0.312	0.302	0.279	**0.363**
Eyes	0.387	0.377	0.372	**0.553**
Nose	0.766	0.728	0.735	**0.789**
Upper lip	0.483	0.482	0.436	**0.531**
Lower lip	0.468	0.424	0.418	**0.614**
All five parts	0.593	0.566	0.560	**0.616**

For single-sample shape adjusting, we use Adam optimizer with a learning rate of 10^{-3}. We run 20 gradient-descent optimization iterations for each shape, and the shape adjusting takes around 31 s for each iteration (on a workstation with one Nvidia Tesla V100-PCIe GPU). On the validation data, the average IoU of five-segment shapes after fitting is 0.683 (discussed in Ablation Study), which is 11% higher than the raw network prediction (0.616 as shown in Table 2). Figure 3 shows some visualization of the alignment performance comparison. For each method, we use one column to show the rendered 3D face on top of the original image, and another column to show the union minus the intersection of P and P′. If the shapes of the face segments are better aligned to the 2D image, the black areas should be less visible.

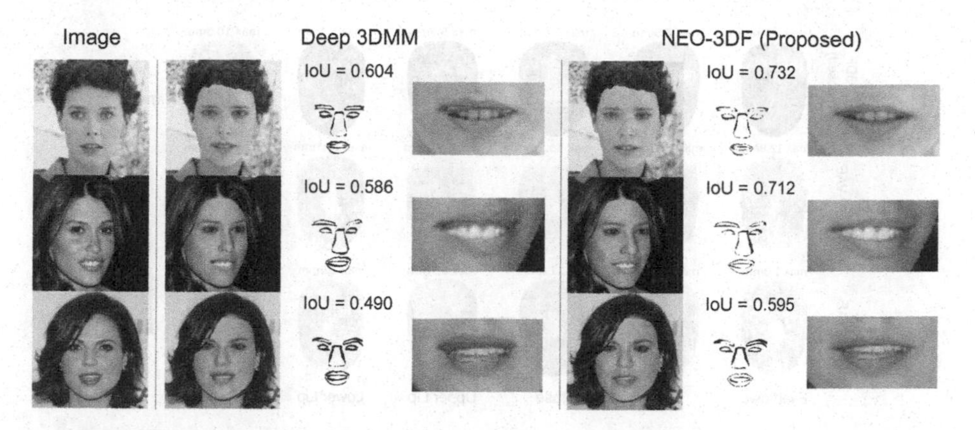

Fig. 3. 3D-to-2D semantic face segments alignment results. Compared with Deep 3DMM, we can see that our method gives better aligned facial segments (yields higher segmentation accuracy, as measured by IoU) and produces more consistent rendered results with original images.

Intuitive Editing. We compare our method with global VAE and LD-VAE [13]. We use the same MLP network structure for both global VAE and the LD-VAE, where the decoder structure is the same as non-linear 3DMM's shape decoder [35]. Our shape decoders have a similar decoder structure (same number of layers and hidden neurons), but each has a smaller output layer because the shape of a segment has fewer vertices than the entire face shape. We group the controllers by segments. In each group, we tune each controller separately to be -3σ and $+3\sigma$, then aggregate all the edited shapes and show the maximum change of each vertex location (Euclidean distance measured in millimeters) from the original mean face shape. The results are visualized as rendered 2D heatmaps (see Fig. 4). Figure 5 shows the results of applying the same set of edits on the nose (*Tip* -3σ, *Breadth* $+3\sigma$, *Bridge Width* -3σ) using different methods. To demonstrate the application of our framework in customizing a reconstructed 3D face, we randomly apply a sequence of edits to the reconstructed faces. Figure 6 shows some example intuitive editing results.

Ablation Study. We first investigate the necessity of employing the proposed offset regressor and blending module. Figure 7 shows the original shape and the editing we made (lift the nose tip) using our unmodified framework, our framework without offset regressor, and our framework without blending module. Note that, in this experiment, only the nose is editable. Without the offset regressor, we need to use the absolute vertex locations to train the five-segments decoders. So the relative location of a five-part shape is still entangled into the latent representation. As a result, after removing the offset regressor, we see that the entire nose moves up when lifting the nose tip. The issue of eliminating the blending module is evident, as we can see the juncture around the nose.

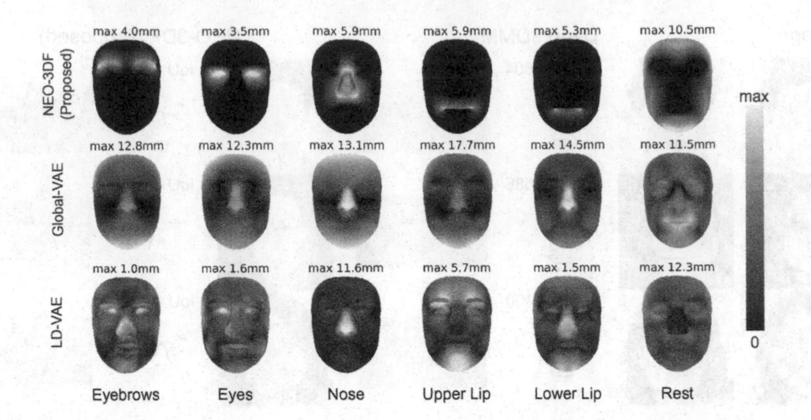

Fig. 4. Vertex location change heatmaps. For visualization purposes, the values in each heatmap are normalized to be from 0 to 1.0. The value above each heatmap indicates the highest value of that heatmap. We can see that our method gives near-perfect disentanglement while the other methods cannot isolate local edits well.

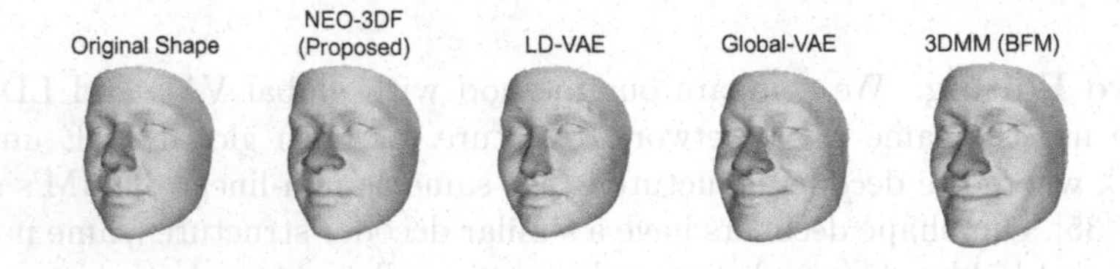

Fig. 5. Nose shape editing comparison.

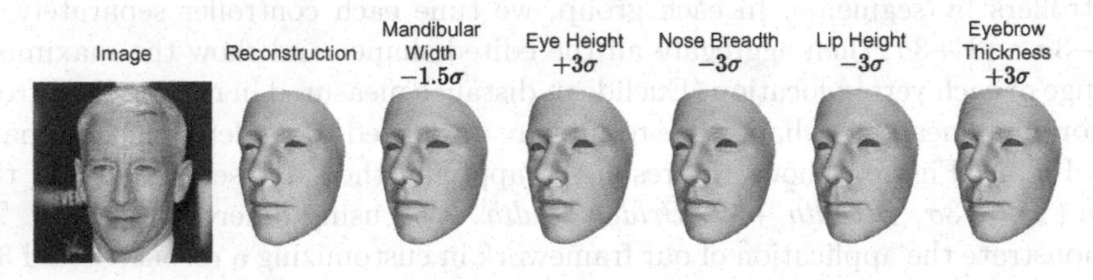

Fig. 6. Reconstruction and further editing examples. The edits are made cumulatively from left to right. Editing via our framework is more intuitive and expressive since subsequent changes (on different segments) will not affect earlier ones.

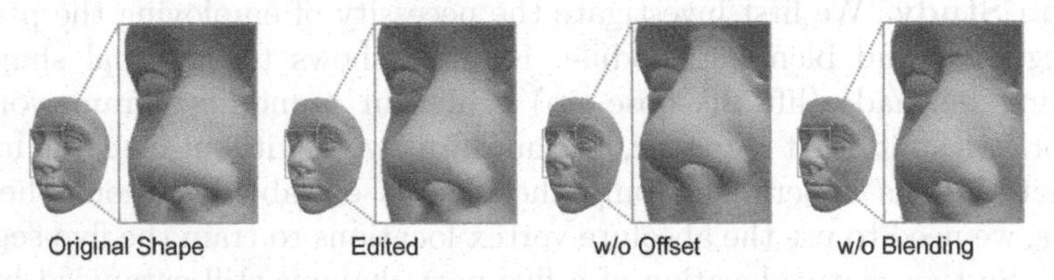

Fig. 7. Ablation study shows that the offset regressor and blending module are necessary for our framework. Without the offsets, alignment between segments is compromised, while visible seams appear around the edited segment without the blending module.

We also investigate different ways of improving the 3D-to-2D alignment in our fitting-based single shape adjusting (see Table 3). We find that with the help of our differentiable ARAP, the best results are obtained by the fine-tuning of the editing controllers and the shape offsets. This yields 7% higher IoU than by directly adjusting the latent encodings and offsets (IoU is 0.632). Our explanation for this finding is that the editing controllers' space is less complex and more interpretable than the latent encoding space, thus reducing the learning complexity. Interestingly, we noticed that it does not work if we do not optimize the offsets and only optimize the editing controllers. We presume that the face segments must be roughly aligned to provide useful loss information.

Table 3. Single-sample automatic 3D part shapes adjustment results. (Metric: IoU averaged on validation data)

	w/o Blending	w/ diff. ARAP Blending
Latent representations only	0.621	0.626
Editing controllers only	0.581	0.566
Latent representations & offsets	0.629	0.632
Editing controllers & offsets	0.649	**0.683**

5 Conclusion

We proposed NEO-3DF, a 3D face creation framework that models the different segments of the 3D face independently. NEO-3DF is the first method that couples single-image 3D face reconstruction and 3D face intuitive editing. To train NEO-3DF, we created a 3D face dataset using existing large 2D face image datasets and an off-the-shelf 3D face reconstruction method named deep 3DMM. The face segment-based model of the 3D face structure makes face editing more intuitive and user-friendly. It also allows fine-tuning the reconstructed 3D face by using 2D semantic face segments as guidance and yields 14% improvement in IoU. We also proposed a differentiable version of ARAP to obtain an end-to-end trainable framework, which enables the automatic adjusting of the variables in the editing controllers. This resulted in an additional 5% improvement in IoU. The main limitation of our current framework is that the differentiable ARAP is computationally expensive and only used to fine-tune the result at the last construction stage. We will explore a less expensive method for blending the face segments in the future. We believe our concept of combining the 3D face editing and reconstruction also sets a new direction in 3D face reconstruction research.

References

1. Allen, B., Curless, B., Popović, Z.: The space of human body shapes: reconstruction and parameterization from range scans. ACM Trans. Graph. (TOG) **22**(3), 587–594 (2003)
2. Bai, Z., Cui, Z., Liu, X., Tan, P.: Riggable 3D face reconstruction via in-network optimization. In: Proceedings of the IEEE/CVF Conference on Computer Vision and Pattern Recognition, pp. 6216–6225 (2021)
3. Blanz, V., Vetter, T.: A morphable model for the synthesis of 3D faces. In: Proceedings of the 26th Annual Conference on Computer Graphics and Interactive Techniques, pp. 187–194 (1999)
4. Bouritsas, G., Bokhnyak, S., Ploumpis, S., Bronstein, M., Zafeiriou, S.: Neural 3D morphable models: spiral convolutional networks for 3D shape representation learning and generation. In: Proceedings of the IEEE/CVF International Conference on Computer Vision, pp. 7213–7222 (2019)
5. Chang, F.J., Tran, A.T., Hassner, T., Masi, I., Nevatia, R., Medioni, G.: Expnet: landmark-free, deep, 3D facial expressions. In: 2018 13th IEEE International Conference on Automatic Face & Gesture Recognition (FG 2018), pp. 122–129. IEEE (2018)
6. Chen, A., Chen, Z., Zhang, G., Mitchell, K., Yu, J.: Photo-realistic facial details synthesis from single image. In: Proceedings of the IEEE/CVF International Conference on Computer Vision, pp. 9429–9439 (2019)
7. Deng, Y., Yang, J., Chen, D., Wen, F., Tong, X.: Disentangled and controllable face image generation via 3D imitative-contrastive learning. In: Proceedings of the IEEE/CVF Conference on Computer Vision and Pattern Recognition, pp. 5154–5163 (2020)
8. Deng, Y., Yang, J., Xu, S., Chen, D., Jia, Y., Tong, X.: Accurate 3D face reconstruction with weakly-supervised learning: from single image to image set. In: Proceedings of the IEEE/CVF Conference on Computer Vision and Pattern Recognition Workshops (2019)
9. Egger, B., et al.: 3D morphable face models-past, present, and future. ACM Trans. Graph. (TOG) **39**(5), 1–38 (2020)
10. Farkas, L.G., Kolar, J.C., Munro, I.R.: Geography of the nose: a morphometric study. Aesthetic Plastic Surg. **10**(1), 191–223 (1986)
11. Feng, Y., Feng, H., Black, M.J., Bolkart, T.: Learning an animatable detailed 3D face model from in-the-wild images. ACM Trans. Graph. (ToG) **40**(4), 88:1–88:13 (2021)
12. Feng, Y., Wu, F., Shao, X., Wang, Y., Zhou, X.: Joint 3D face reconstruction and dense alignment with position map regression network. In: Proceedings of the European Conference on Computer Vision (ECCV), pp. 534–551 (2018)
13. Foti, S., Koo, B., Stoyanov, D., Clarkson, M.J.: 3D shape variational autoencoder latent disentanglement via mini-batch feature swapping for bodies and faces. In: Proceedings of the IEEE/CVF Conference on Computer Vision and Pattern Recognition, pp. 18730–18739 (2022)
14. Ghafourzadeh, D., et al.: Local control editing paradigms for part-based 3D face morphable models. Comput. Anim. Virt. Worlds **32**(6), e2028 (2021)
15. Karras, T., Laine, S., Aittala, M., Hellsten, J., Lehtinen, J., Aila, T.: Analyzing and improving the image quality of stylegan. In: Proceedings of the IEEE/CVF Conference on Computer Vision and Pattern Recognition, pp. 8110–8119 (2020)

16. Kesterke, M.J., et al.: Using the 3D facial norms database to investigate craniofacial sexual dimorphism in healthy children, adolescents, and adults. Biol. Sex Differ. **7**(1), 1–14 (2016)
17. Koizumi, T., Smith, W.A.: Shape from semantic segmentation via the geometric rényi divergence. In: Proceedings of the IEEE/CVF Winter Conference on Applications of Computer Vision, pp. 2312–2321 (2021)
18. Lattas, A., et al.: Avatarme: realistically renderable 3D facial reconstruction "in-the-wild". In: Proceedings of the IEEE/CVF Conference on Computer Vision and Pattern Recognition, pp. 760–769 (2020)
19. Le, B.H., Deng, Z.: Interactive cage generation for mesh deformation. In: Proceedings of the 21st ACM SIGGRAPH Symposium on Interactive 3D Graphics and Games, pp. 1–9 (2017)
20. Lee, C.H., Liu, Z., Wu, L., Luo, P.: Maskgan: towards diverse and interactive facial image manipulation. In: IEEE Conference on Computer Vision and Pattern Recognition (CVPR) (2020)
21. Lee, G.H., Lee, S.W.: Uncertainty-aware mesh decoder for high fidelity 3D face reconstruction. In: Proceedings of the IEEE/CVF Conference on Computer Vision and Pattern Recognition, pp. 6100–6109 (2020)
22. Lewis, J.P., Anjyo, K., Rhee, T., Zhang, M., Pighin, F.H., Deng, Z.: Practice and theory of blendshape facial models. Eurograph. (State Art Rep.) **1**(8), 2 (2014)
23. Li, T., Bolkart, T., Black, M.J., Li, H., Romero, J.: Learning a model of facial shape and expression from 4D scans. ACM Trans. Graph. (TOG) **36**(6), 1–17 (2017)
24. Lin, J., Yuan, Y., Shao, T., Zhou, K.: Towards high-fidelity 3D face reconstruction from in-the-wild images using graph convolutional networks. In: Proceedings of the IEEE/CVF Conference on Computer Vision and Pattern Recognition, pp. 5891–5900 (2020)
25. Martyniuk, T., Kupyn, O., Kurlyak, Y., Krashenyi, I., Matas, J., Sharmanska, V.: Dad-3dheads: a large-scale dense, accurate and diverse dataset for 3D head alignment from a single image. In: Proceedings of the IEEE/CVF Conference on Computer Vision and Pattern Recognition, pp. 20942–20952 (2022)
26. Paysan, P., Knothe, R., Amberg, B., Romdhani, S., Vetter, T.: A 3D face model for pose and illumination invariant face recognition. In: 2009 Sixth IEEE International Conference on Advanced Video and Signal Based Surveillance, pp. 296–301. IEEE (2009)
27. Piao, J., Sun, K., Wang, Q., Lin, K.Y., Li, H.: Inverting generative adversarial renderer for face reconstruction. In: Proceedings of the IEEE/CVF Conference on Computer Vision and Pattern Recognition, pp. 15619–15628 (2021)
28. Ramanathan, N., Chellappa, R.: Modeling age progression in young faces. In: 2006 IEEE Computer Society Conference on Computer Vision and Pattern Recognition (CVPR 2006), vol. 1, pp. 387–394. IEEE (2006)
29. Ranjan, A., Bolkart, T., Sanyal, S., Black, M.J.: Generating 3D faces using convolutional mesh autoencoders. In: Proceedings of the European Conference on Computer Vision (ECCV), pp. 704–720 (2018)
30. Rhee, S.C., Woo, K.S., Kwon, B.: Biometric study of eyelid shape and dimensions of different races with references to beauty. Aesthetic Plastic Surg. **36**(5), 1236–1245 (2012)
31. Schroff, F., Kalenichenko, D., Philbin, J.: Facenet: a unified embedding for face recognition and clustering. In: Proceedings of the IEEE Conference on Computer Vision and Pattern Recognition, pp. 815–823 (2015)

32. Shi, T., Yuan, Y., Fan, C., Zou, Z., Shi, Z., Liu, Y.: Face-to-parameter translation for game character auto-creation. In: Proceedings of the IEEE/CVF International Conference on Computer Vision, pp. 161–170 (2019)
33. Sorkine, O., Alexa, M.: As-rigid-as-possible surface modeling. In: Symposium on Geometry processing, vol. 4, pp. 109–116 (2007)
34. Tena, J.R., De la Torre, F., Matthews, I.: Interactive region-based linear 3D face models. In: ACM SIGGRAPH 2011 Papers, pp. 1–10. ACM (2011)
35. Tran, L., Liu, X.: Nonlinear 3D face morphable model. In: Proceedings of the IEEE Conference on Computer Vision and Pattern Recognition, pp. 7346–7355 (2018)
36. Vetter, T., Blanz, V.: Estimating coloured 3D face models from single images: an example based approach. In: Burkhardt, H., Neumann, B. (eds.) ECCV 1998. LNCS, vol. 1407, pp. 499–513. Springer, Heidelberg (1998). https://doi.org/10.1007/BFb0054761
37. Wood, E., Baltrusaitis, T., Hewitt, C., Dziadzio, S., Cashman, T.J., Shotton, J.: Fake it till you make it: face analysis in the wild using synthetic data alone. In: Proceedings of the IEEE/CVF International Conference on Computer Vision, pp. 3681–3691 (2021)
38. Zhu, W., Wu, H., Chen, Z., Vesdapunt, N., Wang, B.: Reda: reinforced differentiable attribute for 3D face reconstruction. In: Proceedings of the IEEE/CVF Conference on Computer Vision and Pattern Recognition, pp. 4958–4967 (2020)

LSMD-Net: LiDAR-Stereo Fusion with Mixture Density Network for Depth Sensing

Hanxi Yin[1], Lei Deng[2], Zhixiang Chen[3], Baohua Chen[4], Ting Sun[2], Yuseng Xie[2], Junwei Xiao[1], Yeyu Fu[2], Shuixin Deng[2], and Xiu Li[1](✉)

[1] Shenzhen International Graduate School, Tsinghua University, Shenzhen, China
{yhx20,xjw20}@mails.tsinghua.edu.cn, li.xiu@sz.tsinghua.edu.cn
[2] School of Instrument Science and Opto-Electronics Engineering, Beijing Information Science and Technology University, Beijing, China
[3] Department of Computer Science, The University of Sheffield, Sheffield, UK
[4] Department of Automation, Tsinghua University, Beijing, China

Abstract. Depth sensing is critical to many computer vision applications but remains challenge to generate accurate dense information with single type sensor. The stereo camera sensor can provide dense depth prediction but underperforms in texture-less, repetitive and occlusion areas while the LiDAR sensor can generate accurate measurements but results in sparse map. In this paper, we advocate to fuse LiDAR and stereo camera for accurate dense depth sensing. We consider the fusion of multiple sensors as a multimodal prediction problem. We propose a novel end-to-end learning framework, dubbed as LSMD-Net to faithfully generate dense depth. The proposed method has dual-branch disparity predictor and predicts a bimodal Laplacian distribution over disparity at each pixel. This distribution has two modes which captures the information from two branches. Predictions from the branch with higher confidence is selected as the final disparity result at each specific pixel. Our fusion method can be applied for different type of LiDARs. Besides the existing dataset captured by conventional spinning LiDAR, we build a multiple sensor system with a non-repeating scanning LiDAR and a stereo camera and construct a depth prediction dataset with this system. Evaluations on both KITTI datasets and our home-made dataset demonstrate the superiority of our proposed method in terms of accuracy and computation time.

1 Introduction

Real-time, dense and accurate depth sensing is crucial for many computer vision applications, including SLAM, autonomous driving and augmented realities. There are two kinds of sensors, active and passive sensors used to sense depth. However, either active sensors like LiDAR scanner or passive sensors like stereo camera have their limitations. On the one hand, stereo camera can provide dense

---

Supplementary Information The online version contains supplementary material available at https://doi.org/10.1007/978-3-031-26319-4_6.

L. Wang et al. (Eds.): ACCV 2022, LNCS 13841, pp. 89–105, 2023.
https://doi.org/10.1007/978-3-031-26319-4_6

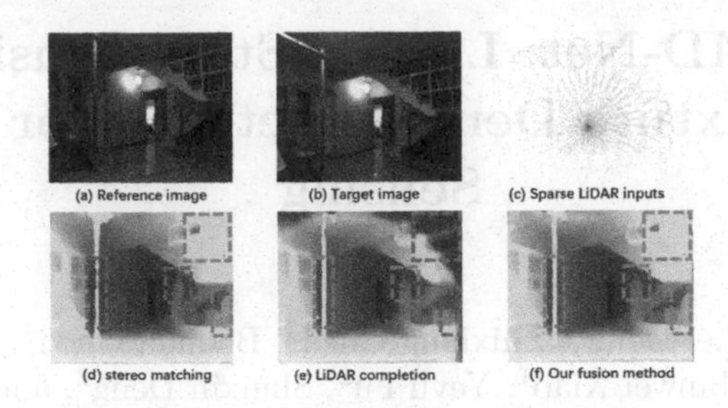

Fig. 1. Illustration of predicted disparity maps (d–f) based on inputs (a–c). Green/red dotted rectangles show the areas where the depth prediction is accurate/inaccurate. The proposed method (f) can take full advantages of different sensors. (Color figure online)

depth estimation but underperforms in texture-less or repetitive areas, occlusion areas, thin structure and poor light conditions. On the other hand, LiDAR scanner often provides precise but relatively sparse depth measurements. These limitations hinder their usages in practical applications. One possible solution to remedy this issue is to combine them by multiple sensor fusion. In terms of fusing LiDAR with RGB camera, there are existing works proposing to fuse LiDAR and monocular camera [16,20,36]. However, the monocular camera setting makes it depend on strong scene priors and is vulnerable to overfitting as monocular depth estimation is inherently unreliable and ambiguous. On the contrary, in this paper, we consider LiDAR-stereo fusion. The stereo camera is more robust as it computes the geometric correspondence between an image pair. The fused depth also benefits from the robustness.

With a stereo camera and a LiDAR sensor, there are two possible ways to generate dense depth prediction: (1) stereo matching from a pair of stereo images, (2) LiDAR completion from sparse LiDAR measurements and a RGB image. The former estimates disparities between image pairs by matching pixels and recovers depth through triangulation, while the latter utilizes a corresponding RGB image to guide the depth interpolation. These two methods exploit information from different modalities with different priori hypotheses and characteristics. The performance of stereo camera depends on image matching, while that of LiDAR completion is limited by the density and quality of point clouds. As illustrated in Fig. 1, stereo matching works well in rich textured areas, but has difficulties in dealing with fine structure and texture-less areas. LiDAR completion performs depth interpolation accurately. However, it has poor extrapolation ability in areas where point clouds are too sparse or missing. Besides, the quality of LiDAR point clouds are poor in reflective surface and distant areas. Based on this analysis, these two methods are expected to complement each other from the perspective of multimodal fusion.

Existing LiDAR-stereo fusion works either use LiDAR information to assist stereo matching [29,37] or simply combine them at the output stage [28,38]. The former one simply injects LiDAR information into cost volume which is the core component of stereo matching. It is confined to the stereo matching architecture and therefore cannot avoid the inherent drawbacks of image matching.

The latter one lacks deep feature fusion, which makes it not fully utilize the intrinsic information of different sources. To take full advantage of the unique characteristics of different sources, we propose a confidence based fusion method by combining stereo matching and LiDAR completion branches. The features of reference image is fed into both branches to fuse with features from different sensor. This breaks the symmetry of stereo image pair and thus gets rid of the limitation of stereo matching pipeline. Besides, the confidence-based fusion can better solve the redundancy and contradiction between heterogeneous sources. Furthermore, our model is built over disparity which is inversely proportional to depth. Inverse depth allows probability distribution to describe depth from nearby to infinity and is more stable to regress with a finite boundary. Specifically, we formulate the task of LiDAR-Stereo fusion as a multimodal prediction problem. Instead of regress disparity directly, we exploit a mixture density network to estimate a bimodal probability distribution over possible disparities for each pixel. Predictions from the branch with higher confidence is selected as the final disparity result at each specific pixel.

To further evaluate our method, we have constructed a dataset based on a solid state Livox LiDAR. Compared with conventional spinning LiDAR, Solid state LiDAR is more suitable for our LiDAR-stereo fusion task in various scenarios for large FOV overlap with RGB camera, advantages in terms of point cloud density and affordable cost.

In summary, the contributions of this paper are summarized as follows:

(I) We propose a novel end-to-end learning dual-branch framework called LSMD-Net (**Li**DAR-**S**tereo fusion with **M**ixture **D**ensity Network) to fuse LiDAR and stereo camera for accurate and dense depth estimation in real time.
(II) We treat multisensor fusion as a multimodal prediction problem. A bimodal distribution is utilized to capture information from different modes and provides a measure of confidence for them at each pixel, which can take full advantage of different sensors for better depth prediction.
(III) We build a data collecting system equipped with a solid state LiDAR and a stereo camera and present a depth prediction dataset.

2 Related Works

Stereo Matching. With the development of convolutional neural networks (CNNs), learning-based stereo matching methods have achieved great success. An end-to-end stereo matching architecture has four trainable components: (a) feature extraction, (b) cost volume, (c) aggregation and (d) regression. Most methods can be categorized into 2D architectures and 3D architectures according to the type of cost volume. The first class [24,27] performs correlation layer to build 2D cost volume and uses 2D CNNs aggregation, which is less accurate than the second. The second class constructs 3D cost volume by concatenating image features [4,19] or using group-wise correlation [13]. Although more accurate, the second class suffers from computational complexities. Stereo matching

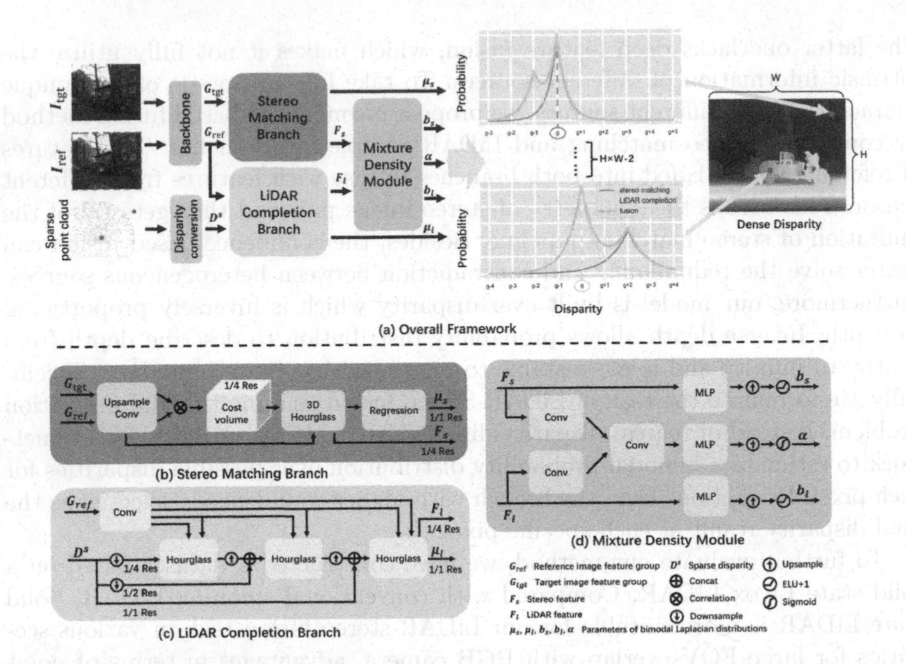

Fig. 2. An overview of our method. Our model has two branches (blue and yellow) which extract features from different sensors. Features from different branches are fused in mixture density module. For each pixel in reference image, a bimodal Laplacian distribution (light orange curve) is predicted. At inference time, we use the expectation of the more confident branch as the final disparity for each specific pixel (as shown by the dotted line). (Color figure online)

performs bad in texture-less or repetitive areas. Fusion with LiDAR is therefore important for obtaining reliable depth estimation.

LiDAR-RGB Fusion. LiDAR-camera fusion is well known for its practicability in 3D perception. There are two types of fusion: LiDAR-monocular and LiDAR-stereo fusion. The former one, also known as depth completion, regresses dense depth from sparse depth cues with the help of monocular image information [16,20,36]. Relying on priors of a particular scene, LiDAR-monocular fusion is inherently unreliable and ambiguous [37]. LiDAR-stereo fusion is less ambiguous in terms of the absolute distance for stereo matching relies on the geometric correspondence across images. Several works [1,9,22,26,34] studied the application of LiDAR-stereo fusion in robotic for the past two decades. Park et al. [28] was the first to implement CNNs in context of LiDAR and stereo fusion. Learning-based methods can be roughly divided into feature level fusion and decision level fusion. Feature level fusion [5,29,37,41] encodes LiDAR information at early stage in stereo matching while decision level fusion [28,38] directly fuses hypotheses generated by different sensors. Our LiDAR-stereo fusion method takes advantage of both feature level fusion and confidence-based decision level fusion.

Multimodal Predictions with CNNs. Standard depth prediction works directly regress a scalar depth at every pixels. However, LiDAR-stereo fusion system is complex for there are multimodal inputs and we can obtain more than one candidate outputs from them. A lot of works have been done for multiple solutions from CNNs. Guzman-Rivera et al. [14] introduced the Winner-Takes-All (WTA) loss for classification tasks while another option is Mixture Density Networks (MDNs) by C. M. Bishop [3]. Instead of using a parametric distribution, MDNs learn parameterization as a part of the neural network. O. Makansi et al. [23] used MDNs for Multimodal Future Prediction. G. Hager [17] proposed a MDNs-based approach to estimate uncertainty in stereo disparity prediction networks. F. Tosi [35] uses a bimodal approach to solve the over-smoothing issue in stereo matching, which inspired us greatly. In contrast to them, we apply bimodal distributions to capture information from two sensors, which can solve the redundancy and contradiction between heterogeneous sources.

3 LSMD-Net

As shown in Fig. 2(a), our proposed model aims to generate an accurate dense disparity map $\boldsymbol{D}^d \in \mathbb{R}^{H\times W}$ given sparse LiDAR measurements $\boldsymbol{S} \in \mathbb{R}^{n\times 3}$ ($n < H \times W$) and a pair of stereo images $\boldsymbol{I}_{ref}, \boldsymbol{I}_{tgt} \in \mathbb{R}^{H\times W\times 3}$. A dense depth map can be further obtained by $\boldsymbol{D}^d$. There are two stages in our model pipeline. At the first stage (Sect. 3.1), we separately estimate dense disparity maps and confidence related feature maps for images and LiDAR measurements. The image based disparity estimation is obtained by stereo matching. The LiDAR based disparity estimation is completed by LiDAR completion with reference image. At the second stage (Sect. 3.2), we employ a mixture density module to fuse these estimations into a final dense depth map. Specifically, we introduce a confidence-based fusion to effectively exploit the information from different sensors.

3.1 Dual-Branch Disparity Predictor

To estimate dense disparity and features from images and LiDAR measurements, we employ separated disparity prediction branches of stereo matching and LiDAR completion. Before passing through the individual branches, we firstly preprocess the images and LiDAR signals. The images are processed by a backbone network to extract meaningful features. Specifically, we adopt the MobileNetV2 model [31] pre-trained on ImageNet [8] to extract image features at scales of 1/2, 1/4, 1/8, 1/16 and 1/32 of the input image resolution. Both reference and target images are passed through the same backbone with shared weight to obtain the corresponding feature groups $\boldsymbol{G}_{ref}, \boldsymbol{G}_{tgt}$. Inspired by the parametrization for monocular SLAM [7], we project sparse LiDAR points $\boldsymbol{S}$ onto the image plane of $\boldsymbol{I}_{ref}$ and convert the projected depth map to a disparity map $\boldsymbol{D}^s$. Note that the disparity is inversely proportional to the depth. Our network directly predicts disparity rather than depth. This conversion from depth to disparity has two advantages. First, it allows us to consider a wider depth range in our model. Second, the model prediction is more stable as the regression target is with finite boundary.

Stereo Matching Branch. This branch takes as input the feature groups of reference and target images, $\boldsymbol{G}_{ref}$ and $\boldsymbol{G}_{tgt}$ to generate a disparity map $\boldsymbol{\mu}_s \in \mathbb{R}^{H \times W}$ and a feature $\boldsymbol{F}_s \in \mathbb{R}^{H/4 \times W/4 \times D/4}$, where D is the maximum disparity value. D is set to 192 in our network. This feature contains matching probabilities along possible disparities and is used in the fusion stage. Denoting this stereo matching branch as ϕ with parameters θ, we can formally write it as,

$$\{\boldsymbol{F}_s, \boldsymbol{\mu}_s\} = \phi_\theta(\boldsymbol{G}_{ref}, \boldsymbol{G}_{tgt}). \tag{1}$$

Our design of this stereo matching branch follows the mainstream learning-based stereo matching framework. It consists of matching cost computation, cost aggregation, and disparity regression. With feature groups $\boldsymbol{G}_{ref}$, $\boldsymbol{G}_{tgt}$ as input, a U-Net [30] style upsampling module with long skip connections at each scale level is built to propagate context information to higher resolution layers. Image features with less than 1/4 of the input image resolution in $\boldsymbol{G}_{ref}$, $\boldsymbol{G}_{tgt}$ are upsampled by this module to a quarter of the input image resolution. The cost volume is then built by computing the correlation between the outputs of the upsample module. We keep the size of cost volume at $H/4 \times W/4 \times D/4$ to reduce cost aggregation computing costs. As for cost aggregation, instead of employing neighborhood aggregation, we first capture geometric features from cost volume by 3D convolutions, and then utilize the guidance weights generated from image features to redistribute this geometric information to local features as in CoEx [2]. The output feature $\boldsymbol{F}_s$ is from the aggregated cost volume. To reduce the computation time of regression, our model regresses disparity at 1/4 of the input image resolution from cost volume and finally upsamples it to the original input image resolution.

LiDAR Completion Branch. This branch takes as input the feature group of reference image $\boldsymbol{G}_{ref}$ and a sparse disparity map $\boldsymbol{D}^s$ to generate a dense disparity map $\boldsymbol{\mu}_l \in \mathbb{R}^{H \times W}$ and a feature map $\boldsymbol{F}_l \in \mathbb{R}^{H/4 \times W/4 \times C}$, where C is set to 64. This feature contains information extracted from LiDAR measurement and reference image and is used in the fusion stage. Denoting this disparity completion branch as ψ with parameters ω, we can formally write it as,

$$\{\boldsymbol{F}_l, \boldsymbol{\mu}_l\} = \psi_\omega(\boldsymbol{G}_{ref}, \boldsymbol{D}^s). \tag{2}$$

This disparity completion can be considered as a disparity map interpolation guided by the reference image feature. Similar to MSG-CHN [20], we use coarse to fine cascade hourglass CNNs to interpolate the disparity features at three levels. The output of the coarse level is upsampled and concatenated with the sparse disparity map at corresponding scale as the input of the fine level. At each level, the hourglass CNNs refine the disparity features according to both the input disparity features and the corresponding scale image features from $\boldsymbol{G}_{ref}$. The disparity features and the image features are fused by concatenation. We expect that this design can exploit the clues from reference image to guide the interpolation of disparities for LiDAR. The output feature $\boldsymbol{F}_l$ is extracted from the last hourglass CNNs. It contains LiDAR and image information and is with the same resolution as $\boldsymbol{F}_s$.

3.2 Mixture Density Module

The mixture density module is used to fuse the disparity information from two branches. We view the estimated disparity at each pixel as a probability distribution over the possible range of disparities. And the fusion of disparity estimations leads to the final probability distribution, from which we can get the output disparity. To be specific, we utilize the Laplacian distribution to model the probability distribution for each branch as

$$\boldsymbol{P}(\boldsymbol{d}) = \frac{1}{2\boldsymbol{b}} e^{(-\frac{|\boldsymbol{\mu}-\boldsymbol{d}|}{\boldsymbol{b}})}, \tag{3}$$

where $\boldsymbol{\mu}$ is the location parameter map and $\boldsymbol{b}$ is the scale parameter map. We opt to the Laplacian distribution rather than the widely used Gaussian distribution. This is because the Gaussian assumption is sensitive to outliers but the Laplacian distribution is more robust as it has a heavier tails than Gaussian.

We take the estimated disparities from each branch, $\boldsymbol{\mu}_s$ and $\boldsymbol{\mu}_l$ as the location parameters. Rather than setting a global scale parameter for each branch, we propose to learn pixel wise parameters $\boldsymbol{b}$ from the features of each branch $\boldsymbol{F}_s$ and $\boldsymbol{F}_l$ with two independent networks, as shown in Fig. 2(d).

$$\boldsymbol{b}_s = \sigma(MLP(\boldsymbol{F}_s)),\ \boldsymbol{b}_l = \sigma(MLP(\boldsymbol{F}_l)), \tag{4}$$

where $\sigma(\cdot)$ is the activation function. An exponential activation function is adopted in traditional mixture density network to predict parameters with positive value. However, the exponential increases to a very large value in case of high variance, which makes the training unstable. In this work, following [12], we choose ELU as the activation function. The ELU activation function shares the same exponential behavior for small activation value but is linear to the input for large activation value.

$$f(\beta, x) = ELU(\beta, x) + 1 = \begin{cases} \beta(e^x - 1) + 1, & x < 0 \\ x + 1, & x \geq 0 \end{cases}, \tag{5}$$

where β is a parameter to control the slope and is set to 1 in our work.

With the Laplacian distributions of two branches, $\boldsymbol{P}_s$ and $\boldsymbol{P}_l$ at hand, we compute the final probability distribution as a weighted sum of these two distributions.

$$\boldsymbol{P}_m(\boldsymbol{d}) = \boldsymbol{\alpha}\boldsymbol{P}_s(\boldsymbol{d}) + (1-\boldsymbol{\alpha})\boldsymbol{P}_l(\boldsymbol{d}) = \frac{\boldsymbol{\alpha}}{2\boldsymbol{b}_s} e^{-\frac{|\boldsymbol{\mu}_s-\boldsymbol{d}|}{\boldsymbol{b}_s}} + \frac{1-\boldsymbol{\alpha}}{2\boldsymbol{b}_l} e^{-\frac{|\boldsymbol{\mu}_l-\boldsymbol{d}|}{\boldsymbol{b}_l}} \tag{6}$$

where $\boldsymbol{\alpha}$ is a parameter map weighting the contributions of different branches. We also design a network to learn $\boldsymbol{\alpha}$ from the branch features $\boldsymbol{F}_s$ and $\boldsymbol{F}_l$ (Fig. 2(d)). Convolutional layers are applied to process and aggregate the branch features followed by MLP and a sigmoid activation function. We can use the fused distribution in Eq. 6 to compute the loss at training stage. However, at inference stage, we aims to predict a single disparity value for each pixel. One possible solution is to use the conditional expectation as the final output. In our case,

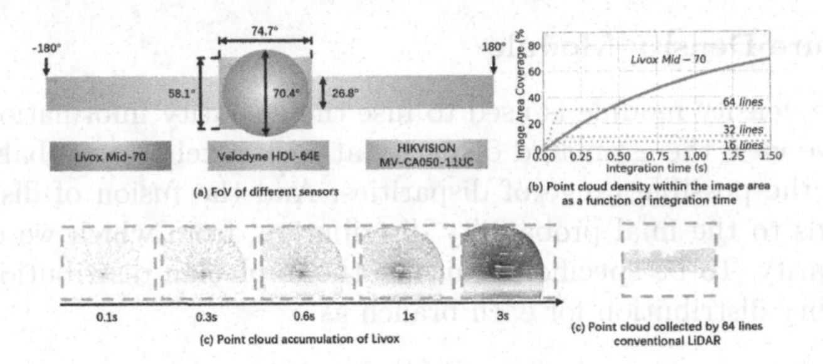

Fig. 3. Illustration of characteristics of two LiDARs. (a) shows that solid state LiDAR (red) fits better with camera (green) than conventional LiDAR (blue) in terms of FoV. (b) and (c) illustrates point cloud accumulation of solid state Livox LiDAR quantitatively and qualitatively. (Color figure online)

one branch may be more confident than the other branch for particular pixels. Simply calculating the conditional expectation will deteriorate the performance as the outliers from the less confident branch are considered. To this end, we propose to use the expectation of the more confident branch as the final disparity prediction $\hat{\boldsymbol{d}}$. And the branch confidence is determined by $\boldsymbol{\alpha}$.

$$\hat{\boldsymbol{d}} = \begin{cases} \boldsymbol{\mu}_s, \ \boldsymbol{\alpha} \geq 0.5 \\ \boldsymbol{\mu}_l, \ \boldsymbol{\alpha} < 0.5 \end{cases}. \tag{7}$$

3.3 Losses

Our model is trained with the supervision on the final output and the intermediate supervisions over two branches. The training objective is to minimize the overall loss

$$\mathcal{L} = \omega_m \mathcal{L}_m + \omega_s \mathcal{L}_s + \omega_l \mathcal{L}_l, \tag{8}$$

where ω_m, ω_s and ω_l are the weighting parameters for the three losses $\mathcal{L}_m$ $\mathcal{L}_s$ $\mathcal{L}_l$. $\mathcal{L}_m$ is the loss over the final output of the fusion method. $\mathcal{L}_s$ and $\mathcal{L}_l$ are the losses over the outputs of stereo matching branch and LiDAR completion branch, respectively. We compute the negative logarithm of the likelihood loss based on the PDFs (P_m, P_s, P_l) in Eq. 6 for each loss which can be expressed as:

$$\mathcal{L}_{NLL}(\theta) = -\mathbb{E}_{d,x,I} \log P(d|x, I_{ref}, I_{tgt}, D^s, \theta), \tag{9}$$

where d is the ground truth disparity at each pixel location x in reference image I_{ref} from the dataset. θ denotes parameters of our model.

4 Livox-Stereo Dataset

There are two widely used kinds of LiDAR sensors: mechanical spinning LiDAR and solid state LiDAR. The mechanical spinning LiDAR uses mechanical rotation to spin the sensor for 360° detection. The density of collected point cloud

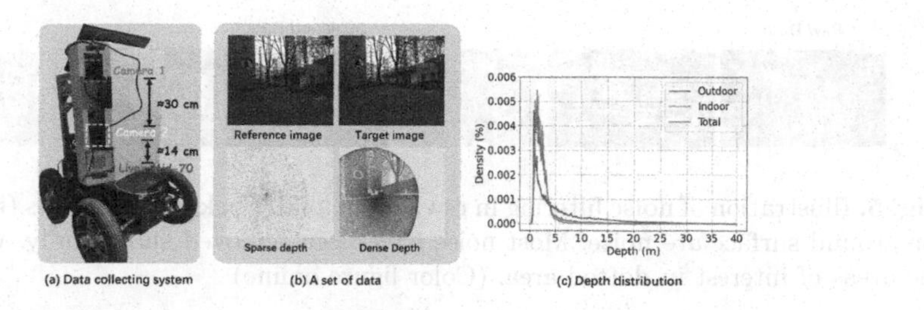

Fig. 4. Illustration of (a) data collecting system hardware, (b) a set of data in dataset and (c) the depth distribution of dataset.

is determined by the number of scanner layer. Most existing depth sensing datasets [36,40] use this kind of sensor to acquire LiDAR information. The solid state LiDAR such as Livox uses prism scanning to acquire depth information in a non-repeating scanning pattern and accumulates point clouds to generate relative dense depth maps [21,42]. Livox LiDAR is promising for LiDAR-stereo fusion task for three reasons. Firstly, as shown in Fig. 3(a), it fits well with cameras because of their large overlapping FoV. Livox collects point cloud which covers a larger image area, rather than focusing on a limited number of scan lines with a narrow FoV like traditional LiDAR, which is illustrated in Fig. 3(c)(d). Secondly, compared to traditional LiDAR, Livox has advantages in terms of point cloud density within the image area when the accumulation time is sufficient as Fig. 3(b) shows. Thirdly, Livox is promising in various scenarios owing to its portability and advantages in terms of cost. To show the feasibility of our proposed depth fusion method on different LiDAR sensors, we further present a Livox-stereo dataset collected by our own system for evaluation.

4.1 Data Collecting System

As shown in Fig. 4(a), our system hardware includes two HIKVISION MV-CA050-11UC color cameras stacked vertically and a Livox Mid-70 LiDAR. The distance between the two cameras is 30 cm, which results in errors in four centimeters for depth within three meters (see Suppl. for the error estimation). The Livox is placed close to the reference camera (the camera below) to increase the overlap between their FoV.

The calibration process of our system can be divided into two steps. First, We follow the binocular calibration process in OpencV [18] to compute the intrinsic and extrinsic parameters of the stereo cameras. After that, We use the calibrated reference camera to calculate the extrinsic parameters of Livox. Specifically, the Livox extrinsic parameters are derived by PnP and RANSAC after the extraction of corresponding key points between the depth map generated by Livox and the remapped image from the reference camera [39].

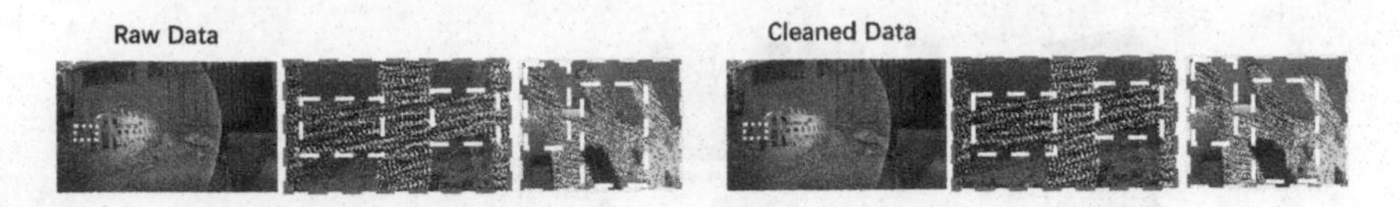

Fig. 5. Illustration of noise filtering in raw point cloud. Background points (blue) on the foreground surface are noise. Most noise have been removed successfully. We zoom-in the areas of interest in dotted area. (Color figure online)

4.2 Livox-Stereo Dataset

We collected 507 sets data in both indoor and outdoor scenes in residential areas. Each set of data consists of a pair of stereo images at the resolution of 1224×1024 and a pair of sparse and dense depth maps. We show a set of data in Fig. 4(b). More examples are in the Suppl. The sparse and dense depth maps are collected by the same Livox sensor on the same scene but with different accumulation time. Specifically, we obtain a sparse depth map with the coverage around 10% by setting the accumulation time to 0.3 s. The dense depth map is obtained by accumulating the point clouds for 3 s to achieve a coverage around 60%. The dense depth maps are used as ground truth to train our fusion model. We split the dataset into train, validation and test subsets at a ratio around 7:1:2. The specific statistics of the subsets is in Suppl.

We present the statistical information of pixel level depth in Fig. 4(b). From the curves in the figure, we can find that the majority pixels are with small depth values. Specifically, the depths of 53.07% pixels fall within 3 m (80.55% and 44.27% for indoor and outdoor) and 68.75% falls within five meters. The depth distribution matches well with the depth range that the stereo cameras can generate depth with low errors.

During data collection, we observed that the difference of projection centers of Livox and stereo cameras leads to noise when point clouds are projected onto the image plane. In order to remove this noise, we identify inconsistent LiDAR points by applying semi-global matching (SGM) [15] and refuse these points. This point cloud filtering is based on the assumption that passive and activate sensors rarely make the same inaccurate prediction in these problematic areas. We show some examples of this filtering in Fig. 5. Emperically, we found that this works well to produce clean point clouds for our task.

In Table 1, we compare our dataset with other relevant datasets. Our dataset is unique in several senses. First, the point clouds in our dataset are collected by Livox LiDAR. Non-repeating scanning Livox LiDAR allows our dataset to provide sparse LiDAR inputs and to be better than other LiDAR-based datasets in term of the density of depth information for supervision. Besides, our dataset includes both indoor and outdoor scenes which is different from the autonomous driving scene in existing datasets. This is promising to improve the generalization the performance of LiDAR-RGB fusion model in practical application.

Table 1. Comparison between our dataset and other published depth sensing datasets.

Datasets	Tools	Real	Sparse LiDAR	SceneType	DataSize		Coverage
					Train	Test	
Middlebury [32]	Structured light scanner	✓		Indoor	15	15	≈ 96%
ETH3D [33]	Structured light scanner	✓		Indoor/ outdoor	27	20	≈ 69%
KITTI stereo [10,25]	Velodyne HDL-64E	✓		Autonomous driving	394	395	≈ 19%
KITTI depth completion [36]	Velodyne HDL-64E	✓	✓	Autonomous driving	43k	1k	16.1%
FlyingThings3D [24]	Software			Animation	22k	42k	100%
DrivingStereo [40]	Velodyne HDL-64E S3	✓		Autonomous driving	174k	8k	≈ 4%
Ours	Livox Mid-70	✓	✓	Indoor/ outdoor	407	100	≈ 60%

5 Experiments

In this section, we demonstrate the effectiveness of our proposed depth fusion method on three datasets, KITTI stereo dataset [36], KITTI depth completion dataset [25] and our new Livox-stereo dataset.

5.1 Datasets and Evaluation Metrics

KITTI Stereo 2015 and KITTI Depth Completion are real-world datasets with street views from a driving car. We follow [9,11,22,37] to evaluated our model on the training set in Stereo 2015 dataset and the validation set in Depth Completion dataset (see Suppl. for more details).

For Stereo 2015 dataset, we report several common metrics in stereo matching tasks: end-point error (EPE, the mean average disparity error in pixels) and the percentage of disparity error that is greater than 1, 2 and 3 pixel(s) away from the ground truth ($> 1px$, $> 2px$ and $> 3px$). For Depth Completion dataset, root mean squared error of depth ($RMSE$, m), mean absolute error of depth (MAE, m), root mean squared error of the inverse depth ($iRMSE$, $1/km$) and mean absolute error of the inverse depth ($iMAE$, $1/km$) are reported.

5.2 Implementation Details

The proposed network is implemented in PyTorch and optimized with Adam ($\beta_1 = 0.9, \beta_2 = 0.999$) and a learning rate of 1e−3. Our model is trained on NVIDIA GeForce RTX 2080 with random change of brightness and contrast, random dropout part of disparity inputs (see Suppl. for more details) and random cropping to 512×256 as data augmentation. We initialize the network with random parameters. A weight decay of 1e−4 is applied for regularization.

For KITTI datasets, the network is trained on Depth Completion dataset for 20 epochs with a batch size of 4 and is tested on two KITTI datasets. The weighting parameter in loss function are set as $\omega_s = 0.8, \omega_m, \omega_l = 0.1$ at first 5

Table 2. Comparison on the KITTI Stereo 2015 dataset.

Methods	Input	$> 3px \downarrow$	$> 2px \downarrow$	$> 1px \downarrow$	$EPE \downarrow$
GC-Net [19]	Stereo	4.24	5.82	9.97	–
CoEx [2]	Stereo	3.82	5.59	10.67	1.06
Prob. Fusion [22]	Stereo + LiDAR	5.91	–	–	–
Park et al. [28]	Stereo + LiDAR	4.84	–	–	–
CCVN [37]	Stereo + LiDAR	3.35	4.38	6.79	–
LSMD-Net(ours)	Stereo + LiDAR	**2.37**	**3.18**	**5.19**	**0.86**

Table 3. Comparison on the KITTI Depth Completion dataset.

Methods	Input	$MAE \downarrow$	$iMAE \downarrow$	$RMSE \downarrow$	$iRMSE \downarrow$
MSG-CHN [20]	Mono + LiDAR	0.2496	1.11	0.8781	2.59
Park et al. [28]	Stereo + LiDAR	0.5005	1.38	2.0212	3.39
SCADC [38]	Stereo + LiDAR	0.4015	1.94	1.0096	3.96
CCVN [37]	Stereo + LiDAR	0.2525	<u>0.81</u>	<u>0.7493</u>	**1.40**
LiStereo [41]	Stereo + LiDAR	0.2839	1.10	0.8322	2.19
VPN [6]	Stereo + LiDAR	**0.2051**	0.99	**0.6362**	1.87
LSMD-Net (ours)	Stereo + LiDAR	<u>0.2100</u>	**0.79**	0.8845	<u>1.85</u>

epochs, $\omega_l = 0.8$, ω_m, $\omega_s = 0.1$ for another 5 epochs and $\omega_m = 0.7$, $\omega_s = 0.2$, $\omega_l = 0.1$ after 10 epochs. Following [37], input images are bottom-cropped to 1216×256 for there is no ground truth on the top. For our Livox-stereo dataset, we fine-tune the network pretrained on KITTI Depth Completion dataset for another 200 epochs. The weighting parameter in loss function are set as $\omega_m = 1.0$, $\omega_s = 0.25$, $\omega_l = 0.125$.

5.3 Results on KITTI Stereo 2015 Dataset

We compared the performance of LSMD-Net with stereo matching methods [2, 19] and other publicly available LiDAR-stereo fusion methods [22,28,37]. Note that CoEx [2] is our baseline stereo matching method. Quantitative results in Table 2 shows that our method outperforms other methods in terms of disparity metrics. This further demonstrates the advantage of our LSMD-Net in depth prediction since disparity is inversely proportional to depth.

5.4 Results on KITTI Depth Completion Dataset

We converted predicted disparity maps into depth maps and compared our LSMD-Net with other depth prediction methods in Table 3. Our method is comparable to other depth prediction methods in terms of depth prediction.

A qualitative comparison on test set is shown in Fig. 6. LiDAR completion (MSG-CHN) is more accurate than stereo matching (CoEx) in depth measurement, but has poor performance at the upper side of maps due to the absence of point clouds. Stereo matching is less precise and performs bad in fine structure, whereas is more stable than LiDAR. Our methods can leverage the unique characteristics of different sensors and provide accurate depth measurements throughout maps.

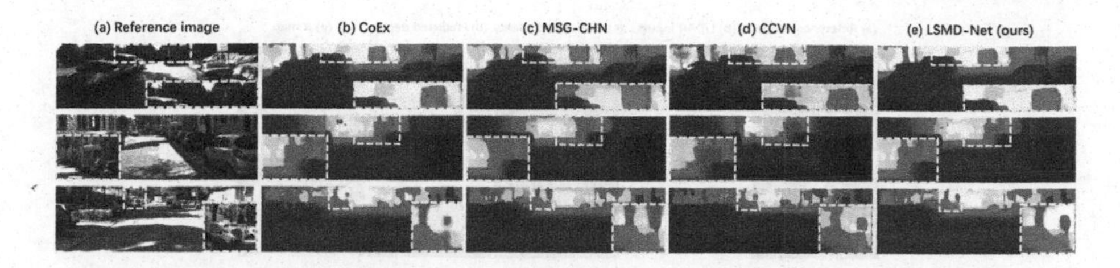

Fig. 6. Qualitative results on KITTI Depth completion test set. Predicted depth map of three scenes from methods based on different sensors are illustrated. We zoom-in the boxes of interest at the bottom on maps.

Table 4. Quantitative results on Livox-stereo test set.

Methods	Input	$> 3px \downarrow$	$EPE \downarrow$	$MAE \downarrow$	$RMSE \downarrow$
CoEx [2]	Stereo	12.72	2.80	–	–
MSG-CHN [20]	Mono + LiDAR	–	–	0.3437	1.08
CCVN [37]	Stereo + LiDAR	6.57	1.86	0.6569	1.85
LSMD-Net (ours)	Stereo + LiDAR	**5.28**	**1.32**	**0.1957**	**0.93**

5.5 Results on Livox-stereo Dataset

The proposed method was further evaluated on home-made Livox-stereo dataset. LSMD-Net is compared with MSG-CHN and CoEx on depth maps and disparity maps respectively in Table 4. Our method has obvious advantage over other depth sensing methods in all indicators.

Qualitative results can be found in Fig. 7. As mentioned in Sect. 3.2, the weight of two modes α determines the branch confidence. The α map in Fig. 7(e) presents unique Livox scanning pattens and discontinuous edges of objects, which indicates that LiDAR completion is less reliable in areas without LiDAR measurements and at the edges of objects. Our method can capture the advantages of different sensors using this map.

5.6 Ablation Study

Ablation study is performed on Livox-stereo dataset to study the effect of using different probability distribution models. Four distribution models are tested and their results are reported in Table 5. The Laplacian distribution over disparity we select outperforms others.

Table 5. Comparison of different probability distribution model.

Models	$>3px \downarrow$	$EPE \downarrow$	$MAE \downarrow$	$RMSE \downarrow$
Gaussian distribution over depth	7.93	1.76	0.3294	1.29
Gaussian distribution over disparity	5.71	1.40	0.2102	0.94
Laplacian distribution over depth	6.15	1.47	0.2750	1.32
Laplacian distribution over disparity	**5.28**	**1.32**	**0.1957**	**0.93**

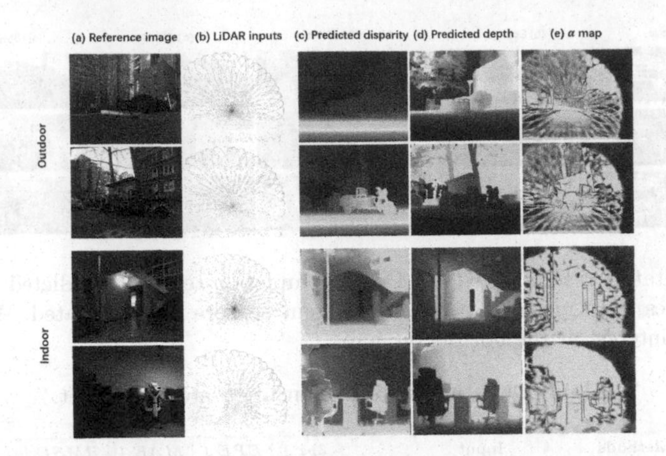

Fig. 7. Qualitative results of our LSMD-Net on Livox-stereo dataset. The brighter part of (e) indicates that LiDAR completion branch is more reliable and the darker part indicates that stereo matching branch is more reliable.

Table 6. Computational time of different methods (unit: millisecond).

Methods	GC-Net [19]	CCVN [37]	VPN [6]	SCADC [38]	CoEx [2]	LSMD-Net(ours)
Time	962	1011	1400	≈ 800	22	27

5.7 Computational Time

We provide a reference for computational time on KITTI in Table 6. The proposed method takes a little bit longer time (5 ms) than baseline method CoEx [2], but provide significant improvement in performance, validating the efficiency of our fusion scheme. Other stereo matching method [19] and LiDAR-Stereo fusion method [6,37,38] take much more times than our method.

6 Conclusion

In this work, we treat multisensor fusion as a multimodal prediction problem and present a real-time dual-branch LiDAR-Stereo fusion method for the task of efficient depth sensing. The proposed method utilizes mixture density network to predict a bimodal Laplacian distribution at each pixel. Each distribution captures information from stereo matching and LiDAR completion and provide a measure of confidence for them. Our method excels in terms of accuracy and computational time on both KITTI and our home-made Livox-stereo datasets.

Acknowledgement. This work was supported by the National Key Research and Development Program of China under Grant 2020AAA0108302.

References

1. Badino, H., Huber, D., Kanade, T.: Integrating lidar into stereo for fast and improved disparity computation. In: 2011 International Conference on 3D Imaging, Modeling, Processing, Visualization and Transmission, pp. 405–412 (2011). https://doi.org/10.1109/3DIMPVT.2011.58
2. Bangunharcana, A., Cho, J.W., Lee, S., Kweon, I.S., Kim, K.S., Kim, S.: Correlate-and-excite: real-time stereo matching via guided cost volume excitation. In: 2021 IEEE/RSJ International Conference on Intelligent Robots and Systems (IROS), pp. 3542–3548 (2021). https://doi.org/10.1109/IROS51168.2021.9635909
3. Bishop, C.M.: Mixture density networks. IEEE Computer Society (1994)
4. Chang, J.R., Chen, Y.S.: Pyramid stereo matching network. In: 2018 IEEE/CVF Conference on Computer Vision and Pattern Recognition, pp. 5410–5418 (2018). https://doi.org/10.1109/CVPR.2018.00567
5. Cheng, X., Zhong, Y., Dai, Y., Ji, P., Li, H.: Noise-aware unsupervised deep lidar-stereo fusion. In: 2019 IEEE/CVF Conference on Computer Vision and Pattern Recognition (CVPR), pp. 6332–6341 (2019). https://doi.org/10.1109/CVPR.2019.00650
6. Choe, J., Joo, K., Imtiaz, T., Kweon, I.S.: Volumetric propagation network: stereo-lidar fusion for long-range depth estimation. IEEE Robot. Automation Lett. **6**(3), 4672–4679 (2021). https://doi.org/10.1109/LRA.2021.3068712
7. Civera, J., Davison, A.J., Montiel, J.M.M.: Inverse depth parametrization for monocular slam. IEEE Trans. Rob. **24**(5), 932–945 (2008). https://doi.org/10.1109/TRO.2008.2003276
8. Deng, J., Dong, W., Socher, R., Li, L.J., Li, K., Fei-Fei, L.: Imagenet: a large-scale hierarchical image database. In: 2009 IEEE Conference on Computer Vision and Pattern Recognition, pp. 248–255 (2009). https://doi.org/10.1109/CVPR.2009.5206848
9. Gandhi, V., Čech, J., Horaud, R.: High-resolution depth maps based on tof-stereo fusion. In: 2012 IEEE International Conference on Robotics and Automation, pp. 4742–4749 (2012). https://doi.org/10.1109/ICRA.2012.6224771
10. Geiger, A., Lenz, P., Urtasun, R.: Are we ready for autonomous driving? the kitti vision benchmark suite. In: 2012 IEEE Conference on Computer Vision and Pattern Recognition, pp. 3354–3361 (2012). https://doi.org/10.1109/CVPR.2012.6248074
11. Godard, C., Aodha, O.M., Brostow, G.J.: Unsupervised monocular depth estimation with left-right consistency. In: 2017 IEEE Conference on Computer Vision and Pattern Recognition (CVPR), pp. 6602–6611 (2017). https://doi.org/10.1109/CVPR.2017.699
12. Guillaumes, A.B.: Mixture density networks for distribution and uncertainty estimation. Ph.D. thesis, Universitat Politècnica de Catalunya. Facultat d'Informàtica de Barcelona (2017)
13. Guo, X., Yang, K., Yang, W., Wang, X., Li, H.: Group-wise correlation stereo network. In: 2019 IEEE/CVF Conference on Computer Vision and Pattern Recognition (CVPR), pp. 3268–3277 (2019). https://doi.org/10.1109/CVPR.2019.00339
14. Guzman-Rivera, A., Batra, D., Kohli, P.: Multiple choice learning: learning to produce multiple structured outputs. In: Advances in Neural Information Processing Systems 25 (2012)
15. Hirschmuller, H.: Stereo processing by semiglobal matching and mutual information. IEEE Trans. Pattern Anal. Mach. Intell. **30**(2), 328–341 (2008). https://doi.org/10.1109/TPAMI.2007.1166

16. Huang, Z., Fan, J., Cheng, S., Yi, S., Wang, X., Li, H.: Hms-net: hierarchical multi-scale sparsity-invariant network for sparse depth completion. IEEE Trans. Image Process. **29**, 3429–3441 (2020). https://doi.org/10.1109/TIP.2019.2960589
17. Häger, G., Persson, M., Felsberg, M.: Predicting disparity distributions. In: 2021 IEEE International Conference on Robotics and Automation (ICRA), pp. 4363–4369 (2021). https://doi.org/10.1109/ICRA48506.2021.9561617
18. Itseez: Open source computer vision library (2015). https://github.com/itseez/opencv
19. Kendall, A., et al.: End-to-end learning of geometry and context for deep stereo regression. In: 2017 IEEE International Conference on Computer Vision (ICCV), pp. 66–75 (2017). https://doi.org/10.1109/ICCV.2017.17
20. Li, A., Yuan, Z., Ling, Y., Chi, W., Zhang, S., Zhang, C.: A multi-scale guided cascade hourglass network for depth completion. In: 2020 IEEE Winter Conference on Applications of Computer Vision (WACV), pp. 32–40 (2020). https://doi.org/10.1109/WACV45572.2020.9093407
21. Liu, Z., Zhang, F., Hong, X.: Low-cost retina-like robotic lidars based on incommensurable scanning. IEEE/ASME Trans. Mechatron. **27**(1), 58–68 (2022). https://doi.org/10.1109/TMECH.2021.3058173
22. Maddern, W., Newman, P.: Real-time probabilistic fusion of sparse 3d lidar and dense stereo. In: 2016 IEEE/RSJ International Conference on Intelligent Robots and Systems (IROS), pp. 2181–2188 (2016). https://doi.org/10.1109/IROS.2016.7759342
23. Makansi, O., Ilg, E., Cicek, Z., Brox, T.: Overcoming limitations of mixture density networks: a sampling and fitting framework for multimodal future prediction. In: 2019 IEEE/CVF Conference on Computer Vision and Pattern Recognition (CVPR), pp. 7137–7146 (2019). https://doi.org/10.1109/CVPR.2019.00731
24. Mayer, N., Ilg, E., Häusser, P., Fischer, P., Cremers, D., Dosovitskiy, A., Brox, T.: A large dataset to train convolutional networks for disparity, optical flow, and scene flow estimation. In: 2016 IEEE Conference on Computer Vision and Pattern Recognition (CVPR), pp. 4040–4048 (2016). https://doi.org/10.1109/CVPR.2016.438
25. Menze, M., Geiger, A.: Object scene flow for autonomous vehicles. In: 2015 IEEE Conference on Computer Vision and Pattern Recognition (CVPR), pp. 3061–3070 (2015). https://doi.org/10.1109/CVPR.2015.7298925
26. Nickels, K., Castano, A., Cianci, C.: Fusion of lidar and stereo range for mobile robots. In: International Conference on Advanced Robotics (2003)
27. Pang, J., Sun, W., Ren, J.S., Yang, C., Yan, Q.: Cascade residual learning: a two-stage convolutional neural network for stereo matching. In: 2017 IEEE International Conference on Computer Vision Workshops (ICCVW), pp. 878–886 (2017). https://doi.org/10.1109/ICCVW.2017.108
28. Park, K., Kim, S., Sohn, K.: High-precision depth estimation with the 3d lidar and stereo fusion. In: 2018 IEEE International Conference on Robotics and Automation (ICRA), pp. 2156–2163 (2018). https://doi.org/10.1109/ICRA.2018.8461048
29. Poggi, M., Pallotti, D., Tosi, F., Mattoccia, S.: Guided stereo matching. In: 2019 IEEE/CVF Conference on Computer Vision and Pattern Recognition (CVPR), pp. 979–988 (2019). https://doi.org/10.1109/CVPR.2019.00107
30. Ronneberger, O., Fischer, P., Brox, T.: U-Net: convolutional networks for biomedical image segmentation. In: Navab, N., Hornegger, J., Wells, W.M., Frangi, A.F. (eds.) MICCAI 2015. LNCS, vol. 9351, pp. 234–241. Springer, Cham (2015). https://doi.org/10.1007/978-3-319-24574-4_28

31. Sandler, M., Howard, A., Zhu, M., Zhmoginov, A., Chen, L.C.: Mobilenetv 2: inverted residuals and linear bottlenecks. In: 2018 IEEE/CVF Conference on Computer Vision and Pattern Recognition, pp. 4510–4520 (2018). https://doi.org/10.1109/CVPR.2018.00474
32. Scharstein, D., Hirschmüller, H., Kitajima, Y., Krathwohl, G., Nešić, N., Wang, X., Westling, P.: High-resolution stereo datasets with subpixel-accurate ground truth. In: Jiang, X., Hornegger, J., Koch, R. (eds.) GCPR 2014. LNCS, vol. 8753, pp. 31–42. Springer, Cham (2014). https://doi.org/10.1007/978-3-319-11752-2_3
33. Schöps, T., Schönberger, J.L., Galliani, S., Sattler, T., Schindler, K., Pollefeys, M., Geiger, A.: A multi-view stereo benchmark with high-resolution images and multi-camera videos. In: 2017 IEEE Conference on Computer Vision and Pattern Recognition (CVPR), pp. 2538–2547 (2017). https://doi.org/10.1109/CVPR.2017.272
34. Shivakumar, S.S., Mohta, K., Pfrommer, B., Kumar, V., Taylor, C.J.: Real time dense depth estimation by fusing stereo with sparse depth measurements. In: 2019 International Conference on Robotics and Automation (ICRA), pp. 6482–6488 (2019). https://doi.org/10.1109/ICRA.2019.8794023
35. Tosi, F., Liao, Y., Schmitt, C., Geiger, A.: Smd-nets: stereo mixture density networks. In: 2021 IEEE/CVF Conference on Computer Vision and Pattern Recognition (CVPR), pp. 8938–8948 (2021). https://doi.org/10.1109/CVPR46437.2021.00883
36. Uhrig, J., Schneider, N., Schneider, L., Franke, U., Brox, T., Geiger, A.: Sparsity invariant CNNs. In: 2017 International Conference on 3D Vision (3DV), pp. 11–20 (2017). https://doi.org/10.1109/3DV.2017.00012
37. Wang, T.H., Hu, H.N., Lin, C.H., Tsai, Y.H., Chiu, W.C., Sun, M.: 3d lidar and stereo fusion using stereo matching network with conditional cost volume normalization. In: 2019 IEEE/RSJ International Conference on Intelligent Robots and Systems (IROS), pp. 5895–5902 (2019). https://doi.org/10.1109/IROS40897.2019.8968170
38. Wu, C.Y., Neumann, U.: Scene completeness-aware lidar depth completion for driving scenario. In: ICASSP 2021–2021 IEEE International Conference on Acoustics, Speech and Signal Processing (ICASSP), pp. 2490–2494 (2021). https://doi.org/10.1109/ICASSP39728.2021.9414295
39. Xie, Y., et al.: A4lidartag: depth-based fiducial marker for extrinsic calibration of solid-state lidar and camera. IEEE Robot. Autom. Lett., 1 (2022). https://doi.org/10.1109/LRA.2022.3173033
40. Yang, G., Song, X., Huang, C., Deng, Z., Shi, J., Zhou, B.: Drivingstereo: a large-scale dataset for stereo matching in autonomous driving scenarios. In: 2019 IEEE/CVF Conference on Computer Vision and Pattern Recognition (CVPR), pp. 899–908 (2019). https://doi.org/10.1109/CVPR.2019.00099
41. Zhang, J., Ramanagopal, M.S., Vasudevan, R., Johnson-Roberson, M.: Listereo: generate dense depth maps from lidar and stereo imagery. In: 2020 IEEE International Conference on Robotics and Automation (ICRA), pp. 7829–7836 (2020). https://doi.org/10.1109/ICRA40945.2020.9196628
42. Zhu, Y., Zheng, C., Yuan, C., Huang, X., Hong, X.: Camvox: a low-cost and accurate lidar-assisted visual slam system. In: 2021 IEEE International Conference on Robotics and Automation (ICRA), pp. 5049–5055 (2021). https://doi.org/10.1109/ICRA48506.2021.9561149

Point Cloud Upsampling via Cascaded Refinement Network

Hang Du, Xuejun Yan, Jingjing Wang, Di Xie, and Shiliang Pu(✉)

Hikvision Research Institute, Hangzhou, China
{duhang,yanxuejun,wangjingjing9,xiedi,pushiliang.hri}@hikvision.com

Abstract. Point cloud upsampling focuses on generating a dense, uniform and proximity-to-surface point set. Most previous approaches accomplish these objectives by carefully designing a single-stage network, which makes it still challenging to generate a high-fidelity point distribution. Instead, upsampling point cloud in a coarse-to-fine manner is a decent solution. However, existing coarse-to-fine upsampling methods require extra training strategies, which are complicated and time-consuming during the training. In this paper, we propose a simple yet effective cascaded refinement network, consisting of three generation stages that have the same network architecture but achieve different objectives. Specifically, the first two upsampling stages generate the dense but coarse points progressively, while the last refinement stage further adjust the coarse points to a better position. To mitigate the learning conflicts between multiple stages and decrease the difficulty of regressing new points, we encourage each stage to predict the point offsets with respect to the input shape. In this manner, the proposed cascaded refinement network can be easily optimized without extra learning strategies. Moreover, we design a transformer-based feature extraction module to learn the informative global and local shape context. In inference phase, we can dynamically adjust the model efficiency and effectiveness, depending on the available computational resources. Extensive experiments on both synthetic and real-scanned datasets demonstrate that the proposed approach outperforms the existing state-of-the-art methods. The code is publicly available at https://github.com/hikvision-research/3DVision.

1 Introduction

Point clouds have been widely adopted in many 3D computer vision studies [13, 15,18,19,28,37,38] in recent years. However, in real-world scenarios, the raw point clouds produced by the 3D sensors are often sparse, noisy, and non-uniform, which have negative impact on the performance of the point cloud analysis and processing tasks. Therefore, in order to facilitate the downstream point cloud

H. Du and X. Yan—Equal contribution.

Supplementary Information The online version contains supplementary material available at https://doi.org/10.1007/978-3-031-26319-4_7.

L. Wang et al. (Eds.): ACCV 2022, LNCS 13841, pp. 106–122, 2023.
https://doi.org/10.1007/978-3-031-26319-4_7

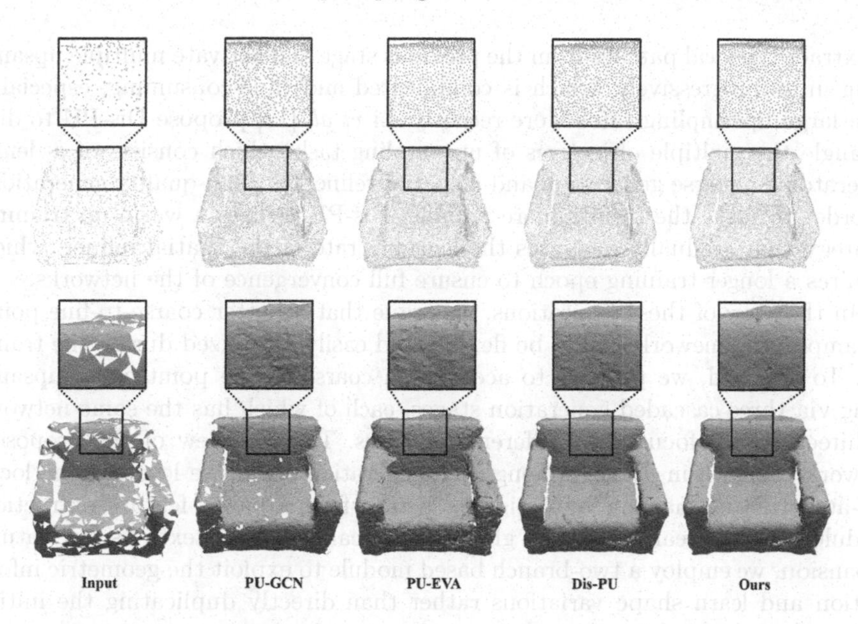

Fig. 1. Upsampling results (×4) on real-scanned point clouds. The top line is the upsampled point clouds, and the bottom line shows the corresponding 3D surface reconstruction results using the ball-pivoting algorithm [2]. Compared with the recent state-of-the-art methods, we can generate high-fidelity point clouds with uniform distribution, resulting in a more smooth reconstruction surface.

tasks, it is necessary to upsample sparse point clouds to a dense, uniform and high-fidelity point set.

In recent years, many deep learning-based methods have been proposed for point cloud upsampling. Compared with the traditional optimization-based methods [1,7,14,29], deep learning-based methods [11,20,21,34,35] are able to handle more complex geometric structures of 3D objects since they can effectively extract deep features from point clouds and learn to generate new points in a data-driven manner. Among them, as a pioneering point cloud upsampling method, Yu *et al.* [34] propose a classic upsampling framework, named PU-Net, which develops a PointNet-based network to learn multi-scale point features and expand the number of points by multi-branch Multi-layer Perceptrons (MLP). Based on the framework of PU-Net, the recent advanced upsampling methods [11,16,20] have made remarkable progress by designing more effective point feature extractor and upsampling unit. However, these methods directly generate the dense and high-fidelity points via a single-stage network, which makes the network challenging to meet multiple objectives and cannot obtain an optimal result through one-stage structure (Fig. 1).

Coarse-to-fine is a widely-used scheme in many point cloud generative models [26,30–32,37]. Certain methods [12,27] also have applied such scheme for point cloud upsampling. Specifically, MPU [27] contains a series of upsampling sub-networks that focus on different level of details. During the training, it needs to extract the local patches from the previous stage and activate multiple upsampling units progressively, which is complicated and time-consuming, especially on a large upsampling rate. More recently, Li *et al.* [12] propose Dis-PU to disentangle the multiple objectives of upsampling task, which consists of a dense generator for coarse generation and a spatial refiner for high-quality generation. In order to make the results more reliable, Dis-PU utilizes a warm-up training strategy that gradually increases the learning rate of the spatial refiner, which requires a longer training epoch to ensure full convergence of the networks.

In the view of these limitations, we argue that a better coarse-to-fine point upsampling framework should be flexible and easily optimized during the training. To this end, we propose to accomplish coarse-to-fine point cloud upsampling via three cascaded generation stages, each of which has the same network architecture but focuses on different purposes. The overview of the proposed network is shown in Fig. 2. Among each generation stage, we leverage the local self-attention mechanism by designing a transformer-based feature extraction module that can learn both the global and local shape context. As for feature expansion, we employ a two-branch based module to exploit the geometric information and learn shape variations rather than directly duplicating the initial point-wise features.

In addition, the learning conflicts between multiple stages is a major challenge to the coarse-to-fine framework. To tackle this problem, we simply adopt a residual learning scheme for point coordinate reconstruction, which predicts the offset of each point and further adjusts the initial point position. In this manner, the entire cascaded refinement network enables to be optimized without extra training strategies.

Resorting to the above practices, we can build a flexible and effective point cloud upsampling framework. Compared with the existing methods, the proposed framework can obtain significant performance improvements. According to the available computational resources, the feed-forward process of refinement stage is optional. Thus, we can dynamically adjust the model efficiency and effectiveness in inference phase. Extensive experiments on both synthetic and real-scanned datasets demonstrate the effectiveness of our proposed method.

The contributions of this paper are summarized as follows:

- We propose a cascaded refinement network for point cloud upsampling. The proposed network can be easily optimized without extra training strategies. Besides, our method has better scalability to dynamically adjust the model efficiency and effectiveness in inference phase.
- We adopt the residual learning scheme in both point upsampling and refinement stages, and develop a transformer-based feature extraction module to learn both global and local shape context.
- We conduct comprehensive experiments on both synthetic and real-scanned point cloud datasets, and achieve the leading performance among the state-of-the-art methods.

2 Related Work

Our work is related to the point cloud upsampling and the networks in point cloud processing. In this section, we provide a brief review of recent advances in these areas.

2.1 Point Cloud Upsampling

Existing point cloud upsampling methods can be roughly divided into traditional optimization-based approaches [1,7,14,29] and the deep learning-based approaches [11,12,16,20–22,27,33–35,39]. The former generally requires the geometric prior information of point clouds (*e.g.,* edges and normal), and achieves poor results on the complex shapes. In contrast, the deep learning-based approaches have been prevailing and dominated the recent state-of-the-arts. Particularly, PU-Net [34] proposes a classic learning-based upsampling framework that involves three components, including feature extraction, feature expansion (upsampling unit), and point coordinate reconstruction. Later on, many works [11,16,20–22,39] follow this classic framework and focus on improving it from many aspects, such as feature extractor, upsampling scheme, training supervision. For example, PU-GAN [11] introduces a upsampling adversarial network with a uniform loss to ensure the point uniformity, and PU-GCN [20] leverages graph convolutional networks for both feature extraction and expansion. The above methods employ a single-stage network to accomplish the multiple objectives of point cloud upsampling. However, it is challenging to achieve all the objectives at the same time, and coarse-to-fine manner is a more appropriate solution. To this end, MPU [27] and Dis-PU [12] propose to divide the upsampling process into multiple steps, however, they require progressive training or warm-up strategy to ensure a better optimization, which is complicated and time-consuming during the training. In this work, instead of using extra training strategies, we develop a more flexible and effective coarse-to-fine framework.

2.2 Point Cloud Processing

There are three major categories of point cloud processing networks, including projection-based, voxel-based, and point-based networks. Since the point clouds have no spatial order and regular structure, projection-based methods [9,24] and voxel-based methods [5,31,40] transform irregular point cloud to 2D pixel or 3D voxel, and then apply 2D/3D convolution to extract the regular representations. In contrast, point-based networks are designed to process the irregular point cloud directly, which is able to avoid the loss of shape context during the projection or voxelization. Among them, some methods [18,19] employ MLP-based structure to extract point-wise features, and others [10,28] utilize graph-based convolutional networks to aggregate the point neighbourhood information. Besides, certain methods [13,23] develop continuous convolutions that can be directly applied to the point cloud. More recently, the success of vision transformer has inspired many 3D point cloud processing approaches [3,6,17,36,38].

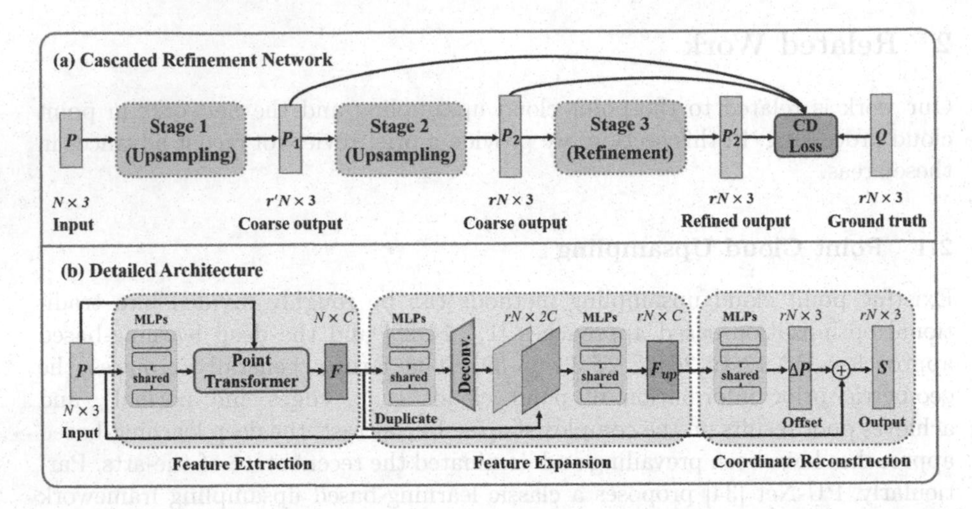

Fig. 2. (a) Overview of our cascaded refinement network for point cloud upsampling, which consists of two upsampling stages and one refinement stage. (b) For each upsampling or refinement stage, the detailed network architecture contains a transformer-based feature extraction module, a two-branch feature expansion module, and a coordinate reconstruction module.

As far as we can see, the transformer-based structure is under-explored in point cloud upsampling. Thus, based on the architecture of point transformer [38], we build a transformer-based feature extraction module in our upsampling network.

3 Proposed Method

Given a sparse point cloud set $\mathcal{P} = \{p_i\}_{i=1}^{N}$, where N is the number of points and p_i is the 3D coordinates, the objective of point clouds upsampling is to generate a dense and high-fidelity point cloud set $\mathcal{S} = \{s_i\}_{i=1}^{rN}$, where r is the upsampling rate. Figure 2 shows the overview of the proposed cascaded refinement network, where the first two upsampling stages generate a coarse dense point cloud set progressively and the last refinement stage servers as a refiner to adjust the coarse points to a better position. During the training, we train the entire framework in an end-to-end manner, and adopt three Chamfer Distance (CD) losses to constrain each stage, simultaneously. In the following sections, we elaborate the detailed network architecture of our framework and the training loss function.

3.1 Network Architecture

As shown in the bottom of Fig. 2, the network architecture of upsampling or refinement stage follows the commonly-used pipeline, which consists of a feature extraction module, a feature expansion module and a coordinate reconstruction

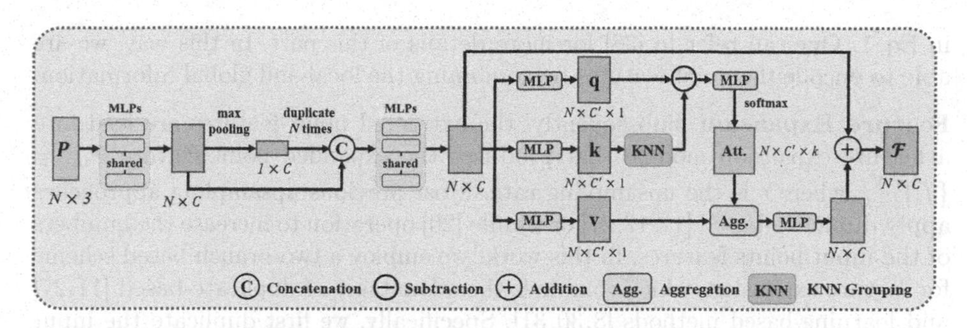

Fig. 3. The detailed network architecture of transform-based feature extraction module. The attention map is calculated locally within the k-nearest neighbors of the current query. We omit the position encoding in value vectors and attention map for the clearness.

module. In the following, we will provide a detailed introduction of each module, respectively.

Feature Extraction. Given a input point cloud $\mathcal{P} = \{p_i\}_{i=1}^{N}$, the feature extraction module aims to encode point features $\mathcal{F} = \{f_i\}_{i=1}^{N}$ of C channel dimensions. Most existing point cloud upsampling methods [11,12,34] fulfill this objective using a MLP-based structure. However, the local details are limited by the insufficient feature representation capability of MLPs. To solve this problem, some upsampling approaches [10,20,28] employ a GCN-based structure to capture the local neighborhood information for point feature extraction.

In the view of the success of transformer-based networks [6,17,38] in the field of 3D computer vision, we explore a transformer-based feature extraction module for the task of point cloud upsampling. Figure 3 illustrates the details on our proposed network architecture. Based on the network architecture in [19], we first extract the point-wise features from the input point clouds via a set of MLPs, and then apply a max pooling operation to obtain the global features. The global features are duplicated N times and concatenated with the initial features, subsequently. After that, a point transformer layer [38] is utilized to refine the local shape context and obtain the output point features. By applying self-attention locally, the learned features can incorporate the point neighborhood information between the point feature vectors. Finally, the encoded point features $\mathcal{F} = \{\mathbf{f}_i\}_{i=1}^{N}, \mathbf{f}_i \in \mathbb{R}^C$ are added with the input features through a residual connection. The locally self-attention operations can be formulated as:

$$\mathbf{f}_i = \mathbf{x}_i \oplus \sum_{j \in \mathcal{N}(i)} softmax\left(\mathbf{q}_i - \mathbf{k}_j\right) \odot \mathbf{v}_j, \tag{1}$$

where $\mathbf{x}_i$ is ith input point feature vector, $\mathbf{q}_i$ is the corresponding query vector, $\mathbf{k}_j$ is the jth key vector from the k-nearest neighbors of $\mathbf{x}_i$, and $\mathbf{v}_j$ is the jth value vector. The attention weights are calculated between the query and its k-nearest neighbors by softmax function. Note that we omit the position encoding in Eq. 1. One can refer to [38] for more details of this part. In this way, we are able to encode the point features by combining the local and global information.

Feature Expansion. Subsequently, the extracted point features are feed into a feature expansion module that produces the expanded point feature $\mathcal{F}_{\text{up}} = \{f_i\}_{i=1}^{rN}$, where r is the upsampling rate. Most previous upsampling approaches apply duplicate-based [11,12,27] or shuffle [20] operation to increase the numbers of the input points features. In this work, we employ a two-branch based scheme for feature expansion, which combines the advantages of duplicate-based [11,27] and learning-based methods [8,30,31]. Specifically, we first duplicate the input features $\mathcal{F} = \{f_i\}_{i=1}^{N}$ with r copies. Meanwhile, a transposed convolution branch is used for feature expansion, which is utilized to learn new point features through a learnable manner. Certain methods [8,30] have discussed the advantages of learning-based feature interpolation for generative models, which can exploit more local geometric structure information with respect to the input shape. Finally, we concatenate the point features from two branches and feed them to a set of MLPs which produces the expanded features as follows:

$$\mathcal{F}_{\text{up}} = \text{MLP}(\text{Concat}[\text{Dup}(\mathcal{F}, r); \text{Deconv}(\mathcal{F})]). \tag{2}$$

The duplicated-based branch is able to preserve the initial shape, and the learnable transposed convolution branch can provide local geometric variations on the input point clouds. The combination of them enables to produce a more expressive upsampled features for reconstructing the point coordinate reconstruction subsequently.

Coordinate Reconstruction. The objective of coordinate reconstruction is to generate a new point set $\mathcal{S} = \{s_i\}_{i=1}^{rN}$ from the expanded feature vectors $\mathcal{F}_{\text{up}} = \{f_i\}_{i=1}^{rN}$. To accomplish this objective, a common way is to regress the 3D point coordinates directly. However, it is difficult to generate high-fidelity points from latent space without noises [12]. In order to solve this problem, several methods [12,27] apply a residual learning strategy that predicts the offset $\Delta\mathcal{P}$ of each point to adjust the initial position, which can be formulated as:

$$\mathcal{P}_{\text{output}} = \Delta\mathcal{P} + \text{Dup}(\mathcal{P}_{\text{input}}, r), \tag{3}$$

where r is the upsampling rate. Moreover, residual learning is a simple yet effective scheme to mitigate the learning conflicts between multiple stages in a coarse-to-fine framework. Thus, we consider the advantages of such scheme are two folds: (i) learning per-point offset can greatly decrease the learning difficulty of a cascaded network and mitigate the conflicts between multiple stages, which results in a stable and better optimization; (ii) residual connection enables to preserve the geometric information of initial shape and provide variations for generating new points. Hence, we adopt residual learning scheme for coordinate reconstruction, which contains two MLP layers to generate the per-point offset $\Delta\mathcal{P}$ from $\mathcal{F}_{\text{up}}$ ($\Delta\mathcal{P} = \text{MLP}(\mathcal{F}_{\text{up}})$), and then adds it on the r times duplicated input points. In this manner, the entire cascaded refinement network can be easily optimized together without extra training strategies (*e.g.*, progressive or warm-up training).

3.2 Compared with Previous Coarse-to-Fine Approaches

The coarse-to-fine scheme has been studied in the previous point cloud upsampling methods [12,27]. Nonetheless, the proposed approach is more efficient and flexible during the training and inference. Specifically, MPU [27] needs to extract the local training patches from the previous stage, and activates the training of each upsampling stage progressively. Dis-PU [12] adopts a warm-up training strategy to control the importance of dense generator and spatial refiner in different training stage. Both of them are complex during the training and cost a longer time to ensure the full convergence of the networks. Moreover, the complicated training strategies rely on the experience of hyper-parameters tuning, which may be sensitive on the new datasets. In contrast, our advantages are two folds. Firstly, the proposed cascaded refinement network can be easily trained without progressive or warm-up training strategies, which enables to require less training consumption. Secondly, our framework is more flexible and scalable. We can dynamically adjust the model efficiency and effectiveness at the stage of inference. For example, we can pursue a faster inference time only using two-stage upsampling structure, and still achieve a leading performance among other competitors (Table 3). Therefore, the proposed coarse-to-fine upsampling framework can be more effective and practical in real-world applications.

3.3 Training Loss Function

During the training, our approach is optimized in an end-to-end manner. To encourage the upsampled points more distributed over the target surface, we employ Chamfer Distance (CD) [4] $\mathcal{L}_{CD}$ for training loss function, which is formulated as:

$$\mathcal{L}_{CD}\left(\mathcal{P}, \mathcal{Q}\right) = \frac{1}{|\mathcal{P}|}\sum_{p\in\mathcal{P}}\min_{q\in\mathcal{Q}}\|p-q\|_2^2 + \frac{1}{|\mathcal{Q}|}\sum_{q\in\mathcal{Q}}\min_{p\in\mathcal{P}}\|p-q\|_2^2, \tag{4}$$

where $\mathcal{P}$ and $\mathcal{Q}$ denote the generated point clouds and their corresponding ground truth, respectively. CD loss aims to calculate the average closest point distance between the two sets of point cloud.

In order to explicitly control the generation process of point clouds, we employ the CD loss on three generation stages, simultaneously. Therefore, the total training loss function $\mathcal{L}_{\text{total}}$ can be defined as:

$$\mathcal{L}_{\text{total}} = \mathcal{L}_{CD}\left(\mathcal{P}_1, \mathcal{Q}\right) + \mathcal{L}_{CD}\left(\mathcal{P}_2, \mathcal{Q}\right) + \mathcal{L}_{CD}(\mathcal{P}_2^{'}, \mathcal{Q}), \tag{5}$$

where $\mathcal{P}_1$, $\mathcal{P}_2$ and $\mathcal{P}_2^{'}$ are the output point clouds from three generation stages, respectively. We validate the effectiveness of the penalty on three stages in the experiment (please refer to supplementary materials). The results show that the simultaneous supervision on three generation stages results in a better performance. Therefore, we adopt CD loss on three generation stages simultaneously.

4 Experiments

In this section, we conduct extensive experiments to verify the effectiveness of the proposed point upsampling network. The section is organized as follows. Section 4.1 introduces the detail experimental settings. Section 4.2 demonstrates the obvious improvements by our approach on two synthetic datasets. Section 4.3 shows the qualitative comparisons on real-scanned data. Section 4.4 provides model complexity analysis with other counterparts. Section 4.5 studies the effect of major components in our approach.

4.1 Experimental Settings

Datasets. We employ two point cloud upsampling datasets, including PU-GAN [11] and PU1K [20]. Among them, PU-GAN dataset contains 24,000 training patches collected from 120 3D models and 27 3D models for testing. In contrast, PU1K dataset is a large-scale point cloud upsampling dataset which is newly released by PU-GCN [20]. PU1K dataset consists of 69,000 training patches collected from 1,020 3D models and 127 3D models for testing. In addition, we also utilize a real-scanned point cloud dataset ScanObjectNN [25] for qualitative evaluation.

Training and Evaluation. Our models are trained with 100 epochs. The batch size is set as 64 for PU1K and 32 for PU-GAN, respectively. The learning rate begins at 0.001 and drops by a decay rate of 0.7 every 50k iterations. The ground truth of each training patch contains 1,024 points and the input contains 256 points that are randomly sampled from the ground truth. In practical implementation, we set the upsampling rate $r = 2$ in the upsampling stage and $r = 1$ in the refinement stage, to achieve $\times 4$ upsampling. More detailed settings of our network can be found in the supplementary material. For inference, we follow the common settings [11,20] to divide the input point clouds into multiple patches based on the seed points. Then, the patches are upsampled with r times. After that, FPS algorithm is used to combine all the upsampled patches as the output point clouds. For quantitative evaluation, we apply three commonly used metrics, including Chamfer Distance (CD), Hausdorff Distance (HD), and Point-to-Surface Distance (P2F). A smaller value of these metrics denotes a better performance.

Comparison Methods. In the following, we will compare the proposed method with five existing point cloud upsampling methods, including PU-Net [34], MPU [27], PU-GAN [11], Dis-PU [12], PU-GCN [20], and PU-EVA [16]. For a fair comparison, we reproduce these methods by their officially released codes and recommended settings on the same experimental environment.

4.2 Results on Synthetic Dataset

PU1K Dataset. As shown in Table 1, we conduct experiments on three different input sizes of point, including sparse (512), medium (1,024), and dense (2,048). From the results, we can observe the proposed method can achieve

Table 1. Quantitative comparison (×4 upsampling) on PU1K dataset with different input sizes of point cloud. The values of CD, HD, and P2F are multiplied by 10^3. A smaller value denotes a better performance.

Methods	Sparse (512) input			Medium (1,024) input			Dense (2,048) input		
	CD	HD	P2F	CD	HD	P2F	CD	HD	P2F
PU-Net [34]	2.990	35.403	11.272	1.920	24.181	7.468	1.157	15.297	4.924
MPU [27]	2.727	30.471	8.190	1.268	16.088	4.777	0.861	11.799	3.181
PU-GAN [11]	2.089	22.716	6.957	1.151	14.781	4.490	0.661	9.238	2.892
Dis-PU [12]	2.130	25.471	7.299	1.210	16.518	4.606	0.731	9.505	2.719
PU-GCN [20]	1.975	22.527	6.338	1.142	14.565	4.082	0.635	9.152	2.590
PU-EVA [16]	1.942	20.980	6.366	1.065	13.376	4.172	0.649	8.870	2.715
Ours	**1.594**	**17.733**	**4.892**	**0.808**	**10.750**	**3.061**	**0.471**	**7.123**	**1.925**

Table 2. Quantitative comparison on PU-GAN [11] dataset. The input size of point cloud is 1,024. The values of CD, HD, and P2F are multiplied by 10^3.

Methods	×4 Upsampling			×16 Upsampling		
	CD	HD	P2F	CD	HD	P2F
PU-Net [34]	0.883	7.132	8.127	0.845	11.225	10.114
MPU [27]	0.589	6.206	4.568	0.365	8.113	5.181
PU-GAN [11]	0.566	6.932	4.281	0.390	8.920	4.988
Dis-PU [12]	0.527	5.706	3.378	0.302	6.939	4.146
PU-GCN [20]	0.584	**5.257**	3.987	0.320	**6.567**	4.381
PU-EVA [16]	0.571	5.840	3.937	0.342	8.140	4.473
Ours	**0.520**	6.102	**3.165**	**0.284**	7.143	**3.724**

the leading performance in all metrics and obvious improvements on the sparse input. Besides, compared the results of left three columns (sparse input) with the right three columns (dense input), we can observe that the sparse input size is a more challenging scenario. Nonetheless, our method can consistently yield evident improvements over the existing state-of-the-art methods, which verifies the robustness of our method to different input sizes. Moreover, the qualitative results in Fig. 4 can also demonstrate that our cascaded refinement network enables to generate more uniform point clouds with better fine-grained details and less outliers, such as the arms of the chair and the wings of the plane.

PU-GAN Dataset. Following the previous settings [12], we also conduct experiments under two different upsampling rates at the input size of 1,024 on PU-GAN dataset. Table 2 reports the quantitative results based on ×4 and ×16 upsampling rates. Since PU-GAN dataset only provides training patches based on ×4 upsampling rate, we apply the model twice for accomplishing ×16 upsampling rates in this experiment. Overall, our method can achieve the best performance in most metrics. For Hausdorff Distance metric, some counterparts slightly outperform our method. We consider the most possible reason is that

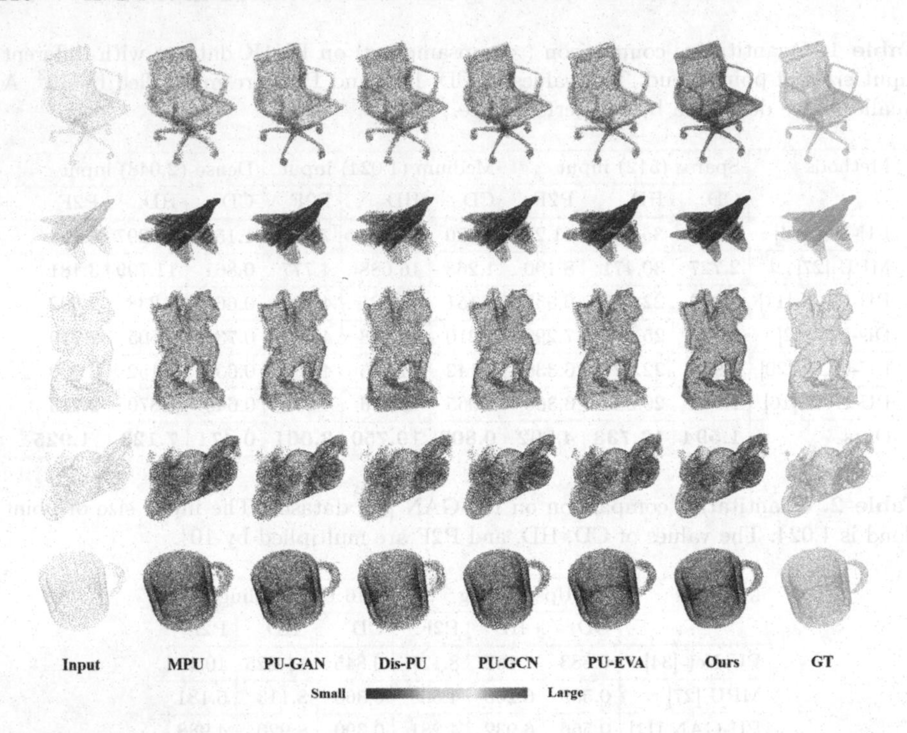

Fig. 4. Point cloud upsampling (×4) results on PU1K dataset. The point clouds are colored by the nearest distance between the ground truth and the prediction of each method. The blue denotes the small errors. One can zoom in for more details. (Color figure online)

our transformer-based feature extraction module require more variations during the training. Compared with the 1,020 3D meshes in PU1K, PU-GAN is a much smaller dataset that only contains 120 meshes for training, which limits the capacity of the proposed method. Nonetheless, as for the results of Chamfer Distance and Point-to-Surface Distance, our method gains evident improvement over other counterparts.

4.3 Results on Real-Scanned Dataset

To verify the effectiveness of our method in real-world scenarios, we utilize the models trained on PU1K and conduct the qualitative experiments on ScanObjectNN [25] dataset. We only visualize the input sparse points and upsampled results, since ScanObjectNN dataset does not provide the corresponding dense points. For clarity, we only choose three recent advanced methods (Dis-PU [12], PU-GCN [20], and PU-EVA [16]) for visualization. As shown in Fig. 5, we can observe upsampling the real-scanned data is more challenging, since the input sparse points are non-uniform and incomplete. Comparing with other counterparts, our method can generate more uniform and high-fidelity point clouds. The

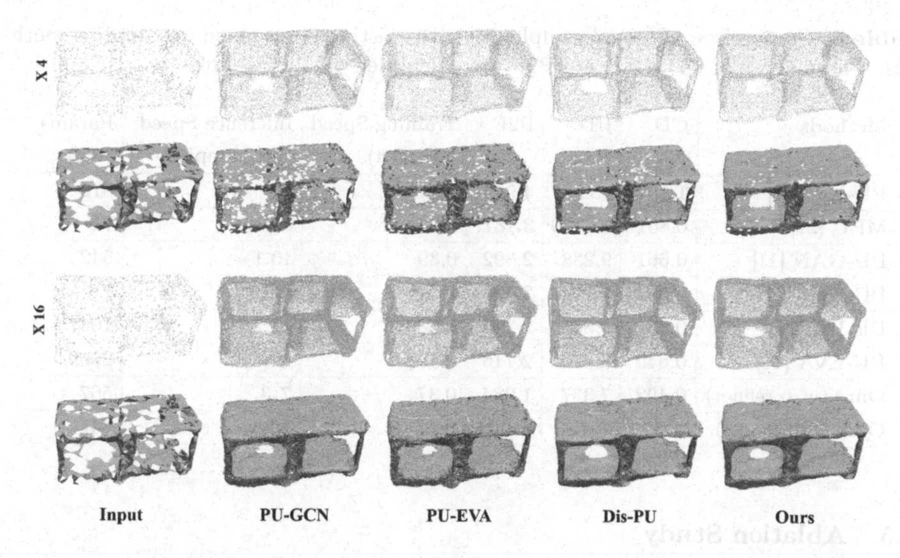

Fig. 5. Point cloud upsampling (×4 and ×16) results on real-scanned sparse inputs. Compared with the other counterparts, our method can generate more uniform with detailed structures. The 3D meshes are reconstructed by ball-pivoting algorithm [2].

visualization of 3D meshes also shows the proposed is able to obtain a better surface reconstruction result. More visualization results on real-scanned data can be found in the supplementary material.

4.4 Model Complexity Analysis

In this section, we investigate the model complexity of our method compared with existing point cloud upsampling methods. The experiments are conducted on a Titan X Pascal GPU. As shown in Table 3, we can find that the proposed method is comparable with other counterparts. In point of training speed, our method is more efficiency than PU-GAN [11], Dis-PU [12], and PU-EVA [16]. Note that Dis-PU [12] employs a warm-up strategy during the training and requires more epoch to ensure the convergence of networks. Thus, the total training time of Dis-PU costs about 48 h on PU-GAN dataset, which is much longer than ours (8 h). As for inference time, we obtain a comparable and affordable inference speed compared with existing methods. In terms of model parameters, the proposed method can also be regarded as a lightweight network for point cloud upsampling. In addition, our method is scalable to achieve a trade-off between model effectiveness and efficiency by removing the refinement stage during the inference. In this manner, we can achieve leading performance as well as the model efficiency among other counterparts (the first bottom line). Overall, our cascaded refinement network can not only obtain the obvious performance improvements but also achieve comparable model complexity with other counterparts.

Table 3. Comparison of model complexity with existing point cloud upsampling methods. The metrics of CD, HD, and P2F are calculated on PU1K dataset.

Methods	CD	HD	P2F	Training Speed (s/batch)	Inference Speed (ms/sample)	Params. (Kb)
PU-Net [34]	1.157	15.297	4.924	**0.13**	9.9	812
MPU [27]	0.861	11.799	3.181	0.23	8.8	**76**
PU-GAN [11]	0.661	9.238	2.892	0.39	13.1	542
PU-GCN [20]	0.635	9.152	2.590	**0.13**	8.6	**76**
Dis-PU [12]	0.731	9.505	2.719	0.58	11.2	1047
PU-EVA [16]	0.649	8.870	2.715	0.57	12.8	2869
Ours (w/o refiner)	0.492	7.337	1.984	0.37	**7.3**	567
Ours (full model)	**0.471**	**7.123**	**1.925**	0.37	10.8	847

4.5 Ablation Study

To verify the effect of major components in our approach, we quantitatively evaluate the contribution of them. In the following, we provide the detailed discussions respectively.

Ablation of Generation Stages. Our full cascaded refinement network consists of two upsampling stage and one refinement stage. In order to analyze the effectiveness of each sub-network, we investigate the setting of the number of generation stages. As shown in Table 4, the "two-stage" refers to remove the refinement stage, and the "four-stage" denotes adding an extra upsampling stage. All the models are trained from scratch. Thus, the "two-stage" is different with merely removing the refiner in inference phase. We can find that the performance consistently increases with the number of generation stages. Our "two-stage" still outperforms the existing SOTA methods with a faster inference speed. In the view of the balance between effectiveness and efficiency, we adopt two upsampling stage and a refinement stage in this work.

Effect of Feature Extraction Module. Feature extraction module plays an important role in point cloud upsampling. As shown in Table 5, we investigate the effect of different network architecture on feature extraction. Specifically, "MLP-based" denotes the same feature extraction unit in MPU [27], and "DenseGCN" refers to replace the point transformer layer with a DenseGCN block in [10]. From the results, we can find that the proposed transform-based feature extraction module outperforms other counterparts, indicating its effectiveness of the locally self-attention operator on shape context learning. Moreover, the comparison of DenseGCN and MLP-based networks proves that an effective feature extraction network can promote obvious improvements on point cloud upsampling.

Study of Residual Learning. For a coarse-to-fine upsampling framework, it is crucial to mitigate the learning conflicts between multiple generation units. As mentioned in Sect. 3.2, MPU needs to progressively activate the training of

Table 4. Ablation of generation stages. The values of CD, HD, and P2F are multiplied by 10^3.

Model	Dense (2048) input			Inference Speed (ms/sample)	Params. (Kb)
	CD	HD	P2F		
Two-stage	0.513	7.808	2.078	**7.3**	**567**
Three-stage	0.471	7.123	1.925	10.8	847
Four-stage	**0.466**	**6.793**	**1.821**	14.7	1127

Table 5. Effect of feature extraction module.

Model	Dense (2,048) input		
	CD	HD	P2F
MLP-based	0.628	9.687	2.721
DenseGCN	0.531	8.462	2.254
Ours	**0.471**	**7.123**	**1.925**

Table 6. Effect of residual learning scheme.

Warm-up	Residual learning	Dense (2,048) input		
		CD	HD	P2F
✓	-	0.511	7.599	2.986
✓	✓	**0.472**	**7.329**	**1.933**
–	–	0.733	9.667	3.380
–	✓	**0.471**	**7.123**	**1.925**

each upsampling unit, and Dis-PU employs a warm-up strategy that gradually increases the learning rate of spatial refiner. In contrast, we only adopt the residual learning in each generation stage without progressive or warm-up training strategy. From the results in Table 6, we can find that residual learning is more effective than warm-up training strategy in our framework. The reason behind is, that the residual connection makes each stage only regress the point offsets of the initial shape and thus mitigate the learning conflicts between multiple stages, resulting in a better optimization. Therefore, we equip the residual learning for point reconstruction, which makes the proposed method be easily optimized without extra training strategies.

5 Conclusion

In this paper, we propose a cascaded refinement network for point cloud upsampling. The proposed method consists of three generation stages, which progressively generates the coarse dense points and refine the coarse points to a better position in the end. We adopt a residual learning scheme for point coordinate reconstruction, which enables to decrease the difficulty for regressing the new points and mitigate the conflicts between multiple stages. Moreover, a transformer-based feature extraction module is designed to aggregate the global and local shape context, and a two-branch based feature expansion module enables to learn a more expressive upsampled features. Compared with the existing coarse-to-fine point upsampling methods, the proposed network can be easily optimized without extra training strategies. Moreover, we can also dynamically adjust the model efficiency and effectiveness at the stage of inference, depending on the available computational resources. Both quantitative and qualitative results demonstrate the superiority of our method over the state-of-the-art methods.

References

1. Alexa, M., Behr, J., Cohen-Or, D., Fleishman, S., Levin, D., Silva, C.T.: Computing and rendering point set surfaces. IEEE Trans. Vis. Comput. Graph. **9**, 3–15 (2003)
2. Bernardini, F., Mittleman, J., Rushmeier, H., Silva, C., Taubin, G.: The ball-pivoting algorithm for surface reconstruction. IEEE Trans. Visual Comput. Graph. **5**(4), 349–359 (1999)
3. Engel, N., Belagiannis, V., Dietmayer, K.C.J.: Point transformer. IEEE Access **9**, 134826–134840 (2021)
4. Fan, H., Su, H., Guibas, L.J.: A point set generation network for 3d object reconstruction from a single image. In: Proceedings of the IEEE Conference on Computer Vision and Pattern Recognition, pp. 605–613 (2017)
5. Graham, B., Engelcke, M., van der Maaten, L.: 3d semantic segmentation with submanifold sparse convolutional networks. In: 2018 IEEE/CVF Conference on Computer Vision and Pattern Recognition, pp. 9224–9232 (2018)
6. Guo, M.H., Cai, J., Liu, Z.N., Mu, T.J., Martin, R.R., Hu, S.: Pct: point cloud transformer. Comput. Vis. Media **7**, 187–199 (2021)
7. Huang, H., Wu, S., Gong, M., Cohen-Or, D., Ascher, U.M., Zhang, H.: Edge-aware point set resampling. ACM Trans. Graph. (TOG) **32**, 1–12 (2013)
8. Hui, L., Xu, R., Xie, J., Qian, J., Yang, J.: Progressive point cloud deconvolution generation network. In: Vedaldi, A., Bischof, H., Brox, T., Frahm, J.-M. (eds.) ECCV 2020. LNCS, vol. 12360, pp. 397–413. Springer, Cham (2020). https://doi.org/10.1007/978-3-030-58555-6_24
9. Lang, A.H., Vora, S., Caesar, H., Zhou, L., Yang, J., Beijbom, O.: Pointpillars: fast encoders for object detection from point clouds. In: 2019 IEEE/CVF Conference on Computer Vision and Pattern Recognition (CVPR), pp. 12689–12697 (2019)
10. Li, G., Müller, M., Thabet, A.K., Ghanem, B.: Deepgcns: can GCNs go as deep as CNNs? In: ICCV, pp. 9266–9275 (2019)
11. Li, R., Li, X., Fu, C.W., Cohen-Or, D., Heng, P.A.: Pu-gan: a point cloud upsampling adversarial network. In: ICCV, pp. 7202–7211 (2019)
12. Li, R., Li, X., Heng, P.A., Fu, C.W.: Point cloud upsampling via disentangled refinement. In: CVPR, pp. 344–353 (2021)
13. Li, Y., Bu, R., Sun, M., Wu, W., Di, X., Chen, B.: Pointcnn: convolution on x-transformed points. In: NeurIPS (2018)
14. Lipman, Y., Cohen-Or, D., Levin, D., Tal-Ezer, H.: Parameterization-free projection for geometry reconstruction. In: SIGGRAPH 2007 (2007)
15. Long, C., Zhang, W., Li, R., Wang, H., Dong, Z., Yang, B.: Pc2-pu: patch correlation and position correction for effective point cloud upsampling. ArXiv abs/2109.09337 (2021)
16. Luo, L., Tang, L., Zhou, W., Wang, S., Yang, Z.X.: Pu-eva: an edge-vector based approximation solution for flexible-scale point cloud upsampling. In: ICCV (2021)
17. Pan, X., Xia, Z., Song, S., Li, L.E., Huang, G.: 3d object detection with pointformer. In: CVPR, pp. 7459–7468 (2021)
18. Qi, C., Su, H., Mo, K., Guibas, L.J.: Pointnet: deep learning on point sets for 3d classification and segmentation. In: CVPR, pp. 77–85 (2017)
19. Qi, C., Yi, L., Su, H., Guibas, L.J.: Pointnet++: deep hierarchical feature learning on point sets in a metric space. In: NIPS (2017)

20. Qian, G., Abualshour, A., Li, G., Thabet, A.K., Ghanem, B.: Pu-gcn: point cloud upsampling using graph convolutional networks. In: CVPR, pp. 11678–11687 (2021)
21. Qian, Y., Hou, J., Kwong, S., He, Y.: PUGeo-Net: a geometry-centric network for 3D point cloud upsampling. In: Vedaldi, A., Bischof, H., Brox, T., Frahm, J.-M. (eds.) ECCV 2020. LNCS, vol. 12364, pp. 752–769. Springer, Cham (2020). https://doi.org/10.1007/978-3-030-58529-7_44
22. Qian, Y., Hou, J., Kwong, S.T.W., He, Y.: Deep magnification-flexible upsampling over 3d point clouds. IEEE Trans. Image Process. **30**, 8354–8367 (2021)
23. Shi, S., Wang, X., Li, H.: Pointrcnn: 3d object proposal generation and detection from point cloud. In: 2019 IEEE/CVF Conference on Computer Vision and Pattern Recognition (CVPR), pp. 770–779 (2019)
24. Tatarchenko, M., Park, J., Koltun, V., Zhou, Q.Y.: Tangent convolutions for dense prediction in 3d. In: Proceedings of the IEEE Conference on Computer Vision and Pattern Recognition, pp. 3887–3896 (2018)
25. Uy, M.A., Pham, Q.H., Hua, B.S., Nguyen, D.T., Yeung, S.K.: Revisiting point cloud classification: a new benchmark dataset and classification model on real-world data. In: ICCV, pp. 1588–1597 (2019)
26. Wang, X., Ang Jr, M.H., Lee, G.H.: Cascaded refinement network for point cloud completion. In: Proceedings of the IEEE/CVF Conference on Computer Vision and Pattern Recognition, pp. 790–799 (2020)
27. Wang, Y., Wu, S., Huang, H., Cohen-Or, D., Sorkine-Hornung, O.: Patch-based progressive 3d point set upsampling. In: CVPR, pp. 5951–5960 (2019)
28. Wang, Y., Sun, Y., Liu, Z., Sarma, S.E., Bronstein, M.M., Solomon, J.M.: Dynamic graph CNN for learning on point clouds. ACM Trans. Graph. (TOG) **38**, 1–12 (2019)
29. WuShihao, H.: GongMinglun, ZwickerMatthias. Deep points consolidation. ACM Transactions on Graphics, Cohen-OrDaniel (2015)
30. Xiang, P., et al.: SnowflakeNet: point cloud completion by snowflake point deconvolution with skip-transformer. In: ICCV (2021)
31. Xie, H., Yao, H., Zhou, S., Mao, J., Zhang, S., Sun, W.: GRNet: gridding residual network for dense point cloud completion. In: Vedaldi, A., Bischof, H., Brox, T., Frahm, J.-M. (eds.) ECCV 2020. LNCS, vol. 12354, pp. 365–381. Springer, Cham (2020). https://doi.org/10.1007/978-3-030-58545-7_21
32. Yan, X., et al.: FBNet: feedback network for point cloud completion. In: Avidan, S., Brostow, G., Cissé, M., Farinella, G.M., Hassner, T. (eds) ECCV 2022. LNCS, vol. 13662. Springer, Cham (2022). https://doi.org/10.1007/978-3-031-20086-1_39
33. Ye, S., Chen, D., Han, S., Wan, Z., Liao, J.: Meta-pu: an arbitrary-scale upsampling network for point cloud. IEEE Trans. Visualization Comput. Graph. (2021)
34. Yu, L., Li, X., Fu, C.W., Cohen-Or, D., Heng, P.: Pu-net: point cloud upsampling network. In: CVPR, pp. 2790–2799 (2018)
35. Yu, L., Li, X., Fu, C.-W., Cohen-Or, D., Heng, P.-A.: EC-Net: an edge-aware point set consolidation network. In: Ferrari, V., Hebert, M., Sminchisescu, C., Weiss, Y. (eds.) ECCV 2018. LNCS, vol. 11211, pp. 398–414. Springer, Cham (2018). https://doi.org/10.1007/978-3-030-01234-2_24
36. Yu, X., Rao, Y., Wang, Z., Liu, Z., Lu, J., Zhou, J.: Pointr: diverse point cloud completion with geometry-aware transformers. In: Proceedings of the IEEE/CVF International Conference on Computer Vision, pp. 12498–12507 (2021)
37. Yuan, W., Khot, T., Held, D., Mertz, C., Hebert, M.: Pcn: point completion network. International Conference on 3D Vision (3DV), pp. 728–737 (2018)

38. Zhao, H., Jiang, L., Jia, J., Torr, P.H.S., Koltun, V.: Point transformer. In: ICCV (2021)
39. Zhao, Y., Hui, L., Xie, J.: Sspu-net: self-supervised point cloud upsampling via differentiable rendering. In: ACMMM (2021)
40. Zhou, Y., Tuzel, O.: Voxelnet: end-to-end learning for point cloud based 3d object detection. In: 2018 IEEE/CVF Conference on Computer Vision and Pattern Recognition, pp. 4490–4499 (2018)

CVLNet: Cross-view Semantic Correspondence Learning for Video-Based Camera Localization

Yujiao Shi[1(✉)], Xin Yu[2], Shan Wang[1], and Hongdong Li[1]

[1] Australian National University, Canberra, Australia
yujiao.shi@anu.edu.au
[2] University of Technology Sydney, Sydney, Australia

Abstract. This paper tackles the problem of Cross-view Video-based camera Localization (CVL). The task is to localize a query camera by leveraging information from its past observations, *i.e.*, a continuous sequence of images observed at previous time stamps, and matching them to a large overhead-view satellite image. The critical challenge of this task is to learn a powerful global feature descriptor for the sequential ground-view images while considering its domain alignment with reference satellite images. For this purpose, we introduce CVLNet, which first projects the sequential ground-view images into an overhead view by exploring the ground-and-overhead geometric correspondences and then leverages the photo consistency among the projected images to form a global representation. In this way, the cross-view domain differences are bridged. Since the reference satellite images are usually pre-cropped and regularly sampled, there is always a misalignment between the query camera location and its matching satellite image center. Motivated by this, we propose estimating the query camera's relative displacement to a satellite image before similarity matching. In this displacement estimation process, we also consider the uncertainty of the camera location. For example, a camera is unlikely to be on top of trees. To evaluate the performance of the proposed method, we collect satellite images from Google Map for the KITTI dataset and construct a new cross-view video-based localization benchmark dataset, KITTI-CVL. Extensive experiments have demonstrated the effectiveness of video-based localization over single image-based localization and the superiority of each proposed module over other alternatives.

1 Introduction

Cross-view image-based localization using ground-to-satellite image matching has attracted significant attention these days [1–11]. It has found many practical applications such as autonomous driving and robot navigation. Prior works have been focused on localizing omnidirectional ground-view images with a 360° Field-of-View (FoV), which helps to provide rich and discriminative features

Supplementary Information The online version contains supplementary material available at https://doi.org/10.1007/978-3-031-26319-4_8.

L. Wang et al. (Eds.): ACCV 2022, LNCS 13841, pp. 123–141, 2023.
https://doi.org/10.1007/978-3-031-26319-4_8

for localization. However, when a regular forward-looking camera with a limited FoV is used, those omnidirectional camera-based algorithms suffer severe performance degradation.

To tackle this challenge, this paper proposes to use a continuous short video, *i.e.*, a sequence of ground-view images, as input for the task of visual localization. Specifically, we localize a camera at the current time stamp t_n by augmenting it with previous observations at time *i.e.*, $t_1 \sim t_{n-1}$, as shown in Fig.1. Compared to using a single query image, a short video provides richer visual and dynamic information about the current location.

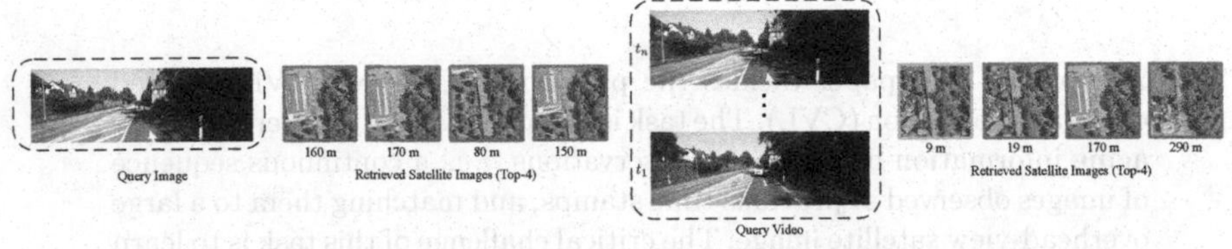

(a) Cross-view image-based localization (b) Cross-view video-based localization

Fig. 1. Single-frame image-based localization (a) Vs. Multi-frame video-based localization (b). The multi-frame video-based localization leverages richer scene context of a query place, increasing the discriminating power of query descriptors compared to single image-based localization. As a result, the matching satellite image, marked by green border, from the database is more likely to be retrieved. Red border indicates non-matching satellite images to the query image. (Color figure online)

We present a Cross-view Video-based Localization Network, named CVLNet, to address the camera localization problem. To the best of our knowledge, our CVLNet is the first vision- and deep-based cross-view geo-localization framework that exploits a continuous video rather than a single image to pinpoint the camera location.

Our CVLNet is composed of two branches that extract deep features from ground and satellite images, respectively. Considering the drastic viewpoint changes between the two-view images, we first introduce a Geometry-driven View Projection (GVP) module to transform ground-view features to the overhead view by explicitly exploring their geometric correspondences. Then, we design a Photo-consistency Constrained Sequence Fusion (PCSF) module to fuse the sequential features. Our PCSF first estimates the reliability of the sequential ground-view features in overhead view by leveraging photo-consistency across them and then aggregates them as a global query descriptor. In this manner, we achieve more discriminative and reliable ground-view feature representation.

Since satellite images in a database are usually pre-cropped and sampled at discretized locations, there would be a misalignment between a query camera location and its matching satellite image center. Furthermore, a query camera is usually impossible in some regions (*e.g.*, on top of a tree), while likely on the other areas (*e.g.*, road). Hence, we propose a Scene-prior driven Similarity Matching (SSM) strategy to estimate the relative displacement between a query camera location and a satellite image center while restricting the search space by scene priors. The scene priors are learned statistically from training rather than pre-defined. With the help of SSM, our CVLNet can eliminate unreasonable localization results.

In order to train and evaluate our method, we curate a new cross-view dataset by collecting satellite images for the KITTI dataset [12] from Google Map [13]. The new dataset combines sequential ground-view images from the original KITTI dataset and the newly collected satellite images. To the best of our knowledge, it is not only the first cross-view video-based localization dataset, but also the first cross-view localization dataset where ground-view images are captured by a perspective pin-hole camera with a restricted FoV (rather than being cropped from Google street-view panoramas [1,8]). Extensive experiments on the newly collected dataset demonstrate that our method effectively localizes camera positions and outperforms the state-of-the-art remarkably.

2 Related Work

Image-Based Localization. The image-based localization problem is initially tackled as a ground-to-ground image matching [14–19], where both the query and database images are captured at the ground level. However, those methods cannot localize query images when there is no corresponding reference image in the database. Thanks to the wide-spread coverage and easy accessibility of satellite imagery, recent works [1–11,20–29] resort to satellite images for city-scale localization.

While recent works on city-scale ground-to-satellite localization have achieved promising results, they mostly focus on localizing isolated omnidirectional ground images. When the query camera has a limited FoV, we propose using a continuous video instead of a single image for camera localization, improving the discriminativeness of the query location representation.

Video-Based Localization. The concept of video-based localization can be divided into three main categories; Visual Odometry (VO) [30–32], Visual-SLAM (vSLAM) [33–38] and Visual Localization [39–45]. VO techniques can be classified according to their camera setup - either monocular or stereoscopic or their processing techniques - either feature-based or appearance-based. VO methods usually use a combination of feature tracking and feature matching [46,47]. vSLAM pertains to simultaneously creating a map of features and localizing the robot in that map, all using visual information [48,49]. Many a time, the map is pre-built, and the robot needs to localize itself using camera-based map-matching, which is referred to as Visual Localization [50]. Even though these methods use a series of image frames to determine the robot's location, they match information from the same viewpoint. In our work, we have developed a cross-view video-based localization approach by leveraging a sequence of images with varied viewpoints and limited FoVs, aiming to improve the representativeness of a query location significantly.

Multi-view Counting/Detection. There are also related methods which project images to the same ground-plane for fusion, such as multi-view counting [51–54], multi-view detection [55–59] methods. We share the similar idea of projecting features on the ground plane, but solve different downstream tasks and have distinct challenges.

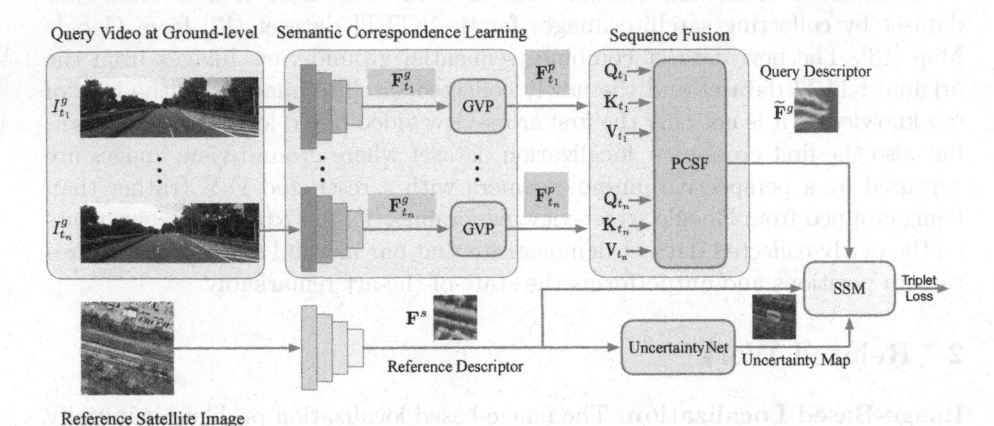

Fig. 2. Overview of our proposed CVLNet. Our Geometry-driven View Projection (GVP) module first aligns the sequential ground-view features in the overhead view and presents them in a unified coordinate system. Next, the Photo-consistency Constrained Sequential Fusion (PCSF) module measures the photo-consistency of an overhead view pixel across the different ground-views and fuses them together, obtaining a global feature representation $\widetilde{\mathbf{F}}^g$ of the query video. The global feature representation is then compared with the satellite feature map $\mathbf{F}^s$ with a Scene-prior driven Similarity Matching (SSM) scheme to determine the relative displacement between the query camera location and the satellite image center, guided by an uncertainty map. After alignment, the feature similarity is then computed for image retrieval.

3 CVLNet: Cross-view Video-Based Localization

This paper tackles the ground-to-satellite localization task. Instead of using a single query image captured at the ground level, we augment the query image with a short video containing previous observations. To solve this task, our motivation is first projecting the images in the ground video to an overhead[1] perspective and then extracting a global description from the projected image sequence for localization. An overview of our pipeline is illustrated in Fig. 2.

3.1 Geometry-Driven View Projection (GVP)

Prior methods often resort to a satellite to ground projection to bridge the cross-view domain gap. This is achieved either by a polar transform [6,8,10] or a projective transform [25,60,61]. However, both transforms need to know the query

[1] For clarity, we use "overhead" throughout the paper to denote the projected features from ground-views, and "satellite" to indicate the real satellite image/features.

camera location with respect to the satellite image center. In the CVUSA and CVACT dataset where polar transform performs excellent, the query images accidentally align with their matching satellite image center, which however does not occur in practice. When there is a large offset between the real camera location and its assumed location with respect to its matching satellite image (*e.g.*, satellite image center in polar transform), the performance will be impeded significntly. Hence, instead of projecting satellite images to ground views, we introduce a Geometry-driven View Projection (GVP) module to transform ground-view images to overhead view.

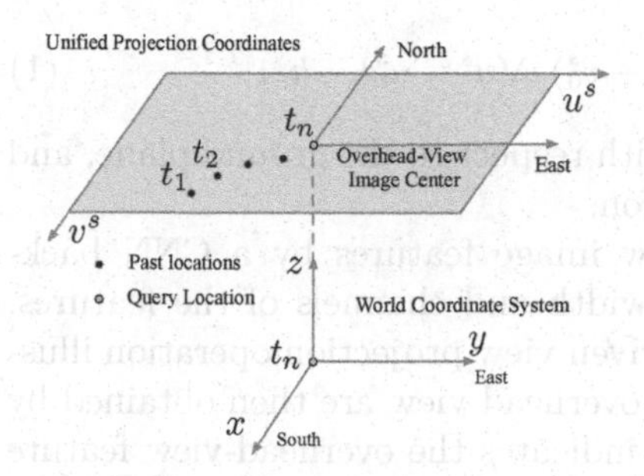

Fig. 3. A unified projection coordinates for the projected overhead-view features. Ground-view observations at timestamps $T = \{t_1, t_2, \ldots, t_n\}$ are projected to the same overhead-view grid with the center corresponding to the query camera location at t_n.

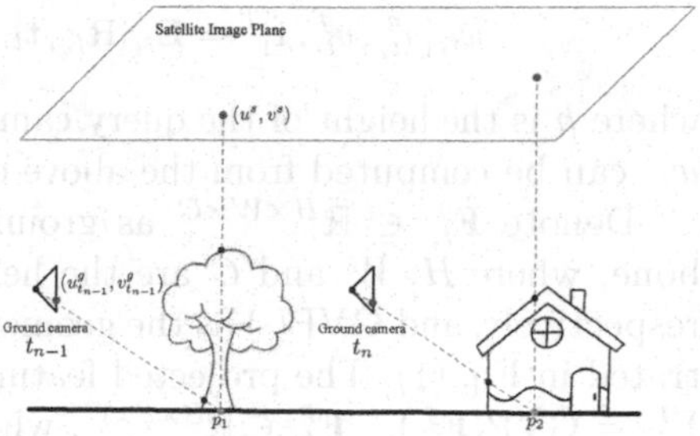

Fig. 4. Geometry-driven cross-view semantic correspondence learning. Both tree canopy and tree trunk are "trees". Building roof and facades are both "buildings".

Starting from a blank canvas in the overhead view with its center corresponds to the geospatial location of the query camera, we aim to fill it with features collected from ground-view images. We set the origin of the world coordinate system to the geo-spatial query camera location as well, with its x-axis pointing to the south direction, y axis pointing to the east direction, and the z-axis vertically upward. Different ground-view images in a video sequence are projected to the same overhead-view coordinate system so that they are geographically aligned after projection. Figure 3 provides a visual illustration of the coordinate systems.

Parallel Projection of a Satellite Camera. The projection between the satellite image coordinate system (u^s, v^s) and the world coordinate system (x, y, z) can be approximated as a parallel projection [60], $[x, y]^T = \lambda[v^s - v_0^s, u^s - u_0^s]^T$, where (u_0^s, v_0^s) indicates the satellite map center, λ indicates the real-world distance between two neighboring pixels in the satellite map.

Perspective Projection of Ground-View Images. Denote $\mathbf{R}_{t_i}$ and $\mathbf{t}_{t_i}$ as the rotation and translation for the camera at time step t_i in the world coordinate, $\mathbf{E}_{t_i}$ as the camera intrinsic, and N as the sequence number. The relative $\mathbf{R}_{t_i}$ and $\mathbf{t}_{t_i}$ can be easily obtained by Structure from Motion [62]. The projection between the world coordinate system (x, y, z) and the ground-view camera coordinate system $(u^g_{t_i}, v^g_{t_i})$ is expressed as $w_{t_i}[u^g_{t_i}, v^g_{t_i}, 1]^T = \mathbf{E}_{t_i}[\mathbf{R}_{t_i}, \mathbf{t}_{t_i}][x, y, z, 1]^T$, where w_{t_i} is a scale factor in the perspective projection.

Ground-to-Satellite Projection. There is a height ambiguity of satellite pixels in the ground-to-satellite projection. Instead of explicitly estimating the heights, we present a simple yet effective solution. Specifically, we project ground-view observations to the overhead view assuming satellite pixels lie on the ground plane. Rather than projecting original image RGB pixels, we project high-level deep features. The geometric projection from the ground-view to the overhead-view is derived as,

$$w_{t_i}[u^g_{t_i}, v^g_{t_i}, 1]^T = \mathbf{E}_{t_i}[\mathbf{R}_{t_i}, \mathbf{t}_{t_i}][\lambda(v^s - v^s_0), \lambda(u^s - u^s_0), -h, 1]^T. \tag{1}$$

where h is the height of the query camera with respect to the ground plane, and w_{t_i} can be computed from the above equation.

Denote $\mathbf{F}^g_{t_i} \in \mathbb{R}^{H \times W \times C}$ as ground-view image features by a CNN backbone, where H, W and C are the height, width and channels of the features, respectively, and GVP$(\cdot)$ as the geometry-driven view projection operation illustrated in Eq. (1). The projected features in overhead view are then obtained by $\mathbf{F}^p_{t_i} = \text{GVP}(\mathbf{F}^g_{t_i})$, $\mathbf{F}^p_{t_i} \in \mathbb{R}^{S \times S \times C}$, where S indicates the overhead-view feature map resolution.

This projection establishes the exact geometric correspondences between the ground and overhead views for scene contents on the ground plane. For scene objects with higher heights, projecting features rather than image pixels can alleviate the strict constraint while providing a cue that corresponding objects exist between the views. As shown in Fig. 4, for pixel (u^s, v^s), the projected feature from the ground-view at t_{n-1} represents the tree trunk, but the feature in the satellite image corresponds to the tree canopy. Both tree canopy and tree trunk indicate there is a tree at location p_2. Then, by applying a matching loss between the two features (tree trunk and tree canopy), the network will be trained to learn viewpoint invariant features, *i.e.*, both tree trunk and tree canopy are mapped to the semantic features of "tree".

The coverage of the canvas for ground-to-satellite projection is set to the reference satellite image coverage, *i.e.*, around 100 m × 100 m, with its center corresponding to the query camera location. When the sequence is too long with some previous image contents exceeding the canvas's pre-set coverage, the exceeded contents will not be collected. This is because scene contents that are too far from the query camera location are less important for localization, and it is better to cover most of the synthetic overhead-view feature map by referencing satellite images.

3.2 Photo-Consistency Constrained Sequence Fusion

We leverage photo consistency among different ground-views for the video sequence fusion. For a satellite pixel, when its corresponding features in several (more than two) ground views are similar, the existence of a scene object at this geographical location is highly reliable for these ground-views. We should highlight these corresponding features when generating descriptors for scene contents. Driven by this, we design a Photo-consistency Constrained Sequence Fusion (PCSF) module. Our PCSF module employs an attention mechanism [63] to emphasize reliable features in fusing a video sequence and obtaining a global descriptor for the video.

Our GVP block has aligned the original ground-view features at different time steps in a unified overhead-view coordinate. When the features of a geographical location observed by different ground views are similar, those features should be more reliable for localization. We leverage the self-attention mechanism [63] to measure the photo-consistency/similarity across different views and find reliable features. Specifically, for each projected feature map $\mathbf{F}^p_{t_i}$ at time step t_i, we compute its query, key and value by two stacked convolutional layers, denoted by $\mathbf{Q}_{t_i}, \mathbf{K}_{t_i}, \mathbf{V}_{t_i} \in \mathbb{R}^{S\times S\times C}$, respectively. The stacked convolutional layers increase the receptive field and the representative ability of the key, query, and value features at each spatial location. Next, we compute the similarities between each projected feature map at t_i and other projected feature maps at $t_j, i, j = 1, ..., N$, and normalize them across all possible j by a softmax operation, expressed as,

$$\mathbf{M}_{i,j} = \text{Softmax}_j\left(\mathbf{Q}_{t_i}^T \mathbf{K}_{t_j}\right), \quad \mathbf{M} \in \mathbb{R}^{N\times N\times S\times S}. \tag{2}$$

The final fused feature is obtained by,

$$\widetilde{\mathbf{F}}^g = \frac{1}{N}\sum\nolimits_i^N \sum\nolimits_j^N \mathbf{M}_{i,j}\mathbf{V}_{t_j}, \quad \widetilde{\mathbf{F}}^g \in \mathbb{R}^{S\times S\times C}. \tag{3}$$

In this way, we highlight the common features across the views and make the global descriptor reliable.

3.3 Scene-Prior Driven Similarity Matching

We want to address the location misalignment between a query camera location and its matching satellite image center by ground-to-satellite projection and spatial correlation between the projected features and the real satellite features. Hence, the satellite feature descriptors should be translational equivariant, which is an inherent property of conventional CNNs. Following most previous works [2,6–8,11], we use VGG16 [64] as our backbone for satellite (and ground) feature extraction. The extracted satellite features, denoted as $\mathbf{F}^s \in \mathbb{R}^{S\times S\times C}$, share the same spatial scale as the global representation of the query video. Next, we adopt a Normalized spatial Cross-Correlation (NCC) to estimate latent alignment between the query location and a satellite image center.

Denote $[\mathbf{F}^s]_{m,n}$ as a shifted version of a satellite feature map with its center at (m, n) in the original satellite feature map, and $m = 0, n = 0$ correspond to the center of the original satellite feature map. The similarity between $\mathbf{F}^s$ and $\widetilde{\mathbf{F}}^g$ aligned at (m, n) computed by NCC is,

$$\mathbf{D}_0(\mathbf{F}^s, \widetilde{\mathbf{F}}^g)_{m,n} = \frac{[\mathbf{F}^s]_{m,n} \cdot \widetilde{\mathbf{F}}^g}{\|[\mathbf{F}^s]_{m,n}\|_2 \|\widetilde{\mathbf{F}}^g\|_2}, \tag{4}$$

where $\mathbf{D}_0(\mathbf{F}^s, \widetilde{\mathbf{F}}^g) \in \mathbb{R}^{h \times w}$ denotes the similarity matrix between $\mathbf{F}^s$ and $\widetilde{\mathbf{F}}^g$ at all possible spatial-aligned locations, $m \in [-\frac{h}{2}, \frac{h}{2}]$, and $n \in [-\frac{w}{2}, \frac{w}{2}]$. A potential spatial-aligned location of the satellite map lies in a region of 10×10 m^2 in our KITTI-CVL dataset, as the database satellite image is collected very ten meters.

To exclude impossible query camera locations, *e.g.*, top of trees, we estimate an uncertainty map from the satellite semantic features, $\mathbf{U}(\mathbf{F}^s) = \mathcal{U}(\mathbf{F}^s)$, $\mathbf{U}(\mathbf{F}^s) \in \mathbb{R}^{h \times w}$, where $\mathcal{U}(\cdot)$ is the uncertainty net, composed of a set of convolutional layers. The value of each element in $\mathbf{U}(\mathbf{F}^s)$ is within the range of $[0, 1]$, forced by a Sigmoid layer. By encoding the uncertainty, The similarity between $\mathbf{F}^s$ and $\widetilde{\mathbf{F}}^g$ aligned at (m, n) is then written as,

$$\mathbf{D}(\mathbf{F}^s, \widetilde{\mathbf{F}}^g)_{m,n} = \frac{\mathbf{D}_0(\mathbf{F}^s, \widetilde{\mathbf{F}}^g)_{m,n}}{\mathbf{U}(\mathbf{F}^s)_{m,n}}. \tag{5}$$

When the uncertainty at (m, n) is large, the similarity between $\mathbf{F}^s$ and $\widetilde{\mathbf{F}}^g$ aligned at this location will be decreased. We do not have explicit supervisions for the uncertainty map. Rather, it is learned statistically from training. The relative displacement between $\mathbf{F}^s$ and $\widetilde{\mathbf{F}}^g$ is obtained by,

$$m^*, n^* = \arg\max_{m,n} \mathbf{D}(\mathbf{F}^s, \widetilde{\mathbf{F}}^g)_{m,n}. \tag{6}$$

During inference, we have no idea which one is the matching reference image for a query image. Thus the uncertainty-guided similarity matching is applied to all reference features (including non-matching ones). Furthermore, it is more challenging when a similarity score between non-matching ground and satellite features is high. Hence, we apply the similarity matching scheme to the pairs of query and non-matching reference images as well during training and minimize their maximum similarity, making the learned features more discriminative.

3.4 Training Objective

We employ the soft-weighted triplet loss [2] to train our network. The loss includes a positive term to maximize the similarity between the matching query and reference pairs and a negative term to minimize the similarity between non-matching pairs. The non-matching term also prevents our view projection module from trivial solutions. Therefore, it is formulated as,

$$\mathcal{L} = \log\left(1 + e^{\alpha\left(d(\widetilde{\mathbf{F}}^g, \mathbf{F}^s) - d(\widetilde{\mathbf{F}}^g, \mathbf{F}^{s^*})\right)}\right), \tag{7}$$

where $\mathbf{F}^s$ is the matching satellite image feature to the ground feature $\mathbf{F}^g$, $\mathbf{F}^{s^*}$ is the non-matching satellite image feature, $d(\cdot, \cdot)$ is the L_2 distance between its two inputs after alignment, and α is set to 10.

4 The KITTI-CVL Dataset

KITTI is one of the widely used benchmark datasets for testing computer vision algorithms for autonomous driving [12]. In this paper, we intend to investigate a method for using a short video sequence for satellite image-based camera localization. For this purpose, we supplement the KITTI drive sequences with corresponding satellite images. This is done by cropping high-definition Google earth satellite images using the KITTI-provided GPS tags for vehicle trajectories. Based on these GPS tags of the ground-view images, we select a large region that covers the vehicle trajectory. We then uniformly partition the region into overlapping satellite image patches. Each satellite image patch has a resolution of 1280 × 1280 pixels, amounting to about 20 cm per pixel.

Training, Validation and Test Sets. The KITTI data contains different trajectories captured at different time. In our Training, Validation and Test set split, the images of Training and Validation set are from the same region. The Validation set is constructed in this way to select the best model during training. In contrast, the images in the test set are captured at different regions from the Training and Validation sets. The test set aims to evaluate the generalization ability of the compared algorithms.

Table 1. Query image numbers in the training, validation and tests sets.

	Training	Validation	Test-1	Test-2
Distractor	✗	✗	✗	✓
Query Num	23,905	2,362	2,473	2,473

Only the nearest satellite image for each ground image in the sampled grids is retained for the Training and Validation set. We use the same method to construct our first test set, Test-1. Furthermore, we construct the second test set, Test-2, where all satellite images in the sampled grids are reserved. In other words, Test-2 contains many distracting satellite images, and it considers the real deployment scenario compared to Test-1. Visual illustrations of the differences between Test-1 and Test-2 are provided in the supplementary material. Table 1 presents the query ground image numbers of the Training, Validation, Test-1, and Test-2 sets.

5 Experiments

Evaluation Metrics. Following the previous cross-view localization work [3], we use the distance and recall at top k ($r@k$) for the performance evaluation. Specifically, when one of the retrieved top k reference images is within 10 meters to the query ground location, it is regarded as a successful localization. The percentage of successfully localized query images is recorded as recall at top k. we set k to 1, 5, 10 and 100, respectively.

Implementation Details. The input satellite image size is 512×512, center cropped from the collected images. The coverage of them is approximately $102\text{m} \times 102\text{m}$. The ground image resolution is 256×1024. The sizes of our global descriptor for query videos and satellite images are both 4096, which is a typical descriptor dimension in image retrieval. We follow prior arts [2–11] to adopt an exhaustive mini-batch strategy [1] with a batch size of $B = 8$ to prepare the training triplets. The Adam optimizer [65] with a learning rate of 10^{-4} is employed, and our network is trained end-to-end with five epochs. Our source code with every detail will be released, and the satellite images will be available for research purposes only and upon request.

5.1 Cross-view Video-Based Localization

Since there are no existing video-based cross-view localization algorithms, we conduct extensive experiments to dissect the effectiveness and necessity of each component in our framework.

Table 2. Performance comparison on different designs for view projection and sequence fusion (sequence = 4)

		Model size	Test-1				Test-2			
			r@1	r@5	r@10	r@100	r@1	r@5	r@10	r@100
View projection	Ours w/o GVP (Unet)	66.4M	0.08	0.61	1.70	26.24	0.00	0.00	0.00	1.09
	Ours w/o GVP	66.2M	1.66	4.33	7.97	36.35	0.04	0.16	0.20	5.22
Direct fusion	Conv2D	66.0M	1.25	5.90	10.80	65.91	8.90	18.44	26.61	76.51
	Conv3D	66.0M	15.08	41.57	53.17	93.09	7.00	20.38	30.33	75.90
	LSTM	66.2M	12.53	32.11	50.42	96.93	5.78	15.89	23.01	70.60
Attention based fusion	Conv2D	66.0M	18.80	47.03	61.75	96.64	11.69	25.03	36.55	81.52
	Conv3D	66.0M	19.65	43.27	58.39	97.41	11.36	24.02	34.45	83.58
	LSTM	66.1M	15.93	47.88	**66.03**	97.61	9.70	24.30	35.26	**85.08**
	Ours	66.2M	**21.80**	**47.92**	64.94	**99.07**	**12.90**	**27.34**	**38.62**	85.00

5.1.1 Geometry-Driven View Projection

Although our GVP module is the basis for the following sequence fusion and similarity matching steps, we investigate whether it can be replaced or removed. We first replace it with an Unet and expect the domain correspondences can be learned implicitly during training, denoted as "Ours w/o GVP (Unet)". Next, we remove it from our pipeline and directly feed the original ground-view features to our sequence fusion module, denoted as "Ours w/o GVP". As indicated by the results in Table 2, the performance of the two baselines is significantly inferior to our whole pipeline, demonstrating the necessity of our geometry-driven view projection module.

Learned Viewpoint-Invariant Semantic Features. To fully understand the capability of our view projection module, we visualize the learned viewpoint-invariant semantic features of our network by using the techniques of Grad-Cam [66]. As seen in Fig. 5, salient features on roads and roads edges are successfully recognized in ground-view images (Fig. 5(a)). The detected salient features in satellite images also concentrate on roads and scene objects along roads edges (Fig. 5(b)). By using our view projection module and the photo-consistency constrained sequence fusion mechanism, the learned global representations of the ground video (Fig. 5(c)) capture similar scene patterns to those of their matching satellite counterparts (Fig. 5(d)).

5.1.2 Photo-Consistency Constrained Sequence Fusion

Our goal is to synthesize an overhead-view feature map from a query ground video. To this end, our PCSF module measures the photo consistency for each overhead view pixel across different ground-view images and fuses them with an attention-based (transformer) architecture. Apart from this design, LSTM (RNNs) and 3D CNNs are also known for their power to handle sequential signals. Hence, we compare with these architectures. For completeness, we also experiment with 2D CNNs.

Direct Fusion. We first replace our PCSF module with Conv2D, Conv3D, and LSTM based networks, respectively. The Conv2D-based fusion network takes

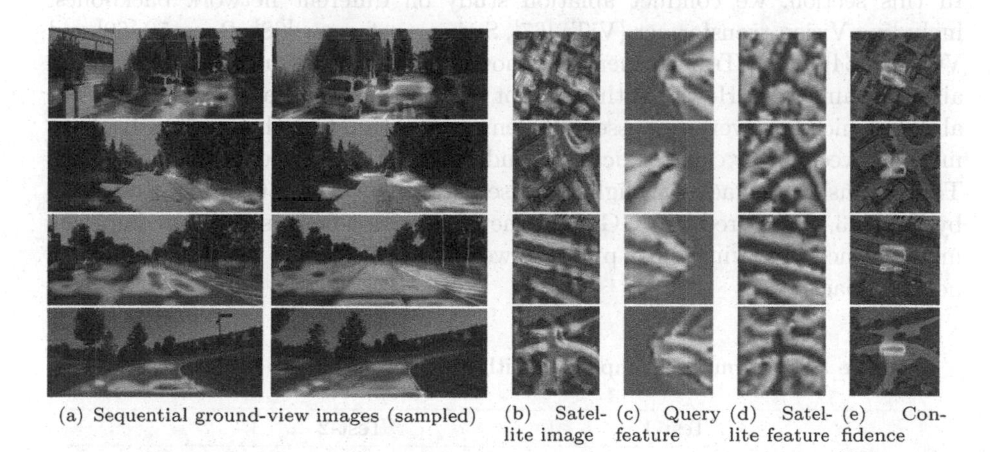

(a) Sequential ground-view images (sampled) (b) Satellite image (c) Query feature (d) Satellite feature (e) Confidence map

Fig. 5. Visualizations of intermediate results of our method. For ground images, the learned activations focus on salient features of the ground and road edges. Interestingly, it automatically ignores the dynamic objects (first row in (a)). The learned activations of satellite images concentrate on the scene objects along the main road (likely visible by a moving vehicle). The fused query video features (c) capture scene objects similar to those of their satellite counterparts (d), and the learned confidence maps attention on the region of road.

the projected sequential ground-view features $\mathbf{F}_{t_i}^p$ separately and computes the average of the outputs of different time steps. The Conv3D-based fusion network uses its third dimension to operate on the temporal dimension. The LSTM-based network includes two bidirectional LSTM layers to enhance the sequential relationship encoding. The outputs of the Conv3D-based and the LSTM-based networks are both directly fused features for the query video. The results are presented in the middle part of Table 2. It can be seen that the performance is significantly inferior to ours.

Attention-Based Fusion. Based on the above observations, we infer that it may be difficult for a network to fuse a sequence of features implicitly. Hence, we employ the Conv2D, Conv3D, and LSTM based network to regress the attention weights for the projected features at different time steps, denoted as $\mathbf{N}_{t_i} \in \mathbb{R}^{S\times S}$. Then, the global query descriptor is obtained by a dot product between the attention weight $\mathbf{N}_{t_i}$ and the features $\mathbf{F}_{t_i}^p$. The results are presented in the bottom part of Table 2. It can be seen that the attention-based fusion methods all outperform the direct fusion methods, indicating that the attention-based decomposition helps to achieve better performance. Among the attention-based fusion ablations, our method achieves the best overall performance. This should be attributed to the explicit photo consistency computation across different ground-views by our PCSF module.

5.1.3 Different Choices for Network Backbone

In this section, we conduct ablation study on different network backbones, including Vision transformer (ViT) [67], Swin transformer [68], Renet50 [69] and VGG16 [64](ours). Transformers are known of their superior feature extraction ability than CNNs. However, they do not preserve the translational equivariance ability, which however is an essential element in estimating the relative displacement between query camera locations and their matching satellite image centers. Thus, transformers achieve slightly worse performance than CNNs, as indicated by Table 3. Compared to VGG16, Resnet50 does not make significant improvement. Hence, following most previous works [2,6–8,11], we use VGG16 as our network backbone.

Table 3. Performance comparison with different backbones (sequence = 4)

	Test-1				Test-2			
	r@1	r@5	r@10	r@100	r@1	r@5	r@10	r@100
ViT [67]	20.05	45.13	60.17	97.53	12.86	27.94	**38.86**	81.64
Swin [68]	18.40	47.80	63.73	**99.11**	12.29	22.31	35.29	80.70
Resnet [69]	**22.68**	**55.16**	**67.69**	97.90	9.75	**28.31**	38.45	73.72
VGG16 [64] (Ours)	21.80	47.92	64.94	99.07	**12.90**	27.34	38.62	**85.00**

Table 4. Effectiveness of the scene-prior driven similarity matching (sequence = 4)

	Test-1				Test-2			
	r@1	r@5	r@10	r@100	r@1	r@5	r@10	r@100
Ours w/o SSM	6.35	25.76	41.97	97.61	3.48	9.42	14.03	63.04
Ours w/o U	13.26	36.76	55.72	97.05	10.47	**27.42**	**39.51**	**88.92**
Ours	**21.80**	**47.92**	**64.94**	**99.07**	**12.90**	27.34	38.62	85.00

5.1.4 Scene-Prior Driven Similarity Matching

Next, we study whether the NCC-based similarity matching can be removed. In this experiment, the distance between the satellite features and the fused ground-view features is directly computed without estimating their potential alignments. Instead, they are assumed to be aligned at the satellite image center. The results are presented in the first row of Table 4. The performance drops significantly compared to our whole baseline, demonstrating that the network does not have the ability to tolerate the spatial shifts between query camera locations, and our explicit alignment strategy (NCC-based similarity matching) is effective.

Furthermore, we investigate the effectiveness of the learned scene prior by the uncertainty map (Eq. 5). To do so, we remove the term of uncertainty map $\mathbf{U}(\mathbf{F}^s)_{m,n}$ in Eq. (5), denoted as "Ours w/o U". The results in the second row of Table 4 indicates the learned uncertainty map boosts the localization performance. Figure 5(e) visualizes the generated confidence maps (inverse of uncertainty) by our method. It can be seen that the higher confidence regions mainly concentrate on roads, indicating that the confidence maps successfully encode the semantic information of satellite images and recognize the correct possible regions for a vehicle location.

5.1.5 Varying Sequence Lengths

One desired property for a video-based localization method is to be robust to various input video lengths after a model is trained. Hence, we investigate the performance of our method on different query video sequences (1–16) using a model trained on sequence 4. Figure 6 shows that the performance increases elegantly with the increase in number of video sequences. This confirms our general intuition that more input images will increase the discriminativeness of the query place and help boost the localization performance. Note that when keep increasing the sequence length until the cameras at previous time steps exceed the pre-set coverage of the projected features, the performance will not increase but stay same because we did not fuse exceeded information. Scene contents too far from the query camera location are also less useful for localization.

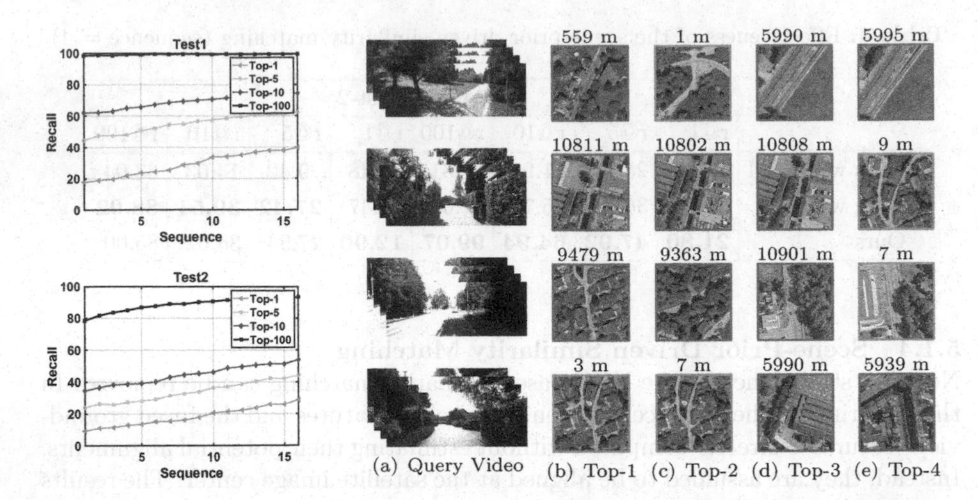

Fig. 6. Recall rates with the increase of input sequence number.

Fig. 7. Qualitative visualization of retrieved results using 4 image frames in a video.

When only using one image for localization, the feature extraction time for a query descriptor is 0.15 s. With the increasing of sequence numbers, the query descriptor extraction time increases linearly. We expect this can be accelerated by parallel computation. The retrieval time for each ground image on Test-2 is around 3 ms, and the coverage of satellite images in Test-2 is about 710, 708 m^2. It takes 8 GB GPU memory when the sequence = 4 and 24 GB when the sequence = 16. We show some qualitative examples of retrieved results in Fig. 7 using sequence number 4.

5.2 Single Image-Based Localization

Single image-based localization is a special case of video-based localization, *i.e.*, when the image frame count in the video is one. In this section, we compare the performance of our method with the recent state-of-the-art (SOTA) that are invented for cross-view single image-based localization, including CVM-NET [2], CVFT [7], SAFA [6], Polar-SAFA [6], DSM [8], Zhu *et al.* [11], and Toker *et al.* [10]. The results are presented in Table 5. It can be seen that our method significantly outperforms the recent SOTA algorithms.

Among the compared algorithms, DSM [8] achieves the best performance, because it explicitly addresses the challenge of limited FoV problem of query images while the others assume that query images are full FoV panoramas. By comparing SAFA and Polar-SAFA, we can observe that the polar transform boosts the performance on Test-1 (one-to-one matching) while impairs the performance on Test-2 (one-to-many matching). This is consistent with the conclusion in Shi *et al.* [6] and Zhu *et al.* [11].

Based on SAFA, Zhu *et al.* [11] proposes two training losses: (1) an IoU loss and (2) a GPS loss. However, we do not found the two items work well

Table 5. Comparison with the recent state-of-the-art on single image based localization

Method	Test-1				Test-2			
	r@1	r@5	r@10	r@100	r@1	r@5	r@10	r@100
CVM-NET [2]	6.43	20.74	32.47	84.07	1.01	4.33	7.52	32.88
CVFT [7]	1.78	7.20	14.40	73.55	0.20	1.29	3.03	16.86
SAFA [6]	4.89	15.77	23.29	87.75	1.62	4.73	7.40	30.13
Polar-SAFA [6]	6.67	17.06	27.62	86.53	1.13	3.76	6.23	28.22
DSM [8]	13.18	41.16	58.67	97.17	5.38	18.12	28.63	75.70
Zhu *et al.* [11]	5.26	17.79	28.22	88.44	0.73	3.28	5.66	27.86
Toker *et al.* [10]	2.79	7.72	11.69	58.92	2.39	5.50	8.90	27.05
Ours	**17.71**	**44.56**	**62.15**	**98.38**	**9.38**	**24.06**	**34.45**	**85.00**

on the KITTI-CVL dataset. We guess that the IoU loss is only suitable for the panorama case. The limited FoV images in the KITTI-CVL dataset have a smaller overlap with satellite images than panoramas, and thus the original IoU loss may not provide correct guidance for training. The GPS loss does not help mainly because of the inaccuracy of the GPS data in our dataset. We provide the GPS accuracy analysis of the KITTI dataset in the supplementary material. In contrast, our method does not rely on the accurate GPS tags of ground or satellite images.

5.3 Limitations

Our method assumes that the north direction is provided by a compass, following previous works [3,6,8,10,11], and the absolute scale of camera translations can be estimated roughly from the vehicle velocity. We have not investigated how significant tilt and roll angle changes will affect the performance, because the tilt and roll angles in the KITTI dataset are very small and we set them to zero. In autonomous driving scenarios, the vehicle-mounted cameras are usually perpendicular to the ground plane. Thus there are only slight changes in tilt and toll during driving.

6 Conclusions

This paper introduced a novel geometry-driven semantic correspondence learning approach for cross-view video-based localization. Our method includes a Geometry-driven View Projection block to bridge the cross-view domain gap, a Photo-consistency Constrained Sequence Fusion module to aggregate the sequential ground-view observations and a Scene-prior driven similarity matching mechanism to determine the location of a ground camera with respect to a satellite image center. Benefiting from the proposed components, we demonstrate that using a video rather than a single image for localization significantly facilitates the localization performance considerably.

Acknowledgments. This research is funded in part by ARC-Discovery grants (DP190102261 and DP220100800 to HL, DP220100800 to XY) and ARC-DECRA grant (DE230100477 to XY). YS is a China Scholarship Council (CSC)-funded Ph.D. student to ANU. We thank all anonymous reviewers and ACs for their constructive suggestions.

References

1. Vo, N.N., Hays, J.: Localizing and orienting street views using overhead imagery. In: Leibe, B., Matas, J., Sebe, N., Welling, M. (eds.) ECCV 2016. LNCS, vol. 9905, pp. 494–509. Springer, Cham (2016). https://doi.org/10.1007/978-3-319-46448-0_30
2. Hu, S., Feng, M., Nguyen, R.M.H., Hee Lee, G.: CVM-Net: cross-view matching network for image-based ground-to-aerial geo-localization. In: The IEEE Conference on Computer Vision and Pattern Recognition (CVPR) (2018)
3. Liu, L., Li, H.: Lending orientation to neural networks for cross-view geo-localization. In: The IEEE Conference on Computer Vision and Pattern Recognition (CVPR) (2019)
4. Regmi, K., Shah, M.: Bridging the domain gap for ground-to-aerial image matching. In: The IEEE International Conference on Computer Vision (ICCV) (2019)
5. Cai, S., Guo, Y., Khan, S., Hu, J., Wen, G.: Ground-to-aerial image geo-localization with a hard exemplar reweighting triplet loss. In: The IEEE International Conference on Computer Vision (ICCV) (2019)
6. Shi, Y., Liu, L., Yu, X., Li, H.: Spatial-aware feature aggregation for image based cross-view geo-localization. In: Advances in Neural Information Processing Systems, pp. 10090–10100 (2019)
7. Shi, Y., Yu, X., Liu, L., Zhang, T., Li, H.: Optimal feature transport for cross-view image geo-localization. Account. Audit. Account. **I**, 11990–11997 (2020)
8. Shi, Y., Yu, X., Campbell, D., Li, H.: Where am I looking at? Joint location and orientation estimation by cross-view matching. In: Proceedings of the IEEE/CVF Conference on Computer Vision and Pattern Recognition. pp. 4064–4072 (2020)
9. Zhu, S., Yang, T., Chen, C.: Revisiting street-to-aerial view image geo-localization and orientation estimation. In: Proceedings of the IEEE/CVF Winter Conference on Applications of Computer Vision, pp. 756–765 (2021)
10. Toker, A., Zhou, Q., Maximov, M., Leal-Taixé, L.: Coming down to earth: Satellite-to-street view synthesis for geo-localization. In: CVPR (2021)
11. Zhu, S., Yang, T., Chen, C.: Vigor: cross-view image geo-localization beyond one-to-one retrieval. In: CVPR (2021)
12. Geiger, A., Lenz, P., Stiller, C., Urtasun, R.: Vision meets robotics: the KITTI dataset. Int. J. Robot. Res. **32**, 1231–1237 (2013)
13. https://developers.google.com/maps/documentation/maps-static/overview
14. Arandjelovic, R., Gronat, P., Torii, A., Pajdla, T., Sivic, J.: Netvlad: CNN architecture for weakly supervised place recognition. In: Proceedings of the IEEE Conference on Computer Vision and Pattern Recognition, pp. 5297–5307 (2016)
15. Kim, H.J., Dunn, E., Frahm, J.M.: Learned contextual feature reweighting for image geo-localization. In: 2017 IEEE Conference on Computer Vision and Pattern Recognition (CVPR), pp. 3251–3260 IEEE (2017)
16. Liu, L., Li, H., Dai, Y.: Stochastic attraction-repulsion embedding for large scale image localization. In: Proceedings of the IEEE International Conference on Computer Vision, pp. 2570–2579 (2019)

17. Noh, H., Araujo, A., Sim, J., Weyand, T., Han, B.: Large-scale image retrieval with attentive deep local features. In: Proceedings of the IEEE International Conference on Computer Vision, pp. 3456–3465 (2017)
18. Ge, Y., Wang, H., Zhu, F., Zhao, R., Li, H.: Self-supervising fine-grained region similarities for large-scale image localization. In: Vedaldi, A., Bischof, H., Brox, T., Frahm, J.-M. (eds.) ECCV 2020. LNCS, vol. 12349, pp. 369–386. Springer, Cham (2020). https://doi.org/10.1007/978-3-030-58548-8_22
19. Zhou, Y., Wan, G., Hou, S., Yu, L., Wang, G., Rui, X., Song, S.: DA4AD: end-to-end deep attention-based visual localization for autonomous driving. In: Vedaldi, A., Bischof, H., Brox, T., Frahm, J.-M. (eds.) ECCV 2020. LNCS, vol. 12373, pp. 271–289. Springer, Cham (2020). https://doi.org/10.1007/978-3-030-58604-1_17
20. Castaldo, F., Zamir, A., Angst, R., Palmieri, F., Savarese, S.: Semantic cross-view matching. In: Proceedings of the IEEE International Conference on Computer Vision Workshops, pp. 9–17 (2015)
21. Lin, T.Y., Belongie, S., Hays, J.: Cross-view image geolocalization. In: Proceedings of the IEEE Conference on Computer Vision and Pattern Recognition, pp. 891–898 (2013)
22. Mousavian, A., Kosecka, J.: Semantic image based geolocation given a map. arXiv preprint arXiv:1609.00278 (2016)
23. Tian, Y., Chen, C., Shah, M.: Cross-view image matching for geo-localization in urban environments. In: Proceedings of the IEEE Conference on Computer Vision and Pattern Recognition, pp. 3608–3616 (2017)
24. Hu, S., Lee, G.H.: Image-based geo-localization using satellite imagery. Int. J. Comput. Vision **128**, 1205–1219 (2020)
25. Shi, Y., Yu, X., Liu, L., Campbell, D., Koniusz, P., Li, H.: Accurate 3-DOF camera geo-localization via ground-to-satellite image matching. arXiv preprint arXiv:2203.14148 (2022)
26. Zhu, S., Shah, M., Chen, C.: Transgeo: transformer is all you need for cross-view image geo-localization. In: Proceedings of the IEEE/CVF Conference on Computer Vision and Pattern Recognition (CVPR), pp. 1162–1171 (2022)
27. Elhashash, M., Qin, R.: Cross-view slam solver: global pose estimation of monocular ground-level video frames for 3d reconstruction using a reference 3d model from satellite images. ISPRS J. Photogramm. Remote. Sens. **188**, 62–74 (2022)
28. Guo, Y., Choi, M., Li, K., Boussaid, F., Bennamoun, M.: Soft exemplar highlighting for cross-view image-based geo-localization. IEEE Trans. Image Process. **31**, 2094–2105 (2022)
29. Zhao, J., Zhai, Q., Huang, R., Cheng, H.: Mutual generative transformer learning for cross-view geo-localization. arXiv preprint arXiv:2203.09135 (2022)
30. Bloesch, M., Omari, S., Hutter, M., Siegwart, R.: Robust visual inertial odometry using a direct ekf-based approach. In,: IEEE/RSJ International Conference on Intelligent Robots and Systems (IROS).pp. 298–304. IEEE (2015)
31. Leutenegger, S., Lynen, S., Bosse, M., Siegwart, R., Furgale, P.: Keyframe-based visual-inertial odometry using nonlinear optimization. Int. J. Robot. Res. **34**, 314–334 (2015)
32. Chien, H.J., Chuang, C.C., Chen, C.Y., Klette, R.: When to use what feature? sift, surf, orb, or a-kaze features for monocular visual odometry. 2016 International Conference on Image and Vision Computing New Zealand (IVCNZ), pp. 1–6 (2016)
33. Cadena, C., Carlone, L., Carrillo, H., Latif, Y., Scaramuzza, D., Neira, J., Reid, I., Leonard, J.J.: Past, present, and future of simultaneous localization and mapping: Toward the robust-perception age. IEEE Trans. Rob. **32**, 1309–1332 (2016)

34. Engel, J., Schöps, T., Cremers, D.: LSD-SLAM: large-scale direct monocular SLAM. In: Fleet, D., Pajdla, T., Schiele, B., Tuytelaars, T. (eds.) ECCV 2014. LNCS, vol. 8690, pp. 834–849. Springer, Cham (2014). https://doi.org/10.1007/978-3-319-10605-2_54
35. Klein, G., Murray, D.: Parallel tracking and mapping for small AR workspaces. In,: 6th IEEE and ACM International Symposium on Mixed and Augmented Reality. pp. 225–234. IEEE (2007)
36. Mur-Artal, R., Montiel, J.M.M., Tardos, J.D.: Orb-slam: a versatile and accurate monocular slam system. IEEE Trans. Rob. **31**, 1147–1163 (2015)
37. Mur-Artal, R., Tardós, J.D.: Orb-slam2: An open-source slam system for monocular, stereo, and RGB-D cameras. IEEE Trans. Rob. **33**, 1255–1262 (2017)
38. Campos, C., Elvira, R., Rodríguez, J.J.G., Montiel, J.M., Tardós, J.D.: Orb-slam3: an accurate open-source library for visual, visual-inertial, and multimap slam. IEEE Trans. Robot. **37**, 1874–1890 (2021)
39. Mur-Artal, R., Tardós, J.D.: Visual-inertial monocular slam with map reuse. IEEE Robot. Autom. Lett. **2**, 796–803 (2017)
40. Wolcott, R.W., Eustice, R.M.: Visual localization within lidar maps for automated urban driving. 2014 IEEE/RSJ International Conference on Intelligent Robots and System, pp. 176–183 (2014)
41. Voodarla, M., Shrivastava, S., Manglani, S., Vora, A., Agarwal, S., Chakravarty, P.: S-BEV: semantic birds-eye view representation for weather and lighting invariant 3-DOF localization (2021)
42. Stenborg, E., Toft, C., Hammarstrand, L.: Long-term visual localization using semantically segmented images. In,: IEEE International Conference on Robotics and Automation (ICRA). pp .6484–6490. IEEE (2018)
43. Stenborg, E., Sattler, T., Hammarstrand, L.: Using image sequences for long-term visual localization. In: 2020 International Conference on 3D Vision (3DV), pp. 938–948 IEEE (2020)
44. Vaca-Castano, G., Zamir, A.R., Shah, M.: City scale geo-spatial trajectory estimation of a moving camera. In: 2012 IEEE Conference on Computer Vision and Pattern Recognition, pp. 1186–1193 IEEE (2012)
45. Regmi, K., Shah, M.: Video geo-localization employing geo-temporal feature learning and GPS trajectory smoothing. In: Proceedings of the IEEE/CVF International Conference on Computer Vision, pp. 12126–12135 (2021)
46. Yousif, K., Bab-Hadiashar, A., Hoseinnezhad, R.: An overview to visual odometry and visual slam: applications to mobile robotics. Intell. Ind. Syst. **1**, 289–311 (2015)
47. Scaramuzza, D., Fraundorfer, F.: Visual odometry [tutorial]. IEEE Robot. Autom. Mag. **18**, 80–92 (2011)
48. Gao, X., Wang, R., Demmel, N., Cremers, D.: Ldso: direct sparse odometry with loop closure. In: 2018 IEEE/RSJ International Conference on Intelligent Robots and Systems (IROS), pp. 2198–2204 IEEE (2018)
49. Kasyanov, A., Engelmann, F., Stückler, J., Leibe, B.: Keyframe-based visual-inertial online slam with relocalization. In,: IEEE/RSJ International Conference on Intelligent Robots and Systems (IROS), pp. 6662–6669. IEEE (2017)
50. Liu, D., Cui, Y., Guo, X., Ding, W., Yang, B., Chen, Y.: Visual localization for autonomous driving: mapping the accurate location in the city maze (2020)
51. Hou, Y., Zheng, L., Gould, S.: Multiview Detection with Feature Perspective Transformation. In: Vedaldi, A., Bischof, H., Brox, T., Frahm, J.-M. (eds.) ECCV 2020. LNCS, vol. 12352, pp. 1–18. Springer, Cham (2020). https://doi.org/10.1007/978-3-030-58571-6_1

52. Hou, Y., Zheng, L.: Multiview detection with shadow transformer (and view-coherent data augmentation). In: Proceedings of the 29th ACM International Conference on Multimedia, pp. 1673–1682 (2021)
53. Vora, J., Dutta, S., Jain, K., Karthik, S., Gandhi, V.: Bringing generalization to deep multi-view detection. arXiv preprint arXiv:2109.12227 (2021)
54. Ma, J., Tong, J., Wang, S., Zhao, W., Zheng, L., Nguyen, C.: Voxelized 3d feature aggregation for multiview detection. arXiv preprint arXiv:2112.03471 (2021)
55. Zhang, Q., Lin, W., Chan, A.B.: Cross-view cross-scene multi-view crowd counting. In: Proceedings of the IEEE/CVF Conference on Computer Vision and Pattern Recognition, pp. 557–567 (2021)
56. Zhang, Q., Chan, A.B.: Wide-area crowd counting via ground-plane density maps and multi-view fusion CNNS. In: Proceedings of the IEEE/CVF Conference on Computer Vision and Pattern Recognition, pp. 8297–8306 (2019)
57. Zhang, Q., Chan, A.B.: 3d crowd counting via multi-view fusion with 3d gaussian kernels. Proceedings of the AAAI Conference on Artificial Intelligence. **34**, 12837–12844 (2020)
58. Zhang, Q., Chan, A.B.: Wide-area crowd counting: Multi-view fusion networks for counting in large scenes. Int. J. Comput Vis. **130**, 1938–1960 (2022)
59. Chen, L., et al.: Persformer: 3D lane detection via perspective transformer and the openlane benchmark. arXiv preprint arXiv:2203.11089 (2022)
60. Shi, Y., Campbell, D.J., Yu, X., Li, H.: Geometry-guided street-view panorama synthesis from satellite imagery. IEEE Trans. Pattern Anal. Mach. Intell. **44**, 10009–10022(2022)
61. Shi, Y., Li, H.: Beyond cross-view image retrieval: Highly accurate vehicle localization using satellite image. In: Proceedings of the IEEE/CVF Conference on Computer Vision and Pattern Recognition, pp. 17010–17020 (2022)
62. Schonberger, J.L., Frahm, J.M.: Structure-from-motion revisited. In: Proceedings of the IEEE Conference on Computer Vision and Pattern Recognition, pp. 4104–4113 . (2016)
63. Vaswani, A., et al.: Attention is all you need. In: Advances in Neural Information Processing Systems, pp. 5998–6008 (2017)
64. Simonyan, K., Zisserman, A.: Very deep convolutional networks for large-scale image recognition. CoRR abs/1409.1556 (2014)
65. Kingma, D.P., Ba, J.: Adam: A method for stochastic optimization. arXiv preprint arXiv:1412.6980 (2014)
66. Selvaraju, R.R., Cogswell, M., Das, A., Vedantam, R., Parikh, D., Batra, D.: Grad-cam: visual explanations from deep networks via gradient-based localization. In: Proceedings of the IEEE International Conference on Computer Vision (ICCV) (2017)
67. Dosovitskiy, A., et al.: An image is worth 16x16 words: Transformers for image recognition at scale. arXiv preprint arXiv:2010.11929 (2020)
68. Liu, Z., et al.: Swin transformer: hierarchical vision transformer using shifted windows. In: Proceedings of the IEEE/CVF International Conference on Computer Vision, pp. 10012–10022 (2021)
69. He, K., Zhang, X., Ren, S., Sun, J.: Deep residual learning for image recognition. In: Proceedings of the IEEE Conference on Computer Vision and Pattern Recognition, pp. 770–778 (2016)

Vectorizing Building Blueprints

Weilian Song[✉], Mahsa Maleki Abyaneh, Mohammad Amin Shabani, and Yasutaka Furukawa

Simon Fraser University, Burnaby, BC, Canada
weilians@sfu.ca

Abstract. This paper proposes a novel vectorization algorithm for high-definition floorplans with construction-level intricate architectural details, namely a blueprint. A state-of-the-art floorplan vectorization algorithm starts by detecting corners, whose process does not scale to high-definition floorplans with thin interior walls, small door frames, and long exterior walls. Our approach 1) obtains rough semantic segmentation by running off-the-shelf segmentation algorithms; 2) learning to infer missing smaller architectural components; 3) adding the missing components by a refinement generative adversarial network; and 4) simplifying the segmentation boundaries by heuristics. We have created a vectorized blueprint database consisting of 200 production scanned blueprint images. Qualitative and quantitative evaluations demonstrate the effectiveness of the approach, making significant boost in standard vectorization metrics over the current state-of-the-art and baseline methods. We will share our code at https://github.com/weiliansong/blueprint-vectorizer.

Keywords: Vectorization · Blueprint · Segmentation

1 Introduction

Blueprints are technical drawings of buildings, conveying rich architectural and engineering information for building maintenance, assessing building code compliance, and remodeling. Unfortunately, blueprints are often stored as scanned raster images, where a professional architect spends hours manually converting to a vector format for down-stream applications. Automated blueprint vectorization will have tremendous impacts on real-estate or construction industries, significantly reducing the amount of human resources in converting hundreds of thousands of blueprints ever scanned and stored as raster images.

Floorplan vectorization has a long history in the domain of document scanning [3,9]. A classical approach relies on heuristics and was never robust enough at a production level. With the advent of deep learning, Liu *et al.*made a breakthrough in combining convolutional neural networks for corner detection and binary integer programming for their connection inference [11]. Their system achieves more than 90% precision and recall for the task of vectorizing consumer-grade floorplan images in the real-estate industry.

Supplementary Information The online version contains supplementary material available at https://doi.org/10.1007/978-3-031-26319-4_9.

L. Wang et al. (Eds.): ACCV 2022, LNCS 13841, pp. 142–157, 2023.
https://doi.org/10.1007/978-3-031-26319-4_9

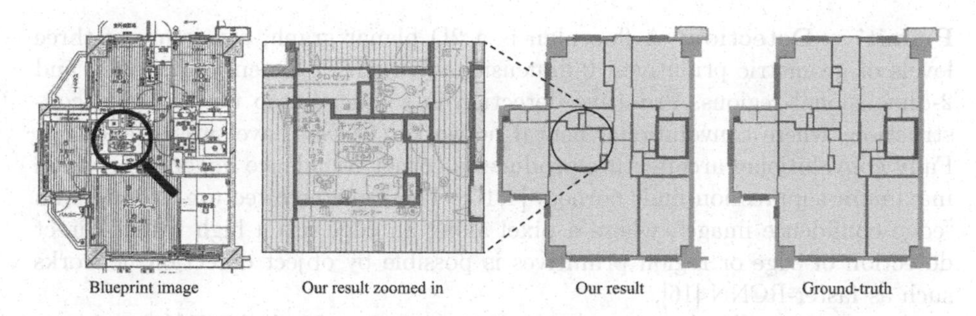

Fig. 1. This paper proposes a novel algorithm for vectorizing building blueprints with intricate architectural details. The left is an input raster blueprint image. Our vectorized blueprint is shown in the middle two columns (the left is a zoomed-in view). The right is the ground-truth.

Fig. 2. Sample floorplans from the R2V [11] dataset (left two) in comparison to our dataset (right two).

In this paper, we propose a novel vectorization algorithm for building blueprints with intricate architectural details. Our system outputs the floorplan in vector-graphics representation, and keeps intermediate representations as raster segmentation. A standard vectorization process starts by corner detection, which fails severely on complex blueprints. Our approach 1) starts by region-detection (which involves higher level primitives and is more robust) by an instance segmentation technique; 2) detects missing or extraneous architectural components by topological reasoning; 3) adds or removes components by a refinement generative adversarial network; and 4) further refines the boundaries by heurisics.

We have annotated 200 raster blueprints as vector-graphics images. We compare against the current state-of-the-art and baseline methods, based on the standard metrics as well as a new one focusing on the topological correctness. The proposed approach makes significant improvements over existing methods.

2 Related Works

We review related literature, namely, primitive detection, floorplan vectorization, and structured reconstruction.

Primitive Detection: A floorplan is a 2D planar graph, consisting of three levels of geometric primitives: 0-dimensional corners, 1-dimensional edges, and 2-dimensional regions. Primitive detection is a crucial step for floorplan construction, where convolutional neural networks (CNNs) have proven effective. Fully convolutional architecture produces a corner confidence image, where non maximum suppression finds corners [4,15]. The same architecture produces an "edge confidence image", where a pixel along an edge has a high value. Direct detection of edge or region primitives is possible by object detection networks such as faster-RCNN [16].

Standard instance segmentation techniques such as Mask-RCNN [8] and metric learning [6] yield primitive segmentation. However, blueprints contain many extremely thin and elongated regions, where these techniques perform poorly. We divide an input image into smaller patches, perform metric learning locally per patch, and merge results.

Floorplan Vectorization: Liu *et al.* [11] presented the first successful floorplan vectorization approach by combining CNNs for corner detection and Integer Programming for edge/region inference. However, their approach is unsuitable for our problem due to two key differences in the data. First, their data are consumer-grade floorplans with much simpler graph structures in comparison to our blueprints. Second, their input are digitally rasterized images (e.g., a jpeg image), while our input are optically scanned images from printed papers, exhibiting severe noise and distortions. Some sample floorplans from Liu *et al.* [11] are shown in Fig. 2. Our approach is to 1) only utilize region detection (i.e., no corner nor edge detection, which are less robust) and 2) learn to add or remove tiny architectural components, which are the hardest to detect, by topological reasoning.

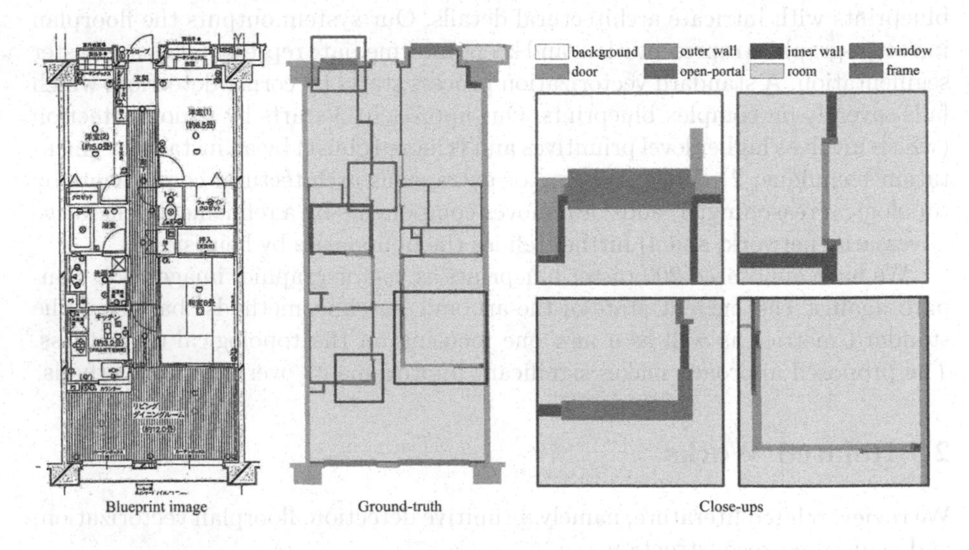

Fig. 3. A sample annotated blueprint with close-ups.

Table 1. Statistics of our dataset, consisting of 200 blueprint images. All averages are rounded to the nearest integer.

	Image		Primitives			Regions by type						
	Height	Width	Corners	Edges	Regions	Outer	Inner	Window	Door	Portal	Room	Frame
Total	N/A	N/A	123,230	139,808	17,077	1,384	2,111	1019	3,203	172	3,756	5,432
Average	1,279	906	616	699	85	7	11	5	16	1	19	27

Structured Reconstruction: Structured reconstruction [12] seeks to turn raw sensor data (e.g., images or point-clouds) into structured geometry representation such as CAD models. The challenge lies in the noisy input sensor data. A general rule of thumb is to utilize higher level primitive detection (i.e., regions) instead of corner/edge detection, which is less robust [5,13]. Our approach also borrows this idea.

The state-of-the-art structured reconstruction algorithms handle outdoor architecture [13,19] or commercial floorplans [5,12], which are much simpler than the building blueprints. Our approach learns topological reasoning to handle tiny architectural components as described above.

3 Blueprint Vectorization Dataset

We collected 200 building blueprints from a production pipeline. Following the annotation protocol by an existing floorplan vectorization paper [11], we use the VIA annotator [7] to annotate a planar graph and associate an architectural component type to each region. There are eight architectural component types: background, outer wall, inner wall, windows, doors, open-portals, rooms, and frames. Table 1 shows various statistics of our dataset, and Fig. 3 illustrates a sample blueprint along with its annotation. We do not have access to previous datasets [11] to compare statistics, but ours is many times more complex.

To expand upon our semantic types: open-portals denote an opening between rooms without a physical door. Frames are door trims that often surround doors within our dataset. From the top-down view, they appear as small rectangles on the ends of a door (small red rectangles in Fig. 3). We make distinction between outer (exterior) and inner (interior) walls, which can be adjacent to each other for architectural reasons.

4 Blueprint Vectorization Algorithm

A high-level view of our system is shown in Fig. 4. Our blueprint vectorization consists of four steps: 1) instance segmentation and type classification; 2) frame connectivity inference; 3) frame correction; and 4) ad-hoc boundary simplification. We first explain our intermediate data representation, then provide details of each step.

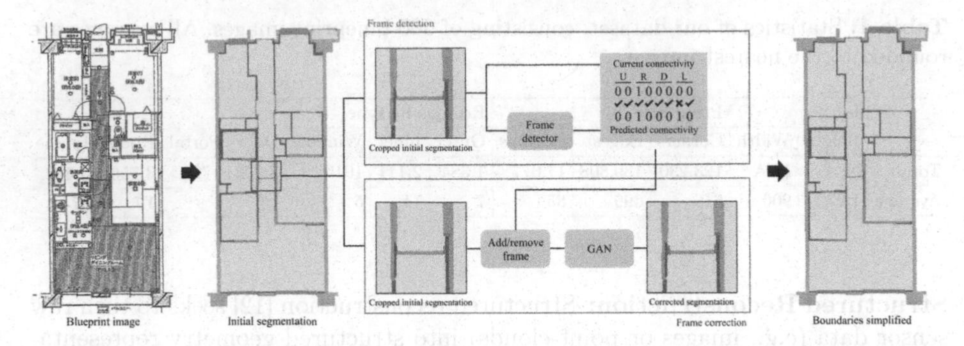

Fig. 4. System overview. After obtaining our initial segmentation through instance segmentation followed by type classification, we detect and correct missing/extraneous building-frames. Segment boundaries are simplified with heuristics to produce the final result.

4.1 Blueprint Representation

While the goal is to reconstruct a blueprint as a planar graph, our algorithm also uses a pixel-wise semantic segmentation image as an intermediate representation, which is an 8-channel image as there exist 8 component types. Note that a semantic segmentation image is instance-unaware, and cannot represent two instances of the same type sharing a boundary. However, such cases are rare and we merge such instances into a single segment for simplicity.

A planar graph is converted to a semantic segmentation image by simple rasterization. A semantic segmentation image is, in turn, converted to a graph representation by tracing the segmentation boundaries while assuming a Manhattan world (i.e., edges are either horizontal or vertical). For example, a single-pixel segmentation is converted to a square with four corners in the vector representation.

4.2 Instance Segmentation and Type Classification

Instance Segmentation: Blueprint images need to be in high resolution to retain intricate geometric structures. We found that standard instance segmentation techniques such as Mask-RCNN [8] or metric learning [6] do not work well. Semantic segmentation [10,18] is also an option (given our semantic segmentation representation), but it does not work well due to the data-imbalance (i.e., the open-portal is the rare type as shown in Table 1).

We divide a scanned blueprint image into a set of overlapping "crops", each of which is a 256×256 grayscale image and overlaps with the other crops by 32 pixels.[1] Since Mark-RCNN is not effective for thin structures, we use a metric-learning based approach [6].

[1] At training, we randomly pick a pixel and extract a patch around it.

First, a standard CNN converts a crop into an embedding of size $256 \times 256 \times 8$ (i.e., 8 dimensional embedding per pixel) with a discriminative loss function [6] (implementation borrowed from [1]), which optimizes pixel embedding from the same instance to be similar and vice-versa. We refer to the supplementary for the architectural specification.

Second, a mean-shift clustering algorithm extracts instances from the embedding (i.e., a python module MeanShift from the scikit-learn library with binseeding and the bandwidth parameter set to 2.0). We also classify an architectural component type for each instance (see the paragraph below), and merge results by simple heuristics: A pixel in an overlapping region has multiple segmentation results from multiple crops. We take the most frequent component type within a 3×3 window centered at the pixel in all the crops.

Type Classification: We train a CNN-based classifier to assign component type to each instance in each crop. For each instance in a crop, we pass the blueprint image of the crop and the binary segmentation mask of the instance as a 4 channel image to a standard CNN-based encoder with a softmax cross-entropy loss. Again, see the supplementary for the full architectural specification. To handle a large instance, we uniformly sample 20 pixel locations from the binary instance mask and use them as centers to extract crops. At test time, we take the average probability distribution from all the crops and pick the type with the highest probability.

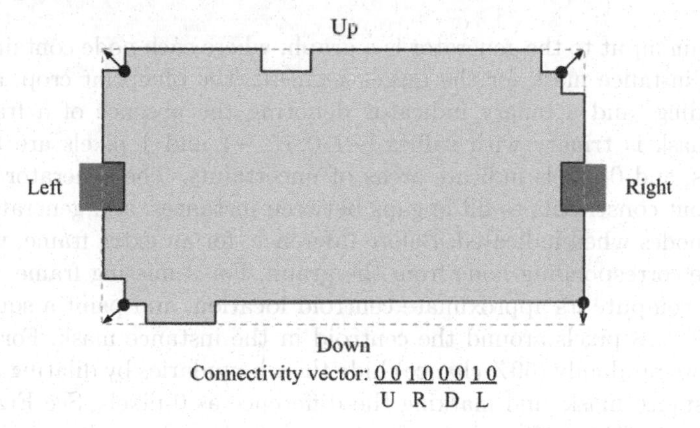

Fig. 5. Figure illustrating the heuristic used to determine the four sides of a segment. Dashed lines indicate the bounding box of a segment. Four disks indicate the vertices closest to the bounding box corners. We then trace the perimeter between the vertices to obtain the four sides.

4.3 Frame Detection Module

Small architectural components tend to be missed by the instance segmentation technique in the first step. We train a classifier that infers the correct "connectivity" of frames. Precisely, we define the connectivity of a segment to be a set of neighboring segments.

To encode the frame connectivity, we first apply a heuristic to determine the segment's four sides. Then, assuming a maximum of two frames on each side[2], we use an 8-bit binary vector to denote the presence of frames on all four sides. Figure 5 illustrates the heuristic. The top-left, top-right, bottom-left, and bottom-right corners of the instance are first determined, which are defined as the four vertices closest to the instance bounding box. The four sides are then obtained as boundary edges in between two corners, and each frame is attached to one side only.

With our representation, we train a classifier to predict a segment's ground-truth frame connectivity. The classifier takes as input three masks: a one-hot encoded semantic segmentation mask around the segment, the corresponding gray-scale blueprint crop, and a binary mask highlighting the segment. The network directly outputs the 8-bit vector, and is optimized with the standard binary cross-entropy loss. We use a mix of ground-truth and predicted data to generate training input-label pairs, where each predicted instance is assigned the frame vector of the corresponding ground-truth instance. See supplementary material for more details on training data generation.

4.4 Frame Correction Module

Once we detect a discrepancy in the current and predicted frame arrangement, we employ a generative adversarial network (GAN) to locally correct the segmentation.

The main input to the generator is a graph, where each node contains a noise vector, an instance mask for the target segment, the blueprint crop, a one-hot type encoding, and a binary indicator denoting the absence of a frame. The instance mask is trinary with values $[-1, 0, 1]$; -1 and 1 pixels are the input constraints, and 0-pixels indicate areas of uncertainty. The generator learns to expand input constraints to fill in gaps between instances, and generate missing frames in nodes when indicated. Before inference, for an extra frame, we simply remove the corresponding node from the graph. For a missing frame, we add a new node, compute its approximate centroid location, and paint a square zero-mask of 18×18 pixels around the centroid in the instance mask. For all other segments, we randomly (50% chance) hide their boundaries by dilating and eroding the instance mask, and marking the difference as 0-pixels. See Fig. 6 for an example input of door with missing frame, and please refer to the supplementary material for heuristics on approximate centroid computation.

Our generator and discriminator designs are inspired by House-GAN++ [14]. They are both convolutional message passing networks (ConvMPN) [19], and considers an input blueprint crop as a graph of instances. For simplicity, we consider the graph to be fully-connected. The generator first encodes input masks down to a lower spatial resolution, then performs a series of four convolutional message passing and up-sampling layers back to the input resolution. The discriminator performs a series of three convolutional message passing and down-

[2] One window in our dataset does not follow this rule, which we ignore for simplicity.

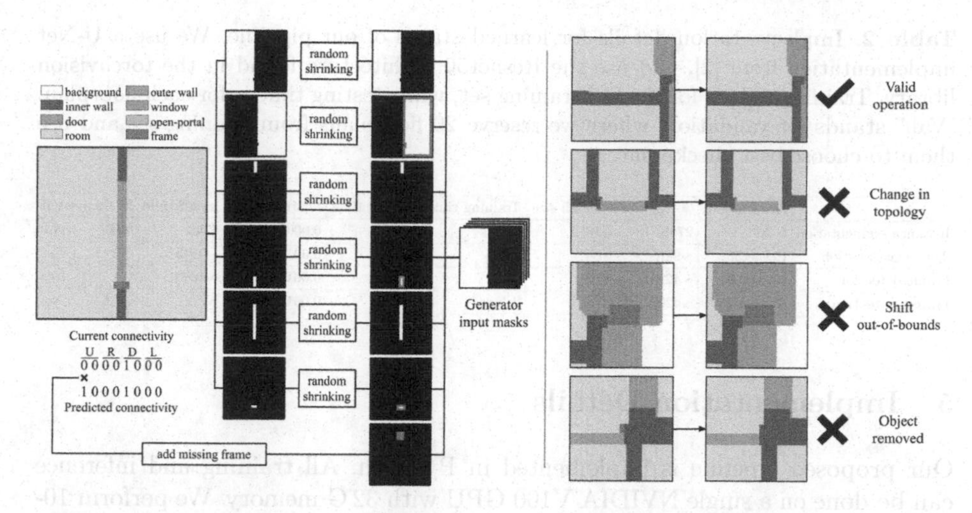

Fig. 6. Visualization of generator preprocessing routine (left) and example valid/invalid edge shift for heuristic simplification (right).

sampling layers, followed by a special pooling and FC module to output a single value for the input graph. During training, we do self-supervised learning on the ground-truth data by randomly hiding frame instances. During testing, following House-GAN++ we do iterative refinement by running the generator a max of 50 steps or until convergence. Please see the supplementary for details.

4.5 Heuristic Simplification

Given our corrected segmentation, we use heuristics to simplify segment boundaries. Concretely, we take an edge shorter than 2 pixels (i.e., a reference edge) and consider neighboring edges that share one of its end-points. We find the neighboring edge that is perpendicular and closest to the reference. We shift the reference perpendicularly until the closest neighboring edge gets contracted. We perform this operation for each edge shorter than 2 pixels, unless one of the following three conditions occur (see Fig. 6).

Change in Topology: If an edge shift changes the segment connectivity (e.g., a door being disconnected from a wall).

Shift Out-of-Bounds: If an edge shift expands the segment too much. Concretely, we dilate the original segment with a 7×7 kernel and check if the new segment is fully contained inside the dilated segment.

Instance Added/Removed: If an edge shift removes a segment or splits a segment into two instances or more.

Table 2. Implementation details for learned stages of our pipeline. We use a U-Net implementation from [2], and use the Resnet50 architecture found in the torchvision library. Training time is for the full training set, while testing time is for each floorplan. "Val." stands for validation, where we reserve 20 floorplans from training set and use them to choose best checkpoint.

	Architecture	# iterations	Batch size	Training time	Testing time	Learning rate	lr. schedule	lr. decay	Val
Instance segmentation	U-Net	270K	16	2 days	~1 h	0.0003	6750	0.96	No
Type classification	Resnet50	~98K	64	5 h	<1 min	0.001	~9800	0.5	Yes
Frame detection	Resnet50	~42K	32	2 h	<1 min	0.001	~4200	0.1	Yes
Frame correction	Conv-MPN	~420K	1	2 days	~1 h	0.001	N/A	N/A	No

5 Implementation Details

Our proposed pipeline is implemented in PyTorch. All training and inference can be done on a single NVIDIA V100 GPU with 32 G memory. We perform 10-fold cross-validation on our dataset to obtain test-time results for each floorplan. Table 2 shows implementation details for pipeline stages that require learning. In addition, for frame detection training we perform random rotation augmentation at 0, 90, 180, and 270 °C.

6 Evaluations

We compare against one baseline (Mask-RCNN [8]), one state-of-the-art method (Raster-to-Vector [11] denoted as R2V), and a few variants of our system. Our method and Mask-RCNN both perform segmentation, but we predict in smaller local crops and merge predictions. R2V performs corner detection first and then edge/region inference, which our method aims to avoid detecting primitives.

Mask-RCNN: We use the detectron2 implemenation [17] using the predefined config of ResNet50+FPN backbone, pretrained on COCO dataset, and fine-tuned for 60k iterations. The learning rate is set to 0.0025. The anchor sizes and ratios are set to (16, 32, 64, 128, 256) and (0.1, 0.5, 1.0, 2.0, 5.0), respectively.

R2V: We downloaded and modified the official implementation [11], while considering every boundary as a wall of 2 pixels. All types are considered to be rooms. For example, a door is a line in the R2V system, but a region with a boundary in our problem. The network was trained for 150 epochs with a batch size of 4 on 512×512 images.

6.1 Quantitative Results

Besides the standard corner, edge, and region metrics in the existing literature [11,13], we introduce a new one that measures the correctness of the segment "connectivity". In the evaluation, we also consider directions in addition to

Table 3. Quantitative evaluations are based on the standard corner/edge/region metrics [11] and a new connectivity correctness accuracy. "I", "F", and "H" refers to the the instance segmentation and type classification (Sect. 4.2), the frame detection and correction modules (Sects. 4.3 and 4.4), and the heuristic simplification (Sect. 4.5), respectively. The "italic" and the "bold" denote the best and the second best results, respectively.

Model	Corner			Edge			Region			Connectivity Acc.	
	Pre.	Rec.	F1	Pre.	Rec.	F1	Pre.	Rec.	F1	All	Frames
R2V	*71.3*	1.9	3.7	**11.5**	0.3	0.5	97.8	4.5	8.7	0.0	0.0
Mask-RCNN	6.9	25.5	10.8	0.5	1.8	0.8	44.0	15.8	23.2	0.0	35.7
Mask-RCNN + H	24.6	17.1	20.2	5.1	3.2	4.0	44.0	15.8	23.2	0.0	35.7
Ours(I)	27.2	*65.4*	38.4	7.6	17.2	10.6	*81.7*	**89.3**	*85.3*	**46.6**	76.7
Ours (I + F)	27.9	**63.7**	**38.8**	8.2	**17.6**	**11.2**	**81.3**	*89.7*	*85.3*	*48.4*	*80.4*
Ours (I + F+ H)	**70.4**	53.1	*60.5*	*38.3*	*30.4*	*33.9*	81.2	*89.7*	**85.2**	*48.4*	**80.3**

being just neighbors. Here, we define the connectivity of a segment to be a set of neighboring segments and their directions, where a direction is given by a ray connecting the centroids of the segments.

We declare that a segment connectivity is correct if it is exactly equal to the connectivity of the corresponding ground-truth segment with a tolerance of 15° in the directions. Note that we only evaluate the connectivity for segments that are matched with ground-truth segments during region evaluation. Therefore, we report only the accuracy number for the connectivity correctness.

Table 3 shows the main results, comparing against one state-of-the-art (R2V [11]) and one baseline method (Mask-RCNN [8]). For our system, we report numbers at three different stages. Mask-RCNN is a "dense" segmentation method where segment boundaries consist of many corners. Therefore, we also apply our heuristic simplification (Sect. 4.5) after Mask-RCNN. For the connectivity, we report the average accuracy across all segment types as well as only for type "Frame". Note that frames are extremely small segments which pose challenges to existing techniques.

The table shows that R2V and Mask-RCNN both fail with significant margins. Our system after the instance segmentation (I) recovers most regions and achieves high region metrics while suffering severely in corner and edge metrics. The frame detection/correction modules (F) improves the connectivity accuracy for frames by 3.7%. The heuristic simplification (H) further improves the corner and edge metrics significantly.

Table 4 shows the performance of the proposed approach while varying the order and the presence of the three steps. After the instance segmentation (I), it is best to perform frame detection/correction first (F), then the heuristic simplification (H).

Fig. 7. Full-blueprint comparison between our full pipeline and competing methods. "H" stands for heuristic simplification (Sect. 4.5). Our results match the ground-truth data very closely, while other methods cannot recover most instances.

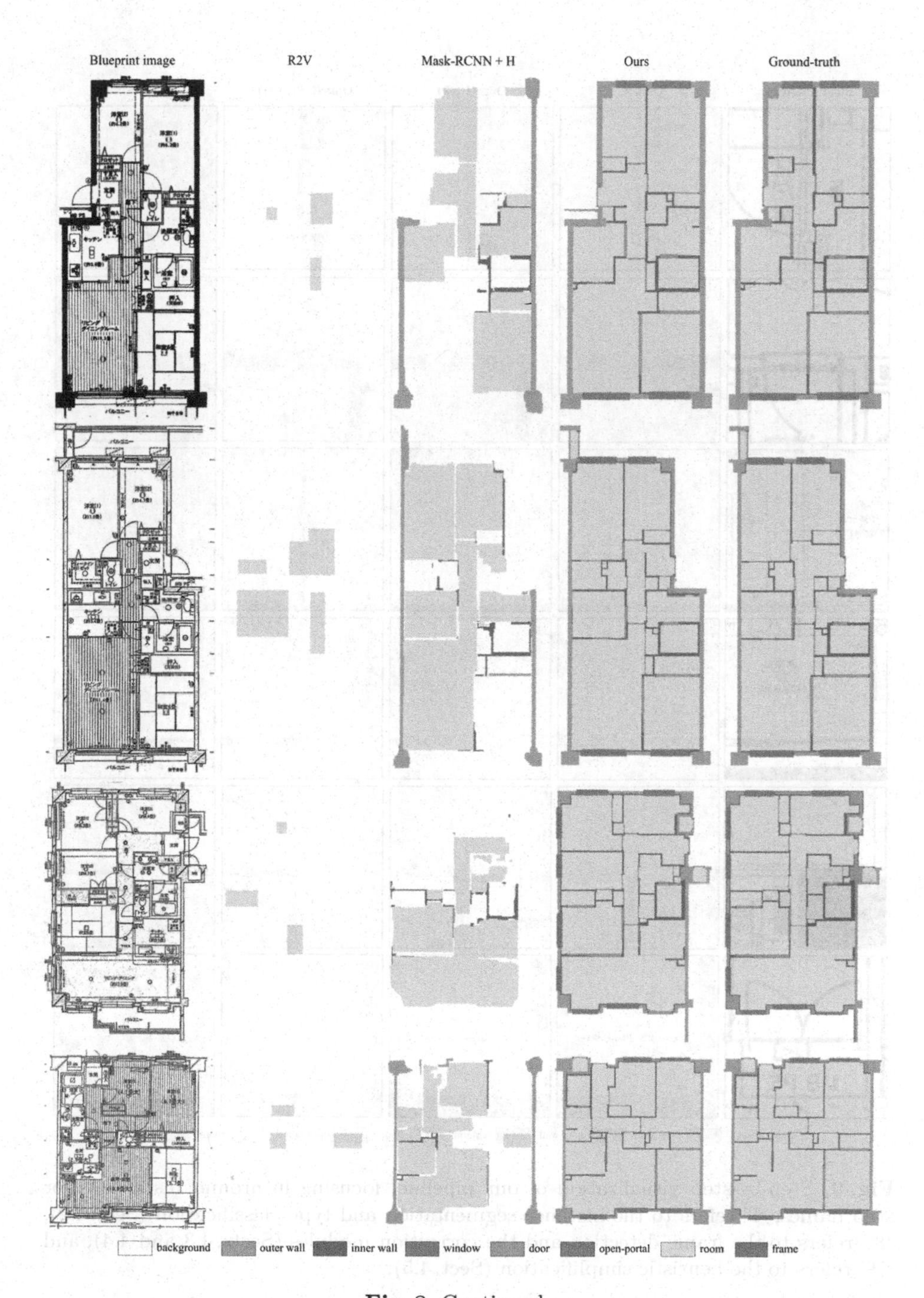

Fig. 8. Continued.

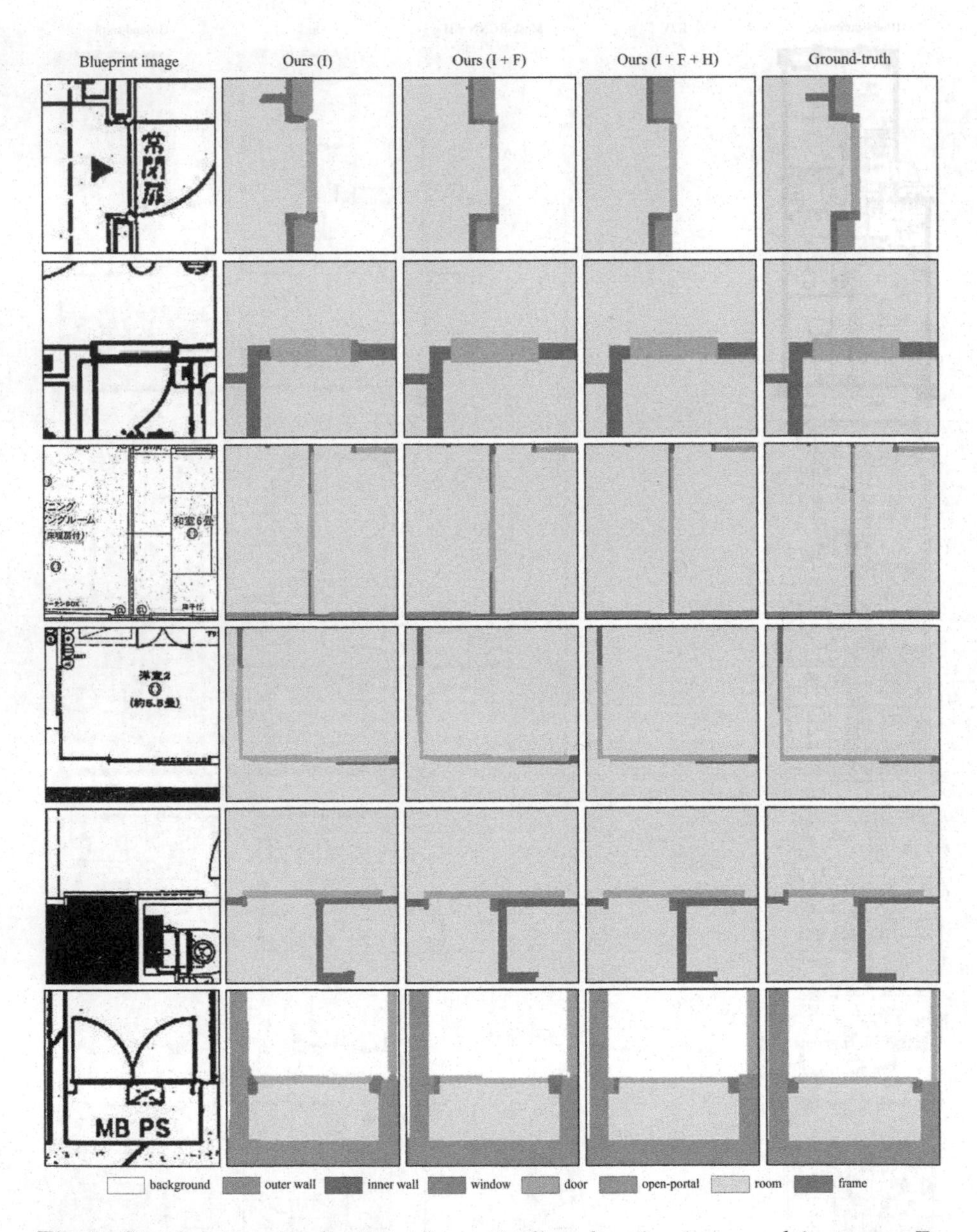

Fig. 9. Step-by-step visualization of our pipeline, focusing in around instances. For step names, "I" refers to the instance segmentation and type classification (Sect. 4.2), "F" refers to the frame detection and the correction modules (Sects. 4.3 and 4.4), and "H" refers to the heuristic simplification (Sect. 4.5).

6.2 Qualitative Results

Figures 7 and 8 compare our pipeline against the competing methods. R2V [11] is not designed for varying wall thickness or small segments, and hence fail to recover most segments. Mask-RCNN with heuristic simplification did not perform well, producing results that are far from complete. Our method clearly outperforms the two.

Table 4. Ablation study. The table shows the performance of the proposed approach while varying the order and the presence of the three steps, namely, "I" the instance segmentation and the type classification (Sect. 4.2), "F" the frame detection and the correction modules (Sects. 4.3 and 4.4), and "H" the heuristic simplification (Sect. 4.5). The "italic" and the "bold" denote the best and the second best results, respectively.

Config	Corner	Edge	Region	Connectivity Acc.	
	F1	F1	F1	All	Frames
Ours (I)	38.4	10.6	85.3	46.6	76.7
Ours (I + H)	*61.1*	34.1	*85.3*	46.6	76.7
Ours (I + F)	38.8	11.2	85.3	*48.4*	*80.4*
Ours (I + H + F)	59.4	**32.5**	**85.2**	**48.1**	79.8
Ours (I + F + H)	**60.5**	*33.9*	**85.2**	*48.4*	**80.3**

Figures 9 evaluate the quality of our frame connectivity. A mix of adding or removing actions and different frame connections are shown, where our system successfully corrects frame connectivity. Refinement GAN based simplification removes many jagged edges, followed by heuristic simplification which produces further visually-pleasing results.

The second row of Fig. 9 is a common case for bathroom doors, where there should be no door frames. Our pipeline removed the extra frame and filled in the open region. The first row is a common example of doors with two frames. Here, we see a failure case of the frame correction module where the top left horizontal blue wall disappears due to the refinement procedure. The last row is interesting where a door frame was segmented, but mistakenly classified as a blue wall. Our system is able to handle this mistake, and also smooth the green door's jagged line with our heuristic simplification.

7 Conclusion

This paper proposes a novel image vectorization algorithm for building blueprints, whose geometric structures are far more complex and detailed than standard floorplan images. Qualitative and quantitative evaluations demonstrate the effectiveness of our approach, outperforming the current state-of-the-art and a baseline method with significant margins. However, the precision and recall are still far from the production quality, where our main future work will be to further robustify the algorithm.

Acknowledgements. The research is supported by NSERC Discovery Grants, NSERC Discovery Grants Accelerator Supplements, and DND/NSERC Discovery Grants. We also thank GA Technologies for providing us with the building blue-print images.

References

1. instance-seg. https://github.com/alicranck/instance-seg. Accessed 16 Apr 2021
2. U-Net: semantic segmentation with PyTorch. https://github.com/milesial/Pytorch-UNet. Accessed 16 Apr 2021
3. Ahmed, S., Liwicki, M., Weber, M., Dengel, A.: Improved automatic analysis of architectural floor plans. In: 2011 International Conference on Document Analysis and Recognition, pp. 864–869. IEEE (2011)
4. Cao, Z., Hidalgo, G., Simon, T., Wei, S.E., Sheikh, Y.: OpenPose: realtime multi-person 2d pose estimation using part affinity fields. IEEE Trans. Pattern Anal. Mach. Intell. **43**(1), 172–186 (2019)
5. Chen, J., Liu, C., Wu, J., Furukawa, Y.: Floor-SP: inverse cad for floorplans by sequential room-wise shortest path. In: The IEEE International Conference on Computer Vision (ICCV) (2019)
6. De Brabandere, B., Neven, D., Van Gool, L.: Semantic instance segmentation with a discriminative loss function. In: CVPR Workshop on "Deep Learning for Robotic Vision" (2017)
7. Dutta, A., Zisserman, A.: The VIA annotation software for images, audio and video. In: Proceedings of the 27th ACM International Conference on Multimedia, MM 2019. ACM, New York (2019). https://doi.org/10.1145/3343031.3350535
8. He, K., Gkioxari, G., Dollár, P., Girshick, R.: Mask R-CNN. In: 2017 IEEE International Conference on Computer Vision (ICCV), pp. 2980–2988 (2017). https://doi.org/10.1109/ICCV.2017.322
9. de las Heras, L.P., Terrades, O.R., Robles, S., Sánchez, G.: CVC-FP and SGT: a new database for structural floor plan analysis and its groundtruthing tool. Int. J. Doc. Anal. Recogn. (IJDAR) **18**(1), 15–30 (2015)
10. Kalervo, A., Ylioinas, J., Häikiö, M., Karhu, A., Kannala, J.: CubiCasa5K: a dataset and an improved multi-task model for floorplan image analysis. In: Felsberg, M., Forssén, P.-E., Sintorn, I.-M., Unger, J. (eds.) SCIA 2019. LNCS, vol. 11482, pp. 28–40. Springer, Cham (2019). https://doi.org/10.1007/978-3-030-20205-7_3
11. Liu, C., Wu, J., Kohli, P., Furukawa, Y.: Raster-to-vector: revisiting floorplan transformation. In: 2017 IEEE International Conference on Computer Vision (ICCV), pp. 2214–2222 (2017). https://doi.org/10.1109/ICCV.2017.241
12. Liu, C., Wu, J., Furukawa, Y.: FloorNet: a unified framework for floorplan reconstruction from 3D scans. In: Ferrari, V., Hebert, M., Sminchisescu, C., Weiss, Y. (eds.) ECCV 2018. LNCS, vol. 11210, pp. 203–219. Springer, Cham (2018). https://doi.org/10.1007/978-3-030-01231-1_13

13. Nauata, N., Furukawa, Y.: Vectorizing world buildings: planar graph reconstruction by primitive detection and relationship inference. In: Vedaldi, A., Bischof, H., Brox, T., Frahm, J.-M. (eds.) ECCV 2020. LNCS, vol. 12353, pp. 711–726. Springer, Cham (2020). https://doi.org/10.1007/978-3-030-58598-3_42
14. Nauata, N., Hosseini, S., Chang, K.H., Chu, H., Cheng, C.Y., Furukawa, Y.: House-GAN++: generative adversarial layout refinement networks. In: Proceedings of the IEEE/CVF Conference on Computer Vision and Pattern Recognition (CVPR) (2021)
15. Newell, A., Yang, K., Deng, J.: Stacked hourglass networks for human pose estimation. In: Leibe, B., Matas, J., Sebe, N., Welling, M. (eds.) ECCV 2016. LNCS, vol. 9912, pp. 483–499. Springer, Cham (2016). https://doi.org/10.1007/978-3-319-46484-8_29
16. Ren, S., He, K., Girshick, R., Sun, J.: Faster R-CNN: towards real-time object detection with region proposal networks. In: Advances in Neural Information Processing Systems, vol. 28 (2015)
17. Wu, Y., Kirillov, A., Massa, F., Lo, W.Y., Girshick, R.: Detectron2 (2019). https://github.com/facebookresearch/detectron2
18. Zeng, Z., Li, X., Yu, Y.K., Fu, C.W.: Deep floor plan recognition using a multi-task network with room-boundary-guided attention. In: Proceedings of the IEEE/CVF International Conference on Computer Vision (ICCV) (October 2019)
19. Zhang, F., Nauata, N., Furukawa, Y.: Conv-MPN: convolutional message passing neural network for structured outdoor architecture reconstruction. In: Proceedings of the IEEE/CVF Conference on Computer Vision and Pattern Recognition (CVPR) (June 2020)

Unsupervised 3D Shape Representation Learning Using Normalizing Flow

Xiang Li[1(✉)], Congcong Wen[2], and Hao Huang[2]

[1] King Abdullah University of Science and Technology, Thuwal, Saudi Arabia
xiangli92@ieee.org
[2] New York University Abu Dhabi, Abu Dhabi, UAE
{cw3437,hh1811}@nyu.edu

Abstract. Learning robust and compact shape representation learning plays an important role in many 3D vision tasks. Existing supervised learning-based methods have achieved remarkable performance, meanwhile requiring large-scale human-annotated datasets for model training. Self-supervised/unsupervised methods provide an attractive solution to this issue that can learn shape representations without the need for ground truth labels. In this paper, we introduce a novel self-supervised method for shape representation learning using normalizing flows. Specifically, we build a model upon a variational normalizing flow framework where a sequence of normalizing flow layers are adopted to model exact posterior latent distribution and enhance the representation power of the learned latent code. To further encourage inter-shape separability and intra-shape compactness among a batch of shapes, we design a contrastive-center loss that performs metric learning on features on a hypersphere. We validate the representation learning ability of our model on downstream classification tasks. Experiments on ModelNet40/10, ScanobjectNN, and ScanNet datasets demonstrate the superior performance of our method compared with current state-of-the-art methods.

Keywords: Shape representation learning · Normalizing flow · Contrastive learning

1 Introduction

With recent advancements in range sensors (i.e. LiDAR and RGBD cameras) and imaging technologies (i.e. 3D MRI), the amount of available 3D geometric data has increased dramatically. It is therefore of great importance to develop methods that can take advantage of the ubiquity of 3D point cloud data for 3D scene understanding. One fundamental problem with 3D geometric data is learning representative and robust feature representations. To handle this problem, existing supervised-learning-based methods have achieved remarkable performance with the help of large-scale human-annotated datasets. However, human

X. Li and C. Wen—Equal contribution.

L. Wang et al. (Eds.): ACCV 2022, LNCS 13841, pp. 158–175, 2023.
https://doi.org/10.1007/978-3-031-26319-4_10

annotations are usually labor-intensive and time-consuming and an inadequate dataset may lead to poor generalization ability of the learned models. Therefore, unsupervised representation learning stands out as an attractive alternative and drew huge research attention in the 3D vision community.

Several studies have been devoted to addressing this challenging problem [1–3]. To train the neural network models without ground truth labels, these methods formulate self-supervision signals from careful-designed generation or reconstruction tasks, including self-reconstruction [1,2,4], transformation equivariant [5–7], local-to-global reconstruction [3,8,9] and distribution approximation [10,11]. Although these methods obtain ever-increasing downstream classification performance on several benchmark datasets, two challenging issues still exist and impede these methods to get better performance than state-of-the-art supervised methods. First, existing methods mostly focus on formulating self-supervision signals from latent representation while failing to regularize latent distribution, and thus the *learned latent representation cannot well characterize the structural distribution of input data.* Second, these methods usually overemphasize global representations while neglecting *semantic local structures and the relationship between local and global representations.*

For the first issue, a direct remedy is to use a simple Gaussian prior over shape representations, like the ones used in VAE models [12]. But it has been shown that a restricted prior tends to limit the performance of VAEs [13]. Inspired by the great success of normalizing flow-based models for unsupervised density modeling [14,15]. In this paper, we introduce a variational normalizing flow-based module to encourage more flexible latent distribution which can potentially *better characterize the global structures of irregular 3D shapes by exact log-likelihood modeling.* To the best of our knowledge, we are the first to use normalizing flows for unsupervised shape representation learning.

For the second issue, we generate our solution based on the observation that local patterns of 3D shapes are highly related to global patterns. The human can recognize an object category from only part of the object and also identify whether a local patch can be a constructive part of a given object. A desirable shape representation model should take into account both local and global structures when designing the feature learning module. To this end, we aim to enhance global shape representations by incorporating a *self-supervised local-global semantic supervision.* Specifically, we formulate a contrastive-center loss on local and global embeddings to encourage inter-shape separability and intra-shape compactness of learned embeddings.

We validate the representation learning ability of our model on downstream classification tasks. Extensive experiments are conducted on three benchmark datasets and results show that the proposed unsupervised method obtains better performance than its supervised counterpart and exhibits robustness to sparse point sampling and input noise. The proposed method also reports new state-of-the-art performance on ModelNet40, ModelNet10, and ScanNet datasets, with a single view classification accuracy of 93.3%, 95.6%, and 90.8% respectively.

2 Related Work

2.1 3D Point Cloud Representation Learning

Supervised-Learning Based Methods. As a pioneering work, PointNet [16] introduce the first deep learning-based method that directly learns point features from unstructured raw point clouds. Although it provides a simple and efficient architecture for point cloud signature learning, it lacks the ability to capture the local structure information. PointNet++ [17] tries to address this issue using hierarchy point sampling and grouping techniques. Subsequent works try to improve the performance by designing new point convolution operations that can better capture local structural information. DGCNN [18] and its following works [19] regard point clouds as undirected graphs and formulate point feature learning a dynamic message passing process on graph data. PointCNN [20] learns an $\mathcal{X}$-transformation to reorder the input points into a canonical order. KPConv [21] build point convolution based on rigid or deformable kernel points. In light of the great success of vision transformers [22], recent works [23–25] develop point convolutions with self-attention networks. In this paper, we build our unsupervised shape representation learning framework using PointNet++ as the backbone network for point feature learning. Other PointNet++-like variants can be easily adapted into our pipeline.

Unsupervised-Learning Based Methods. To learn feature representation for 3D point clouds without access to ground truth labels, previous methods have developed various types of self-supervision signals. The most intuitive self-supervised signal can be formulated in a self-reconstruction process where the global feature representations are first learned from the input point clouds and then a decoder network is used to reconstruct the inputs from the feature representations [1,4]. Similarly, contrastive learning-based methods [7,26] have also been explored for unsupervised pre-training for 3d representation learning. In the light of adversarial networks for various data generation tasks, researchers proposed to use generative adversarial networks (GANs) [27] to learn a probabilistic latent space of 3D objects [28]. Instead of using an explicit encoder network to learn 3D shape representations, recent works also explored auto-decoder networks for shape representation learning [29,30]. Although these methods have obtained ever-increasing performance for unsupervised 3D shape representation learning, they usually fail to capture high-level semantic information thus the performance fall behind state-of-the-art supervised methods. To address this issue, recent works [3,9] incorporated semantic knowledge by simultaneously exploiting local and global self-supervision in order to learn discriminative representations. In this paper, we aim to enhance the learned shape representation by exploiting the semantic relation between local and global structures by a newly designed contrastive-center loss.

2.2 Normalizing Flows

Normalizing Flows (NFs) are a family of generative models based on an invertible mapping between the data distribution and latent distribution. Pioneering work [14] introduced the first flow-based deep learning framework for high-dimensional density estimation using change of variable theory. To enable the tractability of the Jacobian determinant, a coupling layer was proposed with efficient bijective transformation. Recent works have demonstrated the superior performance in many generation tasks, including image generation [13,15], audio synthesis [31,32], video generation [33], and machine translation [34]. Thanks to the attractive merits of exact log-likelihood modeling, normalizing flows have become a powerful technique for unsupervised density modeling.

Recent efforts have full-filled theoretical developments and applications of flow-based methods. In [13], the authors introduced a variational normalizing flow model that combines the merits of VAE and normalizing models in a unified framework where flow layers are used to transform latent variable from a simple diagonal Gaussian distribution to a highly flexible distribution that characterizes the true posterior. Glow [15] introduced a simple but effective generative flow using an invertible 1×1 convolution and demonstrated its effectiveness and efficiency for synthesizing realistic high-resolution natural images. A comprehensive review of normalizing flow can be found [35]. Recent works [11,36–38] have developed numerous flow-based methods for a wide range of 3D tasks, such as point cloud generation, single-view 3D reconstruction. In this paper, a variational normalizing flow module is designed to enhance latent representations by using normalizing flows to characterize the exact latent distribution.

3 Method

In this section, we introduce the normalizing Flow-based method for unsupervised 3D Shape representation learning, named SFlow. Our proposed method is built upon a self-reconstruction framework with normalizing flow modules to ensure the learned latent code can characterize the exact probability distribution of input data. A newly designed contrastive loss is further applied to the semantic embeddings and global representations to encourage the discrimination abilities of the learned features. Figure 1 gives an overview of the proposed method. Our method includes three main components. The first component is **Self-Supervised Reconstruction** module. In this module, our model first leverages an encoder network Q_ϕ to learn global representation z from an input shape X, i.e., $z = Q_\phi(X)$, and then employs a decoder network D to decode the global representation into a reconstructed shape $\hat{X}$, i.e., $\hat{X} = D(z)$. The network architecture will be illustrated in Sect. 3.1. The second component is **Variational Normalizing Flow** module, in which a reparametrization trick is leveraged to generate initial latent code z_0 from a Gaussian distribution $\mathcal{N}(\mu, \sigma)$. Then, we leverage a sequence of normalizing flow layers to learn the exact probability distribution z_K. The initial probability distribution after the encoder network 'flows' through the sequence of invertible mappings and is finally constrained by standard Gaussian prior, see Sect. 3.2. For the third component, **Feature Contrastive** module, we formulate a contrastive-center loss to encourage intra-shape compactness and inter-shape separability of the local and

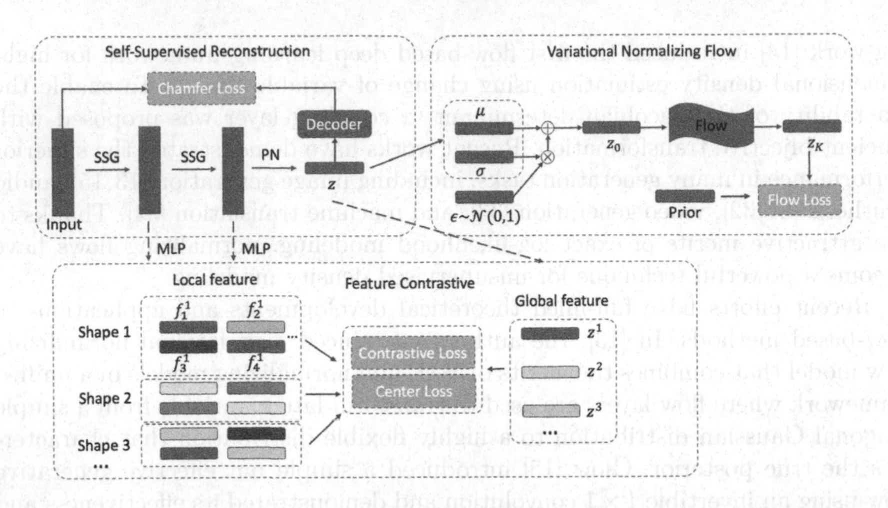

Fig. 1. Overview of the proposed method. Our SFlow model starts with a PointNet++ [16] backbone network to extract global feature z of the input shape. 'PN' denotes a unit PointNet [16], and 'SSG' denotes Single-Scale Grouping (SSG) proposed in PointNet++ [17]. A decoder network is leveraged to recover the input shape with Chamfer loss as a supervision signal. Then, a variational normalizing flow module is developed to transform the latent distribution into a standard Gaussian prior through a sequence of invertible mappings. A feature contrastive module with both contrastive loss and center loss is designed in the embedding space to encourage intra-shape compactness and inter-shape separability of the local and global embeddings.

global embeddings. Both a softmax-based contrastive loss and a center loss are defined to perform metric learning on features on a hypersphere, see Sect. 3.3. A shared multi-layer perceptron (MLP) network is leveraged after each downsampling block to transform the local features into the same dimension as the global feature before calculating contrastive-center loss.

3.1 Self-supervised Reconstruction

Self-supervised reconstruction, or point auto-encoding, is one of the first family of methods for unsupervised 3D shape representation learning [1,2,4]. This type of method starts by using an encoder network to learn global shape representation and then a decoder network to reconstruct input shapes. A self-reconstruction loss, e.g., Chamfer distance [39], can then be used to provide self-supervision signals for model training. In our method, we leverage a hierarchy point feature learning network proposed in PointNet++ [17] as the encoder. Given a 3D point set $X = \{x_1, x_2, ..., x_N\}$, where each point x_i is represented by a 3D coordinate and possibly attributes (e.g., surface normal), and N is the number of points. To directly learn feature representations from raw point sets, pioneering work PointNet [16] proposed to use a shared MLP network to learn per-point feature embeddings followed by a symmetry function, e.g. max-pooling, to get global shape representation. PointNet++ [17] enhances the method by introducing a set abstraction and feature interpolation layer to enable a hierarchy feature learning. Specifically, at each set abstraction layer, a smaller number of points are selected from the previous layer using farthest point sampling, and a unit PointNet is applied to the local neighborhood around each selected point. The global shape representation can be obtained by applying a smaller PointNet on the final abstraction layer.

To perform self-reconstruction, a folding-based [1] decoder network is adopted to transform the global shape representation into a set of 3D coordinates. Specifically, the global shape representation is concatenated with the coordinate of a canonical 2D grid and a multi-level MLP network is used to deform the 2D grid onto an underlying 3D object surface, i.e., $\hat{X} = D(z, \mathcal{G})$, where $\mathcal{G}$ is the coordinates of regular 2D grid. A self-supervised Chamfer loss is adopted to train the self-reconstruction network, defined as:

$$\mathcal{L}_{rec} = \sum_{x \in X} \min_{y \in \hat{X}} ||x - y||_2 + \sum_{x \in \hat{X}} \min_{y \in X} ||x - y||_2. \tag{1}$$

Optionally, a normal estimation network Ψ can be built upon the learned global representation to further encourage high-level semantic feature learning. Unlike previous methods [17] that use normal as additional inputs, our method uses normal information as auxiliary output supervision, thus relieving the need for normal information at the inference stage. Specifically, we concatenate the 3D coordinate of each input point x_i with the global feature vector z and feed it into a shared MLP network to predict the normal estimations. The cosine similarity loss is used to train the network:

$$\mathcal{L}_{nor} = -\frac{1}{N} \sum_{i} cos(\Psi(z, x_i), \mathbf{n}_{x_i}), \tag{2}$$

where $\mathbf{n}_{x_i}$ denotes the ground truth normal for point x_i.

3.2 Variational Normalizing Flow

The above self-supervised point auto-encoding (AE) [40] model can be easily extended to a probabilistic form of variational auto-encoder (VAE) [12] by constraining the latent variable by some underlying probability distributions. Given input data X, a typical VAE model characterize the data distribution via latent variable z with a prior distribution $P_\psi(z)$, and captures the distribution of X given z using a decoder network $P_\theta(X|z)$. An encoder/inference network is typically used to generate the mean and variance of latent distribution $Q_\phi(z|X)$. During training, the parameters of the encoder and decoder networks are jointly optimized to maximize a lower bound on the log-likelihood of the input data,

$$\begin{aligned} \log P_\theta(X) &\geq \log P_\theta(X) - \mathcal{D}_{KL}(Q_\phi(z|X)||p_\theta(z|X)) \\ &= E_{Q_\phi(z|X)}[\log p_\theta(X|z)] - D_{KL}(Q_\phi(z|X)||p_\psi(z)) \\ &= -\mathcal{L}(X), \end{aligned} \tag{3}$$

which is also called the evidence lower bound (ELBO). From the above equation, the ELBO jointly optimizes the negative reconstruction error (the first term) and a latent distribution regularizer (the second term), which is KL divergence between the approximate posterior and the prior distribution. In practice, $Q_\phi(z|x)$ is modeled by a diagonal Gaussian distribution $\mathcal{N}(\mu_\phi(X), \sigma_\phi(X))$ where the mean $\mu_\phi(X)$ and the standard-deviation $\sigma_\phi(X)$ are predicted by a deep neural network $Q_\phi(z|X)$.

One limitation of the VAE model lies in the available choices of posterior approximating families where the true posterior is unknown and is generally more complex than the assumption allows for. Choosing a highly flexible and computationally-feasible approximate posterior distribution stands as one of the bottlenecks of VAE models. To handle this issue, one feasible solution is to use normalizing flows to transform a simple distribution into a highly complex one as the posterior in VAE, which makes the model become variational normalizing flows [13].

A normalizing flow defines the transformation from an initial known distribution to a more complicated one using a sequence of invertible mappings. Let $f_1, ..., f_K$ denotes a sequence of invertible functions, where each $f : \mathbb{R}^d \rightarrow \mathbb{R}^d$ with inverse $f^{-1} = g$, s.t., $g \circ f(x) = x$. Given a latent variable z_0 ($z_0 = z$) with distribution $q(z_0)$, a variable z_K with more complex distribution can be generated by recursively apply the transformation, i.e., $z_K = f_K \circ f_{K-1} \circ f_1(z_0)$. The probability distribution of the resulting variable z_K can be generated by the change of variables formula:

$$\log q(z_K) = \log q(z_0) - \sum_{k=1}^{K} \log |\det \frac{\partial f_k}{\partial z_{k-1}}|. \tag{4}$$

Thanks to the invertible characteristic of each transformation function, z_0 can be computed from z_K using inverse flow: $z_0 = f_1^{-1} \circ f_2^{-1} \circ f_K^{-1}(z_K)$. In practice, $f_1, ..., f_n$ are implemented using neural networks with an architecture that ensures the determinant of the Jacobian det $\frac{\partial f_k}{\partial z_{k-1}}$ can be easily computed. In this paper, we use Glow-like 1×1 invertible convolutions for density transformation, interested readers can refer to [15] for details. After applying the above flow transformations, the marginal log-likelihood in Eq. (3) can be reformulated as:

$$\begin{aligned}
-\mathcal{L}(X) &= \log P_\theta(X) - \mathcal{D}_{KL}(Q_\phi(z|X)||P_\theta(z|X)) \quad \%\textit{The first row of Eq. (4)} \\
&= \log P_\theta(X) - E_{Q_\phi(z|X)}(\log Q_\phi(z|X) - \log P_\theta(z|X)) \\
&= \log P_\theta(X) - E_{Q_\phi(z|X)}(\log Q_\phi(z|X) - \log P_\theta(z, X) + \log P_\theta(X)) \\
&= E_{Q_\phi(z|X)}[\log Q_\phi(z|X) - \log P_\theta(X, z)] \\
&= E_{q(z_0)}[\log q(z_K) - \log P_\theta(X, z_K)] \quad \%\textit{Replace } Q_\phi(z|X) \textit{ with } z_K \\
&= \mathcal{H}(q(z_0)) - E_{q(z_0)}[\sum_{k=1}^{K} \log |\det \frac{\partial f_k}{\partial z_{k-1}}|] \quad \%\textit{Replace } z_K \textit{ using Eq. (5)} \\
&- E_{q(z_0)}[\log p(X, z_K)],
\end{aligned} \tag{5}$$

where $\mathcal{H}$ represents the entropy. The first term is the entropy of the approximated posterior, the second term is prior regularization. We denote the first two terms as $\mathcal{L}_{flow}$ in the following sections. The third term is the reconstruction log-likelihood of the input point set, calculated as Eq. (1).

3.3 Contrastive-Center Loss

The above self-reconstruction process only characterizes input shapes from a global perspective. In this section, we aim to enhance shape representation by exploiting the relation between local structures and global shapes. Specifically, we first design a contrastive loss to encourage inter-shape separability by encouraging semantic embeddings of each point to be closer to the global representation of the same object than other objects. In the light of instance discrimination [41], our method treats the global representation of one object as the positive class and uses the global representation of other objects as the negative class, and formulates a classification loss to encourage separability. Given a bunch of input shapes $\{X^b\}_{b=1}^{B}$, $\mathbf{f}_i^b$ as the embedding of point x_i^b on shape X^b and the global representation z^b,

$$\mathcal{L}_{cont} = -\frac{1}{B*N}\sum_{b=1}^{B}\sum_{i=1}^{N}\log\frac{\exp(s\omega(\mathbf{f}_i^b)^T z^b)}{\sum_{j=1}^{B}\exp(s\omega(\mathbf{f}_i^b)^T z^j)}. \tag{6}$$

The above loss function will maximize the similarity of each point embeddings $\mathbf{f}_i^b$ with the global representation of the same shape z^b meanwhile minimizing the similarity of each point embeddings with the global representation of other shapes $z^j (j \neq b)$. ω is an MLP network that maps $\mathbf{f}$ to the same dimension as z. Similar to the metric losses used for face recognition [42,43], we normalize all feature embeddings onto a hypersphere before computing similarities and use a constant value s = 64 to re-scale the features. Note that Eq. (6) is calculated on local embeddings at all downsampling levels.

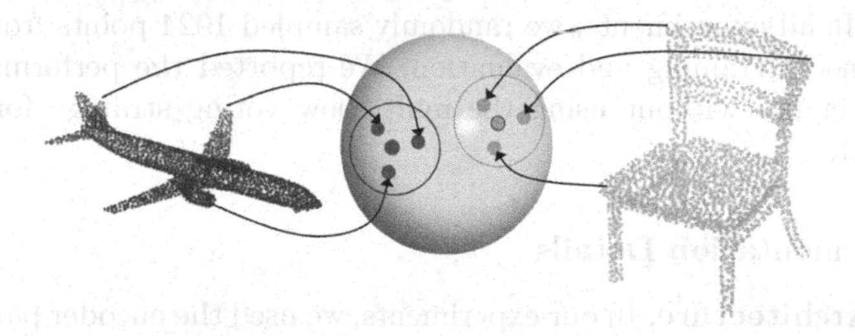

Fig. 2. Illustration of contrastive-center loss on the hypersphere of feature embeddings. The red circle indicates global shape representation, which is treated as the pseudo center point for center loss calculation. (Color figure online)

One may note that the above contrastive loss only encourages inter-class separability. Inspired by the studies on contrastive-center loss for face recognition

[44], we further introduce a center loss to enforce intra-shape compactness. Unlike [44] that dynamically updates class centers using gradient decent, we directly treat the global feature of each input shape as the "class center" and develop the center loss as,

$$\mathcal{L}_{center} = \frac{1}{B * N} \sum_{b=1}^{B} \sum_{i=1}^{N} ||\mathbf{f}_i^b - z^b||. \tag{7}$$

Figure 2 gives an illustration of the proposed contrastive-center loss. The final metric loss used in our method is defined as:

$$\mathcal{L}_{met} = \mathcal{L}_{cont} + \mathcal{L}_{center}. \tag{8}$$

Combining self-reconstruction, normalizing flow loss, and contrastive-center loss, the overall training objective is defined as:

$$\mathcal{L} = \mathcal{L}_{rec} + \mathcal{L}_{flow} + \mathcal{L}_{met} + \mathcal{L}_{nor}. \tag{9}$$

Note that all loss terms in Eq. (9) can be calculated in a purely unsupervised way, without using the ground truth labels.

4 Experiments and Results

4.1 Experimental Datasets

We evaluated the 3D shape representation learning performance of our model on ModelNet [45], ScanObjectNN [46] and ScanNet [47] datasets. The ModelNet40/10 dataset consists of 9832/3991 training shapes and 2468/908 test shapes from 40/10 object classes. All point sets are sampled from CAD models with surface normal information provided. ScanObjectNN is a real-world dataset that consists of 2902 3D objects from 15 categories. We used the "object-only" split in our experiments. ScanNet [47] is another real-world scan dataset with 17 object categories where we followed [20] to get points sets from instance segmentation labels. In all experiments, we randomly sampled 1024 points from each 3D object for model training and evaluation. We reported the performance using single view inputs without using the multi-view voting strategy for potential enhancement.

4.2 Implementation Details

Network Architecture. In our experiments, we used the encoder part of PointNet++ (PN++) as the backbone network for feature learning. Three set abstraction layers were sequentially applied to reduce the number of points to 512, 128, and 64, with a radius of 0.23, 0.32, and 0.46 respectively, followed by a unit PointNet to get the global representations. In each set abstraction layer, we used Single-Scale Grouping (SSG) instead of Multi-Scale Grouping (MSG) [17] for feature aggregation to reduce model capacity. In a self-supervised reconstruction decoder, we adopted a two-level folding process to reconstruct 3D shapes.

Table 1. Classification accuracy (%) on ModelNet40 (MN40.) and ModelNet10 (MN10.) datasets. (L) denotes the model with a large PN++ backbone network. † indicates the mode is trained on the ShapeNet dataset.

Method	Input	Accuracy	
		MN40	MN10
TL Network [50]	Voxel	74.40	–
VConv-DAE [51]	Voxel	75.50	80.50
3DGAN [28]	Voxel	83.30	91.00
VSL [52]	Voxel	84.50	91.00
VIPGAN [53]	Views	91.98	94.05
†LGAN [2]	Points	85.70	95.30
LGAN [2]	Points	87.27	92.18
†FoldingNet [1]	Points	88.40	94.40
FoldingNet [1]	Points	84.36	91.85
MAP-VAE [3]	Points	90.15	94.82
GraphTER [7]	Points	92.02	–
GLR [9]	Points	92.22	94.82
GLR(L) [9]	Points	93.02	95.53
SFlow	Points	92.78	94.82
SFlow (L)	Points	**93.31**	**95.60**

In the variational normalizing flow module, we used a Glow architecture with 16 flow layers, and each has 4 flow steps. The hidden dimension was set to 128 and divided into 8 groups. To evaluate the downstream classification performance, we trained a linear SVM [48] using the feature representations obtained from the training set and evaluated the classification performance on the test split.

Network Optimization. Our model was optimized using the Adam optimizer. The initial learning rate was set to 1e−3 and decayed with a scale of 0.7 every 20 epochs. We used a momentum of 0.9 for Batch Normalization layers [49] and decayed with a rate of 0.5 every 20 epochs. Our model was trained for 300 epochs with a batch size of 32 and it took around 30 h on a single Titan XP GPU.

4.3 Results on ModelNet

To demonstrate the effectiveness of our proposed method for unsupervised shape representation learning, we compared our method with state-of-the-art unsupervised methods in Table 1. We also included the results of our SFlow model with a larger backbone (4× channel width), similar to recent work GLR [9] which reports state-of-the-art performance on the ModelNet dataset. From Table 1, our method achieves new state-of-the-art performance, with a classification accuracy of 93.31% and 95.60% on ModelNet40 and ModelNet10 dataset respectively.

We further compared the performance of our SFlow model with its supervised counterpart. Specifically, we trained supervised PointNet++ models with the same backbone networks, followed by several fully connected layers and a softmax layer to generate the prediction labels. From Table 2, one can see that our SFlow model obtains better performance than its supervised counterpart using both small and large backbone networks. This demonstrates that our unsupervised SFlow model can potentially learn more discriminative representation than its supervised counterpart.

Table 2. Comparison with the supervised counterpart on ModelNet40 dataset.

Model	PN++	PN++ (L)
PN++ (supervised)	91.69	92.01
SFlow (unsupervised)	**92.78**	**93.31**

4.4 Cross-Dataset Evaluation

We conducted downstream classification experiments on the real-world ScanObjectNN and ScanNet datasets. Following the experimental settings of GLR [9], we trained our shape representation learning network on the ModelNet40 dataset and evaluated the downstream classification performance on the ScanObjectNN and ScanNet datasets. Note that we did not fine-tune our model on the target datasets. Even though ScanObjectNN and ScanNet datasets have different object categories with the ModelNet40 dataset, our SFlow model can still produce well-separable representations without training/finetuning on the target datasets, as shown in Table 3. This demonstrates that our SFlow method can successfully *learn generic representations from object structures without labels.* Moreover, our model achieves significant better performance than current STOA method GLR [9] on ScanNet dataset (90.8% vs. 89.2%), and a obtains a comparable performance with GLR [9] on ScanObjectNN dataset (87.0% vs. 87.2%). It should be noted that we only got a classification accuracy of 86.2% on the ScanObjectNN dataset using the official code of GLR [9].

4.5 Robustness Analysis

In this section, we investigate the robustness of our model under different numbers of sampled points and noise levels. To achieve this, we evaluate the downstream classification performance on the ModelNet40 dataset with sparser points of 1024, 512, 256, 128, and 64, while the backbone network is still trained on 1024 points. From Fig. 3(a), our SFlow model is a lot more robust than its supervised counterpart and maintains an accuracy higher than 86.1% with only 128 points. For the latter, we added Gaussian noise of $\mathcal{N}(0, \sigma_r)$ to input point sets and generated feature representation using the backbone trained on the clean dataset. We conducted experiments with σ_r choose from $[0, 0.01, 0.02, 0.03, 0.05]$. From Fig. 3(b), our SFlow shows a smaller performance drop than its supervised counterpart and maintains an accuracy higher than 86.5% with a noise level of 0.05.

Table 3. Transferring accuracy (%) on ScanObjectNN (SON.) and ScanNet (SN.) datasets. (L) denotes a large PN++ backbone network. 'Sup' denotes supervised methods. * denotes our reproduced results.

Method	Sup.	Accuracy	
		SON	SN
PointNet++ [17]	✓	84.3	–
PointCNN [20]	✓	85.5	–
DGCNN [18]	✓	86.2	–
GLR(L) [9]	✗	**87.2**	89.2
*GLR(L) [9]	✗	86.2	89.2
SFlow (L)	✗	87.0	**90.8**

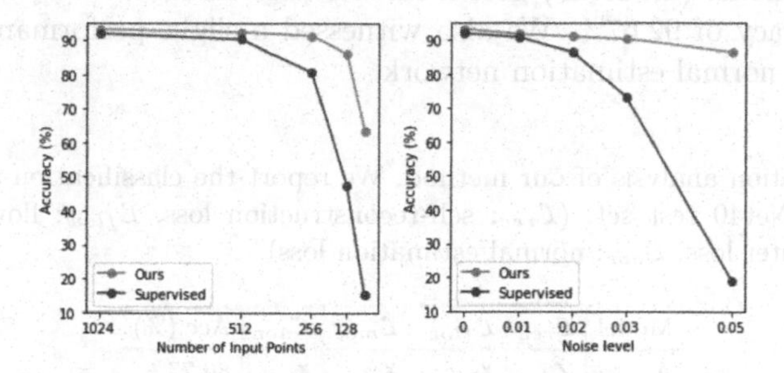

Fig. 3. Robustness test. The classification of our model with different numbers of sampled points and different noise levels.

4.6 Complexity Analysis

In Table 4, we report the model capacity and inference time of our SFlow model and its supervised counterpart. We calculated the total inference time on the whole test set of the ModelNet40 dataset with a batch size of 24 and a single Titan XP GPU. From Table 4, compared to the supervised PN++ model, our SFlow model only brings slightly more computation cost but can get significantly better performance. Moreover, our SFlow model with a small backbone shows a better trade-off in speed and accuracy compared to the one with a larger backbone.

Table 4. Model Complexity and inference time comparison.

Model	#Param	Time	Acc. (%)
PN++	**1.29M**	**2.99** s	91.69
PN++ (L)	12.11M	8.66 s	92.01
SFlow	2.94M	3.29	92.78
SFlow (L)	14.99M	9.03	**93.31**

4.7 Ablation Analysis

In this section, we conduct a detailed ablation study to verify the effectiveness of our model design. We conduct our experiments on the ModelNet40 dataset using a small PointNet++ backbone. In Table 5, the baseline model (model A) is an variant of FoldingNet [1] and trained using reconstruction loss only. By introducing the proposed variational normalizing flow model, our model (model B) got a significant performance boost, from 86.77% to 88.65%. By comparing model A and model C, our newly proposed contrastive-center loss improves the performance by a large margin (+5.43%). This demonstrates our shape representation model benefits a lot by encouraging intra-shape compactness and inter-shape separability. Combining our normalizing flow module and feature contrastive module, our model (model D) gets a further performance boost, with a classification accuracy of 92.67%. We also witnessed a slight performance boost by introducing a normal estimation network.

Table 5. Ablation analysis of our method. We report the classification accuracy (%) on the ModelNet40 test set. ($\mathcal{L}_{rec}$: self-reconstruction loss, $\mathcal{L}_{flow}$: flow loss, $\mathcal{L}_{met}$: contrastive-center loss, $\mathcal{L}_{nor}$: normal estimation loss).

Model	$\mathcal{L}_{rec}$	$\mathcal{L}_{flow}$	$\mathcal{L}_{met}$	$\mathcal{L}_{nor}$	Acc.(%)
A	✓	✗	✗	✗	86.77
B	✓	✓	✗	✗	88.65
C	✓	✗	✓	✗	92.10
D	✓	✓	✓	✗	92.67
E	✓	✓	✓	✓	**92.78**

Self-reconstruction with Normalizing Flows. To better understand the effect of our normalizing flow module, we build a VAE model that directly constrains the latent representation by a Gaussian prior and trained the model by maximizing the lower bound defined in Eq. 3. From Table 6, using a simple Gaussian prior leads to similar reconstruction performance as the baseline AE model, as indicated by Chamfer distance (0.062 vs. 0.063), but can slightly enhance the downstream classification accuracy with a more powerful representation (87.60% vs. 86.77%). In contrast, thanks to a more flexible latent distribution enabled by normalizing flow transformations, our model enhances the latent representation with a better reconstruction performance (0.059 vs. 0.063). The downstream classification performance in Table 6 also supports that our variational normalizing flow module contributes to a more powerful latent representation (88.65% vs. 86.77%). Moreover, we also investigated the effect of different numbers of flow blocks. From Table 6, our SFlow model gets the best performance with 16 flow blocks.

Effect of Different Metric Losses. In the above section, we show the effectiveness of our proposed contrastive-center loss. In this section, we check the

Table 6. Performance comparison with different models and different numbers of flow layers. C.D. denotes Chamfer distance.

Model	C.D	Acc. (%)
Baseline (AE)	0.063	86.77
VAE	0.062	87.60
Ours (k = 16)	**0.059**	**88.65**
Ours (k = 12)	0.060	88.41
Ours (k = 20)	0.058	88.43

effect of each metric separately. Specifically, we conducted experiments using contrastive loss, center loss, and both losses. From Table 7, both metrics can significantly enhance the performance, while softmax-based contrastive loss led to a larger performance boost than center loss (+5.04% vs. +1.40%). Our model got the best performance by using both metrics.

Table 7. Classification accuracy (%) on ModelNet40 test set with different metric losses.

Center loss	Contrastive loss	Acc.(%)
✗	✗	86.77
✓	✗	88.17
✗	✓	91.84
✓	✓	**92.10**

5 Conclusion

In this paper, we introduce an unsupervised method for 3D shape representation learning based on normalizing flow and a newly designed feature discrimination loss. By introducing a variational normalizing flow module to the self-reconstruction process, our model is able to model the exact log-likelihood of latent distribution thus enhancing the representation power of learned latent code. We further designed a feature discrimination loss that combines contrastive loss and center loss to encourage inter-shape separability and intra-shape compactness. We validate the representation learning ability of our model on downstream classification tasks. Experimental results demonstrated our unsupervised method could achieve better performance than its supervised counterpart and our SFlow model obtains new state-of-the-art performance on ModelNet40/10 and ScanNet datasets.

References

1. Yang, Y., Feng, C., Shen, Y., Tian, D.: FoldingNet: point cloud auto-encoder via deep grid deformation. In: Proceedings of the IEEE Conference on Computer Vision and Pattern Recognition, pp. 206–215 (2018)
2. Achlioptas, P., Diamanti, O., Mitliagkas, I., Guibas, L.: Learning representations and generative models for 3d point clouds. In: International Conference on Machine Learning, pp. 40–49. PMLR (2018)
3. Han, Z., Wang, X., Liu, Y.S., Zwicker, M.: Multi-angle point cloud-VAE: unsupervised feature learning for 3d point clouds from multiple angles by joint self-reconstruction and half-to-half prediction. In: 2019 IEEE/CVF International Conference on Computer Vision (ICCV), pp. 10441–10450. IEEE (2019)
4. Deng, H., Birdal, T., Ilic, S.: PPF-FoldNet: unsupervised learning of rotation invariant 3d local descriptors. In: Ferrari, V., Hebert, M., Sminchisescu, C., Weiss, Y. (eds.) ECCV 2018. LNCS, vol. 11209, pp. 620–638. Springer, Cham (2018). https://doi.org/10.1007/978-3-030-01228-1_37
5. Zhang, L., Qi, G.J., Wang, L., Luo, J.: AET vs. AED: unsupervised representation learning by auto-encoding transformations rather than data. In: Proceedings of the IEEE/CVF Conference on Computer Vision and Pattern Recognition, pp. 2547–2555 (2019)
6. Qi, G.J., Zhang, L., Chen, C.W., Tian, Q.: AVT: unsupervised learning of transformation equivariant representations by autoencoding variational transformations. In: Proceedings of the IEEE/CVF International Conference on Computer Vision, pp. 8130–8139 (2019)
7. Gao, X., Hu, W., Qi, G.J.: GraphTER: unsupervised learning of graph transformation equivariant representations via auto-encoding node-wise transformations. In: Proceedings of the IEEE/CVF Conference on Computer Vision and Pattern Recognition, pp. 7163–7172 (2020)
8. Liu, X., Han, Z., Wen, X., Liu, Y.S., Zwicker, M.: L2G Auto-encoder: understanding point clouds by local-to-global reconstruction with hierarchical self-attention. In: Proceedings of the 27th ACM International Conference on Multimedia, pp. 989–997 (2019)
9. Rao, Y., Lu, J., Zhou, J.: Global-local bidirectional reasoning for unsupervised representation learning of 3d point clouds. In: Proceedings of the IEEE/CVF Conference on Computer Vision and Pattern Recognition, pp. 5376–5385 (2020)
10. Li, C.L., Zaheer, M., Zhang, Y., Poczos, B., Salakhutdinov, R.: Point cloud GAN. arXiv preprint arXiv:1810.05795 (2018)
11. Yang, G., Huang, X., Hao, Z., Liu, M.Y., Belongie, S., Hariharan, B.: PointFlow: 3d point cloud generation with continuous normalizing flows. In: Proceedings of the IEEE/CVF International Conference on Computer Vision, pp. 4541–4550 (2019)
12. Kingma, D.P., Welling, M.: Auto-encoding variational Bayes. In: International Conference on Learning Representations (2013)
13. Rezende, D., Mohamed, S.: Variational inference with normalizing flows. In: International Conference on Machine Learning, pp. 1530–1538. PMLR (2015)
14. Dinh, L., Krueger, D., Bengio, Y.: NICE: non-linear independent components estimation. arXiv preprint arXiv:1410.8516 (2014)
15. Kingma, D.P., Dhariwal, P.: Glow: generative flow with invertible 1× 1 convolutions. In: Proceedings of the 32nd International Conference on Neural Information Processing Systems, pp. 10236–10245 (2018)

16. Qi, C.R., Su, H., Mo, K., Guibas, L.J.: PointNet: deep learning on point sets for 3d classification and segmentation. In: Proceedings of the Computer Vision and Pattern Recognition (CVPR), p. 4. IEEE (2017)
17. Qi, C.R., Yi, L., Su, H., Guibas, L.J.: PointNet++: deep hierarchical feature learning on point sets in a metric space. In: Advances in Neural Information Processing Systems, pp. 5099–5108 (2017)
18. Wang, Y., Sun, Y., Liu, Z., Sarma, S.E., Bronstein, M.M., Solomon, J.M.: Dynamic graph CNN for learning on point clouds. ACM Trans. Graph. (TOG) **38**, 1–12 (2019)
19. Wang, L., Huang, Y., Hou, Y., Zhang, S., Shan, J.: Graph attention convolution for point cloud semantic segmentation. In: Proceedings of the IEEE/CVF Conference on Computer Vision and Pattern Recognition, pp. 10296–10305 (2019)
20. Li, Y., Bu, R., Sun, M., Wu, W., Di, X., Chen, B.: PointCNN: convolution on x-transformed points. In: Advances in Neural Information Processing Systems 31, pp. 820–830 (2018)
21. Thomas, H., Qi, C.R., Deschaud, J.E., Marcotegui, B., Goulette, F., Guibas, L.J.: KPConv: flexible and deformable convolution for point clouds. In: Proceedings of the IEEE/CVF International Conference on Computer Vision, pp. 6411–6420 (2019)
22. Vaswani, A., et al.: Attention is all you need. In: Advances in Neural Information Processing Systems, pp. 5998–6008 (2017)
23. Zhao, H., Jiang, L., Jia, J., Torr, P.H., Koltun, V.: Point transformer. In: Proceedings of the IEEE/CVF International Conference on Computer Vision, pp. 16259–16268 (2021)
24. Guo, M.H., Cai, J.X., Liu, Z.N., Mu, T.J., Martin, R.R., Hu, S.M.: PCT: point cloud transformer. Comput. Vis. Media **7**, 187–199 (2021)
25. Engel, N., Belagiannis, V., Dietmayer, K.: Point transformer. IEEE Access **9**, 134826–134840 (2021)
26. Xie, S., Gu, J., Guo, D., Qi, C.R., Guibas, L., Litany, O.: PointContrast: unsupervised pre-training for 3d point cloud understanding. In: Vedaldi, A., Bischof, H., Brox, T., Frahm, J.-M. (eds.) ECCV 2020. LNCS, vol. 12348, pp. 574–591. Springer, Cham (2020). https://doi.org/10.1007/978-3-030-58580-8_34
27. Goodfellow, I., et al.: Generative adversarial nets. In: Advances in Neural Information Processing Systems 27 (2014)
28. Wu, J., Zhang, C., Xue, T., Freeman, W.T., Tenenbaum, J.B.: Learning a probabilistic latent space of object shapes via 3d generative-adversarial modeling. In: Proceedings of the 30th International Conference on Neural Information Processing Systems, pp. 82–90 (2016)
29. Park, J.J., Florence, P., Straub, J., Newcombe, R., Lovegrove, S.: DeepSDF: learning continuous signed distance functions for shape representation. In: Proceedings of the IEEE/CVF Conference on Computer Vision and Pattern Recognition, pp. 165–174 (2019)
30. Duan, Y., Zhu, H., Wang, H., Yi, L., Nevatia, R., Guibas, L.J.: Curriculum DeepSDF. In: Vedaldi, A., Bischof, H., Brox, T., Frahm, J.-M. (eds.) ECCV 2020. LNCS, vol. 12353, pp. 51–67. Springer, Cham (2020). https://doi.org/10.1007/978-3-030-58598-3_4
31. Kim, S., Lee, S.G., Song, J., Kim, J., Yoon, S.: FloWaveNet: a generative flow for raw audio. In: International Conference on Machine Learning, pp. 3370–3378. PMLR (2019)

32. Prenger, R., Valle, R., Catanzaro, B.: WaveGlow: a flow-based generative network for speech synthesis. In: 2019 IEEE International Conference on Acoustics, Speech and Signal Processing (ICASSP), ICASSP 2019, pp. 3617–3621. IEEE (2019)
33. Kumar, M., et al.: VideoFlow: a flow-based generative model for video. arXiv preprint arXiv:1903.01434 2 (2019)
34. Ma, X., Zhou, C., Li, X., Neubig, G., Hovy, E.: FlowSeq: non-autoregressive conditional sequence generation with generative flow. In: Proceedings of the 2019 Conference on Empirical Methods in Natural Language Processing and the 9th International Joint Conference on Natural Language Processing (EMNLP-IJCNLP), pp. 4282–4292 (2019)
35. Kobyzev, I., Prince, S., Brubaker, M.: Normalizing flows: an introduction and review of current methods. IEEE Trans. Pattern Anal. Mach. Intell. **43**, 3964–3979 (2020)
36. Klokov, R., Boyer, E., Verbeek, J.: Discrete point flow networks for efficient point cloud generation. In: Vedaldi, A., Bischof, H., Brox, T., Frahm, J.-M. (eds.) ECCV 2020. LNCS, vol. 12368, pp. 694–710. Springer, Cham (2020). https://doi.org/10.1007/978-3-030-58592-1_41
37. Kim, H., Lee, H., Kang, W.H., Lee, J.Y., Kim, N.S.: SoftFlow: probabilistic framework for normalizing flow on manifolds. In: Advances in Neural Information Processing Systems 33 (2020)
38. Pumarola, A., Popov, S., Moreno-Noguer, F., Ferrari, V.: C-Flow: conditional generative flow models for images and 3d point clouds. In: Proceedings of the IEEE/CVF Conference on Computer Vision and Pattern Recognition, pp. 7949–7958 (2020)
39. Fan, H., Su, H., Guibas, L.J.: A point set generation network for 3d object reconstruction from a single image. In: Proceedings of the IEEE Conference on Computer Vision and Pattern Recognition, pp. 605–613 (2017)
40. Hinton, G.E., Zemel, R.S.: Autoencoders, minimum description length, and Helmholtz free energy. Adv. Neural. Inf. Process. Syst. **6**, 3–10 (1994)
41. Wu, Z., Xiong, Y., Yu, S.X., Lin, D.: Unsupervised feature learning via non-parametric instance discrimination. In: Proceedings of the IEEE Conference on Computer Vision and Pattern Recognition, pp. 3733–3742 (2018)
42. Liu, W., Wen, Y., Yu, Z., Li, M., Raj, B., Song, L.: SphereFace: deep hypersphere embedding for face recognition. In: Proceedings of the IEEE Conference on Computer Vision and Pattern Recognition, pp. 212–220 (2017)
43. Deng, J., Guo, J., Xue, N., Zafeiriou, S.: ArcFace: additive angular margin loss for deep face recognition. In: Proceedings of the IEEE/CVF Conference on Computer Vision and Pattern Recognition, pp. 4690–4699 (2019)
44. Qi, C., Su, F.: Contrastive-center loss for deep neural networks. In: 2017 IEEE International Conference on Image Processing (ICIP), pp. 2851–2855. IEEE (2017)
45. Wu, Z., et al.: 3D ShapeNets: a deep representation for volumetric shapes. In: Proceedings of the IEEE Conference on Computer Vision and Pattern Recognition, pp. 1912–1920 (2015)
46. Uy, M.A., Pham, Q.H., Hua, B.S., Nguyen, T., Yeung, S.K.: Revisiting point cloud classification: a new benchmark dataset and classification model on real-world data. In: Proceedings of the IEEE/CVF International Conference on Computer Vision, pp. 1588–1597 (2019)
47. Dai, A., Chang, A.X., Savva, M., Halber, M., Funkhouser, T., Nießner, M.: ScanNet: richly-annotated 3d reconstructions of indoor scenes. In: Proceedings of the IEEE Conference on Computer Vision and Pattern Recognition, pp. 5828–5839 (2017)

48. Cortes, C., Vapnik, V.: Support-vector networks. Mach. Learn. **20**, 273–297 (1995)
49. Ioffe, S., Szegedy, C.: Batch normalization: accelerating deep network training by reducing internal covariate shift. In: International Conference on Machine Learning, pp. 448–456. PMLR (2015)
50. Girdhar, R., Fouhey, D.F., Rodriguez, M., Gupta, A.: Learning a predictable and generative vector representation for objects. In: Leibe, B., Matas, J., Sebe, N., Welling, M. (eds.) ECCV 2016. LNCS, vol. 9910, pp. 484–499. Springer, Cham (2016). https://doi.org/10.1007/978-3-319-46466-4_29
51. Sharma, A., Grau, O., Fritz, M.: VConv-DAE: deep volumetric shape learning without object labels. In: Hua, G., Jégou, H. (eds.) ECCV 2016. LNCS, vol. 9915, pp. 236–250. Springer, Cham (2016). https://doi.org/10.1007/978-3-319-49409-8_20
52. Liu, S., Giles, L., Ororbia, A.: Learning a hierarchical latent-variable model of 3d shapes. In: 2018 International Conference on 3D Vision (3DV), pp. 542–551. IEEE (2018)
53. Han, Z., Shang, M., Liu, Y.S., Zwicker, M.: View inter-prediction GAN: unsupervised representation learning for 3d shapes by learning global shape memories to support local view predictions. Proc. AAAI Conf. Artif. Intell. **33**, 8376–8384 (2019)

Learning Inter-superpoint Affinity for Weakly Supervised 3D Instance Segmentation

Linghua Tang, Le Hui, and Jin Xie(✉)

Nanjing University of Science and Technology, Nanjing, China
{tanglinghua,le.hui,csjxie}@njust.edu.cn

Abstract. Due to the few annotated labels of 3D point clouds, how to learn discriminative features of point clouds to segment object instances is a challenging problem. In this paper, we propose a simple yet effective 3D instance segmentation framework that can achieve good performance by annotating only one point for each instance. Specifically, to tackle extremely few labels for instance segmentation, we first oversegment the point cloud into superpoints in an unsupervised manner and extend the point-level annotations to the superpoint level. Then, based on the superpoint graph, we propose an inter-superpoint affinity mining module that considers the semantic and spatial relations to adaptively learn inter-superpoint affinity to generate high-quality pseudo labels via semantic-aware random walk. Finally, we propose a volume-aware instance refinement module to segment high-quality instances by applying volume constraints of objects in clustering on the superpoint graph. Extensive experiments on the ScanNet-v2 and S3DIS datasets demonstrate that our method achieves state-of-the-art performance in the weakly supervised point cloud instance segmentation task, and even outperforms some fully supervised methods. Source code is available at https://github.com/fpthink/3D-WSIS.

1 Introduction

Point cloud instance segmentation is a classic task in 3D computer vision, and it can be applied in many fields, including indoor navigation systems, augmented reality, and robotics. The fully supervised instance segmentation methods [2, 12,17] have achieved impressive results, but they rely on numerous manually labeled data. However, annotating a large number of point clouds is extremely time-consuming and expensive. Thus, it is meaningful to segment point clouds in a semi-/weakly supervised manner that requires a small number of annotations. However, how to fully exploit the limited labels to improve the performance of instance segmentation is still a challenging problem.

Few efforts have been dedicated to semi-/weakly supervised point cloud instance segmentation. As a pioneer, Liao *et al.* [18] proposed a semi-supervised

---

Supplementary Information The online version contains supplementary material available at https://doi.org/10.1007/978-3-031-26319-4_11.

L. Wang et al. (Eds.): ACCV 2022, LNCS 13841, pp. 176–192, 2023.
https://doi.org/10.1007/978-3-031-26319-4_11

point cloud instance segmentation method using bounding boxes as supervision, where a network is used to generate bounding box proposals. And instance segmentation is achieved by refining the point cloud within the bounding box proposals. Besides, Tao *et al.* [24] proposed a two-stage seg-level supervision 3D instance and semantic segmentation method, which first leverages a segment grouping network to generate pseudo labels for the whole scenes, and then the generated pseudo point-level labels are used as the ground truth to train the network. However, these simple pseudo label generation strategies cannot effectively generate high-quality pseudo labels, resulting in poor 3D instance segmentation results.

In this paper, we propose a simple yet effective weakly supervised 3D instance segmentation framework, which can achieve impressive results with one point annotation per instance. For weakly supervised point cloud instance segmentation with few annotated labels, our intuition lies in two folds: (1) Under rare annotations, effective label propagation is essential to produce high-quality pseudo labels, especially in 3D instance segmentation. (2) Weakly supervised 3D instance segmentation is more challenging than weakly supervised 3D semantic segmentation, so we consider introducing the object volume constraint to improve the instance segmentation results. Specifically, we first use an unsupervised method [14] to over-segment the point cloud into superpoints and build the superpoint graph. In this way, point-level labels can be extended to superpoint-level labels. Then, we propose an inter-superpoint affinity mining module to generate high-quality pseudo labels based on a few annotated superpoint-level labels. Based on the superpoint graph, we leverage the semantic and spatial information of adjacent superpoints to adaptively learn inter-superpoint affinity, which can be used to propagate superpoint labels along the superpoint graph via semantic-aware random walk. Finally, we propose a volume-aware instance refinement module to improve instance segmentation performance. Based on the trained model using superpoint-level propagation, we can obtain coarse instance segmentation results through superpoint clustering and further infer the object volume information from the instance segmentation results. The object volume information contains the number of voxels and the radius of the object. The inferred object volume information is regarded as the ground truth of the corresponding instance to retrain the network. In the test phase, based on the object volume information, we utilize the predicted object volume information to introduce a volume-aware instance clustering algorithm for segmenting high-quality instances. Extensive experiments on the ScanNet-v2 [6] and S3DIS [1] datasets can demonstrate the effectiveness of our method.

The main contributions of our paper are as follows:

- We present an inter-superpoint affinity mining module that considers the semantic and spatial relation to adaptively learn inter-superpoint affinity for random-walk based label propagation.
- We present a volume-aware instance refinement module, which guides the superpoint clustering on the superpoint graph to segment instances by using the object volume information.
- Our simple yet effective framework achieves state-of-the-art weakly supervised 3D instance segmentation performance on popular datasets ScanNet-v2 and S3DIS.

2 Related Work

2.1 3D Semantic Segmentation

Fully Supervised 3D Semantic Segmentation. Many methods have been proposed to achieve point cloud semantic segmentation. Some methods [11, 15,25] project point clouds into a series of regular 2D images from different views, and then fuse features extracted through 2D convolutional neural networks (CNNs). To apply 3D CNNs on the irregular point cloud and alleviate large memory costs, many efforts [5,8] first voxelize the point cloud into voxels and then utilize the sparse convolutional neural network to extract features of the point cloud. PointNet [20] directly extracts features from points with shared multi-layer perceptrons and max-pooling layer. Inspired by PointNet, different local feature aggregation operators [4,21,26,32] are proposed to work on point cloud, which directly consume point cloud. Besides, various methods [10,30] capture intrinsic spatial and geometric features by constructing the graph on the point cloud. Various approaches exploit different local feature aggregation networks to extract discriminative point features and use multi-layer perceptrons to achieve 3D semantic segmentation.

Semi-/Weakly Supervised 3D Semantic Segmentation. Inspired by class activation map in 2D images, Wei *et al.* [31] introduce a multi-path region mining module to generate pseudo labels, which only requires cloud-level weak labels. Xu *et al.* [33] use three additional losses to constrain on unlabeled points, achieving impressive performance with 10% labels. Cheng *et al.* [3] use a dynamic label propagation strategy to generate pseudo labels, and learn discriminative features with a coupled attention module. Zhang *et al.* [36] exploit the consistency generated by perturbation to obtain additional supervision and propagate implicit labels by constructing the graph topology of the point cloud. Liu *et al.* [19] first build a supervoxel graph on the point cloud and then conduct label propagation by learning the similarity among graph nodes. Li *et al.* [16] utilize a hybrid contrastive regularization strategy with point cloud augmentation to provide additional constraints for network training. To generate pseudo labels for outdoor point cloud scenes, Shi *et al.* [22] design a matching module to propagate pseudo labels in both temporal and spatial spaces.

2.2 3D Instance Segmentation

Fully Supervised 3D Instance Segmentation. Compared with point cloud semantic segmentation, instance segmentation is more challenging because it not only requires predicting semantic scores but also distinguishing instances of the same class. According to the different manners of generating instances, instance segmentation methods can be mainly divided into clustering-based methods and proposal-based methods. Given point clouds as input, clustering-based methods regard instance segmentation as the post-processing task after network inference, and the result is obtained by clustering on point clouds with the predicted features. As a pioneer, Wang *et al.* [28] introduce a similarity matrix to measure the distances between the features of all point pairs, which guides clustering points as proposals. Wang *et al.* [29] integrate semantic and instance segmentation into a parallel framework, which benefits from each other task. Lahoud *et al.* [13] design a multi-task neural network architecture, where instances are simultaneously separated in the feature vector space and direction vector space by a discriminative loss [7] and a directional loss. Jiang *et al.* [12] generate proposals by clustering points on the original and offset-shifted coordinate spaces, which benefits from both advantages. Hou *et al.* [9] jointly learn color and geometry features for instance segmentation from different modalities. Lately, Chen *et al.* [2] introduce a hierarchical aggregation method that iteratively clusters point clouds into instance proposals. Liang *et al.* [17] propose a semantic superpoint tree structure and achieved instance segmentation by tree traversal and splitting. Vu *et al.* [27] design a soft group algorithm to reduce the semantic prediction errors to significantly boost the segmentation performance.

For proposal-based methods, instance segmentation consists of two procedures, first generating rough proposals and then predicting precise instance masks. Yang *et al.* [34] propose an end-to-end trainable network which directly generates 3D bounding boxes as proposals and infers point-wise instance masks for points inside proposals. Instead of getting proposals via 3D bounding box regression, Yi *et al.* [35] introduce an approach to obtain proposals by object generation, and then predict instance masks within proposals.

Semi-/Weakly Supervised 3D Instance Segmentation. Few efforts have been made on semi-/weakly supervised point cloud instance segmentation. Tao *et al.* [24] propose a method to generate pseudo labels for the whole training scene and the generated pseudo point-level labels are used to train existing full supervised methods for point cloud instance segmentation, where one point per instance is clicked as the weak label. Nonetheless, the quality of pseudo labels is limited due to lack of learning discriminative instance features. With bounding boxes as weak labels, Liao *et al.* [18] propose a semi-supervised point cloud instance segmentation method, where a network is leveraged to generate bounding box proposals and instance segmentation is achieved by refining points within bounding box proposals.

3 Method

The overall architecture of our method is depicted in Fig. 1. The backbone network (Sect. 3.1) first takes the point cloud and superpoint graph as input and predicts superpoint-wise semantic labels and offset vectors. Then, the inter-superpoint affinity mining module (Sect. 3.2) propagates labels on the superpoint graph via semantic-aware random walk. Finally, the volume-aware instance refinement module (Sect. 3.3) learns object volume information to improve instance segmentation performance.

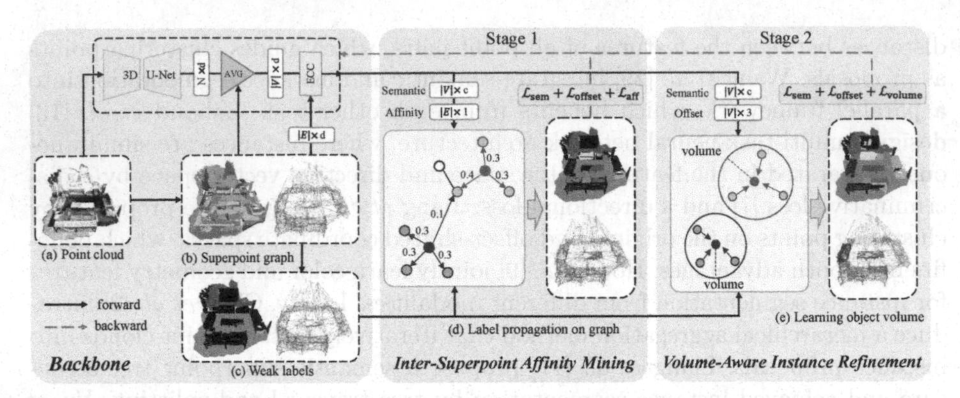

Fig. 1. Overview of our framework for weakly supervised 3D instance segmentation. We first oversegment point cloud (a) to build the superpoint graph (b) and extend weak labels (c) to the corresponding superpoint. Then, based on superpoint graph, the random walk with the predicted affinity and semantic is used for label propagation (d). Finally, combining the predicted semantic and offset, learning pseudo object volume (e) is achieved. c is the number of categories, d is the feature dimension, N is the number of points, $|V|$ is the number of superpoints, and $|E|$ is the number of edges.

3.1 Backbone Network

Superpoint Graph Construction. Following [14,17], we adopt an unsupervised point cloud oversegmentation method to generate the superpoints and construct the superpoint graph. The superpoint graph is a geometry representation of point clouds, defined as $G = (V, E)$. Vertice V means superpoints that are generated by aggregating points with similar geometric characteristics, and edge E indicates the prior connection relationship between adjacency superpoints which are constructed by linking k-nearest superpoints. In weakly supervised 3D instance segmentation, the benefits of the use of superpoints are two-fold. On one hand, the superpoint is a geometrically homogeneous unit, so we can extend annotated point label to the corresponding superpoint, thereby alleviating the sparsity of point-level annotations. On the other hand, the superpoint graph captures the spatial relationship of different instances so that we can utilize it to perform label propagation efficiently.

Superpoint Feature Extraction. Specifically, we first extract point features using 3D U-Net [8] on the point cloud and then aggregate point features into superpoint features by average pooling. After that, based on the superpoint graph, we use the edge-conditioned convolutions (ECC) [23] to extract superpoint features. Finally, we use the superpoint features to predict the semantic label of superpoints.

3.2 Inter-superpoint Affinity Mining

To perform label propagation, we develop an inter-superpoint affinity mining module to learn the superpoint relationship in the semantic and coordinate spaces. By

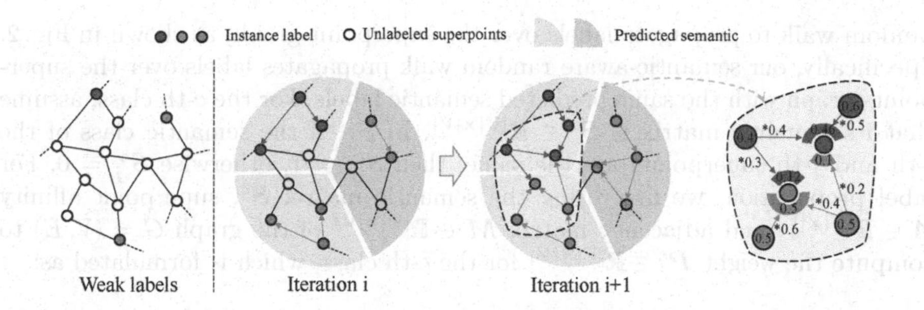

Fig. 2. The process of label propagation with predicted instance affinity and semantics.

using the learned inter-superpoint affinity, we design a simple semantic-aware random walk algorithm for label propagation on the superpoint graph.

Superpoint Affinity Learning. Based on the superpoint graph, we learn the relationship between two adjacent superpoints to characterize their affinity. It is desired that the learned affinity between two adjacent superpoints can guide the label propagation along the edge of the superpoint graph. Assuming the learned superpoint embedding in the backbone network is $\boldsymbol{X} \in \mathbb{R}^{|V| \times d}$, where $|V|$ is the number of superpoints and d is the feature dimension. Given the i-th superpoint embedding $\boldsymbol{X}_i \in \mathbb{R}^d$ and its first-order neighbors $\mathcal{N}_i$, we leverage the semantic and spatial information of the superpoints to adaptively learn inter-superpoint affinity. The affinity A_{ij} between the i-th superpoint and its j-th neighbor is formulated as:

$$A_{ij} = \frac{\exp(\sigma(\phi(\boldsymbol{X}_i), \psi(\boldsymbol{X}_j)) * \gamma(\boldsymbol{p}_i - \boldsymbol{p}_j))}{\sum_{k \in \mathcal{N}_i} \exp(\sigma(\phi(\boldsymbol{X}_i), \psi(\boldsymbol{X}_k)) * \gamma(\boldsymbol{p}_i - \boldsymbol{p}_k))} \tag{1}$$

where $\boldsymbol{X}_i \in \mathbb{R}^d$ and $\boldsymbol{X}_j \in \mathbb{R}^d$ are the superpoint embedding. $\boldsymbol{p}_i \in \mathbb{R}^3$ and $\boldsymbol{p}_j \in \mathbb{R}^3$ are the centroid coordinate of the superpoints. $\phi(\cdot)$ and $\psi(\cdot)$ are linear projections, and $\gamma(\cdot)$ is a multi-layer perceptron. $\sigma(\cdot, \cdot)$ is the dot production for learning the similarity of the i-th and g-th superpoints. In Eq. (1), semantic similarity is measured by dot production while spatial similarity is measured by subtraction. As a result, the affinity A_{ij} considers the semantic and spatial information of the superpoints. After that, we use the learned inter-superpoint affinity to update the superpoint embeddings. For the i-th superpoint, the new superpoint embedding $\widetilde{\boldsymbol{X}}_i \in \mathbb{R}^d$ is written as:

$$\widetilde{\boldsymbol{X}}_i = A_{ij} \cdot \rho(\boldsymbol{X}_j) + \boldsymbol{X}_i \tag{2}$$

where $\rho(\cdot)$ is linear projection. During the training, we employ a discriminative loss (dubbed $\mathcal{L}_{\text{aff}}$) used in [7] to draw $\widetilde{\boldsymbol{X}}$ belonging to the same object towards each other, and make $\widetilde{\boldsymbol{X}}$ in different objects away. It is expected that the affinity A_{ij} between the superpoints of the same instance can be enhanced.

Label Propagation via Semantic-Aware Random Walk. After obtaining the inter-superpoint affinity $\boldsymbol{A} \in \mathbb{R}^{|V| \times |V|}$, we design a simple semantic-aware random walk to propagate labels over the superpoint graph, as shown in Fig. 2. Specifically, our semantic-aware random walk propagates labels over the superpoints graph with the same predicted semantic labels. For the c-th class, assume that its semantic matrix is $\boldsymbol{S}^c \in \mathbb{R}^{|V|\times|V|}$. In $\boldsymbol{S}^c$, if the semantic class of the i-th and j-th superpoints are the same, then $S^c_{ij} = 1$, otherwise $S^c_{ij} = 0$. For label propagation, we first using the semantic matrix $\boldsymbol{S}^c$, superpoint affinity $\boldsymbol{A} \in \mathbb{R}^{|V|\times|V|}$ and adjacency matrix $\boldsymbol{M} \in \mathbb{R}^{|V|\times|V|}$ of the graph $G = (V, E)$ to compute the weight $\boldsymbol{P}^c \in \mathbb{R}^{|V|\times|V|}$ for the c-th class, which is formulated as:

$$\boldsymbol{P}^c = \boldsymbol{M} \odot \boldsymbol{S}^c \odot \boldsymbol{A} \tag{3}$$

where $\odot$ is Hadamard product. Note that the weight $\boldsymbol{P}^c$ considers the semantic information and superpoint affinity simultaneously. Then, we derive the transition probability matrix $\boldsymbol{T}^c \in \mathbb{R}^{|V|\times|V|}$ for the c-th class, and it is defined as:

$$\boldsymbol{T}^c = \boldsymbol{D}^{-1}\boldsymbol{P}^c, \text{ where } D_{ii} = \sum\nolimits_j P^c_{ij} \tag{4}$$

The diagonal matrix $\boldsymbol{D}$ is used for the row normalization of the matrix $\boldsymbol{P}^c$. Finally, the pseudo instance label I_j of the j-th superpoint is propagated by:

$$I_j = I_k, \text{where } k = \underset{i=1,\dots,|V|}{\operatorname{argmax}}\,(\hat{\boldsymbol{T}}^c_{ij}),\ \hat{\boldsymbol{T}}^c = (\boldsymbol{T}^c)^t \tag{5}$$

where $\hat{\boldsymbol{T}}^c_{ij}$ indicates the probability of propagating the instance label of the i-th superpoint to the j-th superpoint, and t is the iteration number. In Eq. (5), the j-th superpoint selects the instance label of the k-th superpoint, which has the highest probability of propagating to the j-th superpoint. In this way, we can propagate the instance label of each annotated superpoint to unlabeled superpoints on the superpoint graph.

3.3 Volume-Aware Instance Refinement

We propose the volume-aware instance refinement module to segment instances by using object volume information. We first introduce how to predict object volume via pseudo instances. Then, we present the volume-aware instance clustering algorithm to generate instances on the superpoint graph.

Object Volume Prediction via Pseudo Instance. In order to predict object volume information, we first train the network with the pseudo labels generated by label propagation in the first stage. As shown in Fig. 1(e), we use the pre-trained model in the first stage to generate the pseudo instances by voting the superpoints to the closest annotated point. Specifically, by adding the predicted offset vector (refer to the first stage in Sect. 3.4) to the corresponding superpoint center, we can shift each superpoint center closer to the center of the corresponding object. To generate instances, the shifted superpoints are assigned the same instance labels as the closest annotated points with the same semantic labels. Here we regard the generated instances as the pseudo instances. According to the generated pseudo instances, we compute its volume information. We consider

Algorithm 1. Volume-Aware Instance Clustering Algorithm.

Input: superpoint shifted coordinate $\{\widetilde{p}_1, \ldots, \widetilde{p}_{|V|}\} \in \mathbb{R}^{|V| \times 3}$; superpoint semantic label $\{s_1, \ldots, s_{|V|}\} \in \mathbb{R}^{|V| \times C}$; the predicted voxel number $\{u_1, \ldots, u_{|V|}\} \in \mathbb{R}^{|V|}$; the predicted radius $\{r_1, \ldots, r_{|V|}\} \in \mathbb{R}^{|V|}$
Output: generated instances $\mathbf{I} \in \{I_1, \ldots, I_m\}$, m is the number of instances.

1: Initialize an array f (visited flag) of length $|V|$ with all zeros
2: Initialize an empty proposal set H, an empty instance set $\mathbf{I}$
3: // Using the predicted radius r to filter superpoints
4: **for** $v = 1$ to $|V|$ **do**
5:   **if** $f_v == 0$ **then**
6:     Initialize an empty queue Q
7:     Initialize a set H
8:     $f_v = 1$; Q.enqueue(v) ; add v to H
9:     **while** Q is not empty **do**
10:       $j = Q$.dequeue()
11:       **for** each $k \in \{k \mid k \in \mathcal{N}_j, s_k == s_j, \|\widetilde{p}_k - \widetilde{p}_j\|_2 < \lambda r_j\}$ **do**
12:         **if** $f_k == 0$ **then**
13:           $f_k = 1$; Q.enqueue(k) ; add k to H
14:     add H to $\mathbf{H}$
15: // Using the predicted voxel numbers u to filter proposals
16: **for** each $H \in \mathbf{H}$ **do**
17:   compute $\bar{w} = \text{avg}(\{u_i \mid i \in H\})$, w for H
18:   **if** $w > \beta\bar{w}$ **then**
19:     add H to $\mathbf{I}$
20: **for** each $H \in \mathbf{H}$ **do**
21:   compute $\bar{w} = \text{avg}(\{u_i \mid i \in H\})$, w for H
22:   **if** $w \leq \beta\bar{w}$ **then**
23:     $I_{closest}$ = findClosestInstance($\{I \mid I \in \mathbf{I}, s_I == s_H\}$)
24:     $I_{closest} = I_{closest} \cup H$
25: **return I**

the number of voxels inside the instance and the instance radius to measure the volume of the instance. The instance radius is defined as the distance between the instance center and the farthest point. Thus, for each pseudo instance, we can obtain its volume information after the first stage.

Volume-Aware Instance Clustering. After predicting the volume of the object in the first stage, we additionally use the predicted volume as the supervision to retrain the network (refer to the second stage in Sect. 3.4), which is regarded as the second stage. Thus, in the second stage, we can additionally predict the instance volume information (the number of voxels and the radius) for each superpoint. For the i-th superpoint, assume that the predicted semantic is $s_i \in \mathbb{R}^{1 \times C}$, offset vector is $o_i \in \mathbb{R}^3$, the number of voxels is u_i, and the radius is r_i. Note that the predicted semantic s_i is the one hot label. We first obtain the shifted coordinate of the superpoint by $\widetilde{\boldsymbol{p}}_i = \boldsymbol{p}_i + \boldsymbol{o}_i$, which makes the superpoint close to the corresponding instance center. Based on the shifted coordinate and the graph structure, the i-th superpoint merges its neighbors $\{j \mid j \in \mathcal{N}_i, s_i = s_j, \|\widetilde{\boldsymbol{p}}_i - \widetilde{\boldsymbol{p}}_j\|_2 < \lambda r_i\}$ into the same cluster, where the hyperparameter λ is set to 0.25 empirically. Note that the radius r_i is used to filter the superpoints far from the object center. Here, we use the breadth-first search on the superpoint graph to group nodes in the same cluster for generating compact instance proposals. After that, we further count the number of voxels w in the proposal to filter the fragmented proposals. The predicted number of voxels $\bar{w}$ of the proposal is computed by averaging the predicted voxel numbers of superpoints within the proposal. If $w > \beta\bar{w}$, the corresponding proposal can be regarded as the instance. The hyperparameter β is empirically set to 0.3. Finally, the remaining proposals are aggregated to the closest instance with the same semantic label. The volume-aware instance clustering algorithm is shown in Algorithm 1.

3.4 Network Training

Our method is a two-stage framework. As shown in Fig. 1, the first stage learns the inter-superpoint affinity to propagate labels via random walk, while the second stage leverage the object volume information to refine the instance.

First Stage. As shown in Fig. 1, the first stage is supervised by the semantic loss $\mathcal{L}_{\text{sem}}$, offset loss $\mathcal{L}_{\text{offset}}$, and affinity loss $\mathcal{L}_{\text{aff}}$. The semantic loss $\mathcal{L}_{\text{sem}}$ is defined as the cross-entropy loss:

$$\mathcal{L}_{\text{sem}} = \frac{1}{\sum_{i=1}^{|V|} \mathbb{I}(v_i)} \sum_{i=1}^{|V|} \text{CE}(s_i, s_i^*) \cdot \mathbb{I}(v_i) \tag{6}$$

where s_i is the predicted label and s_i^* is the ground truth label. Note that the original annotated labels and generated pseudo labels are all regarded as the ground truth labels. If the superpoint v_i has the label or is assigned with the pseudo label, the indicator function $\mathbb{I}(v_i)$ is equal to 1, otherwise 0. In addition, we use MLP to predict the offset vector $\boldsymbol{o}_i \in \mathbb{R}^3$. the offset loss $\mathcal{L}_{\text{offset}}$ is used to minimize the predicted offset of the superpoint to its instance center. $\mathcal{L}_{\text{offset}}$ is defined as:

$$\mathcal{L}_{\text{offset}} = \frac{1}{\sum_{i=1}^{|V|} \mathbb{I}(v_i)} \sum_{i=1}^{|V|} \|\boldsymbol{o}_i - \boldsymbol{o}_i^*\|_1 \cdot \mathbb{I}(v_i) \tag{7}$$

where the $\boldsymbol{o}_i$ is the predicted superpoint offset and the $\boldsymbol{o}_i^*$ is the ground truth offset. Note the $\boldsymbol{o}_i^*$ is computed by coarse pseudo instance labels. Following [7], the affinity loss $\mathcal{L}_{\text{aff}}$ (refer to Sect. 3.2) is formulated as:

$$\mathcal{L}_{\text{var}} = \frac{1}{I}\sum_{i=1}^{I}\frac{1}{\sum_{j=1}^{|V|}\mathbb{I}(v_j,i)}\sum_{j=1}^{|V|}\left[\|\boldsymbol{\mu}_i - \widetilde{\boldsymbol{X}}_j\|_2 - \delta_v\right]_+^2 \cdot \mathbb{I}(v_j,i)$$

$$\mathcal{L}_{\text{dist}} = \frac{1}{I(I-1)}\sum_{i_A=1}^{I}\sum_{\substack{i_B=1 \\ i_A \neq i_B}}^{I}[2\delta_d - \|\boldsymbol{\mu}_{i_A} - \boldsymbol{\mu}_{i_B}\|_2]_+^2 \tag{8}$$

$$\mathcal{L}_{\text{aff}} = \mathcal{L}_{\text{var}} + \mathcal{L}_{\text{dist}} + \alpha \cdot \mathcal{L}_{\text{reg}}, \text{where } \mathcal{L}_{\text{reg}} = \frac{1}{I}\sum_{i=1}^{I}\|\boldsymbol{\mu}_i\|_2$$

where I is the number of instances (equal to the number of the annotated points, *i.e*, one point per instance). $\boldsymbol{\mu}_i$ is the mean embedding of the i-th instance and $\widetilde{\boldsymbol{X}}_j$ is the embedding of the j-th superpoint in Eq. (2). According to [7], the margins δ_v and δ_d are set to 0.1 and 1.5, respectively. The parameter α is set to 0.001, and $[x]_+ = max(0, x)$ denotes the hinge. $\mathbb{I}(v_j, i)$ is the indicator function, and $\mathbb{I}(v_j, i)$ equals to 1 if superpoint v_j is labeled as the i-th instance. Note that we only perform $\mathcal{L}_{\text{aff}}$ on the superpoints with annotated labels or pseudo labels. The final loss function in the first stage is defined as:

$$\mathcal{L}_{\text{stage1}} = \mathcal{L}_{\text{sem}} + \mathcal{L}_{\text{offset}} + \mathcal{L}_{\text{aff}} \tag{9}$$

Second Stage. As shown in Fig. 1, the second stage is supervised by the semantic loss $\mathcal{L}_{\text{sem}}$, offset loss $\mathcal{L}_{\text{offset}}$, and volume loss $\mathcal{L}_{\text{volume}}$. As the affinity loss is used for label propagation, we remove it in the second stage. For the volume loss $\mathcal{L}_{\text{volume}}$, it uses the predicted object volume information as the ground truth to train the network (refer to Sect. 3.3). The $\mathcal{L}_{\text{volume}}$ is formulated as:

$$\mathcal{L}_{\text{volume}} = \frac{1}{K}\sum_{i=1}^{K}\sum_{j=1}^{I}(\|u_i - \hat{u}_j\|_1 + \|r_i - \hat{r}_j\|_1) \cdot \mathbb{I}(i,j) \tag{10}$$

where K is the number of labeled superpoints, including the original annotated labels and the generated pseudo labels. If the i-th superpoint belongs to the j-th instance, the indicator function $\mathbb{I}(i, j)$ is equal to 1, otherwise 0. $\hat{u}_j$ and $\hat{r}_j$ indicate the ground truth voxel numbers and radius counted from the pseudo instances, respectively. The generation of the pseudo instances refers to Sect. 3.3. The final loss function in the second stage is defined as:

$$\mathcal{L}_{\text{stage2}} = \mathcal{L}_{\text{sem}} + \mathcal{L}_{\text{offset}} + \mathcal{L}_{\text{volume}} \tag{11}$$

4 Experiments

4.1 Experimental Settings

Datasets. ScanNet-v2 [6] and S3DIS [1] are used in our experiments to conduct 3D instance segmentation. ScanNet-v2 contains 1,613 indoor RGB-D scans with dense semantic and instance annotations. The dataset is split into 1,201 training scenes, 312 validation scenes, and 100 hidden test scenes. The instance segmentation is evaluated on 18 object categories. S3DIS contains 6 large-scale indoor

Table 1. 3D instance segmentation results on the ScanNet-v2 validation set and online test set. "Baseline" means the model trained with the initial annotated labels only.

Method		Annotation	AP	AP_{50}	AP_{25}	AP	AP_{50}	AP_{25}
			Validation set			Online test set		
Fully Sup.	SGPN [28]	100%	–	11.3	22.2	4.9	14.3	39.0
	3D-SIS [9]	100%	–	18.7	35.7	16.1	38.2	55.8
	MTML [13]	100%	20.3	40.2	55.4	28.2	54.9	73.1
	PointGroup [12]	100%	34.8	56.9	71.3	40.7	63.6	77.8
	HAIS [2]	100%	43.5	64.1	75.6	45.7	69.9	80.3
	SSTNet [17]	100%	**49.4**	64.3	74.0	**50.6**	69.8	78.9
	SoftGroup [27]	100%	46.0	**67.6**	**78.9**	50.4	**76.1**	**86.5**
Weakly Sup.	SPIB [18]	0.16%	–	38.6	61.4	–	–	63.4
	SegGroup [24]	0.02%	23.4	43.4	62.9	24.6	44.5	63.7
	Baseline	0.02%	21.2	39.0	61.3	–	–	–
	3D-WSIS (ours)	0.02%	**28.1**	**47.2**	**67.5**	**25.1**	**47.0**	**67.8**

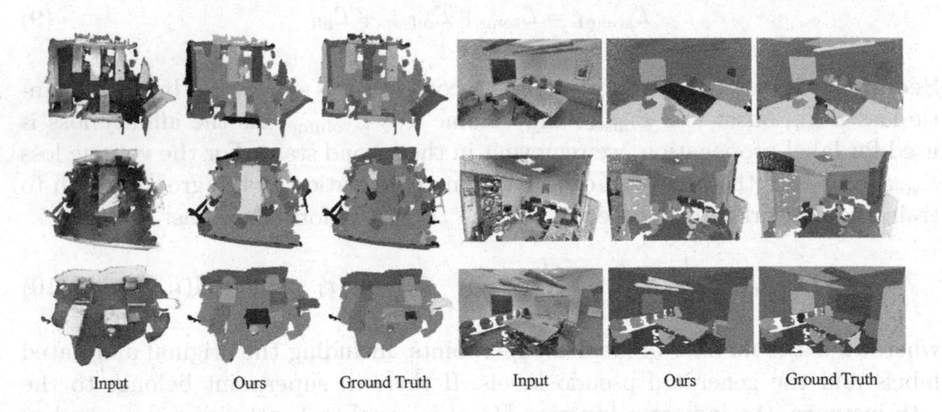

Fig. 3. Visualization of the 3D instance segmentation results on the validation of ScanNet-v2 (left) and S3DIS (right). We randomly select colors for different instances.

areas, which has 272 rooms and 13 categories. For the ScanNet-v2 dataset, we report both validation and online test results. For the S3DIS dataset, we report both Area 5 and the 6-fold cross validation results.

Evaluation Metrics. For the ScanNet-v2 dataset, the mean average precision at the overlap 0.25 (AP_{25}), 0.5 (AP_{50}) and overlaps from 0.5 to 0.95 (AP) are reported. For the S3DIS dataset, we additionally use mean coverage (mCov), mean weighted coverage (mWCov), mean precision (mPrec), and mean recall (mRec) with the IoU threshold of 0.5 as evaluation metrics.

Annotation of Weak Labels. To generate weak labels of point clouds, we randomly click one point of each instance as the ground truth label. Note that our

Table 2. 3D instance segmentation results on the S3DIS dataset. "Baseline" means the model trained with the initial annotated labels only.

Method		Annotation	AP	AP_{50}	AP_{25}	mCov	mWCov	mPrec	mRec
6-fold cross validation									
Fully sup.	SGPN [28]	100%	–	–	–	37.9	40.8	38.2	31.2
	ASIS [29]	100%	–	–	–	51.2	55.1	63.6	47.5
	PointGroup [12]	100%	–	64.0	–	–	–	69.6	69.2
	HAIS [2]	100%	–	–	–	67.0	70.4	73.2	69.4
	SSTNet [17]	100%	54.1	67.8	–	–	–	73.5	**73.4**
	SoftGroup [27]	100%	**54.4**	**68.9**	–	**69.3**	**71.7**	**75.3**	69.8
Weakly Sup.	Baseline	0.02%	19.5	30.5	42.0	41.1	42.3	13.3	37.1
	SegGroup	0.02%	23.1	37.6	48.5	45.5	47.6	56.7	43.3
	3D-WSIS (**ours**)	0.02%	**26.7**	**40.4**	**52.6**	**48.0**	**50.5**	**59.3**	**46.7**
Area 5									
Fully sup.	SGPN [28]	100%	–	–	–	32.7	35.5	36.0	28.7
	ASIS [29]	100%	–	–	–	44.6	47.8	55.3	42.4
	PointGroup [12]	100%	–	57.8	–	–	–	61.9	62.1
	HAIS [2]	100%	–	–	–	64.3	66.0	71.1	65.0
	SSTNet [17]	100%	42.7	59.3	–	–	–	65.5	64.2
	SoftGroup [27]	100%	**51.6**	**66.1**	–	**66.1**	**68.0**	**73.6**	**66.6**
Weakly Sup.	Baseline	0.02%	18.9	26.8	37.7	36.3	37.1	11.1	28.5
	SegGroup	0.02%	21.0	29.8	41.9	39.1	40.8	47.2	34.9
	3D-WSIS (**ours**)	0.02%	**23.3**	**33.0**	**48.3**	**42.2**	**44.2**	**50.8**	**38.9**

annotation strategy is the same as SegGroup [24]. Unlike our method and SegGroup, SPIB [18] adopts 3D box-level annotation, which annotates each instance with bounding box. Compared with time-consuming box-level annotation, clicking one point per instance is faster and more convenient.

4.2 Results

ScanNet-v2. Table 1 reports the quantitative results on the ScanNet-v2 validation set and hidden testing set. Compared with the existing semi-/weakly supervised point cloud instance segmentation methods, our approach achieves state-of-the-art performance and improves the AP_{25} from 62.9% to 67.5% with a gain of about 5% on the ScanNet-v2 validation set. Note that SPIB [18] uses the bounding box of each instance as weak labels, which is different from ours. Although we have the same number of annotated instances, clicking one point per instance provides less information than eight corners of the box. Nonetheless, our method can still achieve higher performance than SPIB.

S3DIS. Table 2 reports the results on Area 5 and 6-fold cross validation of the S3DIS dataset. Compared with the fully supervised 3D instance segmentation methods, our model can still achieve good results, even outperforming the fully

Table 3. The ablation study of different components on the ScanNet-v2 validation set and S3DIS Area 5. "Baseline" means the model trained the initial annotated labels only. Note that "Stage 2" is performed based on "Stage 1".

		ScanNet-v2 Val			S3DIS Area 5						
Settings		AP	AP_{50}	AP_{25}	AP	AP_{50}	AP_{25}	mCov	mWCov	mPrec	mRec
Baseline		21.2	39.0	61.3	18.9	26.8	37.7	36.3	37.1	11.1	28.5
Stage 1	Iter. 1	23.4	42.2	62.8	19.9	27.7	40.2	38.5	39.4	21.7	33.1
	Iter. 2	24.5	43.6	64.4	20.1	28.0	40.7	39.7	40.4	22.2	33.8
	Iter. 3	25.4	45.3	65.8	20.9	28.3	41.9	40.0	40.8	23.1	34.1
Stage 2		**28.1**	**47.2**	**67.5**	**23.3**	**33.0**	**48.3**	**42.2**	**44.2**	**50.8**	**38.9**

Table 4. The ablation study results (proportion/accuracy) of pseudo labels at different iterations on the ScanNet-v2 training set. The proportion and accuracy of pseudo labels are computed at the point level.

Stage 1	Only Random	Random+Affinity	Random+Affinity+Semantic
Iter. 1	33.7/39.9	29.1/52.7	18.2/81.9
Iter. 2	47.7/38.4	35.1/71.6	30.9/82.1
Iter. 3	48.2/38.3	35.2/73.1	31.4/82.5

supervised methods such as SGPN [28]. The quantitative results demonstrate the effectiveness of our method on weakly supervised 3D instance segmentation.

Visualization Results. Figure 3 shows the visualization results of instance segmentation on the ScanNet-v2 validation set and S3DIS. It can be found that although some objects of the same class are close to each other, like chairs, the instances are still segmented properly. Since we additionally predict the size of the objects in the network, our method can effectively use the size information of the objects to guide the clustering, thereby segmenting different instances that are close to each other.

4.3 Ablation Study

Effect of Pseudo Labels. We conduct experiments on the ScanNet-v2 validation set to verify the effectiveness of our label propagation. The quantitative results are reported in Table 3. "Baseline" indicates that our method trained with initial annotated labels, without pseudo labels. In the first stage, we report the mean average precision at different iterations, respectively. It can be observed that the performance gradually increases as the number of iterations increases. Furthermore, based on the stage one (dubbed "Stage 1"), it can be found that the performance is greatly improved after training in the second stage (dubbed "Stage 2"). Since the number of generated pseudo labels in the first stage is still less than that of fully annotated labels, it is difficult to effectively segment

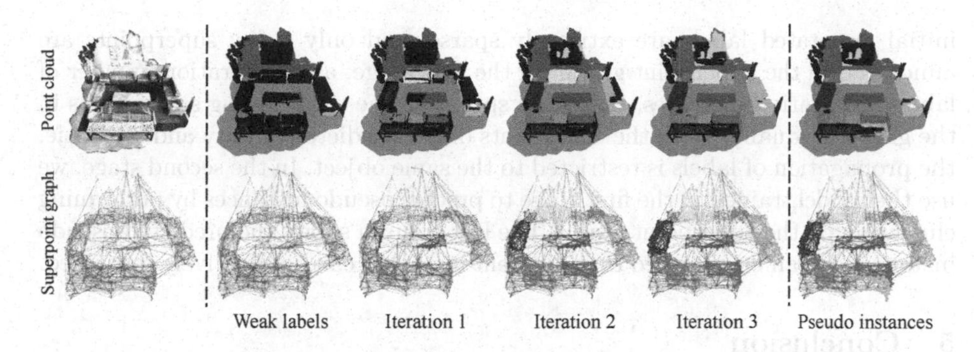

Fig. 4. Visualization results of pseudo label generation. Note that we remove the superpoints on the walls and floor for a better view.

instances. In the second stage, we use the model trained in the first stage to cluster pseudo instances, so we can regard the obtained object volume as the additional supervision to train the network. The quantitative results in the second stage further demonstrate that using the predicted object volume can indeed improve the performance of weakly supervised 3D instance segmentation.

Quality of Pseudo Labels. The pseudo labels are generated on the training set to increase supervision during training, so their quality affects network training. To further study the quality of the generated pseudo labels, we count the proportion and accuracy of pseudo labels on the ScanNet-v2 training set during training. The results are listed in Table 4. When only using random walk (dubbed "Only Random"), the proportion of pseudo labels is high, but the accuracy is low (39.9% at "Iter. 1"). The low-accuracy pseudo labels will affect the training of the network. If we add the extra affinity constraint (dubbed "Random+Affinity"), we can observe that the proportion of pseudo labels is lower, but the accuracy is greatly improved (52.7% at "Iter. 1"). Due to the affinity constraint, the proportion of wrong label propagation is reduced. Therefore, the quality of pseudo labels is improved and high-quality supervision is provided for network training. Furthermore, when we add the semantic constraints (dubbed "Random+Affinity+Semantic"), the accuracy of pseudo labels improves from 52.7% ("Random+Affinity") to 81.9% ("Random+Affinity+Semantic"), which shows that the semantic constraint is useful for the weakly supervised 3D instance segmentation task. As constraints are added gradually, the proportion of the generated pseudo labels decreases, while the accuracy increases.

Label Propagation Times. Different label propagation times influence the quality of pseudo labels. As shown in Table 4, the proportion and accuracy of pseudo labels at three iterations (dubbed "Iter. 3") is comparable to two iterations (dubbed "Iter. 2"), and performing more iterations consumes more resources, thus we choose three iterations for label propagation.

Visualization of Pseudo Labels. Figure 4 shows the pseudo labels at different iterations in the first stage on the superpoint graph. We can observe that the initial annotated labels are extremely sparse, and only a few superpoints are annotated in the superpoint graph. In the first stage, as the iteration number of label propagation increases, the labels spread to the surrounding superpoints in the graph gradually. With the constraints of the predicted affinity and semantic, the propagation of labels is restricted to the same object. In the second stage, we use the model trained in the first stage to predict pseudo instances by performing clustering on the superpoint graph. The last column shows the predicted pseudo instances. It can be observed that different instances can be effectively separated.

5 Conclusion

In this paper, we proposed a simple yet effective method for weakly supervised 3D instance segmentation with extremely few labels. To exploit few point-level annotations, we used an unsupervised point cloud oversegmentation method on the point cloud to generate superpoints and construct the superpoint graph. Based on the constructed superpoint graph, we developed an inter-superpoint affinity mining module to adaptively learn inter-superpoint affinity for label propagation via random walk. We further developed a volume-aware instance refinement module to guide the superpoint clustering on the superpoint graph by learning the object volume information. Experiments on the ScanNet-v2 and S3DIS datasets demonstrate that our method achieves state-of-the-art performance on weakly supervised 3D instance segmentation.

Acknowledgments. The authors would like to thank reviewers for their detailed comments and instructive suggestions. This work was supported by the National Science Fund of China (Grant No. 61876084).

References

1. Armeni, I., et al.: 3D semantic parsing of large-scale indoor spaces. In: CVPR (2016)
2. Chen, S., Fang, J., Zhang, Q., Liu, W., Wang, X.: Hierarchical aggregation for 3D instance segmentation. In: ICCV (2021)
3. Cheng, M., Hui, L., Xie, J., Yang, J.: SSPC-Net: semi-supervised semantic 3D point cloud segmentation network. In: AAAI (2021)
4. Cheng, M., Hui, L., Xie, J., Yang, J., Kong, H.: Cascaded non-local neural network for point cloud semantic segmentation. In: IROS (2020)
5. Choy, C., Gwak, J., Savarese, S.: 4D spatio-temporal ConvNets: Minkowski convolutional neural networks. In: CVPR (2019)
6. Dai, A., Chang, A.X., Savva, M., Halber, M., Funkhouser, T., Nießner, M.: ScanNet: richly-annotated 3D reconstructions of indoor scenes. In: CVPR (2017)
7. De Brabandere, B., Neven, D., Van Gool, L.: Semantic instance segmentation with a discriminative loss function. arXiv (2017)
8. Graham, B., Engelcke, M., Van Der Maaten, L.: 3D semantic segmentation with submanifold sparse convolutional networks. In: CVPR (2018)
9. Hou, J., Dai, A., Nießner, M.: 3D-SIS: 3D semantic instance segmentation of RGB-D scans. In: CVPR (2019)

10. Hui, L., Yuan, J., Cheng, M., Xie, J., Zhang, X., Yang, J.: Superpoint network for point cloud oversegmentation. In: ICCV (2021)
11. Jaritz, M., Gu, J., Su, H.: Multi-view PointNet for 3D scene understanding. In: ICCVW (2019)
12. Jiang, L., Zhao, H., Shi, S., Liu, S., Fu, C.W., Jia, J.: PointGroup: dual-set point grouping for 3D instance segmentation. In: CVPR (2020)
13. Lahoud, J., Ghanem, B., Pollefeys, M., Oswald, M.R.: 3D instance segmentation via multi-task metric learning. In: ICCV (2019)
14. Landrieu, L., Simonovsky, M.: Large-scale point cloud semantic segmentation with superpoint graphs. In: CVPR (2018)
15. Lawin, F.J., Danelljan, M., Tosteberg, P., Bhat, G., Khan, F.S., Felsberg, M.: Deep projective 3d semantic segmentation. In: Felsberg, M., Heyden, A., Krüger, N. (eds.) CAIP 2017. LNCS, vol. 10424, pp. 95–107. Springer, Cham (2017). https://doi.org/10.1007/978-3-319-64689-3_8
16. Li, M., et al.: HybridCR: weakly-supervised 3D point cloud semantic segmentation via hybrid contrastive regularization. In: CVPR (2022)
17. Liang, Z., Li, Z., Xu, S., Tan, M., Jia, K.: Instance segmentation in 3D scenes using semantic superpoint tree networks. In: ICCV (2021)
18. Liao, Y., Zhu, H., Zhang, Y., Ye, C., Chen, T., Fan, J.: Point cloud instance segmentation with semi-supervised bounding-box mining. TPAMI **44**, 10159–10170 (2021)
19. Liu, Z., Qi, X., Fu, C.W.: One thing one click: a self-training approach for weakly supervised 3D semantic segmentation. In: CVPR (2021)
20. Qi, C.R., Su, H., Mo, K., Guibas, L.J.: PointNet: deep learning on point sets for 3D classification and segmentation. In: CVPR (2017)
21. Qi, C.R., Yi, L., Su, H., Guibas, L.J.: PointNet++: deep hierarchical feature learning on point sets in a metric space (2017)
22. Shi, H., Wei, J., Li, R., Liu, F., Lin, G.: Weakly supervised segmentation on outdoor 4D point clouds with temporal matching and spatial graph propagation. In: ICCV (2022)
23. Simonovsky, M., Komodakis, N.: Dynamic edge-conditioned filters in convolutional neural networks on graphs. In: CVPR (2017)
24. Tao, A., Duan, Y., Wei, Y., Lu, J., Zhou, J.: SegGroup: seg-level supervision for 3D instance and semantic segmentation. TIP **31**, 4952–4965 (2022)
25. Tatarchenko, M., Park, J., Koltun, V., Zhou, Q.Y.: Tangent convolutions for dense prediction in 3D. In: CVPR (2018)
26. Thomas, H., Qi, C.R., Deschaud, J.E., Marcotegui, B., Goulette, F., Guibas, L.J.: KPConv: flexible and deformable convolution for point clouds. In: ICCV (2019)
27. Vu, T., Kim, K., Luu, T.M., Nguyen, X.T., Yoo, C.D.: SoftGroup for 3D instance segmentation on 3D point clouds. In: CVPR (2022)
28. Wang, W., Yu, R., Huang, Q., Neumann, U.: SGPN: similarity group proposal network for 3D point cloud instance segmentation. In: CVPR (2018)
29. Wang, X., Liu, S., Shen, X., Shen, C., Jia, J.: Associatively segmenting instances and semantics in point clouds. In: CVPR (2019)
30. Wang, Y., Sun, Y., Liu, Z., Sarma, S.E., Bronstein, M.M., Solomon, J.M.: Dynamic graph CNN for learning on point clouds. TOG (2019)
31. Wei, J., Lin, G., Yap, K.H., Hung, T.Y., Xie, L.: Multi-path region mining for weakly supervised 3D semantic segmentation on point clouds. In: CVPR (2020)
32. Wu, W., Qi, Z., Fuxin, L.: PointConv: deep convolutional networks on 3D point clouds. In: CVPR (2019)

33. Xu, X., Lee, G.H.: Weakly supervised semantic point cloud segmentation: towards 10x fewer labels. In: CVPR (2020)
34. Yang, B., et al.: Learning object bounding boxes for 3D instance segmentation on point clouds. In: NeurIPS (2019)
35. Yi, L., Zhao, W., Wang, H., Sung, M., Guibas, L.J.: GSPN: generative shape proposal network for 3D instance segmentation in point cloud. In: CVPR (2019)
36. Zhang, Y., Qu, Y., Xie, Y., Li, Z., Zheng, S., Li, C.: Perturbed self-distillation: weakly supervised large-scale point cloud semantic segmentation. In: ICCV (2021)

Cross-View Self-fusion for Self-supervised 3D Human Pose Estimation in the Wild

Hyun-Woo Kim[1], Gun-Hee Lee[2], Myeong-Seok Oh[2], and Seong-Whan Lee[1,2](✉)

[1] Department of Artificial Intelligence, Korea University, Seoul, Korea
{kim_hyun_woo,sw.lee}@korea.ac.kr

[2] Department of Computer Science and Engineering, Korea University, Seoul, Korea
{gunhlee,ms_oh}@korea.ac.kr

Abstract. Human pose estimation methods have recently shown remarkable results with supervised learning that requires large amounts of labeled training data. However, such training data for various human activities does not exist since 3D annotations are acquired with traditional motion capture systems that usually require a controlled indoor environment. To address this issue, we propose a self-supervised approach that learns a monocular 3D human pose estimator from unlabeled multi-view images by using multi-view consistency constraints. Furthermore, we refine inaccurate 2D poses, which adversely affect 3D pose predictions, using the property of canonical space without relying on camera calibration. Since we do not require camera calibrations to leverage the multi-view information, we can train a network from in-the-wild environments. The key idea is to fuse the 2D observations across views and combine predictions from the observations to satisfy the multi-view consistency during training. We outperform state-of-the-art methods in self-supervised learning on the two benchmark datasets Human3.6M and MPI-INF-3DHP as well as on the in-the-wild dataset SkiPose. Code and models are available at https://github.com/anonyAcc/CVSF_for_3DHPE.

1 Introduction

Human Pose Estimation (HPE) is widely used in various AI applications such as video analysis, AR/VR, human action recognition, and 3D human reconstruction [1–8]. Owing to the variety of applicability, HPE has received considerable attention in computer vision. Recent methods for 3D HPE have achieved remarkable results in a supervised setting, but they require large amounts of labeled training data. Collecting such datasets is expensive, time-consuming, and mostly limited to fully controlled indoor settings that require a multi-camera motion capture system. Therefore, self-supervised 3D HPE, which does not require 3D annotation, has become an emerging trend in this field.

In this study, we propose a novel self-supervised training procedure that does not require camera calibrations (including camera intrinsic and extrinsic parameters) and any annotations in the multi-view training dataset. Specifically, our model requires at least two temporally synchronized cameras to observe a

L. Wang et al. (Eds.): ACCV 2022, LNCS 13841, pp. 193–210, 2023.
https://doi.org/10.1007/978-3-031-26319-4_12

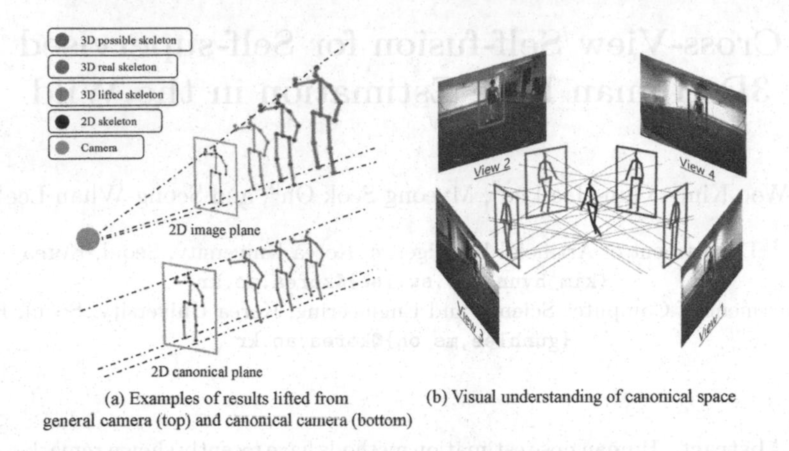

Fig. 1. (a) Examples of results predicted from general and canonical camera. Depth-scale ambiguity in general camera results with numerous 3D candidates of varying scales. In canonical camera, 2D pose is lifted to 3D poses of the same scale irrespective of depth. Red and blue skeletons represent estimated 3D skeletons at varying scales. Scale of green skeletons is the same as that of canonical cameras. (b) Visual understanding of a canonical space. In the space, the scales of all observed human poses are equal, and the relationship between cameras is relative. (Color figure online)

person of interest from different orientations, but no further knowledge regarding the scene and camera parameters is required. Note that multi-view images are used to train a network, but only a single image is used at inference time.

There are only a few comparable methods [9–12] that apply to our self-supervised setting. They require additional knowledge about the observed person such as bone length constraints [11] and 3D human structures [10], or traditional computer vision algorithms to obtain a pseudo ground truth pose [9]. On the other hand, CanonPose [12] learns a monocular 3D human pose estimator using the multi-view images without any prior information. However, the research does not address the 2D pose errors caused by a pretrained 2D pose estimator, which remains fixed during the training. The 2D pose errors not only propagate to the 3D prediction, but also may affect the multi-view consistency requirement during training, which can yield an inaccurate camera rotation estimation.

To address this issue, we refine the 2D pose errors influencing the 3D prediction and then lift the refined 2D pose to a 3D pose. In principle, we train a neural network by satisfying the multi-view consistency between the 2D poses through refining the incorrect 2D pose, as well as the multi-view consistency between the 3D outputs and 2D inputs. However, it is necessary to know the multi-view relationship to refine an incorrect 2D pose in a multi-view setting. The multi-view relationship can be represented using the parameters of each camera. We assume a training setting in which the camera parameters are not given. Also, estimating the camera parameters of each camera is complex and computationally intensive. Therefore, we deploy a canonical form [13] that fixes one camera and represents the remainder with the relative camera parameters based on the fixed camera. According to this form, the relationship between the cameras can be represented as a relative rotation and translation.

On the other hand, every camera position is different for a particular 3D target, so all the scales of the observed 2D poses are different in each view due to perspective. Therefore, there exists an infinite number of 3D poses with multiple scales corresponding to a given 2D pose due to the depth-scale ambiguity, which is illustrated in Fig. 1(a). To address the ambiguity, we transform the 2D poses into a canonical space by normalizing the position and scale of the 2D poses observed with different scales in all views. The transformation allows the distances between the 3D target and each camera to be the same, so we don't need to consider the relative translation. In other words, the relationship between the cameras can be represented only by relative rotation, and all lifted 3D poses bear the same scale. Figure 1(b) illustrates the canonical space in which it has the same scale for all transformed 2D poses and a lifted 3D pose satisfies the multi-view consistency.

The flow of our approach is as follows: First, we transform the estimated 2D pose in an image plane coordinate system into a canonical plane coordinate system. Second, we input the transformed 2D pose into a lifting network. Then, the network predicts a 3D pose in the canonical coordinate system and a camera rotation to rotate the pose to the canonical camera coordinate system. Third, the proposed cross-view self-fusion module takes the 2D poses along with the camera rotations predicted by the lifting network as input. Subsequently, it refines incorrect 2D poses by fusing all the 2D poses with the predicted rotations. Lastly, the refined 2D poses are lifted to 3D poses by the lifting network.

We evaluate our approach on two multi-view 3D human pose estimation datasets, namely Human3.6M [14] and MPI-INF-3DHP [15], and achieve the new state-of-the-art in several metrics for self-supervised manner. Additionally, we present the results for the SkiPose dataset that represents all the challenges arising from outdoor human activities, which can be hard to perform in the limited setting of traditional motion capture systems.

The contributions of our research can be summarized as follows:

- We propose a Cross-view Self-fusion module that refines an incorrect 2D pose using multi-view data without camera calibration. This can be performed in any in-the-wild setting as it does not require camera calibration.
- We improve a self-supervised algorithm to lift a 2D pose to a 3D pose by refining poses across views. Refinement enhances multi-view consistency, and the enhanced consistency enables more accurate refinement.
- We achieve state-of-the-art performance on 3D human pose estimation benchmarks in a self-supervised setting.

2 Related Work

Full Supervision. Recent supervised approaches depend primarily on large datasets with 3D annotations. These approaches can be classified into two categories: image-based and lifting-based 3D human pose estimations. The image-based approaches [16–25] directly estimate the 3D joint locations from images or video frames. Although these approaches generally deliver exceptional performance on similar images, their ability to generalize to other scenes is restricted. In this regard, certain studies [30–33] have attempted to resolve this problem using data augmentation. The lifting-based approaches [34–43] leverage 2D poses from input images or video frames to lift them to the corresponding 3D poses, which is more popular among the state-of-the-art methods in this domain. Martinez et al. [34] showed the prospect of using only 2D joint information for 3D human pose estimation by proposing "a simple and effective baseline for 3D human pose estimation", which uses only 2D information but achieves highly accurate results. Owing to its simplicity, it serves as a baseline for several future studies. However, the main disadvantage of all full-supervised approaches is that they are not appropriately generalized for the unseen poses. Therefore, their application is substantially limited to new environments or in-the-wild scenes.

Self-supervision with Multi-view. Recently, the research interest in self-supervised 3D pose estimation using unlabeled multi-view images has increased, and our research pertains to this category as well. The self-supervised approaches use 2D poses estimated from unlabeled multi-view images. These approaches usually follow a lifting-based pipeline and therefore, they extract the 2D poses from the images using 2D pose estimators [44–47]. In our case, to get 2D joints from the images, we exploit AlphaPose [46] that is pretrained on a MPII dataset [48]. In constraints to the calibrated multi-view supervised approaches [49–52], the self-supervised approaches do not require the camera parameters to use multi-view data and thus, do not use traditional computer algorithms such as triangulation to recover the 3D poses. Kocabas et al. [9] leveraged epipolar geometry to acquire a 3D pseudo ground-truth from multi-view 2D predictions and then used them to train the 3D CNN network. Although this effective and intuitive approach shows promising results, the errors caused by incorrectly estimated joints in 2D estimation lead to an incorrect pseudo ground-truth. Iqbal et al. [11] trained a weakly-supervised network that refines the pretrained 2D pose estimator which predict pixel coordinates of joints and their depth in each view during training. Unfavorably, this method is not robust in environments other than the datasets employed for training, which is a limitation of the method of estimating the 3D pose from the image unit. More recently, Wandt et al. [12] reconstructed the 3D poses in a canonical pose space that was consistent across all views. We take advantage of the canonical space to fuse poses between the multiple views without using any camera parameters to refine the 2D pose incorrectly estimated by the 2D pose estimator.

3 Methods

Our goal is to train a neural network to accurately predict a 3D pose from an estimated 2D pose. At training time, we use 2D poses observed in multi-view images to train the network. At inference time, the network estimates a 3D pose from a 2D pose observed in a single image. The overall process of our framework is as follows. For each view, a neural network takes a 2D pose as an input, and

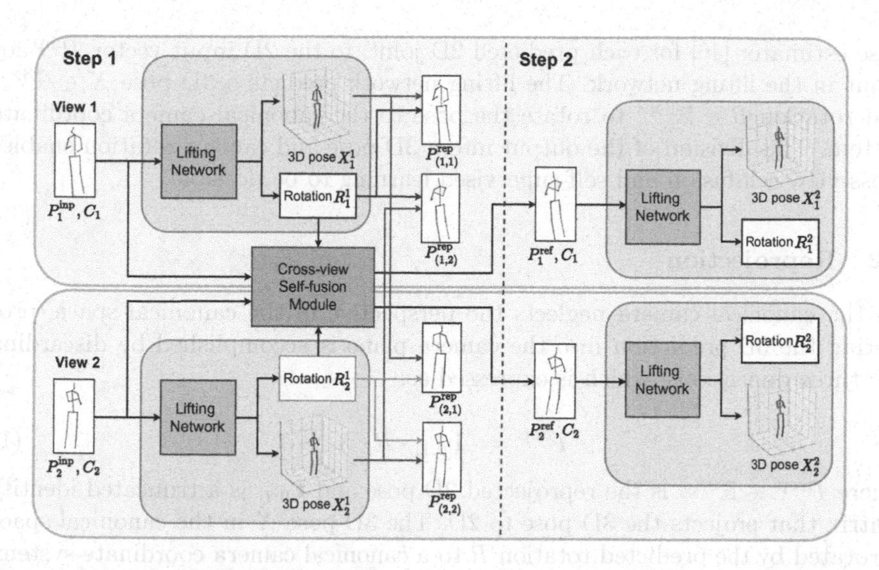

Fig. 2. A framework for learning a single 3D pose estimation from multi-view self-supervision while refining a 2D pose that adversely affects 3D pose estimation. At inference time, only a single view (blue box) is used for estimating a 3D human pose. (Color figure online)

subsequently predict a camera rotation $R^1 \in \mathbb{R}^{3\times3}$ and 3D pose $X^1 \in \mathbb{R}^{3\times J}$ with J joint positions in the first lifting step. Then, the proposed Cross-view Self-fusion module fuses the input 2D poses with the predicted rotations from all views to refine the 2D poses as outputs. In the second lifting step, the refined 2D pose of each view is input into the lifting network to output a second 3D pose X^2 and camera rotation R^2 for each view. We define losses to each step and a total loss using the outputs of multiple weight-sharing neural networks and describe them in Sect. 3.4. The proposed framework with two cameras is illustrated in Fig. 2, which can be conveniently expanded with the availability of more cameras.

3.1 Lifting Network

Before inputting a 2D pose for each view to a lifting network, we normalize the 2D pose by centering it on the root joint and dividing it with its Euclidean norm. As we do not have any 3D annotations, we can train the lifting network by satisfying the multi-view consistency. Although the 3D poses lifted in each view must be identical to satisfy the multi-view consistency, the scales of the lifted 3D poses are different since the scales of the 2D poses are different in each view. Therefore, we transform the estimated 2D pose in each view into a canonical space, where the distances between the 3D target and each camera are the same, by normalizing it. $P^{\text{inp}} \in \mathbb{R}^{2\times J}$ is a transformed 2D pose that is input to the lifting network. We concatenate the confidences $C \in \mathbb{R}^{1\times J}$, provided by the 2D pose estimator [46] for each predicted 2D joint, to the 2D input vector P^{inp} for input in the lifting network. The lifting network predicts a 3D pose $X \in \mathbb{R}^{3\times J}$ and rotation $R \in \mathbb{R}^{3\times 3}$ to rotate the pose to the canonical camera coordinate system. This division of the output into a 3D pose and camera rotation enables cross-view self-fusion and self-supervised learning to be possible.

3.2 Reprojection

As the canonical camera neglects the perspective in the canonical space, projecting the 3D prediction into the camera plane is accomplished by discarding the three dimensions, which is expressed as:

$$P^{\text{rep}} = \mathbf{I}_{[0:1]} \cdot R \cdot X, \tag{1}$$

where $P^{\text{rep}} \in \mathbb{R}^{2\times J}$ is the reprojected 2D pose and $\mathbf{I}_{[0:1]}$ is a truncated identity matrix that projects the 3D pose to 2D. The 3D pose X in the canonical space is rotated by the predicted rotation R to a canonical camera coordinate system. We rotate the m canonical 3D poses into the camera coordinate system of each camera through m rotations, in which the combining provides m^2 combinations. For instance, there are four possible combinations of rotations and poses for two cameras. During training, all possible combinations are reprojected onto the respective cameras. For example, $P^{\text{rep}}_{(2,1)}$ can be obtained by reprojecting a 3D pose X_1 predicted in view-1 to view-2 using the camera rotation R_2 predicted in view-2 as visualized in Fig. 2.

3.3 Cross-View Self-fusion

An inaccurate 2D pose results in incorrect 3D pose estimation. Typically, existing cross-view fusion methods utilize camera parameters to fuse cross-view data for accurate 2D poses. In the case of the canonical space, the relationship between multi-views can be represented only by relative rotation. Therefore, we propose a Cross-view Self-fusion Module (CSM) that refines an incorrect 2D pose using the predicted rotations and other input 2D poses. Figure 3 illustrates our proposed module and a refinement process for a wrist joint that is incorrectly estimated by occlusion. The module takes a set of 2D poses, corresponding confidences and predicted rotations of all views as inputs, and outputs a set of the refined 2D poses: $\mathbf{P}^{\text{ref}} = \text{CSM}(\mathbf{P}^{\text{inp}}, \mathbf{C}, \mathbf{R})$.

Formally, our proposed module is defined as:

$$\begin{aligned}&\text{CSM}\left(\mathbf{P}^{\text{inp}}, \mathbf{C}, \mathbf{R}\right) = \\ &\bigcup_{m\in V}\left\{\left\{argmax\Big(\mathbf{H}\left(p^i_m, c^i_m\right) + \sum_{\substack{n\in V,\\ n\neq m}} \mathbf{H}\left(\mathbf{E}\left(p^i_n, R_{(n,m)}\right), c^i_n\right)\Big) \Big| \forall i \in J\right\}\right\},\end{aligned} \tag{2}$$

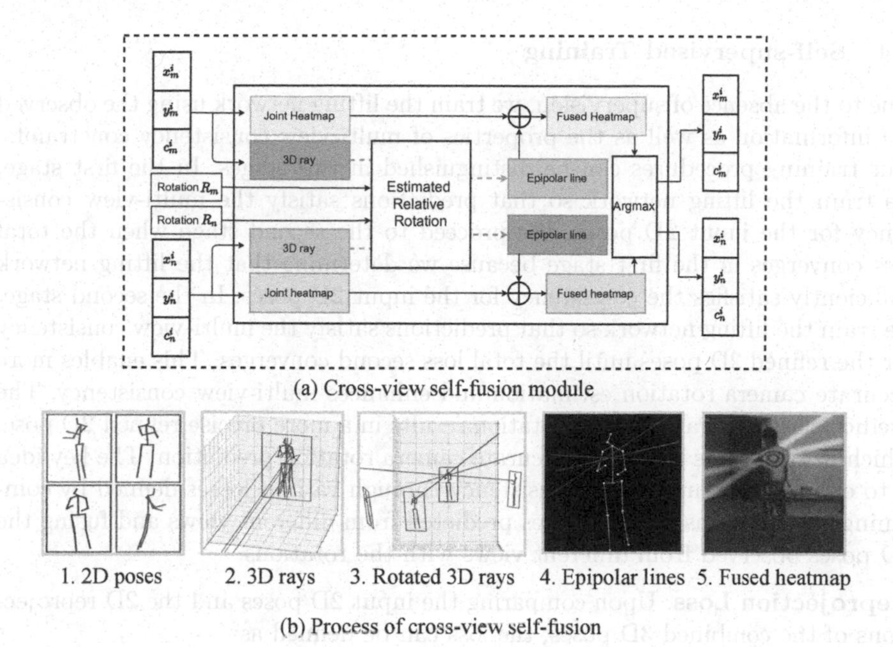

(a) Cross-view self-fusion module

(b) Process of cross-view self-fusion

Fig. 3. Cross-view Self-fusion module. (a) shows our cross-view self-fusion module and (b) provides a process of fusing multi-view information about the wrist joint to refine the joint incorrectly estimated by occlusion.

where $V = \{1, 2, \ldots, v\}$ is a set of views and $J = \{1, 2, \ldots, j\}$ is a set of joints. p_m^i is the i-th joint of 2D pose P_m^{inp} from view-m. It has canonical plane coordinates $\{x_m^i, y_m^i\}$. A set of confidences corresponding to the 2D pose is $C_m = \{c_m^1, \cdots, c_m^j\}$. $R_{(n,m)}$ is the relative rotation between view-n and view-m. The relative rotation $R_{(n,m)}$ by rotating from view-n to view-m using rotation matrices R_n and R_m is defined as $R_{(n,m)} = R_n R_m^T$.

First, a heatmap generator $\mathbf{H}(\cdot)$ takes a joint p_m^i of 2D pose P_m^{inp} from view-m and a confidence c_m^i of the joint as inputs and generates a gaussian heatmap for the joint, in which the maximum value of the heatmap is its confidence. Second, an epipolar line generator $\mathbf{E}(\cdot)$ takes a joint p_n^i of 2D pose P_n^{inp} from view-n and a relative rotation $R_{(n,m)}$ as input and outputs an epipolar line for the joint on the view-m, as illustrated in Fig. 3(b). The input joint is lifted to a 3D ray by simply adding the third dimension since the input is on canonical space where perspective is neglected. It is rotated to view-m by a relative rotation $R_{(n,m)}$ and projected to view-m similar to Sect. 3.2. A rotated and projected 3D ray is represented as a line on a view, which is called an epipolar line. Next, an epipolar line heatmap is generated with the maximum value of the heatmap as the confidence value of the joint. Finally, the position with the maximum value of a heatmap fused with a joint heatmap from view-m and epipolar line heatmaps from the other views becomes the coordinate value of the newly refined 2D joint position. It is repeated for all views.

3.4 Self-supervised Training

Due to the absence of supervision, we train the lifting network using the observed 2D information as well as the properties of multi-view consistency constraints. Our training procedures can be distinguished in two stages. In the first stage, we train the lifting network so that predictions satisfy the multi-view consistency for the input 2D poses. We proceed to the second stage when the total loss converges in the first stage because we determine that the lifting network sufficiently satisfies the consistency for the input 2D poses. In the second stage, we train the lifting network so that predictions satisfy the multi-view consistency for the refined 2D poses until the total loss second converges. This enables more accurate camera rotation estimation and enhances multi-view consistency. The prediction of accurate camera rotation results in a more precise refined 2D pose, which in turn leads to a more accurate camera rotation prediction. The key idea is to enhance the multi-view consistency through various losses defined by combining the rotations and 3D poses predicted from different views and fusing the 2D poses observed from different views with the rotations.

Reprojection Loss. Upon comparing the input 2D poses and the 2D reprojections of the combined 3D poses, the loss can be defined as:

$$\mathcal{L}_{\text{rep}} = \left\| \left(P_m^{\text{inp}} - \frac{P_n^{\text{rep}}}{\|P_n^{\text{rep}}\|_E} \right) \odot C \right\|_1, \tag{3}$$

where $\|\cdot\|_1$ denote the L_1 norm and $\odot$ indicates the Hadamard product. In particular, each deviation between the input and reprojected 2D pose is linearly weighted along with its confidence in order to a strong weight to the predicted joint in certainty and less weight to the predicted joints in uncertainty. Since the global scale of the 3D human pose is not given, the reprojection P^{rep} is scaled by the Euclidean norm. m and n indicate the camera indices.

Refinement Loss. Upon comparing the refined 2D poses and input 2D poses, the loss can be defined as:

$$\mathcal{L}_{\text{ref}} = \left\| \left(P_m^{\text{inp}} - P_n^{\text{ref}} \right) \odot C \right\|_1. \tag{4}$$

According to multi-view geometric consistency, if the 2D joints of all views are accurate and the camera relationship is known correctly, the intersection of epipolar lines on one view from the other views should be on a joint observed in one view. In our initial training, the predicted rotation is not accurate, so the 2D pose is refined to the wrong position. Therefore, we learn the initial refined 2D pose to be equal to the input 2D pose. This loss makes the camera rotation estimation more accurate and ensures that the refined 2D poses are plausible.

Refinement-Reprojection Loss. The loss between the refined 2D poses and 2D reprojections is defined as:

$$\mathcal{L}_{\text{ref-rep}} = \left\| \left(P_m^{\text{ref}} - \frac{P_n^{\text{rep}}}{\|P_n^{\text{rep}}\|_E} \right) \odot C \right\|_1. \tag{5}$$

We learn that the 2D reprojection of the lifted 3D pose is the same as the refined 2D pose, which is the result of the cross-view self-fusion module. This enables the lifting network to learn a 3D pose similar to the 3D poses estimated by other views so that it does not violate multi-view consistency even if it takes an incorrectly estimated 2D pose as an input.

Multi-view 3D Consistency Loss. To ensure multi-view consistency, previous work [12] has attempted to introduce a loss between the 3D poses predicted by multi-view to train a lifting network. As reported, the lifting network learned the 3D poses that were invariant toward the view but were no longer in close correspondence to the input 2D pose, thereby preventing the convergence of the network to plausible solutions. For each view, we lift a refined 2D pose to a 3D pose, as depicted in Fig. 2. More specifically, we enhance the consistency by adding a loss in 3D units using the 3D pose predicted from the refined 2D pose for each view. This loss is defined as the deviation between the 3D poses generated by lifting the refined 2D pose for each view.

$$\mathcal{L}_{3\text{D}} = \left\| X_m^2 - X_n^2 \right\|_1, \tag{6}$$

where X_m^2 is a second 3D pose lifted from a refined 2D pose of view-m and X_n^2 is a second 3D pose lifted from a refined 2D pose of view-n.

Total Loss. We sum up the losses described above for all views. To this end, we can define total loss as follows.

$$\mathcal{L} = \sum_{m=1}^{V} \sum_{n=1}^{V} \left(w_1 \mathcal{L}_{\text{ref}}^{m,n} + w_1 \mathcal{L}_{\text{rep}}^{m,n} + w_2 \mathcal{L}_{\text{ref-rep}}^{m,n} + w_2 \mathcal{L}_{3\text{D}}^{m,n} \right). \tag{7}$$

Until total loss $\mathcal{L}$ first converges, the weight w_1 is set to 1, and the weight w_2 is set to 0.01 in the first stage. Then, until the end of the training, we set the weight w_1 to 0.01, and the weight w_2 to 1 in the second stage.

4 Experiments

4.1 Datasets and Metrics

Dataset. We perform experiments on two standard benchmark datasets: Human3.6M [14] (H36M) and MPI-INF-3DHP [15] (3DHP). We also evaluate our method on the SkiPose dataset [53,54] with six moving cameras to demonstrate the generality of our method to various in-the-wild scenarios. To conform with a setting of self-supervised training for a particular set of activities, we train one network for each dataset without using additional datasets. For each dataset, we follow the self-supervision protocols for training and evaluation [9,12].

Metrics. For quantitative evaluation, we adopt the common protocol, *Normalized Mean Per Joint Position Error* (NMPJPE) and *Procrustes aligned Mean Per Joint Position Error* (PMPJPE) that measure the mean euclidean distance between the ground-truth and the predicted 3D joint positions after applying

Table 1. Per-action PMPJPE of different variants on the Human3.6M dataset.

Method	Act							
	Direct	Discuss	Eating	Greet	Phone	Photo	Pose	Purch
Baseline	50.5	49.0	46.1	54.0	51.2	53.2	59.5	49.1
Ours (S1)	41.9	42.3	41.1	45.3	45.3	45.0	48.8	43.4
Ours (S1+S2)	40.1	40.4	39.9	44.1	44.5	44.2	48.0	42.3
Method	Act							
	Sitting	SittingD	Smoke	Wait	Walk	WalkD	WalkT	**Avg**
Baseline	65.7	83.7	53.7	57.7	48.9	60.1	50.7	55.5
Ours (S1)	59.7	73.0	46.2	46.2	40.6	47.9	40.7	47.1
Ours (S1+S2)	59.0	72.1	45.3	44.3	39.4	46.1	39.9	**45.9**

Table 2. Evaluation of 2D pose refinement accuracy for each dataset. We show JDR (%) for six important joints about each dataset.

Method	Dataset	Hip	Knee	Ankle	Shoulder	Elbow	Wrist
Single	H36M	97.1	97.5	97.5	98.5	96.7	98.2
Ours	H36M	**98.2**	**98.5**	**97.8**	**98.9**	**98.5**	**99.6**
Single	3DHP	97.4	97.8	99.8	96.9	97.0	96.9
Ours	3DHP	**98.8**	**97.8**	**99.9**	**98.4**	**98.4**	**98.3**
Single	Ski	97.0	73.7	**81.2**	90.0	70.0	**60.9**
Ours	Ski	**98.7**	**77.0**	75.1	**91.9**	**71.7**	56.4

the optimal rigid alignment and scale (for NMPJPE), or optimal shift and scale (for PMPJPE) to poses. For 3DHP and SkiPose, we report the N-PCK, which is *Percentage of Correct Keypoints* (PCK) normalized by scale. The N-PCK indicates the percentage of joints whose estimated position is within 150mm of the ground-truth. Lastly, for evaluating our proposed CSM, we measure the refined 2D pose accuracy by *Joint Detection Rate* (JDR), which is the percentage of the successfully detected joints. If the distance between the estimated and ground-truth locations is smaller than a threshold, this joint can be deemed to be successfully detected. The threshold is set to half of the head size.

4.2 Ablation Studies

We analyze the effectiveness of the proposed losses. Specifically, we design several variants of our method, and the details of these variants are shown as follows.
Baseline: The baseline does not consider the Cross-view Self-fusion module. The baseline is trained simply using the reprojection loss.
Step1 (S1): This variant adopts the refinement loss and refinement-reprojection loss by the CSM. It does not consider the lifting of Step 2.
Step2 (S2): It lifts a refined 2D pose to a 3D pose. This variant considers a multi-view 3D consistency loss between second 3D poses of all views.

We train all variants (Baseline, S1, S1+S2) on the H36M, and Table 1 demonstrates the per-action PMPJPE of all variants on the H36M.

(a) Samples of heatmaps (b) Examples of refined 2D poses for SkiPose

Fig. 4. (a) "Detected heatmap" indicated that it is extracted from a image of the target view. "Fused heatmap" is obtained by summing the "Detected heatmap" and the "Epipolar heatmaps" fused from the three remaining views. **(b)** Qualitative results of 2D refined pose for the challenging SkiPose dataset. We compare the visual results of ground-truth, ours and baseline.

Compared with the baseline, our CSM is helpful to obtain better results. We experiment with whether the CSM has the effect of intuitively refining incorrectly estimated 2D poses due to occlusion. Table 2 shows the 2D pose estimation accuracy with JDR (%) for six important joints on the H36M, 3DHP and Skipose datasets. It compares our approach with the 2D pose estimator [46], termed *Single*, which estimates a 2D pose from a single image without performing cross view self-fusion. It can be seen that using CSM improves overall accuracy except for some joints. Figure 4(a) shows examples of the fused heatmap during the cross-view self-fusion process for the H36M. It shows examples of incorrectly estimated wrist joints by occlusions. It can be seen that the 2D pose is correctly refined by fusing the estimated epipolar line heatmaps. If the correct 2D pose is lifted to 3D, a more accurate 3D pose will be estimated. Figure 4(b) shows examples of the refined 2D poses for the SkiPose.

4.3 Comparison with State-of-the-Art Methods

We compare the results of the proposed method with other state-of-the-art approaches. For a fair comparison with [12], we follow the implementation and evaluation performed on [12]. We employ an off-the-shelf detector [46] to extract a 2D human pose required as an input to the proposed method.

Most methods using a lifting network without knowledge of scenes show the large gap between the NMPJPE and PMPJPE as a small error of the 2D pose incorrectly estimated in a particular view among all views leads to a large 3D NMPJPE error. They train their network without addressing the incorrectly estimated 2D pose, which further impacts to estimate the camera rotation because

Table 3. Evaluation results on the Human3.6M and comparison of the 3D pose estimation errors NMPJPE and PMPJPE (*mm*) of previous approaches. The best results are marked in bold. Our model outperforms all self-supervised methods.

Supervision	Method	NMPJPE↓	PMPJPE ↓
Full	Martinez [34]	67.5	52.5
Weak	Rhodin [55]	122.6	98.2
	Rhodin [53]	80.1	65.1
	Wandt [56]	89.9	65.1
	Kolotouros [57]	–	62.0
	Kundu [58]	85.8	–
Self	Kocabas [9]	76.6	67.5
	Jenni [10]	89.6	76.9
	Iqbal [11]	69.1	55.9
	Wandt [12]	74.3	53.0
	Ours (S1)	63.6	46.1
	Ours (S1+S2)	**61.4**	**45.9**

Table 4. Evaluation results on the MPI-INF-3DHP. NMPJPE and PMPJPE are reported in millimeters, and N-PCK is in %. The best results are marked in bold.

Supervision	Method	NMPJPE ↓	PMPJPE ↓	N-PCK ↑
Weak	Rhodin [53]	121.8	–	72.7
	Kolotouros [57]	124.8	–	66.8
	Li [59]	–	–	74.1
	Kundu [58]	103.8	–	**82.1**
Self	Kocabas [9]	125.7	–	64.7
	Iqbal [11]	110.1	68.7	76.5
	Wandt [12]	104.0	70.3	77.0
	Ours (S1)	95.2	57.3	79.3
	Ours (S1+S2)	**94.6**	**56.5**	81.9

Table 5. Evaluation results on the SkiPose. NMPJPE and PMPJPE are given in *mm*, N-PCK is in %. The best results are marked in bold.

Supervision	Method	NMPJPE ↓	PMPJPE ↓	N-PCK ↑
Weak	Rhodin [53]	**85.0**	–	**72.7**
Self	Wandt [12]	128.1	89.6	67.1
	Ours (S1)	118.2	79.3	70.1
	Ours (S1+S2)	115.2	**78.8**	72.4

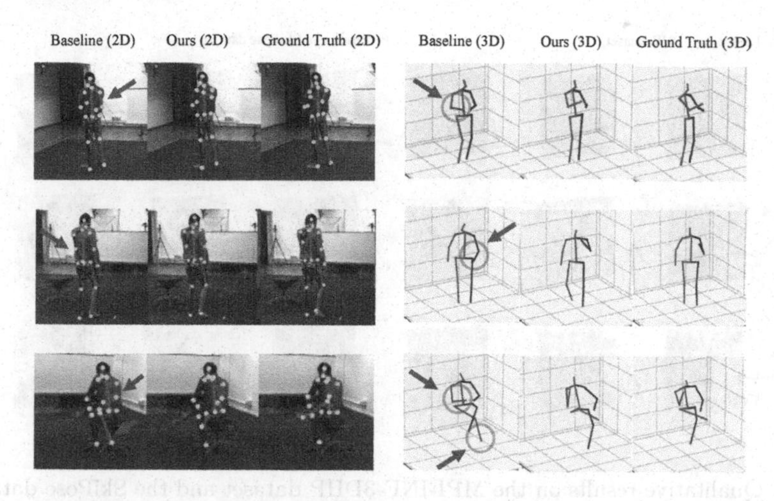

Fig. 5. Qualitative results of our approach on Human3.6M. We present both skeletons of 2D pose on the image and 3D pose in the space by comparing the visual results of baseline, ours, and ground-truth.

the network is trained with the violation of the multi-view consistency constraints. This results in the incorrect 3D pose and rotation estimation. The MPJPE can be considerably enhanced by refining the incorrect 2D pose of any multi-views in training to meet the multi-view consistency constraints.

Results on Human3.6M. A 2D skeleton morphing network introduced by Wandt et al. [12] is used to circumvent the offset between the 2D pose from [46] and the ground-truth 2D pose in the H36M dataset. As illustrated in Table 3, we report the self-supervised pose estimation results in terms of the NMPJPE and PMPJPE. As can be seen, the proposed model outperforms every other comparable approach in terms of the aforementioned metrics. Notably, the achieved performance surpassed our baseline, CanonPose [12], that outperforms the fully supervised method of Martinez et al. [34], which has a lifting network. In Fig. 5, we present some challenging examples on the H36M dataset and qualitatively compare the visualization results. These pictures include some occlusions and show the results of our baseline, ours, and ground-truth.

Our baseline model is already able to output plausible results. However, it does not solve occlusion, so we can visually confirm that an incorrect 2D pose is lifted to an incorrect 3D pose. We demonstrate that our approach solves the problem of occlusion that was not solved in the baseline approach.

Results on MPI-INF-3DHP. We evaluate the proposed approach on the 3DHP dataset [15] following the self-supervised protocols and metrics. The results are presented in Table 4. For a more comprehensive comparison, we report the performance of several recently fully and weakly-supervised methods. The proposed model outperforms all other self-supervised methods. In addition, the

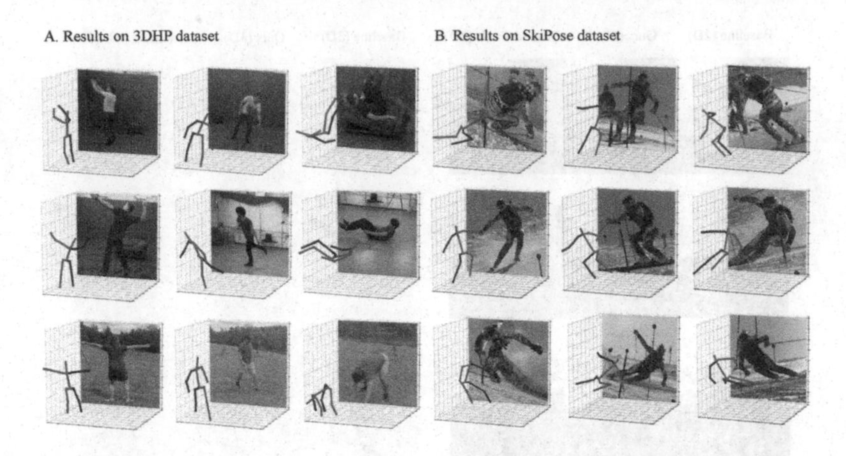

Fig. 6. Qualitative results on the MPI-INF-3DHP dataset and the SkiPose dataset.

visualization results for the test dataset are presented in Fig. 6. Our model yields satisfactory results even for some dynamic action and unseen outdoor scenes.

Results on SkiPose. Our primary motivation is to train 3D human pose estimation in-the-wild with multiple uncalibrated cameras. Moreover, we intend to experiment on in-the-wild human activity data that cannot measure 3D annotation with traditional motion capture systems. The dataset, which is best suited to these conditions, is the SkiPose [53,54]. Our approach can handle all these challenges since it operates without relying on static or calibrated cameras. Table 5 shows our results in comparison to Rhodin et al. [53] and Wandt et al. [12]. Rhodin et al. [53] considers a weakly supervised setting and known camera locations, so direct comparison with ours is impossible. We outperform the baseline approach [12] on the SkiPose and the qualitative results for the dataset are presented in Fig. 6.

5 Conclusion

In this paper, we introduced a novel self-supervised learning method for monocular 3D human pose estimation from unlabeled multi-view images without camera calibration. We exploited multi-view consistency to disentangle 2D estimations into canonical predictions (a 3D pose and camera rotation) that were used to refine the errors of the 2D estimations and reproject the 3D pose on the 2D for self-supervised learning. We conducted quantitative and qualitative experiments on three 3D benchmark datasets and achieve state-of-the-art results. The results demonstrated that our method could be applied to real-world scenarios, including dynamic outdoor human activities like sports.

Acknowledgements. This work was partially supported by the Institute of Information & communications Technology Planning Evaluation (IITP) funded by the Korea government (MSIT) (No. 2019-0-00079, Artificial Intelligence Graduate School Program (Korea University)) and the Technology Innovation Program (No. 20017012, Business Model Development for Golf Putting Simulator using AI Video Analysis and Coaching Service at Home) funded by the Ministry of Trade, Industry & Energy (MOTIE, Korea).

References

1. Liu, W., Mei, T.: Recent advances of monocular 2d and 3d human pose estimation: a deep learning perspective. ACM Comput. Surv. (CSUR) **55**(4), 1–41 (2022)
2. Lim, Y.K., Choi, S.H., Lee, S.W.: Text extraction in mpeg compressed video for content-based indexing. In: Proceedings 15th International Conference on Pattern Recognition, ICPR-2000, vol. 4, pp. 409–412. IEEE (2000)
3. Lee, G.H., Lee, S.W.: Uncertainty-aware human mesh recovery from video by learning part-based 3d dynamics. In: Proceedings of the IEEE/CVF International Conference on Computer Vision, pp. 12375–12384 (2021)
4. Yang, H.D., Lee, S.W.: Reconstruction of 3d human body pose from stereo image sequences based on top-down learning. Pattern Recogn. **40**, 3120–3131 (2007)
5. Ahmad, M., Lee, S.W.: Human action recognition using multi-view image sequences. In: 7th International Conference on Automatic Face and Gesture Recognition (FGR06), pp. 523–528. IEEE (2006)
6. Roh, M.C., Shin, H.K., Lee, S.W.: View-independent human action recognition with volume motion template on single stereo camera. Pattern Recog. Lett. **31**, 639–647 (2010)
7. Ji, X., Fang, Q., Dong, J., Shuai, Q., Jiang, W., Zhou, X.: A survey on monocular 3d human pose estimation. Virtual Reality Intell. Hardware **2**, 471–500 (2020)
8. Roh, M.C., Kim, T.Y., Park, J., Lee, S.W.: Accurate object contour tracking based on boundary edge selection. Pattern Recogn. **40**, 931–943 (2007)
9. Kocabas, M., Karagoz, S., Akbas, E.: Self-supervised learning of 3d human pose using multi-view geometry. In: Proceedings of the IEEE Conference on Computer Vision and Pattern Recognition, pp. 1077–1086 (2019)
10. Jenni, S., Favaro, P.: Self-supervised multi-view synchronization learning for 3d pose estimation. In: Proceedings of the Asian Conference on Computer Vision (2020)
11. Iqbal, U., Molchanov, P., Kautz, J.: Weakly-supervised 3d human pose learning via multi-view images in the wild. In: Proceedings of the IEEE Conference on Computer Vision and Pattern Recognition, pp. 5243–5252 (2020)
12. Wandt, B., Rudolph, M., Zell, P., Rhodin, H., Rosenhahn, B.: CanonPose: selfsupervised monocular 3d human pose estimation in the wild. In: Proceedings of the IEEE Conference on Computer Vision and Pattern Recognition, pp. 13294–13304 (2021)
13. Hartley, R., Zisserman, A.: Multiple View Geometry in Computer Vision. Cambridge University Press (2003)
14. Ionescu, C., Papava, D., Olaru, V., Sminchisescu, C.: Human3.6M: large scale datasets and predictive methods for 3d human sensing in natural environments. IEEE Trans. Pattern Anal. Mach. Intell. **36**, 1325–1339 (2013)

15. Mehta, D., et al.: Monocular 3d human pose estimation in the wild using improved CNN supervision. In: 2017 International Conference on 3D Vision, pp. 506–516 (2017)
16. Tekin, B., Katircioglu, I., Salzmann, M., Lepetit, V., Fua, P.: Structured prediction of 3d human pose with deep neural networks. arXiv preprint arXiv:1605.05180 (2016)
17. Tekin, B., M'arquez-Neila, P., Salzmann, M., Fua, P.: Learning to fuse 2d and 3d image cues for monocular body pose estimation. In: Proceedings of the IEEE International Conference on Computer Vision, pp. 3941–3950 (2017)
18. Sun, X., Shang, J., Liang, S., Wei, Y.: Compositional human pose regression. In: Proceedings of the IEEE International Conference on Computer Vision, pp. 2602–2611 (2017)
19. Pavlakos, G., Zhou, X., Derpanis, K.G., Daniilidis, K.: Coarse-to-fine volumetric prediction for single-image 3d human pose. In: Proceedings of the IEEE Conference on Computer Vision and Pattern Recognition, pp. 7025–7034 (2017)
20. Rogez, G., Weinzaepfel, P., Schmid, C.: LCR-Net: localization-classification regression for human pose. In: Proceedings of the IEEE Conference on Computer Vision and Pattern Recognition, pp. 3433–3441 (2017)
21. Mehta, D., et al.: Single-shot multi-person 3d pose estimation from monocular RGB. In: 2018 International Conference on 3D Vision, pp. 120–130 (2018)
22. Yang, W., Ouyang, W., Wang, X., Ren, J., Li, H., Wang, X.: 3d human pose estimation in the wild by adversarial learning. In: Proceedings of the IEEE Conference on Computer Vision and Pattern Recognition, pp. 5255–5264 (2018)
23. Fang, H.S., Xu, Y., Wang, W., Liu, X., Zhu, S.C.: Learning pose grammar to encode human body configuration for 3d pose estimation. In: Proceedings of the AAAI Conference on Artificial Intelligence (2018)
24. Moon, G., Chang, J.Y., Lee, K.M.: Camera distance-aware top-down approach for 3d multi-person pose estimation from a single RGB image. In: Proceedings of the IEEE International Conference on Computer Vision, pp. 10133–10142 (2019)
25. Wang, C., Li, J., Liu, W., Qian, C., Lu, C.: HMOR: hierarchical multi-person ordinal relations for monocular multi-person 3d pose estimation. In: Vedaldi, A., Bischof, H., Brox, T., Frahm, J.-M. (eds.) ECCV 2020. LNCS, vol. 12348, pp. 242–259. Springer, Cham (2020). https://doi.org/10.1007/978-3-030-58580-8_15
26. Xi, D., Podolak, I.T., Lee, S.W.: Facial component extraction and face recognition with support vector machines. In: Proceedings of 5th IEEE International Conference on Automatic Face Gesture Recognition, pp. 83–88. IEEE (2002)
27. Lee, S.-W., Verri, A. (eds.): SVM 2002. LNCS, vol. 2388. Springer, Heidelberg (2002). https://doi.org/10.1007/3-540-45665-1
28. Lee, S.W., Kim, S.Y.: Integrated segmentation and recognition of handwritten numerals with cascade neural network. IEEE Trans. Syst. Man Cybern. Part C (Appl. Rev.) **29**, 285–290 (1999)
29. Lee, S.W., Kim, J.H., Groen, F.C.: Translation-, rotation-and scale-invariant recognition of hand-drawn symbols in schematic diagrams. Int. J. Pattern Recogn. Artif. Intell. **4**, 1–25 (1990)
30. Rogez, G., Schmid, C.: MoCap-guided data augmentation for 3d pose estimation in the wild. In: Advances in Neural Information Processing Systems (2016)
31. Varol, G., et al.: Learning from synthetic humans. In: Proceedings of the IEEE Conference on Computer Vision and Pattern Recognition, pp. 109–117 (2017)
32. Cheng, Y., Yang, B., Wang, B., Yan, W., Tan, R.T.: Occlusion-aware networks for 3d human pose estimation in video. In: Proceedings of the IEEE International Conference on Computer Vision, pp. 723–732 (2019)

33. Gong, K., Zhang, J., Feng, J.: PoseAug: a differentiable pose augmentation framework for 3d human pose estimation. In: Proceedings of the IEEE Conference on Computer Vision and Pattern Recognition, pp. 8575–8584 (2021)
34. Martinez, J., Hossain, R., Romero, J., Little, J.J.: A simple yet effective baseline for 3d human pose estimation. In: Proceedings of the IEEE International Conference on Computer Vision, pp. 2640–2649 (2017)
35. Xu, T., Takano, W.: Graph stacked hourglass networks for 3d human pose estimation. In: Proceedings of the IEEE Conference on Computer Vision and Pattern Recognition, pp. 16105–16114 (2021)
36. Cai, Y., et al.: Exploiting spatial-temporal relationships for 3d pose estimation via graph convolutional networks. In: Proceedings of the IEEE International Conference on Computer Vision, pp. 2272–2281 (2019)
37. Ci, H., Wang, C., Ma, X., Wang, Y.: Optimizing network structure for 3d human pose estimation. In: Proceedings of the IEEE International Conference on Computer Vision, pp. 2262–2271 (2019)
38. Chen, T., Fang, C., Shen, X., Zhu, Y., Chen, Z., Luo, J.: Anatomy-aware 3d human pose estimation with bone-based pose decomposition. IEEE Trans. Circ. Syst. Video Technol. **32**, 198–209 (2021)
39. Zheng, C., Zhu, S., Mendieta, M., Yang, T., Chen, C., Ding, Z.: 3d human pose estimation with spatial and temporal transformers. In: Proceedings of the IEEE International Conference on Computer Vision, pp. 11656–11665 (2021)
40. Hossain, M.R.I., Little, J.J.: Exploiting temporal information for 3d human pose estimation. In: Ferrari, V., Hebert, M., Sminchisescu, C., Weiss, Y. (eds.) ECCV 2018. LNCS, vol. 11214, pp. 69–86. Springer, Cham (2018). https://doi.org/10.1007/978-3-030-01249-6_5
41. Mehta, D., et al.: XNect: real-time multi-person 3d motion capture with a single RGB camera. ACM Trans. Graph. (TOG) **39**, 82–1 (2020)
42. Cao, X., Zhao, X.: Anatomy and geometry constrained one-stage framework for 3d human pose estimation. In: Proceedings of the Asian Conference on Computer Vision (2020)
43. Liu, K., Zou, Z., Tang, W.: Learning global pose features in graph convolutional networks for 3d human pose estimation. In: Proceedings of the Asian Conference on Computer Vision (2020)
44. Cao, Z., Simon, T., Wei, S.E., Sheikh, Y.: Realtime multi-person 2d pose estimation using part affinity fields. In: Proceedings of the IEEE Conference on Computer Vision and Pattern Recognition, pp. 7291–7299 (2017)
45. Chen, Y., Wang, Z., Peng, Y., Zhang, Z., Yu, G., Sun, J.: Cascaded pyramid network for multi-person pose estimation. In: Proceedings of the IEEE Conference on Computer Vision and Pattern Recognition, pp. 7103–7112 (2018)
46. Fang, H.S., Xie, S., Tai, Y.W., Lu, C.: RMPE: regional multi-person pose estimation. In: Proceedings of the IEEE International Conference on Computer Vision, pp. 2334–2343 (2017)
47. Li, J., Wang, C., Zhu, H., Mao, Y., Fang, H.S., Lu, C.: CrowdPose: efficient crowded scenes pose estimation and a new benchmark. In: Proceedings of the IEEE/CVF Conference on Computer Vision and Pattern Recognition, pp. 10863–10872 (2019)
48. Andriluka, M., Pishchulin, L., Gehler, P., Schiele, B.: 2d human pose estimation: new benchmark and state of the art analysis. In: Proceedings of the IEEE Conference on computer Vision and Pattern Recognition, pp 3686–3693 (2014)
49. Iskakov, K., Burkov, E., Lempitsky, V., Malkov, Y.: Learnable triangulation of human pose. In: Proceedings of the IEEE International Conference on Computer Vision, pp. 7718–7727 (2019)

50. Qiu, H., Wang, C., Wang, J., Wang, N., Zeng, W.: Cross view fusion for 3d human pose estimation. In: Proceedings of the IEEE International Conference on Computer Vision, pp. 4342–4351 (2019)
51. He, Y., Yan, R., Fragkiadaki, K., Yu, S.I.: Epipolar transformers. In: Proceedings of the IEEE Conference on Computer Vision and Pattern Recognition, pp. 7779–7788 (2020)
52. Ma, H., et al.: Transfusion: cross-view fusion with transformer for 3d human pose estimation. In: Proceedings of the British Machine Vision Conference (2021)
53. Rhodin, H., et al.: Learning monocular 3d human pose estimation from multi-view images. In: Proceedings of the IEEE Conference on Computer Vision and Pattern Recognition, pp. 8437–8446 (2018)
54. Spörri, J.: Research dedicated to sports injury prevention - the 'sequence of prevention' on the example of alpine ski racing. Habil. Venia Docendi Biomech. **1**, 7 (2016)
55. Rhodin, H., Salzmann, M., Fua, P.: Unsupervised geometry-aware representation for 3d human pose estimation. In: Ferrari, V., Hebert, M., Sminchisescu, C., Weiss, Y. (eds.) ECCV 2018. LNCS, vol. 11214, pp. 765–782. Springer, Cham (2018). https://doi.org/10.1007/978-3-030-01249-6_46
56. Wandt, B., Rosenhahn, B.: RepNet: weakly supervised training of an adversarial reprojection network for 3d human pose estimation. In: Proceedings of the IEEE Conference on Computer Vision and Pattern Recognition, pp. 7782–7791 (2019)
57. Kolotouros, N., Pavlakos, G., Black, M.J., Daniilidis, K.: Learning to reconstruct 3d human pose and shape via model-fitting in the loop. In: Proceedings of the IEEE International Conference on Computer Vision, pp. 2252–2261 (2019)
58. Kundu, J.N., Seth, S., Jampani, V., Rakesh, M., Babu, R.V., Chakraborty, A.: Self-supervised 3d human pose estimation via part guided novel image synthesis. In: Proceedings of the IEEE Conference on Computer Vision and Pattern Recognition, pp. 6152–6162 (2020)
59. Li, Y., et al.: Geometry-driven self-supervised method for 3d human pose estimation. In: Proceedings of the AAAI Conference on Artificial Intelligence, pp. 11442–11449 (2020)

3D-C2FT: Coarse-to-Fine Transformer for Multi-view 3D Reconstruction

Leslie Ching Ow Tiong[1], Dick Sigmund[2], and Andrew Beng Jin Teoh[3](✉)

[1] Computational Science Research Center, Korea Institute of Science and Technology, 5, Hwarang-ro 14-gil, Seongbuk-gu, Seoul 02792, Republic of Korea
tiongleslie@kist.re.kr

[2] AIDOT Inc., 128, Beobwon-ro, Songpa-gu, Seoul 05854, Republic of Korea
dsigmund@aidot.ai

[3] School of Electrical and Electronic Engineering, Yonsei University, Seoul 120-749, Republic of Korea
bjteoh@yonsei.ac.kr

Abstract. Recently, the transformer model has been successfully employed for the multi-view 3D reconstruction problem. However, challenges remain in designing an attention mechanism to explore the multi-view features and exploit their relations for reinforcing the encoding-decoding modules. This paper proposes a new model, namely 3D coarse-to-fine transformer (3D-C2FT), by introducing a novel coarse-to-fine (C2F) attention mechanism for encoding multi-view features and rectifying defective voxel-based 3D objects. C2F attention mechanism enables the model to learn multi-view information flow and synthesize 3D surface correction in a coarse to fine-grained manner. The proposed model is evaluated by ShapeNet and Multi-view Real-life voxel-based datasets. Experimental results show that 3D-C2FT achieves notable results and outperforms several competing models on these datasets.

Keywords: Multi-view 3D reconstruction · Coarse-to-fine transformer · Multi-scale attention

1 Introduction

Multi-view 3D reconstruction infers geometry and structure of voxel-based 3D objects from multi-view images. The definition of the voxel is essentially 3D pixels in cube size. Theoretically, voxel representation is the perfect modeling technique for replicating reality [9,10,13]. Hence, voxel representation can easily resemble real-world objects into more accurate 3D objects in the virtual environment [14]. Therefore, the multi-view 3D reconstruction plays a crucial role in

L. C. O. Tiong and D. Sigmund—These authors have contributed equally to this work.

Supplementary Information The online version contains supplementary material available at https://doi.org/10.1007/978-3-031-26319-4_13.

L. Wang et al. (Eds.): ACCV 2022, LNCS 13841, pp. 211–227, 2023.
https://doi.org/10.1007/978-3-031-26319-4_13

robot perception [2,22], historical artifact [12,16], dentistry [20,21], etc. Predicaments in estimating 3D structure of an object include solving an ill-posed inverse problem of 2D images, violation of overlapping views, and unconstrained environment conditions [7]. Here, the ill-posed inverse problem refers to the voxel-based 3D objects estimation problem when disordered or limited sources of the inputs are given as inputs [6]. Accordingly, multi-view 3D object reconstruction remains an active topic in computer vision.

Classical approaches to tackle multi-view 3D reconstruction are to introduce feature extraction techniques to map different reconstruction views [5,24,26,31]. However, these approaches hardly reconstruct a complete 3D object when a small number of images, e.g., 1–4, are presented. With the advances of deep learning (DL), most attempts utilize an encoder-decoder network architecture for feature extraction, fusion, and reconstruction. Existing DL models usually rely on the convolutional neural networks (CNN) or recurrent neural networks (RNN) to fuse multiple deep features encoded from 2D multi-view images [3,25]. However, the CNN encoder processes each view image independently, and thus, the relations in different views are barely utilized. This leads to difficulty designing a fusion method that can traverse the relationship between views. Although the RNNs can rectify the fusion problem, the model suffers a long processing time [3].

Recently, transformers [27] have gained exponential attention and proved to be a huge success in vision-related problems [4,18]. In the multi-view 3D reconstruction problem, [28] and [35] propose a unified multi-view encoding, fusion, and decoding framework via a transformer to aggregate features among the patch embeddings that can explore the profound relation between multi-view images and perform the decent reconstruction. However, both studies only consider single-scaled multi-head self-attention (MSA) mechanism to reinforce the view-encoding. Unfortunately, the native single-scaled MSA is not designed to explore the correlation of relevant features between the subsequent layers for object reconstruction.

Considering the limitations above, we propose a new transformer model, namely 3D coarse-to-fine transformer (3D-C2FT). The 3D-C2FT introduces a novel coarse-to-fine (C2F) attention mechanism for encoded features aggregation and decoded object refinement in a multi-scale manner. On top of exploring multi-view images relationship via MSA in the native transformer, the C2F attention mechanism enables aggregating coarse (global) to fine (local) features, which is favored to learn comprehensive and discriminative encoded features. Then, a concatenation operation is used to aggregate multiple sequential C2F features derived from each C2F attention block. In addition, a C2F refiner is devised to rectify the defective decoded 3D objects. Specifically, the refiner leverages a C2F cube attention mechanism to focus on the voxelized object's attention blocks that benefit coarse-to-fine correction of the 3D surface. In general, C2F attention utilizes the coarse and fine-grained features to be paired and promotes the information flow that helps the 3D objects reconstruction task.

Besides that, we compile a new dataset called Multi-view Real-life dataset for evaluation. However, unlike existing dataset such as ShapeNet [32] that is assembled from synthetic data, our dataset is composed of real-life single or multi-view images taken from internet without ground truth. We propose this dataset to support real-life study for multi-view 3D reconstruction evaluation. The contributions of this paper are summarized as follows:

- A novel multi-scale C2F attention mechanism is outlined for the multi-view 3D reconstruction problem. The proposed C2F attention mechanism learns the correlation of the features in a sequential global to the local manner, on top of exploring multi-view images relation via MSA. Accordingly, coarse-grained features draw attention towards the global object structure, and the fine-grained feature pays attention to each local part of the 3D object.
- A novel transformer model, namely 3D-C2FT, for multi-scale multi-view images encoding, features fusion, and coarse-to-fine decoded objects refinement is proposed.
- Extensive experiments demonstrate better reconstruction results on standard benchmark ShapeNet dataset and challenging real-life dataset than several competing models, even under stringent constraints such as occlusion.

This paper is organized as follows: Sect. 2 reviews the related works of the DL-based multi-view 3D reconstruction. Then, Sect. 3 presents our proposed method. Next, the experimental setup, results, and ablation study are presented in Sect. 4. Finally, the conclusions are summarized in Sect. 5.

2 Related Works

This section reviews several relevant works of DL-based multi-view 3D reconstruction. We refer the readers to a more comprehensive survey on this subject [7].

2.1 CNN and RNN-Based Models

Early works [3] and [11] utilize modified RNN to fuse information from multi-view images and reconstruct the 3D objects. However, both works have their limitations; for instance, when given the same set of images but in different orders, the model fails to reconstruct the 3D objects consistently. This is due to both models being permutation-variant and relying on the ordered sequence of the input images for feature extraction.

[33,34] and [36] propose a CNN-based model to address randomly ordered and long sequence forgetting issues of the RNNs. However, these works follow the divide-and-conquer principle by introducing a CNN-based single-view encoder, a single-view decoder, and a fusion model, which work independently. Therefore, the encoder and decoder hardly utilize the relations between different views. Instead, the network relies on the fusion model to integrate the

arbitrary ordered multi-view features for reconstruction. Results suggest that designing a robust fusion method with a CNN-based approach is challenging. Furthermore, these works suffer from the model scaling problem while preserving permutation-invariant capability. For instance, if the input views exceed a specific number, such as 10–16 views, the reconstruction performance level-off, implying the hardness of learning complementary knowledge from a large set of independent inputs.

2.2 Transformer-Based Models

Recently, [28] and [35] put forward the transformer-based models that perceive multi-view 3D reconstruction problems as a sequence-to-sequence prediction problem that permits fully exploiting information from 2D images with different views. [28] leverages native multi-head self-attention (MSA) mechanism for feature encoding, fusing, and decoding to generate a 3D volume for each query token. However, this model only relies on the MSA to explore the relation of multi-view images by fostering different representations of each view. Hence, it falls short in analyzing the low to high-level interactions within multi-view images that signify the global structure and local components of the 3D objects.

Yagubbayli et al. [35] proposes another transformer model simultaneously along with [28] known as Legoformer. This model adopts an encoder with pre-norm residual connections [17] and a non-autoregressive decoder that takes advantage of the decomposition factors without explicit supervision. Although this approach is more effective, the model only focuses on learning 3D objects reconstruction parts by parts. Such a strategy is deemed local-to-local attention, which does not benefit low-level interaction to support information flow across local parts of 3D objects.

Considering the limitations of the previous transformer-based models, we are motivated to design a new attention mechanism by introducing a C2F patch attention mechanism in the encoder to extract multi-scale features of the multi-view images. Furthermore, we also put forward a C2F cube attention mechanism in the refiner to rectify the reconstructed object surface in a coarse to fine-grained manner.

3 3D Coarse-to-Fine Transformer

3D coarse-to-fine transformer (3D-C2FT) consists of an image embedding module, a 2D-view encoder, a 3D decoder, and a 3D refiner, as illustrated in Fig. 1. Specifically, the encoder accepts either single or multi-view images in the embedding form, which are managed by the image embedding module and then C2F patch attention is used to aggregate the different views of inputs by extracting feature representations for the decoder to reconstruct 3D volume. Finally, the refiner with the C2F cube attention mechanism is to rectify the defective reconstructed 3D volume. The 3D-C2FT network is explained in detail in the following subsections.

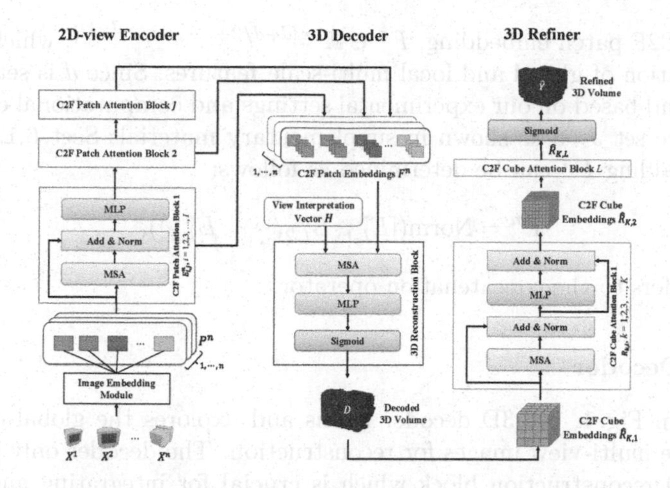

Fig. 1. The architecture of 3D-C2FT. The network comprises an image embedding module - DenseNet121, a 2D-view encoder, a 3D decoder, and a 3D refiner.

3.1 Image Embedding Module

We leverage DenseNet121 [8] as an image embedding module for the proposed 3D-C2FT. For the details, readers are referred to the supplementary materials Sect. 6.1. Given a set of n view images of a 3D object $X = \{X^1, X^2, \cdots, X^n\}$, each view image X^n is fed into the image embedding module to obtain patch embeddings $P^n \in \mathbb{R}^{1 \times d}$, where d is the embedding dimensions. We follow the ViT setting where d is fixed at 768.

3.2 2D-View Encoder

The encoder of 3D-C2FT receives patch embeddings of view n, P^n as well as its associated positional embeddings. These patches and positional embeddings are sent to a C2F patch attention block that consists of I-layer of MSA [27] and multilayer perceptron (MLP), which is simply a fully connected feed forward network, as shown in Fig. 1. The C2F patch attention block is repeated J times. Thus, the output of the C2F patch attention block, $E^n_{I,j}$ is calculated as follows:

$$\hat{E}^n_{i,j} = \text{MSA}(\text{Norm}(P^n_{i-1,j})) + P^n_{i-1,j}, \tag{1}$$

$$E^n_{i,j} = \text{MLP}(\text{Norm}(\hat{E}^n_{i,j})), \tag{2}$$

where Norm$(\cdot)$ denotes layer normalization, i is the layer index of $E^n_{i,j}$ where $i = 1, \cdots, I$ and I is empirically set to 4 in this paper. j is the block index of $E^n_{i,j}$, where $j = 1, \cdots, J$.

Each C2F patch attention block is responsible to extract the features i.e., $E^n_{I,j}$ in a coarse to fine-grained manner by shrinking the dimension of $E^n_{I,j}$ in a ratio of two. Thereafter, we fuse each $E^n_{I,j}$ from J blocks by concatenation to produce a C2F patch embedding, $F^n \in \mathbb{R}^{1\times[d+d/2+,\cdots,+d/2^{(J-1)}]}$, which captures the aggregation of global and local multi-scale features. Since d is set to 768 in this work and based on our experimental settings and computational complexity concerns, we set J=3 as shown in supplementary materials Sect. 6.1. The C2F patch embedding F^n can be determined as follows:

$$F^n = \text{Norm}([E^n_{I,1}, E^n_{I,2}, \cdots, E^n_{I,J}]), \tag{3}$$

where $[\cdot]$ refers to the concatenation operator.

3.3 3D Decoder

As shown in Fig. 1, the 3D decoder learns and explores the global correlation between the multi-view images for reconstruction. The decoder only consists of a single 3D reconstruction block which is crucial for integrating and converting F^n to a 3D volume D with size 32 × 32 × 32. The attention module of the 3D reconstruction block accepts both view interpretation matrix H and F^n to project different views of F^n into the D. Specifically, given F^n and $H \in \mathbb{R}^{g\times[d+d/2+,\cdots,+d/2^{(J-1)}]}$, the decoder can easily aggregate the n-view of F^n together and share across all potential inputs in the network. Here, H is a randomly initialized learnable weight matrix that plays a role in interpreting and addressing F^n, and g denotes the number of cube embeddings that are required to assemble the D. In this paper, we set g=8 × 8 × 8 = 512. The 3D volume output of the decoder, D, is specified as:

$$D = \sigma\,(\text{MLP}(\text{MSA}(H, F^n))) \in \mathbb{R}^{32\times32\times32}, \tag{4}$$

where $\sigma(\cdot)$ is a sigmoid activation function. However, the reconstructed 3D volume, D at this stage is defective (Sect. 4.3) due to simplicity of the decoder. Therefore, a refiner is needed for further rectification.

3.4 3D Refiner

We propose a C2F cube attention mechanism intending to correct and refine the D surface for the refiner. The refiner is composed of L C2F cube attention blocks where each block consists of K layers of ViT-like attention module (Fig. 1). Here, we set K=6, which the experiments determine. In addition, $L=2$ is set in this work as the reconstructed volume resolution is a mere 32 × 32 × 32. Therefore, a larger L can be used for high-resolution volume.

Specifically, D from Eq. 4 is transformed to the C2F cube embeddings $\hat{R}_{k,l}$ by partitioning the D to $\hat{R}_{k,1}$ with a size of 8 × 8 × 8 in the first C2F cube attention block. For the second block, $\hat{R}_{k,2}$ size is further reduced by a factor of two; thus 4 × 4 × 4. For each block, the cube attention block output can be computed from:

$$\hat{R}_{k,l} = \text{MSA}(\text{Norm}(\hat{R}_{k-1,l})) + \hat{R}_{k-1,l}, \tag{5}$$

$$R_{k,l} = \mathrm{MLP}(\mathrm{Norm}(\hat{R}_{k,l})) + \hat{R}_{k,l} \tag{6}$$

where k is the layer index of $R_{k,l}$ where $k = 1, 2, \cdots, K$ and l is the block index of $R_{k,l}$ where $l = 1, 2, \cdots, L$.

By adopting the C2F notion, the proposed refiner leverages the C2F cube attention blocks of different scales of embeddings that can benefit the structure and parts correction of the 3D object. Furthermore, it enables the refiner to iteratively draw attention from coarse to temperate regions by rectifying the spatial surfaces gradually through multi-scale attention.

3.5 Loss Function

For training, we adopt a combination of mean-square-error ($\mathcal{L}_{\text{MSE}}$) loss and 3D structural similarity ($\mathcal{L}_{\text{3D-SSIM}}$) loss as a loss function ($\mathcal{L}_{\text{total}}$) to capture better error information about the quality degradation of the 3D reconstruction. The motivation of adopting $\mathcal{L}_{\text{3D-SSIM}}$ loss [30] is for 3D-C2FT to learn and produce visually pleasing 3D objects by quantifying volume structural differences between a reference and a reconstructed object. Furthermore, the $\mathcal{L}_{\text{MSE}}$ is used to evaluate the voxel loss that measures the L2 differences between the ground-truth (GT) and reconstructed objects. Thus, $\mathcal{L}_{\text{total}}$ is given as follows:

$$\mathcal{L}_{\text{total}}(Y, \hat{Y}) = \mathcal{L}_{\text{MSE}}(Y, \hat{Y}) + \mathcal{L}_{\text{3D-SSIM}}(Y, \hat{Y}), \tag{7}$$

$$\mathcal{L}_{\text{MSE}}(Y, \hat{Y}) = \frac{1}{M^3} \sum_{x=0}^{M} \sum_{y=0}^{M} \sum_{z=0}^{M} (Y_{x,y,z} - \hat{Y}_{x,y,z})^2, \tag{8}$$

$$\mathcal{L}_{\text{3D-SSIM}}(Y, \hat{Y}) = 1 - \frac{(2\mu_Y \mu_{\hat{Y}} + c_1)/(2\sigma_{Y\hat{Y}} + c_2)}{(\mu_Y^2 + \mu_{\hat{Y}}^2 + c_1)(\sigma_Y^2 + \sigma_{\hat{Y}}^2 + c_2)}, \tag{9}$$

where Y is the GT of 3D volume and $\hat{Y}$ is reconstructed 3D volume. Note that, in Eq. 8, M refers to the dimension of 3D volume; x, y and z are defined as the indexes of voxel coordinates, respectively. In Eq. 9, μ_Y and $\mu_{\hat{Y}}$ denote the mean of the voxels; σ_Y^2 and $\sigma_{\hat{Y}}^2$ are the variance of the voxels, respectively. In addition, $\sigma_{Y\hat{Y}}$ denotes the covariance between Y and $\hat{Y}$. We set $c_1 = 0.01$ and $c_2 = 0.03$ to avoid instability when the mean and variances are close to zero.

4 Experiments

4.1 Evaluation Protocol and Implementation Details

Dataset. To have a fair comparison, we follow the protocols specified by [3] and [35] evaluate the proposed model. A subset of the ShapeNet dataset, which comprises 43,783 objects from 13 categories, is adopted. Each object is rendered from 24 different views. We resize the original images from 137×137 to 224×224, and the uniform background color is applied before passing them to the network. For each 3D object category, three subsets with a ratio of 70:10:20 for training, validation, and testing are partitioned randomly. Note that all the 3D object samples are provided with a low-resolution voxel representation of $32 \times 32 \times 32$.

Evaluation Metrics. Intersection over Union (IoU) and F-score are adopted to evaluate the reconstruction performance of the proposed model. A higher IoU score implies a better reconstruction. A higher F-score value suggests a better reconstruction. Therefore, the F-score@1% is adopted in this paper.

Implementation. The proposed model is trained with the batch size 32 with view input images of size 224×224 and the dimension of 3D volume is set to $32 \times 32 \times 32$. For multi-view training, the number of input views is fixed to 8, which is the best obtained from the experiments. The full-fledged results are given in supplementary materials Sect. 6.1.

During training, the views are randomly sampled out of 24 views at each iteration, and the network is trained by an SGD optimizer with an initial learning rate of 0.01. The learning rate decay is used, and it is subsequently reduced by 10^{-1} for every 500 epochs. The minimum learning rate is defined as 1.0×10^{-4}. Our network is implemented by using the PyTorch toolkit [19], and it is performed by an NVidia Tesla V100. The source code is provided at GitHub[1]

4.2 Result

Multi-view 3D Reconstruction. The performance comparisons of 3D-C2FT against several benchmark models, namely 3D-R2N2 (RNN-based) [3], AttSets (RNN-based) [36], Pix2Vox++/F (CNN-based) [34], Pix2Vox++/A (CNN-based) [34], LegoFormer (transformer-based) [35], 3D-RETR (transformer-based) [23], VoIT+ (transformer-based) [28], and EVoIT (transformer-based) [28] are presented in this subsection. To have fair comparisons, we re-implemented several models such as AttSets, Pix2Vox++/F, Pix2Vox++/A, and 3D-RETR that were provided by the respective authors in order to follow the same experimental settings and protocols. Table 1 shows reconstruction performance over a different number of views in terms of IoU and F-score on the ShapeNet dataset.

Compared with RNN-based and CNN-based models, it can be observed that 3D-C2FT achieves the highest IoU and F-score with all views as tabulated in Table 1. As an illustration, we demonstrate several reconstruction instances from the ShapeNet dataset as shown in Fig. 2 and supplementary material Video01. As can be seen in Fig. 2, the objects reconstructed by the 3D-C2FT have complete and smoother surfaces, even compared with the LegoFormer. These results suggest that C2F patch and cube attention mechanisms play a crucial role in multi-view 3D object reconstruction.

Among transformer-based models (see Table 1), 3D-C2FT outperforms LegoFormer, VoIT+, EVoIT, and 3D-RETR significantly in almost all views in terms of IoU score and F-score, even under extreme cases where one view image is used as input. Note VoIT+ and EVoIT are published without code availability; thus, we only report the results for 4, 8, 12, and 20 views inputs [28]. Nevertheless, the results reveal the distinctive advantages of the C2F patch and cube attention over competing transformer models that utilize a plain single-scaled MSA mechanism in both encoder and decoder. However, 3D-C2FT underperforms EVoIT

[1] Source Code URL: https://github.com/tiongleslie/3D-C2FT/.

Table 1. Performance comparisons of single and multi-view 3D reconstruction on ShapeNet with IoU and F-score. The best score for each view is written in bold. * refers to re-implemented models with the same settings of this study. The '-' results of VoIT+ and EVoIT are not provided in [28] and source code is not publicly available.

Model	Number of views								
	1	2	3	4	5	8	12	18	20
Metric: IoU									
RNN & CNN-based									
3D-R2N2 (2016)	0.560	0.603	0.617	0.625	0.634	0.635	0.636	0.636	0.636
AttSets* (2020)	0.607	0.634	0.639	0.658	0.669	0.674	0.677	0.679	0.681
Pix2Vox++/F* (2019)	0.585	0.636	0.658	0.669	0.677	0.686	0.690	0.693	0.694
Pix2Vox++/A* (2020)	0.623	0.674	0.690	0.695	0.699	0.704	0.707	0.709	0.710
Transformer-based									
LegoFormer (2021)	0.519	0.644	0.679	0.694	0.703	0.713	0.717	0.719	0.721
VoIT+ (2021)	–	–	–	0.695	–	0.707	0.714	–	0.715
EVoIT (2021)	–	–	–	0.609	–	0.698	**0.720**	–	**0.738**
3D-RETR* (2021)	0.608	0.661	0.672	0.679	0.682	0.685	0.687	0.688	0.689
3D-C2FT	**0.629**	**0.678**	**0.695**	**0.702**	**0.708**	**0.716**	**0.720**	**0.723**	0.725
Metric: F-score									
RNN & CNN-based									
3D-R2N2 (2016)	0.351	0.368	0.372	0.378	0.382	0.383	0.382	0.382	0.383
AttSets* (2020)	0.358	0.360	0.370	0.379	0.384	0.386	0.389	0.403	0.408
Pix2Vox++/F* (2019)	0.341	0.367	0.388	0.398	0.405	0.417	0.429	0.430	0.432
Pix2Vox++/A* (2020)	0.365	0.419	0.435	0.443	0.447	0.452	0.457	0.460	0.461
Transformer-based									
LegoFormer (2021)	0.282	0.392	0.428	0.444	0.453	0.464	0.470	0.472	0.473
VoIT+ (2021)	–	–	–	0.451	–	0.464	0.469	–	0.474
EVoIT (2021)	–	–	–	0.358	–	0.448	0.475	–	**0.497**
3D-RETR* (2021)	0.355	0.412	0.425	0.432	0.436	0.440	0.442	0.443	0.444
3D-C2FT	**0.371**	**0.424**	**0.443**	**0.452**	**0.458**	**0.468**	**0.476**	**0.477**	0.479

for 20 views. We speculate that the issue is associated with the attention layer in the decoder being less effective for a large number of input views.

Next, we also perform single-view 3D reconstruction against several benchmark models, namely 3D-R2N2, OGN (CNN-based) [25], Pix2Mesh (CNN-based) [29], OccNet (CNN-based) [15], Pix2Vox++F, Pixe2Vox++A, LegoFormer, and 3D-RETR. For reference, we follow the protocols specified by [3] and [15] in this experiment. As shown in Table 2, 3D-C2FT attains the highest IoU value and F-score for most of the categories. By comparison, the C2FT attention mechanism clearly achieves better aggregation performance even with single-view images.

Performance Comparisons with Occlusion Images. In this subsection, the reconstruction performance of 3D-C2FT and the competing models is evaluated with occluded images. Here, we use the testing set from the subset of the ShapeNet dataset by following the same protocols in the previous section. We add occlusion boxes with different sizes to the odd number ordered lists of 2D view images: *20 × 20*, *25 × 25*, *30 × 30*, *35 × 35* and *40 × 40*, which impeded the essential parts of the images intentionally. Readers are referred to the supplementary materials Sect. 6.2 for details.

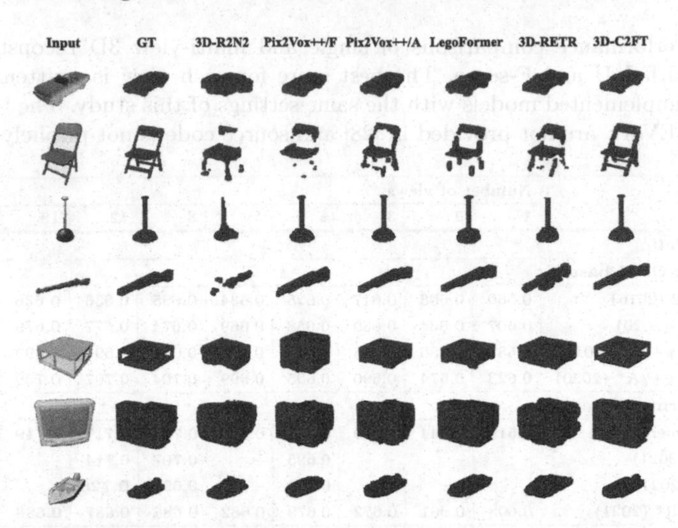

Fig. 2. 3D object reconstruction using 8 views (only 1 are shown) for specific categories: *car*, *chair*, *lamp*, *rifle*, *table*, *display*, and *watercraft*.

Table 2. Performance comparisons of single-view 3D reconstruction results for each category on the ShapeNet dataset. * refers to re-implemented models with the same settings in this study. The best score for each category is written in bold.

Category	3D-R2N2	OCN	Pix2Mesh	OccNet	Pix2Vox++F*	Pix2Vox++A*	3D-RETR*	3D-C2FT
Airplane	0.513	0.587	0.420	0.571	0.533	**0.652**	0.647	0.640
Bench	0.421	0.481	0.323	0.485	0.454	0.547	0.547	**0.549**
Cabinet	0.716	0.729	0.664	0.733	0.726	0.760	0.712	**0.761**
Car	0.798	0.816	0.552	0.737	0.801	0.831	0.818	**0.832**
Chair	0.466	0.483	0.396	0.501	0.482	0.522	0.511	**0.525**
Display	0.468	**0.502**	0.490	0.471	0.463	0.499	0.434	0.489
Lamp	0.381	0.398	0.323	0.371	0.427	0.445	0.436	**0.468**
Loudspeaker	0.662	0.637	0.599	0.647	0.690	0.695	0.632	**0.697**
Rifle	0.544	**0.593**	0.474	0.402	0.528	0.583	0.567	0.583
Sofa	0.628	0.646	0.613	0.680	0.637	**0.695**	0.672	0.668
Table	0.513	0.536	0.395	0.506	0.541	0.568	0.536	**0.571**
Telephone	0.661	0.702	0.661	**0.720**	0.702	**0.720**	0.624	**0.720**
Watercraft	0.513	**0.632**	0.397	0.530	0.514	0.570	0.547	0.558
Overall	0.560	0.596	0.480	0.571	0.585	0.623	0.607	**0.629**

As shown in Table 3, the proposed 3D-C2FT achieves the best IoU scores for all sizes of occlusion boxes. Among the benchmark models, LegoFormer performs second-best over various occlusion boxes. Unlike the LegoFormer that draws attention to the fine-grained features only to explore the correlation between multi-view images, 3D-C2FT focuses on both coarse and fine-grained features so that they can assist the model in exploiting global and local interactions within the occluded regions, which make the model more robust against predicting unknown elements in the multi-view images.

Table 3. Performance comparisons of 12-view 3D reconstruction on ShapeNet using different sizes of occlusion box. The best score for each size is highlighted in bold.

Model	Occlusion box				
	20 × 20	25 × 25	30 × 30	35 × 35	40 × 40
Metric: IoU					
Pix2Vox++/F	0.672	0.661	0.651	0.636	0.621
Pix2Vox++/A	0.698	0.695	0.690	0.685	0.678
LegoFormer	0.701	0.699	0.695	0.691	0.686
3D-RETR	0.667	0.651	0.623	0.603	0.584
3D-C2FT	**0.717**	**0.715**	**0.712**	**0.706**	**0.698**
Metric: F-score					
Pix2Vox++/F	0.416	0.410	0.405	0.396	0.388
Pix2Vox++/A	0.451	0.449	0.446	0.442	0.437
LegoFormer	0.453	0.451	0.448	0.443	0.439
3D-RETR	0.414	0.392	0.361	0.340	0.319
3D-C2FT	**0.471**	**0.468**	**0.465**	**0.459**	**0.451**

Experiment on Multi-view Real-Life Dataset. This subsection presents the performance comparisons on Multi-view Real-life Dataset. All models were only trained with the ShapeNet dataset. To create our dataset, we randomly searched the images by following the 3D object categories of the ShapeNet database from the Google image search engine. After that, all the images were manually verified to ensure they were correctly labeled. However, collecting many different input views for real images is difficult. Therefore, we manually combined the same samples with similar images from different views. The dataset is designed as a testing set that only contains 60 samples across 13 categories with three cases:

- *Case I*: samples with at least 1 to 5 views;
- *Case II*: samples with at least 6 to 11 views;
- *Case III*: samples with at least 12 or above views;

The experiment is designed to verify the 3D objects based on human perception without the GT of 3D volume. In Fig. 3 and supplementary materials Sect. 6.3, we show the qualitative results for single and multi-view 3D reconstruction. It is observed that 3D-C2FT performs substantially better than the benchmark models in terms of the refined 3D volume and surface quality even in *Case I.* Interestingly, most benchmark models fail to reconstruct the 3D objects, or the surface is not accurately generated. This is likely because the benchmark models lack in utilizing multi-scale feature extraction mechanisms, which causes these models to fail to capture relevant components more richly.

Note that [35] advocates a part-by-part composition notion by means of fine-to-fine attention, which is opposed to our global-to-local attention notion. As depicted in Fig. 3, LegoFormer underperforms the proposed model. This could be due to LegoFormer does not utilize 3D structure information essential in interpreting global interaction between semantic components. As a result, the model ignores the importance of the structure and orientation information of the 3D objects in guiding the model to perform reconstruction in challenging environments. In summary, the advantage of the C2F attention mechanism is that

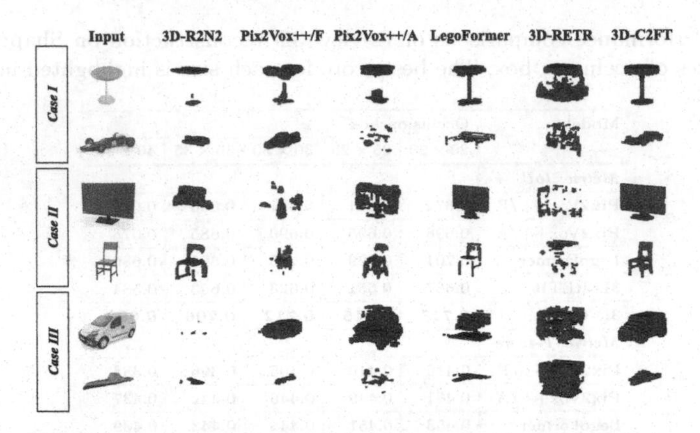

Fig. 3. 3D object reconstruction for *Case I*, *Case II* and *Case III* with Multi-view Real-life dataset

Table 4. Comparisons on parameter sizes of competing models

Model	Param. (M)	Backbone	Speed per View (s)
3D-R2N2	36.0	–	–
AttSets	53.1	–	0.011
Pix2Vox++/F	114.2	ResNet50	0.010
Pix2Vox++/A	96.3	ResNet50	0.011
LegoFormer	168.4	VGG16	0.013
3D-RETR	168.0	RETR	0.012
3D-C2FT	90.1	DenseNet121	0.013

it fully utilizes the coarse and fine-grained features to be paired and represents the specific information flow that benefits the 3D object reconstruction task.

Computational Cost. Table 4 tabulates the number of parameters in various benchmark models. Note that the backbone refers to the pre-trained image embedding module. 3D-C2FT has a smaller number of parameters than most of the competing models. Although 3D-R2N2 has significantly smaller parameter sizes, the reconstruction performances are below par, as revealed in Table 1. Besides, we also compare reconstruction time (speed per view) in Table 4. As can be seen in this table, 3D-C2FT has a well-balanced of the best balance of speed (see Table 4) and performance (IoU score, see Table 1).

4.3 Ablation Study

In this section, all the experiments are conducted with the ShapeNet dataset described in Sect. 4.1.

Visualization on C2F Attention. To better understand the benefit of C2F attention, we present visualization results of three C2F patch attention blocks separately, labeled as $C2F_1$, $C2F_2$ and $C2F_3$ with attention rollout [1]. Figure 4

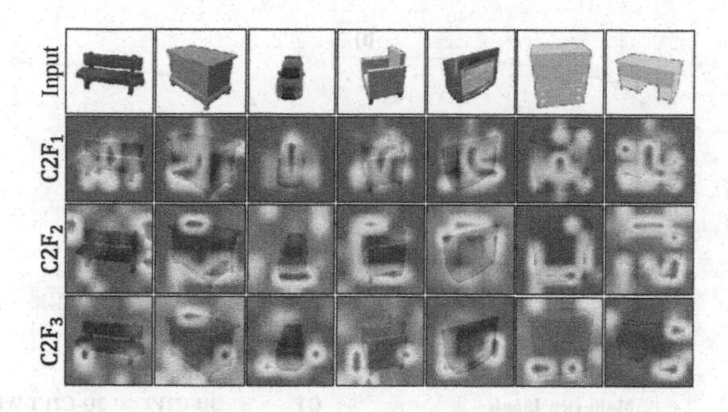

Fig. 4. Visualization of attention maps for each C2F patch attention block.

depicts the attention maps of each stage of the C2F patch attention block, which interpret and focus semantically relevant regions that result in a coarse to fine-grained manner.

As shown in Fig. 4, the $C2F_1$ attention is drawn to the central parts of the given images, which represent coarse or global features of the 3D objects. Similarly, the $C2F_2$ shows attention is drawn toward the edges of the objects, which can assist the model to distinguish the view orientations for reconstruction. The final block ($C2F_3$) suggests that the attention is drawn to the specific parts of the objects, which can be considered as fine-grained features. The results indicate that the proposed coarse-to-fine mechanism pays specific attention to the relevant components in the 3D objects, which is essential for the decoder to predict and reconstruct accurate 3D volume.

Reconstruction With and Without Refiner. In Fig. 5a and 5b, we evaluate the influence of the refiner on 3D reconstruction results using 3D-C2FT and 3D-C2FT without refiner (3D-C2FT-WR). We observe that 3D-C2FT can significantly achieve the best IoU score and F-score in all views. We also show several qualitative 3D reconstruction results in Fig. 5c. Without refiner, the decoder generates defective 3D surfaces due to a lack of drawing attention to rectifying wrongly reconstructed parts within the 3D objects. Therefore, our analysis indicates that the refiner plays a crucial role in improving the 3D surface quality and can remove the noise of the reconstructed 3D volume from the decoder.

Loss Functions. We also show the performance of 3D-C2FT trained with different loss functions, namely $\mathcal{L}_{\text{total}}$, $\mathcal{L}_{\text{MSE}}$ and $\mathcal{L}_{\text{3D-SSIM}}$, as shown in Table 5. It is noticed that 3D-C2FT trained with $\mathcal{L}_{\text{total}}$ achieves the best IoU and F-score compared to either $\mathcal{L}_{\text{MSE}}$ or $\mathcal{L}_{\text{3D-SSIM}}$ alone. $\mathcal{L}_{\text{MSE}}$ minimizes the voxel grid error between the predicted and GT of 3D objects. In contrast, $\mathcal{L}_{\text{3D-SSIM}}$ reduces the structural difference, which benefits in improving the surface of predicted 3D objects. As a result, both loss functions drive the model training towards more reliable and better surface quality reconstruction.

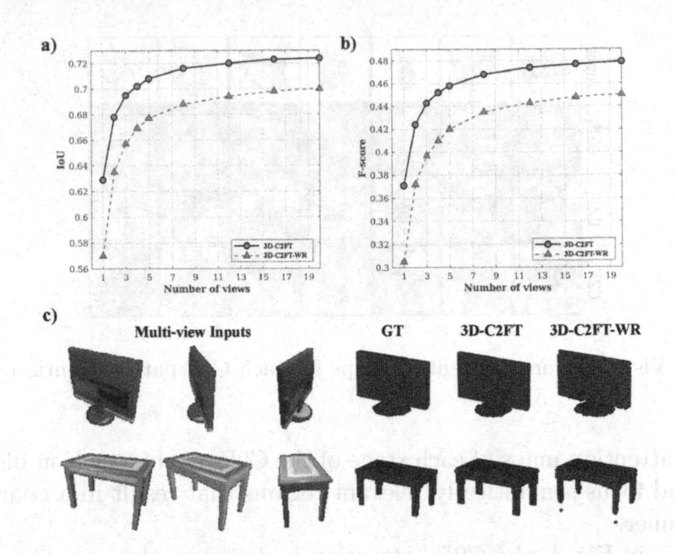

Fig. 5. Performance comparisons of multi-view 3D reconstruction results on 3D-C2FT and 3D-C2FT-WR. a) IoU score; b) F-score; c) 3D reconstruction results using 8 views (only 3 are shown) between 3D-C2FT and 3D-C2FT-WR.

Table 5. Performance comparisons of multi-view 3D reconstruction results on 3D-C2FT using different loss functions. The best score for each view is written in bold.

Loss function	Number of views								
	1	2	3	4	5	8	12	18	20
Metric: IoU									
$\mathcal{L}_{\text{MSE}}$	0.623	0.673	0.683	0.696	0.701	0.708	0.713	0.716	0.716
$\mathcal{L}_{\text{3D-SSIM}}$	0.617	0.668	0.683	0.691	0.696	0.705	0.709	0.712	0.712
$\mathcal{L}_{\text{total}}$	**0.629**	**0.678**	**0.695**	**0.702**	**0.708**	**0.716**	**0.720**	**0.723**	**0.724**
Metric: F-score									
$\mathcal{L}_{\text{MSE}}$	0.369	0.421	0.437	0.451	0.457	0.466	0.471	0.474	0.476
$\mathcal{L}_{\text{3D-SSIM}}$	0.365	0.418	0.437	0.446	0.452	0.462	0.467	0.470	0.472
$\mathcal{L}_{\text{total}}$	**0.371**	**0.424**	**0.443**	**0.452**	**0.458**	**0.468**	**0.475**	**0.477**	**0.479**

5 Conclusion

This paper proposed a multi-view 3D reconstruction model that employs a C2F patch attention mechanism in the 2D encoder, a 3D decoder, and a C2F cube attention mechanism in the refiner. Our experiments showed that 3D-C2FT could achieve significant results compared to the several competing models on the ShapeNet dataset. Further study with the Multi-view Real-life dataset showed that 3D-C2FT is far more robust than other models. For future works, we will consider improving the attention mechanism and the scaling of 3D-C2FT. Finally, we will explore other mechanisms to enhance the 3D reconstruction task, which we hope will facilitate the practical usage of robotics, historical artifacts, and other related domains.

References

1. Abnar, S., Zuidema, W.: Quantifying attention flow in transformers. arXiv e-prints (2020). https://arxiv.org/abs/2005.00928
2. Burchfiel, B., Konidaris, G.: Bayesian eigenobjects: a unified framework for 3D robot perception. In: Robotics: Science and Systems, vol. 13 (2017)
3. Choy, C.B., Xu, D., Gwak, J.Y., Chen, K., Savarese, S.: 3D-R2N2: a unified approach for single and multi-view 3d object reconstruction. In: Leibe, B., Matas, J., Sebe, N., Welling, M. (eds.) ECCV 2016. LNCS, vol. 9912, pp. 628–644. Springer, Cham (2016). https://doi.org/10.1007/978-3-319-46484-8_38
4. Dosovitskiy, A., et al.: An image is worth 16 × 16 words: transformers for image recognition at scale. In: International Conference on Learning Representations (ICLR) (2021)
5. Gao, Y., Luo, J., Qiu, H., Wu, B.: Survey of structure from motion. In: Proceedings of 2014 International Conference on Cloud Computing and Internet of Things, pp. 72–76 (2014)
6. Groen, I.I.A., Baker, C.I.: Previews scenes in the human brain: comparing 2D versus 3D representations. Neuron **101**(1), 8–10 (2019)
7. Han, X.F., Laga, H., Bennamoun, M.: Image-based 3D object reconstruction: state-of-the-art and trends in the deep learning era. IEEE Trans. Pattern Anal. Mach. Intell. **43**(5), 1578–1604 (2021)
8. Huang, G., Liu, Z., van der Maaten, L., Weinberger, K.Q.: Densely connected convolutional networks. In: Proceedings of the IEEE Conference on Computer Vision and Pattern Recognition (CVPR), pp. 4700–4708 (2017)
9. Jabłoński, S., Martyn, T.: Real-time voxel rendering algorithm based on screen space billboard voxel buffer with sparse lookup textures. In: 24th Conference on Computer Graphics, Visualization and Computer Vision, pp. 27–36 (2016)
10. Kanzler, M., Rautenhaus, M., Westermann, R.: A voxel-based rendering pipeline for large 3d line sets. IEEE Trans. Visual Comput. Graph. **25**(7), 2378–2391 (2019)
11. Kar, A., Häne, C., Malik, J.: Learning a multi-view stereo machine. In: Proceedings of the 31st International Conference on Neural Information Processing Systems (NIPS), pp. 364–375. Curran Associates, Inc. (2017)
12. Kargas, A., Loumos, G., Varoutas, D.: Using different ways of 3D reconstruction of historical cities for gaming purposes: the case study of Nafplio. Heritage **2**(3), 1799–1811 (2019)
13. Kniaz, V.V., Knyaz, V.A., Remondino, F., Bordodymov, A., Moshkantsev, P.: Image-to-voxel model translation for 3d scene reconstruction and segmentation. In: Vedaldi, A., Bischof, H., Brox, T., Frahm, J.-M. (eds.) ECCV 2020. LNCS, vol. 12352, pp. 105–124. Springer, Cham (2020). https://doi.org/10.1007/978-3-030-58571-6_7
14. Malik, J., et al.: HandVoxNet: deep voxel-based network for 3d hand shape and pose estimation from a single depth map. In: Proceedings of the IEEE Conference on Computer Vision and Pattern Recognition (CVPR), pp. 7111–7120 (2020)
15. Mescheder, L., Oechsle, M., Niemeyer, M., Nowozin, S., Geiger, A.: Occupancy networks: learning 3d reconstruction in function space. In: Proceedings of the IEEE Conference on Computer Vision and Pattern Recognition (CVPR) (2019)
16. Nabil, M., Saleh, F.: 3D reconstruction from images for museum artefacts: a comparative study. In: International Conference on Virtual Systems and Multimedia (VSMM), pp. 257–260. IEEE (2014)

17. Nguyen, T.Q., Salazar, J.: Transformers without tears: improving the normalization of self-attention. In: Proceedings of the 16th International Conference on Spoken Language Translation, Hong Kong (2019)
18. Park, N., Kim, S.: How do vision transformers work? In: International Conference on Learning Representations (ICLR) (2022)
19. Paszke, A., et al.: PyTorch: an imperative style, high-performance deep learning library. In: Proceedings of the 34th International Conference on Neural Information Processing Systems (NIPS), pp. 8024–8035 (2019)
20. Păvăloiu, I.B., Vasilăţeanu, A., Goga, N., Marin, I., Ilie, C., Ungar, A., Pătraşcu, I.: 3D dental reconstruction from CBCT data. In: International Symposium on Fundamentals of Electrical Engineering (ISFEE), pp. 4–9 (2014)
21. Roointan, S., Tavakolian, P., Sivagurunathan, K.S., Floryan, M., Mandelis, A., Abrams, S.H.: 3D dental subsurface imaging using enhanced truncated correlation-photothermal coherence tomography. Sci. Rep. **9**(1), 1–12 (2019)
22. Shi, Q., Li, C., Wang, C., Luo, H., Huang, Q., Fukuda, T.: Design and implementation of an omnidirectional vision system for robot perception. Mechatronics **41**, 58–66 (2017)
23. Shi, Z., Meng, Z., Xing, Y., Ma, Y., Wattenhofer, R.: 3D-RETR: end-to-end single and multi-view 3D reconstruction with transformers. In: British Machine Vision Conference (BMVC), pp. 1–14 (2021)
24. Silveira, G., Malis, E., Rives, P.: An efficient direct approach to visual SLAM. IEEE Trans. Rob. **24**(5), 969–979 (2008)
25. Tatarchenko, M., Dosovitskiy, A., Brox, T.: Octree generating networks: efficient convolutional architectures for high-resolution 3D outputs. In: IEEE International Conference on Computer Vision (ICCV), pp. 2088–2096 (2017)
26. Tron, R., Vidal, R.: Distributed 3-D localization of camera sensor networks from 2-D image Measurements. IEEE Trans. Autom. Control **59**(12), 3325–3340 (2014)
27. Vaswani, A., et al.: Attention is all you need. In: Proceedings of the 31st International Conference on Neural Information Processing Systems (NIPS), vol. 30, pp. 6000–6010 (2017)
28. Wang, D., et al.: Multi-view 3D reconstruction with transformer. In: International Conference on Computer Vision (ICCV), pp. 5722–5731 (2021)
29. Wang, N., Zhang, Y., Li, Z., Fu, Y., Liu, W., Jiang, Y.-G.: Pixel2Mesh: generating 3d mesh models from single RGB images. In: Ferrari, V., Hebert, M., Sminchisescu, C., Weiss, Y. (eds.) ECCV 2018. LNCS, vol. 11215, pp. 55–71. Springer, Cham (2018). https://doi.org/10.1007/978-3-030-01252-6_4
30. Wang, Z., Bovik, A., Sheikh, H., Simoncelli, E.: Image quality assessment: from error visibility to structural similarity. IEEE Trans. Image Process. **13**(4), 600–612 (2004)
31. Wilson, K., Snavely, N.: Robust global translations with 1DSfM. In: Fleet, D., Pajdla, T., Schiele, B., Tuytelaars, T. (eds.) ECCV 2014. LNCS, vol. 8691, pp. 61–75. Springer, Cham (2014). https://doi.org/10.1007/978-3-319-10578-9_5
32. Wu, Z., et al.: 3D ShapeNets: a deep representation for volumetric shapes. In: Proceedings of the IEEE Conference on Computer Vision and Pattern Recognition (CVPR), pp. 1912–1920 (2015)
33. Xie, H., Yao, H., Sun, X., Zhou, S., Zhang, S.: Pix2Vox: context-aware 3D reconstruction from single and multi-view images. In: IEEE International Conference on Computer Vision (ICCV), pp. 2690–2698 (2019)
34. Xie, H., Yao, H., Zhang, S., Zhou, S., Sun, W.: Pix2Vox++: multi-scale context-aware 3D object reconstruction from single and multiple images. Int. J. Comput. Vis. **128**(12), 2919–2935 (2020)

35. Yagubbayli, F., Tonioni, A., Tombari, F.: LegoFormer: transformers for block-by-block multi-view 3D reconstruction. arXiv e-prints (2021). http://arxiv.org/abs/2106.12102
36. Yang, B., Wang, S., Markham, A., Trigoni, N.: Robust attentional aggregation of deep feature sets for multi-view 3D reconstruction. Int. J. Comput. Vis. **128**(1), 53–73 (2020)

SymmNeRF: Learning to Explore Symmetry Prior for Single-View View Synthesis

Xingyi Li[1], Chaoyi Hong[1], Yiran Wang[1], Zhiguo Cao[1], Ke Xian[2](✉), and Guosheng Lin[2]

[1] Key Laboratory of Image Processing and Intelligent Control, Ministry of Education, School of AIA, Huazhong University of Science and Technology, Wuhan, China
{xingyi_li,cyhong,wangyiran,zgcao}@hust.edu.cn
[2] S-lab, Nanyang Technological University, Singapore, Singapore
{ke.xian,gslin}@ntu.edu.sg

Abstract. We study the problem of novel view synthesis of objects from a single image. Existing methods have demonstrated the potential in single-view view synthesis. However, they still fail to recover the fine appearance details, especially in self-occluded areas. This is because a single view only provides limited information. We observe that man-made objects usually exhibit symmetric appearances, which introduce additional prior knowledge. Motivated by this, we investigate the potential performance gains of explicitly embedding symmetry into the scene representation. In this paper, we propose SymmNeRF, a neural radiance field (NeRF) based framework that combines local and global conditioning under the introduction of symmetry priors. In particular, SymmNeRF takes the pixel-aligned image features and the corresponding symmetric features as extra inputs to the NeRF, whose parameters are generated by a hypernetwork. As the parameters are conditioned on the image-encoded latent codes, SymmNeRF is thus scene-independent and can generalize to new scenes. Experiments on synthetic and real-world datasets show that SymmNeRF synthesizes novel views with more details regardless of the pose transformation, and demonstrates good generalization when applied to unseen objects. Code is available at: https://github.com/xingyi-li/SymmNeRF.

Keywords: Novel view synthesis · NeRF · Symmetry · HyperNetwork

1 Introduction

Novel view synthesis is a long-standing problem in computer vision and graphics [4,9,20]. The task is to synthesize novel views from a set of input views or even a single input view, which is challenging as it requires comprehensive 3D understanding [36]. Prior works mainly focus on explicit geometric representations, such as voxel grids [14,21,32,42], point clouds [1,7], and triangle

---

Supplementary Information The online version contains supplementary material available at https://doi.org/10.1007/978-3-031-26319-4_14.

L. Wang et al. (Eds.): ACCV 2022, LNCS 13841, pp. 228–244, 2023.
https://doi.org/10.1007/978-3-031-26319-4_14

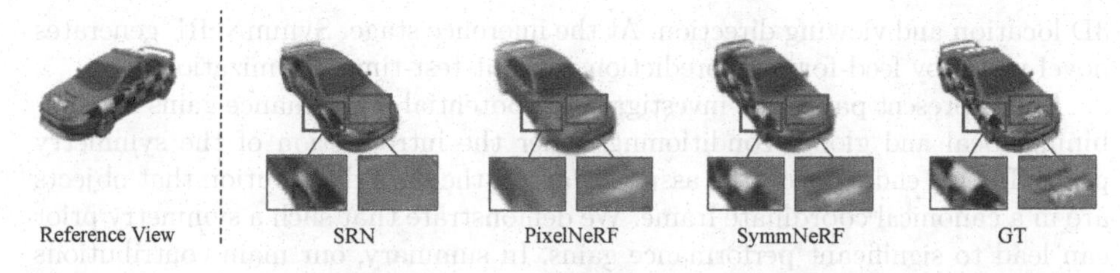

Fig. 1. Novel views from a single image synthesized by SRN [33], PixelNeRF [44] and our SymmNeRF. The competitive methods are prone to miss some texture details, especially when the pose difference between the reference view and target view is large. By contrast, SymmNeRF augmented with symmetry priors recovers more appearance details.

meshes [16,29,39]. However, these methods suffer from limited spatial resolution and representation capability because of the discrete properties. Recently, differentiable neural rendering methods [25,27,28,31,33,37] have shown great progress in synthesizing photo-realistic novel views. For example, neural radiance fields (NeRFs) [26], which implicitly encode volumetric density and color via multi-layer perceptrons (MLPs), show an impressive level of fidelity on novel view synthesis. However, these methods usually require densely captured views as input and test-time optimization, leading to poor generalization across objects and scenes. To reduce the strong dependency on dense inputs and enable better generalization to unseen objects, in this paper, we explore novel view synthesis of object categories from only a single image.

Novel view synthesis from a single image is challenging, because a single view cannot provide sufficient information. Recent NeRF-based methods [13,44] learn scene priors for reconstruction by training on multiple scenes. Although they have shown the potential in single-view view synthesis, it is particularly challenging when the pose difference between the reference and target view is large (see Fig. 1). We observe that man-made objects in real world usually exhibit symmetric appearances. Based on this, a question arises: *can symmetry priors benefit single-view view synthesis?*

To answer this question, we explore how to take advantage of symmetry priors to introduce additional information for reconstruction. To this end, we present SymmNeRF, a NeRF-based framework that is augmented by symmetry priors. Specifically, we take the pixel-aligned image features and the corresponding symmetric features as extra inputs to NeRF. This allows reasonable recovery of occluded geometry and missing texture. During training, given a set of posed input images, SymmNeRF simultaneously optimizes a convolutional neural network (CNN) encoder and a hypernetwork. The former encodes image features, and generates latent codes which represent the coarse shape and appearance of unseen objects. The latter maps specific latent codes to the weights of the neural radiance fields. Therefore, SymmNeRF is scene-independent and can generalize to unseen objects. Unlike the original NeRF [26], for a single query point, we take as input its original and symmetric pixel-aligned image features besides its

3D location and viewing direction. At the inference stage, SymmNeRF generates novel views by feed-forward prediction without test-time optimization.

In the present paper, we investigate the potential performance gains by combining local and global conditioning under the introduction of the symmetry prior. To this end, we add the assumptions on the data distribution that objects are in a canonical coordinate frame. We demonstrate that such a symmetry prior can lead to significant performance gains. In summary, our main contributions are:

- We propose SymmNeRF, a NeRF-based framework for novel view synthesis from a single image. By introducing symmetry priors into NeRF, SymmNeRF can synthesize high-quality novel views with fine details regardless of pose transformation.
- We combine local features with global conditioning via hypernetworks and demonstrate significant performance gains. Note that we perform inference via a CNN instead of auto-decoding, *i.e.*, without test-time optimization, which is different from SRN [33].
- Given only a single input image, SymmNeRF demonstrates significant improvement over state-of-the-art methods on synthetic and real-world datasets.

2 Related Work

Novel View Synthesis. Novel view synthesis is the task of synthesizing novel camera perspectives of a scene, given source images and their camera poses. The key challenges are understanding the 3D structure of the scene and inpainting of invisible regions of the scene [11]. The research of novel view synthesis has a long history in computer vision and graphics [4,9,20]. Pioneer works typically synthesize novel views by warping, resampling, and blending reference views to target viewpoints, which can be classified as image-based rendering methods [4]. However, they require densely captured views of the scene. When only a few observations are available, ghosting-like artifacts and holes may appear [36]. With the advancement of deep learning, a few learning-based methods have been proposed, most of which focus on explicit geometric representations such as voxel grids [14,21,32,42], point clouds [1,7], triangle meshes [16,29,39], and multiplane images (MPIs) [8,38,46].

Recent works [3,25,27,28,37] show that neural networks can be used as an implicit representation for 3D shapes and appearances. DeepSDF [28] maps continuous spatial coordinates to signed distance and proves the superiority of neural implicit functions. SRN [33] proposes to represent 3D shapes and appearances implicitly as a continuous, differentiable function that maps a spatial coordinate to the local features of the scene properties at that point. Recently, NeRF [26] shows astonishing results for novel view synthesis, which is an implicit MLP-based model that maps 3D coordinates plus 2D viewing directions to opacity and color values. However, NeRF requires enormous posed images and must be independently optimized for every scene. PixelNeRF [44] tries to address this issue by conditioning NeRF on image features, which are extracted by an image encoder. This enables its ability to render novel views from only a single image and its generalization to new scenes. Rematas *et al.* [30] propose ShaRF, a generative model aiming at estimating neural representation of objects from a single image, combining the benefits of explicit and implicit representations, which is capable of generalizing to unseen objects. CodeNeRF [13] learns to disentangle shape and texture by learning separate embeddings from a single image, allowing single view reconstruction and shape/texture editing. However, these methods usually struggle to synthesize reasonable novel views from a single image when self-occlusion occurs. In contrast, SymmNeRF first estimates coarse representations and then takes reflection symmetry as prior knowledge to inpaint invisible regions. This allows reasonable recovery of occluded geometry and missing texture.

HyperNetworks. A hypernetwork [10] refers to a small network that is trained to predict the weights of a large network, which has the potential to generalize previous knowledge and adapt quickly to novel environments. Various methods resort to hypernetworks in 3D vision. Littwin *et al.* [22] recover shape from a single image using hypernetworks in an end-to-end manner. SRN [33] utilizes hypernetworks for single-shot novel view synthesis with neural fields. In this work, we condition the parameters of NeRF on the image-encoded latent codes via the hypernetwork, which allows SymmNeRF to be scene-independent and generalize to new scenes.

Reflection Symmetry. Reflection symmetry plays a significant role in the human visual system and has already been exploited in the computer vision community. Wu *et al.* [41] infer 3D deformable objects given only a single image, using a symmetric structure to disentangle depth, albedo, viewpoint and illumination. Ladybird [43] assigns occluded points with features from their symmetric points based on the reflective symmetry of the object, allowing recovery of occluded geometry and texture. NeRD [47] learns a neural 3D reflection symmetry detector, which can estimate the normal vectors of objects' mirror planes. They focus on the task of detecting the 3D reflection symmetry of a symmetric object from a 2D image. In this work, we focus on exploring the advantages of explicitly embedding symmetry into the scene representation for single-view view synthesis.

3 SymmNeRF

3.1 Overview

Here we present an overview of our proposed method. We propose to firstly estimate holistic representations as well as symmetry planes, followed by fulfilling details, and to explicitly inject symmetry priors into single-view view synthesis of object categories. In particular, we design SymmNeRF to implement the ideas above. Figure 2 shows the technical pipeline of SymmNeRF.

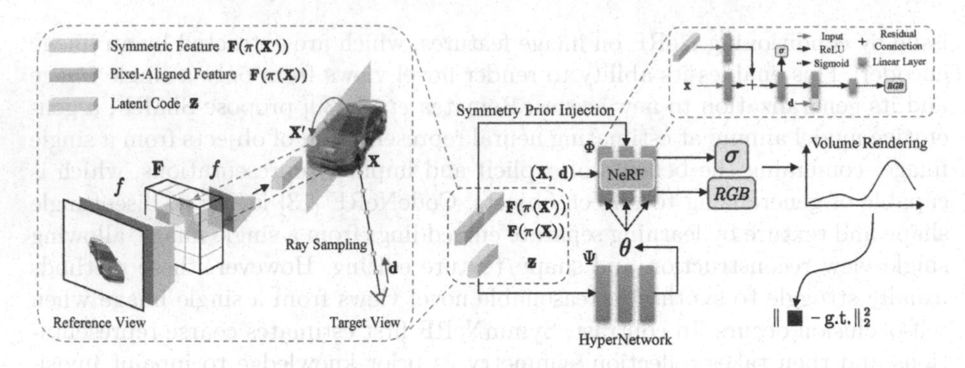

Fig. 2. An overview of our SymmNeRF. Given a reference view, we first encode holistic representations by estimating the latent code $\mathbf{z}$ through the image encoder network f. We then obtain the pixel-aligned feature $\mathbf{F}(\pi(\mathbf{X}))$ and the symmetric feature $\mathbf{F}(\pi(\mathbf{X}'))$, by projecting the query point $\mathbf{X}$ and the symmetric point $\mathbf{X}'$ to the 2D location on the image plane using camera parameters, followed by bilinear interpolation between the pixelwise features on the feature volume $\mathbf{F}$. The hypernetwork transforms the latent code $\mathbf{z}$ to the weights θ of the corresponding NeRF Φ. For a query point $\mathbf{X}$ along a target camera ray with viewing direction $\mathbf{d}$, NeRF takes the spatial coordinate $\mathbf{X}$, ray direction $\mathbf{d}$, pixel-aligned feature $\mathbf{F}(\pi(\mathbf{X}))$ and symmetric feature $\mathbf{F}(\pi(\mathbf{X}'))$ as input, and outputs the color and density.

Given a set of M instances training datasets $\mathcal{D} = \{\mathcal{C}_j\}_{j=1}^{M}$, where $\mathcal{C}_j = \{(\mathcal{I}_i^j, \mathbf{E}_i^j, \mathbf{K}_i^j)\}_{i=1}^{N}$ is a dataset of an instance object, $\mathcal{I}_i^j \in \mathbb{R}^{H\times W\times 3}$ refers to an input image, $\mathbf{E}_i^j = [\mathbf{R}|\mathbf{t}] \in \mathbb{R}^{3\times 4}$ and $\mathbf{K}_i^j \in \mathbb{R}^{3\times 3}$ are the corresponding extrinsic and intrinsic camera matrix respectively, and N denotes the number of the input images, SymmNeRF first encodes a holistic representation and regresses the symmetry plane for the input view. We then extract symmetric features along with pixel-aligned features for the sake of preserving fine-grained details observed in the input view. Subsequently, we transform the holistic representation to the weights of the corresponding NeRF, and inject symmetry priors (*i.e.*, symmetric features as well as pixel-aligned features) to predict colors and densities. Finally, we adopt the classic volume rendering technique [15,26] to synthesize novel views.

3.2 Encoding Holistic Representations

Humans usually understand the 3D shapes and appearances by first generating a profile, then restoring details from observations. Emulating the human visual system, we implement coarse depictions of objects by an image encoder network.

The image encoder network f is responsible for mapping the input image $\mathcal{I}_i$ into the latent code $\mathbf{z}_i$ which characterizes holistic information of the object's shape and appearance:

$$f: \mathbb{R}^{H\times W\times 3} \to \mathbb{R}^{k}, \quad \mathcal{I}_i \mapsto f(\mathcal{I}_i) = \mathbf{z}_i, \tag{1}$$

where k is the dimension of $\mathbf{z}_i$, and the parameters of f are denoted by Ω. Here, we denote the feature volume extracted by f during the encoding of holistic representations as $\mathbf{F}$ (*i.e.*, the concatenation of upsampled feature maps outputted by ResNet blocks). The image encoder network contains four ResNet blocks of ResNet-34 [12], followed by an average pooling layer and a fully connected layer.

3.3 Extracting Symmetric Features

The holistic representations introduced in the previous section coarsely describe objects. To synthesize detailed novel views, we follow PixelNeRF [44] and adopt pixel-aligned image features to compensate for the details. However, simply using pixel-aligned image features ignores the underlying 3D structure. In contrast, humans can infer the 3D shape and appearance from a single image, despite the information loss and self-occlusion that occurs during the imagery capture. This can boil down to the fact that humans usually resort to prior knowledge, *e.g.*, symmetry. Motivated by the above observation, we propose to inpaint invisible regions and alleviate the ill-posedness of novel view synthesis from a single image via symmetry priors. In the following, we briefly introduce the properties of 3D reflection symmetry [47], followed by how symmetry priors are applied.

3D Reflection Symmetry. When two points on an object's surface are symmetric, they share identical surface properties of the object. Formally, we define the symmetry regrading a rigid transformation $\mathbf{M} \in \mathbb{R}^{4\times 4}$ as

$$\forall \mathbf{X} \in \mathbb{S} : \begin{cases} \mathbf{MX} \in \mathbb{S}, \\ \mathcal{F}(\mathbf{X}) = \mathcal{F}(\mathbf{MX}), \end{cases} \tag{2}$$

where $\mathbf{X}$ is the homogeneous coordinate of a point on the object's surface, $\mathbb{S} \subset \mathbb{R}^4$ is the set of points that are on the surface of an object, $\mathbf{MX}$ is the symmetric point of $\mathbf{X}$, and $\mathcal{F}(\cdot)$ stands for the surface properties at a given point.

The 2D projections $\mathbf{x} = [x, y, 1, 1/d]^T$ and $\mathbf{x}' = [x', y', 1, 1/d']^T$ of two 3D points $\mathbf{X}, \mathbf{X}' \in \mathbb{S}$ satisfy

$$\begin{cases} \mathbf{x} = \mathbf{KR_t X}/d, \\ \mathbf{x}' = \mathbf{KR_t X}'/d', \end{cases} \tag{3}$$

where $\mathbf{K} \in \mathbb{R}^{4\times 4}$ and $\mathbf{R_t} \in \mathbb{R}^{4\times 4}$ are respectively the camera intrinsic matrix and extrinsic matrix. The latter transforms the coordinate from the world coordinate system to the camera coordinate system. d, d' are the depth in the camera space. When these two points are symmetric w.r.t. a rigid transformation, *i.e.*, $\mathbf{X}' = \mathbf{MX}$, the following constraint can be derived:

$$\mathbf{x}' = \frac{d}{d'} \mathbf{KR_t M R_t}^{-1} \mathbf{K}^{-1} \mathbf{x}, \tag{4}$$

where $\mathbf{x}$ and $\mathbf{x}'$ are 2D projections of these two 3D points. This suggests that given a 2D projection $\mathbf{x}$, we can obtain its symmetric counterpart $\mathbf{x}'$ if $\mathbf{M}$ and camera parameters are known. In this paper, we focus on exploring the benefits of explicitly embedding symmetry into our representation. To this end, we add the assumptions on the data distribution that objects are in a canonical coordinate frame, and that their symmetry axis is known.

Applying Symmetry Prior. To inpaint invisible regions, we apply the symmetry property introduced above and extract symmetric features $\mathbf{F}(\pi(\mathbf{X}'))$. This can be achieved by projecting the symmetric point $\mathbf{X}'$ to the 2D location $\mathbf{x}'$ on the image plane using camera parameters, followed by bilinearly interpolating between the pixelwise features on the feature volume $\mathbf{F}$ extracted by the image encoder network f. In addition, we follow PixelNeRF [44] and adopt pixel-aligned features $\mathbf{F}(\pi(\mathbf{X}))$, which share the same acquisition approach with symmetric features. They are subsequently concatenated together to form the local image features corresponding to $\mathbf{X}$.

3.4 Injecting Symmetry Prior into NeRF

In this section, we inject symmetry priors into the neural radiance field for single-view view synthesis. Technically, the weights of the neural radiance field are conditioned on the latent code $\mathbf{z}_i$ introduced in Sect. 3.2, which represents a coarse but holistic depiction of the object. To preserve fine-grained details, during the color and density prediction, we also take the pixel-aligned image features and the corresponding symmetric features as extra inputs to fulfill details observed in the input view.

Generating Neural Radiance Fields. We generate a specific neural radiance field by mapping a latent code $\mathbf{z}_i$ to the weights θ_i of the neural radiance field using the hypernetwork Ψ, which can be defined as follows:

$$\Psi : \mathbb{R}^k \to \mathbb{R}^l, \quad \mathbf{z}_i \mapsto \Psi(\mathbf{z}_i) = \theta_i, \tag{5}$$

where, l stands for the dimension of the parameter space of neural radiance fields. We parameterize Ψ as an MLP with parameters ψ. This can be interpreted as a simulation of the human visual system. Specifically, humans first estimate the holistic shape and appearance of the unseen object when given a single image, then formulate a sketch in their mind to represent the object. Similarly, SymmNeRF encodes overall information of the object as a latent code from a single image, followed by generating a corresponding neural radiance field to describe the object.

Color and Density Prediction. Given a reference image with known camera parameters, for a single query point location $\mathbf{X} \in \mathbb{R}^3$ on a ray $\mathbf{r} \in \mathbb{R}^3$ with unit-length viewing direction $\mathbf{d} \in \mathbb{R}^3$, SymmNeRF predicts the color and density at that point in 3D space, which is defined as:

$$\begin{aligned} &\Phi : \mathbb{R}^{m_{\mathbf{X}}} \times \mathbb{R}^{m_{\mathbf{d}}} \times \mathbb{R}^{2n} \to \mathbb{R}^3 \times \mathbb{R}, \\ &(\gamma_{\mathbf{X}}(\mathbf{X}), \gamma_{\mathbf{d}}(\mathbf{d}), \mathbf{F}(\pi(\mathbf{X})), \mathbf{F}(\pi(\mathbf{X}'))) \mapsto \\ &\Phi(\gamma_{\mathbf{X}}(\mathbf{X}), \gamma_{\mathbf{d}}(\mathbf{d}), \mathbf{F}(\pi(\mathbf{X})), \mathbf{F}(\pi(\mathbf{X}'))) = (\mathbf{c}, \sigma), \end{aligned} \tag{6}$$

where Φ represents a neural radiance field, an MLP network whose weights are given by the hypernetwork Ψ, $\mathbf{X}' \in \mathbb{R}^3$ is the corresponding symmetric 3D point of $\mathbf{X}$, $\gamma_{\mathbf{X}}(\cdot)$ and $\gamma_{\mathbf{d}}(\cdot)$ are positional encoding functions for spatial locations and viewing directions, n, $m_{\mathbf{X}}$ and $m_{\mathbf{d}}$ are respectively the dimensions of pixel-aligned features (symmetric features), $\gamma_{\mathbf{X}}(\mathbf{X})$ and $\gamma_{\mathbf{d}}(\mathbf{d})$, π denotes the process of projecting the 3D point onto the image plane using known intrinsics, and $\mathbf{F}$ is the feature volume extracted by the image encoder network f.

3.5 Volume Rendering

To render the color of a ray $\mathbf{r}$ passing through the scene, we first compute its camera ray $\mathbf{r}$ using the camera parameters and sample K points $\{\mathbf{X}_k\}_{k=1}^{K}$ along the camera ray $\mathbf{r}$ between near and far bounds, and then perform classical volume rendering [15,26]:

$$\tilde{C}(\mathbf{r}) = \sum_{k=1}^{K} T_k(1 - \exp(-\sigma_k \delta_k))\mathbf{c}_k, \tag{7}$$

$$\text{where} \quad T_k = \exp(-\sum_{j=1}^{k-1} \sigma_j \delta_j), \tag{8}$$

where $\mathbf{c}_k$ and σ_k denote the color and density of the k-th sample on the ray, respectively, and $\delta_k = \|\mathbf{X}_{k+1} - \mathbf{X}_k\|_2$ is the interval between adjacent samples.

3.6 Training

To summarize, given a set of M instances training datasets $\mathcal{D} = \{\mathcal{C}_j\}_{j=1}^{M}$, where $\mathcal{C}_j = \{(\mathcal{I}_i^j, \mathbf{E}_i^j, \mathbf{K}_i^j)\}_{i=1}^{N}$ is a dataset of an instance object, we optimize SymmNeRF to minimize the rendering error of observed images:

$$\min_{\Omega,\psi} \sum_{j=1}^{M} \sum_{i=1}^{N} \mathcal{L}(\mathcal{I}_i^j, \mathbf{E}_i^j, \mathbf{K}_i^j; \Omega, \psi), \tag{9}$$

$$\mathcal{L} = \sum_{\mathbf{r} \in \mathcal{R}} \left\| \tilde{C}(\mathbf{r}) - C(\mathbf{r}) \right\|_2^2, \tag{10}$$

where Ω and ψ are respectively the parameters of the image encoder network f and the hypernetwork Ψ, $\mathcal{R}$ is the set of camera rays passing through image pixels, and $C(\mathbf{r})$ denotes the ground truth pixel color.

4 Experiments

4.1 Datasets

Synthetic Renderings. We evaluate our approach on the synthetic ShapeNet benchmark [2] for single-shot reconstruction. 1) We mainly focus on the

Table 1. Quantitative comparisons against state-of-the-art methods on "Cars" and "Chairs" classes of the ShapeNet-SRN dataset. The best performance is in **bold**, and the second best is underlined.

Methods	Chairs		Cars		Average	
	PSNR↑	SSIM↑	PSNR↑	SSIM↑	PSNR↑	SSIM↑
GRF [37] (ICCV'21)	21.25	0.86	20.33	0.82	20.79	0.84
TCO [35] (ECCV'16)	21.27	0.88	-	-	-	-
dGQN [6] (Science'18)	21.59	0.87	-	-	-	-
ENR [5] (ICML'20)	22.83	-	22.26	-	22.55	-
SRN [33] (NeurIPS'19)	22.89	0.89	22.25	0.89	22.57	0.89
PixelNeRF [44] (CVPR'21)	23.72	0.91	23.17	0.90	23.45	0.91
ShaRF [30] (ICML'21)	23.37	**0.92**	22.53	0.90	22.90	0.91
CodeNeRF [13] (ICCV'21)	23.66	0.90	**23.80**	**0.91**	23.74	0.91
Ours	**24.32**	**0.92**	23.44	**0.91**	**23.88**	**0.92**

ShapeNet-SRN dataset, following the same protocol adopted in [33]. This dataset includes two object categories: 3,514 "Cars" and 6,591 "Chairs". The train/test split is predefined across object instances. There are 50 views per object instance in the training set. For testing, 251 novel views in an Archimedean spiral are used for each object instance in the test set. All images are at 128×128 pixels; 2) Similar to PixelNeRF [44], we also test our method on the ShapeNet-NMR dataset [17] under two settings: category-agnostic single-view reconstruction and generalization to unseen categories, following [17,23,27]. This dataset contains the 13 largest categories of ShapeNet and provides 24 fixed elevation views for each object instance. All images are of 64×64 resolution.

Real-World Renderings. We also generalize our model, trained only on the ShapeNet-SRN dataset, directly to two complex real-world datasets. One is the Pix3D [34] dataset containing various image-shape pairs with 2D-3D alignment. The other is the Stanford Cars [19] dataset which contains various real images of 196 classes of cars. All images of the two datasets are cropped and resized to 128×128 pixels during testing.

4.2 Implementation Details

SymmNeRF is trained using the AdamW optimizer [18,24]. The learning rate follows the warmup [12] strategy: linearly growing from 0 to 1×10^{-4} during the first 2k iterations and then decaying exponentially close to 0 over the optimization. The network parameters are updated with around 400–500k iterations. We use a batch size of 4 objects and a ray batch size of 256, each queried at 64 samples. Experiments are conducted on 2 NVIDIA GeForce RTX 3090 GPUs.

4.3 Comparisons

Here we compare SymmNeRF against the existing state-of-the-art methods, among which CodeNeRF, SRN and ShaRF require test-time optimization at inference, and ShaRF entails 3D ground truth voxel grids besides 2D supervision. To evaluate the quality of renderings, we adopt two standard image quality metrics: the Peak Signal-to-Noise Ratio (PSNR) and the Structure Similarity

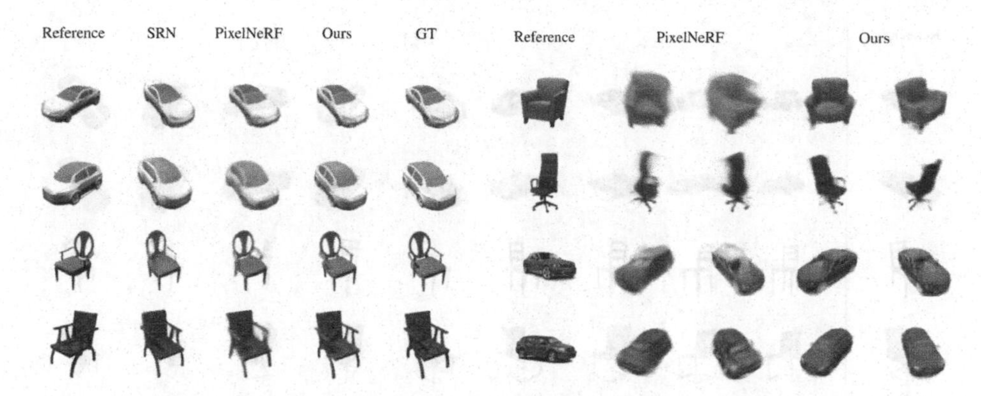

Fig. 3. Qualitative comparisons on "Cars" and "Chairs" classes. SymmNeRF can produce high-quality renderings with fine-grained details, proper geometry and reasonable appearance.

Fig. 4. Qualitative comparisons with PixelNeRF [44] on real-world Pix3D [34] and Stanford Cars [19] datasets. Compared with PixelNeRF, SymmNeRF yields better generalization.

Index Measure (SSIM) [40]. We also include LPIPS [45] in all experiments except the ShapeNet-SRN dataset. The better approach favors the higher PSNR and SSIM, and the lower LPIPS. Please refer to the supplementary material for more visualization.

Evaluations on the ShapeNet-SRN Dataset. In general, as shown in Table 1, our method outperforms or at least is on par with state-of-the-art methods. For the "Cars" category, SymmNeRF outperforms its competitors including PixelNeRF, SRN and ShaRF, and achieves comparable performance with CodeNeRF. Note that *our SymmNeRF solves a much harder problem than SRN and CodeNeRF*. In particular, SymmNeRF directly infers the unseen object representation in a single forward pass, while SRN and CodeNeRF need to be retrained on all new objects to optimize the latent codes. In addition, we observe that most cars from the "Cars" category share similar 3D shapes and simple textures. As a result, the experiment on the "Cars" category is in favor of CodeNeRF. In contrast, for the "Chairs" category, SymmNeRF significantly outperforms all baselines across all metrics by a large margin. This result implies that our model can generalize well on new objects, as the shapes and textures of chairs in the "Chairs" category vary considerably. This implies that SymmNeRF indeed captures the underlying 3D structure of objects with the help of symmetry priors and the hypernetwork, rather than simply exploiting data biases.

Here we compare SymmNeRF qualitatively with SRN and PixelNeRF in Fig. 3. One can observe that: i) SRN is prone to generate overly smooth renderings and is unable to capture the accurate geometry, leading to some distortions; ii) PixelNeRF performs well when the query view is close to the reference one, but fails to recover the details invisible in the reference, especially when the rendered view is far from the reference; iii) SymmNeRF, by contrast, can synthesize

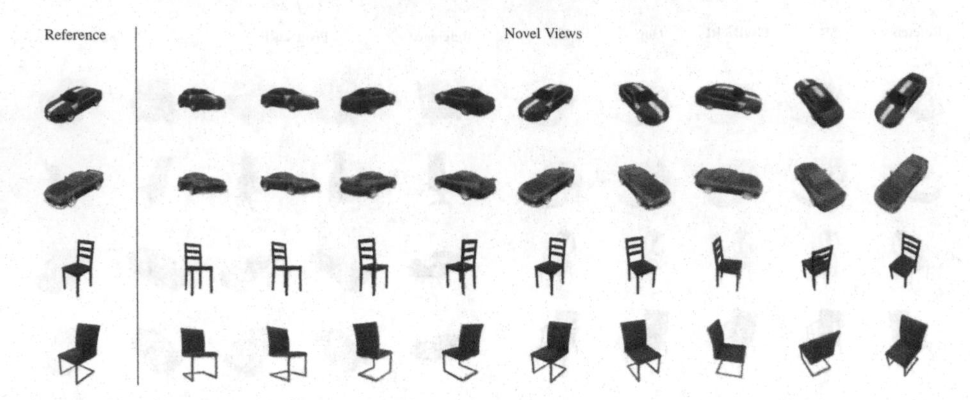

Fig. 5. Novel view synthesis on "Cars" and "Chairs" of ShapeNet-SRN dataset.

photo-realistic, reasonable novel views with fine-grained details close to ground truths.

We further demonstrate the high-quality results of SymmNeRF by providing more novel view synthesis visualization in Fig. 5. As can be seen, SymmNeRF can always synthesize photo-realistic and reasonable novel renderings from totally different viewpoints.

Generalization on Real-World Datasets. To further investigate the generalization of SymmNeRF, we evaluate SymmNeRF on two real-world datasets, *i.e.*, the Pix3D [34] and the Stanford Cars [19]. Note that for the lack of ground truth, we only show the qualitative results on the two datasets. Here we apply SymmNeRF trained on the synthetic chairs and cars directly on the real-world images without any finetuning. As shown in Fig. 4, PixelNeRF [44] fails to synthesize reasonable novel views, because it only notices the use of pixel-aligned image features, ignoring the underlying 3D structure the reference view provides. Compared with PixelNeRF [44], SymmNeRF can effectively infer the geometry and appearance of real-world chairs and cars. Please also note that there are no camera poses for real-world objects from Pix3D and Stanford Cars. Our model assumes that objects are at the center of the canonical space and once trained, can estimate camera poses for each reference view similar to CodeNeRF [13].

Evaluations on the ShapeNet-NMR Dataset. Although the experimental results of two common categories have demonstrated that including symmetry is simple yet effective, we further explore our approach on the ShapeNet-NMR dataset under two settings: category-agnostic single-view reconstruction and generalization to unseen categories. 1) Category-agnostic single-view reconstruction: only a single model is trained across the 13 largest categories of ShapeNet. We show in Table 2 and Fig. 6 that SymmNeRF outperforms other state-of-the-art methods [23,27,33,44]. This also implies that *symmetry priors benefit the reconstruction of almost all symmetric objects*; 2) Generalization to unseen categories: we reconstruct ShapeNet categories which are not involved in training.

Table 2. Quantitative comparisons against state-of-the-art methods on the 13 largest categories of the ShapeNet-NMR dataset.

	Methods	plane	bench	cbnt.	car	chair	disp.	lamp	spkr.	rifle	sofa	table	phone	boat	mean
PSNR↑	DVR	25.29	22.64	24.47	23.95	19.91	20.86	23.27	20.78	23.44	22.35	21.53	24.18	25.09	22.70
	SRN	26.62	22.20	23.42	24.40	21.85	19.07	22.17	21.04	24.95	23.65	22.45	20.87	25.86	23.28
	PixelNeRF	29.76	26.35	27.72	27.58	23.84	24.22	28.58	24.44	30.60	26.94	25.59	27.13	29.18	26.80
	Ours	**30.57**	**27.44**	**29.34**	**27.87**	**24.29**	**24.90**	**28.98**	**25.14**	**30.64**	**27.70**	**27.16**	**28.27**	**29.71**	**27.57**
SSIM↑	DVR	0.905	0.866	0.877	0.909	0.787	0.814	0.849	0.798	0.916	0.868	0.840	0.892	0.902	0.860
	SRN	0.901	0.837	0.831	0.897	0.814	0.744	0.801	0.779	0.913	0.851	0.828	0.811	0.898	0.849
	PixelNeRF	0.947	0.911	0.910	0.942	0.858	0.867	0.913	0.855	0.968	0.908	0.898	0.922	0.939	0.910
	Ours	**0.955**	**0.925**	**0.922**	**0.945**	**0.865**	**0.875**	**0.917**	**0.862**	**0.970**	**0.915**	**0.917**	**0.929**	**0.943**	**0.919**
LPIPS↓	DVR	0.095	0.129	0.125	0.098	0.173	0.150	0.172	0.170	0.094	0.119	0.139	0.110	0.116	0.130
	SRN	0.111	0.150	0.147	0.115	0.152	0.197	0.210	0.178	0.111	0.129	0.135	0.165	0.134	0.139
	PixelNeRF	0.084	0.116	0.105	0.095	0.146	0.129	0.114	0.141	0.066	0.116	0.098	0.097	0.111	0.108
	Ours	**0.062**	**0.085**	**0.068**	**0.082**	**0.120**	**0.104**	**0.096**	**0.108**	**0.054**	**0.086**	**0.067**	**0.068**	**0.089**	**0.084**

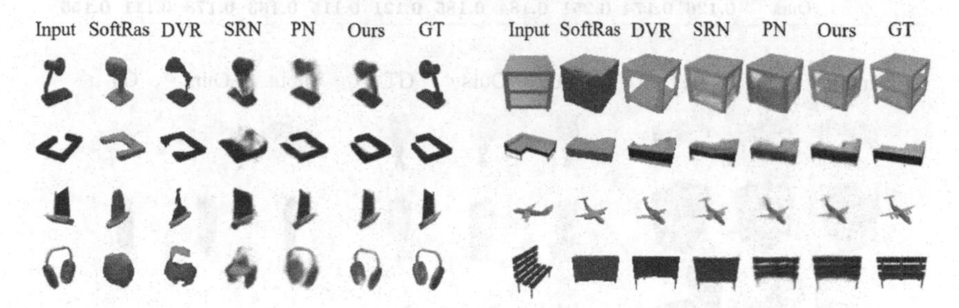

Fig. 6. Qualitative comparisons on the ShapeNet-NMR dataset under the category-agnostic single-view reconstruction setting.

The results in Table 3 and Fig. 7 suggest that our method performs comparably to PixelNeRF. This means that *our method can also handle out-of-distribution categories with the help of symmetry priors.*

Asymmetric Objects. As shown in Fig. 6 (Row 2), *our method can also deal with objects that are not perfectly symmetric.* This is because a few asymmetric objects are also included in the training dataset. Our model can perceive and recognize asymmetry thanks to the global latent code and hypernetwork. SymmNeRF therefore adaptively chooses to utilize local features to reconstruct asymmetric objects.

4.4 Ablation Study

To validate the design choice of SymmNeRF, we conduct ablation studies on the synthetic "Chairs" and "Cars" from the ShapeNet-SRN dataset. Table 4 shows the results corresponding to the effectiveness of the pixel-aligned, symmetric features and the hypernetwork. One can observe: i) *The symmetry priors injection benefits novel view synthesis.* Compared with (b) in average performance, our full model (c) with the symmetric priors injection yields a relative improvement of 6.0% PSNR and 2.3% SSIM. This finding highlights the importance of the

Table 3. Quantitative comparisons against state-of-the-art methods on 10 unseen categories of ShapeNet-NMR dataset. The models are trained on only planes, cars and chairs.

	Methods	bench	cbnt.	disp.	lamp	spkr.	rifle	sofa	table	phone	boat	mean
PSNR↑	DVR	18.37	17.19	14.33	18.48	16.09	20.28	18.62	16.20	16.84	22.43	17.72
	SRN	18.71	17.04	15.06	19.26	17.06	23.12	18.76	17.35	15.66	24.97	18.71
	PixelNeRF	23.79	**22.85**	**18.09**	**22.76**	**21.22**	23.68	24.62	**21.65**	**21.05**	26.55	**22.71**
	Ours	**23.87**	21.36	16.83	22.68	19.98	**23.77**	**25.10**	21.10	20.48	**26.80**	22.36
SSIM↑	DVR	0.754	0.686	0.601	0.749	0.657	0.858	0.755	0.644	0.731	0.857	0.716
	SRN	0.702	0.626	0.577	0.685	0.633	0.875	0.702	0.617	0.635	0.875	0.684
	PixelNeRF	0.863	**0.814**	**0.687**	0.818	**0.778**	0.899	0.866	**0.798**	0.801	0.896	**0.825**
	Ours	**0.873**	0.780	0.663	**0.824**	0.751	**0.902**	**0.881**	0.792	**0.802**	**0.909**	0.823
LPIPS↓	DVR	0.219	0.257	0.306	0.259	0.266	0.158	0.196	0.280	0.245	0.152	0.240
	SRN	0.282	0.314	0.333	0.321	0.289	0.175	0.248	0.315	0.324	0.163	0.280
	PixelNeRF	0.164	0.186	0.271	0.208	0.203	0.141	0.157	0.188	0.207	0.148	0.182
	Ours	**0.126**	**0.174**	**0.251**	**0.184**	**0.185**	**0.121**	**0.115**	**0.163**	**0.178**	**0.111**	**0.155**

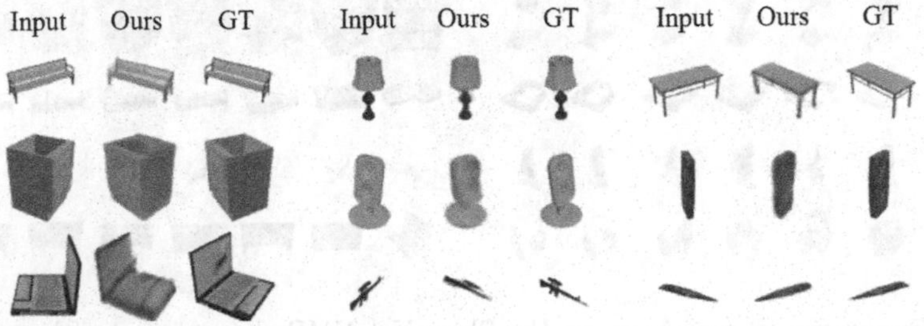

Fig. 7. Qualitative visualization on the ShapeNet-NMR dataset under the generalization to unseen categories setting.

symmetry priors on novel view synthesis when only a single image is provided; ii) *The hypernetwork matters.* Compared with our full model (c), the rendering quality of (d) deteriorates if we do not adopt the hypernetwork. This may lie in the fact that simply conditioning on local features ignores the underlying 3D structure of objects. In contrast, combining local and global conditioning via the hypernetwork module not only enables recovery of rendering details, but also improves generalization to unseen objects in a coarse-to-fine manner. We also visualize the comparative results in Fig. 8. The baseline model (a) tends to render smoothly. Simply using pixel-aligned image features (b) still fails to fully understand 3D structure. In contrast, our full model (c) reproduces photo-realistic details from most viewpoints. The rendering quality of (d) deteriorates as the hypernetwork is not adopted. We have to emphasize that, *only including both the symmetry priors and the hypernetwork can accurately recovers the geometry information and texture details despite the occlusions.*

Table 4. Ablation study on each component of SymmNeRF.

		Image encoder network	Hypernetwork	Local features	Symm features	PSNR(Δ)↑	SSIM(Δ)↑
Chairs	(a)	✓	✓			21.26 (–)	0.87 (–)
	(b)	✓	✓	✓		23.09 (8.6%)	0.91 (4.6%)
	(c)	✓	✓	✓	✓	**24.32 (14.4%)**	**0.92 (5.7%)**
	(d)	✓		✓	✓	19.76 (−7.1%)	0.85 (−2.3%)
Cars	(a)	✓	✓			20.65 (–)	0.87 (–)
	(b)	✓	✓	✓		22.15 (7.3%)	0.89 (2.3%)
	(c)	✓	✓	✓	✓	**23.44 (13.5%)**	**0.91 (4.6%)**
	(d)	✓		✓	✓	21.15 (2.4%)	0.86 (−1.2%)
Average	(a)	✓	✓			20.96 (–)	0.87 (–)
	(b)	✓	✓	✓		22.62 (7.9%)	0.90 (3.4%)
	(c)	✓	✓	✓	✓	**23.88 (13.9%)**	**0.92 (5.7%)**
	(d)	✓		✓	✓	20.46 (−2.4%)	0.86 (−1.1%)

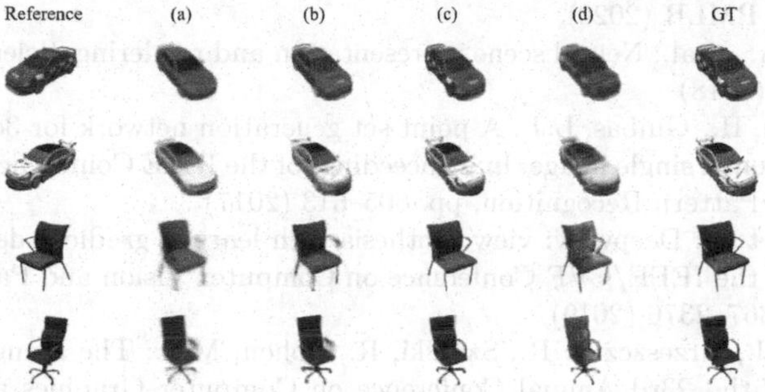

Fig. 8. Qualitative evaluation of different configurations on ShapeNet-SRN.

5 Conclusion

Existing methods [13,44] fail to synthesize fine appearance details of objects, especially when the target view is far away from the reference view. They focus on learning scene priors, but ignore fully exploring the attributes of objects, *e.g.*, symmetry. In this paper, we investigate the potential performance gains of explicitly injecting symmetry priors into the scene representation. In particular, we combine hypernetworks [33] with local conditioning [31,37,44], embedded with the symmetry prior. Experimental results demonstrate that such a symmetry prior can boost our model to synthesize novel views with more details regardless of the pose transformation, and show good generalization when applied to unseen objects.

Acknowledgement. This work is supported in part by the National Natural Science Foundation of China (Grant No. U1913602). This study is also supported under the RIE2020 Industry Alignment Fund - Industry Collaboration Projects (IAF-ICP) Funding Initiative, as well as cash and in-kind contribution from the industry partner(s).

References

1. Achlioptas, P., Diamanti, O., Mitliagkas, I., Guibas, L.: Learning representations and generative models for 3D point clouds. In: International Conference on Machine Learning, pp. 40–49. PMLR (2018)
2. Chang, A.X., et al.: Shapenet: an information-rich 3D model repository. arXiv preprint arXiv:1512.03012 (2015)
3. Chen, Z., Zhang, H.: Learning implicit fields for generative shape modeling. In: Proceedings of the IEEE/CVF Conference on Computer Vision and Pattern Recognition, pp. 5939–5948 (2019)
4. Debevec, P.E., Taylor, C.J., Malik, J.: Modeling and rendering architecture from photographs: a hybrid geometry-and image-based approach. In: Proceedings of the 23rd Annual Conference on Computer Graphics and Interactive Techniques, pp. 11–20 (1996)
5. Dupont, E., Martin, M.B., Colburn, A., Sankar, A., Susskind, J., Shan, Q.: Equivariant neural rendering. In: International Conference on Machine Learning, pp. 2761–2770. PMLR (2020)
6. Eslami, S.A., et al.: Neural scene representation and rendering. Science **360**(6394), 1204–1210 (2018)
7. Fan, H., Su, H., Guibas, L.J.: A point set generation network for 3d object reconstruction from a single image. In: Proceedings of the IEEE Conference on Computer Vision and Pattern Recognition, pp. 605–613 (2017)
8. Flynn, J., et al.: Deepview: view synthesis with learned gradient descent. In: Proceedings of the IEEE/CVF Conference on Computer Vision and Pattern Recognition, pp. 2367–2376 (2019)
9. Gortler, S.J., Grzeszczuk, R., Szeliski, R., Cohen, M.F.: The lumigraph. In: Proceedings of the 23rd Annual Conference on Computer Graphics and Interactive Techniques, pp. 43–54 (1996)
10. Ha, D., Dai, A., Le, Q.V.: Hypernetworks. arXiv preprint arXiv:1609.09106 (2016)
11. Häni, N., Engin, S., Chao, J.J., Isler, V.: Continuous object representation networks: novel view synthesis without target view supervision. arXiv preprint arXiv:2007.15627 (2020)
12. He, K., Zhang, X., Ren, S., Sun, J.: Deep residual learning for image recognition. In: Proceedings of the IEEE Conference on Computer Vision and Pattern Recognition, pp. 770–778 (2016)
13. Jang, W., Agapito, L.: CodeNerf: disentangled neural radiance fields for object categories. In: Proceedings of the IEEE/CVF International Conference on Computer Vision, pp. 12949–12958 (2021)
14. Jimenez Rezende, D., Eslami, S., Mohamed, S., Battaglia, P., Jaderberg, M., Heess, N.: Unsupervised learning of 3D structure from images. Adv. Neural. Inf. Process. Syst. **29**, 4996–5004 (2016)
15. Kajiya, J.T., Von Herzen, B.P.: Ray tracing volume densities. ACM SIGGRAPH Comput. Graph. **18**(3), 165–174 (1984)
16. Kanazawa, A., Tulsiani, S., Efros, A.A., Malik, J.: Learning category-specific mesh reconstruction from image collections. In: Ferrari, V., Hebert, M., Sminchisescu, C., Weiss, Y. (eds.) ECCV 2018. LNCS, vol. 11219, pp. 386–402. Springer, Cham (2018). https://doi.org/10.1007/978-3-030-01267-0_23
17. Kato, H., Ushiku, Y., Harada, T.: Neural 3D mesh renderer. In: The IEEE Conference on Computer Vision and Pattern Recognition (CVPR) (2018)

18. Kingma, D.P., Ba, J.: Adam: a method for stochastic optimization. arXiv preprint arXiv:1412.6980 (2014)
19. Krause, J., Stark, M., Deng, J., Fei-Fei, L.: 3D object representations for fine-grained categorization. In: Proceedings of the IEEE International Conference on Computer Vision Workshops, pp. 554–561 (2013)
20. Levoy, M., Hanrahan, P.: Light field rendering. In: Proceedings of the 23rd Annual Conference on Computer Graphics and Interactive Techniques, pp. 31–42 (1996)
21. Liao, Y., Donne, S., Geiger, A.: Deep marching cubes: learning explicit surface representations. In: Proceedings of the IEEE Conference on Computer Vision and Pattern Recognition, pp. 2916–2925 (2018)
22. Littwin, G., Wolf, L.: Deep meta functionals for shape representation. In: Proceedings of the IEEE/CVF International Conference on Computer Vision, pp. 1824–1833 (2019)
23. Liu, S., Li, T., Chen, W., Li, H.: Soft rasterizer: a differentiable renderer for image-based 3D reasoning. In: The IEEE International Conference on Computer Vision (ICCV), October 2019
24. Loshchilov, I., Hutter, F.: Decoupled weight decay regularization. In: International Conference on Learning Representations (2018)
25. Mescheder, L., Oechsle, M., Niemeyer, M., Nowozin, S., Geiger, A.: Occupancy networks: learning 3D reconstruction in function space. In: Proceedings of the IEEE/CVF Conference on Computer Vision and Pattern Recognition, pp. 4460–4470 (2019)
26. Mildenhall, B., Srinivasan, P.P., Tancik, M., Barron, J.T., Ramamoorthi, R., Ng, R.: NeRF: representing scenes as neural radiance fields for view synthesis. In: Vedaldi, A., Bischof, H., Brox, T., Frahm, J.-M. (eds.) ECCV 2020. LNCS, vol. 12346, pp. 405–421. Springer, Cham (2020). https://doi.org/10.1007/978-3-030-58452-8_24
27. Niemeyer, M., Mescheder, L., Oechsle, M., Geiger, A.: Differentiable volumetric rendering: learning implicit 3d representations without 3D supervision. In: Proceedings of the IEEE/CVF Conference on Computer Vision and Pattern Recognition, pp. 3504–3515 (2020)
28. Park, J.J., Florence, P., Straub, J., Newcombe, R., Lovegrove, S.: Deepsdf: learning continuous signed distance functions for shape representation. In: Proceedings of the IEEE/CVF Conference on Computer Vision and Pattern Recognition, pp. 165–174 (2019)
29. Ranjan, A., Bolkart, T., Sanyal, S., Black, M.J.: Generating 3D faces using convolutional mesh autoencoders. In: Ferrari, V., Hebert, M., Sminchisescu, C., Weiss, Y. (eds.) ECCV 2018. LNCS, vol. 11207, pp. 725–741. Springer, Cham (2018). https://doi.org/10.1007/978-3-030-01219-9_43
30. Rematas, K., Martin-Brualla, R., Ferrari, V.: Sharf: shape-conditioned radiance fields from a single view. In: ICML (2021)
31. Saito, S., Huang, Z., Natsume, R., Morishima, S., Kanazawa, A., Li, H.: Pifu: pixel-aligned implicit function for high-resolution clothed human digitization. In: Proceedings of the IEEE/CVF International Conference on Computer Vision, pp. 2304–2314 (2019)
32. Sitzmann, V., Thies, J., Heide, F., Nießner, M., Wetzstein, G., Zollhofer, M.: Deepvoxels: learning persistent 3D feature embeddings. In: Proceedings of the IEEE/CVF Conference on Computer Vision and Pattern Recognition, pp. 2437–2446 (2019)

33. Sitzmann, V., Zollhoefer, M., Wetzstein, G.: Scene representation networks: Continuous 3D-structure-aware neural scene representations. Adv. Neural. Inf. Process. Syst. **32**, 1121–1132 (2019)
34. Sun, X., et al.: Pix3D: dataset and methods for single-image 3D shape modeling. In: Proceedings of the IEEE Conference on Computer Vision and Pattern Recognition, pp. 2974–2983 (2018)
35. Tatarchenko, M., Dosovitskiy, A., Brox, T.: Multi-view 3D models from single images with a convolutional network. In: Leibe, B., Matas, J., Sebe, N., Welling, M. (eds.) ECCV 2016. LNCS, vol. 9911, pp. 322–337. Springer, Cham (2016). https://doi.org/10.1007/978-3-319-46478-7_20
36. Tewari, A., et al.: State of the art on neural rendering. In: Computer Graphics Forum, vol. 39, pp. 701–727. Wiley Online Library (2020)
37. Trevithick, A., Yang, B.: GRF: learning a general radiance field for 3D representation and rendering. In: Proceedings of the IEEE/CVF International Conference on Computer Vision (ICCV), pp. 15182–15192, October 2021
38. Tucker, R., Snavely, N.: Single-view view synthesis with multiplane images. In: Proceedings of the IEEE/CVF Conference on Computer Vision and Pattern Recognition, pp. 551–560 (2020)
39. Wang, N., Zhang, Y., Li, Z., Fu, Y., Liu, W., Jiang, Y.-G.: Pixel2Mesh: generating 3D mesh models from single RGB images. In: Ferrari, V., Hebert, M., Sminchisescu, C., Weiss, Y. (eds.) ECCV 2018. LNCS, vol. 11215, pp. 55–71. Springer, Cham (2018). https://doi.org/10.1007/978-3-030-01252-6_4
40. Wang, Z., Bovik, A.C., Sheikh, H.R., Simoncelli, E.P.: Image quality assessment: from error visibility to structural similarity. IEEE Trans. Image Process. **13**(4), 600–612 (2004)
41. Wu, S., Rupprecht, C., Vedaldi, A.: Unsupervised learning of probably symmetric deformable 3D objects from images in the wild. In: Proceedings of the IEEE/CVF Conference on Computer Vision and Pattern Recognition, pp. 1–10 (2020)
42. Xie, H., Yao, H., Sun, X., Zhou, S., Zhang, S.: Pix2vox: context-aware 3D reconstruction from single and multi-view images. In: Proceedings of the IEEE/CVF International Conference on Computer Vision, pp. 2690–2698 (2019)
43. Xu, Y., Fan, T., Yuan, Y., Singh, G.: Ladybird: Quasi-Monte Carlo sampling for deep implicit field based 3D reconstruction with symmetry. In: Vedaldi, A., Bischof, H., Brox, T., Frahm, J.-M. (eds.) ECCV 2020. LNCS, vol. 12346, pp. 248–263. Springer, Cham (2020). https://doi.org/10.1007/978-3-030-58452-8_15
44. Yu, A., Ye, V., Tancik, M., Kanazawa, A.: PixelNerf: neural radiance fields from one or few images. In: Proceedings of the IEEE/CVF Conference on Computer Vision and Pattern Recognition, pp. 4578–4587 (2021)
45. Zhang, R., Isola, P., Efros, A.A., Shechtman, E., Wang, O.: The unreasonable effectiveness of deep features as a perceptual metric. In: CVPR (2018)
46. Zhou, T., Tucker, R., Flynn, J., Fyffe, G., Snavely, N.: Stereo magnification: learning view synthesis using multiplane images. ACM Trans. Graph. (TOG) **37**(4), 1–12 (2018)
47. Zhou, Y., Liu, S., Ma, Y.: Nerd: neural 3D reflection symmetry detector. In: Proceedings of the IEEE/CVF Conference on Computer Vision and Pattern Recognition, pp. 15940–15949 (2021)

Meta-Det3D: Learn to Learn Few-Shot 3D Object Detection

Shuaihang Yuan[1,2,3,4], Xiang Li[1,2,3,4], Hao Huang[1,2,3,4], and Yi Fang[1,2,3,4(✉)]

[1] NYU Multimedia and Visual Computing Lab, New York, USA
{sy2366,xl1845,hh1811,yfang}@nyu.edu
[2] NYUAD Center for Artificial Intelligence and Robotics (CAIR), Abu Dhabi, United Arab Emirates
[3] NYU Tandon School of Engineering, New York University, New York, USA
[4] New York University Abu Dhabi, Abu Dhabi, United Arab Emirates

Abstract. This paper addresses the problem of few-shot indoor 3D object detection by proposing a meta-learning-based framework that only relies on a few labeled samples from novel classes for training. Our model has two major components: a *3D meta-detector* and a *3D object detector*. Given a query 3D point cloud and a few support samples, the 3D meta-detector is trained over different 3D detection tasks to learn task distributions for different object classes and dynamically adapt the 3D object detector to complete a specific detection task. The 3D object detector takes task-specific information as input and produces 3D object detection results for the query point cloud. Specifically, the 3D object detector first extracts object candidates and their features from the query point cloud using a point feature learning network. Then, a class-specific re-weighting module generates class-specific re-weighting vectors from the support samples to characterize the task information, one for each distinct object class. Each re-weighting vector performs channel-wise attention to the candidate features to re-calibrate the query object features, adapting them to detect objects of the same classes. Finally, the adapted features are fed into a detection head to predict classification scores and bounding boxes for novel objects in the query point cloud. Several experiments on two 3D object detection benchmark datasets demonstrate that our proposed method acquired the ability to detect 3D objects in the few-shot setting.

Keywords: 3D object detection · Indoor scene · Few-shot learning · Meta-learning · Channel-wise attention

1 Introduction

In door 3D object detection is one of the key components in various real-world vision applications, such as indoor navigation [1] and visual SLAM [2], and has

S. Yuan and X. Li—Equal contribution.

Supplementary Information The online version contains supplementary material available at https://doi.org/10.1007/978-3-031-26319-4_15.

L. Wang et al. (Eds.): ACCV 2022, LNCS 13841, pp. 245–261, 2023.
https://doi.org/10.1007/978-3-031-26319-4_15

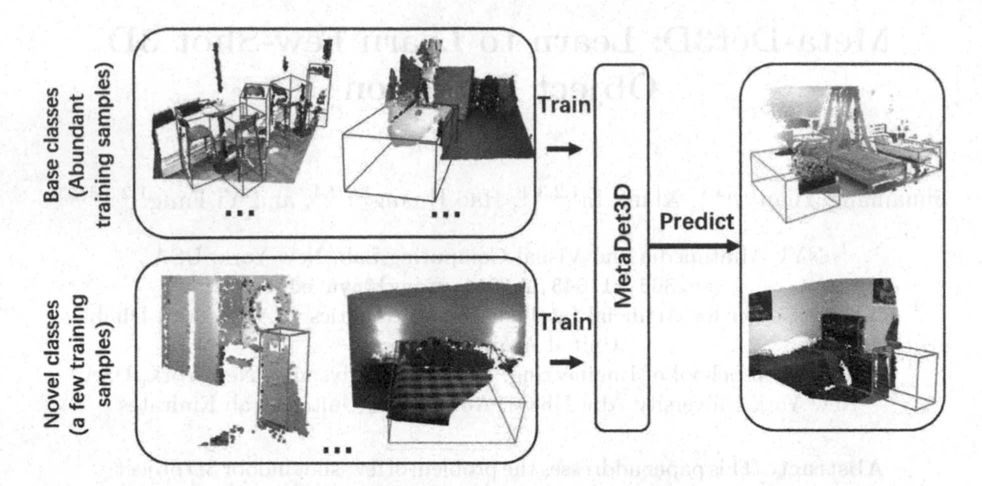

Fig. 1. Given abundant annotated samples of base classes and a few annotated samples of novel classes, our proposed model aims to learn from limited novel class samples to detect unseen 3D objects from the novel classes. (Top) Annotated chairs and sofas in base classes. (Bottom) Beds in the novel classes. Note that the objects are incomplete in the scanned 3D point cloud scenes.

been attracting numerous research attention in the last decades [3,4]. Recent progress in deep-learning-based methods has greatly advanced the performance of 3D object detection. However, existing deep-learning-based 3D object detection methods require ground truth supervision. When only a limited number of training data is provided, existing deep-learning-based methods suffer from severe performance degradation. In contrast, humans can easily learn new concepts and identify new objects with only a few given examples. This observation motivates us to design a novel few-shot 3D object detection framework that leverages the abundant annotated base classes to quickly learn new patterns or concepts from only a few given samples from unseen or novel classes (Fig. 1).

Here, we focus on the problem of indoor 3D object detection under the few-shot setting. Specifically, given a sufficient labeled training sample from the base classes and a few labeled samples from the novel classes, our model learns to detect an object from the novel classes. To achieve this goal, we design our detection model based on meta-learning [5,6]. 3D object detection requires predicting both object category and localization. However, 3D point clouds are unordered and unstructured in nature, which makes it hard to adopt a neural network to learn robust and representative point features. Therefore, 3D object detection under a few-shot scenario is even more challenging. Despite that there are a few previous literatures focusing on meta-learning for 3D point cloud classification [7–9] and segmentation [10,11], meta-learning for few-shot 3D point cloud detection has seldom been explored, and this work paves a path for few-shot 3D point cloud detection using the meta-learning scheme.

Specifically, our proposed method adopts a meta-learning-based paradigm for few-shot 3D object detection based on 3D object detectors, *e.g.*, VoteNet [4]. In VoteNet, a task-specific 3D object detector is designed for object detection with massive training data and thus is inapplicable in few-shot scenarios. To address the problem of 3D object detection in the few-shot scenario, we design a *3D meta-detector* to guide a *3D object detector* to perform few-shot 3D object detection. Specifically, we introduce a class-specific re-weighting module as the 3D meta-detector to guide the 3D object detector to detect objects from a specific class. Given a query 3D point cloud scene and a few support objects from novel classes, the backbone network of the 3D object detector is first adopted to extract candidates and their features from the input scene. Then, the class-specific re-weighting module receives few-shot objects from novel classes with bounding box annotations and produces re-weighting vectors, one for each novel class. Each vector takes channel-wise attention to all features to re-weight and re-calibrate the query object features. In this manner, the candidate features fuse with support information from novel classes and are adapted to detect objects of the same classes. Finally, the adapted features are fed into a detection head to predict classification scores and bounding boxes for novel objects in the query point cloud scene. We validate our method on two public indoor 3D object detection datasets. Experimental results demonstrate that our proposed meta-learning-based 3D few-shot detection method outperforms several baselines. The contributions of our paper are summarized as follows.

- To the best of our knowledge, we are the first to tackle the problem of few-shot 3D object detection using meta-learning, which is of great importance in real-world applications and has seldom been explored in previous literature. Our method learns meta-knowledge through the training process on base classes and transfers the learned meta-knowledge to unseen classes with only a few labeled samples.
- We design a novel few-shot 3D detection framework that consists of a meta-detector to guide a 3D object detector for object detection. To effectively guide the 3D detector to complete a specific task, we propose a class-specific re-weighting module as the 3D meta-detector that receives a few support samples from novel classes and produces class-attentive re-weighting vectors. These re-weighting vectors are used to re-calibrate the cluster features of the same class and strengthen those informative features for detecting novel objects in the query point cloud scene.
- We demonstrate the effectiveness of our proposed model on two 3D object detection benchmark datasets, and our model achieves superior performance than the well-established baseline methods by a large margin.

2 Related Work

2.1 3D Object Detection

Many existing works have been proposed to tackle the task of object detection from 3D point clouds. In this section, we focus on reviewing indoor 3D object detection. A 3D scene can be represented by three-dimensional point clouds. One intuitive way to detect 3D objects from a 3D scene is to employ a 3D deep learning network for object detection. However, this approach requires exhaustive computation, which slows the inference speed when handling large indoor scenes. Moreover, in 3D indoor datasets [12,13], the complex configuration of the layout of indoor 3D objects also introduces challenges. Various learning-based methods have been proposed to address the challenges above. Those learning-based indoor 3D object detection methods can be generally categorized into three classes. The first class is the sliding-window-based method, which divides the whole 3D scene into multiple patches and further adopts a classifier to detect 3D objects. The second type of indoor 3D object detection method extracts the per-point feature in a latent space and clusters corresponding points to produce the detection results. Qi *et al.* [4] propose a method that generates seed points from the input 3D point cloud and then predicts votes from the generated seed points. Then, Qi *et al.* cluster the votes by gathering the corresponding features and predicting 3D bounding boxes. The third type of 3D indoor detection method is proposal-based. Vote3D [14] uses a grid and predicts bounding boxes to detect 3D indoor objects. In addition, [15] uses MLP to directly predict the 3D bounding boxes from extracted point cloud features. Although the aforementioned methods achieve promising results in various 3D object detection benchmarks, they require the ground truth label as supervision during the training to ensure the model's generalization ability. However, this is not practical in real-world applications. To maintain the model's generalization ability for unseen object categories with only a limited amount of training data, *a.k.a.*, few-shot scenario, we design a meta-learning-based method for 3D object detection for point clouds.

2.2 Meta-learning

Meta-learning aims to handle the scenario where only a few labeled training data are available. Meta-learning can also be categorized into three classes as follows.

Metric-Based Methods. Directly training deep neural networks with a few training data leads the networks to over-fit on the training data. To avoid this issue, metric-learning methods adopt a non-parametric distance measurement as a classifier to directly compare the features of input data and predict labels. In addition, a parametric network is often adopted to extract representative and discriminative features that have large inter-class variations and small intra-class variations. Typical metric-based frameworks include Siamese Networks [16], Matching Networks [17], Prototypical Networks [18], and Relation Networks [19].

Model-Based Methods. Distinct from metric-based methods, model-based methods [20,21] adopt two different parametric networks (often termed as *learners*) where the first one serves as a meta-learner and the second one works as a classifier. The classifier only needs a single feed-forward pass to predict data labels, and its parameters are directly estimated by the meta-learner. Model-based methods only require to calculate gradients of the meta-learner for back-propagation, yielding efficient learning for few-shot problems. However, the generalization ability of the model-based methods on out-of-distribution data is not satisfactory.

Gradient-Based Methods. The key idea of gradient-based methods is to train a meta-learner to learn meta-knowledge from a dataset, thus to help a classifier to achieve a quick convergence on a few labeled data from unseen classes. In the pioneer work MAML [22], a meta-learner is used to learn a proper parameter initialization of a classifier from training data such that after a small number of back-propagation steps on the classifier, it can achieve good performance on few-shot tasks with limited labeled data. In addition to learning network parameter initialization, this method can also be utilized to update network hyper-parameters [23].

3 Review of VoteNet

We give a brief review of VoteNet [4] that is incorporated as the building block of our proposed method. VoteNet is a 3D object detection model that takes raw 3D point clouds as input and produces 3D bounding boxes without using 2D detectors. The object detection process in VoteNet can be divided into three steps. The first step is to generate seed points from the input 3D point cloud; the second step is to predict votes from the generated seed points, and the final step is to cluster votes by gathering their features and predicting 3D box proposals for each object.

To generate accurate votes for 3D object detection, VoteNet leverage PointNet++ [24], a hierarchical point signature learning model, as the backbone network to learn representative point features for geometric reasoning. The backbone network takes raw point clouds as input and produces a subset of the input point clouds with the coordinates of each point and the corresponding feature vector. Each output point is called a *seed* point and will produce one vote. Inspired by the generalized Hough voting for object detection [25], VoteNet leverages the power of a deep neural network to generate votes from seed points. Given the seed set $\{seed_i\}_{i=1}^{P}$, consisting P seed points, where each $seed_i$ is represented by the 3D coordinate x_i and feature vector f_i, a shared voting module is used to generate votes from each seed point independently. The voting module is simply a multi-layer perceptron (MLP). The MLP takes $[x_i, f_i]$ as input and outputs the displacement of each seed point in Euclidean space as well as the feature offset in feature space represented as $[\Delta x_i, \Delta f_i]$. The final vote is then represented as $v_i = [y_i = x_i + \Delta x_i, g_i = f_i + \Delta f_i]$. This voting process is supervised by the ℓ_2 distance that penalizes the offsets between the predicted and ground-truth distances from the seed points on the object surface to the object centers. During training, the MLP learns to drive the seed points to move towards the associating object centers and making it easier for the following vote clustering process.

Then, VoteNet adopts farthest point sampling (FPS) [24] to select Q votes based on the vote coordinates in Euclidean space and group votes by finding the nearest votes to the cluster centers. A cluster $\{C_q\}_{q=1}^{Q}$ is represented by $[y_{qi}, g_{qi}]$ where y_{qi} is the 3D coordinate of the ith vote in this cluster and g_{qi} is the corresponding feature vector. After that, each cluster is transformed into a local coordinate by subtracting each vote's coordinate from its associating cluster center. Finally, a detection head based on PointNet [26] is introduced to generate 3D bounding boxes from each of the vote clusters. The head network takes the normalized vote clusters as input and predicts the bounding boxes and classification scores. However, directly applying VoteNet in few-shot scenarios yields inferior performance as VoteNet requires a large amount of labeled data for training. In the next section, we describe our model utilizing VoteNet to overcome the challenges of few-shot 3D object detection with limited training labels.

4 Method

4.1 Few-Shot 3D Object Detection

In this section, we introduce the basic setting of the meta-learning-based few-shot 3D object detection. Different from conventional 3D object detection, in the few-shot scenario, the whole dataset $\mathscr{D}$ is separated into two parts, *i.e.*, base classes $\mathscr{D}_{base}$ and novel classes $\mathscr{D}_{novel}$ where $\mathscr{D}_{base} \cap \mathscr{D}_{novel} = \emptyset$. For each of the object categories from the base classes, abundant labeled data is available and $\mathscr{D}_{base}$ is divided into training and testing set $\mathcal{D}^{train}_{base}$ and $\mathcal{D}^{test}_{base}$. Similarly, $\mathscr{D}_{novel}$ is also divided into training and testing sets $\mathcal{D}^{train}_{novel}$ or $\mathcal{D}^{test}_{novel}$. However, in the few-shot setting, only a few labeled samples from the novel class can be used for training. Few-shot 3D object detection aims to predict 3D bounding boxes and class labels for each object from the *novel* classes with a few labeled data during training.

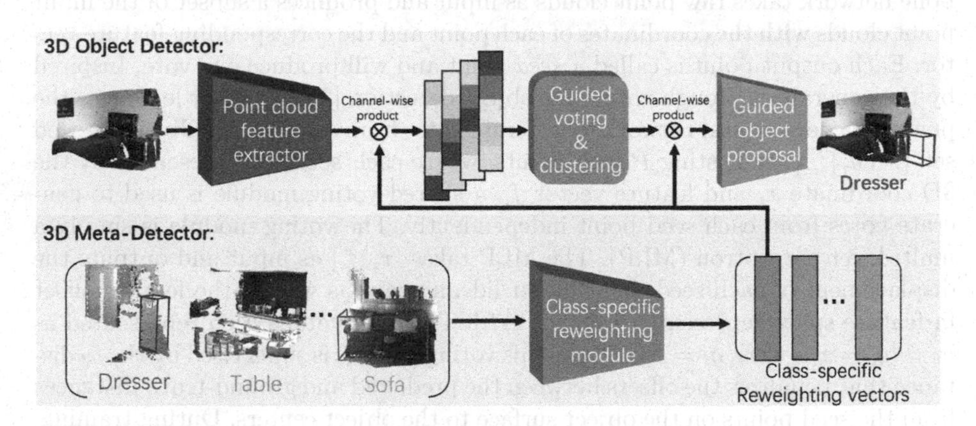

Fig. 2. Our MetaDet3D contains two components. The first component (bottom) is the 3D meta-detector which provides task/class-specific object detection guidance through a group of class-specific re-weighting vectors. The second component (top) is the 3D object detector that takes class-specific re-weighting vectors as input and predicts object bounding boxes and labels using three sub-components. A detailed description of each component and the three sub-components are depicted in Sect. 4.3.

To facilitate the training and evaluation of a few-shot 3D object detection model, we construct several episodes from the training and testing set for $\mathscr{D}_{base}$ and $\mathscr{D}_{novel}$ following the settings proposed in [27]. Each episode $\mathcal{E}$ is constructed from a set of support point clouds $\mathbf{S}$ with annotations and a set of query point clouds $\mathbf{Q}$. In a *N-way-K-shot* setting, the support set $\mathbf{S}$ has K instances with bounding box annotations for each of the N classes, *i.e.*, $\mathbf{S} = \{(S_{n,k}, r_{n,k})\}$, where $c = 1, 2, ..., N$ and $k = 1, 2, ..., K$, $S_{n,k}$ denotes the input point cloud, and $r_{n,k}$ denotes the bounding box annotation of the support set instances. We denote the query set as $\mathbf{Q}$, which contains N_q instances from the object classes that are the same as the support set. Our few-shot detection model aims to perform 3D object detection over the query point clouds with the support set as guidance.

Such a few-shot 3D object detection setting is practical in real-world scenarios. One may need to develop an object detection model, but collecting a large-scale well-annotated dataset for the target classes is time-consuming. An alternative would be deploying a detection model pre-trained on some existing large-scale object detection datasets (*e.g.*, Microsoft COCO [28] and SUN RGB-D [13]). However, one may only hope to detect several specific object categories of which these datasets only cover a very limited number of samples. These real-world applications pose a challenge and demand for few-shot object detection models.

In the following sections, we introduce our novel design of a few-shot 3D object detection framework, *MetaDet3D*, short for **Meta-Det**ection-**3D**. The illustration of our model is shown in Fig. 2. Our detection process is divided into two components: 1) a 3D meta-detector, and 2) a 3D object detector. We treat the object detection problem for each class as a learning task. 3D meta-detector is designed to learn the class-specific task distribution over different tasks and dynamically adapt the 3D object detector to complete a specific detection task. The 3D meta-detector learns class-specific re-weighting vectors that are used to guide the 3D object detector. The 3D object detector aims to complete a specific 3D object detection task by leveraging the class-specific information from the 3D meta-detector to detect objects from the same classes. The details of the proposed 3D meta-detector and 3D object detector are described as follows.

4.2 3D Meta-Detector

In this section, we introduce a novel 3D meta-detector, which is designed to learn the class-specific task distribution from different 3D object detection tasks.

We propose a class-specific re-weighting module $\mathcal{M}$ to produce the guidance for the 3D object detector. The 3D meta-detector takes 3D point cloud objects of different classes as input and learns a group of compact and robust re-weighting vectors in the embedding space. Formally, we first crop support object instances from support set $\mathbf{S}$ by using the ground-truth bounding box $r_{n,k}$ to construct supporting instance set $\mathbf{I} = \{I_{n,k}\}$, where $I_{n,k}$ denotes the kth cropped instance from the nth class.

As deep neural networks have been proved effective and efficient in 3D feature learning, we adopt PointNet++ [24] to produce a set of compact re-weighting vectors, one for each class. Specifically, given $N \times K$ support instances, our re-weighting module $\mathcal{M}$ generates N re-weighting vectors $\{z_n\}_{n=1}^{N}$ where $z_n \in \mathbb{R}^W$, one for each class. This step is formulated as:

$$z_n = \frac{1}{K} \sum \mathcal{M}_\theta(I_{n,k}), \quad n = 1, 2, ..., N \ . \tag{1}$$

The output z_n represents the re-weighting vector for the nth class. We use θ to represent the network parameters of our re-weighting module $\mathcal{M}$. The generated z_n is responsible for tuning object features in the 3D object detector for an accurate object candidate generation for the corresponding n^{th} class, which is detailed in the following section.

4.3 3D Object Detector

In this section, we introduce a 3D object detector that is used to complete a 3D object detection task for the given query set $\mathbf{Q}$ with the guidance provided by the 3D meta-detector. Inspired by the previous 3D object detection works [3,4,29], the 3D object detector can be divided into three steps. The first step is to extract point features from the input 3D point cloud. The second step is to generate object candidates. The final step is to predict 3D box proposals for each object.

Point Feature Extraction. To extract point features from a point cloud, we leverage PointNet++ [24], a hierarchical point feature learning model, as the backbone network to learn representative point features for geometric reasoning. The backbone network takes raw point clouds as input and produces a subset of the input point clouds with the coordinates of each point and the corresponding feature vectors. Each output point is called a *seed* and will produce one vote to generate a candidate. Specifically, for a given query scan Q_i from $\mathbf{Q}$, the network (denotes as $\mathcal{F}$) outputs seeds represented by a combination of seeds' coordinates and point features, *i.e.*, $[x, f]$ where $x \in \mathbb{R}^3$ and $f \in \mathbb{R}^W$:

$$[x, f] = \mathcal{F}_\delta(q_i) \ , \tag{2}$$

where δ denotes the network parameters.

Guided Voting and Clustering. Inspired by the voting-based 3D object detection works [3,4,29], we leverage the power of a deep neural network to generate candidates from seeds. Given a group of seeds, we apply channel-wise multiplication $\otimes$ between seeds' feature f and the learned class-specific re-weighting vectors z_n to generate re-weighted feature f' where $f' = z_n \otimes f$.

A shared voting module is then utilized to generate candidates from the seeds. The module is implemented as a multi-layer perceptron (MLP) network with ϕ to denote the network parameters. The MLP takes $[x, f']$ as input and outputs the displacement of the seeds in Euclidean space as well as the feature offset in feature space. The guided voting for object candidates generation is formulated as:

$$[\Delta x, \Delta f'] = MLP_\phi(f'). \tag{3}$$

$$[y, g] = [x + \Delta x, f' + \Delta f']. \tag{4}$$

The final candidates are then represented by $[y, g]$. This voting process is supervised by the Euclidean distance from the seeds on the object surface to the object centers. We denote this voting module as $\mathcal{V}$:

$$\begin{aligned} \mathcal{V}_\phi([x, f], z_n) &= [x, f \otimes z_n] + MLP_\phi(f \otimes z_n) \\ &= [y, g]. \end{aligned} \tag{5}$$

After object candidates generation, we adopt farthest point sampling [24] to select T candidates as center points in Euclidean space and generate clusters $\{\mathcal{C}_t\}_{t=1}^T$ by searching the nearest neighboring points for each center candidates, *i.e.*, $\mathcal{C}_t = \{[y_i, g_i] \| \|y_i - y_t\| \leq d\}$, where y_t is the center coordinates of cluster t, y_i and g_i denote the coordinates and features of the ith candidates, and d is the searching radius for each cluster. We denote the farthest point sampling and grouping process as $\mathcal{G}(\cdot)$.

Guided Object Proposal. We apply a 3D point feature learning network to extract cluster features and generate 3D object proposals. Specifically, for each cluster $\mathcal{C}_t$ represented by $\{[y_i, g_i]\}$, we normalize candidate locations to a local normalized coordinate system by $y'_i = (y_i - y_t)/d$. A shared PointNet [26], represented by $\mathcal{H}$, is adopted to produce cluster features. To generate object proposals with the guidance of 3D meta-detectors, we perform a channel-wise multiplication between cluster features and z_n to generate the re-calibrated cluster features. The guided object proposal module for the nth class is defined as:

$$\mathcal{P}_\psi(\mathcal{C}) = MLP(\mathcal{H}([y'_i, g_i]) \otimes z_c) \ , \tag{6}$$

where MLP is a multi-layer perceptron network serving as a prediction head to predict parameters of 3D bounding boxes and class labels. The whole detection head network is denoted by $\mathcal{P}$ with network parameters ψ.

4.4 Few-Shot Learning Strategy

In this section, we introduce our learning strategy for N-way-K-shot detection. Our model takes support instances $\mathbf{S} = \{S_{n,k}\}$ and query scan q_i as input. We denote the ground-truth bounding boxes of query scan q_i as $\mathcal{R}_{q_i}$. The learning of this task can be reorganized as jointly optimizing network parameters θ, δ, ϕ, and ψ using back-propagation by minimizing the following loss:

$$\min_{\theta,\delta,\phi,\psi} \mathcal{L} = \sum_i \mathcal{L}(\mathcal{P}_\psi(\mathcal{V}_\phi(\mathcal{F}_\delta(q_i), \mathcal{M}_\theta(S_{n,k}))), \mathcal{R}_{q_i}) , \quad (7)$$

It is of importance to employ an appropriate learning strategy such that the 3D meta-detector $\mathcal{M}$ can produce representative re-weighting vectors to guide the 3D object detector $\{\mathcal{F}, \mathcal{V}, \mathcal{P}\}$ to generate the correct object proposals.

To achieve this goal, we adopt a two-phase learning strategy to train our proposed model. We refer to the first phase as *base-training.* During base-training, only episode data from the base classes are used, and we jointly train $\mathcal{F}$, $\mathcal{V}$, $\mathcal{P}$ and $\mathcal{M}$ using ground-truth annotations. This phase guarantees the learner, composed of $\mathcal{F}$, $\mathcal{V}$, and $\mathcal{P}$, to learn to detect class-specific objects by referring to re-weighting vectors. We denote the second phase as *fine-tuning.* In this step, we train our model using episode data from both base classes and novel classes. Similar to the first base-training phase, we jointly optimize θ,δ,ϕ and ψ by minimizing Eq. 7.

Loss Function. Our model simultaneously estimates category labels of the bounding boxes, objectiveness scores, and bounding box parameters. We adopt a similar loss function as in VoteNet [4]. Specifically, We use the cross-entropy loss $\mathcal{L}_{sem}$ and $\mathcal{L}_{obj}$ to supervise the bounding box's label prediction as well as the objective score for candidates. Proposals generated from positive candidates that are close to the object center are further regressed to bounding parameters which include box center, heading angle, and box size. We adopt Huber loss $\mathcal{L}_{box}$ to supervise the bounding box prediction. Hence, the overall detection loss function is $\mathcal{L} = \mathcal{L}_{sem} + \mathcal{L}_{obj} + \mathcal{L}_{box}$.

5 Experiments and Results

In this section, we conduct experiments on two 3D object detection benchmarks to validate the object detection ability in the few-shot setting. In Sect. 5.1, we first give a description of the benchmarks used in our experiments and how we adapt them to few-shot settings. In Sect. 5.2, we describe the detailed implementation and the network architecture of our proposed model. In Sect. 5.3, we compare our model with several baselines approaches for few-shot 3D object detection in an indoor environment. In Sect. 5.4, we further analyze the performance of our model. In Sect. 5.5, we validate the effectiveness of our proposed modules by comparing different designs.

5.1 Datasets

SUN RGB-D. We conduct our experiments on SUN RGB-D [13] dataset. It is a widely used benchmark dataset for object detection. The dataset consists of around 10k RGB-D images captured from indoor scenes. Half of them are used for training, and the rest are used for testing. Each sample in the dataset is annotated with amodal-oriented 3D bounding boxes and category labels. The dataset includes a total of 37 object categories. In the common 3D object detection setting, the training and evaluation process is based on the ten most common categories, including nightstand, sofa, and table. In our few-shot experiment setting, we divide the whole dataset into $\mathscr{D}_{novel}$ and $\mathscr{D}_{base}$ according to the categories, where $\mathscr{D}_{base} \cap \mathscr{D}_{novel} = \emptyset$. More specifically, we select N out of the ten categories as novel categories, and we refer to the other $10-N$ categories as base categories. We further split each category into support/training and query/test sets.

ScanNet (v2). In addition to the SUN RGB-D dataset, we also conduct our experiment on the ScanNet (v2) [12] benchmark, which is a richly-annotated 3D indoor scene dataset that consists of 1,513 real-world scans. The whole dataset is divided into 1,201 training scenes, 312 validation scenes, and 100 test scenes. In contrast to the SUN RGB-D dataset, neither amodal bounding boxes nor their orientations are available in ScanNet (v2). Moreover, different from the SUN RGB-D dataset, ScanNet (v2) contains 18 object categories and provides complete reconstructed meshes of indoor scenes. We select N out of the 18 categories as novel categories and others as base categories. We also split each category into support/training and query/test sets for the SUN RGB-D dataset.

5.2 Implementation Details

Few-Shot Input and Data Augmentation. Our model takes point clouds from depth images (SUN RGB-D) or 3D indoor scans (ScanNet (v2)) as input. We uniformly sample 20k points from the source data, and each point is represented by 3D coordinates and the 1D height feature, which is the distance from the point to the floor. Following the VoteNet [4], we set the 1% percentile of all heights as the floor to calculate the height features. After the sampling, we apply random scaling from 0.8 to 1.2 on the data, followed by the random rotation for $-3°$ to $3°$ and random flipping.

Table 1. Quantitative results for few-shot 3D object detection on **SUN RGB-D**. Experiments are conducted on three different novel classes with $N = 3$, $K = 10$.

Method	Novel Set 1				Novel Set 2				Novel Set 3			
	bathtub	bed	chair	mAP	dresser	sofa	table	mAP	chair	desk	table	mAP
VoteNet-JT	0.0	5.1	6.7	3.9	0.0	1.0	5.6	2.2	12.4	0.8	0.8	4.7
VoteNet-FT	0.0	16.1	4.8	7.0	0.1	4.2	6.9	3.7	13.7	2.5	2.2	6.1
VoteNet-2	0.0	16.6	5.5	7.4	0.1	4.8	7.5	4.1	16.5	2.0	2.1	6.9
Ours	3.5	21.8	10.3	11.9	10.4	9.9	11.0	10.4	10.2	12.7	12.1	11.6

Table 2. Quantitative results for few-shot 3D object detection on **ScanNet (v2)**. Experiments are conducted on three different novel classes with $N = 3$, $K = 10$. "shower cur." is the abbreviation for "shower curtain", and "garb. bin" is the abbreviation for "garbage bin".

Method	Novel Set 1				Novel Set 2				Novel Set 3			
	curtain	shower cur.	desk	mAP	garb. bin	bed	sink	mAP	sofa	table	chair	mAP
VoteNet-JT	1.9	2.2	1.0	1.7	2.4	1.0	2.0	1.8	0.3	0.5	0.0	0.3
VoteNet-FT	3.4	4.9	2.2	3.5	6.1	1.7	4.4	4.1	1.6	1.8	0.5	1.3
VoteNet-2	4.8	5.2	2.9	4.3	7.5	2.1	5.1	4.9	2.2	2.6	1.1	1.9
Ours	26.5	11.0	3.0	13.5	19.4	10.8	1.1	10.4	9.2	8.0	8.4	8.5

Network Architecture. We use PointNet++ [24] as backbone for the class-specific re-weighting module. The feature extractor network contains three set abstraction (SA) layers followed by a max-pooling layer. The radius of three SA layers are [0.3, 0.5, 1] with the output channel sizes of [1024, 512, 256]. The 256-dimensional feature vector is fed into the weight learning network which is an MLP with two hidden layers of size [512, 256]. We fuse the class-specific re-weighting module with VoteNet [4] as our 3D object detector.

Network Training. We follow the two-phase learning strategy to train our model from scratch with an Adam optimizer. In the base-training phase, we set the initial learning rate to 0.001 on $\mathcal{D}_{base}^{train}$. We decrease the learning rate by ten times after 50 epochs and 100 epochs. In the fine-tuning phase, we reset the initial learning rate to 0.001, and we decrease the learning rate every 30 epochs on $\mathcal{D}_{base}^{train}$ and $\mathcal{D}_{novel}^{train}$. Our model is implemented using PyTorch [30] and runs on NVIDIA GTX 2080 GPUs.

Table 3. Results on **SUN RGB-D**. The table shows the evaluation results for different methods on three novel classes with different numbers of shots $K = 30, 50$. The evaluation metric is mAP with IoU threshold 0.25.

Method	Novel Set 1		Novel Set 2		Novel Set 3	
	30	50	30	50	30	50
VoteNet-JT	5.8	6.8	3.4	4.8	5.3	5.9
VoteNet-FT	8.1	9.8	4.6	5.6	7.3	8.1
VoteNet-2	9.0	11.3	5.1	6.1	8.0	8.9
Ours	12.1	14.0	11.7	13.6	13.2	13.8

5.3 Comparison with the Baselines

Experiment Settings. To the best of our knowledge, this is the first work focusing on the few-shot 3D object detection task. In order to explore the generalization ability of our proposed model on few-shot 3D object detection settings, we conduct experiments on SUN RGB-D, and ScanNet (v2) benchmarks to compare our model against various baseline models built on VoteNet [4]. Specifically,

Table 4. Results on **ScanNet (v2)**. The table shows the evaluation results for different methods on three novel classes with different numbers of shots $K = 30, 50$. The evaluation metric is mAP with IoU threshold 0.25.

Method	Novel Set 1		Novel Set 2		Novel Set 3	
	30	50	30	50	30	50
VoteNet-JT	2.2	4.1	2.4	3.1	0.6	0.8
VoteNet-FT	4.8	6.0	5.9	7.0	1.9	2.7
VoteNet-2	6.0	6.9	7.3	8.8	2.5	4.0
Ours	14.7	16.1	11.4	12.7	9.9	10.8

we adopt three baseline models for the comparison: 1) VoteNet-JT. In this baseline method, VoteNet is alternately trained on the base classes and novel classes without base-training and fine-tuning phases. 2) VoteNet-FT. In this baseline method, VoteNet first uses a base-training phase to train on the data from base classes, and then it only uses data from novel classes to fine-tune the model. 3) VoteNet-2. In this method, we train the VoteNet following the two-phase learning strategy until full convergence. Both our model and the three baseline models are trained in the few-shot setting where $N = 3$, $K = \{10, 30, 50\}$. During inference, we use the class-specific re-weighting vectors learned from the whole supporting set to guide 3D object detection on the query set.

Result Analysis. The quantitative results on SUN RGB-D and ScanNet (v2) benchmarks are presented in Table 1 and Table 2 respectively. In both tables, the average precision values with the IoU threshold 0.25 are reported. Results presented in Table 1 and Table 2 show that our model outperforms the baseline models. The comparison results for different numbers of shots on two benchmarks are presented in Table 3 and Table 4. As shown in the tables, our model improves the mAP by at least 3.1% on SUN RGB-D benchmark when $K = 30$, and 2.7% when $K = 50$. Moreover, our method improves the mAP by at least 4.1% on ScanNet (v2) benchmark when $K = 30$, and 3.9% when $K = 50$. Refer to the supplementary material for the per-instance mAP. The results also demonstrate that directly adopting conventional object detectors leads to limited generalization abilities for novel classes. In Fig. 4 and Fig. 5, we show several qualitative examples of the 3D detection results on SUN RGB-D and ScanNet (v2) benchmarks respectively.

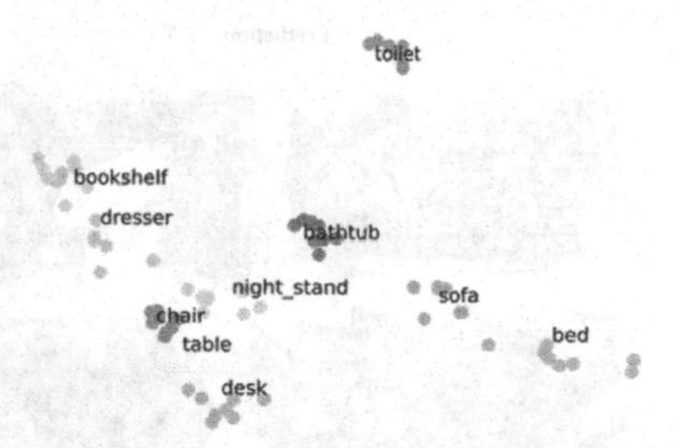

Fig. 3. The t-SNE visualization of class-specific re-weighting vectors from SUN-RGB D.

5.4 Performance Analysis

Class-Specific Re-weighting Vector. The re-weighting vectors learned from the 3D meta-detector serve as important task-specific information to guide the detection process of the 3D object detector. In this section, we project the re-weighting vectors for different object categories onto a 2D plane to verify that the

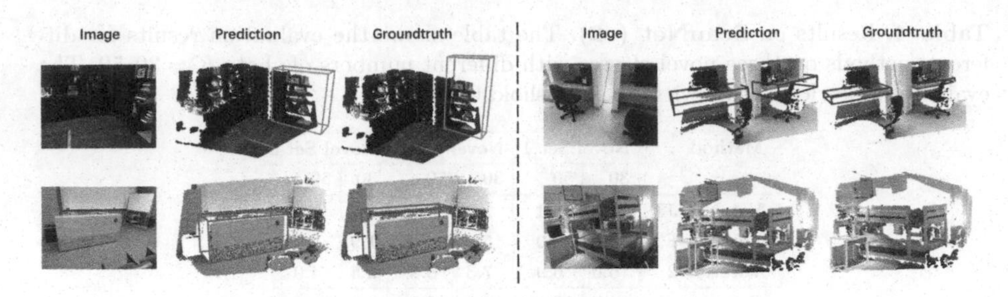

Fig. 4. Qualitative results on **SUN RGB-D**.

re-weighting vectors are learned to contain class-specific information. We conduct experiments on the SUN RGB-D dataset. Figure 3 shows the projections of re-weight vectors by using t-SNE [31]. It is obvious that the vectors that belong to the same class tend to cluster with each other. Moreover, we also observe that classes sharing similar visual features are projected close to each other. Specifically, the desk, table, and chair classes are projected to the bottom-left in the figure, while the sofa and bed classes are projected to the bottom-right.

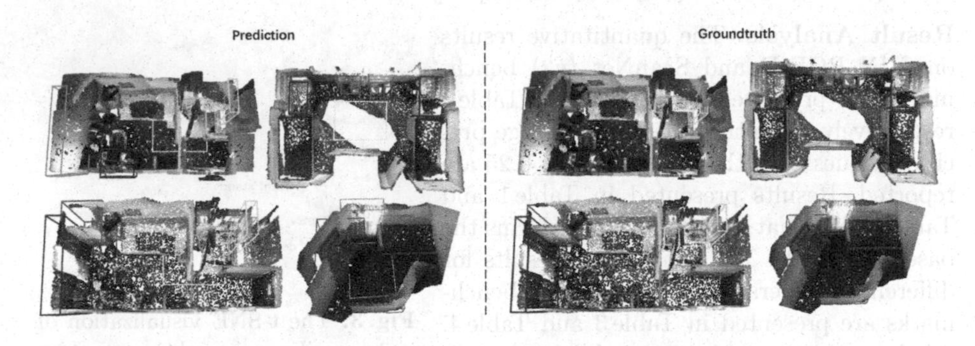

Fig. 5. Qualitative results on **ScanNet (v2)**.

Table 5. The comparison for the model size and the inference time for the object detection in one query scan.

Method	Model size	SUN RGB-D	ScanNet (v2)
VoteNet	11.2 MB	0.10 s	0.14 s
Ours	11.6 MB	0.11 s	0.16 s

Speed and Size. We also analyze the inference speed and model size on different benchmark datasets. In Table 5, the comparison of our model and two baseline models is presented. Our model takes around 0.16 s to detect objects in one query scan in ScanNet (v2) dataset, which is almost the same as the VoteNet. In addition, our proposed model has a size of 11.7 MB, which is only 0.5 MB larger than VoteNet. This observation demonstrates that our model can effectively detect novel objects with almost the same memory cost as VoteNet.

5.5 Where to Re-weight

We analyze the effect of different re-weighting locations on the final 3D object detection performance. In this experiment, we compare the object detection performance on ScanNet (v2) using three different designs of our proposed model. In the first design, we only use re-weighting in the "Guided Voting" module, and we denote this design as **V.R.**. In the second design, we only employ re-weighting in the object proposal module, which is referred as **O.R.**, and the last design corresponds to our proposed method that applies re-weighting in both object proposal modules. Table 6 shows the mAP of the three different designs. We notice that our proposed model that applies re-weighting in both object proposal modules achieves the best performance.

Table 6. The object detection results on SUN RGB-D with different model designs.

	V.R.	O.R.	Ours
Base classes	52.1	52.5	59.0
Novel classes	10.9	11.1	11.3

6 Conclusion

To the best of our knowledge, this is the first work to tackle the few-shot 3D object detection problem. Our proposed method generalizes its ability from the meta-training process to infer 3D object proposals and predict 3D bounding boxes for unseen 3D objects in novel classes. A novel class-specific re-weighting vector is introduced with the goal of facilitating the meta-detector to learn task distributions for different object classes and then dynamically adapt the 3D object detector to complete a specific detection task. Our method can be easily adapted to other two-stage detection methods that contain the object proposal stage. Experiments on two public 3D object detection benchmarks demonstrate that our model can effectively detect 3D objects from novel classes with superior performance over the well-established baseline models. In-depth analyses of our model further indicate the effectiveness and efficiency of each proposed component.

Acknowledgements. The authors appreciate the generous support provided by Inception Institute of Artificial Intelligence (IIAI) in the form of NYUAD Global Ph.D. Student Fellowship. This work was also partially supported by the NYUAD Center for Artificial Intelligence and Robotics (CAIR), funded by Tamkeen under the NYUAD Research Institute Award CG010.

References

1. Afif, M., Ayachi, R., Said, Y., Pissaloux, E., Atri, M.: An evaluation of retinanet on indoor object detection for blind and visually impaired persons assistance navigation. Neural Process. Lett. **51**, 2265–2279 (2020)
2. Yang, S., Scherer, S.: Cubeslam: monocular 3-D object slam. IEEE Trans. Rob. **35**, 925–938 (2019)
3. Qi, C.R., Chen, X., Litany, O., Guibas, L.J.: Imvotenet: boosting 3D object detection in point clouds with image votes. In: Proceedings of the IEEE/CVF Conference on Computer Vision and Pattern Recognition, pp. 4404–4413 (2020)
4. Qi, C.R., Litany, O., He, K., Guibas, L.J.: Deep hough voting for 3D object detection in point clouds. In: Proceedings of the IEEE International Conference on Computer Vision, pp. 9277–9286 (2019)
5. Hospedales, T.M., Antoniou, A., Micaelli, P., Storkey, A.J.: Meta-learning in neural networks: a survey. Trans. Pattern Anal. Mach. Intell. (2021)
6. Huisman, M., van Rijn, J.N., Plaat, A.: A survey of deep meta-learning. Artif. Intell. Rev. **54**(6), 4483–4541 (2021). https://doi.org/10.1007/s10462-021-10004-4
7. Nie, J., Xu, N., Zhou, M., Yan, G., Wei, Z.: 3D model classification based on few-shot learning. Neurocomputing **398**, 539–546 (2020)
8. Zhang, B., Wonka, P.: Training data generating networks: linking 3D shapes and few-shot classification. arXiv preprint arXiv:2010.08276 (2020)
9. Huang, H., Li, X., Wang, L., Fang, Y.: 3D-metaconnet: meta-learning for 3D shape classification and segmentation. In: International Conference on 3D Vision, pp. 982–991. IEEE (2021)
10. Yuan, S., Fang, Y.: Ross: Robust learning of one-shot 3D shape segmentation. In: The IEEE Winter Conference on Applications of Computer Vision, pp. 1961–1969 (2020)
11. Wang, L., Li, X., Fang, Y.: Few-shot learning of part-specific probability space for 3D shape segmentation. In: Proceedings of the IEEE/CVF Conference on Computer Vision and Pattern Recognition, pp. 4504–4513 (2020)
12. Dai, A., Chang, A.X., Savva, M., Halber, M., Funkhouser, T., Nießner, M.: Scannet: richly-annotated 3D reconstructions of indoor scenes. In: Proceedings of the IEEE Conference on Computer Vision and Pattern Recognition, pp. 5828–5839 (2017)
13. Song, S., Lichtenberg, S.P., Xiao, J.: Sun RGB-D: a RGB-D scene understanding benchmark suite. In: Proceedings of the IEEE Conference on Computer Vision and Pattern Recognition, pp. 567–576 (2015)
14. Wang, D.Z., Posner, I.: Voting for voting in online point cloud object detection. In: Robotics: Science and Systems, vol. 1, pp. 10–15, Rome, Italy (2015)
15. Yang, B., et al.: Learning object bounding boxes for 3D instance segmentation on point clouds. In: Advances in Neural Information Processing Systems, vol. 32 (2019)
16. Koch, G., Zemel, R., Salakhutdinov, R.: Siamese neural networks for one-shot image recognition. In: ICML Deep Learning Workshop, vol. 2, Lille (2015)
17. Vinyals, O., Blundell, C., Lillicrap, T., Wierstra, D., et al.: Matching networks for one shot learning. In: Advances in Neural Information Processing Systems, pp. 3630–3638 (2016)
18. Snell, J., Swersky, K., Zemel, R.: Prototypical networks for few-shot learning. In: Advances in Neural Information Processing Systems, pp. 4077–4087 (2017)
19. Sung, F., Yang, Y., Zhang, L., Xiang, T., Torr, P.H., Hospedales, T.M.: Learning to compare: relation network for few-shot learning. In: Proceedings of the IEEE Conference on Computer Vision and Pattern Recognition, pp. 1199–1208 (2018)

20. Bertinetto, L., Henriques, J.F., Valmadre, J., Torr, P., Vedaldi, A.: Learning feed-forward one-shot learners. In: Advances in Neural Information Processing Systems, pp. 523–531 (2016)
21. Graves, A., Wayne, G., Danihelka, I.: Neural turing machines. arXiv preprint arXiv:1410.5401 (2014)
22. Finn, C., Abbeel, P., Levine, S.: Model-agnostic meta-learning for fast adaptation of deep networks. arXiv preprint arXiv:1703.03400 (2017)
23. Li, Z., Zhou, F., Chen, F., Li, H.: Meta-SGD: learning to learn quickly for few-shot learning. arXiv preprint arXiv:1707.09835 (2017)
24. Qi, C.R., Yi, L., Su, H., Guibas, L.J.: Pointnet++: Deep hierarchical feature learning on point sets in a metric space. In: Advances in Neural Information Processing Systems, pp. 5099–5108 (2017)
25. Leibe, B., Leonardis, A., Schiele, B.: Combined object categorization and segmentation with an implicit shape model. In: Workshop on Statistical Learning in Computer Vision, ECCV, vol. 2, p. 7 (2004)
26. Qi, C.R., Su, H., Mo, K., Guibas, L.J.: Pointnet: deep learning on point sets for 3D classification and segmentation. In: Proceedings of the IEEE Conference on Computer Vision and Pattern Recognition, pp. 652–660 (2017)
27. Santoro, A., Bartunov, S., Botvinick, M., Wierstra, D., Lillicrap, T.: Meta-learning with memory-augmented neural networks. In: International Conference on Machine Learning, pp. 1842–1850. PMLR (2016)
28. Lin, T.-Y., et al.: Microsoft COCO: common objects in context. In: Fleet, D., Pajdla, T., Schiele, B., Tuytelaars, T. (eds.) ECCV 2014. LNCS, vol. 8693, pp. 740–755. Springer, Cham (2014). https://doi.org/10.1007/978-3-319-10602-1_48
29. Xie, Q., et al.: Mlcvnet: multi-level context votenet for 3D object detection. In: Proceedings of the IEEE/CVF Conference on Computer Vision and Pattern Recognition, pp. 10447–10456 (2020)
30. Paszke, A., et al.: Automatic differentiation in PyTorch (2017)
31. Van der Maaten, L., Hinton, G.: Visualizing data using t-SNE. J. Mach. Learn. Res. **9**, 2579–2605 (2008)

ReAGFormer: Reaggregation Transformer with Affine Group Features for 3D Object Detection

Chenguang Lu[1], Kang Yue[1,2], and Yue Liu[1(✉)]

[1] Beijing Engineering Research Center of Mixed Reality and Advanced Display, School of Optics and Photonics, Beijing Institute of Technology, Beijing, China
liuyue@bit.edu.cn

[2] Institute of Software, Chinese Academy of Sciences, Beijing, China

Abstract. Direct detection of 3D objects from point clouds is a challenging task due to sparsity and irregularity of point clouds. To capture point features from the raw point clouds for 3D object detection, most previous researches utilize PointNet and its variants as the feature learning backbone and have seen encouraging results. However, these methods capture point features independently without modeling the interaction between points, and simple symmetric functions cannot adequately aggregate local contextual features, which are vital for 3D object recognition. To address such limitations, we propose ReAGFormer, a reaggregation Transformer backbone with affine group features for point feature learning in 3D object detection, which can capture the dependencies between points on the aligned group feature space while retaining the flexible receptive fields. The key idea of ReAGFormer is to alleviate the perturbation of the point feature space by affine transformation and extract the dependencies between points using self-attention, while reaggregating the local point set features with the learned attention. Moreover, we also design multi-scale connections in the feature propagation layer to reduce the geometric information loss caused by point sampling and interpolation. Experimental results show that by equipping our method as the backbone for existing 3D object detectors, significant improvements and state-of-the-art performance are achieved over original models on SUN RGB-D and ScanNet V2 benchmarks.

Keywords: 3D object detection · Transformer · Point cloud

1 Introduction

3D object detection from point clouds is a fundamental task in 3D scene understanding and has wide applications in robotics, augmented reality, etc. However, most of the latest progress in 2D object detection cannot be directly applied to 3D object detection due to the sparsity and irregularity of point clouds.

Supplementary Information The online version contains supplementary material available at https://doi.org/10.1007/978-3-031-26319-4_16.

L. Wang et al. (Eds.): ACCV 2022, LNCS 13841, pp. 262–279, 2023.
https://doi.org/10.1007/978-3-031-26319-4_16

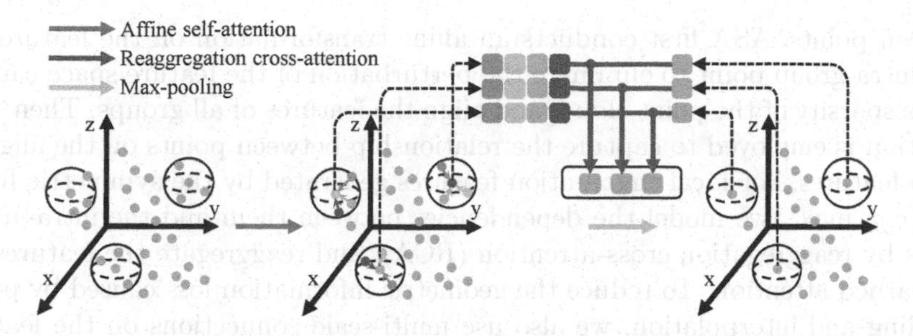

Fig. 1. Illustration of the ReAGF Transformer block. Affine self-attention is introduced to align the group feature space while capturing the dependencies between points within each group. Compared to using only symmetric functions (*e.g.* max), we reaggregate point features via dependencies learned from reaggregation cross-attention, thus improve feature aggregation efficiency.

Prior works first convert the point cloud into the regular data format [1–4] and then use convolutional neural networks for feature extraction and 3D object detection. However, the conversion process always leads to geometric information loss due to quantization errors. PointNet and its variants [5,6] alleviate this issue by extracting features directly from the raw point clouds, which can preserve the spatial structure and geometric information of the point cloud. As a result, PointNet and its variants are widely used as the feature learning backbone in 3D object detection [7–11]. However, these methods cannot adequately consider the dependencies between points during the capture of point features, and the simple symmetric function (*e.g.* max) cannot effectively utilize the dependencies between points to aggregate local contextual features, which are vital for 3D object detection.

Recently, Transformer [12] has achieved great success in computer vision [13–16]. Thanks to its long-range dependencies modeling capability, the Transformer is an ideal way to address the above limitations. However, how to integrate the advantages of PointNet-like backbone and Transformer to boost 3D object detection is still an open problem. One effort is to combine Transformer with PointNet++ and its variants, such as PCT [17] and PT [18], which focus on the classification and segmentation of point clouds, and the resulting architecture may be suboptimal for other tasks such as 3D object detection. Other solutions introduce the sampling and grouping in PointNet++ into Transformer and design a pure Transformer model for 3D object detection, such as Pointformer [19], but such solutions still use simple symmetric functions (*e.g.* max) to aggregate point features, which limits the representation of the model.

In this paper, we propose a plug-and-play reaggregation Transformer backbone with affine group features for 3D object detection, named as ReAGFormer, which utilizes the ability of the Transformer to model the dependencies between points and reaggregate point features through learned attention, while retaining the flexible receptive fields. Specifically, we propose a reaggregation Transformer block with affine group features (ReAGF Transformer block) to form the down-sampling stage of the backbone. As shown in Fig. 1, in the ReAGF Transformer block, we introduce affine self-attention (ASA) to interact on the relationship between points. ASA first conducts an affine transformation on the features of each intra-group point to eliminate the perturbation of the feature space caused by the sparsity of the point cloud, and align the features of all groups. Then self-attention is employed to capture the relationship between points on the aligned group features. For local aggregation features generated by the symmetric function (*e.g.* max), we model the dependencies between them and the intra-group points by reaggregation cross-attention (RCA), and reaggregate the features by the learned attention. To reduce the geometric information loss caused by point sampling and interpolation, we also use multi-scale connections on the feature propagation layer [6] in the upsampling stage.

To validate the effectiveness and generalization of our method, we replace the backbone of three different state-of-the-art methods of VoteNet [7], BRNet [10] and Group-Free [11] with our proposed ReAGFormer while not changing the other network structures. Experimental results show that when using our proposed ReAGFormer as the feature extraction backbone, all three methods achieve significant improvements, and the modified BRNet and Group-Free achieve state-of-the-art results on ScanNet V2 [20] and SUN RGB-D [21] datasets, respectively.

Our main contributions can be summarized as follows:

- We introduce the reaggregation Transformer block with affine group features (ReAGF Transformer block), which alleviates the perturbation of the local feature space by affine transformation and models the dependencies between points, while reaggregating the point set features with the learned attention.
- Based on ReAGF Transformer block, we build reaggregation Transformer backbone with affine group features, named as ReAGFormer, which can align different groups of feature space and efficiently capture the relationship between points for 3D object detection. Our ReAGFormer can be served as a plug-and-play replacement features learning backbone for 3D object detection.
- Experiments demonstrate the effectiveness and generalization of our backbone network. Our proposed method enables different state-of-the-art methods to achieve significant performance improvements.

2 Related Work

2.1 Point Cloud Representation Learning

Grid-based methods such as projection-based methods [22,23] and voxel-based methods [2,24,25] were frequently used in early point cloud representations. Such methods can effectively solve the problem of difficult point cloud feature extraction caused by irregular point clouds. However, the quantification process in the projection-based methods suffers from information loss, while voxel-based methods require careful consideration of computational effort and memory cost.

Recently, the method of learning features directly from the raw point cloud has received increasing attention. Prior works include MLP-based [5,6,26] methods, convolution-based methods [27–33] and graph-based methods [34–37]. As representative methods, PointNet and its variants [5,6] are widely used for point feature learning in 3D object detection. However, these methods lack the ability to capture dependencies, in addition, the symmetric function in PointNet and its variants cannot adequately aggregate local point set features. In this work, we address the above limitations with Transformer to boost 3D object detection.

2.2 3D Object Detection in Point Clouds

Due to the sparsity and irregularity of point clouds, early 3D object detection methods usually transformed point clouds into regular data structures. One class of methods [1,4,38] projects point cloud to the bird's eye view. Another class of methods [2,3,39,40] converts the point cloud into voxels. There are also methods that use templates [41] or clouds of oriented gradients [42] for 3D object detection.

With the rapid progress of deep learning on point clouds, a series of networks represented by PointNet [5] and PointNet++ [6] that directly processes point clouds are proposed and gradually serve as the backbone of 3D object detectors. PointRCNN [43] introduces a two-stage object detection method that generates 3D proposals directly from the raw point cloud. PV-RCNN [44] combines the advantages of both PointNet++ and voxel-based methods. VoteNet [7] introduces deep hough voting to design an end-to-end 3D object detector, and subsequently derives a series of methods. MLCVNet [8] and HGNet [45] use attention mechanism and hierarchical graph network, respectively, to boost the detection performance. To address the issues of outlier points on detection performance, H3DNet [9] and BRNet [10] introduce hybrid geometric primitives and back-tracing representative points strategy to generate more robust results, respectively. DisARM [46] designs a displacement aware relation module to capture the contextual relationships between carefully selected anchor. In contrast to these methods, we focus on feature learning backbone in 3D object detection. We show that our ReAGFormer can serve as the point feature learning backbone for most of the above methods.

2.3 Transformers in Computer Vision

Transformer [12] has been successfully applied to computer vision and has seen encouraging results in such tasks as image classification [13,47], detection [14,48] and segmentation [49,50]. Transformer is inherently permutation invariant and therefore also well suited for point cloud data. PCT [17] and PT [18] construct transformer on point clouds for classification and segmentation. Stratified Transformer [51] proposes a stratified transformer architecture to capture long-range contexts for point cloud segmentation. DCP [52] is the first method to introduce transformer to the point cloud registration task. In point cloud video understanding, P4Transformer [53] introduces point 4D convolution and transformer to embed local features and capture information about the entire video. Transformer also shows great potential for low-level tasks in point clouds, such as point cloud upsampling [54], denoising [55] and completion [56]. Transformer is

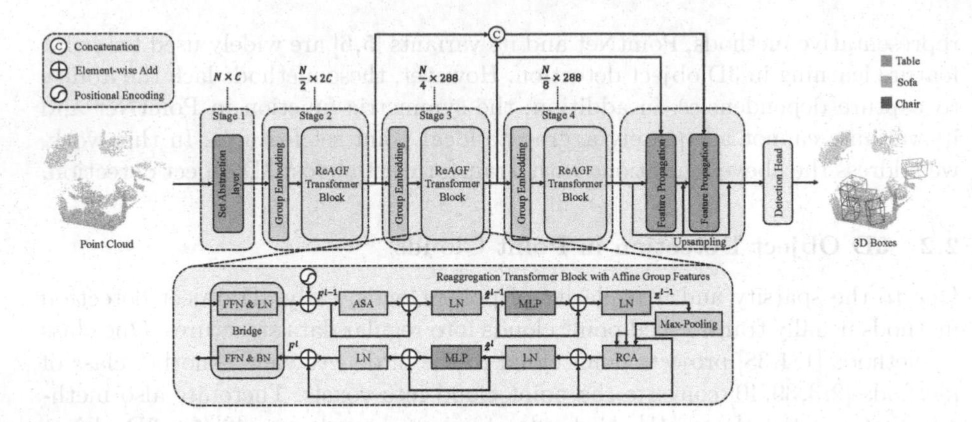

Fig. 2. The architecture of our ReAGFormer backbone for 3D object detection. ReAGFormer has four downsampling stages. The first stage is a set abstraction layer [6], and all other stages consist of group embedding and ReAGF Transformer block. The upsampling stage is a feature propagation layer with multi-scale connection.

also used for 3D object detection such as Pointformer [19], Group-Free [11] and 3DETR [57]. These methods have seen great progress, however, they neglect the perturbation of the local point set feature space caused by the sparsity and irregularity of point clouds, while still using simple symmetric functions to aggregate point set features. In contrast, we propose a reaggregation Transformer block with affine group features that can alleviate the perturbation of the feature space, while reaggregating the point set features with the learned attention to boost the symmetric function.

3 Proposed Method

In this work, we proposed ReAGFormer, a reaggregation Transformer backbone with affine group features for point feature learning in 3D object detection. As shown in Fig. 2, the proposed ReAGFormer involves four stages of downsampling to generate point sets with different resolutions, and an upsampling stage to recover the number of points. Each downsampling stage involves two main components: group embedding and reaggregation Transformer block with affine group features (ReAGF Transformer block). The group embedding is used to generate the suitable input for the Transformer block. In the ReAGF Transformer block, we introduce affine self-attention (ASA) to apply an affine transformation on the group feature space, while modeling the dependencies between points. For group aggregation point features generated by symmetric functions (*e.g.* max), we introduce reaggregation cross-attention (RCA) to boost the efficiency of symmetric function aggregation by reaggregating group features using the captured dependencies. Moreover, we also utilize feature propagation layers [6] with multi-scale connections in the upsampling stage to reduce the information loss due to point sampling and interpolation. In this section, we describe each part in detail.

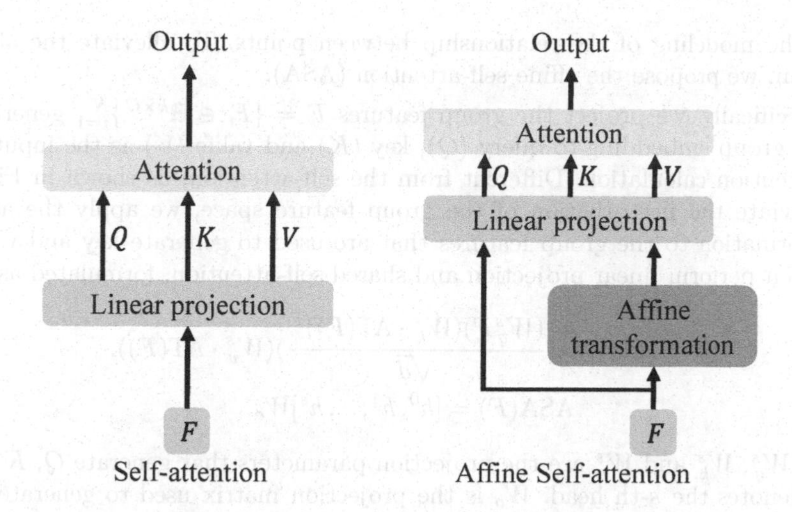

Fig. 3. A comparison between self-attention and our affine self-attention. By a simple affine transformation, we align the feature spaces of different groups and therefore better capture the dependencies between points within a group.

3.1 Group Embedding

The way in which the point cloud data is fed into the ReAGF Transformer block is vital to the overall architecture. To efficiently capture the fine-grained spatial geometric features of the point cloud scene and obtain the flexible receptive fields, we follow the sampling and grouping strategy in PointNet++ [6] to generate local point sets.

Specifically, we use farthest point sampling (FPS) to sample N' points from the input point cloud $P = \{x_i\}_{i=1}^{N}$. Taking each sampling point as the centroid, a ball query is used to generate N' groups according to the specified radius r, in which each group contains k points. Groups are denoted as $\{G_i\}_{i=1}^{N'}$ and feature learning is performed on groups using the shared MLP layer to extract group features $F = \{F_i \in \mathbb{R}^{k\times C}\}_{i=1}^{N'}$, where C is the feature dimension of each point in the group. The group features F is served as the input sequence for the subsequent Transformer block.

3.2 Reaggregation Transformer Block with Affine Group Features

Affine Self-attention (ASA). To extract the dependencies between each intra-group point, we resort to Transformer and self-attention [12]. However, the feature space of each group consisting of intra-group point features may be unaligned due to the sparsity and irregularity of point clouds, as well as the variety of structures within each group, which may lead to perturbations in the group feature space. We argue that the perturbation in the feature space can limit the modeling of the relationship between points. To alleviate the above problem, we propose the affine self-attention (ASA).

Specifically, we project the group features $F = \{F_i \in \mathbb{R}^{k \times C}\}_{i=1}^{N'}$ generated by the group embedding to query (Q), key (K) and value (V) as the input for the attention calculation. Different from the self-attention, as shown in Fig. 3, to alleviate the perturbation of the group feature space, we apply the affine transformation to the group features that are used to generate key and value, and then perform linear projection and shared self-attention, formulated as:

$$h^s = \text{softmax}(\frac{(W_q^s F)(W_k^s \cdot \text{AT}(F))^T}{\sqrt{d}})(W_v^s \cdot \text{AT}(F)), \tag{1}$$

$$\text{ASA}(F) = [h^0, h^1, ..., h^s]W_o, \tag{2}$$

where W_q^s, W_k^s and W_v^s are the projection parameters that generate Q, K and V. s denotes the s-th head. W_o is the projection matrix used to generate the output. $[\cdot]$ is the concatenation and d is the feature dimension of the s-th head. AT is the affine transformation.

For affine transformation (AT) module, inspired by [58], we utilize a simple transformation method. Specifically, for the group feature $F = \{F_i \in \mathbb{R}^{k \times C}\}_{i=1}^{N'}$ generated by the group embedding, we formulate the following operation:

$$\hat{F}_i = \delta(\frac{F_i - f_i}{\sigma + \epsilon}), \quad \sigma = \sqrt{\frac{1}{N' \times k \times C}\sum_{i=1}^{N'}\sum_{j=1}^{k}(F_i - f_i)^2}, \tag{3}$$

$$\hat{F} = \{\hat{F}_i\}_{i=1}^{N'} = \text{AT}(F), \tag{4}$$

where $F_i = \{f_j \in \mathbb{R}^C\}_{j=1}^k$ is the point features of the i-th group. $f_i \in \mathbb{R}^C$ is centroid feature of the i-th group and $\delta(\cdot)$ is shared MLP. $\hat{F}$ denotes the group feature after affine transformation and ϵ is set to 1e-5 to ensure the correctness of the calculation.

Reaggregation Cross-Attention (RCA). Although the simple symmetric function (*e.g.* max) can satisfy the permutation invariance of the point cloud, it cannot utilize dependencies between points to aggregate the local point set features, which limits the representation of the model. To alleviate this issue, we propose the reaggregation cross-attention. Specifically, for the group features F extracted by ASA, we first apply a symmetric function to aggregate the point features of each group, and then perform cross-attention on the aggregated features and their corresponding intra-group point features, which can be formulated as:

$$h^s = \text{softmax}(\frac{(W_q^s \cdot \text{MAX}(F))(W_k^s F)^T}{\sqrt{d}})(W_v^s F), \tag{5}$$

$$\text{RCA}(\text{MAX}(F), F) = [h^0, h^1, ..., h^s]W_o, \tag{6}$$

where W_q^s, W_k^s, W_v^s, W_o, d and $[\cdot]$ have the same meaning as Eq. (1) and Eq. (2). MAX denotes the symmetric function and we use the max-pooling.

Based on ASA and RCA, the reaggregation Transformer block with affine group features (ReAGF Transformer block) can be summarized as:

$$\begin{aligned} \hat{z}^{l-1} &= \mathrm{LN}(\mathrm{ASA}(F^{l-1}) + F^{l-1}), \\ z^{l-1} &= \mathrm{LN}(\mathrm{MLP}(\hat{z}^{l-1}) + \hat{z}^{l-1}), \\ \hat{z}^{l} &= \mathrm{LN}(\mathrm{RCA}(\mathrm{MAX}(z^{l-1}), z^{l-1}) + \mathrm{MAX}(z^{l-1})), \\ F^{l} &= \mathrm{LN}(\mathrm{MLP}(\hat{z}^{l}) + \hat{z}^{l}) + \mathrm{MAX}(z^{l-1}), \end{aligned} \tag{7}$$

where ASA and RCA are affine self-attention and reaggregation cross-attention, respectively. F^l and z denote the output group features of stage l and temporary variables, respectively. LN denotes layer normalization.

Normalized Relative Positional Encoding. The positional encoding has a vital role in the Transformer. Since the coordinates of points naturally express positional information, such methods of applying transformer on point clouds as PCT [17] do not use positional encoding. However, we find that adding positional encoding helps to improve detection performance. We argue that the reason is that the detection task requires explicit position information to help with object localization. In this work, we use learnable normalized relative positional encoding. Specifically, we compute the normalized relative position between each intra-group point and the corresponding centroid, then we map it to the group feature dimension by shared MLP layer, which can be formulated as:

$$\mathrm{NRPE} = \mathrm{MLP}(\frac{P_i - p_i}{r}), \tag{8}$$

where $P_i = \{p_j \in \mathbb{R}^3\}_{j=1}^{k}$ denotes the point coordinates of the i-th group. $p_i \in \mathbb{R}^3$ is the centroid coordinates of the i-th group. NRPE is the relative positional encoding and r indicates the radius of the ball query in the group embedding. The positional encoding is added to the input F^{l-1} before performing the attention calculation shown in Eq. (7).

Feature Connection Bridge. The computational process of group embedding and ReAGF Transformer block have a large difference, so their output features may have semantic gap. To connect these two modules more naturally, inspired by [59], we use a simple bridge module, as shown in Fig. 2. Specifically, we use linear projection and normalization between the group embedding and ReAGF Transformer block to eliminate their semantic gap.

3.3 Feature Propagation Layer with Multi-scale Connection

Previous methods [7,8,10,11] often utilize hierarchical feature propagation layers [6] for upsampling of points. Point sampling in the group embedding and interpolation in the feature propagation layer inevitably suffer from information loss. To alleviate the above limitations, inspired by [60,61], we introduce a multi-scale connection based on the feature propagation layer, as shown in the

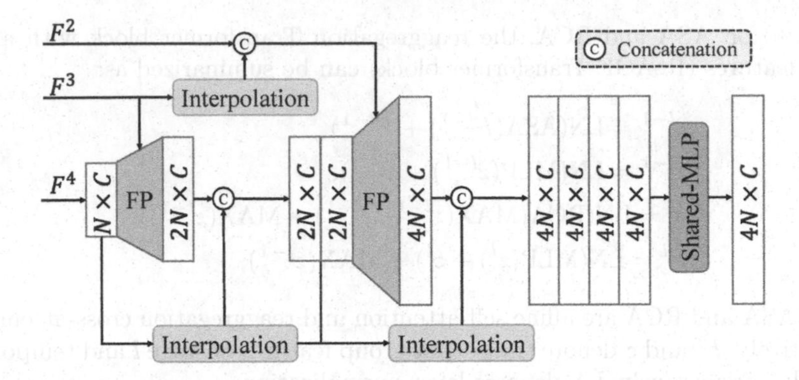

Fig. 4. Upsampling stage consisting of feature propagation (FP) layer with multi-scale connection. F^l denotes output group features of downsampling stage l. N is the number of points and C is the point feature dimension.

Fig. 4. Specifically, in the upsampling stage, we perform point interpolation and feature mapping on the input point features of each layer using the feature propagation layer. The output of all previous layers is concatenated to the output of the current layer, and the concatenated features are served as input to the subsequent layer. Moreover, we fuse the features from the downsampling on the skip connection and concatenate them to the corresponding upsampling layer.

4 Experiments

4.1 Datasets and Evaluation Metrics

We validate our approach using two large-scale indoor scene datasets: SUN RGB-D [21] and ScanNet V2 [20]. SUN RGB-D is a 3D scene understanding dataset with 10,335 monocular RGB-D images and oriented 3D bounding box annotations for 37 categories. We follow VoteNet [7] and divide ~5K samples for training, while using 10 common categories for evaluation. ScanNet V2 is a large-scale 3D reconstructed indoor dataset consisting of 1513 scenes, containing 18 categories of axis-aligned 3D bounding box annotations, and point clouds obtained from reconstructed mesh. Following the setup of VoteNet, we use about 1.2K training samples. For both datasets, we follow a standard evaluation protocol [7], which is mean Average Precision (mAP) with IoU thresholds of 0.25 and 0.5.

4.2 Implementation Details

We apply our ReAGFormer on three state-of-the-art models (*i.e.* VoteNet [7], BRNet [10], and Group-Free [11]) by replacing the backbone of these models with our ReAGFormer, and the replaced models are named as *ReAGF-VoteNet*, *ReAGF-BRNet*, and *ReAGF-Group-Free*, respectively. The number of input points and the data augmentation follow the corresponding baseline [7,10, 11]. Our model is divided into four stages in the downsampling part. Except for

Table 1. Performance comparison by applying our backbone to state-of-the-art models on SUN RGB-D and ScanNet V2. VoteNet* denotes that the result is implemented in MMDetection3D [63], which has better results than the original paper [7]. For Group-Free, we reports the results for 6-layer decoder and 256 object candidates.

Method	SUN RGB-D		ScanNet V2	
	mAP@0.25	mAP@0.5	mAP@0.25	mAP@0.5
VoteNet* [7]	59.1	35.8	62.9	39.9
+Ours (ReAGF-VoteNet)	**62.3(↑3.2)**	**40.7(↑4.9)**	**66.1(↑3.2)**	**45.4(↑5.5)**
BRNet [10]	61.1	43.7	66.1	50.9
+Ours (ReAGF-BRNet)	**61.5(↑0.4)**	**44.8(↑1.1)**	**67.4(↑1.3)**	**52.2(↑1.3)**
Group-Free [11]	**63.0**	45.2	**67.3**	48.9
+Ours (ReAGF-Group-Free)	62.9(↓0.1)	**45.7(↑0.5)**	67.1(↓0.2)	**50.0(↑1.1)**

stage 1, each stage contains a group embedding and a reaggregation Transformer block with affine group features. Note that we use the standard set abstraction layer [6] in stage 1, because using transformer in the early stages does not help the results. We argue that the point feature extraction is not complete in the shallower layers, and thus the dependencies between points cannot be built effectively. For the upsampling stage, we use 2 feature propagation layers with multi-scale connections. The ball query radius of the group embedding is set to $\{0.2, 0.4, 0.8, 1.2\}$ and the number of sampling points is $\{2048, 1024, 512, 256\}$. The upsampling stage interpolates the points to $\{512, 1024\}$. The feature dimension of the points generated by the backbone is set to 288. For ASA and RCA, the number of head is set to 8, and a dropout of 0.1 is used. The initial learning rate of the Transformer block is 1/20 of the other parts, and the model is optimized with the AdamW optimizer [62]. More implementation details are described in the supplementary material.

4.3 Evaluation Results

Evaluation on Different State-of-the-Art Models. We apply our proposed ReAGFormer on three existing state-of-the-art models: VoteNet [7], BRNet [10] and Group-Free [11]. We replace the backbone of these three methods with our proposed ReAGFormer and evaluate them on SUN RGB-D and ScanNet V2 datasets, and the results are shown in Table 1. Our proposed ReAGFormer enables all three methods to achieve performance improvements. In particular, ReAGF-VoteNet gets 4.9% and 5.5% improvement on mAP@0.5 on both datasets. Similarly, ReAGF-BRNet outperforms BRNet with gains of 1.1% mAP@0.5 and 1.3% mAP@0.5 on SUN RGB-D and ScanNet V2, respectively. For both datasets, ReAGF-Group-Free also achieves improvement of 0.5% and 1.1% on mAP@0.5, respectively. Note that by applying our approach to the baseline model, the performance improvement on the more challenging mAP@0.5 is better than mAP@0.25, which demonstrates that our ReAGFormer adequately models the interaction between points and improves object localization accuracy.

Table 2. Performance comparison on SUN RGB-D (left) and ScanNet V2 (right). VoteNet* indicates that the resulting implementation is based on the MMDetection3D [63] toolbox, which has better results than the original paper [7]. 4×PointNet++ denotes 4 individual PointNet++. - indicates that the corresponding method does not report results under this condition or dataset. For Group-Free, we report the results for 6-layer decoder and 256 object candidates.

SUN RGB-D	backbone	mAP@0.25	mAP@0.5	ScanNet V2	backbone	mAP@0.25	mAP@0.5
VoteNet* [7]	PointNet++	59.1	35.8	VoteNet* [7]	PointNet++	62.9	39.9
MLCVNet [8]	PointNet++	59.8	-	MLCVNet [8]	PointNet++	64.7	42.1
HGNet [45]	GU-Net	61.6	-	HGNet [45]	GU-Net	61.3	34.4
SPOT [64]	PointNet++	60.4	36.3	SPOT [64]	PointNet++	59.8	40.4
H3DNet 1BB [9]	PointNet++	-	-	H3DNet 1BB [9]	PointNet++	64.4	43.4
H3DNet 4BB [9]	4×PointNet++	60.1	39.0	H3DNet 4BB [9]	4×PointNet++	67.2	48.1
Pointformer+VoteNet [19]	Pointformer	61.1	36.6	Pointformer+VoteNet [19]	Pointformer	64.1	42.6
3DETR [57]	PointNet++	59.1	32.7	3DETR [57]	PointNet++	65.0	47.0
BRNet [10]	PointNet++	61.1	43.7	BRNet [10]	PointNet++	66.1	50.9
Group-Free [11]	PointNet++	**63.0**	45.2	Group-Free [11]	PointNet++	67.3	48.9
CaVo [65]	U-Net	61.3	44.3	CaVo[65]	U-Net	-	-
DisARM+VoteNet [46]	PointNet++	61.5	41.3	DisARM+VoteNet [46]	PointNet++	66.1	49.7
DisARM+Group-Free [46]	PointNet++	-	-	DisARM+Group-Free [46]	PointNet++	67.0	50.7
ReAGF-VoteNet (Ours)	ReAGFormer (Ours)	62.3	40.7	ReAGF-VoteNet (Ours)	ReAGFormer (Ours)	66.1	45.4
ReAGF-BRNet (Ours)	ReAGFormer (Ours)	61.5	44.8	ReAGF-BRNet (Ours)	ReAGFormer (Ours)	**67.4**	**52.2**
ReAGF-Group-Free (Ours)	ReAGFormer (Ours)	62.9	**45.7**	ReAGF-Group-Free (Ours)	ReAGFormer (Ours)	67.1	50.0

Table 3. Ablation study on ASA and RCA of the ReAGF Transformer block. If ASA and RCA are not used, each layer well be a standard set abstraction layer [6].

ASA	RCA	mAP@0.25	mAP@0.5
–	–	64.1	42.6
✓	–	65.0	45.2
–	✓	66.0	44.9
✓	✓	**66.1**	**45.4**

Comparisons with the State-of-the-Art Methods. In order to verify the effectiveness of our proposed ReAGFormer, we compare ReAGF-VoteNet, ReAGF-BRNet and ReAGF-Group-Free with previous state-of-the-art methods on SUN RGB-D and ScanNet V2 datasets. Table 2 shows the comparison results. By replacing the original backbone with our ReAGFormer, VoteNet achieves competitive results on both datasets. For ScanNet V2, ReAGF-BRNet achieves 67.4% on mAP@0.25 and 52.2% on mAP@0.5, which outperforms all previous state-of-the-art methods. On the SUN RGB-D dataset, ReAGF-Group-Free achieves 62.9% on mAP@0.25 and 45.7% on mAP@0.5, which is better than previous state-of-the-art methods on the more challenging mAP@0.5.

4.4 Ablation Study

In this section, we conduct ablation experiments to verify the effectiveness of each module. If not specified, the models used in all experiments are trained on ReAGF-VoteNet, and evaluated on ScanNet V2 validation set.

ReAGF Transformer Block. We investigate the effects of the ReAGF Transformer block consisting of ASA and RCA, and the results are summarized in

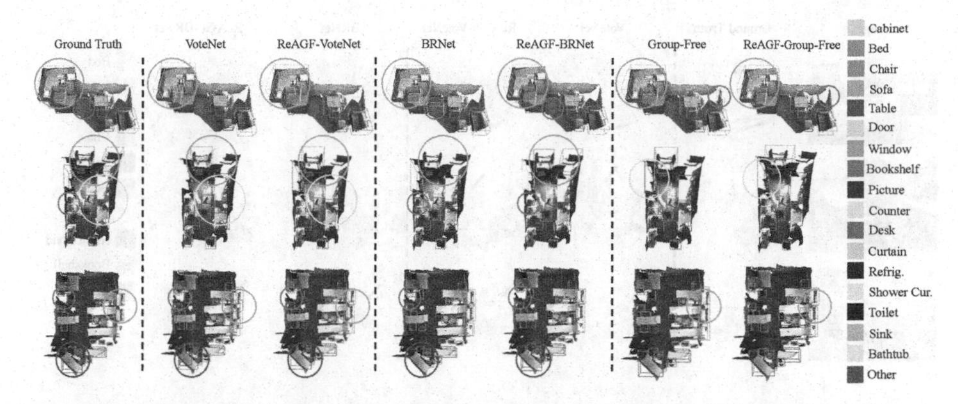

Fig. 5. Qualitative comparison results of 3D object detection on ScanNet V2. ReAGF-VoteNet, ReAGF-BRNet and ReAGF-Group-Free denote the replacement of the baseline original backbone with our ReAGFormer. With the help of ReAGFormer, VoteNet and BRNet achieve more reliable results (orange circles, blue circles and purple circles). Objects with similar shapes (*e.g.* Table and Desk, Bookshelf and Door) can be easily confused, but our method can alleviate such problem (yellow circles). Color is used for better illustration purpose, and it is not used in the experiment. (*Best viewed in color.*)

Table 4. Ablation study on affine transformation of the ASA. If the affine transformation as shown in Eq. (4) is not used, ASA will be the standard self-attention [12].

Downsampling method	Affine transformation	mAP@0.25	mAP@0.5
Set abstraction layer	–	64.1	42.6
	✓	64.6	43.7
ReAGF transformer block	–	65.9	44.5
	✓	**66.1**	**45.4**

Table 3. If the ReAGF Transformer block consisting of ASA and RCA is not used, each layer will be a standard set abstraction layer [6]. We can observe that by applying ASA and RCA separately, performance is improved by 2.6% and 2.3% on mAP@0.5, respectively. If both ASA and RCA are used, we can achieve the best performance improvement. Table 4 ablates the affine transformation (AT) module in the ASA. The best result is achieved using our complete ReAGF transformer block, and our AT also improves the performance of the set abstraction layer, which demonstrates the effectiveness of our AT module.

Positional Encoding. To investigate whether positional encoding is effective and normalized relative positional encoding is better, we conduct comparison without positional encoding and with absolute or normalized relative positional encoding. As shown in Table 5, using normalized relative positional encoding brings 2.1% mAP@0.25 improvement and 0.4% mAP@0.5 improvement compared to not using positional encoding. We argue that the reason is that the detection task requires explicit position information to help object localization. We also find that normalized relative positional encoding outperforms absolute

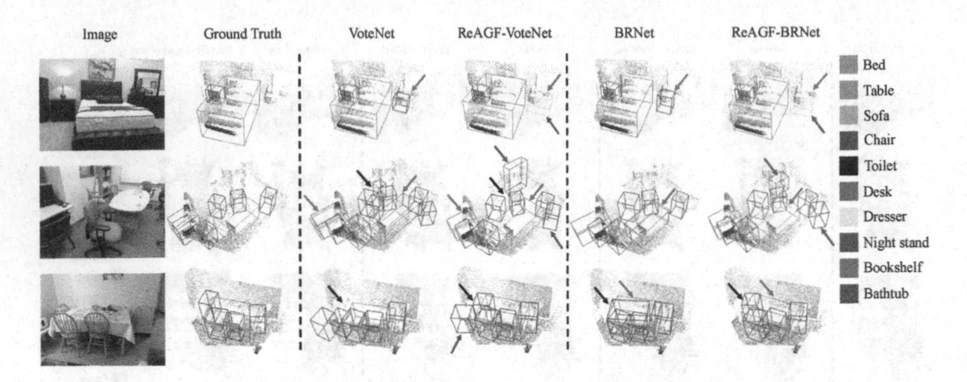

Fig. 6. Qualitative results of 3D object detection on SUN RGB-D. Our method generates more reasonable boxes (see black arrows) and can distinguish between objects with similar shapes (see blue arrows). Moreover, our method can even detect objects that are not annotated in ground truth (see red arrows). Images and colors are only used for better illustration, and they are not used in our network. (*Best viewed in color.*)

Table 5. Ablation study on the effectiveness of positional encoding and performance of different positional encoding.

Positonal encoding	mAP@0.25	mAP@0.5
None	64.0	45.0
Absolute	64.8	43.9
Normalized relative	**66.1**	**45.4**

positional encoding, and even the network without positional encoding is 1.1% better than that using absolute positional encoding on mAP@0.5.

Feature Connection Bridge. In Table 6, we compare the impact of with and without features connection bridge on the 3D object detection performance. With the feature connection bridge, we eliminate the semantic gap between the group embedding and ReAGF Transformer block, thus achieving the improvement of 0.5% on mAP@0.25 and 0.9% on mAP@0.5.

Multi-scale Connection. We investigate the effect of multi-scale connection by replacing it with cascade connection [66,67] and residual connection [61], and the results are summarized in Table 7. We can see that multi-scale connection achieves the best results compared to the other methods. This demonstrates that multi-scale connection can more fully aggregate multi-scale contextual information and reduce information loss caused by point sampling and interpolation.

4.5 Qualitative Results and Discussion

Figure 5 illustrates the qualitative comparison of the results on ScanNet V2. These results show that applying our ReAGFormer to the baseline achieves

Table 6. Ablation study on the feature connection bridge.

Feature connection bridge	mAP@0.25	mAP@0.5
–	65.6	44.5
✓	**66.1**	**45.4**

Table 7. Ablation study on the different connection methods for feature propagation layer.

Connection method	mAP@0.25	mAP@0.5
Cascade	65.2	44.9
Residual	65.0	44.3
Multi-scale	**66.1**	**45.4**

more reliable results. Specifically, ReAGF-VoteNet, ReAGF-BRNet and ReAGF-Group-Free can detect more reasonable and accurate results (orange circles, blue circles and purple circles), despite the challenges of cluttered scenes or fewer points. In addition, our method achieves better results for similarly shaped objects (yellow circles). For example, the desk in the second row of the scene is treated as a table by VoteNet [7] in Fig. 5, but ReAGF-VoteNet successfully detects a desk.

Figure 6 visualizes the qualitative results on SUN RGB-D scenes. Our model generate more reasonable boxes even in cluttered and occluded scenes (see black arrows). In addition, our method can also better distinguish between similarly shaped objects on SUN RGB-D. For example, in the first row of Fig. 6, we successfully solve the problem of different categories generated by the same object (see blue arrows). In the second row, we can detect the table and the desk correctly (see blue arrows). Besides, our method can even detect objects that are not annotated in the ground truth (see red arrows).

5 Conclusion

In this paper, we present ReAGFormer, a reaggregation Transformer backbone with affine group features for 3D object detection. We introduce affine self-attention to align different groups of feature spaces while modeling the dependencies between points. To improve the efficiency of feature aggregation, we utilize reaggregation cross-attention to reaggregate group features based on learned attention. Moreover, we also introduce a multi-scale connection in the feature propagation layer to reduce the information loss caused by point sampling and interpolation. We apply our ReAGFormer to existing state-of-the-art detectors and achieve significant performance improvements on the main benchmarks. Experiments demonstrate the effectiveness and generalization of our method.

Acknowledgements. This work was supported by the Key-Area Research and Development Program of Guangdong Province (No. 2019B010139004) and the National Natural Science Foundation Youth Fund No. 62007001.

References

1. Chen, X., Ma, H., Wan, J., Li, B., Xia, T.: Multi-view 3D object detection network for autonomous driving. In: CVPR, pp. 1907–1915 (2017)
2. Zhou, Y., Tuzel, O.: Voxelnet: end-to-end learning for point cloud based 3D object detection. In: CVPR, pp. 4490–4499 (2018)
3. Song, S., Xiao, J.: Deep sliding shapes for amodal 3D object detection in RGB-D images. In: CVPR, pp. 808–816 (2016)
4. Yang, B., Luo, W., Urtasun, R.: Pixor: real-time 3D object detection from point clouds. In: CVPR, pp. 7652–7660 (2018)
5. Qi, C.R., Su, H., Mo, K., Guibas, L.J.: Pointnet: deep learning on point sets for 3D classification and segmentation. In: CVPR, pp. 652–660 (2017)
6. Qi, C.R., Yi, L., Su, H., Guibas, L.J.: Pointnet++: deep hierarchical feature learning on point sets in a metric space. NeurIPS **30**, 5099–5108 (2017)
7. Qi, C.R., Litany, O., He, K., Guibas, L.J.: Deep Hough voting for 3D object detection in point clouds. In: ICCV, pp. 9277–9286 (2019)
8. Xie, Q., et al.: Mlcvnet: multi-level context votenet for 3D object detection. In: CVPR, pp. 10447–10456 (2020)
9. Zhang, Z., Sun, B., Yang, H., Huang, Q.: H3DNet: 3D object detection using hybrid geometric primitives. In: Vedaldi, A., Bischof, H., Brox, T., Frahm, J.-M. (eds.) ECCV 2020. LNCS, vol. 12357, pp. 311–329. Springer, Cham (2020). https://doi.org/10.1007/978-3-030-58610-2_19
10. Cheng, B., Sheng, L., Shi, S., Yang, M., Xu, D.: Back-tracing representative points for voting-based 3D object detection in point clouds. In: CVPR, pp. 8963–8972 (2021)
11. Liu, Z., Zhang, Z., Cao, Y., Hu, H., Tong, X.: Group-free 3D object detection via transformers. In: ICCV (2021)
12. Vaswani, A., et al.: Attention is all you need. In: NeurIPS, pp. 5998–6008 (2017)
13. Dosovitskiy, A., et al.: An image is worth 16x16 words: transformers for image recognition at scale. arXiv preprint arXiv:2010.11929 (2020)
14. Carion, N., Massa, F., Synnaeve, G., Usunier, N., Kirillov, A., Zagoruyko, S.: End-to-end object detection with transformers. In: Vedaldi, A., Bischof, H., Brox, T., Frahm, J.-M. (eds.) ECCV 2020. LNCS, vol. 12346, pp. 213–229. Springer, Cham (2020). https://doi.org/10.1007/978-3-030-58452-8_13
15. Liu, Z., et al.: Swin transformer: hierarchical vision transformer using shifted windows. In: ICCV (2021)
16. Liu, Y., et al.: A survey of visual transformers. arXiv preprint arXiv:2111.06091 (2021)
17. Guo, M.H., Cai, J.X., Liu, Z.N., Mu, T.J., Martin, R.R., Hu, S.M.: PCT: point cloud transformer. Comput. Vis. Media **7**(2), 187–199 (2021)
18. Zhao, H., Jiang, L., Jia, J., Torr, P., Koltun, V.: Point transformer. arXiv preprint arXiv:2012.09164 (2020)
19. Pan, X., Xia, Z., Song, S., Li, L.E., Huang, G.: 3D object detection with pointformer. In: CVPR, pp. 7463–7472 (2021)

20. Dai, A., Chang, A.X., Savva, M., Halber, M., Funkhouser, T., Nießner, M.: Scannet: richly-annotated 3D reconstructions of indoor scenes. In: CVPR, pp. 5828–5839 (2017)
21. Song, S., Lichtenberg, S.P., Xiao, J.: Sun RGB-D: a RGB-D scene understanding benchmark suite. In: CVPR, pp. 567–576 (2015)
22. Su, H., Maji, S., Kalogerakis, E., Learned-Miller, E.: Multi-view convolutional neural networks for 3D shape recognition. In: ICCV, pp. 945–953 (2015)
23. Lang, A.H., Vora, S., Caesar, H., Zhou, L., Yang, J., Beijbom, O.: Pointpillars: fast encoders for object detection from point clouds. In: CVPR, pp. 12697–12705 (2019)
24. Riegler, G., Osman Ulusoy, A., Geiger, A.: Octnet: learning deep 3D representations at high resolutions. In: CVPR, pp. 3577–3586 (2017)
25. Maturana, D., Scherer, S.: Voxnet: a 3D convolutional neural network for real-time object recognition. In: IROS, pp. 922–928 (2015)
26. Jiang, L., Zhao, H., Liu, S., Shen, X., Fu, C.W., Jia, J.: Hierarchical point-edge interaction network for point cloud semantic segmentation. In: ICCV, pp. 10433–10441 (2019)
27. Li, Y., Bu, R., Sun, M., Wu, W., Di, X., Chen, B.: PointCNN: convolution on x-transformed points. NeurIPS **31**, 820–830 (2018)
28. Xu, Y., Fan, T., Xu, M., Zeng, L., Qiao, Yu.: SpiderCNN: Deep Learning on Point Sets with Parameterized Convolutional Filters. In: Ferrari, V., Hebert, M., Sminchisescu, C., Weiss, Y. (eds.) ECCV 2018. LNCS, vol. 11212, pp. 90–105. Springer, Cham (2018). https://doi.org/10.1007/978-3-030-01237-3_6
29. Wang, S., Suo, S., Ma, W.C., Pokrovsky, A., Urtasun, R.: Deep parametric continuous convolutional neural networks. In: CVPR, pp. 2589–2597 (2018)
30. Wu, W., Qi, Z., Fuxin, L.: PointConv: deep convolutional networks on 3D point clouds. In: CVPR, pp. 9621–9630 (2019)
31. Thomas, H., Qi, C.R., Deschaud, J.E., Marcotegui, B., Goulette, F., Guibas, L.J.: KPConv: flexible and deformable convolution for point clouds. In: ICCV, pp. 6411–6420 (2019)
32. Xu, M., Ding, R., Zhao, H., Qi, X.: PaConv: position adaptive convolution with dynamic kernel assembling on point clouds. In: CVPR, pp. 3173–3182 (2021)
33. Boulch, A., Puy, G., Marlet, R.: FKAConv: feature-kernel alignment for point cloud convolution. In: Ishikawa, H., Liu, C.-L., Pajdla, T., Shi, J. (eds.) ACCV 2020. LNCS, vol. 12622, pp. 381–399. Springer, Cham (2021). https://doi.org/10.1007/978-3-030-69525-5_23
34. Wang, Y., Sun, Y., Liu, Z., Sarma, S.E., Bronstein, M.M., Solomon, J.M.: Dynamic graph CNN for learning on point clouds. ACM Trans. Graph. (TOG) **38**(5), 1–12 (2019)
35. Wang, L., Huang, Y., Hou, Y., Zhang, S., Shan, J.: Graph attention convolution for point cloud semantic segmentation. In: CVPR, pp. 10296–10305 (2019)
36. Xu, Q., Sun, X., Wu, C.Y., Wang, P., Neumann, U.: Grid-GCN for fast and scalable point cloud learning. In: CVPR, pp. 5661–5670 (2020)
37. Zhao, H., Jiang, L., Fu, C.W., Jia, J.: Pointweb: enhancing local neighborhood features for point cloud processing. In: CVPR, pp. 5565–5573 (2019)
38. Liang, M., Yang, B., Wang, S., Urtasun, R.: Deep continuous fusion for multi-sensor 3D object detection. In: Ferrari, V., Hebert, M., Sminchisescu, C., Weiss, Y. (eds.) ECCV 2018. LNCS, vol. 11220, pp. 663–678. Springer, Cham (2018). https://doi.org/10.1007/978-3-030-01270-0_39
39. Hou, J., Dai, A., Nießner, M.: 3D-sis: 3D semantic instance segmentation of RGB-D scans. In: CVPR, pp. 4421–4430 (2019)

40. Yan, Y., Mao, Y., Li, B.: Second: sparsely embedded convolutional detection. Sensors **18**(10), 3337 (2018)
41. Yi, L., Zhao, W., Wang, H., Sung, M., Guibas, L.J.: GSPN: generative shape proposal network for 3D instance segmentation in point cloud. In: CVPR, pp. 3947–3956 (2019)
42. Ren, Z., Sudderth, E.B.: Three-dimensional object detection and layout prediction using clouds of oriented gradients. In: CVPR, pp. 1525–1533 (2016)
43. Shi, S., Wang, X., Li, H.: PointRCNN: 3D object proposal generation and detection from point cloud. In: CVPR, pp. 770–779 (2019)
44. Shi, S., et al.: Pv-RCNN: point-voxel feature set abstraction for 3D object detection. In: CVPR, pp. 10529–10538 (2020)
45. Chen, J., Lei, B., Song, Q., Ying, H., Chen, D.Z., Wu, J.: A hierarchical graph network for 3D object detection on point clouds. In: CVPR, pp. 392–401 (2020)
46. Duan, Y., Zhu, C., Lan, Y., Yi, R., Liu, X., Xu, K.: Disarm: displacement aware relation module for 3D detection. In: CVPR, pp. 16980–16989 (2022)
47. Touvron, H., Cord, M., Douze, M., Massa, F., Sablayrolles, A., Jégou, H.: Training data-efficient image transformers & distillation through attention. In: ICML, pp. 10347–10357. PMLR (2021)
48. Zhu, X., Su, W., Lu, L., Li, B., Wang, X., Dai, J.: Deformable DETR: deformable transformers for end-to-end object detection. arXiv preprint arXiv:2010.04159 (2020)
49. Zheng, S., et al.: Rethinking semantic segmentation from a sequence-to-sequence perspective with transformers. In: CVPR, pp. 6881–6890 (2021)
50. Strudel, R., Garcia, R., Laptev, I., Schmid, C.: Segmenter: transformer for semantic segmentation. arXiv preprint arXiv:2105.05633 (2021)
51. Lai, X., et al.: Stratified transformer for 3D point cloud segmentation. In: Proceedings of the IEEE/CVF Conference on Computer Vision and Pattern Recognition, pp. 8500–8509 (2022)
52. Wang, Y., Solomon, J.M.: Deep closest point: learning representations for point cloud registration. In: Proceedings of the IEEE/CVF International Conference on Computer Vision, pp. 3523–3532 (2019)
53. Fan, H., Yang, Y., Kankanhalli, M.: Point 4D transformer networks for spatio-temporal modeling in point cloud videos. In: Proceedings of the IEEE/CVF Conference on Computer Vision and Pattern Recognition, pp. 14204–14213 (2021)
54. Qiu, S., Anwar, S., Barnes, N.: Pu-transformer: point cloud upsampling transformer. arXiv preprint arXiv:2111.12242 (2021)
55. Xu, X., Geng, G., Cao, X., Li, K., Zhou, M.: TDnet: transformer-based network for point cloud denoising. Appl. Opt. **61**(6), C80–C88 (2022)
56. Yu, X., Rao, Y., Wang, Z., Liu, Z., Lu, J., Zhou, J.: Pointr: diverse point cloud completion with geometry-aware transformers. In: ICCV, pp. 12498–12507 (2021)
57. Misra, I., Girdhar, R., Joulin, A.: An end-to-end transformer model for 3D object detection. In: ICCV (2021)
58. Ma, X., Qin, C., You, H., Ran, H., Fu, Y.: Rethinking network design and local geometry in point cloud: a simple residual MLP framework. In: ICLR (2021)
59. Peng, Z., et al.: Conformer: local features coupling global representations for visual recognition. arXiv preprint arXiv:2105.03889 (2021)
60. Huang, G., Liu, Z., Van Der Maaten, L., Weinberger, K.Q.: Densely connected convolutional networks. In: CVPR, pp. 4700–4708 (2017)
61. Chen, B., Liu, Y., Zhang, Z., Lu, G., Zhang, D.: Transattunet: multi-level attention-guided u-net with transformer for medical image segmentation. arXiv preprint arXiv:2107.05274 (2021)

62. Loshchilov, I., Hutter, F.: Decoupled weight decay regularization. arXiv preprint arXiv:1711.05101 (2017)
63. Contributors, M.: MMDetection3D: OpenMMLab next-generation platform for general 3D object detection. https://github.com/open-mmlab/mmdetection3d (2020)
64. Du, H., Li, L., Liu, B., Vasconcelos, N.: SPOT: selective point cloud voting for better proposal in point cloud object detection. In: Vedaldi, A., Bischof, H., Brox, T., Frahm, J.-M. (eds.) ECCV 2020. LNCS, vol. 12356, pp. 230–247. Springer, Cham (2020). https://doi.org/10.1007/978-3-030-58621-8_14
65. You, Y., et al.: Canonical voting: towards robust oriented bounding box detection in 3D scenes. In: CVPR, pp. 1193–1202 (2022)
66. Qin, X., Zhang, Z., Huang, C., Dehghan, M., Zaiane, O.R., Jagersand, M.: U2-net: going deeper with nested U-structure for salient object detection. Pattern Recogn. **106**, 107404 (2020)
67. Cai, Y., Wang, Y.: Ma-unet: an improved version of unet based on multi-scale and attention mechanism for medical image segmentation. arXiv preprint arXiv:2012.10952 (2020)

Adaptive Range Guided Multi-view Depth Estimation with Normal Ranking Loss

Yikang Ding[1], Zhenyang Li[1], Dihe Huang[1], Kai Zhang[1], Zhiheng Li[1(✉)], and Wensen Feng[2(✉)]

[1] Tsinghua University, Beijing, China
zhhli@mail.tsinghua.edu.cn
[2] Huawei Company, Shenzhen, China
fengwensen@huawei.com

Abstract. Deep learning algorithms for Multi-view Stereo (MVS) have surpassed traditional MVS methods in recent years, due to enhanced reconstruction quality and runtime. Deep-learning based methods, on the other side, continue to generate overly smoothed depths, resulting in poor reconstruction. In this paper, we aim to Boost Depth Estimation (BDE) for MVS and present an approach, termed as **BDE-MVSNet**, for reconstructing high-quality point clouds with precise depth prediction. We present a non-linear strategy that derives an adaptive depth range (ADR) from the estimated probability, motivated by distinctive differences in estimated probability between foreground and background pixels. ADR also tends to decrease fuzzy boundaries via upsampling low-resolution depth maps between stages. Additionally, we provide a novel structure-guided normal ranking (SGNR) loss that imposes geometrical consistency in boundary areas by using the surface normal vector. Extensive experiments on *DTU* dataset, *Tanks and Temples* benchmark, and *BlendedMVS* dataset demonstrate that our method outperforms known methods and achieves state-of-the-art performance.

1 Introduction

Multi-view Stereo (MVS) is the process of reconstructing the dense 3D geometry of an observed scene using posed images and camera parameters. It is a key problem in computer vision, with applications to various domains such as augmented and virtual reality, robotics, and 3D modeling. Although MVS has been studied for several decades, computing a high-quality 3D reconstruction in the presence of occlusions, low-textured regions, and blur remains a challenge [1]. Convolutional Neural Networks (CNNs) were adopted by MVS methods as an alternative for hand-crafted matching metrics and regularization schemes [2–6] following the success of deep learning in many fields of computer vision, offering a significant improvement to the completeness of the reconstructed model and the runtime required for generating it [7–13]. The basic learning-based MVS

Y. Ding and Z. Li—Equal contribution.

L. Wang et al. (Eds.): ACCV 2022, LNCS 13841, pp. 280–295, 2023.
https://doi.org/10.1007/978-3-031-26319-4_17

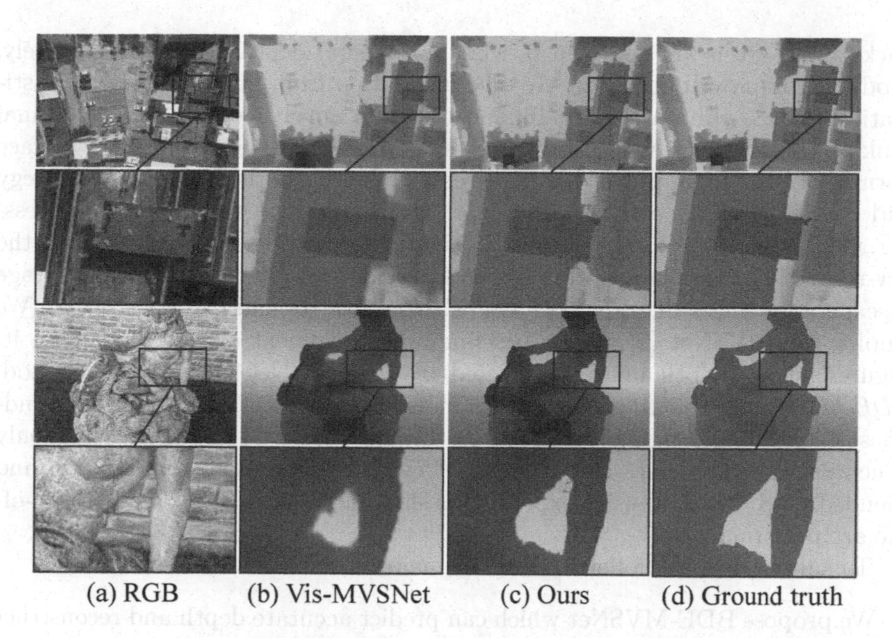

Fig. 1. Comparison of depth maps. (a) RGB images. (b)–(d) Depth maps predicted by Vis-MVSNet [9] and our approach, and the respective ground truth depth. Our method predicts much more accurate depth compared to Vis-MVSNet [9].The resolution of input images is 640×512.

approach [7] begins with the extraction of deep features using a 2D CNN. The corresponding cost maps, each generated by considering the variance of warped activation maps, are then stacked to form a cost volume over different depth hypotheses. Before applying Softmax and regressing the depth of the reference image, a 3D CNN is applied to finalize the cost volume normalization. Recent state-of-the-art MVS approaches have recommended various optimizations, such as using coarse-to-fine processing of activation maps [8–12], since runtime and memory expand cubically with spatial resolution and have a complexity of $O(n^3)$. The above mentioned methods using depth range estimation in pixels [10,11] and imposing pixel-wise visibility constraints [9]. Despite these recent advancements, learning-based MVS methods still produce overly smoothed depth and imprecise boundaries, which are harmful to the reconstruction results.

In this work, we propose a MVS method, named BDE-MVSNet, which aims at boosting depth estimation, especially on the boundaries and the background areas in multi-view stereo tasks. The uncertainty associated with depth estimation in the foreground and background/boundary regions motivated our method. After a single estimating step, we observe that the depth at foreground pixels is assigned with high probability, whereas the depth at background pixels is commonly assigned with low probability. Thus, we propose an adaptive depth range (ADR) method that computes the depth range per pixel in a non-linear way from its probability. As a consequence, depth ranges for boundary and background pixels are kept broad, while foreground depth is sampled precisely, producing improved accuracy. We take inspiration from Monocular Depth Estimation (MDE) approaches [14–16] and present a novel structure-guided normal ranking (SGNR) loss, which promotes geometrical consistency using the surface normal vector, to improve depth regression. With the help of ADR strategy and SGNR loss, we estimate the depth with only two stages while processing relative high-resolution images. Without the requirement to up-sample the low-resolution depth maps, this method provides an alternative to three-stage cascade approaches like Vis-MVSNet [9], UCSNet [10] and CasMVSNet [8]. We employ Vis-MVSNet [9] as our baseline and implement our method upon it. Figure 1 shows depth images predicted by Vis-MVSNet [9] and our method. *BDE-MVSNet* is able to produce much more accurate depth and sharper boundaries, even in challenging scenarios. We evaluate BDE-MVSNet on commonly benchmarked MVS datasets, namely DTU [17], Tanks and Temples [18] and BlendedMVS [19]. Extensive experiments show BDE-MVSNet achieves state-of-the-art performances.

To summarize, the following are our main contributions.

- We propose BDE-MVSNet which can predict accurate depth and reconstruct high-quality point clouds.
- We introduce ADR strategy to derive a per-pixel depth interval in a non-linear manner, which helps predict accurate depth in only two stages.
- We propose SGNR loss and show it can help in predicting sharp boundary and in decreasing tailing errors in the reconstructed point cloud.
- We qualify the performance of our method on multiple MVS datasets and show it achieves state-of-the-art performance.

2 Related Work

2.1 Learning-Based MVS

The accuracy and efficiency of learning-based methods are directly affected by the number of depth hypotheses, the interval from which they are sampled and the spatial resolution of the activation maps, which determine the dimensions of the regularized cost volume [8]. Recent state-of-the-art learning-based MVS methods have suggested different strategies for extending the basic learned MVS paradigm (MVSNet [7]) and optimizing the aforementioned trade-offs. Different recurrent models were suggested for regularizing the cost volume in a sequential manner [20–22]. However, while decreasing memory consumption, sequential processing does not scale well with spatial resolution. Cascade methods [8–13] proposed instead to form a feature pyramid and regress depth maps in a coarse-to-fine manner. Typically, at the coarsest stage, depth is regressed as in the MVSNet paradigm. As the resolution increases, the number of depth hypotheses is decreased, resulting in improved efficiency. The depth range at a given stage is often centered around the depth estimated at the previous coarser stage resulting in more accurate depth ranges to sample from [8–11]. One drawback of

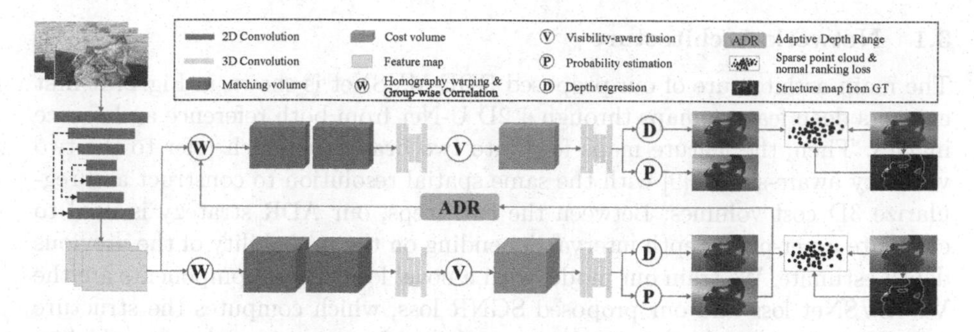

Fig. 2. Overview of BDE-MVSNet. We use a visibility-aware architecture, similar to [9], but apply only two processing stages using the same resolution feature maps. Our ADR strategy determines the per-pixel depth range at the second stage based on the estimated probability. We further apply our SGNR loss to enforce geometrical consistency.

coarse-to-fine methods is the need to up-sample depth maps from stage to stage, which can yield fuzzy boundaries.

2.2 MVS Loss Functions

Learning-based MVS methods are typically optimized to minimize the L1 loss between the predicted depth map and ground truth (GT) depth [7,8,10,12,21] or cross-entropy loss between the predicted probability volume and the GT probability, which is generated by one-hot encoding [20–22]. A recent extension to this formulation is introduced by Vis-MVSNet [9], where the per-pixel visibility is explicitly addressed by computing the cost volume per pair of reference and source images before constructing and regularizing the joint cost volume. The common L1 loss is also extended to minimize the depth maps estimated in a pairwise and joint manner. Other extensions are also proposed to improve geometrical consistency through constraints on the surface normal vector [23,24]. Besides multi-view depth estimation, MDE approaches [14,15,25–27] also offered novel strategies for improving depth prediction. For example, ranking loss [14,15] and structure-guided point sampling [16] were shown to improve depth accuracy. In this work, we extend propose a novel SGNR loss based on MDE strategies and surface normal vector constraints.

3 Method

Given a collection of images and camera parameters (intrinsic matrices K, rotation matrices R and translation vector t), our proposed BDE-MVSNet aims to predict a dense depth map $d \in \mathbb{R}^{h\times w}$ for each reference image $I_0 \in \mathbb{R}^{h\times w\times 3}$, using respective m source images $\{I_i\}_{i=1}^{m}$ with highest co-visibility.

3.1 Network Architecture

The main architecture of our proposed BDE-MVSNet is shown in Fig. 2. It first extracts deep feature maps through a 2D U-Net from both reference and source images. Then, the feature maps feed into two branches, which refer to the two visibility aware-stages [9] with the same spatial resolution to construct and regularize 3D cost volumes. Between the two steps, our ADR strategy is used to calculate a per-pixel depth interval depending on the probability of the previous depth estimate. We train our model with a novel loss, whose components are the Vis-MVSNet loss and our proposed SGNR loss, which computes the structure map from the ground truth depth map and samples pair-wise points to calculate the normal ranking loss.

Feature Extraction. We use a 2D U-Net to extract deep features from reference I_0 and source $\{I_i\}_{i=1}^m$ and process the finest-resolution feature maps, of size $\frac{h}{2} \times \frac{w}{2} \times 32$.

Cost Volume Construction. Following [7,9], we construct pair-wise and joint cost volumes using differentiable homography warping and group-wise correlation. The warping process from I_i to I_0 can be described as:

$$H_{i,j} = K_i R_i (I - \frac{(t_0 - t_i) a_0^T}{d_j}) R_0^T K_0^{-1}, \tag{1}$$

where $H_{i,j}$ refers to the homography matrix at depth d_j and a_0 denotes the principle axis of the reference image. We first compute pairwise cost volumes and respective probability volumes [9] using group-wise correlation [28] and 3D CNN, which will then be fused to construct the final cost volume. Then, we apply a depth-wise 3D CNN with shape $1 \times 1 \times 1$ and a Softmax function to compute the probability volume $P \in \mathbb{R}^{N \times \frac{h}{2} \times \frac{w}{2}}$. The final depth with its probability map can be obtained from P using regression or winner-take-all. The generation of cost volume is identical for both stages, e.g. uses same spatial resolution and same sampling strategy. The main difference between the two stages lies in the prior depth range derivation. For the first stage, we use a fixed prior depth range for all pixels as in previous methods. While for the second stage, we update the depth range for each pixel using our ADR strategy (Sect. 3.2).

Depth Regression. Given the probability volume P, we regress the predicted depth $\overline{d}(p)$ of each pixel p in the reference image by taking the probability-weighted mean of all N hypotheses:

$$\overline{d}(p) = \sum_{j=1}^{N} d_j \cdot P(p, j). \tag{2}$$

Depth Fusion. At inference time, we apply our model to regress the depth maps of all images and then filter and fuse them to reconstruct the 3D point cloud as in [8], based on photometric and geometric consistency.

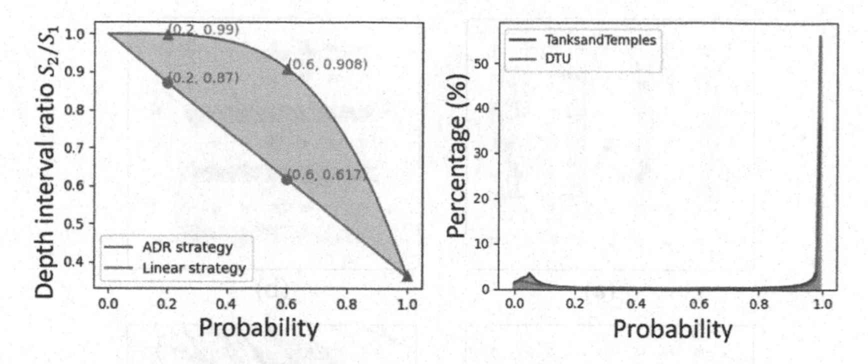

Fig. 3. Illustration of ADR. The left column shows comparison of ADR and linear strategy [10,11]. The right column shows the distribution of the first stage's probability on DTU validation set [17].

3.2 Adaptive Depth Range

At the end of the first stage, the network produces a probability volume P and an estimated depth map $\overline{d}$. Given these two outputs, we can obtain an estimated probability $\overline{P}$ for each pixel, as in MVSNet [7]. The probability $\overline{P}(p)$ at pixel p provides an measurement for the uncertainty. We leverage this information for adapting the depth range at the next stage. We propose a non-linear strategy motivated by distinctive differences in pixel uncertainty between foreground and background areas. As shown in Fig. 3, when analyzing the distribution of estimated probability, we find foreground pixels are typically assigned with a probability of 0.9–0.999, while pixels at boundary and background regions present a probability of 0.3–0.7 and <0.3 respectively. For depth estimations with high uncertainty, we would like to sample from a relatively wide range to decrease error rate. On the other hand, when the estimation is made with high certainty, a narrow range can help in achieving improved accuracy. Following this intuition, we propose our novel Adaptive Depth Range (ADR) method.

Given a depth range $[d_{min,i}, d_{max,i}]$, a fixed depth interval δ and a fixed number of depth hypotheses N_i for stage i, we can obtain the *j-th* depth hypothesis $d_{j,i}$ at the *i-th* stage,

$$\begin{cases} d_{j,i} = d_{min,i} + (j-1) \cdot \delta \\ d_{max,i} = d_{min,i} + N_i \cdot \delta \end{cases}, \tag{3}$$

We simply take a fixed $[d_{min,i}, d_{max,i}]$ for all pixels in the first stage ($i = 1$). While for the second stage, our ADR strategy computes a pixel-specific depth range $[\overline{d}_1(p) - ADR(p),\ \overline{d}_1(p) + ADR(p)]$ based on the estimated depth $\overline{d}_1(p)$ and probability $\overline{P}_1(p)$, where $ADR(p)$ can be written as,

$$ADR(p) = \frac{1}{2} \cdot N_2 \cdot s(p) \cdot \delta. \tag{4}$$

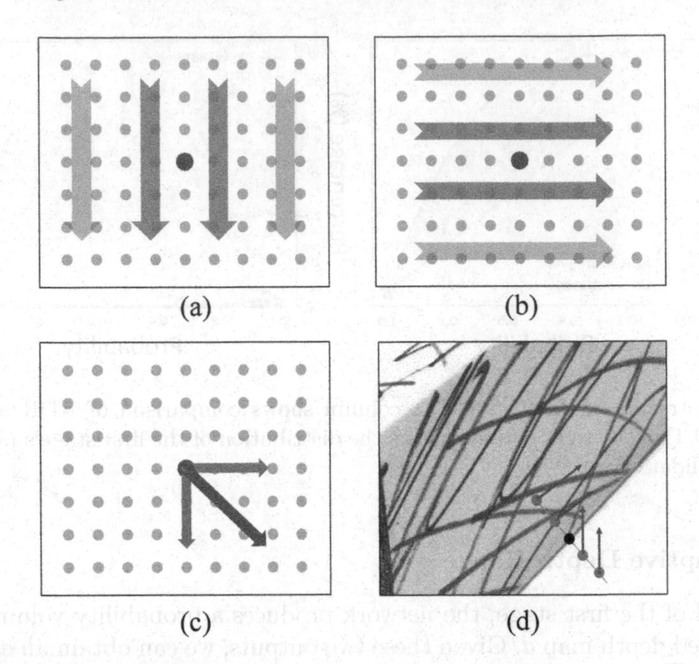

Fig. 4. Illustration of our structure-guided sampling scheme. (a)–(b): Sobel operator in y and x direction. (c): Gradient of center pixel (red point). (d): Four sampled points (in green) around a center point (in black) at a boundary area. (Color figure online)

$s(p)$ is a pixel-specific scaling factor which can be obtained by,

$$s(p) = \cos(k \cdot \overline{P}_1^2(p)), \tag{5}$$

where k is a hyper-parameter (we set it to 1.2) which controls the range scaling. We illustrate the differences between the linear strategy employed by [10,11] and our ADR strategy in Fig. 3. ADR tries to maintain a large depth range for low-probability pixels and strictly narrow the depth range for high-probability pixels, which helps to assign proper depth range and predict accurate depth in fewer stages.

3.3 Structure-Guided Normal Ranking Loss

Inspired by recent advancements in MDE, we propose to modify the *pairwise ranking loss* [14] for MVS, and suggest imposing geometric consistency by updating the ordinal label using the *surface normal*. Our proposed SGNR loss mainly consists of two steps: structure-guided sampling and pairwise normal ranking loss.

Structure-Guided Sampling. In order to accommodate for the typical MVS setting (dataset), where the background is usually masked in both RGB and depth images, we propose a modified version of four-point sampling scheme [16] to sample the point pairs $S = \{(p_{i,0}, p_{i,1})\}_{i=1}^{N}$. Given a reference image I, we first convert it to a gray-scale image I^* and use the *Sobel* operator to get the gradient maps G_x, G_y and gradient magnitude G (Fig. 4(a)–(c)). Before we get the final edge map, we compute a solid-region mask to avoid sampling points in masked background regions. We execute this operation by reducing 16 pixels along the orthogonal line crossing the background-mask to get the solid-region mask $M_{solid} \in \mathbb{R}^{h \times w}$. Then the final edge map E can be obtained by applying a solid-region mask to avoid sampling points in masked background regions by

$$E = I^*[G(p) \geq \alpha \cdot max(G)]_{p \in M_{solid}}, \tag{6}$$

where α is a threshold which controls the density of E, and we set it to 0.05 in our experiments.

For each edge point $e = (x, y) \in E$, we sample four points $\{(x_k, y_k)\}_{k=0}^{3}$ by

$$\begin{cases} x_k = x + \delta_k G_x(e)/G(e) \\ y_k = y + \delta_k G_y(e)/G(e) \end{cases}, \tag{7}$$

where we sample $\delta_0 < \delta_1 < 0 < \delta_2 < \delta_3$ within a small distance range β from the edge point. Given four sampled points p_0, p_1, p_2, p_3, we form three pairs of points for pairwise ranking: $[(p_0, p_1), (p_1, p_2), (p_2, p_3)]$. Similar to [16], we also sample some points using a random sampling scheme in order to preserve global structures. Specifically, we sample $3n$ pairs points through the four-point sampling scheme and n pairs through the random sampling scheme, where n is set to 1000 in our experiments.

Pairwise Normal Ranking Loss. The pair-wise ranking loss performs well in the majority of cases, however it loses geometry consistency in plane or boundary regions. Take a look at the box-like object in Fig. 4(d), where a sample of four points is chosen from the area around an edge point at the object's boundary. Contrary to the [14], we believe it is irrational to suppose nearby points have the same depth estimation while ignoring their crucial geometrical constraints. In our work, we consider the inner point pair (p_0, p_1) and outer point pair (p_2, p_3) have the same surface normal vector, due to the pairs are relatively close in spatial. To better leverage the geometrical constraints, we propose to incorporate the surface normal vector in the loss formulation. After we have the GT depth map D^* with its corresponding camera intrinsic matrix K and the sampled points set S from our 4-point sampling scheme, we first select eight neighboring points $\{(x_{n,i}, y_{n,i})\}_{i=0}^{7}$ for each point p in S. In order to calculate the surface normal $n_p \in \mathbb{R}^{3 \times 1}$of p, we unproject the center point p and its neighboring points into 3D space as below,

$$P^* = K^{-1} D^*(p) p \tag{8}$$

Then, we compute the surface normal vector using three points for eight times as below,

$$\overrightarrow{N_{p,i}^*} = \overrightarrow{P^* P_{i,0}^*} \times \overrightarrow{P^* P_{i,1}^*}, \tag{9}$$

where $\{(P^*_{i,0}, P^*_{i,1})\}_{i=0}^{7}$ indicates two neighboring points of center point P^*. The final surface normal of p in GT depth map is presented as,

$$\overrightarrow{N^*_p} = \frac{1}{8}\sum_{i=0}^{7}\overrightarrow{N^*_{p,i}} \tag{10}$$

We can compute the surface normal map N^* and N with respect to GT depth map D^* and predicted depth map D in this way. After that, we calculate the surface normal vectors at each point and assign an ordinal label to each pair of points $(p_0, p_1) \in S$ based on their *cosine similarity*,

$$l = \begin{cases} 1, & |\cos(n^*_0, n^*_1)| \leq \frac{1}{1+\tau} \\ 0, & otherwise \end{cases}, \tag{11}$$

where n^*_0 and n^*_1 are the GT depth-derived surface normal vectors corresponding to p_0 and p_1, and τ is a tolerance threshold. The *SGNR* loss for (p_0, p_1) is then given by,

$$\phi_{sgnr}(p_0, p_1) = \begin{cases} \log(1 + \exp(-\tan(\frac{|n_0, n_1| + \epsilon}{2}))), & l = 1 \\ (\tan(\frac{|n_0, n_1| + \epsilon}{2}))^2, & l = 0 \end{cases}, \tag{12}$$

where n_0 and n_1 are the surface normal vectors computed using the predicted depth map for p_0 and p_1 respectively, $|n_0, n_1|$ refers to the angle between n_0 and n_1, ϵ is a perturbation which we set to 1×10^{-4}.

In a word, the SGNR loss enforces the model to predict similar normal vectors for pairs with similar surface normals and dissimilar them otherwise. Our total SGNR loss is given by,

$$L_{sgnr} = \frac{1}{|S|}\sum_{(p_0,p_1)\in S}\phi_{sgnr}(p_0, p_1). \tag{13}$$

where $|S|$ refers to the number of point pairs.

3.4 Total Loss Function

The total loss function of BDE-MVSNet is composed of SGNR loss and the losses in Vis-MVSNet: the pair-wise L1 loss L_1^{pair}, the pair-wise joint loss L^{joint} and the L1 loss of the final depth map L^{final} for each stage. The final loss formulation is thus given by,

$$L_{total} = \sum_{k=1}^{2}\lambda_k[L^{final}_{1,k} + \frac{1}{m}\sum_{i=1}^{m}(L^{pair}_{1,k,i} + L^{joint}_{k,i}) + \alpha L_{sgnr,k}] \quad , \tag{14}$$

where λ_k is the weight for k-th stage, m is the number of source images and α is the weight of L_{sgnr}, which we set to be 0.5.

Table 1. Accuracy, completeness and overall scores of different MVS methods on the DTU test set. The resolution of input images is 864 × 1152.

Method	Acc. (mm) ↓	Comp. (mm) ↓	Overall (mm) ↓
Gipuma [30]	**0.283**	0.873	0.758
COLMAP [31]	0.400	0.664	0.532
MVSNet [7]	0.396	0.527	0.462
Vis-MVSNet [9]	0.369	0.361	0.365
PatchmatchNet [12]	0.427	**0.277**	0.352
CVP-MVSNet [32]	0.296	0.406	0.351
CasMVSNet [8]	0.346	0.351	0.348
UCSNet [10]	0.338	0.346	0.344
DDR-Net [11]	0.339	0.320	0.329
BDE-MVSNet	0.338	0.302	**0.320**

4 Experiments

4.1 Datasets

We evaluate our method using three commonly benchmarked MVS datasets: DTU [17], BlendedMVS [19] and Tanks and Temples [18]. DTU is an indoor-scene dataset, consisting of 124 scenes, scanned from 49 or 64 views under 7 different lighting conditions. BlendedMVS is a large-scale dataset that contains 17,000 MVS training samples covering a variety of 113 scenes. Tanks and Temples contains multiple realistic scenes. All the settings of evaluation datasets follow Vis-MVSNet [9].

4.2 Implementation Details

Our method is implemented with PyTorch [29] and trained on eight NVIDIA Tesla V100 cards. We optimize our network with Adam using a batch size of 16 and an initial learning rate of 0.001. We train for $160K$ iterations and decrease the learning rate by half at the $100K$, $120K$ and $140K$ iterations. During training, we use an image resolution of o 640 × 512 and set the number of source images to 3. For depth sampling strategy, we set $D_{max} = 128$ and the initial depth interval $\delta = 1$. We set $N_{d,1}, N_{d,2} = 32, 16$ for the number of depth hypotheses for stage 1 and 2 respectively. The ADR factor k is set to 1.2 and the weight of loss λ_k is set to 1.0 for all experiments.

4.3 Evaluation on DTU

Our proposed method benchmarked on the DTU [17] evaluation set. DTU dataset is divided into training, validation and evaluation sets. We train our model on the DTU training set and test on the evaluation set. We set the depth range to [425 mm, 905 mm] and use 5 source views per image at a resolution of 1152×864 for depth estimation. As shown in Table 1, the Gipuma [30] and PatchmatchNet [12] methods achieve the best accuracy and completeness respectively,

Fig. 5. Comparison of reconstructed point clouds on DTU validation set [17] between Vis-MVSNet and ours.

while BDE-MVSNet outperforms other learning-based methods and traditional methods in terms of overall performance. Some qualitative results are shown in Fig. 5, compared with baseline method [9], BDE-MVSNet is able to reconstruct better point cloud results.

Table 2. The F-score of MVS methods on the Tanks and Temples intermediate test set (higher is better). The best method is highlighted in bold for each scene. The resolution of input images is 1920 × 1080.

Method	Pub.	Mean	Family	Francis	Horse	L.H.	M60	Panther	P.G.	Train
MVSNet [7]	ECCV18'	43.48	55.99	28.55	25.07	50.79	53.96	50.86	47.90	34.69
R-MVSNet [20]	CVPR19'	48.40	69.96	46.65	32.59	42.95	51.88	48.80	52.00	42.38
CVP-MVSNet [32]	CVPR20'	54.03	76.50	47.74	36.34	55.12	57.28	54.28	57.43	47.54
UCSNet [10]	CVPR20'	54.83	76.09	53.16	43.03	54.00	55.60	51.49	57.38	47.89
DDR-Net [11]	Arxiv20'	54.91	76.18	53.36	43.43	55.20	55.57	52.28	56.04	47.17
CasMVSNet [8]	CVPR20'	56.42	76.36	58.45	46.20	55.53	56.11	54.02	58.17	46.56
D2HC-RMVSNet [21]	ECCV20'	59.20	74.69	56.04	49.42	60.08	59.81	**59.61**	60.04	53.92
Vis-MVSNet [9]	BMVC20'	60.03	77.40	60.23	47.07	63.44	62.21	57.28	**60.54**	52.07
AA-RMVSNet [33]	ICCV21'	61.51	77.77	59.53	51.53	64.02	64.05	59.47	60.85	54.90
EPP-MVSNet [34]	ICCV21'	61.68	77.86	60.54	52.96	62.33	61.69	60.34	62.44	55.30
BDE-MVSNet (ours)	-	**62.30**	**79.71**	**67.33**	**49.52**	**64.68**	**62.43**	58.28	58.15	**58.32**

4.4 Evaluation on Tanks and Temples

We evaluate our method on the Tanks and Temples dataset's intermediate set, [18]. To train BDE-MVSNet, we utilize the training set of the BlendedMVS dataset [19]. We set the number of source views to 7 and tested on images with a resolution of 1920 × 1080. All other hyper-parameters are set to the same values as in the training stage. The Table 2 reports the F-score of our method

Fig. 6. 3D model of a challenging scene from BlendedMVS [19], reconstructed by Vis-MVSNet [9] and ours.

Fig. 7. More 3D model result of challenging scenes from BlendedMVS [19], reconstructed by Vis-MVSNet [9] and ours.

as well as other state-of-the-art learning-based MVS algorithms. BDE-MVSNet outperforms existing approaches in almost every scene due to superior depth prediction.

4.5 Evaluation on BlendedMVS

While MVS benchmarking typically concentrates only on the final output (the point cloud), we also report the quality of the predicted depth maps.

We follow the depth evaluation protocol of BlendedMVS [19] and report three metrics: the mean absolute error between the predicted and the ground truth depth maps, denoted as end point error (EPE), and the proportion in % of pixels with an error >1 and >3 in the scaled depth maps, denoted as e_1 and e_3, respectively.

We set the number of source images $m = 5$ with a resolution of 640×512 and set all other hyper-parameters as in the training phase. We train our model on the BlendedMVS dataset and evaluate using its validation set (Table 4). Our approach outperforms other state-of-the-art methods in terms of depth quality by a large margin. A qualitative comparison on a challenging scene between Vis-MVSNet and

Table 3. The effect of depth range update strategy on the quality of the depth estimation. Linear strategy is used in [10,11]. The resolution of input images is 640 × 512.

Depth Range strategy	EPE ↓	e_1 ↓	e_3 ↓
None	1.35	17.82	6.55
Linear	1.24	16.81	5.67
ADR (ours)	**1.09**	**15.63**	**5.61**

Table 4. Accuracy (*EPE*, *e*1, *e*3), memory and runtime for depth map estimation results obtained with different MVS methods and ours. Results are reported for the BlendedMVS [19] validation set, using image resolution of 640 × 512.

Method	EPE ↓	e_1 ↓	e_3 ↓	Memory	Runtime
MVSNet [7]	1.49	21.98	8.32	5.50G	1.18 s
CasMVSNet [8]	1.43	19.01	9.77	2.71G	0.44 s
CVP-MVSNet [32]	1.90	19.73	10.24	–	–
DDR-Net [11]	1.41	18.08	8.32	–	–
Vis-MVSNet [9]	1.47	18.47	7.59	1.85G	0.56 s
Ours	**1.06**	**15.14**	**5.13**	1.81G	0.47 s

ours is shown in Fig. 6. Thanks to the predicted accurate depth, BDE-MVSNet is able to reduce the tailing error in the final point cloud result (Fig. 7).

4.6 Runtime and Memory Analysis

We measure the runtime and memory cost of depth estimation using our method and several state-of-the-art methods on a Tesla V100 GPU as shown in Table 4. Our method achieves an improved memory and runtime cost compared to VisMVSNet [9], MVSNet [7]. Also provides a better runtime-memory trade-off compared to CasMVSNet.

4.7 Ablation Study

We carry out ablation experiments in order to evaluate the contribution of our proposed ADR strategy and SGNR loss. As shown in Table 5, ADR is the main contributor for the observed reduction in EPE, e_1 and e_3 values. When employing it together with the SGNR loss we achieve a further improvement in depth accuracy.

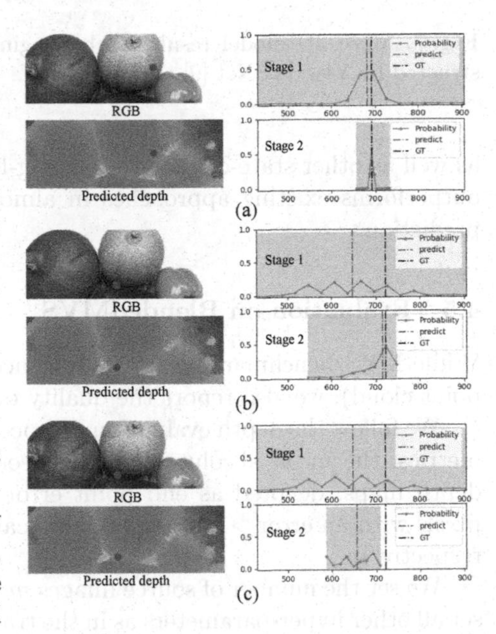

Fig. 8. Comparison between proposed ADR strategy with linear strategy [10,11] in 2 examples.

Table 5. Ablation study on the BlendedMVS validation set. The resolution of input images is 640×512.

ADR	SGNR	EPE ↓	e_1 ↓	e_3 ↓
–	–	1.35	17.82	6.55
–	✓	1.30	16.98	5.99
✓	–	1.09	15.63	5.61
✓	✓	**1.06**	**15.14**	**5.13**

We further compare the linear strategy employed by [10,11] to our proposed ADR strategy. The Table 3 reports the EPE, e_1 and e_3 of our model, when trained without updating the depth range between stages and when doing so either with a linear strategy or with ADR. While the linear depth update improves performance, ADR achieves better depth accuracy.

As shown in Fig. 8. On the left, we show the RGB images and the corresponding predicted depth maps. On the right, we show the probability details of a pixel (red point in images) with depth intervals (pink). (a): For a point with high probability, ADR and linear strategy both narrow the depth range and predict accurate depth in the second stage. (b): For a point with medium probability at the boundary edge, ADR keeps the depth interval large and predicts accurate depth. (c): For the same point with (b), linear strategy narrows the depth interval and gets the wrong depth.

5 Conclusion

In this paper, we present BDE-MVSNet to boost depth estimation for multi-view stereo. We propose a non-linear method for deriving per-pixel depth range and a novel structure-guided normal ranking loss. Together these optimizations yield more accurate depth prediction and 3D reconstruction. To the best of our knowledge, our work is the first to directly tackle boundary prediction and can be used to improve the performance of learning-based MVS methods. As a result, our proposed BDE-MVSNet achieves state-of-the-art performance on multiple datasets.

References

1. Zhu, Q., Min, C., Wei, Z., Chen, Y., Wang, G.: Deep learning for multi-view stereo via plane sweep: a survey (2021)
2. Hirschmuller, H.: Stereo processing by semiglobal matching and mutual information. IEEE Trans. Pattern Anal. Mach. Intell. **30**, 328–341 (2007)
3. Campbell, N.D.F., Vogiatzis, G., Hernández, C., Cipolla, R.: Using multiple hypotheses to improve depth-maps for multi-view stereo. In: Forsyth, D., Torr, P., Zisserman, A. (eds.) ECCV 2008. LNCS, vol. 5302, pp. 766–779. Springer, Heidelberg (2008). https://doi.org/10.1007/978-3-540-88682-2_58

4. Furukawa, Y., Ponce, J.: Accurate, dense, and robust multiview stereopsis. IEEE Trans. Pattern Anal. Mach. Intell. **32**, 1362–1376 (2009)
5. Tola, E., Strecha, C., Fua, P.: Efficient large-scale multi-view stereo for ultra high-resolution image sets. Mach. Vis. Appl. **23**, 903–920 (2012)
6. Galliani, S., Lasinger, K., Schindler, K.: Massively parallel multiview stereopsis by surface normal diffusion. In: Proceedings of the IEEE International Conference on Computer Vision, pp. 873–881 (2015)
7. Yao, Y., Luo, Z., Li, S., Fang, T., Quan, L.: MVSNet: depth inference for unstructured multi-view stereo. In: Ferrari, V., Hebert, M., Sminchisescu, C., Weiss, Y. (eds.) ECCV 2018. LNCS, vol. 11212, pp. 785–801. Springer, Cham (2018). https://doi.org/10.1007/978-3-030-01237-3_47
8. Gu, X., Fan, Z., Zhu, S., Dai, Z., Tan, F., Tan, P.: Cascade cost volume for high-resolution multi-view stereo and stereo matching. In: IEEE Conference on Computer Vision and Pattern Recognition, pp. 2492–2501 (2020)
9. Zhang, J., Yao, Y., Li, S., Luo, Z., Fang, T.: Visibility-aware multi-view stereo network. In: British Machine Vision Conference (2020)
10. Cheng, S., et al.: Deep stereo using adaptive thin volume representation with uncertainty awareness. In: IEEE Conference on Computer Vision and Pattern Recognition, pp. 2521–2531 (2020)
11. Yi, P., Tang, S., Yao, J.: DDR-net: learning multi-stage multi-view stereo with dynamic depth range (2021)
12. Wang, F., Galliani, S., Vogel, C., Speciale, P., Pollefeys, M.: PatchmatchNet: Learned multi-view patchmatch stereo. In: IEEE Conference on Computer Vision and Pattern Recognition. (2021) 14194–14203
13. Ding, Y., et al.: Transmvsnet: global context-aware multi-view stereo network with transformers. arXiv preprint arXiv:2111.14600 (2021)
14. Chen, W., Fu, Z., Yang, D., Deng, J.: Single-image depth perception in the wild. In: Advances in Neural Information Processing Systems, vol. 29 (2016)
15. Xian, K., et al.: Monocular relative depth perception with web stereo data supervision. In: IEEE Conference on Computer Vision and Pattern Recognition, pp. 311–320 (2018)
16. Xian, K., Zhang, J., Wang, O., Mai, L., Lin, Z., Cao, Z.: Structure-guided ranking loss for single image depth prediction. In: IEEE Conference on Computer Vision and Pattern Recognition, pp. 608–617 (2020)
17. Jensen, R., Dahl, A., Vogiatzis, G., Tola, E., Aanaes, H.: Large scale multi-view stereopsis evaluation. In: IEEE Conference on Computer Vision and Pattern Recognition, pp. 406–413 (2014)
18. Knapitsch, A., Park, J., Zhou, Q.Y., Koltun, V.: Tanks and temples: benchmarking large-scale scene reconstruction. ACM Trans. Graph. **36**, 1–13 (2017)
19. Yao, Y., Luo, Z., Li, S., Zhang, J., Quan, L.: BlendedMVS: a large-scale dataset for generalized multi-view stereo networks. In: IEEE Conference on Computer Vision and Pattern Recognition, pp. 1787–1796 (2020)
20. Yao, Y., Luo, Z., Li, S., Shen, T., Fang, T., Quan, L.: Recurrent MVSNet for high-resolution multi-view stereo depth inference. In: IEEE Conference on Computer Vision and Pattern Recognition, pp. 5525–5534 (2019)
21. Yan, J., et al.: Dense hybrid recurrent multi-view stereo net with dynamic consistency checking. In: Vedaldi, A., Bischof, H., Brox, T., Frahm, J.-M. (eds.) ECCV 2020. LNCS, vol. 12349, pp. 674–689. Springer, Cham (2020). https://doi.org/10.1007/978-3-030-58548-8_39

22. Wei, Z., Zhu, Q., Min, C., Chen, Y., Wang, G.: AA-RMVSNet: adaptive aggregation recurrent multi-view stereo network. IEEE International Conference on Computer Vision (2021)
23. Yin, W., Liu, Y., Shen, C.: Virtual normal: enforcing geometric constraints for accurate and robust depth prediction. IEEE Trans. Pattern Anal. Mach. Intell. 1 (2021)
24. Long, X., Liu, L., Theobalt, C., Wang, W.: Occlusion-aware depth estimation with adaptive normal constraints. In: Vedaldi, A., Bischof, H., Brox, T., Frahm, J.-M. (eds.) ECCV 2020. LNCS, vol. 12354, pp. 640–657. Springer, Cham (2020). https://doi.org/10.1007/978-3-030-58545-7_37
25. Chakrabarti, A., Shao, J., Shakhnarovich, G.: Depth from a single image by harmonizing overcomplete local network predictions. In: Advances in Neural Information Processing Systems, vol. 29 (2016)
26. Eigen, D., Fergus, R.: Predicting depth, surface normals and semantic labels with a common multi-scale convolutional architecture. In: IEEE International Conference on Computer Vision, pp. 2650–2658 (2015)
27. Fu, H., Gong, M., Wang, C., Batmanghelich, K., Tao, D.: Deep ordinal regression network for monocular depth estimation. In: IEEE Conference on Computer Vision and Pattern Recognition, pp. 2002–2011 (2018)
28. Guo, X., Yang, K., Yang, W., Wang, X., Li, H.: Group-wise correlation stereo network. In: IEEE Conference on Computer Vision and Pattern Recognition, pp. 3268–3277 (2019)
29. Paszke, A., et al.: PyTorch: an imperative style, high-performance deep learning library. In: Advances in Neural Information Processing Systems, vol. 32 (2019)
30. Galliani, S., Lasinger, K., Schindler, K.: Massively parallel multiview stereopsis by surface normal diffusion. In: IEEE International Conference on Computer Vision, pp. 873–881 (2015)
31. Schönberger, J.L., Zheng, E., Frahm, J.-M., Pollefeys, M.: Pixelwise view selection for unstructured multi-view stereo. In: Leibe, B., Matas, J., Sebe, N., Welling, M. (eds.) ECCV 2016. LNCS, vol. 9907, pp. 501–518. Springer, Cham (2016). https://doi.org/10.1007/978-3-319-46487-9_31
32. Yang, J., Mao, W., Alvarez, J.M., Liu, M.: Cost volume pyramid based depth inference for multi-view stereo. In: IEEE Conference on Computer Vision and Pattern Recognition, pp. 4877–4886 (2020)
33. Wei, Z., Zhu, Q., Min, C., Chen, Y., Wang, G.: AA-RMVSNET: adaptive aggregation recurrent multi-view stereo network. In: Proceedings of the IEEE/CVF International Conference on Computer Vision, pp. 6187–6196 (2021)
34. Ma, X., Gong, Y., Wang, Q., Huang, J., Chen, L., Yu, F.: Epp-mvsnet: epipolar-assembling based depth prediction for multi-view stereo. In: Proceedings of the IEEE/CVF International Conference on Computer Vision, pp. 5732–5740 (2021)

Training-Free NAS for 3D Point Cloud Processing

Ping Zhao(✉), Panyue Chen, and Guanming Liu

Tongji University, Shanghai, China
{zhaoping,chenpanyue,2130776}@tongji.edu.cn

Abstract. Deep neural networks for 3D point cloud processing have exhibited superior performance on many tasks. However, the structure and computational complexity of existing networks are relatively fixed, which makes it difficult for them to be flexibly applied to devices with different computational constraints. Instead of manually designing the network structure for each specific device, in this paper, we propose a novel training-free neural architecture search algorithm which can quickly sample network structures that satisfy the computational constraints of various devices. Specifically, we design a cell-based search space that contains a large number of latent network structures. The computational complexity of these structures varies within a wide range to meet the needs of different devices. We also propose a multi-objective evolutionary search algorithm. This algorithm scores the candidate network structures in the search space based on multiple training-free proxies, encourages high-scoring networks to evolve, and gradually eliminates low-scoring networks, so as to search for the optimal network structure. Because the calculation of training-free proxies is very efficient, the whole algorithm can be completed in a short time. Experiments on 3D point cloud classification and part segmentation demonstrate the effectiveness of our method (Codes will be available).

Keywords: 3D point cloud processing · Training-free proxies · Neural architecture search

1 Introduction

Deep neural networks (DNNs) for 3D point cloud processing have achieved superior performance on many tasks and have received more and more extensive attention. However, most of the existing DNNs rely on researchers to manually design the network structures, which places high demands on the researchers' experience and expertise. In addition, the structure and computational complexity of existing networks are relatively fixed, which makes it difficult for them to be flexibly applied to devices with different computational constraints. For example, complex networks [20,30] in academia often cannot be deployed on mobile devices. This restriction creates a split between academia and industry, limiting the development of downstream tasks such as autonomous driving [12,27]

L. Wang et al. (Eds.): ACCV 2022, LNCS 13841, pp. 296–310, 2023.
https://doi.org/10.1007/978-3-031-26319-4_18

Table 1. Two categories of existing training-free proxies. We classify them according to the properties that they can reflect.

Pruning-Based Proxies for Network Trainability	**Gradnorm** [1]: Euclidean norm of the parameter gradients.
	Synflow [23]: absolute Hadamard product of the parameter gradients and parameters.
Linear-Region-Based Proxies for Network Expressivity	**Naswot** [16]: Hamming distance of binary activation patterns between mini-batch samples
	Zen-score [13]: gradients of deep layer feature maps to mini-batch samples

and robotics [22]. To address the above issues, some studies [8,9,14] proposed to use one-shot neural architecture search (NAS) to quickly sample structures. However, these methods require careful training of a relatively large supernet and suffer from the degenerate search-evaluation correlation problem [8], i.e., the performance of the sampled network at search time does not match its performance at evaluation time. This problem further limits the development of one-shot NAS methods.

In this paper, we focus on bridging the gap between academia and industry for different needs of deep neural networks. Specifically, we propose a novel training-free neural architecture search algorithm which can quickly sample network structures that satisfy the computational constraints of various devices. We first design a cell-based search space in which the network structure has dynamic layer number and feature dimensions. The computational complexity of structures in this search space varies within a wide range to meet the needs of different devices, e.g., we can sample very complex structures for cloud device, relative simple structures for mobile device. Then we propose a multi-objective evolutionary search algorithm. This algorithm scores the candidate network structures in the search space based on multiple training-free proxies, encourages high-scoring networks to evolve, and gradually eliminates low-scoring networks, so as to search for the optimal network structure.

Training-free proxies are numerical metrics that can evaluate the performance of a neural network before training. Previous studies [1,13,16] usually use a single proxy to guide the search. However, a single proxy can only reflect a single property[1] of the network structure, cannot fully reflect different properties of a complex network structure. Based on such single proxy, we can only encounter a small part of network structures in the search space. Therefore, in this paper, we propose to use multiple training-free proxies jointly. The objective of our search algorithm is to find network structures that perform well across multiple training-free proxies. We believe that in this way similar to ensemble learning, the search algorithm can sample more potential structures from the search space. To

[1] A network has various properties, e.g., trainability (how effective a network can be optimized via gradient descent), expressivity (how complex the function a network can represent).

jointly use proxies that reflect various properties, we divide them into different categories, as shown in Table 1.

To sum up, there are two stages in our method. In the search stage, based on multiple training-free proxies, we use an evolutionary search algorithm to find the potential network structures under a specific computational constraints in the designed search space. In the evaluation stage. We train these structures from scratch on different tasks to obtain their final performance. Experiments on 3D point cloud classification and part segmentation show that, compared with previous methods, our method can find better network structures under different computational constraints, which makes our method widely applicable to various devices. Furthermore, compared to traditional NAS methods, our search stage only takes hours, which further reduces the usage bottleneck of our method.

Our contributions are as follows:

1. We design a novel search space adapted to training-free proxies for 3D point cloud processing neural networks. This search space allows us to explore potential structures with varying amounts of computation;
2. We propose a multi-objective evolutionary search algorithm, which enables us to search with multiple proxies in an ensemble learning manner.
3. Our method largely bridges the gap between academia and industry, experiments on 3D point cloud classification and part segmentation demonstrate the effectiveness of our method.

2 Related Works

2.1 Point Cloud Processing

PointNet [18] first proposes to use a multi-layer perceptron with shared weights to process each point and use a symmetric aggregation function to obtain the global feature. PointNet++ [19] uses ball query for each point to obtain features of its neighbor points. PointCNN [11] aligns the points in a certain order by predicting the transformation matrix of the local point set. PCNN [2] proposes a parameterized continuous convolution operation. DGCNN [28] uses features from k-Nearest Neighbors (k-NN) and proposes an edge convolution operator for feature extraction. AdaptConv [35] proposes to jointly use feature relationship and coordinate relationship to achieve efficient graph convolution. These studies have proved the effectiveness of deep neural networks in 3D point cloud processing. However, they have relatively fixed network structures, which makes them difficult to be flexibly applied to devices with different computational constraints.

2.2 Training-Free Neural Architecture Search

Training-free neural architecture search algorithms first come from network pruning. Some studies [7,23,26] try to prune networks in the initialization stage. SNIP [7] proposes an importance metric to approximate the change of network loss after a specific parameter is removed. SynFlow [23] proposes `synflow` metric that can avoid layer collapse during pruning. Abdelfattah et al. [1] extend these indicators to neural architecture search and use them to evaluate the performance of the whole neural network. Other studies [13,16] are motivated by recent theory advances in deep networks and can also be calculated before network training. NASWOT [16] calculates Hamming distance of binary activation patterns between mini-batch samples to estimate linear regions of RELU networks. Zen-NAS [13] proves that the expressivity of the network can be efficiently measured by its expected Gaussian complexity. These works have achieved competitive performance. However, few of them use multiple training-free proxies jointly. In addition, they are all applied to 2D image classification and have not been verified on more tasks such as 3D point cloud classification and segmentation.

2.3 Neural Architecture Search in 3D Point Cloud Processing

Some studies [8,9,14,17,24,24] apply one-shot NAS to 3D point cloud processing and achieve relatively good performance. LC-NAS [9] implements a latency constraint formulation to trade-off between accuracy and latency in 3D NAS. PolyConv [14] introduces a poly-convolutional feature encoder that comprises multiple feature aggregation functions. SPVNAS [24] implements a sparse point-voxel convolution network to effectively process large-scale 3D scenes. SGAS [8] proposes a greedy algorithm and designs multiple numerical metrics such as edge importance for efficient structure selection. PointSeaNet [17] proposes a differentiable convolution search paradigm to create a group of suitable convolutions for 3D point cloud processing. In contrast to these one-shot NAS studies, our training-free method does not require training a relatively large supernet and does not suffer from the degenerate search-evaluation correlation problem.

3 Methods

3.1 Preliminaries and Notations

To quickly sample network structures that satisfy the computational constraints of various devices, we propose a novel training-free neural architecture search algorithm. In Sect. 3.2, we analyze the existing training-free proxies and classify them based on the properties they can reflect. In Sect. 3.3, we propose a search space for 3D point cloud processing network. In Sect. 3.4, we introduce a novel multi-objective evolutionary search algorithm based on multiple training-free proxies.

Let $\boldsymbol{X} = \{\boldsymbol{x}_i\}_{i=1}^N$ denote a minibatch point cloud data, $\boldsymbol{x}_i \in \mathbb{R}^{3\times G}$ denote a point cloud with G points, $\boldsymbol{\theta}$ denote network parameters which is initialized by a Gaussian distribution $\mathcal{N}(0,1)$, $\mathcal{L}(\boldsymbol{x}_i, \boldsymbol{\theta})$ denote network loss to the input $\boldsymbol{x}_i$.

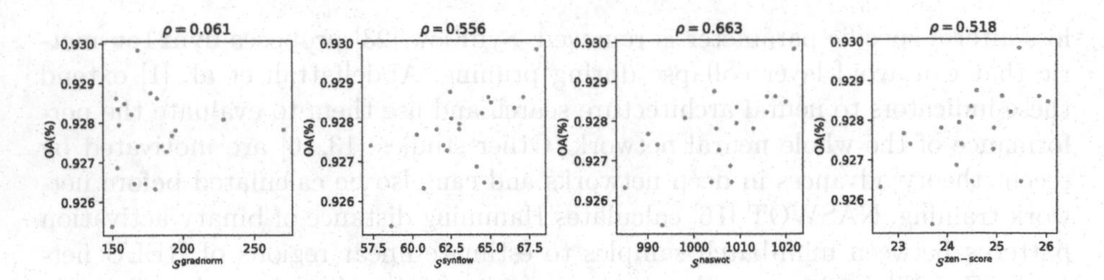

Fig. 1. Correlation between the score of different training-free proxies (horizontal axis) and the overall accuracy (vertical axis). ρ stands for the Pearson coefficient.

3.2 Training-Free Proxies

We divide existing training-free proxies into two categories according to the properties that they can reflect: (1) Pruning-based proxies for trainability of neural networks. These kinds of proxies come from network pruning studies and indicate the performance of the neural network by calculating an indicator related to network parameter gradient at initialization; (2) Linear-region-based proxies for the expressivity of neural network. The linear region [4,31] is an essential indicator of the expressivity of the RELU neural network. These kinds of proxies indicate the performance of the network by estimating the number of linear regions of neural networks.

(1) Pruning-Based Proxies for Trainability

Gradnorm [1] is one of the simplest proxies. It sums the Euclidean norm of the parameter gradients after a single backward of minibatch data and uses this proxy to represent the trainability of a network.

$$\texttt{gradnorm}: \mathcal{S}^{\text{gradnorm}} = \sum_{\theta} \left\| \frac{\partial \mathcal{L}(\boldsymbol{X}, \boldsymbol{\theta})}{\partial \theta} \right\| \tag{1}$$

Synflow [1,23] proposes iterative magnitude pruning to avoid layer-collapse in premature pruning. Formally, it is the absolute Hadamard product of the gradient of loss to each parameter and the parameter itself in a single minibatch. Following [1], when calculating `synflow`, we compute a loss which is simply the product of all parameters in the network, and sum up `synflow` of each parameter to represent the trainability of a network.

$$\texttt{synflow}: \mathcal{S}^{\text{synflow}} = \sum_{\theta} \left| \frac{\partial \mathcal{L}(\boldsymbol{X}, \boldsymbol{\theta})}{\partial \theta} \odot \theta \right| \tag{2}$$

(2) Linear-Region-Based Proxies for Expressivity

Naswot [16] estimates the linear regions of RELU networks by calculating the Hamming distance of binary activation patterns between mini-batch samples. Suppose there are M rectified linear units in the network. For a specific input

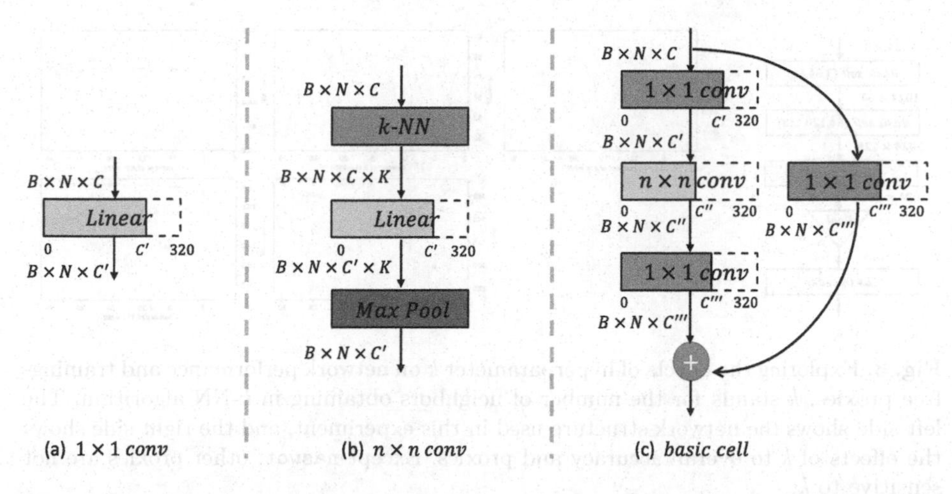

Fig. 2. Illustration of 1×1 convolution (a), $n \times n$ convolution (b) and basic cell (c). 'Linear' means a linear layer shared by all points. Note that the output feature channels of each 'Linear' layer are dynamic (The parts in solid lines mean they are chosen while those in dashed lines are not.), leading to various basic cells.

$\boldsymbol{x}_i$, we can arrange its activation patterns into a pattern string $\boldsymbol{p}_i$. In this way, the number of linear regions can be estimated by calculating the Hamming distance d_H of different inputs in a minibatch.

$$\texttt{naswot} : \mathcal{S}^{\text{naswot}} = \log \left| \begin{pmatrix} M - d_H(\boldsymbol{p}_1 - \boldsymbol{p}_1) & \dots & M - d_H(\boldsymbol{p}_1 - \boldsymbol{p}_N) \\ \vdots & \ddots & \vdots \\ M - d_H(\boldsymbol{p}_N - \boldsymbol{p}_1) & \dots & M - d_H(\boldsymbol{p}_N - \boldsymbol{p}_N) \end{pmatrix} \right| \tag{3}$$

Zen-score [13] calculates the change of deep layer feature maps $f(\boldsymbol{X}, \boldsymbol{\theta})$ in the network after adding disturbance to the input $\boldsymbol{X}$. This result is equal to the Gaussian complexity of the neural network, which considers not only the distribution of linear regions but also the Gaussian complexity of the linear classifier in each linear region.

$$\texttt{zen-score} : \mathcal{S}^{\text{zen-score}} = \log \left\| \frac{\partial f(\boldsymbol{X}, \boldsymbol{\theta})}{\partial \boldsymbol{X}} \right\| \tag{4}$$

Through the classification of existing training-free proxies, we can jointly use proxies with different properties for neural architecture search. We believe that NAS can achieve better performance through this ensemble learning-like way. Experiments have also proved this view.

3.3 Point Cloud Search Space

Similar to many classical neural architecture search algorithms [3,21,34,36], we design a cell-based search space, the macro architecture of the network is stacked

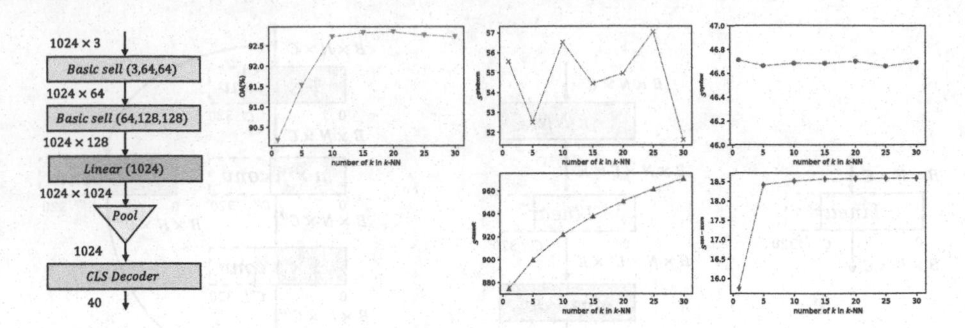

Fig. 3. Exploring the effects of hyperparameter k on network performance and training-free proxies. k stands for the number of neighbors obtaining in k-NN algorithm. The left side shows the network structure used in this experiment, and the right side shows the effects of k to overall accuracy and proxies. Except `naswot`, other proxies are not sensitive to k.

by several basic cells, only the inner architecture of each cell needs to be searched. In our paper, as shown in Fig. 2, a **basic cell** is a bottleneck block [5] consisting of two 1×1 convolutions and one $n \times n$ convolution. However, 3D point cloud data is disordered, and it is difficult to use regular convolutions in 2D image processing to process it, so we first define the convolutions for 3D point cloud processing based on previous studies.

1×1 convolution comes from PointNet [18]. Considering the permutation invariance of 3D point cloud data, PointNet first proposes to use a linear layer with shared weights to process each point. This operation well maintains the permutation invariance but can not mine the relationship between points. We take this linear layer with shared weights as a basic component which corresponds to 1×1 convolution in 2D image processing.

$n \times n$ convolution comes from DGCNN [28]. During a $n \times n$ convolution, the k-NN algorithm is firstly used to obtain the features of k neighbors of each point in the point cloud based on the feature distance between points. Then a linear layer with shared weights is used to deal with these features (maintaining permutation invariance). Finally, a max pooling operation is used to aggregate features from neighbors. We take these operations as a basic component which corresponds to the convolution with a size greater than 1×1 in 2D image processing.

It should be noted that in 2D image processing, the receptive field of $n \times n$ convolution is determined by the size of n, while in 3D point cloud processing, the receptive field of $n \times n$ convolution is determined by the size of k in k-NN. In our experiments (Fig. 3), we fix the network structure and the number of feature channels, and only change the value of k in the $n \times n$ convolution to observe the change of overall accuracy (training from scratch on ModelNet40 [29] dataset) and proxies, we can find that except `naswot`, other training-free proxies are not sensitive to the value of k. This may be because the parameter amount of $n \times n$ convolution in 3D point cloud processing is completely determined by its linear layer. This operation is similar to 1×1 convolution in the case of random

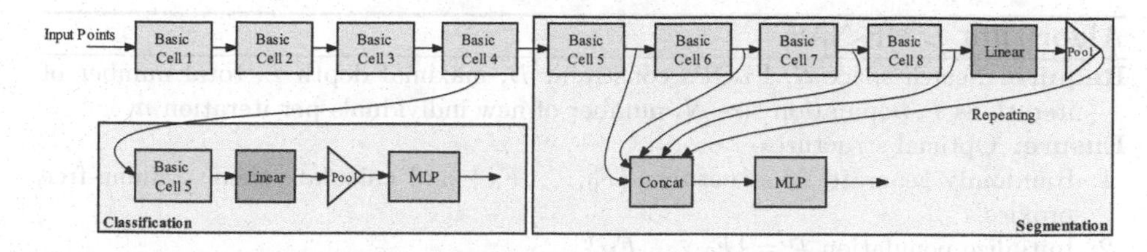

Fig. 4. Examples of the network structure for classification with 5 basic cells and the network structure for segmentation with 8 basic cells, respectively. Note that we only search structures of network encoder, and the decoder part is a multi-layer perceptron.

initialization. Therefore, in this paper, we fix the k value as the empirical value of 20.

As shown in Fig. 4, we stack multiple different basic cells to form the encoder of the network, the decoder is a multi-layer perceptron. In the search stage, The number of output feature channels of three convolutions in each cell varies in [32,64..., 320], which means that there are 1000 different cells in our search space.

We randomly sample 20 classification networks with 3 cells in this search space, calculate their values on different training-free proxies, and train them from scratch on ModelNet40 [29] dataset to obtain the overall accuracy. Then we report the correlation between the proxies and the accuracy. As shown in Fig. 1, `gradnorm` does not perform well in this search space. Therefore, in the subsequent experiments, we will not consider `gradnorm`. The other three proxies can achieve high correlation coefficients.

3.4 Training-Free NAS

We use the standard single-objective evolutionary algorithm in [13] as a baseline. In this paper, we call it SE-NAS (**S**ingle-objective **E**volutionary **N**eural **A**rchitecture **S**earch). It uses one single training-free proxy as objective to rank candidate structures and randomly mutate structures in population. In [13], they only use `zen-score` as objective. We also use `synflow` and `naswot` for comparison. After multiple iterations of SE-NAS, we select the three highest-scoring structures in the population and retrain these structures to report the optimal performance they can achieve.

To use multiple proxies jointly in an ensemble learning way, we design ME-NAS (Algorithm 1): a **m**ulti-objective **e**volutionary **n**eural **a**rchitecture **s**earch algorithm. Different from SE-NAS, since each structure corresponds to multiple proxies, it is difficult for us to compare two structures directly through the value of proxies. Therefore, we introduce the concept of Pareto dominance, that is, when the values of a structure $\mathcal{A}$ on all proxies are better than those of structure $\mathcal{B}$, then structure $\mathcal{A}$ dominates structure $\mathcal{B}$, or structure $\mathcal{A}$ is better than structure $\mathcal{B}$. Through this method, we can divide all structures in the population into multiple non-dominated fronts, the structure of the upper front dominates the structure of the lower front, and there is no dominance relationship

Algorithm 1. ME-NAS

Require: Search space $\mathcal{S}$, FLOPs constraint B, maximal depth L, total number of iterations T, population size N, number of new individuals per iteration n.

Ensure: Optimal structures.

1: Randomly generate N structures $\{F_0, \ldots, F_N\}$ and calculate their training-free proxies.
2: Initialize population $\mathcal{P} = \{F_0, \ldots, F_N\}$.
3: **for** $t = 1, 2, \cdots, T$ **do**
4: Devide $\mathcal{P}$ into M non-dominated fronts $(\mathcal{R}_0, \mathcal{R}_1, \ldots, \mathcal{R}_M)$ by fast non-dominated sort on training-free proxies and calculate crowding-distance in each non-dominated front.
5: Define new population $\hat{\mathcal{P}}$.
6: **for** $m = 0, 1, \cdots, M$ **do**
7: **if** $|\hat{\mathcal{P}}| + |\mathcal{R}_m| > N$ **then**
8: break.
9: **end if**
10: Add all individuals in $\mathcal{R}_m$ to $\hat{\mathcal{P}}$.
11: **end for**
12: Fill $\hat{\mathcal{P}}$ to $|\hat{\mathcal{P}}| = N$ by adding individuals from $\mathcal{R}_m$ according to crowding-distance.
13: $\mathcal{P} = \hat{\mathcal{P}}$.
14: Repeat steps 4 once.
15: Select n individuals and MUTATE (Algorithm 2) them to generate n new individuals.
16: Calculate training-free proxies of new individuals and add them to $\mathcal{P}$.
17: **end for**
18: Return all architectures in $\mathcal{R}_0$.

between the structures of the same front. When the population evolves, we select the structure to be mutated by its front. The higher the level, the more likely it is to mutate. Finally, ME-NAS returns multiple top-front structures (usually 2 to 4), which we individually retrain and report the optimal performance they can achieve.

4 Experiments

4.1 Implementation Details

Our experiments are mainly carried out on the 3D point cloud classification and part segmentation task. First, we compare SE-NAS and ME-NAS under different computation constraints and compare our results with existing methods on 3D point cloud classification. Then, we give the quantitative and qualitative results of 3D point cloud part segmentation.

Algorithm 2. MUTATE

Require: Structure F_t, Cell number M of F_t, Search space $\mathcal{S}$.
Ensure: Randomly mutated architecture $\hat{F}_t$.
1: Uniformly select an integer i from $[0, M]$.
2: **if** $i = M$ **then**
3: Generate a new cell $\hat{b}$ from $\mathcal{S}$.
4: Add $\hat{b}$ to the end of F_t.
5: **else**
6: Uniformly alternate channel width of i-th cell in F_t within some range or even delete this cell.
7: **end if**
8: Return the mutated architecture $\hat{F}_t$.

During neural architecture search, we set the maximum number of cells to 5 and 8 for classification and part segmentation, respectively. The number of input channels of each cell is determined by the previous cell, and the number of middle-layer channels and output channels are selected from $[32, 64, ..., 320]$. The population size of the evolutionary algorithm is 50, and the number of iterations is 30000. To show that our algorithm can be flexibly applied to various devices, we set several different computation constraints in the search process. For classification tasks, we set the computation constraint to 0.5G MACs (Tiny), 1G (Small), 2G (Middle), 3G (Large), 4G (Huge). For segmentation tasks, we set the computation constraint to 4G (Tiny), 8G (Small), 12G (Middle), 16G (Large), 20G (Huge).

Our classification experiments are carried out on the ModelNet40 [29] dataset. This dataset contains 12311 meshed CAD models from 40 categories, where 9843 models are used for training and 2468 models for testing. We randomly sample 1024 points from each object as network inputs and only use the $(\boldsymbol{x}, \boldsymbol{y}, \boldsymbol{z})$ coordinates of the sampled points as input. The training strategy is similar to [35]. It should be noted that the randomness of 3D point cloud classification is relatively large, so in all classification experiments, we train the same structure three times with different seeds and report the average accuracy.

Table 2. SE-NAS and ME-NAS with different training-free proxies under different computational constraints.

Method	`synflow`	`naswot`	`zen-score`	Tiny	Small	Middle	Large	Huge
SE-NAS	✓			92.82	93.01	93.05	93.29	93.31
SE-NAS		✓		92.92	93.06	93.22	93.33	93.46
SE-NAS [13]			✓	92.85	92.99	93.12	93.24	93.37
ME-NAS	✓	✓		93.03	**93.20**	**93.40**	**93.53**	93.66
ME-NAS		✓	✓	93.00	93.07	93.27	93.38	93.49
ME-NAS	✓		✓	93.02	93.17	93.37	93.50	93.68
ME-NAS	✓	✓	✓	**93.08**	93.19	93.38	**93.53**	**93.71**

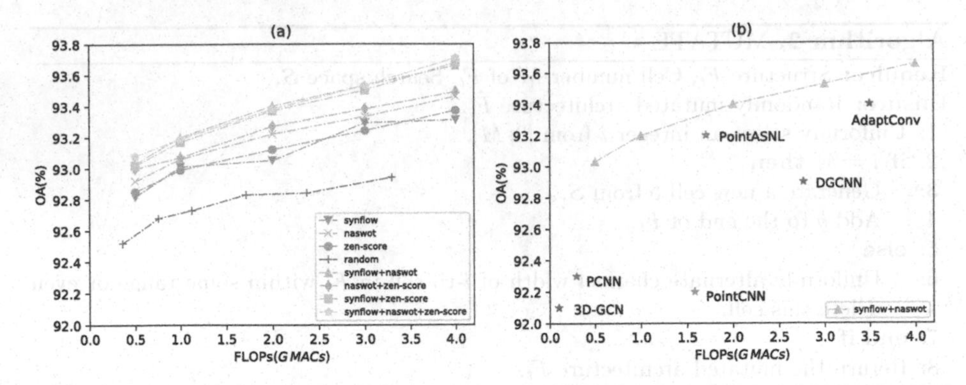

Fig. 5. Exploring the relationships between computational constraints during NAS (horizontal axis) and network performance (vertical axis) on the ModelNet40 dataset. (a) shows the comparisons of SE-NAS and ME-NAS with different training-free proxies. `random` means randomly selecting structures. (b) shows the comparisons of ME-NAS (synflow+naswot) and other studies.

Our part segmentation experiments are carried out on the ShapeNetPart [33] dataset. This dataset contains 16,881 shapes from 16 categories, with 14,006 for training and 2,874 for testing. Each point is annotated with one label from 50 parts, and each point cloud contains 2-6 parts. We randomly sample 2048 points from each shape for segmentation. The training strategy is similar to [35].

4.2 3D Point Cloud Classification

SE-NAS and ME-NAS Under Different Computational Constraints We report the results of SE-NAS and ME-NAS under different computational constraints and different training-free proxies in Table 2 and Fig. 5 (a), where `random` means randomly selecting structures within the computational constraints. As we can see, all training-free proxies can achieve better results than `random`, and jointly using multiple proxies that reflect different properties in an ensemble learning way can further improve performance.

Comparisons with Other Works. We compare the results[2] reported by ME-NAS (synflow+naswot) with other studies in 3D point cloud classification. In Fig. 5 (b), we draw the FLOPs-OA diagram[3]. In Table 3, we provide more comparison results. Experiments show that with a large computational constraints, the performance of the proposed ME-NAS surpasses existing manual methods and one-shot NAS methods, with a small computational constraints, ME-NAS can also achieve relatively good performance.

[2] Note that we do not use the voting strategy during testing in all experiments.
[3] We just test a few networks' FLOPs, cause some papers are not source-opened.

Table 3. Classification results of ME-NAS (synflow+naswot) and other studies on the ModelNet40 dataset. 'Design' means how this work is designed, 'MD' means manual design. 'OS' means one-shot NAS methods, 'ZS' means zero-shot NAS methods.

Method	Design	Input	#points	OA(%)
PointNet [18]	MD	xyz	1 k	89.2
PointNet++ [19]	MD	xyz, normal	5 k	91.9
3D-GCN [15]	MD	xyz, normal	1 k	92.1
PointCNN [11]	MD	xyz	1 k	92.2
PCNN [2]	MD	xyz	1 k	92.3
LC-NAS [9]	OS	xyz	1 k	92.8
DGCNN [28]	MD	xyz	1 k	92.9
KPConv [25]	MD	xyz	6.8 k	92.9
PointASNL [32]	MD	xyz, normal	1 k	93.2
SGAS [8]	OS	xyz	1 k	93.2
AdaptConv [35]	MD	xyz	1 k	93.4
PolyConv [14]	OS	xyz	1 k	93.5
ME-NAS (synflow+naswot)-Tiny	ZS	xyz	1 k	93.03
ME-NAS (synflow+naswot)-Small	ZS	xyz	1 k	93.20
ME-NAS (synflow+naswot)-Middle	ZS	xyz	1 k	93.40
ME-NAS (synflow+naswot)-Large	ZS	xyz	1 k	93.53
ME-NAS (synflow+naswot)-Huge	ZS	xyz	1 k	93.66

Table 4. Part segmentation results of ME-NAS (synflow+naswot) and other studies on the ShapeNetPart dataset evaluated as the mean class IoU (mcIoU) and mean instance IoU (mIoU).

Method	mcIoU	mIoU	air plane	bag	cap	car	chair	ear phone	guitar	knife	lamp	laptop	motor bike	mug	pistol	rocket	skate board	table
Kd-Net [6]	77.4	82.3	80.1	74.6	74.3	70.3	88.6	73.5	90.2	87.2	81.0	94.9	87.4	86.7	78.1	51.8	69.9	80.3
PointNet [18]	80.4	83.7	83.4	78.7	82.5	74.9	89.6	73.0	91.5	85.9	80.8	95.3	65.2	93.0	81.2	57.9	72.8	80.6
PointNet++ [19]	81.9	85.1	82.4	79.0	87.7	77.3	90.8	71.8	91.0	85.9	83.7	95.3	71.6	94.1	81.3	58.7	76.4	82.6
SO-Net [10]	81.0	84.9	82.8	77.8	88.0	77.3	90.6	73.5	90.7	83.9	82.8	94.8	69.1	94.2	80.9	53.1	72.9	83.0
DGCNN [28]	82.3	85.2	84.0	83.4	86.7	77.8	90.6	74.7	91.2	87.5	82.8	95.7	66.3	94.9	81.1	63.5	74.5	82.6
PCNN [2]	–	85.1	82.4	80.1	85.5	79.5	90.8	73.2	91.3	86.0	85.0	95.7	73.2	94.8	83.3	51.0	75.0	81.8
PointCNN [11]	–	86.1	84.1	86.4	86.0	80.8	90.6	79.7	92.3	88.4	85.3	96.1	77.2	95.3	84.2	64.2	80.0	83.0
PointASNL [32]	–	86.1	84.1	84.7	87.9	79.7	92.2	73.7	91.0	87.2	84.2	95.8	74.4	95.2	81.0	63.0	76.3	83.2
3D-GCN [15]	82.1	85.1	83.1	84.0	86.6	77.5	90.3	74.1	90.9	86.4	83.8	95.6	66.8	94.8	81.3	59.6	75.7	82.8
AdaptConv [35]	83.4	86.4	84.8	81.2	85.7	79.7	91.2	80.9	91.9	88.6	84.8	96.2	70.7	94.9	82.3	61.0	75.9	84.2
ME-NAS (synflow+naswot)-Tiny	83.1	85.9	84.1	84.4	88.0	79.2	91.1	74.7	91.5	88.2	85.7	96.0	69.8	94.0	82.0	59.5	76.2	83.3
ME-NAS (synflow+naswot)-Small	83.5	86.1	84.5	85.3	88.6	80.2	91.0	76.1	92.2	88.4	85.7	95.4	71.0	94.3	82.4	60.5	76.2	82.9
ME-NAS (synflow+naswot)-Middle	83.7	86.3	85.0	85.6	87.3	80.6	91.2	74.3	92.1	88.6	86.0	96.1	72.1	94.2	82.3	61.2	74.6	83.1
ME-NAS (synflow+naswot)-Large	83.9	86.5	85.1	86.0	88.3	80.4	91.2	77.6	91.3	88.6	85.4	96.1	71.6	94.6	82.2	62.1	75.4	83.6
ME-NAS (synflow+naswot)-Huge	84.1	86.6	85.3	86.2	88.6	80.5	91.5	77.8	91.7	88.7	86.0	96.0	71.8	94.4	82.3	62.6	75.4	83.3

4.3 3D Point Cloud Part Segmentation

Comparisions with Other Works

3D point cloud part segmentation is a more challenging task, we compare the results reported by ME-NAS (synflow+naswot) with other studies on the ShapeNetPart dataset in Table 4, the proposed ME-NAS (synflow+naswot) can also surpass existing methods.

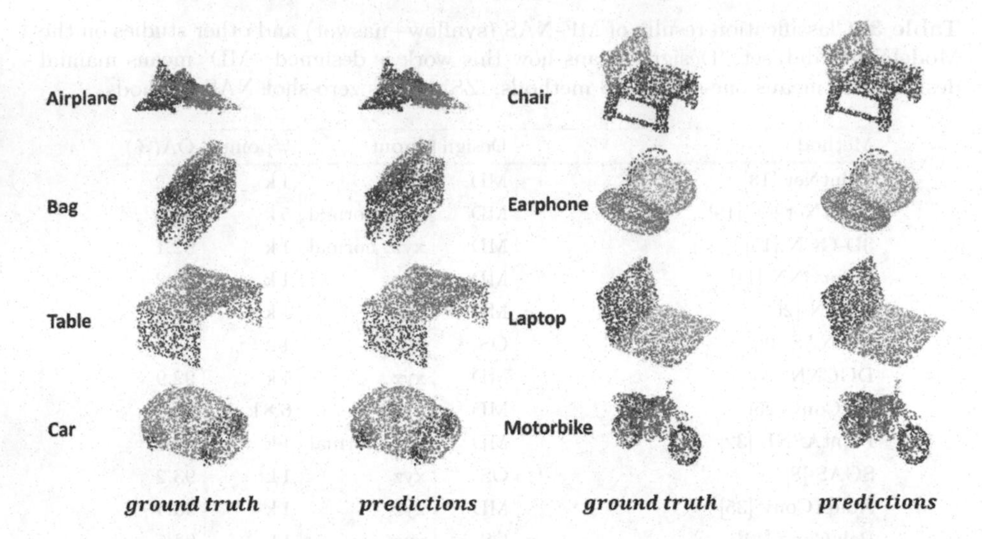

Fig. 6. Segmentation results on the ShapeNetPart dataset. We select 8 categories for visualization.

Visualization. Finally, we report the visualization results of ME-NAS (synflow+naswot)-Tiny on the ShapeNetPart dataset. As shown in Fig. 6, our model can achieve results similar to ground truth annotations on multiple categories.

5 Conclusions

In this paper, we propose a novel training-free neural architecture search algorithm for 3D point cloud processing. Our method can flexibly adjust the computation constraint of the searched structures to meet the needs of various devices. Experiments on 3D point cloud classification and part segmentation show that our method can achieve competitive results under different computation constraints, which will promote the development of downstream tasks. Furthermore, our method is highly scalable, future research can add new cells to the search space or come up with new training-free proxies for a better performance.

Acknowledgements. The work is partially supported by the National Nature Science Foundation of China (No. 61976160, 61906137, 61976158, 62076184, 62076182) and Shanghai Science and Technology Plan Project (No.21DZ1204800) and Technology research plan project of Ministry of Public and Security (Grant No.2020JSYJD01).

References

1. Abdelfattah, M.S., Mehrotra, A., Dudziak, Ł., Lane, N.D.: Zero-cost proxies for lightweight NAS. arXiv preprint arXiv:2101.08134 (2021)
2. Atzmon, M., Maron, H., Lipman, Y.: Point convolutional neural networks by extension operators. arXiv preprint arXiv:1803.10091 (2018)
3. Elsken, T., Metzen, J.H., Hutter, F.: Efficient multi-objective neural architecture search via Lamarckian evolution. arXiv preprint arXiv:1804.09081 (2018)
4. Hanin, B., Rolnick, D.: Complexity of linear regions in deep networks. In: International Conference on Machine Learning, pp. 2596–2604. PMLR (2019)
5. He, K., Zhang, X., Ren, S., Sun, J.: Deep residual learning for image recognition. In: Proceedings of the IEEE Conference on Computer Vision and Pattern Recognition, pp. 770–778 (2016)
6. Klokov, R., Lempitsky, V.: Escape from cells: Deep KD-networks for the recognition of 3D point cloud models. In: Proceedings of the IEEE International Conference on Computer Vision, pp. 863–872 (2017)
7. Lee, N., Ajanthan, T., Torr, P.H.: Snip: single-shot network pruning based on connection sensitivity. arXiv preprint arXiv:1810.02340 (2018)
8. Li, G., Qian, G., Delgadillo, I.C., Muller, M., Thabet, A., Ghanem, B.: SGAS: sequential greedy architecture search. In: Proceedings of the IEEE/CVF Conference on Computer Vision and Pattern Recognition, pp. 1620–1630 (2020)
9. Li, G., Xu, M., Giancola, S., Thabet, A., Ghanem, B.: LC-NAS: latency constrained neural architecture search for point cloud networks. arXiv preprint arXiv:2008.10309 (2020)
10. Li, J., Chen, B.M., Lee, G.H.: SO-Net: self-organizing network for point cloud analysis. In: Proceedings of the IEEE Conference on Computer Vision and Pattern Recognition, pp. 9397–9406 (2018)
11. Li, Y., Bu, R., Sun, M., Wu, W., Di, X., Chen, B.: PointCNN: convolution on X-transformed points, vol. 31 (2018)
12. Liang, M., Yang, B., Wang, S., Urtasun, R.: Deep continuous fusion for multi-sensor 3D object detection. In: Proceedings of the European Conference on Computer Vision, pp. 641–656 (2018)
13. Lin, M., et al.: Zen-NAS: a zero-shot NAS for high-performance image recognition. In: Proceedings of the IEEE/CVF International Conference on Computer Vision, pp. 347–356 (2021)
14. Lin, X., Chen, K., Jia, K.: Object point cloud classification via poly-convolutional architecture search. In: Proceedings of the 29th ACM International Conference on Multimedia, pp. 807–815 (2021)
15. Lin, Z.H., Huang, S.Y., Wang, Y.C.F.: Convolution in the cloud: learning deformable kernels in 3d graph convolution networks for point cloud analysis. In: Proceedings of the IEEE/CVF Conference on Computer Vision and Pattern Recognition, pp. 1800–1809 (2020)
16. Mellor, J., Turner, J., Storkey, A., Crowley, E.J.: Neural architecture search without training. In: International Conference on Machine Learning, pp. 7588–7598 (2021)
17. Nie, X., et al.: Differentiable convolution search for point cloud processing. In: 2021 IEEE/CVF International Conference on Computer Vision (ICCV) (2021)
18. Qi, C.R., Su, H., Mo, K., Guibas, L.J.: PointNet: deep learning on point sets for 3D classification and segmentation. In: Proceedings of the IEEE Conference on Computer Vision and Pattern Recognition, pp. 652–660 (2017)

19. Qi, C.R., Yi, L., Su, H., Guibas, L.J.: PointNet++: deep hierarchical feature learning on point sets in a metric space, vol. 30 (2017)
20. Ran, H., Zhuo, W., Liu, J., Lu, L.: Learning inner-group relations on point clouds. In: Proceedings of the IEEE/CVF International Conference on Computer Vision, pp. 15477–15487 (2021)
21. Real, E., Aggarwal, A., Huang, Y., Le, Q.V.: Aging evolution for image classifier architecture search. In: AAAI Conference on Artificial Intelligence, vol. 3 (2019)
22. Rusu, R.B., Marton, Z.C., Blodow, N., Dolha, M., Beetz, M.: Towards 3D point cloud based object maps for household environments. Robot. Auton. Syst. **56**(11), 927–941 (2008)
23. Tanaka, H., Kunin, D., Yamins, D.L., Ganguli, S.: Pruning neural networks without any data by iteratively conserving synaptic flow, vol. 33, pp. 6377–6389 (2020)
24. Tang, H., et al.: Searching efficient 3D architectures with sparse point-voxel convolution. In: Vedaldi, A., Bischof, H., Brox, T., Frahm, J.-M. (eds.) ECCV 2020. LNCS, vol. 12373, pp. 685–702. Springer, Cham (2020). https://doi.org/10.1007/978-3-030-58604-1_41
25. Thomas, H., Qi, C.R., Deschaud, J.E., Marcotegui, B., Goulette, F., Guibas, L.J.: KPConv: flexible and deformable convolution for point clouds. In: Proceedings of the IEEE/CVF International Conference on Computer Vision, pp. 6411–6420 (2019)
26. Wang, C., Zhang, G., Grosse, R.: Picking winning tickets before training by preserving gradient flow. arXiv preprint arXiv:2002.07376 (2020)
27. Wang, S., Suo, S., Ma, W.C., Pokrovsky, A., Urtasun, R.: Deep parametric continuous convolutional neural networks. In: Proceedings of the IEEE Conference on Computer Vision and Pattern Recognition, pp. 2589–2597 (2018)
28. Wang, Y., Sun, Y., Liu, Z., Sarma, S.E., Bronstein, M.M., Solomon, J.M.: Dynamic graph CNN for learning on point clouds. ACM Trans. Graph. **38**(5), 1–12 (2019)
29. Wu, Z., et al.: 3D shapeNets: a deep representation for volumetric shapes. In: Proceedings of the IEEE Conference on Computer Vision and Pattern Recognition, pp. 1912–1920 (2015)
30. Xiang, T., Zhang, C., Song, Y., Yu, J., Cai, W.: Walk in the cloud: learning curves for point clouds shape analysis. In: Proceedings of the IEEE/CVF International Conference on Computer Vision, pp. 915–924 (2021)
31. Xiong, H., Huang, L., Yu, M., Liu, L., Zhu, F., Shao, L.: On the number of linear regions of convolutional neural networks. In: International Conference on Machine Learning, pp. 10514–10523. PMLR (2020)
32. Yan, X., Zheng, C., Li, Z., Wang, S., Cui, S.: PointASNL: robust point clouds processing using nonlocal neural networks with adaptive sampling. In: Proceedings of the IEEE/CVF Conference on Computer Vision and Pattern Recognition, pp. 5589–5598 (2020)
33. Yi, L., et al.: A scalable active framework for region annotation in 3D shape collections. ACM Trans. Graph. **35**(6), 1–12 (2016)
34. Zhong, Z., Yan, J., Wu, W., Shao, J., Liu, C.L.: Practical block-wise neural network architecture generation. In: Proceedings of the IEEE Conference on Computer Vision and Pattern Recognition, pp. 2423–2432 (2018)
35. Zhou, H., Feng, Y., Fang, M., Wei, M., Qin, J., Lu, T.: Adaptive graph convolution for point cloud analysis. In: Proceedings of the IEEE/CVF International Conference on Computer Vision, pp. 4965–4974 (2021)
36. Zoph, B., Vasudevan, V., Shlens, J., Le, Q.V.: Learning transferable architectures for scalable image recognition. In: Proceedings of the IEEE Conference on Computer Vision and Pattern Recognition, pp. 8697–8710 (2018)

Re-parameterization Making GC-Net-Style 3DConvNets More Efficient

Takeshi Endo[1(✉)], Seigo Kaji[1], Haruki Matono[1], Masayuki Takemura[2], and Takeshi Shima[2]

[1] Hitachi, Ltd., Research & Development Group, Chiyoda City, Japan
{takeshi.endo.cw,seigo.kaji.vc,haruki.matono.dm}@hitachi.com
[2] Hitachi Astemo, Ltd., Sunnyvale, USA
{masayuki.takemura.gx,takeshi.shima.rb}@hitachiastemo.com

Abstract. For depth estimation using a stereo pair, deep learning methods using 3D convolution have been proposed. While the estimation accuracy is high, 3D convolutions on cost volumes are computationally expensive. Hence, we propose a method to reduce the computational cost of 3D convolution-based disparity networks. We apply kernel re-parameterization, which is used for constructing efficient backbones, to disparity estimation. We convert learned parameters, and these values are used for inference to reduce the computational cost of filtering cost volumes. Experimental results on the KITTI 2015 dataset show that our method can reduce the computational cost by 31–61% from those of trained models without any performance loss. Our method can be used for any disparity network that uses 3D convolution for cost volume filtering.

Keywords: Stereo matching · Re-parameterization · Efficient architecture

1 Introduction

Depth estimation using a stereo pair is the core problem in computer vision and is applicable to autonomous driving. The depth can be obtained by calculating the disparity of a stereo pair, where disparity d is the horizontal displacement between a pair of corresponding pixels in the left and right images of the pair. If the position (x, y) in the left image corresponds to the position $(x - d, y)$ in the right image, then we can calculate the depth from $\frac{fB}{d}$, where f is the camera's focal length and B is the distance between the left and right cameras.

In recent years, deep learning has become the mainstream method for disparity estimation. Since DispNetC [12] was proposed, many methods have directly estimated disparities from stereo pairs. Specifically, DispNetC computes the correlation between left and right features to construct a cost volume which represents the matching cost. Then, it filters the cost volume with 2D convolutions to aggregate the cost. This method is efficient, but it loses rich information for channels. On the other hand, GC-Net [7] forms a concatenated cost volume to keep

L. Wang et al. (Eds.): ACCV 2022, LNCS 13841, pp. 311–325, 2023.
https://doi.org/10.1007/978-3-031-26319-4_19

channel information and aggregates the cost with 3D convolutions. This method offers improved accuracy by incorporating geometric and contextual information for disparity estimation. Subsequent 3D convolution-based methods have been proposed, such as PSMNet [2] and GWCNet [5]. These methods can easily be implemented on existing frameworks such as TensorFlow [1] because they have few network-specific operations. They can also be implemented on custom hardware, such as ASICs and FPGAs, with little additional module development. However, these methods filter the cost for each disparity d, which is computationally expensive. CRL [14] and HitNet [15] were thus proposed to speed up disparity estimation. While these 2D convolution-based approaches enable fast disparity estimation, they involve network-specific operations such as warping of feature maps to refine disparities. Accordingly, their implementation on custom hardware requires the development of new additional modules, which increases the implementation cost.

In this work, we aim to reduce the computational cost of 3D convolution-based methods because they can easily be implemented in various environments. For these methods, the most computationally expensive layer in the model is the first 3D convolution for the cost volume. Hence, we propose a method to re-parameterize the learned 3D convolution kernels for the cost volume and then use them during inference. Figure 1 gives an overview of the proposed method. Our method reduces the computational cost without any performance loss, in contrast to network compression methods such as pruning [9], which degrade performance. Our method can easily be applied to any disparity network that uses 3D convolutions for cost volume filtering.

Our contributions are summarized below.

- We propose a re-parameterization method that can reduce the computational cost of any 3D convolution-based network for disparity estimation.
- We show experimentally that our method is effective for the networks of various model sizes.

2 Related Work

2.1 Disparity Estimation

Disparity estimation methods based on deep learning can be divided into two categories, depending on whether they are based on 2D or 3D convolution. 2D convolution-based methods compute the correlation between left and right feature maps to generate a cost volume, which is aggregated using 2D convolutions. In general, 2D methods are computationally inexpensive but have lower performance than 3D methods. Accordingly, structures that refine disparity maps have been proposed to achieve a favorable speed-accuracy tradeoff [14,15]. In CRL [14] and HitNet [15], feature maps are warped to calculate error information, and the disparity map is updated by using that information. As these methods include network-specific operations, running them on custom hardware requires support for those unique operations.

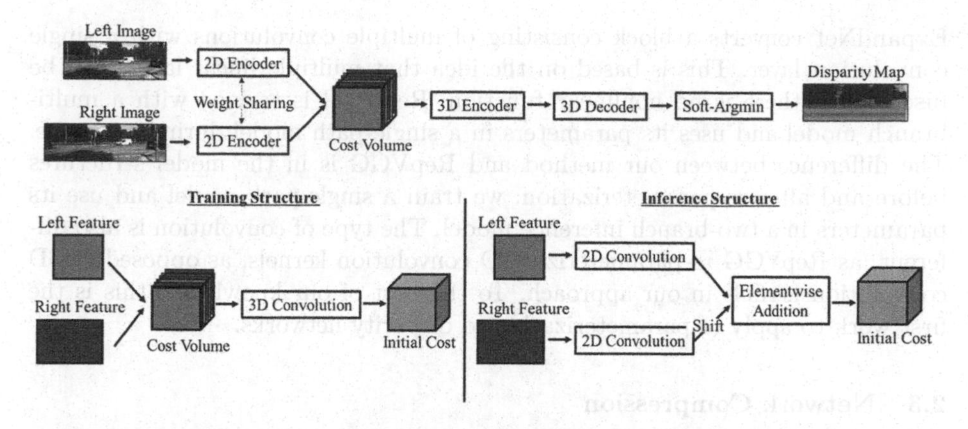

Fig. 1. Overview of the proposed method. (top) The 2D encoder extracts features by 2D convolution. Left and right features are concatenated to form the cost volume, which is then aggregated through the 3D encoder and decoder. Finally, a soft-argmin regresses the disparity. (bottom) Training and inference have different structures for computing the initial cost. During training, the initial cost is formed by 3D convolution on the cost volume. On the other hand, during inference, two 2D convolutions are used and these output values are added to form the initial cost.

In contrast, 3D convolution-based methods concatenate left and right feature maps to form a cost volume, which is filtered by using 3D convolutions to aggregate the cost. GC-Net [7] was the first such 3D method. It aggregates the cost with 3D convolutions and regresses the disparity map with a soft-argmin process. PSMNet [2] uses a spatial pyramid module for feature extraction, which enables the cost volume to incorporate multi-scale information. The problem with GC-Net and PSMNet is the increased computational cost due to the 3D convolutions. As alternatives, a method forming a low-resolution cost volume [8] and a method using group-wise correlation to construct a cost volume [5] were proposed. These network architectures of 3D methods are simpler than those of 2D methods. In addition, their architectures comprise generic operations that are used in other networks. On the other hand, the 3D methods are more computationally expensive than the 2D methods. Here, we aim to reduce the computational cost of 3D convolution-based methods from the viewpoint of ease of implementation.

2.2 Kernel Re-parameterization

To reduce the computational cost, we convert the learned parameters and use the values for inference. This idea is similar to research on kernel re-parameterization. [16] parameterizes a kernel as $\hat{\mathrm{W}}=\mathrm{diag(a)I+diag(b)W_{norm}}$, where $\mathrm{W_{norm}}$ is the normalized weight, and a and b are learnable vectors. The model uses implicit skip-connections, and it has the same structures during learning and inference, whereas we use different model structures during training and inference. Our approach is closely related to ExpandNet [4] and RepVGG [3].

ExpandNet converts a block consisting of multiple convolutions with a single convolution layer. This is based on the idea that multiple linear layers can be fused when there is no nonlinear function. RepVGG is trained with a multi-branch model and uses its parameters in a single-path model during inference. The difference between our method and RepVGG is in the model structures before and after re-parameterization: we train a single-path model and use its parameters in a two-branch inference model. The type of convolution is also different, as RepVGG re-parameterizes 2D convolution kernels, as opposed to 3D convolution kernels in our approach. To the best of our knowledge, this is the first work to apply re-parameterization to disparity networks.

2.3 Network Compression

Network compression methods such as pruning also reduce the computational cost. [9] proposes a structured pruning method. They remove filters that are identified as having a small effect on the output accuracy, and then finetune the network. Another pruning method [11] forces channel-level sparsity and leverages learnable batch normalization parameters for pruning. [10] explores the rank of feature maps and measures the information richness of feature maps by using the rank. All of these network compression methods change the network outputs before and after compression. In contrast, our method matches the outputs before and after re-parameterization, which places it along a different research direction from network compression.

3 Re-parameterization for 3DConvNets

In this section, we first describe the architecture of 3D convolution-based networks such as GC-Net. Then, we describe our re-parameterization approach in detail, followed by the implementation tricks required for re-parameterization.

3.1 GC-Net-Style 3DConvNet

We first describe a general 3DConvNet (3D Convolution-based Network) as represented by GC-Net [7]. The model shown in the upper part of Fig. 1 represents the basic structure of a 3DConvNet. A stereo pair of left and right images is input, and shared 2D convolutions are used to extract the left feature map $\mathbf{f}_l$ and the right feature map $\mathbf{f}_r$. Given $\mathbf{f}_l$ and $\mathbf{f}_r$, the cost volume is computed, as

$$\mathbf{C}_{\text{concat}}(d, x, y) = \text{Concat}\{\mathbf{f}_l(x, y), \mathbf{f}_r(x - d, y)\}, \tag{1}$$

where d denotes the disparity level. We omit the channel dimension here for simplicity.

The cost volume $\mathbf{C}_{\text{concat}}$ is formed from $\mathbf{f}_l$ and $\mathbf{f}_r$ shifted relative to $\mathbf{f}_l$. The 3D encoder aggregates the cost with 3×3×3 3D convolutions, while the 3D decoder uses 3×3×3 3D deconvolutions to upsample the aggregated cost. The disparity map is regressed by a soft-argmin process [7].

Table 1 lists the details of the model structures to which our method applies. H and W denote the height and width of an input image, respectively. F denotes the number of channels for the feature map, and D indicates the maximum disparity to be estimated. The structures described in Table 1 are almost the same as those in GC-Net: the only difference is that we add a Stem2 layer to the 2D encoder to adjust the model's computational cost. In our experiments, we varied the stride values of the layers to build models having different computational costs.

The layers of the 3D encoder are denoted as follows. The output of the first convolution on the cost volume $\mathbf{C}_{\text{concat}}$ is denoted as $\mathbf{C}_{\text{initial}}$, because it represents the initial matching cost. The layers that correspond to $\mathbf{C}_{\text{initial}}$ in Table 1 are the InitialCost1 and InitialCost2 layers of the 3D encoder. Next, we denote the output of the 3D encoder's subsequent convolutions as $\mathbf{C}_{\text{aggregated}}$, because it represents the aggregated cost.

The computational cost of 3D convolution is proportional to the output size, which is similar to the property of 2D convolution. In the 3D encoder described in Table 1, the bottom layers such as InitialCost1 are computationally expensive. Hence, in this work, we propose a re-parameterization method to reduce the computational cost of constructing the initial cost $\mathbf{C}_{\text{initial}}$.

3.2 Re-parameterization for Efficient Inference

In our approach, we re-parameterize the 3×3×3 3D convolution kernels used to construct the initial cost $\mathbf{C}_{\text{initial}}$. Specifically, the 3D kernels are converted to 2D convolution kernels.

The initial cost $\mathbf{C}_{\text{initial}}$ is computed as follows,

$$\mathbf{C}_{\text{initial}}(d, x, y) = \sum_{-1 \leq l,m,n \leq 1} \mathbf{W}_{3d}(l, m, n) \cdot \mathbf{C}_{\text{concat}}(d + l, x + m, y + n), \quad (2)$$

where $\mathbf{W}_{3d}$ is the 3D convolution kernel. From (1), (2) is equivalent to the following:

$$\begin{aligned}\mathbf{C}_{\text{initial}}(d, x, y) = & \sum_{-1 \leq l,m,n \leq 1} \mathbf{W}_{3d,l}(l, m, n) \cdot \mathbf{f}_l(x + m, y + n) \\ & + \sum_{-1 \leq l,m,n \leq 1} \mathbf{W}_{3d,r}(l, m, n) \cdot \mathbf{f}_r(x - d + m - l, y + n), \quad (3)\end{aligned}$$

where $\mathbf{W}_{3d,l}$ and $\mathbf{W}_{3d,r}$ are 3D convolution kernels and parts of $\mathbf{W}_{3d}$. $\mathbf{W}_{3d,l}$ and $\mathbf{W}_{3d,r}$ are used for filtering $\mathbf{f}_l$ and $\mathbf{f}_r$, respectively. If the number of channels in $\mathbf{f}_l$ and $\mathbf{f}_r$ is F, then $\mathbf{W}_{3d}$ has 2F channels. The kernel corresponding to channels $0 \cdots (\text{F} - 1)$ is $\mathbf{W}_{3d,l}$, and the one corresponding to channels $\text{F} \cdots (2\text{F} - 1)$ is $\mathbf{W}_{3d,r}$. The first term in (3) is the operation on the left feature map $\mathbf{f}_l$ and the second term is the operation on the right feature map $\mathbf{f}_r$.

In the first term of (3), the distributive law holds. Therefore, $\mathbf{W}_{3d,l}$ can be fused in the disparity direction l. This yields the following equation for the first

Table 1. Summary of our 3DConvNet architecture, which has a 2D encoder, cost volume layer, 3D encoder, and 3D decoder. Each convolutional layer comprises convolution, batch normalization, and ReLU activation. The layer properties consist of the kernel size, convolution type, number of kernels, and stride, in this order.

(a) 2D encoder

Name	Layer Property	Output Size
Stem1	5×5, conv2d, 32, 2	$\frac{H}{2}\times\frac{W}{2}\times F$
Stem2	3×3, conv2d, 32, 1	$\frac{H}{2}\times\frac{W}{2}\times F$
Resblocks	(3×3, conv2d, 32, 1)×2 skip connect }×8	$\frac{H}{2}\times\frac{W}{2}\times F$
Conv1	3×3, conv2d, 32, 1 (no ReLU and BN)	$\frac{H}{2}\times\frac{W}{2}\times F$

(b) Cost volume layer and 3D encoder

Name	Layer Property	Output Size
	Cost Volume	
ConcatCost	Concat left & right features	$\frac{D}{2}\times\frac{H}{2}\times\frac{W}{2}\times 2F$
	3D Encoder	
InitialCost1	From ConcatCost: 3×3×3, conv3d, 32, 1	$\frac{D}{2}\times\frac{H}{2}\times\frac{W}{2}\times F$
AggCost1	3×3×3, conv3d, 32, 1	$\frac{D}{2}\times\frac{H}{2}\times\frac{W}{2}\times F$
InitialCost2	From ConcatCost: 3×3×3, conv3d, 64, 2	$\frac{D}{4}\times\frac{H}{4}\times\frac{W}{4}\times 2F$
AggCost2-1	(3×3×3, conv3d, 64, 1)×2	$\frac{D}{4}\times\frac{H}{4}\times\frac{W}{4}\times 2F$
AggCost3-1	From InitialCost2: 3×3×3, conv3d, 64, 2	$\frac{D}{8}\times\frac{H}{8}\times\frac{W}{8}\times 2F$
AggCost3-2	(3×3×3, conv3d, 64, 1)×2	$\frac{D}{8}\times\frac{H}{8}\times\frac{W}{8}\times 2F$
AggCost4-1	From AggCost3-1: 3×3×3, conv3d, 64, 2	$\frac{D}{16}\times\frac{H}{16}\times\frac{W}{16}\times 2F$
AggCost4-2	(3×3×3, conv3d, 64, 1)×2	$\frac{D}{16}\times\frac{H}{16}\times\frac{W}{16}\times 2F$
AggCost5-1	From AggCost4-1: 3×3×3, conv3d, 128, 2	$\frac{D}{32}\times\frac{H}{32}\times\frac{W}{32}\times 4F$
AggCost5-2	(3×3×3, conv3d, 128, 1)×2	$\frac{D}{32}\times\frac{H}{32}\times\frac{W}{32}\times 4F$

(c) 3D decoder

Name	Layer Property	Output Size
AggCost6	3×3×3, deconv3d, 64, 2 skip connect (from AggCost4-2)	$\frac{D}{16}\times\frac{H}{16}\times\frac{W}{16}\times 2F$
AggCost7	3×3×3, deconv3d, 64, 2 skip connect (from AggCost3-2)	$\frac{D}{8}\times\frac{H}{8}\times\frac{W}{8}\times 2F$
AggCost8	3×3×3, deconv3d, 64, 2 skip connect (from AggCost2-1)	$\frac{D}{4}\times\frac{H}{4}\times\frac{W}{4}\times 2F$
AggCost9	3×3×3, deconv3d, 32, 2 skip connect (from AggCost1)	$\frac{D}{2}\times\frac{H}{2}\times\frac{W}{2}\times F$
AggCost10	3×3×3, deconv3d, 1, 2 (no ReLU and BN)	D×H×W×1

term of (3):

$$\sum_{-1\le l,m,n\le 1} \mathbf{W}_{3d,l}(l,m,n)\cdot \mathbf{f}_l(x+m,y+n) = \sum_{-1\le m,n\le 1} \mathbf{W}'_{2d,l}(m,n)\cdot \mathbf{f}_l(x+m,y+n), \quad (4)$$

where $\mathbf{W}'_{2d,l}(m,n)$ is defined as

$$\mathbf{W}'_{2d,l}(m,n) = \sum_{-1\le l\le 1} \mathbf{W}_{3d,l}(l,m,n). \quad (5)$$

Next, for the second term in (3), we let the 3×5 2D kernel $\mathbf{W}'_{2d,r}$ be given by,

$$\mathbf{W}'_{2d,r}(m,n) = \sum_{-1\le l\le 1} \mathbf{W}'_{3d,r}(l,m+l,n), \quad (6)$$

where

$$\mathbf{W}'_{3d,r}(l,m,n) = \begin{cases} \mathbf{W}_{3d,r}(l,m,n), & -1\le m\le 1 \\ 0, & \text{otherwise.} \end{cases} \quad (7)$$

Substituting (6) into the second term in (3), we obtain the following equation:

$$\sum_{-1\leq l,m,n\leq 1} \mathbf{W}_{3d,r}(l,m,n)\cdot \mathbf{f}_r(x-d+m-l,y+n) = \sum_{\substack{-2\leq m\leq 2\\ -1\leq n\leq 1}} \mathbf{W}'_{2d,r}(m,n)\cdot \mathbf{f}_r(x-d+m,y+n). \quad (8)$$

From (4) and (8), $\mathbf{C}_{\text{initial}}(d,x,y)$ is given by,

$$\mathbf{C}_{\text{initial}}(d,x,y) = \sum_{-1\leq m,n\leq 1} \mathbf{W}'_{2d,l}(m,n)\cdot \mathbf{f}_l(x+m,y+n) + \sum_{\substack{-2\leq m\leq 2\\ -1\leq n\leq 1}} \mathbf{W}'_{2d,r}(m,n)\cdot \mathbf{f}_r(x-d+m,y+n). \quad (9)$$

Convolution is not affected by translation. As the order of shift and convolution operations can be interchanged, we have

$$\mathbf{W} * \text{Shift}(\mathbf{f},d) = \text{Shift}(\mathbf{W} * \mathbf{f},d), \quad (10)$$

where $\text{Shift}(\mathbf{X},d)$ denotes shifting a tensor $\mathbf{X}$ by d pixels. Hence, we derive the following:

$$\mathbf{C}_{\text{initial}}(d) = \mathbf{W}'_{2d,l} * \mathbf{f}_l + \text{Shift}(\mathbf{W}'_{2d,r} * \mathbf{f}_r, d), \quad (11)$$

where $*$ is the convolution operator.

(11) indicates that the initial cost $\mathbf{C}_{\text{initial}}$ can be computed only from 2D convolutions, without 3D convolutions. It is obvious that we only need to calculate $\mathbf{W}'_{2d,l} * \mathbf{f}_l$ and $\mathbf{W}'_{2d,r} * \mathbf{f}_r$ once to form $\mathbf{C}_{\text{initial}}$. Hence, we compute $\mathbf{C}_{\text{initial}}$ as shown in Fig. 1. We first filter the left feature map $\mathbf{f}_l$ and the right feature map $\mathbf{f}_r$ by using the respective 2D convolution kernels $\mathbf{W}'_{2d,l}$ and $\mathbf{W}'_{2d,r}$. The filtered right feature map is then shifted and added to the filtered left feature map, which allows us to construct the initial cost $\mathbf{C}_{\text{initial}}$.

Here, we re-parameterize the trained model's 3D convolution kernels into 2D convolution kernels, as shown in Fig. 2. The 3D kernels are decomposed to filter the left and right feature maps. As seen in (5), the kernels for left feature maps are re-parameterized by element addition, while those for right feature maps are re-parameterized according to (6) and (7). The re-parameterized 2D convolution kernels are then used for inference.

The re-parameterized 2D convolutions reduce the computational cost. We measure the computational cost in terms of FLOPs, the number of multiply-adds required for convolutions. For the model described in Table 1, the computation of the InitialCost1 layer requires $3^3 \cdot \frac{1}{2}\text{D} \cdot \frac{1}{2}\text{W} \cdot \frac{1}{2}\text{H} \cdot 2\text{F}^2$ FLOPs. On the other hand, the re-parameterized model requires $3^2 \cdot \frac{1}{2}\text{W} \cdot \frac{1}{2}\text{H} \cdot \text{F}^2 + 3 \cdot 5 \cdot \frac{1}{2}\text{W} \cdot \frac{1}{2}\text{H} \cdot \text{F}^2$ FLOPs. Thus, the computational cost is reduced to $\frac{8}{9\cdot\text{D}}$. As D is 192 in general, the layer can be computed with 0.5% of the original computational cost in FLOPs.

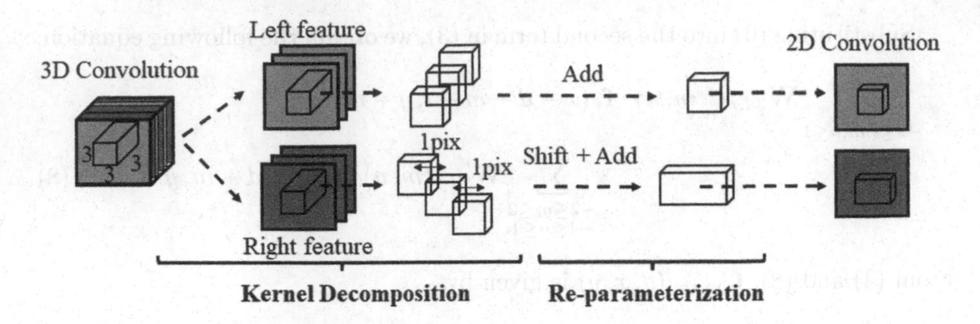

Fig. 2. Kernel re-parameterization, in which 3D convolution kernels are converted into 2D kernels. Each 3D kernel is decomposed into left and right kernels. The left kernels corresponding to each disparity are fused by elementwise addition, while the right kernels corresponding to each disparity are fused by a shift operation and elementwise addition. The re-parameterized 2D kernels are then used for inference.

3.3 Implementation Tricks

Because of the effect of padding, the outputs before and after re-parameterization do not match. Hence, we describe implementation tricks to avoid this issue, namely, left image cropping and initial cost cropping. Because these tricks reduce the output size of the disparity map, our approach cannot estimate disparities at left and right sides of an image. However, in the case of a real application, disparity estimation is performed on a partial region of interest. Accordingly, we assume that disparity calculations at the left and right sides are not necessarily required.

Left Image Cropping. To construct the cost volume $\mathbf{C}_{\text{concat}}$, the left and (shifted) right feature maps are concatenated in the channel direction, as seen in (1). The left side of the right feature map is padded with zeros when the feature map is shifted to the right. The number of padded columns is equal to the number of shifts. In the left feature map, the same number of columns is also padded. Thus, the left feature map $\mathbf{f}_l$ is padded differently at each disparity level d. This is also true for the right feature map $\mathbf{f}_r$.

In our method, however, $\mathbf{f}_l$ must not be padded differently at different disparity levels d. This is because $\mathbf{W}'_{2d,l} * \mathbf{f}_l$ is repeatedly used for constructing the initial cost $\mathbf{C}_{\text{initial}}(d)$ at any disparity level. This is also true for $\mathbf{f}_r$. To satisfy these conditions, we use stereo pairs with different widths. Given the shift of the right feature map, the width of the right image is set to be D pixels larger than that of the left image. Thus, we crop the left image so that its width is $\mathrm{W_l} = \mathrm{W_r} - \mathrm{D}$, where $\mathrm{W_r}$ is that of the right image. This gives the difference between the widths of the left and right feature maps. Hence, we crop the right feature $\mathbf{f}_r$ to the same width as the left feature map $\mathbf{f}_l$. Then, we concatenate them in the channel direction.

Initial Cost Cropping. In general, a cost volume is padded with zeros in the spatial and disparity directions when a 3D convolution is performed on it. Accordingly, we must handle the effect of padding in the x and d directions so that the outputs before and after re-parameterization match. During training, we use the 3D convolution to filter the cost volume. The convolution outputs at position $x = 0$, W and $d = 0$, D are affected by padding. During inference, we use the shift operation and elementwise addition to construct the initial cost, as seen in (11). Because these operations lead to a different padding effect, we crop the cost volume to eliminate pixels that do not match in training and inference, as shown in Fig. 3.

During training, we first crop the left and right sides of the left feature map $\mathbf{f}_l$. This operation is necessary to keep the original disparity range ($d = 0 \cdots \mathrm{D}$). Then, the left and right feature maps are concatenated, and the 3D convolution is performed to construct the initial cost. Finally, we crop the initial cost in the x and d directions to remove the pixels that are affected by padding.

During inference, we use a similar procedure. First, the left feature map is cropped. Then, the re-parameterized 2D convolutions are performed on the left and right feature maps, and the shift operation and elementwise addition are used to construct the initial cost. Finally, we crop the initial cost in the x and d directions.

3.4 Disparity Regression

We use the disparity regression method proposed in [7] to estimate disparities. The soft-argmin operation enables us to estimate disparities with sub-pixel accuracy. The probability of each disparity level d is used for estimation. The negative of the predicted cost c_d is converted to the probability via the softmax operation. The predicted disparity $\hat{d}$ is then calculated as the sum of each disparity level d weighted by its probability:

$$\hat{d} = \sum_{d=0}^{D} d \times \sigma(-c_d), \tag{12}$$

where σ denotes the softmax operation.

3.5 Loss Function

We use the loss function proposed in [7], which is defined as follows:

$$Loss = \frac{1}{N} \sum_{n=0}^{N} \|d - \hat{d}\|_1, \tag{13}$$

where N is the number of labeled pixels, $\hat{d}$ is the predicted disparity, and d is the ground-truth disparity.

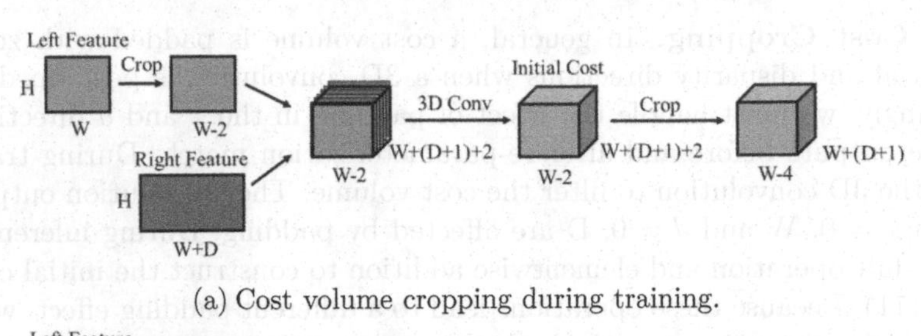

(a) Cost volume cropping during training.

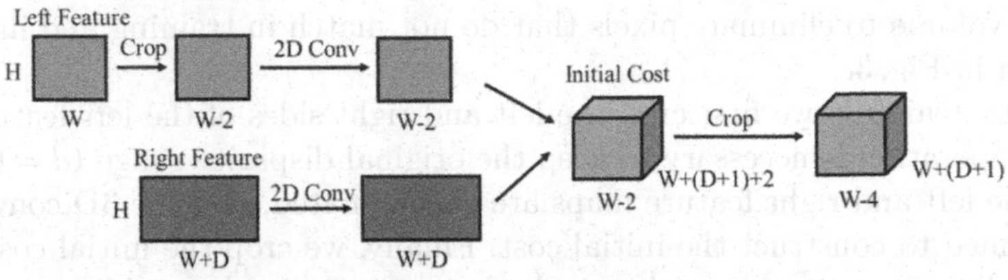

(b) Cost volume cropping during inference.

Fig. 3. Cost volume cropping. (a) During training, the left and right sides of the left feature map are cropped. Then, the initial cost is constructed by a 3D convolution, and it is finally cropped in the x and d directions. (b) During inference, the cropping is almost the same as during training; the difference is the way to construct the initial cost.

4 Experiments

In this section, we describe our experimental conditions and our results on the KITTI 2015 dataset [13]. In these experiments, we compared several models having different computational costs. We also performed ablation studies to further evaluate our method.

4.1 Experimental Conditions

During training, we used the Monkaa dataset [12] for pre-training and the KITTI 2015 dataset [13] for finetuning. For evaluation, we used the KITTI 2015 dataset. As we evaluated several models, we did not use the KITTI 2015 official evaluation dataset. Instead, we split the original 200 training images into 160 training images and 40 evaluation images.

For training on these datasets, we cropped the left and right images to input sizes of 256×512 and 256×704, respectively. We set $\mathrm{D} = 192$ and $\mathrm{F} = 32$. RMSprop [6] was used as the optimizer, with a learning rate of 1×10^{-3}. These settings are the same as in [7]. For evaluation, we padded zeros at the top and on the right side of an image. The input sizes were 384×1056 for left images and 384×1248 for right images.

Table 2. Model structures of the (a) 3DConvNet-B1, (b) 3DConvNet-B0, and (c) Rep3DConvNet-B0 models.

(a) 3DConvNet-B1

Name	Layer Property	Output Size
	3D Encoder	
InitialCost1	From ConcatCost: 3×3×3, conv3d, 64, 1	$\frac{D}{4}\times\frac{H}{4}\times\frac{W}{4}\times 2F$
AggCost1	3×3×3, conv3d, 64, 1	$\frac{D}{4}\times\frac{H}{4}\times\frac{W}{4}\times 2F$
InitialCost2	From ConcatCost: 3×3×3, conv3d, 64, 2	$\frac{D}{8}\times\frac{H}{8}\times\frac{W}{8}\times 2F$
AggCost2-1	(3×3×3, conv3d, 64, 1)×2	$\frac{D}{8}\times\frac{H}{8}\times\frac{W}{8}\times 2F$
AggCost3-1	From InitialCost2: 3×3×3, conv3d, 64, 2	$\frac{D}{16}\times\frac{H}{16}\times\frac{W}{16}\times 2F$
AggCost3-2	(3×3×3, conv3d, 64, 1)×2	$\frac{D}{16}\times\frac{H}{16}\times\frac{W}{16}\times 2F$
AggCost4-1	From AggCost3-1: 3×3×3, conv3d, 128, 2	$\frac{D}{32}\times\frac{H}{32}\times\frac{W}{32}\times 4F$
AggCost4-2	(3×3×3, conv3d, 128, 1)×2	$\frac{D}{32}\times\frac{H}{32}\times\frac{W}{32}\times 4F$
	3D Decoder	
AggCost5	3×3×3, deconv3d, 64, 2 skip connect (from AggCost3-2)	$\frac{D}{16}\times\frac{H}{16}\times\frac{W}{16}\times 2F$
AggCost6	3×3×3, deconv3d, 64, 2 skip connect (from AggCost2-1)	$\frac{D}{8}\times\frac{H}{8}\times\frac{W}{8}\times 2F$
AggCost7	3×3×3, deconv3d, 64, 2 skip connect (from AggCost1)	$\frac{D}{4}\times\frac{H}{4}\times\frac{W}{4}\times 2F$
AggCost8	3×3×3, deconv3d, 1, 2 (no ReLU and BN)	$\frac{D}{2}\times\frac{H}{2}\times\frac{W}{2}\times 1$

(b) 3DConvNet-B0

Name	Layer Property	Output Size
	3D Encoder	
InitialCost1	From ConcatCost: 3×3×3, conv3d, 64, 1	$\frac{D}{4}\times\frac{H}{4}\times\frac{W}{4}\times 2F$
AggCost2-1	From InitialCost1: 3×3×3, conv3d, 64, 2	$\frac{D}{8}\times\frac{H}{8}\times\frac{W}{8}\times 2F$
AggCost2-2	(3×3×3, conv3d, 64, 1)×2	$\frac{D}{8}\times\frac{H}{8}\times\frac{W}{8}\times 2F$
AggCost3-1	From AggCost2-1: 3×3×3, conv3d, 64, 2	$\frac{D}{16}\times\frac{H}{16}\times\frac{W}{16}\times 2F$
AggCost3-2	(3×3×3, conv3d, 64, 1)×2	$\frac{D}{16}\times\frac{H}{16}\times\frac{W}{16}\times 2F$
	3D Decoder	
AggCost4	3×3×3, deconv3d, 64, 2 skip connect (from AggCost2-2)	$\frac{D}{8}\times\frac{H}{8}\times\frac{W}{8}\times 2F$
AggCost5	3×3×3, deconv3d, 1, 2 (no ReLU and BN)	$\frac{D}{4}\times\frac{H}{4}\times\frac{W}{4}\times 1$

(c) Rep3DConvNet-B0

Name	Layer Property	Output Size
Conv2D-L	From 2D encoder's left feature: 3×3, conv2d, 64, 1	$\frac{H}{4}\times\frac{W}{4}\times 2F$
Conv2D-R	From 2D encoder's right feature: 3×5, conv2d, 64, 1	$\frac{H}{4}\times\frac{W+D}{4}\times 2F$
InitialCost1	For each disparity d: Conv2D-L+Shift(Conv2D-R,d)	$\frac{D}{4}\times\frac{H}{4}\times\frac{W}{4}\times 2F$
	3D Encoder	
	AggCost∗ layers in 3D Encoder of 3DConvNet-B0	
	3D Decoder	
	AggCost∗ layers in 3D Decoder of 3DConvNet-B0	

4.2 Experimental Results

We trained four models with different computational costs. We refer to the model described in Sect. 3.1 as 3DConvNet-B3. We created three other models with reduced computational costs: 3DConvNet-B2, 3DConvNet-B1, and 3DConvNet-B0. First, 3DConvNet-B2 limited the number of convolutions of the 3D encoder, as listed in Table 1. Specifically, the AggCost1 is removed, and the number of convolution layers was reduced to one in the AggCost2-1, AggCost3-2, AggCost4-2, and AggCost5-2 layers. Next, 3DConvNet-B1 and 3DConvNet-B0 used a stride of two for the 2D encoder's Stem2 layer to reduce the spatial resolution. The resulting structures of 3DConvNet-B1 and 3DConvNet-B0 are given in Table 2. The structure of 3DConvNet-B1 was similar to that of 3DConvNet-B3, but the 3D encoder's input size and the 3D decoder's output size were different. 3DConvNet-B0 was the lightest model: its 3D encoder and 3D decoder were shallow, and the output disparity size was one-fourth of the input size.

As described in Sect. 3, the proposed method uses 3D convolutions to compute the initial cost during training. The 3D convolutions are re-parameterized, and the re-parameterized kernels are then used for inference. We refer to the re-parameterized model that uses 2D convolutions to compute the initial cost as Rep3DConvNet. In Table 2(c), we show the structure of Rep3DConvNet-B0. The difference between the model and 3DConvNet-B0 is in the initial cost construction process. As described in Sect. 3.3, our methods cannot estimate disparities

Table 3. Experimental results on 40 evaluation images from the KITTI 2015 dataset. The abbreviations 'bg' and 'fg' refer to the results on background and dynamic object pixels, respectively. Each model was evaluated before and after re-parameterization. The column 'FLOPs' gives the computational cost of the entire network, while the column 'FLOPs†' shows the cost for the initial cost construction. The runtime and memory consumption were measured on an NVIDIA Tesla V100 GPU (16 GB).

Model	Inference Structure		D1 error (%)			FLOPs (B)	FLOPs† (G)	Runtime (sec)	Memory (GiB)
	3DConv	2DConv	D1-bg	D1-fg	D1-all				
3DConvNet-B0	✓		3.19	9.15	4.01	0.23	141.9	0.059	3.1
Rep3DConvNet-B0		✓				0.09	1.4	0.041	2.8
3DConvNet-B1	✓		2.58	7.76	3.30	0.52	160.8	0.102	4.9
Rep3DConvNet-B1		✓				0.36	1.7	0.077	3.1
3DConvNet-B2	✓		2.48	4.69	2.79	1.68	695.8	0.378	15.3
Rep3DConvNet-B2		✓				0.98	4.2	0.225	9.9
3DConvNet-B3	✓		2.20	5.18	2.62	2.10	695.8	0.462	15.3
Rep3DConvNet-B3		✓				1.41	4.2	0.311	10.4

(a) Input image

(b) Rep3DConvNet-B3

(c) Rep3DConvNet-B0

Fig. 4. Example of qualitative results on KITTI 2015, showing (a) the input image, and the output results from (b) Rep3DConvNet-B3 and (c) Rep3DConvNet-B0. In (b) and (c), the top and bottom images are the disparity and error maps, respectively.

at the left and right sides of an image. Thus, in our evaluation, the accuracy was measured for pixels excluding those at the left and right sides. For evaluation, we used the D1 error metric, which counts the number of pixels that satisfy $|d - \hat{d}| > 3$pix, or $\frac{|d-\hat{d}|}{d} > 5\%$, for the ground-truth d and estimated disparity $\hat{d}$. Better models have lower scores for the D1 error. Note that our models estimate disparity maps at different scales. If the estimated disparity maps are smaller than the input images, we upsample the outputs by bilinear interpolation.

The evaluation results on the KITTI dataset are summarized in Table 3. Specifically, we give the results for 3DConvNet and Rep3DConvNet. The results show that our method could reduce the computational cost without any performance loss. In particular, the re-parameterization reduced the computational cost of the lightest model, 3DConvNet-B0, by 61%, and that of 3DConvNet-B3, the largest model, by 33%. The reduction rate depends on the computational cost for the initial cost construction. Overall, our experiment showed that the proposed method was effective for various models with different computational costs. We also measured the actual run time and memory consumption. As seen in the table, the re-parameterization accelerated the inference and reduced the memory footprint.

Table 4. Ablation study results. (a) D1 error for 3DConvNet-B3 in various training settings, where 'Trick1' and 'Trick2' denote left image cropping and initial cost cropping, respectively. (b) Accuracy comparison between 2DConvNet-B1, which is trained with the same structure as the re-parameterized model, and Rep3DConvNet-B1.

(a) Ablation 1

		D1 error(%)		
Trick1	Trick2	D1-bg	D1-fg	D1-all
		2.59	7.43	3.26
✓		2.33	5.31	2.74
✓	✓	2.20	5.18	2.62

(b) Ablation 2

	D1 error(%)		
Model	D1-bg	D1-fg	D1-all
2DConvNet-B1	2.77	8.19	3.52
Rep3DConvNet-B1	2.58	7.76	3.30

Figure 4 shows an example of the output results for Rep3DConvNet-B0 and Rep3DConvNet-B3. Rep3DConvNet-B3 had more layers and a higher resolution for disparity maps. Thus, it could suppress depth mixing, which means the errors between foreground and background pixels, better than Rep3DConvNet-B0 could. On the other hand, the overall disparity was similar to that for Rep3DConvNet-B3, even though Rep3DConvNet-B0 had 6% of the convolution operations in Rep3DConvNet-B3. The required accuracy depends on the application for disparity estimation. The results demonstrate that our method can reduce the computational cost from computationally expensive models to much smaller models.

4.3 Ablation Study

Table 4(a) lists the results of this ablation study, where 'trick1' and 'trick2' denote left image cropping and initial cost cropping, respectively. It can be seen that the performance was improved by using trick1. This was because trick1 eliminates occlusion pixels, enabling the model to find better correspondences. We can also see that trick2 did not degrade the performance. Thus, these tricks are unlikely to have negative impacts on the accuracy.

Next, although we use different model structures during training and inference, it is also possible to train a network with the same structure as the re-parameterized model. We refer to the resulting 2D convolution-based model, which is not re-parameterized, as 2DConvNet. Table 4(b) lists the results of a comparison between 2DConvNet-B1 and Rep3DConvNet-B1, which shows that the re-parameterized model had better accuracy. In our re-parameterization process, the number of parameters during training is larger than the number during inference. This can be viewed as learning an over-parameterized model and then using a compact model during inference. In ExpandNet [4], over-parameterization stabilizes the gradients during training. It is also known that the loss landscape becomes flat, which gives better generalization performance. Accordingly, it is important to train models by using 3D convolution and then re-parameterize them during inference, as in our proposed method.

5 Conclusion

We have proposed a kernel re-parameterization method to reduce the computational cost of 3D convolution-based disparity networks. Our method re-parameterizes learned 3D convolution kernels and uses them as 2D convolution kernels. Our experimental results show that the proposed method can reduce the computational cost without degrading the trained model's performance. Specifically, the computational cost could be reduced by 31–61% for models with different sizes. In addition, the proposed method dramatically reduces the computational cost for constructing the initial cost, which makes it feasible to construct many initial costs. Hence, a future work will be to develop a network structure in which our method is effective.

Acknowledgements. We want to thank Atsushi Yokoyama and Kumud Shishir for their helpful comments on this paper.

References

1. Abadi, M., et al.: TensorFlow: large-scale machine learning on heterogeneous systems. In: Software available from tensorflow.org (2015)
2. Chang, J., Chen, Y.: Pyramid stereo matching network. In: The IEEE International Conference on Computer Vision and Pattern Recognition (CVPR), pp. 5410–5418 (2018)
3. Ding, X., Zhang, X., Ma, N., Han, J., Ding, G., Sun, J.: RepVGG: making VGG-style convnets great again. In: The IEEE International Conference on Computer Vision and Pattern Recognition (CVPR), pp. 13733–13742 (2021)
4. Guo, S., Alvarez, J., Salzmann, M.: ExpandNets: linear over-parameterization to train compact convolutional networks. In: Advances in Neural Information Processing Systems (NeurIPS) 33, pp. 1298–1310 (2020)
5. Guo, X., Yang, K., Yang, W., Wang, X., Li, H.: Group-wise correlation stereo network. In: The IEEE International Conference on Computer Vision and Pattern Recognition (CVPR), pp. 3273–3282 (2019)
6. Hinton, G., Srivastava, N., Swersky, K.: Neural networks for machine learning lecture 6a overview of mini-batch gradient descent. Cited on **14**(8), 2 (2012)
7. Kendall, A., et al.: End-to-end learning of geometry and context for deep stereo regression. In: The IEEE International Conference on Computer Vision (ICCV), pp. 66–75 (2017)
8. Khamis, S., Fanello, S., Rhemann, C., Kowdle, A., Valentin, J., Izadi, S.: StereoNet: guided hierarchical refinement for real-time edge-aware depth prediction. In: Ferrari, V., Hebert, M., Sminchisescu, C., Weiss, Y. (eds.) ECCV 2018. LNCS, vol. 11219, pp. 596–613. Springer, Cham (2018). https://doi.org/10.1007/978-3-030-01267-0_35
9. Li, H., Kadav, A., Durdanovic, I., Samet, H., Graf, H.P.: Pruning filters for efficient convnets. In: arXiv pre-print arXiv: 1608.08710 (2016)
10. Lin, M., et al.: HRank: filter pruning using high-rank feature map. In: The IEEE International Conference on Computer Vision and Pattern Recognition (CVPR), pp. 1529–1538 (2020)

11. Liu, Z., Li, J., Shen, Z., Huang, G., Yan, S., Zhang, C.: Learning efficient convolutional networks through network slimming. In: The IEEE International Conference on Computer Vision (ICCV), pp. 2736–2744 (2017)
12. Mayer, N., et al.: A large dataset to train convolutional networks for disparity, optical flow, and scene flow estimation. In: The IEEE International Conference on Computer Vision and Pattern Recognition (CVPR), pp. 4040–4048 (2016)
13. Menze, M., Geiger, A.: Object scene flow for autonomous vehicles. In: The IEEE International Conference on Computer Vision and Pattern Recognition (CVPR), pp. 3061–3070 (2015)
14. Pang, J., Sun, W., Ren, J.S., Yang, C., Yan, Q.: Cascade residual learning: a two-stage convolutional neural network for stereo matching. In: The IEEE International Conference on Computer Vision (ICCV), pp. 887–895 (2017)
15. Tankovich, V., Hane, C., Zhang, Y., Kowdle, A., Fanello, S., Bouaziz, S.: HITNet: hierarchical iterative tile refinement network for real-time stereo matching. In: The IEEE International Conference on Computer Vision and Pattern Recognition (CVPR), pp. 14362–14372 (2021)
16. Zagoruyko, S., Komodakis, N.: DiracNets: training very deep neural networks without skip-connections. In: arXiv pre-print arXiv: 1706.00388 (2018)

PU-Transformer: Point Cloud Upsampling Transformer

Shi Qiu[1,2(✉)], Saeed Anwar[1,2], and Nick Barnes[1]

[1] Australian National University, Canberra, Australia
{shi.qiu,saeed.anwar,nick.barnes}@anu.edu.au
[2] Data61-CSIRO, Sydney, Australia

Abstract. Given the rapid development of 3D scanners, point clouds are becoming popular in AI-driven machines. However, point cloud data is inherently sparse and irregular, causing significant difficulties for machine perception. In this work, we focus on the point cloud upsampling task that intends to generate dense high-fidelity point clouds from sparse input data. Specifically, to activate the transformer's strong capability in representing features, we develop a new variant of a multi-head self-attention structure to enhance both point-wise and channel-wise relations of the feature map. In addition, we leverage a positional fusion block to comprehensively capture the local context of point cloud data, providing more position-related information about the scattered points. As the first transformer model introduced for point cloud upsampling, we demonstrate the outstanding performance of our approach by comparing with the state-of-the-art CNN-based methods on different benchmarks quantitatively and qualitatively.

1 Introduction

3D computer vision has been attracting a wide range of interest from academia and industry since it shows great potential in many fast-developing AI-related applications such as robotics, autonomous driving, augmented reality, *etc.*. As a basic representation of 3D data, point clouds can be easily captured by 3D sensors [1,2], incorporating the rich context of real-world surroundings.

Unlike well-structured 2D images, point cloud data has inherent properties of *irregularity* and *sparsity*, posing enormous challenges for high-level vision tasks such as point cloud classification [3–5], segmentation [6–8], and object detection [9–11]. For instance, Uy *et al.* [12] fail to classify the real-world point clouds while they apply a pre-trained model of synthetic data; and recent 3D segmentation and detection networks [8,13,14] achieve *worse* results on the distant/smaller objects (*e.g.*, bicycles, traffic-signs) than the closer/larger objects (*e.g.*, vehicles, buildings). If we mitigate point cloud data's *irregularity* and *sparsity*, further improvements in visual analysis can be obtained (as verified in [15]). Thus, point cloud upsampling deserves a deeper investigation.

Supplementary Information The online version contains supplementary material available at https://doi.org/10.1007/978-3-031-26319-4_20.

L. Wang et al. (Eds.): ACCV 2022, LNCS 13841, pp. 326–343, 2023.
https://doi.org/10.1007/978-3-031-26319-4_20

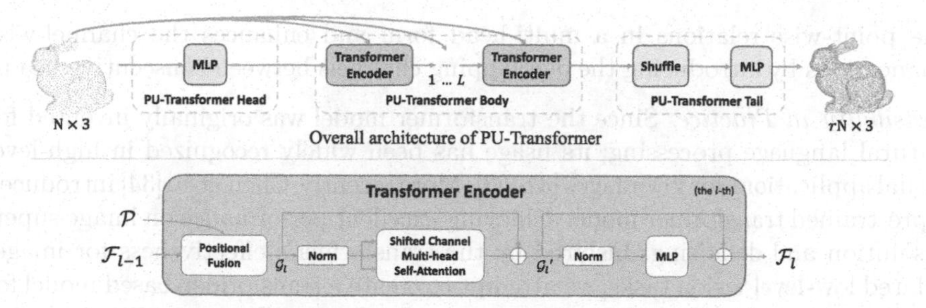

Fig. 1. The details of PU-Transformer. The upper chart shows the overall architecture of the PU-Transformer model containing three main parts: the PU-Transformer head (Sect. 4.1), body (Sect. 4.2), and tail (Sect. 4.3). The PU-Transformer body includes a cascaded set of Transformer Encoders (*e.g.*, L in total), serving as the core component of the whole model. Particularly, the detailed structure of each Transformer Encoder is shown in the lower chart, where all annotations are consistent with Line 3–5 in Algorithm 1.

As a basic 3D low-level vision task, point cloud upsampling aims to generate dense point clouds from sparse input, where the generated data should recover the fine-grained structures at a higher resolution. Moreover, the upsampled points are expected to lie on the underlying surfaces in a uniform distribution, benefiting downstream tasks for both 3D visual analysis [16,17] and graphic modeling [18,19]. Following the success of Convolution Neural Networks (CNNs) in image super-resolution [20–22] and Multi-Layer-Perceptrons (MLPs) in point cloud analysis [3,6], previous methods tended to upsample point clouds via complex network designs (*e.g.*, Graph Convolutional Network [23], Generative Adversarial Network [24]) and dedicated upsampling strategies (*e.g.*, progressive training [25], coarse-to-fine reconstruction [26], disentangled refinement [27]). As far as we are concerned, these methods share a key to point cloud upsampling: learning the representative features of given points to estimate the distribution of new points. Considering that regular MLPs have limited-expression and generalization capability, we need a more powerful tool to extract fine-grained point feature representations for high-fidelity upsampling. To this end, we introduce a succinct transformer model, PU-Transformer, to effectively upsample point clouds following a simple pipeline as illustrated in Fig. 1. The main reasons for adopting transformers to point cloud upsampling are as follows:

Plausibility in Theory. As the core operation of transformers, self-attention [28] is a set operator [29] calculating long-range dependencies between elements regardless of data order. On this front, self-attention can easily estimate the point-wise dependencies without any concern for the inherent *unorderedness*. However, to comprehensively represent point cloud features, channel-wise information is also shown to be a crucial factor in attention mechanisms [5,11]. Moreover, such channel-wise information enables an efficient upsampling via a simple periodic shuffling [30] operated on the channels of point features, saving complex designs [24–27] for upsampling strategy. Given these facts, we propose a Shifted Channel Multi-head Self-Attention (SC-MSA) block, which strengthens the point-wise relations in a multi-head form and enhances the channel-wise connections by introducing the overlapping channels between consecutive heads.

Feasibility in Practice. Since the transformer model was originally invented for natural language processing; its usage has been widely recognized in high-level visual applications for 2D images [31–33]. More recently, Chen *et al.* [34] introduced a pre-trained transformer model achieving excellent performance on image super-resolution and denoising. Inspired by the transformer's effectiveness for image-related low-level vision tasks, we attempt to create a transformer-based model for point cloud upsampling. Given the mentioned differences between 2D images and 3D point clouds, we introduce the Positional Fusion block as a replacement for positional encoding in conventional transformers: on the one hand, local information is aggregated from both the *geometric* and *feature* context of the points, implying their 3D positional relations; on the other hand, such *local* information can serve as complementary to subsequent self-attention operations, where the point-wise dependencies are calculated from a *global* perspective.

Adaptability in Various Applications. Transformer-based models are considered as a luxury tool in computer vision due to the huge consumption of data, hardware, and computational resources. However, our PU-Transformer can be easily trained with a *single* GPU in a few hours, retaining a similar model complexity to regular CNN-based point cloud upsampling networks [25,27,35]. Moreover, following a patch-based pipeline [25], the trained PU-Transformer model can effectively and flexibly upsample different types of point cloud data, including but not limited to regular object instances or large-scale LiDAR scenes (as shown in Fig. 3, 4 and 6). Starting with the upsampling task in low-level vision, we expect our approach to transformers will be affordable in terms of resource consumption for more point cloud applications. Our main contributions are:

- To the best of our knowledge, we are the first to introduce a transformer-based model[1] for point cloud upsampling.
- We quantitatively validate the effectiveness of the PU-Transformer by significantly outperforming the results of state-of-the-art point cloud upsampling networks on two benchmarks using three metrics.
- The upsampled visualizations demonstrate the superiority of PU-Transformer for diverse point clouds.

2 Related Work

Point Cloud Networks: In early research, the projection-based methods [36, 37] used to project 3D point clouds into multi-view 2D images, apply regular 2D convolutions and fuse the extracted information for 3D analysis. Alternatively, discretization-based approaches [38] tended to convert the point clouds to voxels [39] or lattices [40], and then process them using 3D convolutions or sparse tensor convolutions [41]. To avoid context loss and complex steps during data conversion, the point-based networks [3,4,6] directly process point cloud data via

[1] The project page is: https://github.com/ShiQiu0419/PU-Transformer.

MLP-based operations. Although current mainstream approaches in point cloud upsampling prefer utilizing MLP-related modules, in this paper, we focus on an advanced transformer structure [28] in order to further enhance the point-wise dependencies between known points and benefit the generation of new points.

Point Cloud Upsampling: Despite the fact that current point cloud research in low-level vision [35,42] is less active than that in high-level analysis [3,8,9], there exists many outstanding works that have contributed significant developments to the point cloud upsampling task. To be specific, PU-Net [35] is a pioneering work that introduced CNNs to point cloud upsampling based on a PointNet++ [6] backbone. Later, MPU [25] proposed a patch-based upsampling pipeline, which can flexibly upsample the point cloud patches with rich local details. In addition, PU-GAN [24] adopted the architecture of Generative Adversarial Networks [43] for the generation problem of high-resolution point clouds, while PUGeo-Net [44] indicated a promising combination of discrete differential geometry and deep learning. More recently, Dis-PU [27] applies disentangled refinement units to gradually generate the high-quality point clouds from coarse ones, and PU-GCN [23] achieves good upsampling performance by using graph-based network constructions [4]. Moreover, there are some papers exploring *flexible-scale* point cloud upsampling via meta-learning [15], self-supervised learning [45], decoupling ratio with network architecture [46], or interpolation [47], *etc.*. As the first work leveraging transformers for point cloud upsampling, we focus on the effectiveness of PU-Transformer in performing the fundamental *fixed-scale* upsampling task, and expect to inspire more future work in relevant topics.

Transformers in Vision: With the capacity in parallel processing as well as the scalability to deep networks and large datasets [48], more visual transformers have achieved excellent performance on image-related tasks including either low-level [34,49] or high-level analysis [31–33,50]. Due to the inherent gaps between 3D and 2D data, researchers introduce the variants of transformer for point cloud analysis [51–55], using vector-attention [29], offset-attention [56], and grid-rasterization [57], *etc.*. However, since these transformers still operate on an overall classical PointNet [3] or PointNet++ architecture [6], the improvement is relatively limited while the computational cost is too expensive for most researchers to re-implement. To simplify the model's complexity and boost its adaptability in point cloud upsampling research, we only utilize the general structure of transformer encoder [32] to form the body of our PU-Transformer.

3 Methodology

3.1 Overview

As shown in Fig. 1, given a sparse point cloud $\mathcal{P} \in \mathbb{R}^{N\times 3}$, our proposed PU-Transformer can generate a dense point cloud $\mathcal{S} \in \mathbb{R}^{rN\times 3}$, where r denotes the upsampling scale. Firstly, the PU-Transformer head extracts a preliminary feature map from the input. Then, based on the extracted feature map and the inherent 3D coordinates, the PU-Transformer body gradually encodes a more comprehensive feature map via the cascaded Transformer Encoders. Finally, in

Algorithm 1: PU-Transformer Pipeline

```
input: a sparse point cloud P ∈ R^{N×3}
output: a dense point cloud S ∈ R^{rN×3}
  # PU-Transformer Head
1 F_0 = MLP(P)
  # PU-Transformer Body
2 for each Transformer Encoder do
      # l = 1 ... L
      # the l-th Transformer Encoder
3     G_l = PosFus(P, F_{l-1});
4     G_l' = SC-MSA(Norm(G_l)) + G_l;
5     F_l = MLP(Norm(G_l')) + G_l';
6 end for
  # PU-Transformer Tail
7 S = MLP(Shuffle(F_L))
```

the PU-Transformer tail, we use the shuffle operation [30] to form a dense feature map and reconstruct the 3D coordinates of $\mathcal{S}$ via an MLP.

In Algorithm 1, we present the basic operations that are employed to build our PU-Transformer. As well as the operations ("MLP" [3], "Norm" [58], "Shuffle" [30]) that have been widely used in image and point cloud analysis, we propose two novel blocks targeting a transformer-based point cloud upsampling model *i.e.*, the Positional Fusion block ("**PosFus**" in Algorithm 1), and the Shifted-Channel Multi-head Self-Attention block ("**SC-MSA**" in Algorithm 1). In the rest of this section, we introduce these two blocks in detail. Moreover, for a compact description, we only consider the case of an *arbitrary* Transformer Encoder; thus, in the following, we discard the subscripts that are annotated in Algorithm 1 denoting a Transformer Encoder's specific index in the PU-Transformer body.

3.2 Positional Fusion

Usually, a point cloud consisting of N points has two main types of context: the 3D coordinates $\mathcal{P} \in \mathbb{R}^{N\times 3}$ that are explicitly sampled from synthetic meshes or captured by real-world scanners, showing the original geometric distribution of the points in 3D space; and the feature context, $\mathcal{F} \in \mathbb{R}^{N\times C}$, that is implicitly encoded by convolutional operations in C-dimensional embedding space, yielding rich latent clues for visual analysis. Older approaches [24,25,35] to point cloud upsampling generate a dense point set by heavily exploiting the encoded features $\mathcal{F}$, while recent methods [23,44] attempt to incorporate more geometric information. As the core module of the PU-Transformer, the proposed Transformer Encoder leverages a Positional Fusion block to encode and combine both the given $\mathcal{P}$ and $\mathcal{F}$[2] of a point cloud, following the local geometric relations between the scattered points.

Based on the metric of *3D-Euclidean distance*, we can search for neighbors $\forall p_j \in Ni(p_i)$ for each point $p_i \in \mathbb{R}^3$ in the given point cloud $\mathcal{P}$, using the k-nearest-neighbors (knn) algorithm [4]. Coupled with a grouping operation, we thus obtain a matrix $\mathcal{P}_j \in \mathbb{R}^{N\times k\times 3}$, denoting the 3D coordinates of the

[2] Equivalent to "$\mathcal{F}_{l-1}$" in Algorithm 1.

neighbors for all points. Accordingly, the relative positions between each point and its neighbors can be formulated as:

$$\Delta\mathcal{P} = \mathcal{P}_j - \mathcal{P}, \quad \Delta\mathcal{P} \in \mathbb{R}^{N\times k\times 3}; \tag{1}$$

where k is the number of neighbors. In addition to the neighbors' relative positions showing each point's local detail, we also append the centroids' positions in 3D space, indicating the global distribution for all points. By duplicating $\mathcal{P}$ in a dimension expanded k times, we concatenate the local *geometric* context:

$$\mathcal{G}_{geo} = \text{concat}\big[\underset{k}{\text{dup}}(\mathcal{P}); \Delta\mathcal{P}\big] \in \mathbb{R}^{N\times k\times 6}. \tag{2}$$

Further, for the feature matrix $\mathcal{F}_j \in \mathbb{R}^{N\times k\times C}$ of all searched neighbors, we conduct similar operations (Eq. 1 and 2) as on the counterpart $\mathcal{P}_j$, computing the relative features as:

$$\Delta\mathcal{F} = \mathcal{F}_j - \mathcal{F}, \quad \Delta\mathcal{F} \in \mathbb{R}^{N\times k\times C}; \tag{3}$$

and representing the local *feature* context as:

$$\mathcal{G}_{feat} = \text{concat}\big[\underset{k}{\text{dup}}(\mathcal{F}); \Delta\mathcal{F}\big] \in \mathbb{R}^{N\times k\times 2C}. \tag{4}$$

After the local *geometric* context $\mathcal{G}_{geo}$ and local *feature* context $\mathcal{G}_{feat}$ are constructed, we then fuse them for a comprehensive point feature representation. Specifically, $\mathcal{G}_{geo}$ and $\mathcal{G}_{feat}$ are encoded via two MLPs, $\mathcal{M}_\Phi$ and $\mathcal{M}_\Theta$, respectively; further, we comprehensively aggregate the local information, $\mathcal{G} \in \mathbb{R}^{N\times C'}$[3], using a concatenation between the encoded two types of local context, followed by a max-pooling function operating over the neighborhoods. The above operations can be summarized as:

$$\mathcal{G} = \max_k\Big(\text{concat}\big[\mathcal{M}_\Phi(\mathcal{G}_{geo}); \mathcal{M}_\Theta(\mathcal{G}_{feat})\big]\Big). \tag{5}$$

Unlike the local graphs in DGCNN [4] that need to be updated in every encoder based on the *dynamic* relations in embedding space, both of our $\mathcal{G}_{geo}$ and $\mathcal{G}_{feat}$ are constructed (*i.e.*, Eqs. 2 and 4) and encoded (*i.e.*, $\mathcal{M}_\Phi$ and $\mathcal{M}_\Theta$ in Eq. 5) in the same way, following *fixed* 3D geometric relations (*i.e.*, $\forall p_j \in Ni(p_i)$ defined upon *3D-Euclidean distance*). The main benefits of our approach can be concluded from two aspects: (i) it is practically efficient since the expensive knn algorithm just needs to be conducted once, while the searching results can be utilized in all Positional Fusion blocks of the PU-Transformer body; and (ii) the local *geometric* and *feature* context are represented in a similar manner following the same metric, contributing to *fairly fusing* the two types of context. A detailed behavior analysis of this block is provided in the supplementary material.

Overall, the Positional Fusion block can not only encode the positional information about a set of unordered points for the transformer's processing, but also aggregate comprehensive local details for accurate point cloud upsampling.

[3] Equivalent to "$\mathcal{G}_l$" in Algorithm 1.

Algorithm 2: Shifted Channel Multi-head Self-Attention (**SC-MSA**)

input: a point cloud feature map: $\mathcal{I} \in \mathbb{R}^{N \times C'}$

output: the refined feature map: $\mathcal{O} \in \mathbb{R}^{N \times C'}$

others: channel-wise split width: w; channel-wise shift interval: d, $d < w$; the number of heads: M

1. $\mathcal{Q} = \text{Linear}(\mathcal{I})$ # Query Mat $\mathcal{Q} \in \mathbb{R}^{N \times C'}$
2. $\mathcal{K} = \text{Linear}(\mathcal{I})$ # Key Mat $\mathcal{K} \in \mathbb{R}^{N \times C'}$
3. $\mathcal{V} = \text{Linear}(\mathcal{I})$ # Value Mat $\mathcal{V} \in \mathbb{R}^{N \times C'}$
4. **for** $m \in \{1, 2, ..., M\}$ **do**
5. $\quad \mathcal{Q}_m = \mathcal{Q}[\,:\,, (m-1)d : (m-1)d + w]$;
6. $\quad \mathcal{K}_m = \mathcal{K}[\,:\,, (m-1)d : (m-1)d + w]$;
7. $\quad \mathcal{V}_m = \mathcal{V}[\,:\,, (m-1)d : (m-1)d + w]$;
8. $\quad \mathcal{A}_m = \text{softmax}(\mathcal{Q}_m \mathcal{K}_m{}^T)$;
9. $\quad \mathcal{O}_m = \mathcal{A}_m \mathcal{V}_m$;
10. **end for**
11. **obtain:** $\{\mathcal{O}_1, \mathcal{O}_2, ..., \mathcal{O}_M\}$
12. $\mathcal{O} = \text{Linear}\Big(\text{concat}\big[\{\mathcal{O}_1, \mathcal{O}_2, ..., \mathcal{O}_M\}\big]\Big)$

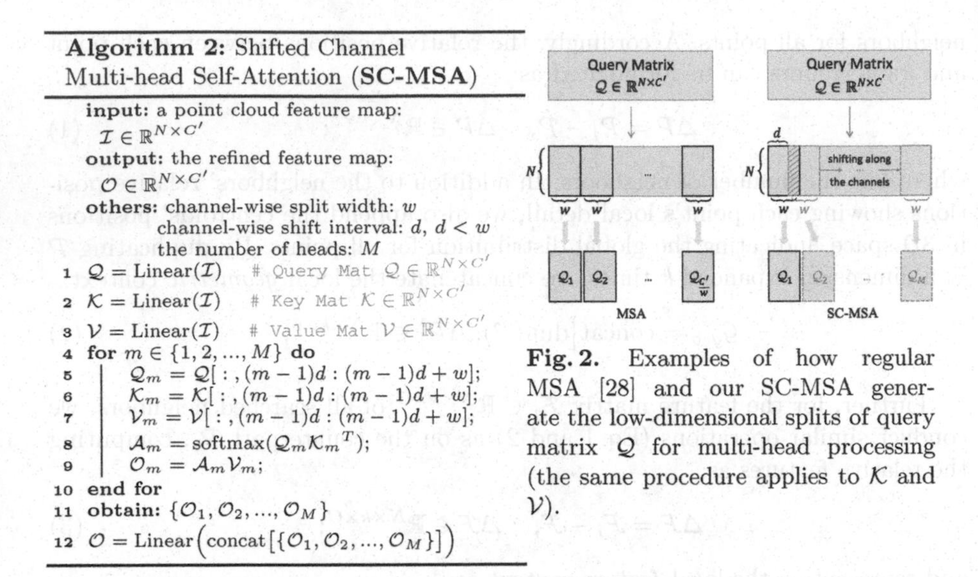

Fig. 2. Examples of how regular MSA [28] and our SC-MSA generate the low-dimensional splits of query matrix $\mathcal{Q}$ for multi-head processing (the same procedure applies to $\mathcal{K}$ and $\mathcal{V}$).

3.3 Shifted Channel Multi-head Self-attention

Different from previous works that applied complex upsampling strategies (*e.g.*, GAN [24], coarse-to-fine [26], task-disentangling [27]) to estimate new points, we prefer generating dense points in a simple way. Particularly, PixelShuffle [30] is a periodic shuffling operation that efficiently reforms the *channels* of each point feature to represent new points without introducing additional parameters. However, with regular multi-head self-attention (MSA) [28] serving as the main calculation unit in transformers, only *point-wise* dependencies are calculated in each independent head of MSA, lacking integration of *channel-related* information for shuffling-based upsampling. To tackle this issue, we introduce a Shifted Channel Multi-head Self-Attention (SC-MSA) block for the PU-Transformer.

As Algorithm 2 states, at first, we apply linear layers (denoted as "Linear", and implement as a 1×1 convolution) to encode the query matrix $\mathcal{Q}$, key matrix $\mathcal{K}$, and value matrix $\mathcal{V}$. Then, we generate low-dimensional splits of $\mathcal{Q}_m, \mathcal{K}_m, \mathcal{V}_m$ for each head. Particularly, as shown in Fig. 2, regular MSA generates the *independent* splits for the self-attention calculation in corresponding heads. In contrast, our SC-MSA applies a window (dashed square) shift along the channels to ensure that any two consecutive splits have an overlap of $(w-d)$ channels (slashed area), where w is the channel dimension of each split and d represents the channel-wise shift interval each time. After generating the $\mathcal{Q}_m, \mathcal{K}_m, \mathcal{V}_m$ for each head in the mentioned manner, we employ self-attention (Alg. 2 steps 8–9) to estimate the point-wise dependencies as the output $\mathcal{O}_m$ of each head. Considering the fact that any two consecutive heads have part of the input in common (*i.e.*, the overlap channels), thus the connections between the outputs $\{\mathcal{O}_1, \mathcal{O}_2, ..., \mathcal{O}_M\}$ (Algorithm 2 step 11) of multiple heads are established. There are two major benefits of such connections: (i) it is easier to integrate the information between the *connected* multi-head outputs (Algorithm 2 step 12), compared to using the *independent* multi-head results of regular MSA; and (ii) as the overlapping context is captured from the channel dimension, our SC-MSA can further enhance the channel-wise relations in the final output $\mathcal{O}$, better fulfilling an efficient and effective shuffling-based upsampling strategy than only using regular MSA's point-wise information. These benefits contribute to a faster training convergence and a better upsampling performance, especially when we deploy fewer Transformer Encoders. More practical evidence is provided in the supplementary material.

It is worth noting that SC-MSA requires the shift interval to be smaller than the channel-wise width of each split (*i.e.*, $d < w$ as in Algorithm 2) for a shared area between any two consecutive splits. Accordingly, the number of heads in our SC-MSA is higher than regular MSA (*i.e.*, $M > C'/w$ in Fig. 2). More implementation detail and the choices of parameters are provided in Sect. 4.2.

4 Implementation

4.1 PU-Transformer Head

As illustrated in Fig. 1, our PU-Transformer model begins with the head to encode a preliminary feature map for the following operations. In practice, we only use a single layer MLP (*i.e.*, a single 1×1 convolution, followed by a batch normalization layer [59] and a ReLU activation [60]) as the PU-Transformer head, where the generated feature map size is $N \times 16$.

4.2 PU-Transformer Body

To balance the model complexity and effectiveness, empirically, we leverage *five* cascaded Transformer Encoders (*i.e.*, $L = 5$ in Algorithm 1 and Fig. 1) to form the PU-Transformer body, where the channel dimension of each output follows: $32 \rightarrow 64 \rightarrow 128 \rightarrow 256 \rightarrow 256$. Particularly, in each Transformer Encoder, we only use the Positional Fusion block to encode the corresponding channel dimension (*i.e.*, C' in Eq. 5), which remains the same in the subsequent operations. For all Positional Fusion blocks, the number of neighbors is empirically set to $k = 20$ as used in previous works [4, 23].

In terms of the SC-MSA block, the primary way of choosing the shift-related parameters is inspired by the Non-local Network [61] and ECA-Net [62]. Specifically, a reduction ratio ψ [61] is introduced to generate the low-dimensional matrices in self-attention; following a similar method, the channel-wise width (*i.e.*, channel dimension) of each split in SC-MSA is set as $w = C'/\psi$. Moreover, since the channel dimension is usually set to a power of 2 [62], we simply set the channel-wise shift interval $d = w/2$. Therefore, the number of heads in SC-MSA becomes $M = 2\psi - 1$. In our implementation, $\psi = 4$ is adopted in all SC-MSA blocks of PU-Transformer.

4.3 PU-Transformer Tail

Based on the practical settings above, the input to the PU-Transformer tail (*i.e.*, the output of the last Transformer Encoder) has a size of $N \times 256$. Then, the periodic shuffling operation [30] reforms the channels and constructs a dense feature map of $rN \times 256/r$, where r is the upsampling scale. Finally, another MLP is applied to estimate the upsampled point cloud's 3D coordinates ($rN \times 3$).

5 Experiments

5.1 Settings

Training Details: In general, our PU-Transformer is implemented using Tensorflow [63] with a single GeForce 2080 Ti GPU running on the Linux OS. In terms of the hyperparameters for training, we heavily adopt the settings from PU-GCN [23] and Dis-PU [27] for the experiments in Table 1 and Table 2, respectively. For example, we have a batch size of 64 for 100 training epochs, an initial learning rate of 1×10^{-3} with a 0.7 decay rate, *etc.*. Moreover, we only use the modified Chamfer Distance loss [25] to train the PU-Transformer, minimizing the average closest point distance between the input set $\mathcal{P} \in \mathbb{R}^{N\times 3}$ and the output set $\mathcal{S} \in \mathbb{R}^{rN\times 3}$ for efficient and effective convergence.

Datasets: Basically, we apply two 3D benchmarks for our experiments:

- **PU1K:** This is a new point cloud upsampling dataset introduced in PU-GCN [23]. In general, the PU1K dataset incorporates 1,020 3D meshes for training and 127 3D meshes for testing, where most 3D meshes are collected from ShapeNetCore [64] covering 50 object categories. To fit in with the patch-based upsampling pipeline [25], the training data is generated from patches of 3D meshes via Poisson disk sampling. Specifically, the training data includes 69,000 samples, where each sample has 256 input points (low resolution) and a ground-truth of 1,024 points (4× high resolution).
- **PU-GAN Dataset:** This is an earlier dataset that was first used in PU-GAN [24] and generated in a similar way as PU1K but on a smaller scale. To be concrete, the training data comprises 24,000 samples (patches) collected from 120 3D meshes, while the testing data only contains 27 meshes. In addition to the PU1K dataset consisting of a large volume of data targeting the basic 4× upsampling experiment, we conduct both 4× and 16× upsampling experiments based on the compact data of the PU-GAN dataset.

Evaluation Metrics: As for the testing process, we follow common practice that has been utilized in previous point cloud upsampling works [23–25,27]. To be specific, at first, we cut the input point cloud into multiple seed patches covering all the N points. Then, we apply the trained PU-Transformer model to upsample the seed patches with a scale of r. Finally, the farthest point sampling algorithm [3] is used to combine all the upsampled patches as a dense output point cloud with rN points. For the 4× upsampling experiments in this paper, each testing sample has a low-resolution point cloud with 2,048 points, as well as a high-resolution one with 8,196 points. Coupled with the original 3D meshes, we quantitatively evaluate the upsampling performance of our PU-Transformer

Table 1. Quantitative comparisons (4× Upsampling) to state-of-the-art methods on the *PU1K* dataset [23]. ("**CD**": Chamfer Distance; "**HD**": Hausdorff Distance; "**P2F**": Point-to-Surface Distance. "**Model**": model size; "**Time**": average inference time per sample; "**Param.**": number of parameters. *: self-reproduced results, –: unknown data.)

Methods	Model (MB)	Time ($\times 10^{-3}$s)	Param. ($\times 10^{3}$)	Results ($\times 10^{-3}$) CD ↓	HD ↓	P2F ↓
PU-Net [35]	10.1	8.4	812.0	1.155	15.170	4.834
MPU [25]	6.2	8.3	76.2	0.935	13.327	3.551
PU-GACNet [66]	–	–	**50.7**	0.665	9.053	2.429
PU-GCN [23]	**1.8**	**8.0**	76.0	0.585	7.577	2.499
Dis-PU* [27]	13.2	10.8	1047.0	0.485	6.145	1.802
Ours	18.4	9.9	969.9	**0.451**	**3.843**	**1.277**

Table 2. Quantitative comparisons to state-of-the-art methods on the *PU-GAN* dataset [24]. (All metric units are 10^{-3}. The best results are denoted in **bold**.)

Methods	4× Upsampling			16× Upsampling		
	CD ↓	HD ↓	P2F ↓	CD ↓	HD ↓	P2F ↓
PU-Net [35]	0.844	7.061	9.431	0.699	8.594	11.619
MPU [25]	0.632	6.998	6.199	0.348	7.187	6.822
PU-GAN [24]	0.483	5.323	5.053	0.269	7.127	6.306
PU-GCN* [23]	0.357	5.229	3.628	0.256	5.938	3.945
Dis-PU [27]	0.315	4.201	4.149	**0.199**	4.716	4.249
Ours	**0.273**	**2.605**	**1.836**	0.241	**2.310**	**1.687**

based on three widely used metrics: (i) Chamfer Distance (CD), (ii) Hausdorff Distance [65] (HD), and (iii) Point-to-Surface Distance (P2F). A lower value under these metrics denotes better upsampling performance.

5.2 Point Cloud Upsampling Results

PU1K: Table 1 shows the quantitative results of our PU-Transformer on the PU1K dataset. It can be seen that our approach outperforms other state-of-the-art methods on all three metrics. In terms of the Chamfer Distance metric, we achieve the best performance among all the tested networks, since the reported values of others are all higher than ours of 0.451. Under the other two metrics, the improvements of PU-Transformer are particularly significant: compared to the performance of the recent PU-GCN [23], our approach can almost *halve* the values assessed under both the Hausdorff Distance (HD: 7.577 → 3.843) and the Point-to-Surface Distance (P2F: 2.499 → 1.277).

PU-GAN Dataset: We also conduct point cloud upsampling experiments using the dataset introduced in PU-GAN [24]. Under more upsampling scales. As shown in Table 2, we achieve best performance under all three evaluation metrics for the 4× upsampling experiment. However, in the 16× upsampling test, we (CD: 0.241) are slightly behind the latest Dis-PU network [27] (CD: 0.199) evaluated under the Chamfer Distance metric: the Dis-PU applies two CD-related items as its loss function, hence getting an edge for CD metric only. As for the results under Hausdorff Distance and Point-to-Surface Distance metrics, our

Table 3. Ablation study of the PU-Transformer's components tested on the *PU1K* dataset [23]. Specifically, models A_1-A_3 investigate the effects of the Positional Fusion block, models B_1-B_3 compare the results of different self-attention approaches, and models C_1-C_3 test the upsampling methods in the tail.

Models	PU-transformer body		PU-transformer tail	Results ($\times 10^{-3}$)		
	Positional fusion	Attention type		CD ↓	HD ↓	P2F ↓
A_1	None	SC-MSA	Shuffle	0.605	6.477	2.038
A_2	$\mathcal{G}_{geo}$	SC-MSA	Shuffle	0.558	5.713	1.751
A_3	$\mathcal{G}_{feat}$	SC-MSA	Shuffle	0.497	4.164	1.511
B_1	$\mathcal{G}_{geo}$ & $\mathcal{G}_{feat}$	SA [61]	Shuffle	0.526	4.689	1.492
B_2	$\mathcal{G}_{geo}$ & $\mathcal{G}_{feat}$	OSA [56]	Shuffle	0.509	4.823	1.586
B_3	$\mathcal{G}_{geo}$ & $\mathcal{G}_{feat}$	MSA [28]	Shuffle	0.498	4.218	1.427
C_1	$\mathcal{G}_{geo}$ & $\mathcal{G}_{feat}$	SC-MSA	MLPs [35]	1.070	8.732	2.467
C_2	$\mathcal{G}_{geo}$ & $\mathcal{G}_{feat}$	SC-MSA	DupGrid [25]	0.485	3.966	1.380
C_3	$\mathcal{G}_{geo}$ & $\mathcal{G}_{feat}$	SC-MSA	NodeShuffle [23]	0.505	4.157	1.404
Full	$\mathcal{G}_{geo}$ & $\mathcal{G}_{feat}$	SC-MSA	Shuffle	**0.451**	**3.843**	**1.277**

PU-Transformer shows significant improvements again, where some values (*e.g.*, P2F in 4×, HD and P2F in 16×) are even lower than *half* of Dis-PU's results.

Overall Comparison: The experimental results in Table 1 and 2 indicate the great effectiveness of our PU-Transformer. Moreover, given quantitative comparisons to CNN-based (*e.g.*, GCN [67], GAN [43]) methods under different metrics, we demonstrate the superiority of transformers for point cloud upsampling by only exploiting the fine-grained feature representations of point cloud data.

5.3 Ablation Studies

Effects of Components: Table 3 shows the experiments that replace PU-Transformer's major components with different options. Specifically, we test three simplified models (A_1–A_3) regarding the Positional Encoding block output (Eq. 5), where employing both local *geometric* $\mathcal{G}_{geo}$ and *feature* $\mathcal{G}_{feat}$ context (model "Full") provides better performance compared to the others. As for models B_1–B_3, we apply different self-attention approaches to the Transformer Encoder, where our proposed SC-MSA (Sect. 3.3) block shows higher effectiveness on point cloud upsampling. In terms of the upsampling method used in the PU-Transformer tail, some learning-based methods are evaluated as in models C_1–C_3. Particularly, with the help of our SC-MSA design, the simple yet efficient periodic shuffling operation (*i.e.*, PixelShuffle [30]) indicates good effectiveness in obtaining a high-resolution feature map.

Robustness to Noise: As the PU-Transformer can upsample different types of point clouds, including real scanned data, it is necessary to verify our model's robustness to noise. Concretely, we test the pre-trained models by adding some random noise to the sparse input data, where the noise is generated from a standard normal distribution $\mathcal{N}(0, 1)$ and multiplied with a factor β. In practice, we conduct the experiments under three noise levels: $\beta = 0.5\%$, 1% and 2%. Table 4 quantitatively compares the testing results of state-of-the-art methods.

Table 4. The model's robustness to random noise tested on the *PU1K* dataset [23], where the noise follows a normal distribution of $\mathcal{N}(0,1)$ and β is the noise level.

Methods	$\beta = 0.5\%$			$\beta = 1\%$			$\beta = 2\%$		
	CD ↓	HD ↓	P2F ↓	CD ↓	HD ↓	P2F ↓	CD ↓	HD ↓	P2F ↓
PU-Net [35]	1.006	14.640	5.253	1.017	14.998	6.851	1.333	19.964	10.378
MPU [25]	0.869	12.524	4.069	0.907	13.019	5.625	1.130	16.252	9.291
PU-GCN [23]	0.621	8.011	3.524	0.762	9.553	5.585	1.107	13.130	9.378
Dis-PU [27]	0.496	6.268	2.604	**0.591**	7.944	4.417	**0.858**	10.960	7.759
Ours	**0.453**	**4.052**	**2.127**	0.610	**5.787**	**3.965**	1.058	**9.948**	**7.551**

Table 5. Model Complexity of PU-Transformer using different numbers of Transformer Encoders. (Tested on the *PU1K* dataset [23] with a single GeForce 2080 Ti GPU.)

# Transformer Encoders	# Parameters	Model size	Training speed (per batch)	Inference speed (per sample)	Results ($\times 10^{-3}$)		
					CD ↓	HD ↓	P2F ↓
$L = 3$	438.3k	8.5M	12.2 s	6.9 ms	0.487	4.081	1.362
$L = 4$	547.3k	11.5M	15.9 s	8.2 ms	0.472	4.010	1.284
$L = 5$	969.9k	18.4M	23.5 s	9.9 ms	0.451	**3.843**	1.277
$L = 6$	2634.4k	39.8M	40.3 s	11.0 ms	**0.434**	3.996	**1.210**

In most tested noise cases, our proposed PU-Transformer achieves the best performance, while Dis-PU [27] shows robustness under the CD metric as explained in Sect. 5.2.

Model Complexity: Generally, our PU-Transformer is a light (<1M parameters) transformer model compared to image transformers [32,33,48] that usually have more than 50M parameters. In particular, we investigate the complexity of our PU-Transformer by utilizing different numbers of the Transformer Encoders. As shown in Table 5, with more Transformer Encoders being applied, the model complexity increases rapidly, while the quantitative performance improves slowly. For a better balance between effectiveness and efficiency, we adopt the model with *five* Transformer Encoders ($L = 5$) in this work. Overall speaking, the PU-Transformer is a powerful and affordable transformer model for the point cloud upsampling task.

5.4 Visualization

Qualitative Comparisons: The qualitative results of different point cloud upsampling models are presented in Figs. 3 and 4. Since we utilize the self-attention based structure to capture the point-wise dependencies from a global perspective, the PU-Transformer's output can better illustrate the overall contours of input point clouds producing fewer outliers (as shown in the zoom-in views of Fig. 3). Particularly, based on the rich local context encoded by our Positional Fusion block, the PU-Transformer precisely upsamples the real point clouds (compared in Fig. 4), retaining a uniform distribution and much structural detail.

Upsampling Different Input Sizes: Figure 5 shows the results of upsampling different sizes of point cloud data using PU-Transformer. Given a relatively low-resolution point cloud (*e.g.*, 256 or 512 input points), our proposed model is still

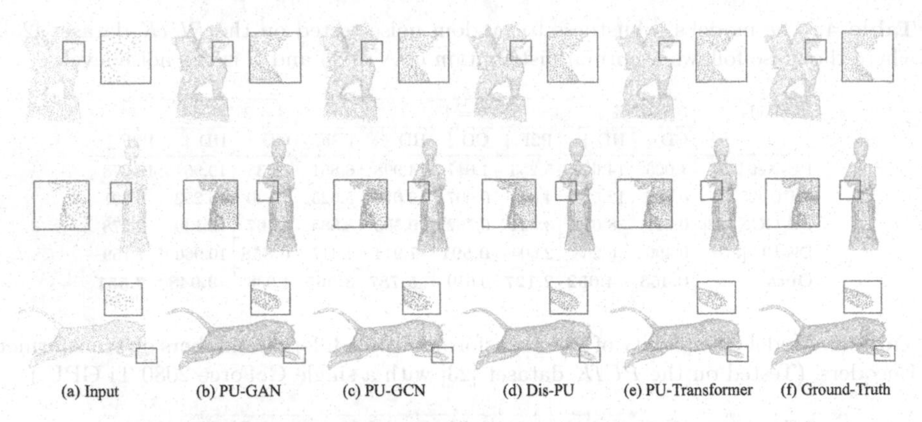

Fig. 3. Comparisons to state-of-the-art methods (PU-GAN [24], PU-GCN [23], Dis-PU [27]) in (4×) upsampling *synthetic* point cloud data using 2048 input points.

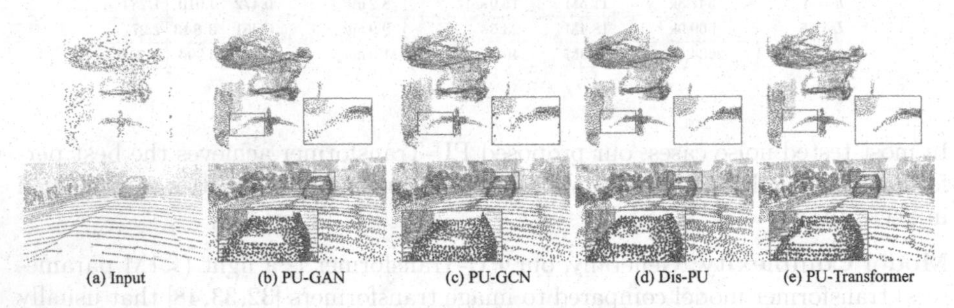

Fig. 4. Comparisons to state-of-the-art methods (PU-GAN [24], PU-GCN [23], Dis-PU [27]) in (4×) upsampling *real* point cloud data from ScanObjectNN [12] dataset and SemanticKITTI [68] dataset.

able to generate dense output with high-fidelity context (*e.g.*, the head/foot of "Panda"). As the input size increases, the new points are uniformly distributed, covering the main flat areas (*e.g.*, the body of "Panda").

Upsampling Real Point Clouds: In addition to Fig. 4, we provide more upsampling results (4× and 16×) on real point cloud samples (*i.e.*, "chair", "office", "room", "street") from *ScanObjectNN* [12], *S3DIS* [69], *ScanNet* [70], and *SemanticKITTI* [68], respectively. As Fig. 6 clearly illustrates, by addressing the sparsity and non-uniformity of raw inputs, not only is the overall quality of point clouds significantly improved, but also the representative features of object instances are enhanced. Particularly, the contours of upsampled object instances (*e.g.*, *tables* in "office/room", *cars* in "street") are clearly distinct from the complex surroundings, obtaining high-fidelity details for visual analysis. More examples for visualization are included in the supplementary material.

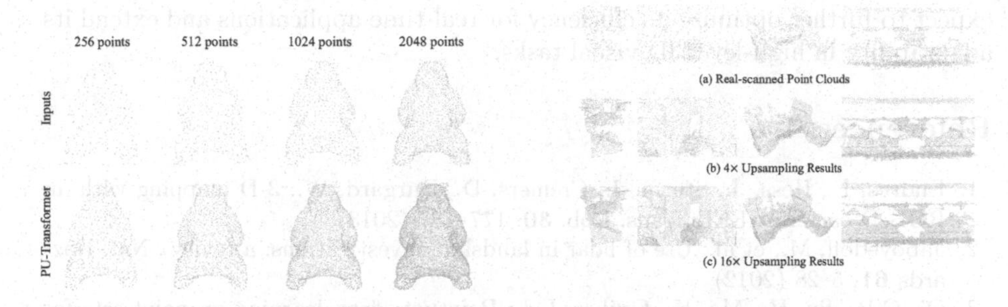

Fig. 5. PU-Transformer's 4× upsampling results, given different sizes of input point cloud data.

Fig. 6. PU-Transformer's 4× and 16× upsampling results, given different real point clouds.

6 Limitations and Future Work

Upsampling Efficiency: Compared to the recent works such as Point Transformer [29] (~7.76M parameters) or PoinTr [55] (~22.7M), PU-Transformer (~0.97M) is an efficient transformer for point clouds. However, it still consumes more parameters than some CNN-based counterparts [9,23,27,35] shown in Table 1. As for inference speed, our approach is very close to others due to the succinct pipeline design, while methods that exploit complex network [24], upsampling strategy [27] or geometric calculations [44] will be a bit slower.

Upsampling Flexibility: To generate different resolutions of output, our PU-Transformer may require some post-processing such as multiple inference iterations and farthest point sampling [3]. For flexible point cloud upsampling, in future work, we will improve the adaptability of the PU-Transformer's body.

Future Work: As a light-weight transformer targeting point clouds, our PU-Transformer has great potential in practice. For example, we could design a *multi-functional* tail to solve different low-level vision problems such as upsampling, completion, and denoising. Moreover, we could further optimize the efficiency of the PU-Transformer in learning fine-grained point feature representations, benefiting the high-level visual analysis of large-scale point clouds.

7 Conclusions

This paper focuses on low-level vision for point cloud data in order to tackle its inherent *sparsity* and *irregularity*. Specifically, we propose a novel transformer-based model, PU-Transformer, targeting the fundamental point cloud upsampling task. Our PU-Transformer shows significant quantitative and qualitative improvements on different point cloud datasets compared to state-of-the-art CNN-based methods. By conducting related ablation studies and visualizations, we also analyze the effects and robustness of our approach. In the future, we expect to further optimize its efficiency for real-time applications and extend its adaptability in high-level 3D visual tasks.

References

1. Endres, F., Hess, J., Sturm, J., Cremers, D., Burgard, W.: 3-D mapping with an RGB-D camera. IEEE Trans. Rob. **30**, 177–187 (2013)
2. Jaboyedoff, M., et al.: Use of lidar in landslide investigations: a review. Nat. Hazards **61**, 5–28 (2012)
3. Qi, C.R., Su, H., Mo, K., Guibas, L.J.: Pointnet: deep learning on point sets for 3D classification and segmentation. In: Proceedings of the IEEE Conference on Computer Vision and Pattern Recognition, pp. 652–660 (2017)
4. Wang, Y., Sun, Y., Liu, Z., Sarma, S.E., Bronstein, M.M., Solomon, J.M.: Dynamic graph CNN for learning on point clouds. ACM Trans. Graph. (TOG) **38**, 146 (2019)
5. Qiu, S., Anwar, S., Barnes, N.: Geometric back-projection network for point cloud classification. IEEE Trans. Multimedia (2021)
6. Qi, C.R., Yi, L., Su, H., Guibas, L.J.: Pointnet++: deep hierarchical feature learning on point sets in a metric space. In: Advances in Neural Information Processing Systems, pp. 5099–5108 (2017)
7. Qiu, S., Anwar, S., Barnes, N.: Dense-resolution network for point cloud classification and segmentation. In: Proceedings of the IEEE/CVF Winter Conference on Applications of Computer Vision (WACV), pp. 3813–3822 (2021)
8. Hu, Q., et al.: Randla-net: efficient semantic segmentation of large-scale point clouds. In: Proceedings of the IEEE/CVF Conference on Computer Vision and Pattern Recognition, pp. 11108–11117 (2020)
9. Qi, C.R., Litany, O., He, K., Guibas, L.J.: Deep Hough voting for 3D object detection in point clouds. In: Proceedings of the IEEE/CVF International Conference on Computer Vision, pp. 9277–9286 (2019)
10. Qi, C.R., Chen, X., Litany, O., Guibas, L.J.: Imvotenet: boosting 3D object detection in point clouds with image votes. In: Proceedings of the IEEE/CVF Conference on Computer Vision and Pattern Recognition, pp. 4404–4413 (2020)
11. Qiu, S., Wu, Y., Anwar, S., Li, C.: Investigating attention mechanism in 3D point cloud object detection. In: International Conference on 3D Vision (3DV). IEEE (2021)
12. Uy, M.A., Pham, Q.H., Hua, B.S., Nguyen, T., Yeung, S.K.: Revisiting point cloud classification: a new benchmark dataset and classification model on real-world data. In: Proceedings of the IEEE/CVF International Conference on Computer Vision, pp. 1588–1597 (2019)
13. Qiu, S., Anwar, S., Barnes, N.: Semantic segmentation for real point cloud scenes via bilateral augmentation and adaptive fusion. In: Proceedings of the IEEE/CVF Conference on Computer Vision and Pattern Recognition (CVPR), pp. 1757–1767 (2021)
14. Park, D., Ambrus, R., Guizilini, V., Li, J., Gaidon, A.: Is pseudo-lidar needed for monocular 3D object detection? In: Proceedings of the IEEE/CVF International Conference on Computer Vision (ICCV), pp. 3142–3152 (2021)
15. Ye, S., Chen, D., Han, S., Wan, Z., Liao, J.: Meta-PU: an arbitrary-scale upsampling network for point cloud. IEEE Trans. Visual. Comput. Graph. (2021)
16. Liu, Y., Fan, B., Xiang, S., Pan, C.: Relation-shape convolutional neural network for point cloud analysis. In: Proceedings of the IEEE Conference on Computer Vision and Pattern Recognition, pp. 8895–8904 (2019)

17. Qiu, S., Anwar, S., Barnes, N.: PNP-3D: a plug-and-play for 3d point clouds. arXiv preprint arXiv:2108.07378 (2021)
18. Mitra, N.J., Nguyen, A.: Estimating surface normals in noisy point cloud data. In: Proceedings of the Nineteenth Annual Symposium on Computational Geometry, pp. 322–328. ACM (2003)
19. Mitra, N.J., Gelfand, N., Pottmann, H., Guibas, L.: Registration of point cloud data from a geometric optimization perspective. In: Proceedings of the 2004 Eurographics/ACM SIGGRAPH Symposium on Geometry Processing, pp. 22–31. ACM (2004)
20. Dong, C., Loy, C.C., He, K., Tang, X.: Image super-resolution using deep convolutional networks. IEEE Trans. Pattern Anal. Mach. Intell. **38**, 295–307 (2015)
21. Kim, J., Lee, J.K., Lee, K.M.: Accurate image super-resolution using very deep convolutional networks. In: Proceedings of the IEEE Conference on Computer Vision and Pattern Recognition, pp. 1646–1654 (2016)
22. Anwar, S., Khan, S., Barnes, N.: A deep journey into super-resolution: a survey. ACM Comput. Surv. (CSUR) **53**, 1–34 (2020)
23. Qian, G., Abualshour, A., Li, G., Thabet, A., Ghanem, B.: PU-GCN: point cloud upsampling using graph convolutional networks. In: Proceedings of the IEEE/CVF Conference on Computer Vision and Pattern Recognition, pp. 11683–11692 (2021)
24. Li, R., Li, X., Fu, C.W., Cohen-Or, D., Heng, P.A.: PU-GAN: a point cloud upsampling adversarial network. In: Proceedings of the IEEE International Conference on Computer Vision, pp. 7203–7212 (2019)
25. Yifan, W., Wu, S., Huang, H., Cohen-Or, D., Sorkine-Hornung, O.: Patch-based progressive 3D point set upsampling. In: Proceedings of the IEEE/CVF Conference on Computer Vision and Pattern Recognition, pp. 5958–5967 (2019)
26. Liu, X., Liu, X., Han, Z., Liu, Y.S.: SPU-net: self-supervised point cloud upsampling by coarse-to-fine reconstruction with self-projection optimization. arXiv preprint arXiv:2012.04439 (2020)
27. Li, R., Li, X., Heng, P.A., Fu, C.W.: Point cloud upsampling via disentangled refinement. In: Proceedings of the IEEE/CVF Conference on Computer Vision and Pattern Recognition, pp. 344–353 (2021)
28. Vaswani, A., et al.: Attention is all you need. In: Advances in Neural Information Processing Systems, pp. 5998–6008 (2017)
29. Zhao, H., Jiang, L., Jia, J., Torr, P.H., Koltun, V.: Point transformer. In: Proceedings of the IEEE/CVF International Conference on Computer Vision, pp. 16259–16268 (2021)
30. Shi, W., et al.: Real-time single image and video super-resolution using an efficient sub-pixel convolutional neural network. In: Proceedings of the IEEE Conference on Computer Vision and Pattern Recognition, pp. 1874–1883 (2016)
31. Carion, N., Massa, F., Synnaeve, G., Usunier, N., Kirillov, A., Zagoruyko, S.: End-to-end object detection with transformers. In: Vedaldi, A., Bischof, H., Brox, T., Frahm, J.-M. (eds.) ECCV 2020. LNCS, vol. 12346, pp. 213–229. Springer, Cham (2020). https://doi.org/10.1007/978-3-030-58452-8_13
32. Dosovitskiy, A., et al.: An image is worth 16x16 words: transformers for image recognition at scale. arXiv preprint arXiv:2010.11929 (2020)
33. Liu, Z., et al.: Swin transformer: hierarchical vision transformer using shifted windows. In: Proceedings of the IEEE/CVF International Conference on Computer Vision (ICCV), pp. 10012–10022 (2021)
34. Chen, H., et al.: Pre-trained image processing transformer. In: Proceedings of the IEEE/CVF Conference on Computer Vision and Pattern Recognition, pp. 12299–12310 (2021)

35. Yu, L., Li, X., Fu, C.W., Cohen-Or, D., Heng, P.A.: PU-net: point cloud upsampling network. In: Proceedings of the IEEE Conference on Computer Vision and Pattern Recognition, pp. 2790–2799 (2018)
36. Su, H., Maji, S., Kalogerakis, E., Learned-Miller, E.: Multi-view convolutional neural networks for 3D shape recognition. In: Proceedings of the IEEE International Conference on Computer Vision, pp. 945–953 (2015)
37. Lawin, F.J., Danelljan, M., Tosteberg, P., Bhat, G., Khan, F.S., Felsberg, M.: Deep projective 3D semantic segmentation. In: Felsberg, M., Heyden, A., Krüger, N. (eds.) CAIP 2017. LNCS, vol. 10424, pp. 95–107. Springer, Cham (2017). https://doi.org/10.1007/978-3-319-64689-3_8
38. Guo, Y., Wang, H., Hu, Q., Liu, H., Liu, L., Bennamoun, M.: Deep learning for 3D point clouds: a survey. IEEE Trans. Pattern Anal. Mach. Intell. (2020)
39. Huang, J., You, S.: Point cloud labeling using 3D convolutional neural network. In: 2016 23rd International Conference on Pattern Recognition (ICPR), pp. 2670–2675 IEEE (2016)
40. Su, H., et al.: Splatnet: sparse lattice networks for point cloud processing. In: Proceedings of the IEEE Conference on Computer Vision and Pattern Recognition, pp. 2530–2539 (2018)
41. Choy, C., Gwak, J., Savarese, S.: 4D spatio-temporal convnets: Minkowski convolutional neural networks. In: Proceedings of the IEEE/CVF Conference on Computer Vision and Pattern Recognition, pp. 3075–3084 (2019)
42. Yuan, W., Khot, T., Held, D., Mertz, C., Hebert, M.: PCN: point completion network. In: 2018 International Conference on 3D Vision (3DV), pp. 728–737. IEEE (2018)
43. Goodfellow, I., et al.: Generative adversarial nets. In: Advances in Neural Information Processing Systems, vol. 27 (2014)
44. Qian, Y., Hou, J., Kwong, S., He, Y.: PUGeo-net: a geometry-centric network for 3D point cloud upsampling. In: Vedaldi, A., Bischof, H., Brox, T., Frahm, J.-M. (eds.) ECCV 2020. LNCS, vol. 12364, pp. 752–769. Springer, Cham (2020). https://doi.org/10.1007/978-3-030-58529-7_44
45. Zhao, Y., Hui, L., Xie, J.: SSPU-net: self-supervised point cloud upsampling via differentiable rendering. In: Proceedings of the 29th ACM International Conference on Multimedia, pp. 2214–2223 (2021)
46. Luo, L., Tang, L., Zhou, W., Wang, S., Yang, Z.X.: PU-EVA: an edge-vector based approximation solution for flexible-scale point cloud upsampling. In: Proceedings of the IEEE/CVF International Conference on Computer Vision, pp. 16208–16217 (2021)
47. Qian, Y., Hou, J., Kwong, S., He, Y.: Deep magnification-flexible upsampling over 3D point clouds. IEEE Trans. Image Process. **30**, 8354–8367 (2021)
48. Khan, S., Naseer, M., Hayat, M., Zamir, S.W., Khan, F.S., Shah, M.: Transformers in vision: a survey. arXiv preprint arXiv:2101.01169 (2021)
49. Yang, F., Yang, H., Fu, J., Lu, H., Guo, B.: Learning texture transformer network for image super-resolution. In: Proceedings of the IEEE/CVF Conference on Computer Vision and Pattern Recognition, pp. 5791–5800 (2020)
50. Zhu, X., Su, W., Lu, L., Li, B., Wang, X., Dai, J.: Deformable DETR: deformable transformers for end-to-end object detection. arXiv preprint arXiv:2010.04159 (2020)
51. Yew, Z.J., Lee, G.H.: REGTR: end-to-end point cloud correspondences with transformers. In: Proceedings of the IEEE/CVF Conference on Computer Vision and Pattern Recognition, pp. 6677–6686 (2022)

52. Fan, H., Yang, Y., Kankanhalli, M.: Point 4D transformer networks for spatio-temporal modeling in point cloud videos. In: Proceedings of the IEEE/CVF Conference on Computer Vision and Pattern Recognition, pp. 14204–14213 (2021)
53. Yu, X., Tang, L., Rao, Y., Huang, T., Zhou, J., Lu, J.: Point-BERT: pre-training 3D point cloud transformers with masked point modeling. In: Proceedings of the IEEE/CVF Conference on Computer Vision and Pattern Recognition, pp. 19313–19322 (2022)
54. Fan, H., Yang, Y., Kankanhalli, M.: Point spatio-temporal transformer networks for point cloud video modeling. IEEE Trans. Pattern Anal. Mach. Intell. (2022)
55. Yu, X., Rao, Y., Wang, Z., Liu, Z., Lu, J., Zhou, J.: PointR: diverse point cloud completion with geometry-aware transformers. In: Proceedings of the IEEE/CVF International Conference on Computer Vision, pp. 12498–12507 (2021)
56. Guo, M.H., Cai, J.X., Liu, Z.N., Mu, T.J., Martin, R.R., Hu, S.M.: PCT: point cloud transformer. Comput. Vis. Media **7**, 187–199 (2021)
57. Mazur, K., Lempitsky, V.: Cloud transformers: a universal approach to point cloud processing tasks. In: Proceedings of the IEEE/CVF International Conference on Computer Vision, pp. 10715–10724 (2021)
58. Ba, J.L., Kiros, J.R., Hinton, G.E.: Layer normalization. arXiv preprint arXiv:1607.06450 (2016)
59. Ioffe, S., Szegedy, C.: Batch normalization: accelerating deep network training by reducing internal covariate shift. arXiv preprint arXiv:1502.03167 (2015)
60. Nair, V., Hinton, G.E.: Rectified linear units improve restricted Boltzmann machines. In: ICML (2010)
61. Wang, X., Girshick, R., Gupta, A., He, K.: Non-local neural networks. In: Proceedings of the IEEE Conference on Computer Vision and Pattern Recognition, pp. 7794–7803 (2018)
62. Wang, Q., Wu, B., Zhu, P., Li, P., Zuo, W., Hu, Q.: ECA-net: efficient channel attention for deep convolutional neural networks. In: IEEE/CVF Conference on Computer Vision and Pattern Recognition (CVPR) (2020)
63. Abadi, M., et al.: Tensorflow: a system for large-scale machine learning. In: 12th {USENIX} Symposium on Operating Systems Design and Implementation ({OSDI} 16) (2016) 265–283
64. Chang, A.X., et al.: Shapenet: an information-rich 3D model repository. arXiv preprint arXiv:1512.03012 (2015)
65. Berger, M., Levine, J.A., Nonato, L.G., Taubin, G., Silva, C.T.: A benchmark for surface reconstruction. ACM Trans. Graph. (TOG) **32**, 1–17 (2013)
66. Han, B., Zhang, X., Ren, S.: PU-GACnet: graph attention convolution network for point cloud upsampling. Image Vision Comput. 104371 (2022)
67. Li, G., Muller, M., Thabet, A., Ghanem, B.: DeepGCNs: can GCNs go as deep as CNNs? In: Proceedings of the IEEE/CVF International Conference on Computer Vision, pp. 9267–9276 (2019)
68. Behley, J., et al.: Semantickitti: a dataset for semantic scene understanding of lidar sequences. In: Proceedings of the IEEE International Conference on Computer Vision, pp. 9297–9307 (2019)
69. Armeni, I., Sax, S., Zamir, A.R., Savarese, S.: Joint 2D-3D-semantic data for indoor scene understanding. arXiv preprint arXiv:1702.01105 (2017)
70. Dai, A., Chang, A.X., Savva, M., Halber, M., Funkhouser, T., Nießner, M.: Scannet: richly-annotated 3d reconstructions of indoor scenes. In: Proceedings of the IEEE Conference on Computer Vision and Pattern Recognition, pp. 5828–5839 (2017)

DIG: Draping Implicit Garment over the Human Body

Ren Li[1](✉), Benoît Guillard[1], Edoardo Remelli[2], and Pascal Fua[1]

[1] CVLab, EPFL, Zurich, Switzerland
{ren.li,benoit.guillard,pascal.fua}@epfl.ch
[2] Meta Reality Labs Research, Zurich, Switzerland
edoremelli@fb.com

Abstract. Existing data-driven methods for draping garments over human bodies, despite being effective, cannot handle garments of arbitrary topology and are typically not end-to-end differentiable. To address these limitations, we propose an end-to-end differentiable pipeline that represents garments using implicit surfaces and learns a skinning field conditioned on shape and pose parameters of an articulated body model. To limit body-garment interpenetrations and artifacts, we propose an interpenetration-aware pre-processing strategy of training data and a novel training loss that penalizes self-intersections while draping garments. We demonstrate that our method yields more accurate results for garment reconstruction and deformation with respect to state of the art methods. Furthermore, we show that our method, thanks to its end-to-end differentiability, allows to recover body and garments parameters jointly from image observations, something that previous work could not do. Our code is available at https://github.com/liren2515/DIG.

1 Introduction

Modeling clothed humans has applications in industries such as fashion design, moviemaking, and video gaming. Many professional tools that rely on Physics-Based Simulation (PBS) [10,24,25,35] can be used to model cloth deformations realistically. However, they are computationally expensive, which precludes real-time use. Some of these can operate in near real-time using an incremental approach in motion sequences. However, these methods remain too slow for static cloth draping over a body in an arbitrary pose.

In recent years, there has therefore been considerable interest in using data-driven techniques to overcome these difficulties. They fall into two main categories. First there are those that use a single model to jointly represent the person and their clothes [6,7,21,32,37]. They produce visually appealing results but,

This work was supported in part by the Swiss National Science Foundation.

Supplementary Information The online version contains supplementary material available at https://doi.org/10.1007/978-3-031-26319-4_21.

L. Wang et al. (Eds.): ACCV 2022, LNCS 13841, pp. 344–359, 2023.
https://doi.org/10.1007/978-3-031-26319-4_21

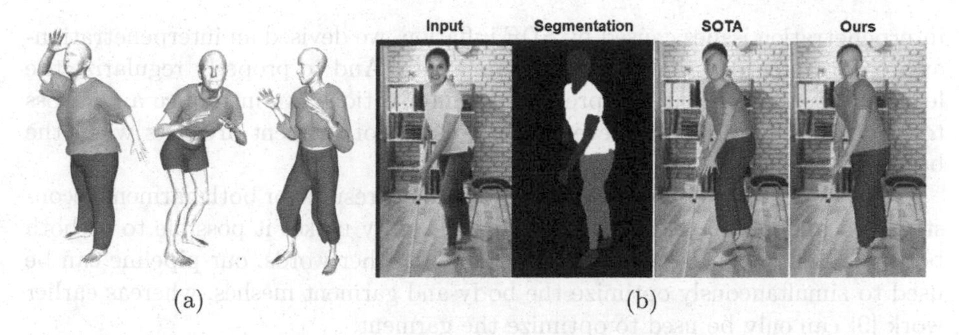

Fig. 1. We introduce a pipeline for (a) generating and draping garments with various topology plausibly and (b) recovering garments from image observations (e.g. segmentation masks). Unlike prior works, our method allows for joint optimization of garment and body meshes, resulting in more faithful reconstruction.

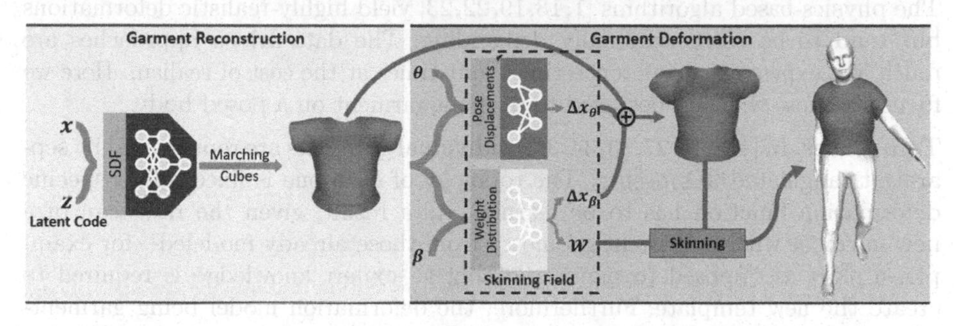

Fig. 2. The pipeline of our approach. The garment in the canonical space is first reconstructed from SDF. Given the shape β and pose θ of the target body, we add the shape and pose displacements (Δx_β and Δx_θ) to the reconstructed garment and drape it to the target body by the skinning function.

because the body and garment are bound together, they do not make it easy to mix and match different bodies and clothing articles. Second, there are methods that represent the body and clothes separately. For example, in [2,14,27,33,36], deep learning is used to define skinning functions that can be used to deform the garments according to body motion. In [9], the explicit representation of clothes is replaced by an implicit one that relies on an inflated SDF surrounding the garment surface. It makes it possible to represent garments with many different topologies using a single model. To this end, it relies on the fact that garments follow the underlying body pose predictably. Hence, for each garment vertex, it uses the blending weights of the closest body vertex in the SMPL model [20]. Unfortunately, this step involves a search, which makes it both computationally expensive and non-differentiable.

In this paper, we propose the novel data-driven approach to skinning depicted by Fig. 2. As in [9], we represent the garments in terms of an inflated SDF but, instead of using the SMPL skinning model, we learn a garment-specific one. This makes our approach both more expressive and fully-differentiable. To address the interpenetration issues caused by SDF inflation, we devised an interpenetration-aware data preprocessing for our training data. And to properly regularize the learned skinning field and to prevent self-intersections, we introduce a new loss term whose minimization prevents the creation of garment artifacts when the body deforms.

As a result, our method yields state-of-the-art results for both garment reconstruction and deformation. Its full differentiability makes it possible to fit both body and garments to partial observations. In other words, our pipeline can be used to simultaneously optimize the body and garment meshes, whereas earlier work [9] can only be used to optimize the garment.

2 Related Work

Most garment deformation approaches are either physics-based or data-driven. The physics-based algorithms [1,18,19,22,23] yield highly-realistic deformations but tend to be computationally demanding. The data-driven approaches are much less expensive at inference-time, sometimes at the cost of realism. Here we focus on those that are designed to drape a garment on a posed body.

Templates. In [4,5,15,27,33,34,36], individual garments are represented by separate triangulated 3D meshes. The topology of each one is fixed and a specific deformation function has to be learned. As a result, given the raw scan of a new garment with a different geometry from those already modeled—for example, a skirt as opposed to pants and shorts—expert knowledge is required to create the new template. Furthermore, the deformation model being garment-dependent makes these approaches impractical on large arrays of garments and, hence, ill-suited to real-world applications.

Point-Clouds. In [14] and DeePSD [2], the meshes are replaced by clouds of 3D points. The deformation is estimated for each point separately, making it possible to animate outfits of arbitrary topology and geometric complexity. However, the garment topology of these work is still non-differentiable because they rely on vertex connections from the template, which are fixed and pre-designed. This is addressed in [39] by using a point-cloud template with a fixed number of points densely sampled from the body mesh. This yields differentiability but the lack of point connections makes the reconstructed garments a group of unordered points instead of a surface with concrete physical properties.

Implicit Functions. Deep implicit functions [8,26] are good at representing surfaces whose topology can change while preserving differentiability [13,30]. SMPLicit [9] is the only work we know of that takes advantage of this to drape garments over bodies. As a result, the model can be fitted to real-world images. However, SMPLicit suffers several limitations. First, it does not handle the interpenetration between the body and garment. Second, it directly uses the blending weights of the closest vertices in the body model [20], which oversimplifies the dynamics and yields over-smoothed results. Finally, the optimization routines used to solve the fitting problem include approximations that produce inaccuracies and prevent the fitting result from being optimal. Our approach is in a similar spirit but overcomes these limitations.

3 Method

We start from an implicit surface model of the garment in a canonical pose, that is, draped over an average body in a T-pose, which we then deform to fit different body shapes and poses. This yields a fully differentiable pipeline that can be used for animation and modeling from images.

3.1 Garment Representation

Watertight surfaces of arbitrary topology can be represented very effectively by the zero crossings of a signed distance function (SDF)

$$f_\Theta(\mathbf{x}, \mathbf{z}) \longrightarrow \mathbb{R}\,, \tag{1}$$

where f is implemented by a neural network with weights Θ, $\mathbf{x} \in \mathbb{R}^3$ is a point in space, and $\mathbf{z}$ is a latent vector that parameterizes the surface shape [26]. However, clothes have openings in them and are not watertight. To nevertheless represent them in this manner, we can first compute unsigned distances to the surfaces, subtract a small ϵ value and treat the result as a signed distance function. This amounts to *inflating* the garments and representing them as watertight thin surfaces of thickness 2ϵ, as in [9,12]. Note that ϵ cannot be too small and must be larger than marching cube's step size, introducing an undesirable dependency between the field and how it is meshed.

Given a database of garments fitted to a body in a T-pose shape and whose vertices coordinates have been normalized to be between -1 and 1, we use an auto-decoding approach to learning the weights Θ and the latent vectors $\mathbf{z}$ associated to specific garments. To this end, for each sample garment and its associated latent vector $\mathbf{z}$, we minimize a loss function

$$Loss = L_{SDF} + \lambda_{grad} L_{grad} + \lambda_{reg} \|\mathbf{z}\|^2\,, \tag{2}$$

$$L_{SDF} = \sum_{\mathbf{x} \in X_v} \|f_\Theta(\mathbf{x}, \mathbf{z}) - s^{gt}(\mathbf{x})\|\,, \tag{3}$$

$$L_{grad} = \sum_{\mathbf{x} \in X_s} \|\nabla_x f_\Theta(\mathbf{x}, \mathbf{z}) - \mathbf{n}^{gt}(\mathbf{x})\|^2 + \sum_{\mathbf{x} \notin X_s} (\|\nabla_x f_\Theta(\mathbf{x}, \mathbf{z})\| - 1)^2\,, \tag{4}$$

where s^{gt} and $\mathbf{n}^{gt}$ are ground-truth values of the signed distance function and normal, X_v and X_s represent points sampled in the $[-1,1]^3$ volume and the garment surface respectively, and λ_{grad} and λ_{reg} are scalars that control the influence of the different terms. Minimizing L_{SDF} ensures that the SDF estimated by f_Θ is close to the ground-truth one in the whole volume while minimizing L_{grad} gives additional emphasis to it producing the right normals close to the surface and being a true SDF with unit gradients elsewhere, as in [11]. We present an ablation study in the results section that shows that both are necessary to produce smooth and accurate surfaces.

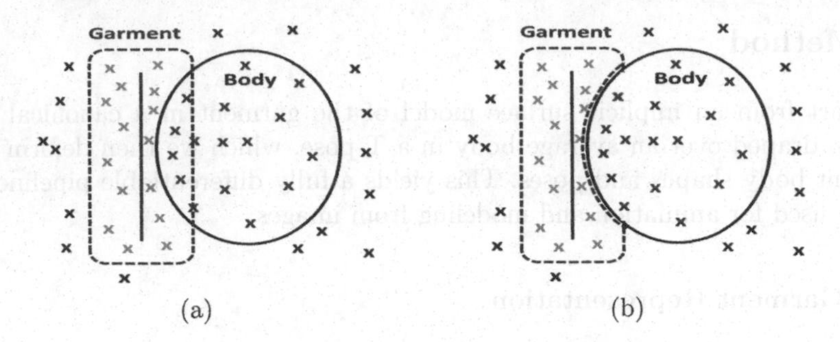

Fig. 3. The illustration of inflation processing for the garment surface (blue solid lines). (a) The inflation strategy of [9,12] will cause interpenetration between the inflated mesh (blue dashed line) and the body mesh, while (b) our proposed interpenetration-aware inflation will not. (Color figure online)

One difficulty with this scheme arises from the fact that the garment is usually close to the underlying body mesh and inflating it by ϵ results in interpenetrations between garment and body, as shown in Fig. 3(a). Intersections between garments and the human body are problematic because they do not allow to employ the reconstructed meshes for downstream tasks such as e.g. physics simulations. Furthermore, in the experiment section, we show that learning a physically correct representation of garments where there are no interpenetration results in more accurate clothing deformations. To address this, we perform the interpenetration-aware pre-processing illustrated by Fig. 3(b) when sampling the surface points in the X_s set of Eq. 4. Given a garment mesh G, we sample a $256 \times 256 \times 256$ grid in $[-1,1]^3$ to produce a set of points X and compute their signed distance to G. We then run Marching Cubes to recover the watertight mesh $M_{initial}$ as the dashed line of Fig. 3(a). For any vertex of $M_{initial}$ whose signed distance to the body is negative—meaning that it is inside it—we find the closest body vertex v_c and replace its position by $v_c + \mu \mathbf{n}_{v_c}$, where $\mathbf{n}_{v_c}$ is the surface normal at v_c and μ is a small positive value, which finally gives us the mesh M_{clean} without interpenetrations depicted by the blue dashed line of Fig. 3(b). In this example, X_s consists of points sampled from M_{clean} located on that dashed line. X_v comprises the points randomly sampled from $[-1,1]^3$. Their position is not affected but their ground-truth signed distance is computed with respect to M_{clean}.

3.2 Modeling Garment Deformations

SMPL is a statistical parametric model that uses Linear Blend Skinning to deform a rigged body template $\mathbf{T} \in \mathbb{R}^{N_B \times 3}$ with N_B vertices. Given the parameters of shape β and pose θ, SMPL can generate the body mesh $M_B(\beta, \theta)$ by

$$M_B(\beta, \theta) = W(T_B(\beta, \theta), J(\beta), \theta, \mathcal{W}) , \quad (5)$$

$$T_B(\beta,\theta) = \mathbf{T} + B_s(\beta) + B_p(\theta) , \tag{6}$$

where W is the skinning function with weight $\mathcal{W} \in \mathbb{R}^{N_B \times 24}$ and joint locations $J(\beta) \in \mathbb{R}^{24\times 3}$. $B_s(\beta) \in \mathbb{R}^{N_B\times 3}$ and $B_p(\theta) \in \mathbb{R}^{N_B\times 3}$ are the shape and pose displacements. The SMPLicit algorithm [9] exploits the fact that the garment follows the pose of the underlying body in a predictable way by using for each garment vertex the blending weights of the closest body vertex. This step involves a search, which makes it both computationally expensive and non-differentiable.

To remedy this, we instead learn a specific blending model for the garment, which is different from that of the body. More specifically, we write

$$\begin{aligned} M_G(\mathbf{x},\beta,\theta) &= W(\mathbf{x}_{(\beta,\theta)}, J(\beta), \theta, \mathcal{W}(\mathbf{x})) , \\ \mathbf{x}_{(\beta,\theta)} &= \mathbf{x} + \Delta x_\beta(\mathbf{x}) + \Delta x_\theta(\mathbf{x}) , \end{aligned} \tag{7}$$

where $W(\cdot)$ is the SMPL skinning function with learned skinning weights $\mathcal{W}(x) \in \mathbb{R}^{24}$, $\Delta x_\beta(\mathbf{x})$ and $\Delta x_\theta(\mathbf{x})$ are shape and pose displacements, and $\mathbf{x} \in \mathbb{R}^3$ denotes a generic 3D point instead of on a template. $\Delta x_\beta(\mathbf{x})$ models the shape offset conditioned on body shape β, while $\Delta x_\theta(\mathbf{x})$ represents a deformation field conditioned on body pose θ.

More specifically, $\mathcal{W}(\mathbf{x})$ and $\Delta x_\beta(\mathbf{x})$ are computed using the skinning weight $\mathcal{W}$ and shape displacement $B_s(\beta)$ from SMPL as base priors. They are extended to the whole 3D volume by writing

$$\mathcal{W}(\mathbf{x}) = w(\mathbf{x})\mathcal{W}, \quad \Delta x_\beta(\mathbf{x}) = w(\mathbf{x}) B_s(\beta), \tag{8}$$

where $w(\mathbf{x}) \in \mathbb{R}^{N_B}$ are shared weights. Since $w(\cdot)$ is implemented by a neural network and $W(\cdot)$ is a differentiable function, M_G is fully differentiable, unlike the SMPLicit model [9]. The approach of [33] does something similar but in a more complex manner because it needs to learn separate models for blending weights and shape displacement, whereas we need only one. Furthermore, because $\mathbf{x}$ can be *any* 3D point, we can deform garments of arbitrary topology, instead of being restricted to a single garment template as in [15,27,33].

3.3 Training the Model

To train the network that implements the function w of Eq. 8, we use the same sampling strategy as in [33] to collect target $\bar{w}(x)$ values. For each $\mathbf{x} \in \mathbb{R}^3$, we sample N points $\mathcal{P} = \{\mathbf{p} : \mathbf{p} \sim \mathcal{N}(\mathbf{x}, d)\}$, where d is the distance from $\mathbf{x}$ to the body. We take $\bar{w}(\mathbf{x})$ to be

$$\bar{w}(\mathbf{x}) = \frac{1}{N} \sum_{\mathbf{p}\in\mathcal{P}} w_{bary}(\phi(\mathbf{p})), \tag{9}$$

where $\phi(\cdot)$ denotes the closest point on the body surface and $w_{bary}(\cdot)$ is a N_B-vector that uses the barycentric coordinate of the closest point as the weight for each body vertex. Since $\bar{w}(\mathbf{x})$ can be regarded as the weight distribution of body vertices, at training time, we introduce the loss

$$L_{KL} = \sum_x KL(w(\mathbf{x}) || \bar{w}(\mathbf{x})) , \tag{10}$$

where KL is the KL-divergence. After the training of $w(x)$, we fix its parameter weights, plug it into our skinning model (Eq. 7), and then minimize the following loss for the training of Δx_θ

$$Loss = \lambda_{deform} L_{deform} + \lambda_{interp} L_{interp} + \lambda_{order} L_{order}. \tag{11}$$

where L_{interp} and L_{order} are regularization terms described below and λ_{deform}, λ_{interp}, and λ_{order} are scalar weights.

Dynamics. To capture detailed dynamics induced by pose changing, we define the deformation loss

$$L_{deform} = \sum_{\mathbf{x} \in X_s} |\bar{x}_d - \hat{x_d}(\mathbf{x})| + \sum_{\mathbf{x} \notin X_s} |\Delta x_\theta(\mathbf{x}) - \Delta x_\theta(\mathbf{x}_c)| \,, \tag{12}$$

where X_s denotes vertices of the ground-truth garment that forms an open surface, $\hat{x}_d$ and $\bar{x}_d$ are the point deformed according to Eq. 7 and the corresponding ground-truth position, respectively. $\mathbf{x}_c = \arg\min_{\mathbf{x}' \in X_s} d(\mathbf{x}', \mathbf{x})$ denotes the surface point closest to $\mathbf{x}$. As there are no correspondences in the training data for $\mathbf{x} \notin X_s$, the second term in Eq. 12 allows them be learned under the guidance of the closest surface points in the garment.

Interpenetrations. To prevent them, we utilize the SDF of the body mesh $M_B(\beta, \theta)$ to penalize the presence of deformed points inside the body. We write

$$L_{interp} = \sum_{\mathbf{x}} max(0, \epsilon_{SDF} - SDF_B(\hat{x_d}(\mathbf{x}))) \,, \tag{13}$$

where ϵ_{SDF} is a small value chosen to prevent $\hat{x_d}(\mathbf{x})$ from overlapping with the body surface.

Self-intersections. Minimizing L_{deform} and L_{interp} usually suffices to deform open surfaces realistically. Unfortunately, when deforming the inflated watertight meshes we use, self-intersections can appear as shown on the left of Fig. 4(b). This can be understood as follows. Let us assume there are two points $\mathbf{x}_1$ and $\mathbf{x}_2$ on the inflated mesh whose closest surface point $\mathbf{x}_0$ is the same, as illustrated by Fig. 4(a). Let us further assume that $\mathbf{x}_2$ is initially farther from the body than $\mathbf{x}_1$. After deformation, nothing prevents $\mathbf{x}_1$ from ending up farther than $\mathbf{x}_2$ and yielding a self-intersection. To prevent this, we introduce the ordering loss

$$L_{order} = \sum_{(\mathbf{x}_1, \mathbf{x}_2) \in O} max(0, SDF_B(\mathbf{x}_2 + \Delta x_\theta(\mathbf{x}_2)) - SDF_B(\mathbf{x}_1 + \Delta x_\theta(\mathbf{x}_1))) \,,$$

$$O = \{(x_1, x_2) | \psi(x_2) = \psi(x_1) \text{ and } SDF_B(x_1) > SDF_B(x_2)\} \,, \tag{14}$$

where $\psi(\cdot)$ denotes the closest garment vertex. Its minimization maintains the spatial relationship between points like $\mathbf{x}_1$ and $\mathbf{x}_2$ because it ensures that points, close to the body before deformation are still close after deformation.

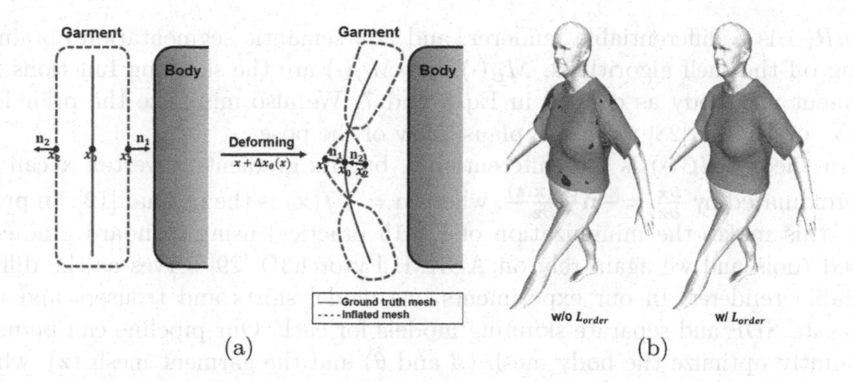

(a) (b)

Fig. 4. (a) The illustration of how artifacts are produced by deformation. (b) Shirts deformed by the models trained w/ and w/o L_{Order}. Without it, the inner face of the t-shirt can intersect the outer one. This is shown in dark blue.(Color figure online)

3.4 Implementation Details

The SDF $f_\Theta(\mathbf{x}, \mathbf{z})$ of Eq. 1 is implemented by a 9-layer multilayer perceptron (MLP) with a skip connection from the input layer to the middle. We use Softplus as the activation function. The Θ weights and the latent code $z \in \mathbb{R}^{12}$ for each garment are optimized jointly during the training using a learning rate of 5e−4.

We use the architecture of [4] to implement the pose displacement network Δx_θ of Eq. 7. It comprises two MLP's with ReLU activation in-between. One encodes the pose θ to an embedding, and the other one predicts the blend matrices for the input point. The pose displacement is computed as the matrix product of the embedding and the blend matrices. The weight distribution $w(\cdot)$ of Eq. 8 is implemented by an MLP with an extra Softmax layer at the end to normalize the output. $N = 1000$ points are sampled to obtain the ground-truth $\bar{w}$ used to train w. We use the ADAM [16] optimizer with a learning rate of 1e-3 for the training of w and Δx_θ.

4 Experiments and Results

Our models can operate in several different ways. First, they can serve as generative models. By varying the latent code of our SDF f_Θ and using Marching Cubes, we can generate triangulated surfaces for garments of different topologies that can then be draped over bodies of changing shapes and poses using the deformation model M_G of Eq. 7, as shown in Fig. 1(a). Second, they can be used to recover both body *and* cloth shapes from images by minimizing

$$L(\beta, \theta, \mathbf{z}) = L_{\text{IoU}}(NeuR(M_G(\mathbf{G}, \beta, \theta), M_B(\beta, \theta)), \mathbf{S}) + L_{prior}(\theta) , \quad (15)$$
$$\mathbf{G} = MC(f_\Theta(\mathbf{x}, \mathbf{z})) ,$$

where L_{IoU} is the IoU loss [17] that measures the difference between segmentation masks, $\mathbf{G}$ is the garment surface reconstructed by Marching Cubes $MC(\cdot)$,

$NeuR(\cdot)$ is a differentiable renderer, and $\mathbf{S}$ a semantic segmentation obtained using off-the-shelf algorithms. $M_B(\cdot)$ and $M_G(\cdot)$ are the skinning functions for garment and body as defined in Eq. 5 and 7. We also minimize the prior loss L_{prior} of VPoser [28] to ensure plausibility of the pose.

In theory $MC(\cdot)$ is not differentiable, but its gradient at vertex $\mathbf{x}$ can be approximated by $\frac{\partial \mathbf{x}}{\partial \mathbf{z}} = -\mathbf{n}\frac{\partial f(\mathbf{x},\mathbf{z})}{\partial \mathbf{z}}$, where $\mathbf{n} = \nabla f(\mathbf{x})$ is the normal [12]. In practice, this makes the minimization of Eq. 15 practical using standard gradient-based tools and we again rely on ADAM. Pytorch3D [29] serves as the differentiable renderer. In our experiments, we model shirts and trousers and use separate SDF and separate skinning models for each. Our pipeline can be used to jointly optimize the body mesh (β and θ) and the garment mesh ($\mathbf{z}$), while previous work [9] can only be used to optimize the garment.

In this section, we demonstrate both uses of our model. To this end, we first introduce the dataset and metrics used for our experiments. We then evaluate our method and compare its performance with baselines for garment reconstruction and deformation. Finally, we demonstrate the ability of our method to model people and their clothes from synthetic and real images.

4.1 Dataset and Evaluation Metrics

We train our models on data from CLOTH3D [3]. It contains over 7k sequences of different garments draped on animated 3D human SMPL models. Each garment has a different template and a single motion sequence that is up to 300 frames long. We randomly select 100 shirts and 100 trousers, and transform them to a body with neutral-shape and T-pose by using displacement of the closest SMPL body vertex, which yields meshes in the canonical space. For each garment sequence, we use the first 90% frames as the training data and the rest as the test data (denoted as TEST EASY). We also randomly select 30 unseen sequences (denoted as TEST HARD) to test the generalization ability of our model. Chamfer Distance (CD), Euclidean Distance (ED), Normal Consistency (NC) and Interpenetration Ratio (IR) are reported as the evaluation metrics. NC is implemented as in [12]. IR is computed as the area ratio of garment faces inside the body to the overall garment faces.

4.2 Garment Reconstruction

The insets of Fig. 5(a) contrast our reconstruction results against those of SMPLicit [9]. The latter yields large interpenetrations while the former does not. Figure 5(b) showcases the role of the L_{grad} term of Eq. 4 in producing smooth surfaces.

In Table 1, we report quantitative results for both the shirt and the trousers. We outperform SMPLicit (the first row - w/o *proc.*, w/o L_{grad}) in all three metrics. The margin in IR is over 18%, which showcases the ability of the interpenetration-aware processing of Sect. 3.1.

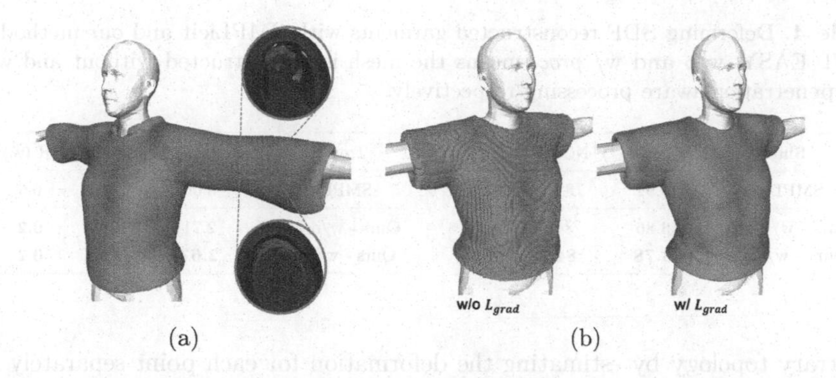

Fig. 5. Reconstruction results. (a) The inside of the same garment reconstructed by SMPLicit (upper inset) and our method (bottom inset). The first features interpenetrations whereas the second does not. (b) Reconstructed garment by a model trained without and with L_{grad}. The latter is smoother and preserves details better.

Table 1. Comparative reconstruction results. *proc.* indicates our proposed interpenetration-aware pre-processing.

Shirt	CD ($\times 10^{-4}$)	NC (%)	IR (%)
Ours w/o *proc.*, w/o L_{grad}	1.88	92.1	18.1
Ours w/o L_{grad}	1.58	90.3	**0.0**
Ours	**1.48**	**92.3**	**0.0**

Trousers	CD ($\times 10^{-4}$)	NC (%)	IR (%)
Ours w/o *proc.*, w/o L_{grad}	1.65	92.0	18.6
Ours w/o L_{grad}	**1.22**	91.8	**0.0**
Ours	1.34	**92.3**	**0.0**

Table 2. Deforming unposed ground truth garments with DeePSD, SMPLicit and our method on TEST EASY.

Shirt	ED (mm)	NC (%)	IR (%)
DeePSD	26.1	82.3	5.8
SMPLicit	35.9	84.0	13.3
Ours	**19.0**	**85.3**	**1.6**

Trousers	ED (mm)	NC (%)	IR (%)
DeePSD	17.5	85.4	1.5
SMPLicit	27.0	85.6	6.3
Ours	**14.8**	**86.7**	**0.2**

Table 3. Deforming unposed ground truth garments with DeePSD, SMPLicit and our method on TEST HARD.

Shirt	ED (mm)	NC (%)	IR (%)
DeePSD	95.6	72.6	46.4
SMPLicit	35.4	83.9	12.9
Ours	**26.5**	**85.1**	**3.0**

Trousers	ED (mm)	NC (%)	IR (%)
DeePSD	37.8	79.5	27.8
SMPLicit	31.9	84.9	8.8
Ours	**24.8**	**85.8**	**0.7**

4.3 Garment Deformation

In this section, we compare our deformation results against those of SMPLicit [9] and DeePSD [2]. The input to DeePSD is the point cloud formed by the vertices of ground-truth mesh so that, like our algorithm, it can deform garments of

Table 4. Deforming SDF reconstructed garments with SMPLicit and our method on TEST EASY. w/o and w/ *proc.* means the mesh is reconstructed without and with interpenetration-aware processing respectively.

Shirt	CD ($\times 10^{-4}$)	NC (%)	IR (%)
SMPLicit	7.91	83.7	16.2
Ours - w/o *proc.*	3.86	84.4	1.6
Ours - w/ *proc.*	**3.78**	**84.7**	**1.5**

Trousers	CD ($\times 10^{-4}$)	NC (%)	IR (%)
SMPLicit	3.66	84.1	6.6
Ours - w/o *proc.*	2.71	85.3	**0.2**
Ours - w/ *proc.*	**2.67**	**85.4**	**0.2**

arbitrary topology by estimating the deformation for each point separately. To skin the garment, it learns functions to predict the blending weight and pose displacement. It also includes a self-consistency module to handle body-garment interpenetration. Hence, for a fair comparison, we retrain DeePSD using the same training data as before.

To test the deformation behavior of our model, we use the SMPL parameters β and θ provided by the test data as the input of our skinning model. As to the garment mesh to be deformed, we either use the ground-truth unposed mesh from the data, which is an open surface, or the corresponding watertight mesh reconstructed by our SDF model.

In Fig. 4(b), we presented a qualitative result that shows the importance of the ordering term of L_{order} in Eq. 14. We report quantitative results with the ground-truth mesh in Table 2 on TEST EASY. Our model performs substantially better than both baselines with the lowest ED and IR and the highest NC. For example, comparing to SMPLicit, the ED and IR of our model drop by more than 15 mm and 10% for the deformation of shirt. In Table 3, we report similar results on TEST HARD, which is more challenging since it resembles less the training set, and we can draw the same conclusions. Since the learning of blending weights in DeePSD does not exploit the prior of the body model as us (Eq. 8), it suffers a huge performance deterioration in this case where its ED even goes up to 95.6 mm and 37.8 mm for the shirt and trousers respectively. Table 4 reports the results with SDF reconstructed mesh. Again, our method performs consistently better than SMPLicit in all metrics (row 2 vs row 4). It is also noteworthy that our interpenetration-aware pre-processing can help reduce deformation error and interpenetration ratio as indicated by the results of row 3 and 4. This demonstrates that learning a physically accurate model of garment interpenetrations results in more accurate clothing deformations.

In the qualitative results of Fig. 6, we can observe that SMPLicit cannot generate realistic dynamics and its results tend to be over-smoothed due to its simple skinning strategy. DeePSD can produce results that are better but too noisy. Besides, neither of them is able to address the body-garment interpenetration. Figure 7 visualizes the level of interpenetrations happening different body region. We can notice that interpenetrations occur on almost everywhere in the body for SMPLicit. DeePSD shows less but still not as good as ours.

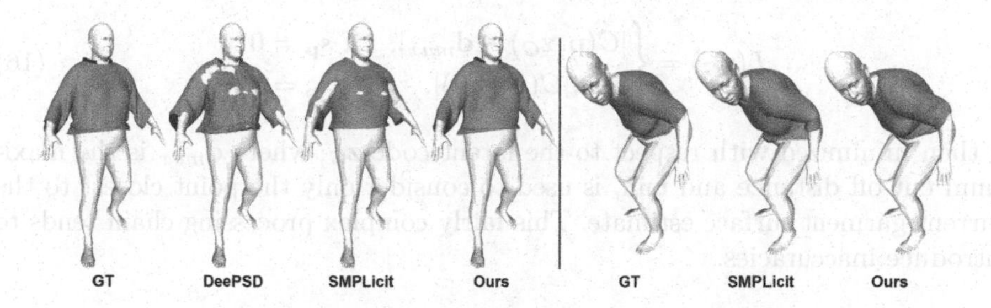

Fig. 6. The skinning results for the ground-truth shirt (left) and the SDF reconstructed shirt (right). Since the input of DeePSD should be the point cloud of the mesh template, we only evaluate it with the unposed ground-truth mesh. Compared to DeePSD and SMPLicit, our method can produce more realistic details and have less body-garment interpenetration.

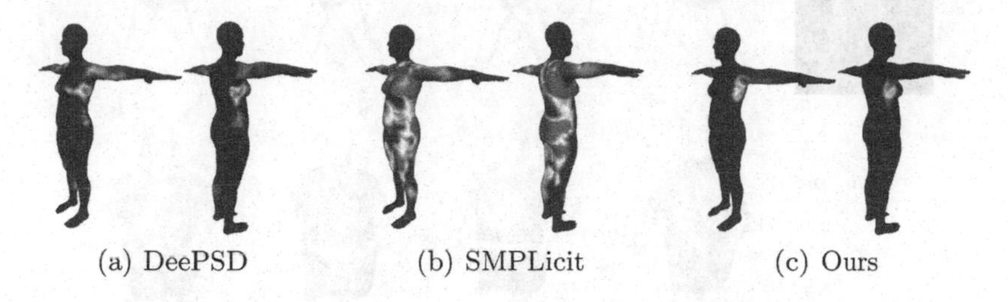

(a) DeePSD (b) SMPLicit (c) Ours

Fig. 7. The visualization of the body region having interpenetrations (marked in red). (Color figure online)

Table 5. The evaluation results of SMPLicit-raw (w/o smoothing), SMPLicit (w/ smoothing) and our method for garment fitting on the synthetic data.

Shirt	CD ($\times 10^{-4}$)	NC (%)	IR (%)	Trousers	CD ($\times 10^{-4}$)	NC (%)	IR (%)
SMPLicit-raw	17.77	82.1	41.5	SMPLicit-raw	4.22	81.2	35.7
SMPLicit	18.73	82.8	37.3	SMPLicit	4.50	82.2	29.2
Ours	**4.69**	**87.3**	**3.9**	Ours	**2.23**	**89.2**	**0.7**

4.4 From Images to Clothed People

Our model can be used to recover the body and garment shapes of clothed people from images by minimizing L of Eq. 15 with respect to β, θ, and $\mathbf{z}$. To demonstrate this, we use both synthetic and real images and compare our results to those of SMPLicit. Our optimizer directly uses the posed garment to compute the loss terms. In contrast, SMPLicit performs the optimization on the unposed garment. It first samples 3D points $\mathbf{p}$ in the canonical space. i.e. on the unposed body, and uses the weights of the closest body vertices to project these points into posed space and into 2D image space to determine if semantic label, 1 if inside the garment, 0 otherwise. The loss

$$L(\mathbf{z}_G) = \begin{cases} |C(\mathbf{p}, \mathbf{z}_G) - \mathbf{d}_{max}|, & \text{if } s_\mathbf{p} = 0 \\ \min_i |C(\mathbf{p}^i, \mathbf{z}_G)|, & \text{if } s_\mathbf{p} = 1 \end{cases}, \tag{16}$$

is then minimized with respect to the latent code $\mathbf{z}_G$, where $\mathbf{d}_{max}$ is the maximum cut-off distance and $\min_i$ is used to consider only the point closest to the current garment surface estimate. This fairly complex processing chain tends to introduce inaccuracies.

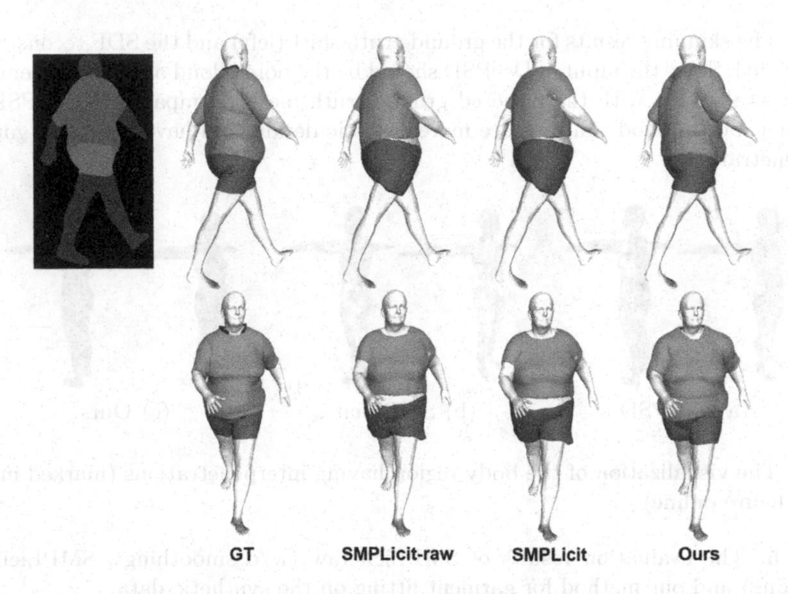

Fig. 8. Fitting results on a synthetic image. Left to right: the ground-truth segmentation and garment meshes, SMPLicit w/o smoothing (SMPLicit-raw), SMPLicit w/ smoothing (SMPLicit) and ours. Note that SMPLicit requires post-processing to remove artifacts, while our method does not.

Synthetic Images. We use the body and garment meshes from CLOTH3D as the synthetic data. Since the ground-truth SMPL parameters are available, we only optimize the latent code $\mathbf{z}$ for the garment and drop the pose prior term L_{prior} from Eq. 15. Image segmentation such as the one of Fig. 8 are obtained by using Pytorch3D to render meshes under specific camera configurations. Given the ground-truth β, θ and segmentation, we initialize $\mathbf{z}$ as the mean of learned codes and then minimize the loss. Figure 8 shows qualitative results in one specific case. The quantitative results reported in Table 5 confirm the greater accuracy and lesser propensity to produce interpenetrations of our approach.

Real Images. In real-world scenarios such as those depicted by Fig. 9, there are no ground-truth annotations but we can get the required information from single images from off-the-shelf algorithms. As in SMPLicit, we use [31] to estimate the SMPL parameters $\hat{\beta}$ and $\hat{\theta}$ and the algorithm of [38] to produce a segmentation. In SMPLicit, $\hat{\beta}$ and $\hat{\theta}$ are fixed and only the garment model is updated. In

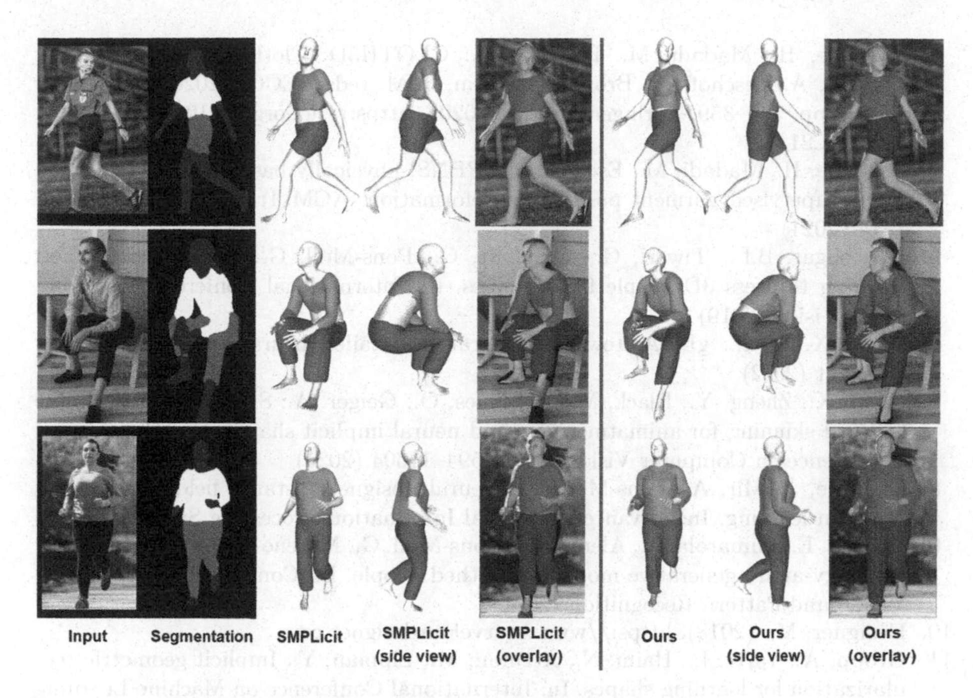

Fig. 9. Fitting results on images in-the-wild. Left to right: the input images and their segmentation, SMPLicit and ours. Note that SMPLicit recovers garments based on body meshes estimated from [31], while we can optimize the body and garment parameters jointly for more accurate results.

contrast, in our approach, $\hat{\beta}$, $\hat{\theta}$, and the latent vector $\mathbf{z}$ are all optimized. As can be seen in Fig. 9, this means that inaccuracies in the $\hat{\beta}$ and $\hat{\theta}$ initial values can be corrected, resulting in an overall better fit of both body and garments.

5 Conclusion

We have presented a fully differentiable approach to draping a garment on a body so that both body and garment parameters can be jointly optimized. At its heart is a skinning model that learns to prevent self-penetration. We have demonstrated its effectiveness both for animation purposes and to recover body and cloth shapes from real images. In future work, we will incorporate additional physics-based constraints to increase realism and to reduce the required amount of training data.

References

1. Baraff, D., Witkin, A.: Large steps in cloth simulation. In: ACM SIGGRAPH, pp. 43–54 (1998)
2. Bertiche, H., Madadi, M., Tylson, E., Escalera, S.: DeePSD: automatic deep skinning and pose space deformation for 3D garment animation. In: International Conference on Computer Vision (2021)

3. Bertiche, H., Madadi, M., Escalera, S.: CLOTH3D: Clothed 3D humans. In: Vedaldi, A., Bischof, H., Brox, T., Frahm, J.-M. (eds.) ECCV 2020. LNCS, vol. 12365, pp. 344–359. Springer, Cham (2020). https://doi.org/10.1007/978-3-030-58565-5_21
4. Bertiche, H., Madadi, M., Escalera, S.: PBNS: physically based neural simulation for unsupervised garment pose space deformation. ACM Trans. Graphics **40**(6), 1–14 (2021)
5. Bhatnagar, B.L., Tiwari, G., Theobalt, C., Pons-Moll, G.: Multi-garment net: learning to dress 3D people from images. In: International Conference on Computer Vision (2019)
6. Chen, X., et al.: gDNA: towards generative detailed neural avatars. In: arXiv Preprint (2022)
7. Chen, X., Zheng, Y., Black, M.J., Hilliges, O., Geiger, A.: SNARF: differentiable forward skinning for animating non-rigid neural implicit shapes. In: International Conference on Computer Vision, pp. 11594–11604 (2021)
8. Chibane, J., Mir, A., Pons-Moll, G.: Neural unsigned distance fields for implicit function learning. In: Advances in Neural Information Processing Systems (2020)
9. Corona, E., Pumarola, A., Alenya, G., Pons-Moll, G., Moreno-Noguer, F.: Smplicit: topology-aware generative model for clothed people. In: Conference on Computer Vision and Pattern Recognition (2021)
10. Designer, M.: (2018). https://www.marvelousdesigner.com
11. Gropp, A., Yariv, L., Haim, N., Atzmon, M., Lipman, Y.: Implicit geometric regularization for learning shapes. In: International Conference on Machine Learning (2020)
12. Guillard, B., et al.: DeepMesh: differentiable ISO-surface extraction. In: arXiv Preprint (2021)
13. Guillard, B., Stella, F., Fua, P.: MeshUDF: fast and differentiable meshing of unsigned distance field networks. In: arXiv Preprint (2021)
14. Gundogdu, E., et al.: Garnet++: improving fast and accurate static 3D cloth draping by curvature loss. IEEE Trans. Pattern Anal. Mach. Intell. **22**(1), 181–195 (2022)
15. Jiang, B., Zhang, J., Hong, Y., Luo, J., Liu, L., Bao, H.: BCNet: learning body and cloth shape from a single image. In: Vedaldi, A., Bischof, H., Brox, T., Frahm, J.-M. (eds.) ECCV 2020. LNCS, vol. 12365, pp. 18–35. Springer, Cham (2020). https://doi.org/10.1007/978-3-030-58565-5_2
16. Kingma, D.P., Ba, J.: Adam: a method for stochastic optimization. In: International Conference on Learning Representations (2015)
17. Li, R., Zheng, M., Karanam, S., Chen, T., Wu, Z.: Everybody is unique: towards unbiased human mesh recovery. In: British Machine Vision Conference (2021)
18. Li, Y., Habermann, M., Thomaszewski, B., Coros, S., Beeler, T., Theobalt, C.: Deep physics-aware inference of cloth deformation for monocular human performance capture. In: International Conference on 3D Vision, pp. 373–384 (2021)
19. Liang, J., Lin, M., Koltun, V.: Differentiable cloth simulation for inverse problems. In: Advances in Neural Information Processing Systems (2019)
20. Loper, M.M., Black, M.J.: OpenDR: an approximate differentiable renderer. In: Fleet, D., Pajdla, T., Schiele, B., Tuytelaars, T. (eds.) ECCV 2014. LNCS, vol. 8695, pp. 154–169. Springer, Cham (2014). https://doi.org/10.1007/978-3-319-10584-0_11
21. Ma, Q., Saito, S., Yang, J., Tang, S., Black, M.J.: SCALE: modeling clothed humans with a surface codec of articulated local elements. In: Conference on Computer Vision and Pattern Recognition (2021)

22. Narain, R., Pfaff, T., O'Brien, J.F.: Folding and crumpling adaptive sheets. ACM Trans. Graphics **32**(4), 1–8 (2013)
23. Narain, R., Samii, A., O'brien, J.F.: Adaptive anisotropic remeshing for cloth simulation. ACM Trans. Graphics **31**(6), 1–10 (2012)
24. Nvidia: Nvcloth (2018)
25. Nvidia: NVIDIA Flex (2018). https://developer.nvidia.com/flex
26. Park, J.J., Florence, P., Straub, J., Newcombe, R.A., Lovegrove, S.: DeepSDF: learning continuous signed distance functions for shape representation. In: Conference on Computer Vision and Pattern Recognition (2019)
27. Patel, C., Liao, Z., Pons-Moll, G.: Tailornet: predicting clothing in 3D as a function of human pose, shape and garment style. In: Conference on Computer Vision and Pattern Recognition, pp. 7365–7375 (2020)
28. Pavlakos, G., et al.: Expressive body capture: 3D hands, face, and body from a single image. In: Conference on Computer Vision and Pattern Recognition, pp. 10975–10985 (2019)
29. Ravi, N., et al.: PyTorch3D (2020). https://github.com/facebookresearch/pytorch3d
30. Remelli, E., et al.: MeshSDF: differentiable ISO-surface extraction. In: Advances in Neural Information Processing Systems (2020)
31. Rong, Y., Shiratori, T., Joo, H.: FrankMOCAP: fast monocular 3D hand and body motion capture by regression and integration. In: arXiv Preprint (2020)
32. Saito, S., Yang, J., Ma, Q., Black, M.J.: SCANimate: weakly supervised learning of skinned clothed avatar networks. In: Conference on Computer Vision and Pattern Recognition, pp. 2886–2897 (2021)
33. Santesteban, I., Thuercy, N., Otaduy, M.A., Casas, D.: Self-supervised collision handling via generative 3D garment models for virtual try-on. In: Conference on Computer Vision and Pattern Recognition (2021)
34. Santesteban, I., Otaduy, M.A., Casas, D.: SNUG: self-Supervised Neural Dynamic Garments. In: arXiv Preprint (2022)
35. Software, O.F.D.: (2018). https://optitex.com/
36. Tiwari, G., Bhatnagar, B.L., Tung, T., Pons-Moll, G.: SIZER: a dataset and model for parsing 3D clothing and learning size sensitive 3D clothing. In: Vedaldi, A., Bischof, H., Brox, T., Frahm, J.-M. (eds.) ECCV 2020. LNCS, vol. 12348, pp. 1–18. Springer, Cham (2020). https://doi.org/10.1007/978-3-030-58580-8_1
37. Tiwari, G., Sarafianos, N., Tung, T., Pons-Moll, G.: Neural-GIF: neural generalized implicit functions for animating people in clothing. In: International Conference on Computer Vision, pp. 11708–11718 (2021)
38. Yang, L., et al.: Renovating parsing R-CNN for accurate multiple human parsing. In: Vedaldi, A., Bischof, H., Brox, T., Frahm, J.-M. (eds.) ECCV 2020. LNCS, vol. 12357, pp. 421–437. Springer, Cham (2020). https://doi.org/10.1007/978-3-030-58610-2_25
39. Zakharkin, I., Mazur, K., Grigorev, A., Lempitsky, V.: Point-based modeling of human clothing. In: International Conference on Computer Vision, pp. 14718–14727 (2021)

Pyramidal Signed Distance Learning for Spatio-Temporal Human Shape Completion

Boyao Zhou[1(✉)], Jean-Sébastien Franco[1], Martin de La Gorce[2], and Edmond Boyer[1]

[1] Inria-University Grenoble Alpes-CNRS-Grenoble INP-LJK, Grenoble, France
{boyao.zhou,jean-sebastien.franco,edmond.boyer}@inria.fr
[2] Microsoft, Cambridge, UK
martin.delagorce@microsoft.com

Abstract. We address the problem of completing partial human shape observations as obtained with a depth camera. Existing methods that solve this problem can provide robustness, with for instance model-based strategies that rely on parametric human models, or precision, with learning approaches that can capture local geometric patterns using implicit neural representations. We investigate how to combine both properties with a novel pyramidal spatio-temporal learning model. This model exploits neural signed distance fields in a coarse-to-fine manner, this in order to benefit from the ability of implicit neural representations to preserve local geometry details while enforcing more global spatial consistency for the estimated shapes through features at coarser levels. In addition, our model also leverages temporal redundancy with spatio-temporal features that integrate information over neighboring frames. Experiments on standard datasets show that both the coarse-to-fine and temporal aggregation strategies contribute to outperform the state-of-the-art methods on human shape completion.

1 Introduction

Completing shape models is a problem that arises in many applications where perception devices provide only partial shape observations. This is the case with single depth cameras which only perceive the front facing geometric information. Completing the 3D geometry in such a situation while retaining the observed geometric details is the problem we consider in this work. We particularly focus on human shapes and take benefit of the camera ability to provide series of depth images over time.

The task is challenging since different and potentially conflicting issues must be addressed. Strong priors on human shapes and their clothing are required as large shape parts are usually occluded in depth images. However, such priors

Supplementary Information The online version contains supplementary material available at https://doi.org/10.1007/978-3-031-26319-4_22.

L. Wang et al. (Eds.): ACCV 2022, LNCS 13841, pp. 360–377, 2023.
https://doi.org/10.1007/978-3-031-26319-4_22

can in practice conflict with the capacity to preserve geometric details present in the observations. Another issue lies in the ability to leverage information over time with local geometric patterns that can be either temporally consistent or time varying, as with folds on human clothing.

Existing methods fall into two main categories with respect to their global or local approach to the human shape completion problem. On the one hand, model-based methods build on parametric human models which can be fitted to incomplete observations, for instance SMPL [25] Dyna [39] or NPMs [34]. Approaches in this category can leverage strong priors on human shapes and hence often yield robust solutions to the shape completion problem. However, parametric models usually impose that the estimated shapes lie in a low dimensional space, which limits the ability to generalize over human shapes, in particular when considering clothing. Another consequence of the low dimensional shape space is that local geometric details tend to be filtered out when encoding shapes into that space.

On the other hand, neural network based approaches, *e.g.* NDF [13], IF-Net [12] or STIF [51], use implicit representations to model the observed surfaces. Such representations give access to a larger set of shapes, although this set is still bounded by the coverage in the training data. They also present better abilities to preserve geometric details as they are, by construction in these approaches, local representations. Besides, in contrast to model based methods, the local aspect of the representation impacts the robustness with estimated shapes that can often be spatially inconsistent.

In order to retain the benefit of local representations while providing better robustness and spatial coherence we propose a novel hierarchical coarse-to-fine strategy that builds on implicit neural representations [30,35]. This strategy applies to the spatial domain as well as to the temporal domain. We investigate the temporal dimension since redundancy over time can contribute to shape completion, as demonstrated in [51]. This dimension naturally integrates into the pyramidal spatio-temporal model we present. Our approach considers as input consecutive depth images of humans in a temporal sequence and estimates in turn a sequence of 3D distance functions that maps any point in the space time cube covered by the input frames to its distance to the observed human shape. This distance mapping function is learned over temporally coherent 3D mesh sequences and through a MLP decoder that is fed with multi-scale spatio-temporal features, as encoded from the input depth images. Our experiments demonstrate the spatial and temporal benefit of the hierarchical strategy we introduce with both local and global shape properties that are substantially better preserved with respect to the state of the art.

The main contributions of this paper are, first, to show the benefit of a pyramidal architecture that aggregates a residual pyramid of features in the context of implicit surface regression, and second, to propose a tailored training strategy that exploits this residual architecture by using losses that are distributed at each scale of the feature map both spatially and temporally. Code is released at https://gitlab.inria.fr/bzhou/PyramidSpatioTempoHumanShapeComplete.git.

2 Related Work

The set of methods proposed in the literature to estimate full human body surfaces from a sequence of depth images can be broadly divided in 4 main categories: model-based methods, regression-based methods, dynamic fusion based methods and hybrid methods. We detail here these different methods and discuss their strengths and limitations.

Model-Based Methods build on learned parametric human models that are fitted to incomplete observations. The parametric model generally represents a 3D undressed human with global shape and pose parameters. The surface can be represented explicitly using a triangulated surface similarly to SMPL [4,25,36,46] or Dyna [39], or implicitly using either a parameterized occupancy or a signed distance function [5,6,18,34]. Model based methods that decouple the body shape parameters from the pose like [25,34] can be used to efficiently exploit temporal coherence in a sequence by estimating a single shape parameter vector for the whole sequence and different pose for each frame. The low dimensionality of the parameterized space inherently limits the ability of these models to represent details such as clothing or hair present in the data. To model the details beyond the parameterized space [2,3,8,29,38] add a cloth displacement map to represent 3D dressed humans, which significantly improves the accuracy of the reconstructed surface but often still impose some constraint on the topology of the surface and thus do not allow to fully explain complex observed data.

Regression-Based Methods directly predict the shapes from the incomplete data. This allows to predict less constrained surfaces and keep more of the details from the input data than model-based methods. The representation of the predicted surface can be explicit [19,40,52] or implicit which allows for changes in the topology [11,35]. This approach has been used for non-articulated objects [13,24,30,33,37,47] and articulated human body [12,16,20,40,42,43,51]. The implicit surface can be represented using a (truncated) signed distance function, an unsigned distance function [13] or an occupancy function as in IF-Net [12]. Learning truncated signed distance functions tends to lead to more precise surfaces than learning occupancy, as the signed distance provides a richer supervision signal for points that are sampled near but not on the surface. One main challenge with these approaches is to efficiently exploit the temporal coherence to extract surface detail from previous frames.

Dynamic Fusion approaches [17,31] are another classical set of approaches that are neither model-based nor regression-based and which consist in tracking point cloud correspondences, aligning point cloud in a non-rigid manner and fusing the aligned point clouds into a single surface representation. This allows to exploit temporal coherence to complete the surface in regions that are observed in other frames in the sequence. Omitting strong priors on the human body surface, either in the form of a model or a trained regressor (although human prior can be used to help the tracking), these methods are able to retain specific details of the observed surface but are also, consequently, not able to predict the surface in regions that are not observed in any frame. These methods are also subject to drift and error accumulation when used on challenging long sequences with fast motion and topological changes.

Hybrid Methods that mix various of these approaches have been proposed to combine their advantages. IP-Net [7] generates the implicit surface with [12] and register the surface with SMPL+D [1,21] and thus inherits the limited ability of model based methods to represent geometric details. Function4D [50] combines a classical dynamic fusion method with a post processing step to repair holes in the surface using a learned implicit surface regression. While these methods improve results w.r.t each of the combined methods, they still tend to retain part of their drawbacks.

Exploiting Temporal Coherence is a main challenge for regression based methods, as mentioned above. Using a pure feed-forward approach where a set of frames is fed into the regressor becomes quickly unmanageable as the size of the temporal window increases. Oflow [32] exploits the temporal coherence by estimating a single surface for the sequence initial frame which is deformed over all frames using a dense correspondence field conditioned on the whole sequence inputs. This strategy enforces by construction temporal consistency, but it also prevents temporal information to benefit to the shape model. STIF [51] addresses the problem of temporal integration using a recurrent GRU [14,15] layer to aggregate information. Such layer can however only be used at the low-resolution features and hence surface details do not propagate from one frame to the next.

In this paper we suggest a pyramidal approach that can exploit both spatial and temporal dimensions. Taking inspiration from pyramidal strategies applied successfully to other prominent vision problems such as object detection [23], multi-view stereo [48], 3D reconstruction [45] and optical flow [44] we devise a method that aggregates spatio-temporal information in a coarse to fine manner, propagating features from low to high resolution through up-sampling, concatenation with higher resolution features and the addition of residuals.

3 Network Architecture

We wish to compute the completion of a sequence of shapes observed from noisy depth maps. These maps, noted $\{\mathcal{D}\} = \{\mathcal{D}\}_{t\in\{1,\ldots,N\}}$, provide a truncated front point cloud of the human shape in motion. Our output, by contrast, is a corresponding sequence of *complete* shapes encoded as a set of per-frame Truncated Signed Distance functions, using neural implicit functions $\mathrm{SDF}_t(p)$ with $t \in \{1,\ldots,N\}$, which can be each queried for arbitrary 3D points $p = (x, y, z)$. The surface can then be extracted using standard algorithms [22,26].

In the following, we explain how the $\mathrm{SDF}_t(p)$ functions can be coded with a network inspired from point cloud SDF networks [13,35], but which also jointly leverages the analysis of small temporal frame subgroups of size K. Without loss of generality we expose the framework for a given subgroup of frames $\{1,\ldots,K\}$ and drop the more general index N. We also propose a more general pyramidal coarse-to-fine architecture to balance global information and local detail in the inference, which was previously only demonstrated with explicit TSDF encodings

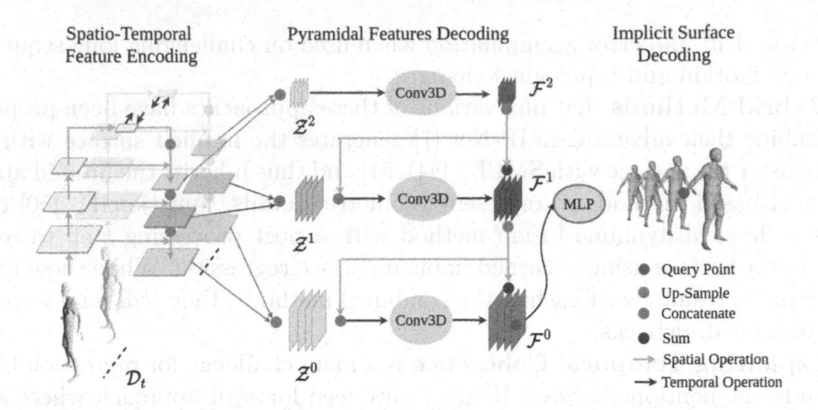

Fig. 1. Network Architecture. Data is processed in 3 phases: depth images are processed with Spatio-Temporal Feature encoding backbone in Sect. 3.1 into three-level-feature maps. These feature maps are latter aggregated with 3D convolution considering the temporal dimension in the coarse-to-fine manner in Sect. 3.2. The aggregated feature map is used to predict the signed distance field with MLP in Sect. 3.3.

for a different incremental reconstruction problem [45]. Our network is articulated around three main phases represented in Fig. 1, a feature pyramid extraction phase (Sect. 3.1), a pyramidal feature decoding phase (Sect. 3.2), and an implicit surface decoding phase (Sect. 3.3), detailed as follows.

3.1 Spatio-Temporal Feature Encoding

To build a hierarchical inference structure, a time-tested pattern (*e.g.* [44,48]) is to build a feature hierarchy or feature pyramid [23] to extract various feature scale levels. The recent literature [51] shows that a measurable improvement to this strategy, in the context of temporal input images sequences, is to link the coarsest layer in each input image's pyramid with a bidirectional recurrent GRU component [14,15], instead of building independent per-frame pyramids. This allows the model to learn common global characteristics and their mutual updates at a modest computational cost. We modify the backbone encoder U-GRU [51] in order to hierarchically yield a set of per-frame feature maps $\mathcal{Z}_t^0, \mathcal{Z}_t^1, \mathcal{Z}_t^2$ at three levels of respectively high, mid and low resolutions, for each input depth map $\mathcal{D}_t$ with $t \in \{1, \dots, K\}$. In the rest of the discussion we will only use temporal aggregates of each feature level $\mathcal{Z}^0 = \{\mathcal{Z}_t^0\}, \mathcal{Z}^1 = \{\mathcal{Z}_t^1\}$, and $\mathcal{Z}^2 = \{\mathcal{Z}_t^2\}$ with $t \in \{1, \dots, K\}$, such that we can denote their extraction:

$$\{\mathcal{Z}^0, \mathcal{Z}^1, \mathcal{Z}^2\} = \text{Encode}(\mathcal{D}_1, \dots, \mathcal{D}_K), \tag{1}$$

The feature map here contains 3 dimensions: x, y and t in the three different levels of the pyramid. However, the coarsening is only done in the spatial domain at this stage so that all scale feature maps contain the same number of frames K. More details on these feature map's dimensions are provided in Sect. 3.4.

3.2 Pyramidal Feature Decoding

In the next stage, our goal is to allow the network to progressively decode and refine spatio-temporal features of the sequence that will be used for TSDF decoding. To this goal, we first process the coarsest feature aggregate $\mathcal{Z}^2$ with a full 3D (2D + t) convolution to extract a sequence feature $\mathcal{F}^2$:

$$\mathcal{F}^2 = \text{Conv3D}(\mathcal{Z}^2). \tag{2}$$

We then subsequently process the finer feature levels $l \in \{1, 0\}$, first by up-sampling the previous-level sequence feature map in the spatial domain using bilinear interpolation, noted $\mathcal{U}^l$, then concatenating it with the aggregate feature $\mathcal{Z}_l$ as input to 3D convolutions, as shown in Fig. 1. The output of the 3D convolutions is the residual correction of the previous-level aggregated feature, echoing successful coarse-to-fine architectures [44,48]:

$$\mathcal{U}^l = \text{Up-Sample}(\mathcal{F}^l + 1), \tag{3}$$

$$\mathcal{F}^l = \mathcal{U}^l + \text{Conv3D}(\{\mathcal{U}^l, \mathcal{Z}^l\}). \tag{4}$$

3.3 Implicit Surface Decoding

The final high-resolution feature $\mathcal{F}^0$ obtained as output of the previous process serves as a latent vector map, from which we sample a latent vector at a given query point's image coordinates. The selected latent vector is then decoded using an MLP, and provides the TSDF value for that point. More formally, given a query point (x, y, z) and a time frame $t \in \{1, \ldots, K\}$ this signed distance function is computed by first sampling bi-linearly at the corresponding 2D location (x, y) the slice $\mathcal{F}_t^0$ of the high resolution 3D feature map that corresponds to frame t of the temporal window. We can note this sampler f_t^0:

$$f_t^0(x, y) = B(\mathcal{F}_t^0; x, y), \tag{5}$$

After obtaining the queried feature vector in the high resolution scale, we concatenate it with depth feature z. The signed distance function can then be computed by

$$\text{SDF}_t^0(x, y, z) = \text{MLP}(f_t^0(x, y), z), \tag{6}$$

where an MLPis used to decode the TSDF value for a given pixel x, y, and frame t of the high-resolution feature map, and an implicitly accounted z. Operational details and constants are discussed in the next section and supplementary.

3.4 Implementation Details

We adapt U-Net [41]-like encoder [51] as backbone, for the purpose of hierarchical learning, combined with GRU [14,15] for pyramidal feature extraction in Sect. 3.1. Each 3D convolution block contains 3 convolutions layers with $3 \times 3 \times 3$ kernels in Sect. 3.2. We use zero padding spatially and temporally. In Sect. 3.3, the depth feature z is multiplied by 128 in order to get a similar activation ranges as the values in the feature vector $f_t^0(x, y)$. The final output layer of the MLP is activated with *tanh* in order to bound the signed distance in the range $[-1, 1]$. The pre-computed SDF is scaled with the factor of 75 and we set the truncate value as 1 in the normalized space. In practice, this multiplication improves also the numerical stability, otherwise the loss would be too small during the training. While it would be possible to compute the inference results over an input sequence in a sliding window fashion, for computational trade-off with quality we compute the results on consecutive groups of $K = 4$ frames. A training batch is composed of 4×512^2 depth images and 4×2800 query points per scale for the SDF evaluation. It takes 2.86 seconds with GPU memory footprint about 7.86GB for a batch training.

4 Training Strategies

In the previous section, we discussed the network used and inference path for a given set of Kinput frames. The same path can be used for supervised training, but pyramidal architectures are usually trained with intermediate loss objectives that guide the coarser levels. This is easy to do with a supervised loss when the coarser objectives are just a down-sampled version of the expected result, *e.g.* for object detection, depth map or optical flow inference [44,48]. A main contribution of this paper is to show the benefit of such schemes with an implicit representation of the surface reconstructed from a residual pyramid of features. One of the key difficulties lies in characterizing the intermediate, lower resolution contribution to the final result of coarser layers of the implicit decoder. To this goal we present four training strategies, with several temporal and pyramidal combinations, and discuss their ablation in the experiments. A visual summary of these strategies is provided in Fig. 2.

4.1 Common Training Principles

We render depth maps of 3D raw scans from a training set to associate depth maps observations to ground truth computed SDF values for a given set of query points in the 3D observation space. Following general supervision principles of implicit networks [42,51], we provide training examples for three groups: on-surface points, points in the vicinity of the surface, and other examples uniformly sampled over the whole observed 3D volume. On-surface training examples are obtained simply by sampling k_s vertices from ground truth 3D model and associating them with SDF value 0. For surface vicinity points at any given time t, as shown in Fig. 2, for each of the k_s vertices, we randomly draw 4 training points with their ground truth SDF values among points on a 26-point 3D grid centered on a ground truth surface point, in which the closest projected grid edge length matches the observed feature map pixel size, as illustrated in the right most column in Fig. 2. For the third, uniform point group, we randomly draw k_u sample training points covering the whole acquisition space, for which

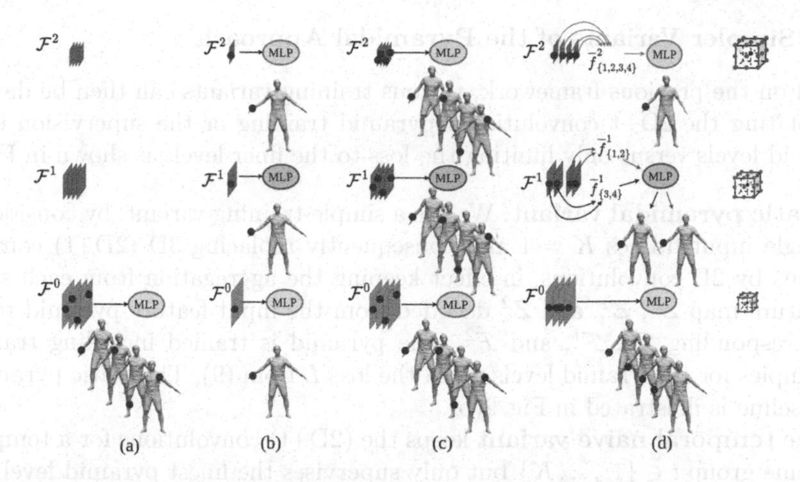

Fig. 2. Training strategy variants, from left to right: (a) temporal naive, (b) static pyramidal, (c) temporal pyramidal and (d) spatio-temporal pyramidal. More detailed explanations can be found in Sect. 4.3.

we provide the SDF to the ground truth surface. In our experiments, we keep $k_s = 400$ and $k_u = 800$ constant regardless of training scenario.

4.2 Pyramidal Training Framework

For explicit pyramidal training, we need to supply ground truth SDFs for all levels of the pyramid, including the coarser ones. To this goal, for a set of sampled ground truth surface points at time t, we extend the set of surface-vicinity training samples by randomly drawing them from three different 26-point-grid cubes centered on the ground truth surface point, instead of only one. Each of these grid cubes corresponds to a network pyramid level feature $\mathcal{F}_t^l$ with $l \in \{0, 1, 2\}$ and its closest projected grid edge length matches the finest feature map pixel size $\mathcal{F}_t^0$. We still draw 4 samples at each feature level l, among the 26 possibilities.

For each of the m^l query points $p_i = (x_i, y_i, z_i)$ in the training batch, noting its frame t_i and pyramid level l, we generalize Eqs. 5 and 6 to provide an MLP decoding and training path of SDFs in the temporal window:

$$f_{t_i}^l(x_i, y_i) = B(\mathcal{F}_{t_i}^l; x_i, y_i) \tag{7}$$

$$\mathrm{SDF}_{t_i}^l(x_i, y_i, z_i) = \mathrm{MLP}(f_{t_i}^l(x_i, y_i), z_i) \tag{8}$$

Given the ground truth SDF value s_i^l computed from ground truth 3D models for every training point p_i for level l, the loss can then be defined as:

$$L = \sum_l \lambda^l \frac{1}{m^l} \sum_{i=1}^{m^l} \|\mathrm{SDF}_{t_i}^l(x_i, y_i, z_i) - s_i^l\|^2 \tag{9}$$

where λ is the weight to balance the final loss, in practice we set $\lambda^0 = 4$ for the finest level, $\lambda^1 = 1$ and $\lambda^2 = 0.1$ for the coarsest level.

4.3 Simpler Variants of the Pyramidal Approach

Based on the previous framework, various training variants can then be devised by ablating the 2D+t convolutional pyramid training or the supervision of all pyramid levels versus only limiting the loss to the finer level, as shown in Fig. 2.

- **Static pyramidal variant.** We use a simple training variant, by considering single input frames $K = 1$ and consequently replacing 3D (2D+t) convolutions by 2D convolutions, in effect keeping the aggregation from each single feature map $\mathcal{Z}^0$, $\mathcal{Z}^1$, and $\mathcal{Z}^2$ decoded from the input feature pyramid to the corresponding $\mathcal{F}^0$, $\mathcal{F}^1$, and $\mathcal{F}^2$. The pyramid is trained including training samples for all pyramid levels $\mathcal{F}^l$ in the loss L from (9). The static pyramidal baseline is illustrated in Fig. 2(b).
- The **temporal naive variant** keeps the (2D+t) convolutions for a temporal frame group $t \in \{1, ..., K\}$ but only supervises the finest pyramid level.
- The **temporal pyramidal variant** includes the loss terms for all 3 pyramid levels, and uses unmatched, untracked randomized sample training points in each frame of the processed temporal frame group. The temporal naive and temporal pyramidal variants are illustrated respectively in Fig. 2(a) and (c).

4.4 Full Spatio-Temporal Pyramidal Training

While the previous pyramidal variants account for spatial hierarchical aspects in training, no component in the previous training schemes accounts for hierarchical temporal aggregation or weak surface motion priors in the training losses. One would expect such terms to help the training balance global spatio-temporal aspects against local spatio-temporal details for the underlying surface in motion.

Here we propose a temporal coarse-to-fine spatio-temporal training supervision for $K = 4$ time frames, extending the previous temporal pyramidal variant. In this training strategy we add gradual pooling of the queried features from feature maps $\mathcal{F}_t^l$ to provide 4 supervision signals at the finer level with map $\mathcal{F}^0$, 2 supervision signals at the mid-level $\mathcal{F}^1$ from frame groups 1, 2 and 3, 4, and one signal global to the sequence for the coarsest level $\mathcal{F}^2$, averaging the feature over all 4 frames. At training time, for a point $p = (x, y, z)$ among the uniform group of training points, this can be done simply by retrieving the four features in the four temporal maps corresponding to point p in the coarse map $\mathcal{F}^2$ and the two averages by retrieving the two features from temporal frames 1, 2 and 3, 4 respectively. To perform this in a meaningful way for points on and at the vicinity of the surface however, the features pooled temporally should be aggregated from a coherent point trajectory in the temporal sequence. For this we leverage training datasets for which a temporal template fitting is provided (*e.g.* with SMPL [25]) to work with temporally registered vertices. Intuitively, this allows the proposed network to account for a weak prior on underlying point trajectory coherence when producing estimated SDF sequences at inference time.

To formalize this sampling strategy we can first denote v_i^t the 3D position of the ith vertex in the temporal registered model in frame t. Then, similarly to what is done in Sect. 4.2, each of the query points p_i^1 generated in the vicinity to the surface in the first frame is obtained by choosing a point on a 26-point 3D grid centered around a vertex of index v_i^1 on the fitted ground truth model in the first frame. However instead of sampling the query point independently for each of the 4 frames in the temporal window, we reuse the same vertex indices and the same offset w.r.t to that vertex throughout the 4 frames to obtain query points $(p_i^t)_{t=1}^4$ along a trajectory that follows the body motion. i.e. $p_i^t = v_i^t + p_i^1 - v_i^1$. We can then formalize the averaging of the features extracted across several frames in a set S using the point trajectories $(p_i^t)_{t \in S}$ as:

$$\overline{f}^l_{\{S\}} = \frac{1}{|S|} \sum_{t \in S} f^l_t(x^t_i, y^t_i) \tag{10}$$

with $p_i^t = (x_i^t, y_i^t, z_i^t)$. As shown in Fig. 2(d), for each point sample we use features $\overline{f}^2_{\{1,2,3,4\}}$ to compute a low-level loss and the features $\overline{f}^1_{\{1,2\}}$ and $\overline{f}^1_{\{3,4\}}$ to compute two mid-level losses. For each of these pooled features, we use the mean of the point's ground truth SDFs as supervision signal after MLP decoding.

5 Experiments

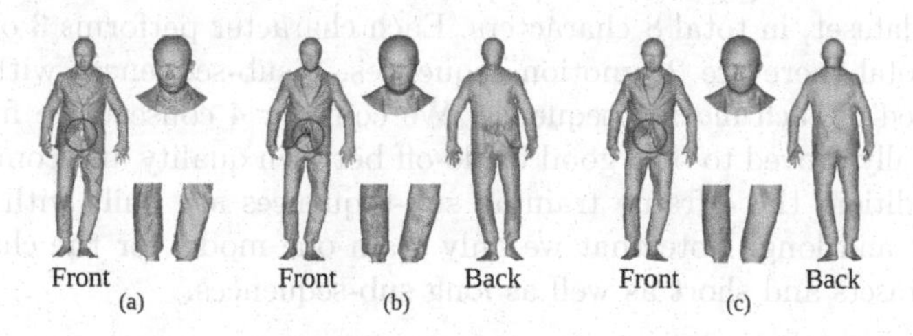

Fig. 3. Shape completion on CAPE. (a) Front-view partial scan. Completion with: (b) our static pyramidal method, (c) our spatio-temporal method.

We provide quantitative and qualitative comparisons on real-world scan data. In particular, we discuss the results of the different training strategies introduced in Sect. 4. We also show numerical comparisons, on raw scan data, between our method and both learning-based method and model-based methods.

5.1 Dataset

Focusing on dynamic human shape completion, we collect data from the CAPE [27] dataset for dressed humans with different clothing styles for characters, and the DFAUST [9] dataset for undressed humans, as in competitive methods [34,51]. Both are captured at 60fps and provide temporally coherent mesh sequences alongside the raw scan data. We render depth images in resolution 512^2 from the raw

Table 1. Pyramidal approach variants. Spatial completion with IoU and Chamfer-L1 distances ($\times 10^{-1}$) in the real 3D space.

Data	CAPE		DFAUST	
Method	IoU ↑	Chamfer-L1 ↓	IoU ↑	Chamfer-L1 ↓
(i) Naive temporal (a)	0.777	0.174	0.858	0.115
(ii) Static pyramidal (b)	0.788	0.172	0.871	0.112
(iii) Temporal pyramidal (c)	0.808	0.182	0.873	0.118
(iv) Spatio-temporal pyramidal (d)	**0.839**	**0.161**	**0.898**	**0.103**
(v) Spatio-temporal occupancy	0.800	0.168	0.872	0.109

scans in order to preserve the measurement noise. We also pre-compute pseudo ground-truth signed distances using the mesh models to avoid topological artifacts present in the raw data. Meshes in these datasets are actually obtained by fitting the SMPL [25] model to the raw scan data. However, thanks to the local reasoning, our network tries to capture the partwise space spanned by the local geometric patterns from the fitted models and is agnostic to the associated shapewise parametric space, here SMPL.

5.2 Training Protocol

We follow the training protocol of [51]. We take 2 male and 2 female characters from each dataset, in total 8 characters. Each character performs 3 or 4 motion styles, in total there are 28 motion sequences. 6 sub-sequences with 4 frames are extracted in each motion sequence. We consider 4 consecutive frames as it experimentally proved to be a good trade-off between quality and computational cost. In addition, the 4-frame training sub-sequences are built with two durations: short and long. Note that we only train one model for the characters in the two datasets and short as well as long sub-sequences.

5.3 Ablation and Variants Comparison

We validate our method by testing the different variants of Sect. 4 on 152 frames from new released CAPE data and 100 frames of two unseen identities from DFAUST in Table 1. Our evaluation metrics are Intersection over Union (IoU) and Chamfer-L1 distance. These results show that the spatio-temporal pyramidal training strategy in Sect. 4.4 and illustrated in Fig. 2(d) yields the best results. In Table 1, from row (i) to row (iii), the pyramidal training is more effective than the naive training. In row (iv), we leverage the coarse-to-fine strategy not only spatially but also temporally and obtain the improvement from row (ii) to row (iv), especially for the more difficult case of clothed humans with the CAPE dataset. Comparing row (iv) and (v) highlights the effectiveness of our neural signed distance representation with respect to occupancy. In order to better illustrate the contribution of the temporal information, Fig. 3 shows how the spatio-temporal pyramidal model can correct artefacts still present with the static pyramidal.

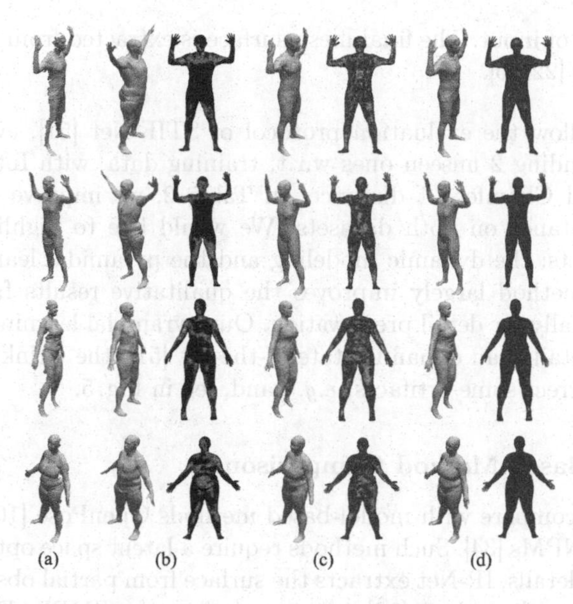

Fig. 4. Static Shape completion, top two identities are from CAPE and bottom two are from DFAUST. From left to right, (a) front-view partial scan, completion with (b) 3D-CODED [19], (c) IF-Net [12] and (d) our method. We set the maximum reconstruction to ground truth Chamfer-L1 distance as 2 cm for heatmaps.

5.4 Learning-Based Method Comparisons

Baselines. We first compare our method with learning-based baselines in the following categories: 1. *Static methods* as 3D-CODED [19], ONet [30], BPS [40], IF-Net [12] and 2. *Dynamic methods* as OFlow [32] and STIF [51]. 3D-CODED and BPS are point-based methods which take point cloud as input and output the template-aligned mesh. ONet, IF-Net, OFlow and STIF-Net are implicit function learning methods as ours. IF-Net relies on the 3D convolution so it pre-processes the partial observation input into voxel grid. And STIF-Net inputs the depth image as ours. To fairly compare with IF-Net and STIF-Net, we use the

Table 2. Comparison with learning-based methods with IoU and Chamfer-L1 distances ($\times 10^{-1}$) in the real 3D space on [51] benchmarck.

Data	CAPE		DFAUST	
Method	IoU ↑	Chamfer-L1 ↓	IoU ↑	Chamfer-L1 ↓
3D-CODED [19]	0.455	0.591	0.578	0.347
ONet [30]	0.488	0.476	0.604	0.340
IF-Net [12]	0.853	0.121	0.876	0.107
BPS [40]	–	–	0.761	0.197
OFlow [32]	–	–	0.740	0.231
STIF [51]	0.834	0.113	0.865	0.104
Ours	**0.880**	**0.100**	**0.914**	**0.092**

same resolution of input. The final mesh surface is extracted from 256^3 grid with marching cubes [22,26].

Results. We follow the evaluation protocol of STIF-Net [51], evaluating on 9 characters, including 2 unseen ones w.r.t. training data, with Intersection over Union(IoU) and Chamfer-L1 distance. In Table 2, we improve both IoU and Chamfer-L1 distance on both datasets. We would like to highlight the benefit from 2 aspects: the dynamic modeling and the pyramidal learning strategy. In Fig. 4, our method largely improves the qualitative results from the static methods, especially for detail preservation. Our pyramidal learning strategy still retains more detail than dynamic state-of-the-art [51], the wrinkle, face details and it could correct some artifacts, *e.g.* hand, leg in Fig. 5.

5.5 Model-Based Method Comparisons

Baselines. We compare with model-based methods OpenPose [10]+SMPL [25], IP-Net [7] and NPMs [34]. Such methods require a latent space optimization process. For more details, IP-Net extracts the surface from partial observation input with learning-based method [12], then optimizes the SMPL+D [1,21] model parameters to fit the surface. NPMs learns the neural parameter model from real-world data and synthetic data and optimize such parameters to fit the partial observation during inference. Note that IP-Net [7] and NPMs [34] train on 45000 frames from not only CAPE, but also from the synthetic data DeformingThings4D [49] and motion capture data AMASS [28], while we train on 1344 frames from the real scan data of CAPE and DFAUST only.

Results. We follow the evaluation protocol of NPMs [34] on 4 identities in 2 cloth styles with IoU and Chamfer-L2 distance. In Table 3, our method improves the numeric result on IoU but not on the Chamfer-L2 loss. One may note that since our method is devoid of a parametric or latent control space, it has more freedom to reconstruct details, see Fig. 5. NPMs improves over a first inference phase by using a latent-space optimization which specifically minimizes for an L2 surface loss. Thus, NPMs is slightly better than ours on Chamfer-L2 metric but increases the inference time. While lacking such a stage, the results produced by our method are still observed to be generally competitive with the latter optimization methods.

Table 3. Comparison with model-based methods with IoU and Chamfer-L2 distances ($\times 10^{-3}$) for CAPE in the normalized space on [34] benchmark. Reuse the table of [34].

Method	IoU ↑	Chamfer-L2 ↓
OpenPose+SMPL	0.68	0. 243
IP-Net [7]	0.82	0. 034
NPMs [34]	0.83	**0.022**
Ours	**0.89**	0.029

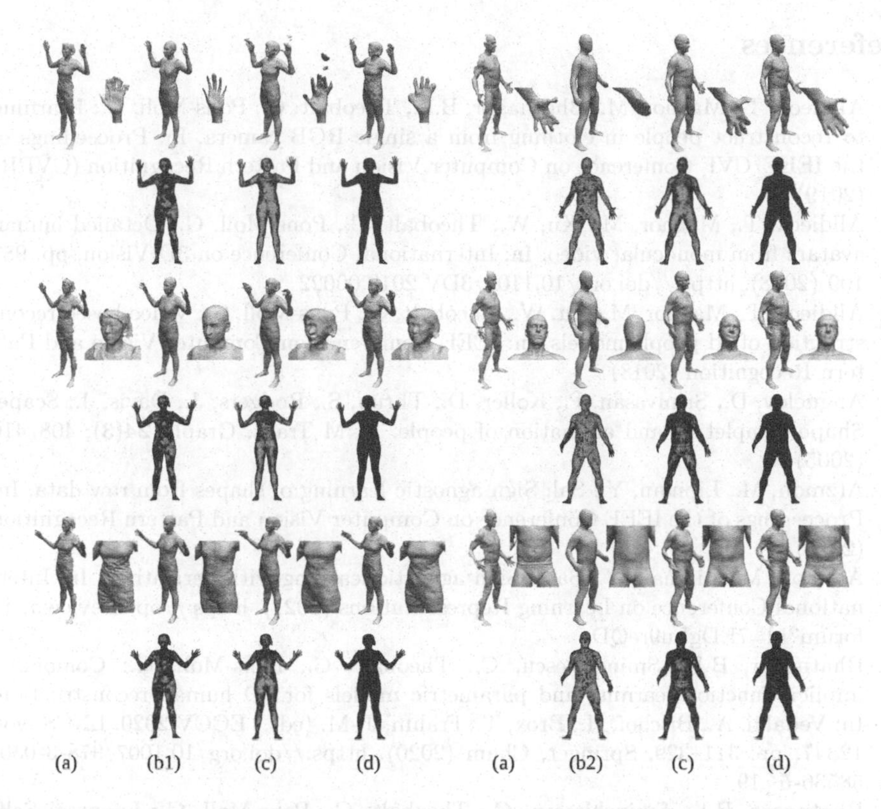

Fig. 5. Motion completion on CAPE(left) and DFAUST(right) datasets. From left to right, (a) front-view partial scan, completion with (b1) NPMs [34], (b2) OFlow [32], (c) STIF [51] and (d) our method. We set the maximum reconstruction to ground truth Chamfer-L1 distance as 2 cm for heatmaps.

6 Conclusion

We propose in this paper a complete framework for coarse-to-fine treatment and training of implicit depth completion from partial depth maps. We demonstrate the substantial quality improvement that can be obtained by using an architecture that proposes a residual pyramid of features in the context of implicit surface regression and a corresponding tailored training strategy which distributes losses at each scale of the feature map both spatially and temporally. The proposed architecture benefits from the temporal consistency and allows to preserve the high-frequency details of the surface. This result suggests the applicability of such implicit pyramid schemes for other problems in the realm of 3D vision and reconstruction.

Acknowledgements. This work has been partially supported by MIAI@Grenoble Alpes (ANR-19-P3IA-0003).

References

1. Alldieck, T., Magnor, M., Bhatnagar, B.L., Theobalt, C., Pons-Moll, G.: Learning to reconstruct people in clothing from a single RGB camera. In: Proceedings of the IEEE/CVF Conference on Computer Vision and Pattern Recognition (CVPR) (2019)
2. Alldieck, T., Magnor, M., Xu, W., Theobalt, C., Pons-Moll, G.: Detailed human avatars from monocular video. In: International Conference on 3D Vision, pp. 98–109 (2018). https://doi.org/10.1109/3DV.2018.00022
3. Alldieck, T., Magnor, M., Xu, W., Theobalt, C., Pons-Moll, G.: Video based reconstruction of 3d people models. In: IEEE Conference on Computer Vision and Pattern Recognition (2018)
4. Anguelov, D., Srinivasan, P., Koller, D., Thrun, S., Rodgers, J., Davis, J.: Scape: Shape completion and animation of people. ACM Trans. Graph. **24**(3), 408–416 (2005)
5. Atzmon, M., Lipman, Y.: Sal: Sign agnostic learning of shapes from raw data. In: Proceedings of the IEEE Conference on Computer Vision and Pattern Recognition (2020)
6. Atzmon, M., Lipman, Y.: Sald: Sign agnostic learning with derivatives. In: International Conference on Learning Representations (2021). https://openreview.net/forum?id=7EDgLu9reQD
7. Bhatnagar, B.L., Sminchisescu, C., Theobalt, C., Pons-Moll, G.: Combining implicit function learning and parametric models for 3D human reconstruction. In: Vedaldi, A., Bischof, H., Brox, T., Frahm, J.-M. (eds.) ECCV 2020. LNCS, vol. 12347, pp. 311–329. Springer, Cham (2020). https://doi.org/10.1007/978-3-030-58536-5_19
8. Bhatnagar, B.L., Sminchisescu, C., Theobalt, C., Pons-Moll, G.: Loopreg: Self-supervised learning of implicit surface correspondences, pose and shape for 3d human mesh registration. In: Advances in Neural Information Processing Systems (NeurIPS) (2020)
9. Bogo, F., Romero, J., Pons-Moll, G., Black, M.J.: Dynamic faust: Registering human bodies in motion. In: Proceedings of the IEEE Conference on Computer Vision and Pattern Recognition, pp. 6233–6242 (2017)
10. Cao, Z., Hidalgo Martinez, G., Simon, T., Wei, S., Sheikh, Y.A.: Openpose: Real-time multi-person 2d pose estimation using part affinity fields. IEEE Trans. Pattern Anal. Mach. Intell. (2019)
11. Chen, Z., Zhang, H.: Learning implicit fields for generative shape modeling. In: Proceedings of the IEEE/CVF Conference on Computer Vision and Pattern Recognition, pp. 5939–5948 (2019)
12. Chibane, J., Alldieck, T., Pons-Moll, G.: Implicit functions in feature space for 3d shape reconstruction and completion. In: Proceedings of the IEEE Conference on Computer Vision and Pattern Recognition. IEEE (2020)
13. Chibane, J., Mir, A., Pons-Moll, G.: Neural unsigned distance fields for implicit function learning. In: Advances in Neural Information Processing Systems (NeurIPS) (2020)
14. Cho, K., Van Merriënboer, B., Bahdanau, D., Bengio, Y.: On the properties of neural machine translation: Encoder-decoder approaches. arXiv preprint arXiv:1409.1259 (2014)
15. Chung, J., Gulcehre, C., Cho, K., Bengio, Y.: Empirical evaluation of gated recurrent neural networks on sequence modeling. arXiv preprint arXiv:1412.3555 (2014)

16. Deng, B., et al.: NASA neural articulated shape aproximation. In: Vedaldi, A., Bischof, H., Brox, T., Frahm, J.-M. (eds.) ECCV 2020. LNCS, vol. 12352, pp. 612–628. Springer, Cham (2020). https://doi.org/10.1007/978-3-030-58571-6_36
17. Dou, M., et al.: Fusion4d: Real-time performance capture of challenging scenes. ACM Transactions on Graphics (TOG) - Proceedings of ACM SIGGRAPH 2016 35 (2016). https://www.microsoft.com/en-us/research/publication/fusion4d-real-time-performance-capture-challenging-scenes-2/
18. Gropp, A., Yariv, L., Haim, N., Atzmon, M., Lipman, Y.: Implicit geometric regularization for learning shapes. In: Proceedings of Machine Learning and Systems 2020, pp. 3569–3579 (2020)
19. Groueix, T., Fisher, M., Kim, V.G., Russell, B., Aubry, M.: 3d-coded : 3d correspondences by deep deformation. In: Proceedings of the European Conference on Computer Vision, pp. 235–251 (2018)
20. Huang, Z., Xu, Y., Lassner, C., Li, H., Tung, T.: Arch: Animatable reconstruction of clothed humans. In: Proceedings of the IEEE/CVF Conference on Computer Vision and Pattern Recognition, pp. 3093–3102 (2020)
21. Lazova, V., Insafutdinov, E., Pons-Moll, G.: 360-degree textures of people in clothing from a single image. In: International Conference on 3D Vision (3DV) (2019)
22. Lewiner, T., Lopes, H., Vieira, A.W., Tavares, G.: Efficient implementation of marching cubes' cases with topological guarantees. J. Graph. Tools **8**(2), 1–15 (2003)
23. Lin, T.Y., Dollár, P., Girshick, R., He, K., Hariharan, B., Belongie, S.: Feature pyramid networks for object detection. In: IEEE Conference on Computer Vision and Pattern Recognition, pp. 936–944 (2017)
24. Liu, S., Saito, S., Chen, W., Li, H.: Learning to infer implicit surfaces without 3d supervision (2019)
25. Loper, M., Mahmood, N., Romero, J., Pons-Moll, G., Black, M.J.: SMPL: A skinned multi-person linear model. ACM Trans. Graph. **34**(6), 248:1-248:16 (2015)
26. Lorensen, W.E., Cline, H.E.: Marching cubes: A high resolution 3d surface construction algorithm. ACM Siggraph Comput. Graph. **21**(4), 163–169 (1987)
27. Ma, Q., et al.: Learning to dress 3d people in generative clothing. In: Proceedings of the IEEE Conference on Computer Vision and Pattern Recognition (2020)
28. Mahmood, N., Ghorbani, N., Troje, N.F., Pons-Moll, G., Black, M.J.: AMASS: Archive of motion capture as surface shapes. In: International Conference on Computer Vision, pp. 5442–5451 (2019)
29. von Marcard, T., Henschel, R., Black, M., Rosenhahn, B., Pons-Moll, G.: Recovering accurate 3d human pose in the wild using imus and a moving camera. In: Proceedings of the European Conference on Computer Vision (2018)
30. Mescheder, L., Oechsle, M., Niemeyer, M., Nowozin, S., Geiger, A.: Occupancy networks: Learning 3d reconstruction in function space. In: Proceedings of the IEEE Conference on Computer Vision and Pattern Recognition (2019)
31. Newcombe, R.A., Fox, D., Seitz, S.M.: Dynamicfusion: Reconstruction and tracking of non-rigid scenes in real-time. In: Proceedings of the IEEE Conference on Computer Vision and Pattern Recognition, pp. 343–352 (2015)
32. Niemeyer, M., Mescheder, L., Oechsle, M., Geiger, A.: Occupancy flow: 4d reconstruction by learning particle dynamics. In: Proceedings of the IEEE/CVF International Conference on Computer Vision (2019)
33. Niemeyer, M., Mescheder, L., Oechsle, M., Geiger, A.: Differentiable volumetric rendering: Learning implicit 3d representations without 3d supervision. In: Proceedings of the IEEE/CVF Conference on Computer Vision and Pattern Recognition, pp. 3504–3515 (2020)

34. Palafox, P., Božič, A., Thies, J., Nießner, M., Dai, A.: Npms: Neural parametric models for 3d deformable shapes. In: Proceedings of the International Conference on Computer Vision (2021)
35. Park, J.J., Florence, P., Straub, J., Newcombe, R., Lovegrove, S.: Deepsdf: Learning continuous signed distance functions for shape representation. In: Proceedings of the IEEE/CVF Conference on Computer Vision and Pattern Recognition, pp. 165–174 (2019)
36. Pavlakos, G., et al.: Expressive body capture: 3d hands, face, and body from a single image. In: Proceedings IEEE Conference on Computer Vision and Pattern Recognition (CVPR) (2019)
37. Peng, S., Niemeyer, M., Mescheder, L., Pollefeys, M., Geiger, A.: Convolutional occupancy networks. In: European Conference on Computer Vision (ECCV) (2020)
38. Pons-Moll, G., Pujades, S., Hu, S., Black, M.J.: Clothcap: Seamless 4d clothing capture and retargeting. ACM Trans. Graph. **36**(4), 1–15 (2017)
39. Pons-Moll, G., Romero, J., Mahmood, N., Black, M.J.: Dyna: A Model of Dynamic Human Shape in Motion, vol. 34, pp. 120:1–120:14 (2015)
40. Prokudin, S., Lassner, C., Romero, J.: Efficient learning on point clouds with basis point sets. In: Proceedings of the IEEE/CVF International Conference on Computer Vision, pp. 4332–4341 (2019)
41. Ronneberger, O., Fischer, P., Brox, T.: U-Net: Convolutional networks for biomedical image segmentation. In: Navab, N., Hornegger, J., Wells, W.M., Frangi, A.F. (eds.) MICCAI 2015. LNCS, vol. 9351, pp. 234–241. Springer, Cham (2015). https://doi.org/10.1007/978-3-319-24574-4_28
42. Saito, S., Huang, Z., Natsume, R., Morishima, S., Kanazawa, A., Li, H.: Pifu: Pixel-aligned implicit function for high-resolution clothed human digitization. In: Proceedings of the IEEE International Conference on Computer Vision, pp. 2304–2314 (2019)
43. Saito, S., Simon, T., Saragih, J., Joo, H.: Pifuhd: Multi-level pixel-aligned implicit function for high-resolution 3d human digitization. In: Proceedings of the IEEE Conference on Computer Vision and Pattern Recognition (2020)
44. Sun, D., Yang, X., Liu, M.Y., Kautz, J.: PWC-Net: CNNs for optical flow using pyramid, warping, and cost volume (2018)
45. Sun, J., Xie, Y., Chen, L., Zhou, X., Bao, H.: NeuralRecon: Real-time coherent 3D reconstruction from monocular video. In: Proceedings of the IEEE Conference on Computer Vision and Pattern Recognition (2021)
46. Xu, H., Bazavan, E.G., Zanfir, A., Freeman, W.T., Sukthankar, R., Sminchisescu, C.: Ghum & ghuml: Generative 3d human shape and articulated pose models. In: Proceedings of the IEEE/CVF Conference on Computer Vision and Pattern Recognition, pp. 6184–6193 (2020)
47. Xu, Q., Wang, W., Ceylan, D., Mech, R., Neumann, U.: Disn: Deep implicit surface network for high-quality single-view 3d reconstruction. Adv. Neural Inf. Process. Syst. **32**, 492–502 (2019)
48. Yang, J., Mao, W., Alvarez, J.M., Liu, M.: Cost volume pyramid based depth inference for multi-view stereo. In: Proceedings of the IEEE/CVF Conference on Computer Vision and Pattern Recognition (CVPR) (2020)
49. Li, Y., Hikari Takehara, T.T.B.Z., Nießner, M.: 4D complete: Non-rigid motion estimation beyond the observable surface (2021)
50. Yu, T., Zheng, Z., Guo, K., Liu, P., Dai, Q., Liu, Y.: Function4d: Real-time human volumetric capture from very sparse consumer RGBD sensors. In: Proceedings of the IEEE Conference on Computer Vision and Pattern Recognition (2021)

51. Zhou, B., Franco, J.S., Bogo, F., Boyer, E.: Spatio-temporal human shape completion with implicit function networks. In: Proceedings of the International Conference on 3D Vision (2021)
52. Zhou, B., Franco, J.S., Bogo, F., Tekin, B., Boyer, E.: Reconstructing human body mesh from point clouds by adversarial GP network. In: Proceedings of the Asian Conference on Computer Vision (ACCV) (2020)

Layered-Garment Net: Generating Multiple Implicit Garment Layers from a Single Image

Alakh Aggarwal[1], Jikai Wang[1], Steven Hogue[1], Saifeng Ni[2], Madhukar Budagavi[2], and Xiaohu Guo[1(✉)]

[1] The University of Texas at Dallas, Richardson, TX, USA
{alakh.aggarwal,jikai.wang,ditzley,xguo}@utdallas.edu
[2] Samsung Research America, Plano, TX, USA
{saifeng.ni,m.budagavi}@samsung.com

Abstract. Recent research works have focused on generating human models and garments from their 2D images. However, state-of-the-art researches focus either on only a single layer of the garment on a human model or on generating multiple garment layers without any guarantee of the intersection-free geometric relationship between them. In reality, people wear multiple layers of garments in their daily life, where an inner layer of garment could be partially covered by an outer one. In this paper, we try to address this multi-layer modeling problem and propose the Layered-Garment Net (LGN) that is capable of generating intersection-free multiple layers of garments defined by implicit function fields over the body surface, given the person's near front-view image. With a special design of garment indication fields (GIF), we can enforce an implicit covering relationship between the signed distance fields (SDF) of different layers to avoid self-intersections among different garment surfaces and the human body. Experiments demonstrate the strength of our proposed LGN framework in generating multi-layer garments as compared to state-of-the-art methods. To the best of our knowledge, LGN is the first research work to generate intersection-free multiple layers of garments on the human body from a single image.

Keywords: Image-based reconstruction · Multi-layered garments · Neural implicit functions · Intersection-free

1 Introduction

Extracting 3D garments from visual data such as images enables the generation of digital wardrobe datasets for the clothing and fashion industry, and is useful in

A. Aggarwal, J. Wang, S. Hogue and X. Guo are partially supported by National Science Foundation (OAC-2007661).

Supplementary Information The online version contains supplementary material available at https://doi.org/10.1007/978-3-031-26319-4_23.

L. Wang et al. (Eds.): ACCV 2022, LNCS 13841, pp. 378–395, 2023.
https://doi.org/10.1007/978-3-031-26319-4_23

Virtual Try-On applications. With the limitation on certain classes of garments, it is already possible to generate explicit upper and lower garment meshes from a single image or multi-view images [1,2], to introduce different styles to the garments, such as length, along with varying poses and shapes [3–5], and to transfer the garments from one subject to another [1].

However, to the best of our knowledge, none of the existing approaches have the capability of generating multiple *intersection-free* layers of clothing on a base human model where an inner layer of garment could be partially covered by an outer one without any intersection or protrusion. This does not conform to reality because people wear multiple layers of garments in their daily life. The existing techniques either generate a single layer of upper-body cloth (e.g., T-shirt, jacket, etc.) and a single layer of lower-body cloth (e.g., pants, shorts, etc.) without any overlap in their covering regions [1,2], or generate multiple garment layers, but without any guarantee on their intersection-free geometry [5].

The fundamental challenge here is to ensure intersection-free between multiple garment layers when they overlap. Existing approaches to garment representation are based on *explicit* models, by using either displacement fields over SMPL surface (SMPL+D) [1,6] or skinned meshes on top of SMPL [2]. However, with explicit mesh representations, it is very difficult to ensure intersection-free between multiple garment layers. SMPLicit [5] is an implicit approach that generated multiple layers of garments but does not handle intersections among multiple layers. In this paper, we propose to use a set of *implicit* functions – signed distance fields (SDF), to represent different layers of garments. The benefit is that the intersection-free condition can be easily enforced by requiring the SDF of the inner layer to be greater than the SDF of the outer one. We call this the *Implicit Covering Relationship* (Sect. 3.1) for modeling multi-layer garments.

There are two challenges associated with the such implicit representation of garments as well as the enforcement of implicit covering relationship: (1) Most of the garments are *open* surfaces with boundaries, while SDF can represent *closed* surfaces only. (2) The implicit covering relationship should only be enforced in those regions where two layers overlap, but how can we define such overlapping regions? In this paper, we solve these two challenges by proposing an implicit function called *Garment Indication Field* (GIF, Sect. 3.2) which successfully identifies those regions where the garment has "holes" – the open regions where the garment does not cover. With such garment indication fields, we not only can enforce the implicit covering relationships between layers but also can extract the open meshes of garments by trimming the closed marching cubes surfaces.

We propose a *Layered-Garment Net* (LGN), which consists of a parallel SDF subsystem and GIF subsystem, that can take an image of the person as input, and output the corresponding SDF and GIF for each garment layer. Specifically, based on the projection of the query point in image space, we obtain its local image features from the encoded features given by a fully convolutional encoder. Using the local image features and other spatial features of the query point, we train different decoder networks for different layers of garments to predict their SDF and GIF, respectively. The network is trained end-to-end, utilizing a covering inconsistency loss given by GIFs and SDFs of different layers, along with other loss functions to regress the predictions to the ground-truth values. The contributions of this paper can be summarized as follows:

- We present a Layered-Garment Net, the first method that can model and generate multiple intersection-free layers of garments and the human body, from a single image.
- We enforce an implicit covering relationship among different layers of garments by using multiple signed distance fields to represent different layers, which guarantees that multiple layers of garments are intersection-free on their overlapping regions.
- We design garment indication fields that can be used to identify the open regions where the garments do not cover, which can be used to identify the overlapping regions between different layers of garments, as well as to extract open meshes of garments out of the closed surfaces defined by SDF.

2 Related Works

In this section, we will review the recent works in two areas of research that are related to our work. We consider **Full Human Body Reconstruction** where the focus is on generating a good quality clothed human model and **Individual Garment Surface Reconstruction** where the focus is on obtaining individual garments for a human model.

Full Human Body Reconstruction. Many recent works generated explicit representations of human body mesh using parametric models for naked human models to handle varying geometry [7–9]. This allows them to modify the shape and pose of the generated model according to shape parameters β and pose parameters θ. The underlying idea is to obtain the parameters β and θ that closely defines the target human body, and apply linear blend skinning using the blend shapes and blending weights to generate the final human body geometry. Bogo et al. [10] obtained these parameters and fitted a human body model from single unconstrained images. Many deep learning-based methods [11,12] have since then come up, that estimate the shape and pose parameters of a human model. Smith et al. [13] employed the use of silhouettes from different viewpoints to generate the human body. Subsequently, some research works [14,15] also used semantic segmentation of human parts to ensure more accurate parameter estimation. However, the above-mentioned works only generate naked human models and do not reconstruct clothed human models.

To address this issue, several recent research works have focused on the displacements of a naked body. Alldieck et al. [6] used frames at some continuous interval from a video of a subject rotating in front of the camera to ensure accurate parameter estimation from different viewpoints and used SMPL+D for clothed human body reconstruction. Such SMPL+D representation uses a displacement vector for the vertices of the naked human body model to represent clothing details and was later used for single image reconstruction [16]. Tex2Shape [17] was able to obtain better displacement details by predicting the displacement map for a model that aligns with the texture map of the model. Several recent works [18] generate explicit dynamic human models. However, since all the above methods are only based on a naked human body model, they cannot generate a human body wearing complex garments like skirts, dresses, long hair, etc. To address these issues, some research works [19,20] used a volumetric representation of the human body with voxelized grids. Ma et al. [21] obtain the point clouds of clothed humans with varying garment topology. Some recent works [22,23] also focus on generative approaches for 3D clothed human reconstruction.

There have been some recent works focusing on the implicit clothed body surface representation. Mescheder et al. [24] used the occupancy field to determine if a point is inside or outside a surface of any object from the ShapeNet [25] dataset, and then used a classifier to generate a surface dividing the 3D space into inside/outside occupancy values. They calculated occupancy values for each point of the voxel grid and used marching cubes [26] to generate the surface. They do not have to store voxel grid representation or any other mesh information for all the data instances. Different from the occupancy field, Chibane et al. [27] predicted an unsigned distance field using a neural network, and projected the points back to the surface to generate a point cloud-based surface using the gradient of the distance field at that point, and could be used to further generate a complete mesh surface. Several recent works [28–30] predicted a Signed Distance Field and used marching cubes [26] to generate a mesh surface. This ensures more accurate geometry because of the implicit field's dependence on distance. Based on the above works on implicit fields, PIFu [31], PIFuHD [32], StereoPIFu [33], GeoPIFu [34], PaMIR [35] take a 2D image or depth data of human as input, and after extracting the local encoded image features for a point, they predict the occupancy field of the dressed body. MetaAvatar [36] represent cloth-specific neural SDFs for clothed human body reconstruction. Other recent works [37–39] aim to dynamically handle the reconstruction of animatable clothed human models via implicit representation. Several other works [5,40–42] also use implicit fields for 3D human reconstruction. Bhatnagar et al. [43] combine use base explicitly defined SMPL model to implicitly register scans and point clouds. The method identifies the region between garment and body, however, it does not reconstruct different garment layers. Handling individual garment regions like Garment Indication Field is more complicated. Scanimate [42] reconstructs a dynamic human model and utilizes an implicit field for fine-tuning their reconstruction. Instead of supervision, they utilize Implicit Geometric Regularization [44] to reconstruct surfaces using implicit SDF in a semi-supervised approach.

Individual Garment Surface Reconstruction. Instead of simply generating a human body model with displacements, Multi-Garment Net (MGN) [1] generates an explicit representation of parametric garment models with SMPL+D. Using single or multiple images, it predicts different upper and lower garments that are parameterized for varying shapes and poses. However, MGN cannot produce garments that do not comply with naked human models, like skirts and dresses. TailorNet [3] uses the wardrobe dataset from MGN and applies different style transforms like sleeve-length to obtain different styles of garments. DeepCloth [4] enable deep-learning based styling of garments. SIZER [45] provides a dataset enabling resizing of the garment on the human body. Deep Fashion3D [46] generates a wardrobe dataset, consisting of complex garment shapes like skirts and dresses. BCNet [2] uses a deformable mesh on top of SMPL to represent garments and proposes a skinning weights generating network for the garments to support garments with different topologies. SMPLicit [5] obtains shape and style features for each garment layer from the image and uses these parameters to obtain multiple layers of garments, and uses a distance threshold to reconstruct overlapping garment layers. However, they do not guarantee intersection-free reconstruction. GarmentNets [47] reconstructs dynamic garments utilizing Generalized Winding Number [48] for occupancy and correct trimming of openings in garment meshes. Their approach, however, does not provide a garment's indicator field.

To the best of our knowledge, none of the existing works can generate overlapping intersection-free multiple layers of garments where an inner layer could be partially covered by an outer one. All existing works on individual garment generation [1–3,46] use explicit mesh representation, making them difficult to ensure intersection-free between different layers. SMPLicit [5] does not guarantee intersection-free reconstruction among different layers, especially in overlapping regions. In this paper, we resort to implicit representation and model the multiple layers of garments with signed distance fields (SDF) which makes it easy to enforce the implicit covering relationship among different layers of garment surfaces with the help of a carefully designed implicit garment indication field (GIF). The combination of these two implicit functions, SDF and GIF, makes the modeling and learning of multi-layer intersection-free garments possible.

3 The Method

Given a near-front-facing image of a posed human, we aim to generate the different intersection-free garment surface layers. The reconstructed surfaces should follow a covering relationship between each other and the body. Our proposed Layered-Garment Net (LGN) can generate implicit functions of Signed Distance Field (SDF) and Garment Indication Field (GIF) for different layers of garments over varying shapes and poses. An overview of our approach is given in Fig. 1.

3.1 Implicit Covering Relationship

For two layers of garments i and j, let layer i be partially covered by layer j. If a point p belongs to their overlapping regions, the SDF values $s_i(p)$ and $s_j(p)$ for the two layers should follow the covering relationship:

$$s_j(p) < s_i(p). \tag{1}$$

This is illustrated in Fig. 2(left). The inequality does not hold for all the points in 3D space but only holds for the overlapping region between the two layers. We are only interested in the points near the surface of the garment layer. Let us consider an example where layer i is a pant and layer j is a shirt. The inequality

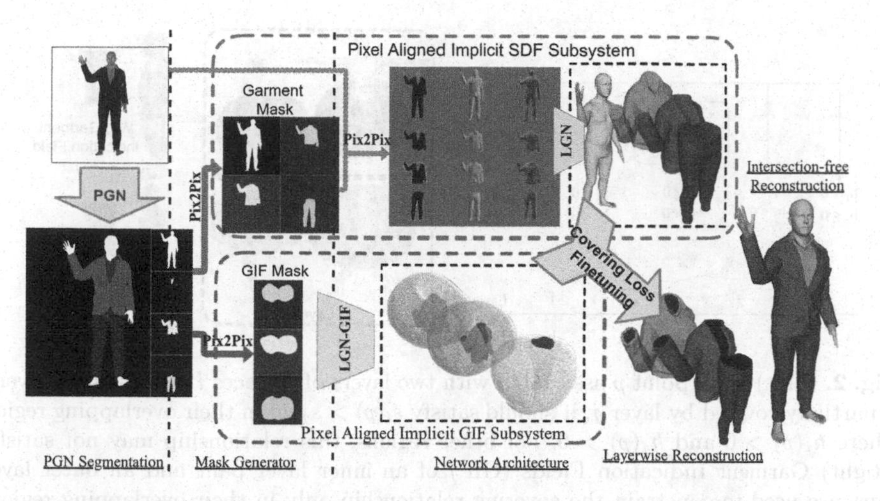

Fig. 1. Given an input image, our method first obtains PGN [49] segmentation and incomplete masks for each garment from segmentation. Then the complete masks of garments are generated by Pix2Pix-HD Garment Mask generator. Similarly the indicator masks for garments are generated by Pix2Pix-HD Indicator Mask generator. Using the masked input image, the front and back normals are obtained using Pix2Pix-HD Normal Subnetwork. Then an Encoder and Decoders of LGN network use masked images and normals and predicts SDF value $s_i(p)$ of layer i for any point p in 3D space. LGN-GIF further uses Indicator Masks ind_i to obtain GIF value $h_i(p)$ of layer i for any point p in 3D space. Finally, LGN is fine-tuned with the covering loss in Eq. (2) to avoid intersection among different layers.

Eq. (1) should not be satisfied in the leg region of the human body, otherwise, this would result in the generation of a shirt layer on top of the pant layer in the leg region, where the shirt originally does not exist. This problem is shown in Fig. 2(right). Hence, we need an indicator function for both layers, and only ensure that the implicit covering relationship Eq. (1) holds on points that are related to both layers i and j. We call this indicator function for layer i the *Garment Indication Field* (GIF), and denote it as $h_i(p)$ (Sec 3.2).

To ensure the network's SDF predictions follow the Implicit Covering Relationship inequality for relevant points p, we can define the covering loss for all layers of surfaces for our network as follows:

$$\mathcal{L}_{cov}(p) = \sum_{j=1}^{N} \sum_{i \in C(j)} h_j(p) * h_i(p) * [max(s_j(p) - s_i(p), 0) + \lambda(s_j(p) - s_i(p))^2], \quad (2)$$

where $C(j)$ is the set of layers partially covered by layer j. The multiplication with $h_j(p)$ and $h_i(p)$ guarantees that the covering loss only applies to the points in the overlapping region between two layers. The last term regularizes the difference between the two SDFs. We choose $\lambda = 0.2$ in all our experiments.

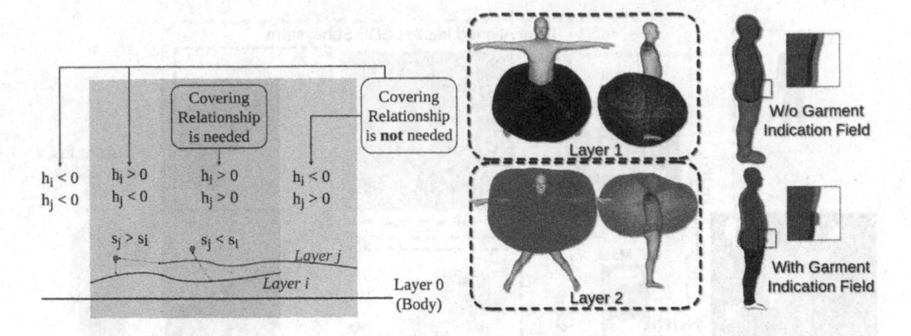

Fig. 2. (Left) For a point p associated with two layers of surfaces i and j, where layer i is partially covered by layer j, it should satisfy $s_i(p) > s_j(p)$ in their overlapping region where $h_i(p) > 0$ and $h_j(p) > 0$. For other regions, this relationship may not satisfy. (Right) Garment Indication Fields (GIF) of an inner layer pant and an outer layer shirt are used to constrain the covering relationship only in their overlapping region. Without GIF, the outer layer would completely cover the inner layer since $s_i(p) > s_j(p)$ would be enforced everywhere.

3.2 Garment Indication Field

For a garment and a query point p, we use its generalized winding number [48], denoted as $W(p)$, to distinguish the open regions from the regions concerned with garment surfaces. Since all the garments are open surfaces, $W(p)$ is equal to 0.5 at the opening regions. $W(p) > 0.5$ for a point inside the surface, and keeps increasing as the point gets farther inside. Similarly, $W(p) < 0.5$ for a point outside the surface, and keeps decreasing as it goes further away from the open regions. In far-off regions and outside the surfaces, $W(p) \leq 0$. Using different field functions as a function of winding number, we can have different observations as shown in Fig. 3.

Observation 1: $o(p) = W(p) - 0.5$ gives the occupancy field for a garment. This has been shown in Fig. 3(b). We call $o(p)$ the *winding occupancy.* This helps us in obtaining the sign of SDF for a non-watertight mesh. Since all garments, in particular, are non-watertight open mesh, for any query point p in 3D space, the distance $d(p)$ to its nearest surface point is essentially an *unsigned* distance because there is no inside/outside for the open surface. Thus we use $o(p)$ to obtain a watertight surface mesh with marching cubes first, then compute a *ground-truth SDF* $s'(p)$ for the watertight garment surface.

Observation 2: $h'(p) = W(p) * (W(p) - w_h)$ gives an indication field of the garment opening region, where $0.5 < w_h < 1$. As previously discussed, $W(p)$ is greater than 0.5 inside the opening region and the surface mesh, and it keeps on increasing inside the mesh. Similarly, $W(p)$ is less than 0.5 outside the opening region and keeps decreasing away from the region outside the mesh. In this paper, we choose $w_h = 0.75$ for all garments. For any point that is inside the mesh and away from the garment opening region, $W(p) > 0.75$, so $h'(p)$ is

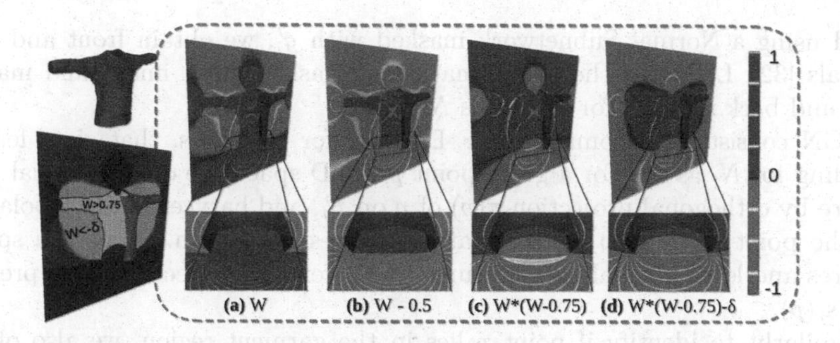

Fig. 3. For a given garment mesh (non-watertight), we show the cross-section views of the following fields: (a) Generalized Winding Number W, (b) Winding Occupancy $W - 0.5$, (d) "Hole" Region Indication $W * (W - 0.75)$, and (e) Garment Indication Field $W * (W - 0.75) - \delta$. Transition from (c) to (d) allows concise bound for GIF, which will not intersect with nearby body surfaces.

positive. Similarly, if it is outside the mesh and away from the garment opening region, $W(p) < 0$, so $h'(p)$ is positive too. However, for any point that is located close to the 0.5-level isosurface, $0 < W(p) < 0.75$, so $h'(p)$ is negative. In this way, $h'(p)$ indicates the open region of the garment. This can be observed from Fig. 3(c).

Furthermore, it also follows that $h'(p) - \delta$, for some $\delta \to 0^+$ gives a bound region of the garment closer to the mesh. This has been shown in Fig. 3(d). We observe that, for $\delta = 0.01$, we get a good quality bound for this indication field. Thus we define the following function as Garment Indication Field (GIF) for the garment surface:

$$\hat{h}(p) = (sign[W(p) * (W(p) - 0.75) - \delta] + 1) * 0.5. \tag{3}$$

Here $\hat{h}(p) = 1$ means the point is in the region close to the garment surface, otherwise $\hat{h}(p) = 0$. Such ground-truth GIF values will be used for enforcing the covering relationship in Eq. (2).

3.3 Layered-Garment Net

Given an input image I of a person, we first obtain the garment segmentation P on the image using Part-Grouping Network [49]. It is possible to obtain different garment masks g_i for garment layer i using the corresponding pixel color. However, the mask g_i may not be complete because of overlap with outer layers. Hence, we train a Garment Mask generator, that takes the incomplete mask g_i and PGN segmentation P as input, and outputs a corrected garment mask g'_i.

Like PiFU-HD [32], we follow a similar pipeline, however, with no requirement for a Fine-Level network, but only a Coarse network for each garment layer. We also use semi-supervised Implicit Geometric Regularization (IGR) [44] for fine-tuning SDF prediction on the surface. For layer i, we use mask g'_i on input image

I and using a Normal Subnetwork masked with g_i', we obtain front and back normals [32]. Let's call the concatenation of masked input image and masked front and back normals for layer i as N_i.

LGN consists of a common SDF Encoder for all layers, that gives feature encoding for N_i as F_i. For a given point p in 3D space, we obtain a local pixel feature by orthogonal projection $\pi(p)$ of p on F_i, and barycentric interpolation. For the point p, we also obtain spatial features like depth. Using the spatial features and local pixel-aligned features, the layer's SDF Decoder $\mathcal{S}_i(.)$ predicts SDF $s_i(p)$.

Similarly, to identify if point p lies in the garment region, we also obtain Indicator Mask ind_i of layer i from PGN P and incomplete mask g_i by training an Indicator Mask generator. Then we train a common GIF Encoder that gives encoding F_i' and the layer's GIF Decoder $\mathcal{H}_i(.)$ to obtain GIF value for layer i as $h_i(p)$.

Mask generators for each garment follow the same architecture as front and back Normal Subnetworks in [32], i.e. Pix2PixHD network [50]. We have a common SDF Encoder and front and back Normal Subnetworks among all garment layers and body layers. A common GIF Encoder is defined for all garment layers. However, we separately define SDF Decoders, GIF Decoders, Garment Mask generators, and Indicator Mask generators for each garment - shirt, pant, coat, skirt, dress.

$$s_i(p) = \mathcal{S}_i(F_i(\pi(p)), \phi(p)),\ h_i(p) = \mathcal{H}_i(F_i'(\pi(p)), \phi(p)), \tag{4}$$

where the spatial feature $\phi(p)$ here is depth.

The L1 loss for the generated SDF is formulated as the following L^1 norm:

$$\mathcal{L}_{sdf}(p) = \sum_{i=0}^{N} |s_i(p) - \hat{s_i}(p)|, \tag{5}$$

where $\hat{s_i}(p)$ is the ground-truth SDF value for point p from layer i, and N is the number of garment layers.

Similarly, the L1 loss for the predicted GIF is formulated as follows:

$$\mathcal{L}_{gif}(p) = \sum_{i=1}^{N} |h_i(p) - \hat{h}_i(p)|, \tag{6}$$

where $\hat{h}_i(p)$ is the ground-truth GIF for the garment layer i and query point p.

We fine-tune network parameters for SDF prediction using Implicit Geometric Regularization (IGR) [44]. Loss for IGR is given as follows:

$$\begin{aligned} \mathcal{L}_{igr}(p) &= \tau(p)\ell_{\mathcal{X}}(p) + \lambda(||\nabla_p s_i(p)|| - 1)^2, \\ \ell_{\mathcal{X}}(p) &= |s_i(p)| + ||\nabla_p s_i(p) - n_p||, \end{aligned} \tag{7}$$

where $s_i(p)$ is the SDF value at p, $\tau(p)$ is an indicator of a point on surface $\mathcal{X}$ and n_p is the surface normal at point p.

3.4 Training and Inference

We first pre-train the Garment Mask generator and Indicator Mask generators on PGN segmentation and incomplete mask as inputs for each garment category - shirt, pant, coat, skirt, dress. To train the network, we sample 20,480 points on the surface of each layer. We add normal perturbation $\mathcal{N}(0, \sigma = 5cm)$ on these points to generate the near-surface samples. We then add random points in 3D space using a ratio of 1 : 16 for the randomly sampled points w.r.t. the near-surface samples. These sampled points are used to optimize the SDF prediction of all garment layers and covering loss between each layer of the garment. We similarly sample points from the 0.5 level iso-surface of the ground-truth Garment Indication Fields (GIFs) of each layer garment layer and add normal perturbation $\mathcal{N}(0, \sigma = 5cm)$ on these points to generate garment indications. We add random points in 3D space using a ratio of 1 : 16 for the randomly sampled points w.r.t. the garment indicating samples. For GIFs, we add additional points along the edges of ground truth mesh to obtain accurate trimming.

Given an input image I, we first obtain PGN segmentation image P which contains different garments in the image. For each garment, we obtain their incomplete masks g_i. Using P and g_i for each layer, we obtain garment masks g_i' and ind_i. We leave out the indicator mask prediction for the body layer since it is not required to obtain GIF for the body. It is assumed that the GIF value for the body layer is 1 at any point. For all the near-surface sampled points, we calculate the ground-truth SDF values $\hat{s}_i$ for each layer i as explained in Sect. 3.2. The encoder and decoder are warmed-up by training with the loss $\mathcal{L}_{sdf}$ and $\mathcal{L}_{igr}$ as defined in Eq. (5) and Eq. (7). We also calculate ground-truth GIF values $\hat{h}_i$ for each layer $i > 0$ with the loss $\mathcal{L}_{gif}$ as defined in Eq. (3). Using the predicted h_i values and the predicted s_i values for all the sampled points, the network is fine-tuned with the covering loss as defined in Eq. (2). This ensures that the output SDF values follow the covering relationship inequality as defined in Eq. (1). For all the garments indicating sampled points, we calculate their ground-truth GIF values for each layer of the garment. Using the ground-truth GIF values for each layer, the GIF value prediction is optimized.

During the inference, after obtaining the SDF values for each layer, we use marching cubes to obtain its triangle mesh. Then, we apply a trivial post-processing step using predictions, to update SDF values. For a given point p where GIF of both layers i and j overlap: if $s_j > s_i - \epsilon$, $s_j = s_i - \epsilon$ where ϵ is a very small number. Experimentally, we use ϵ to be $1e - 3$. Finally, all the triangular meshes obtained for each layer are trimmed by the predicted GIF values on the vertices of the mesh. To trim the garment opening regions, the triangles which have different signs of GIF values for its three vertices are selected, and the triangle is trimmed by linearly interpolating GIF values over each edge. Thus, we finally obtain multiple layers of garments along with the reconstructed Layer-0 body that follows the covering relationship.

For both training and inference, we rely on covering the relationship manually specified with the input image. Different garment layers are then obtained from the output of LGN by satisfying the covering relationship.

4 Experiments

4.1 Dataset and Implementation Details

Dataset Preparation. Our multi-layer garment dataset is constructed from 140 purchased rigged human models from AXYZ [51]. For each rigged model, we first perform SMPL [7] fitting to obtain its body shape and pose parameters. We generate eight images from different views for each human model and run semantic segmentation on each image with Part Grouping Network (PGN) [49]. Using the fitted SMPL, we obtain those segmentations on the SMPL surface and map them to the UV texture space of SMPL. This enables us to perform texture stitching [52] to generate the segmentation texture map. By projecting the texture segmentation onto the 3D human model, we obtain the segmentation of different 3D garment meshes, followed by minor manual corrections on some garment boundaries. Our processed garments include the categories of Shirt, Coat, Dress, Pant (long and short), and Skirt, while Shirt/Coat/Dress all contain three subcategories of no-sleeve, short-sleeve, and long-sleeve. Detailed statistics of the processed garments are provided in the supplementary document.

Table 1. Comparison results (in *cm*) for **A** per-garment Point-to-Surface on (i) Digital Wardrobe [1], (ii) SIZER [45], and **B** Full body reconstruction on BUFF Dataset [53]

	Model	*P2S*		Model	*P2S*		Model	*Chamfer*	*P2S*
A(i)	BCNet	9.75	A(ii)	BCNet	**3.84**	B	PiFU-HD	**1.22**	**1.19**
	SMPLicit	9.12		SMPLicit	6.01		BCNet	1.93	1.96
	Ours	**9.09**		Ours	4.04		Ours	2.75	2.6

Using different garments, we synthesize around 12,000 different combinations of multi-layer garments on top of a layer-0 SMPL body, in 7 different poses. When combining different garment types, we follow the assumption that the length of sleeves for the inner layer should NOT be shorter than that of the outer layer. Otherwise, the sleeves of the inner layer are covered by the outer garment and there is no visual clue to tell its length. We then use this combination of generated garment models with a layer-0 body to train our LGN. We use the synthesized combinations of multi-layer garments as the training set. The geometries of garments are corrected to make sure no intersections exist and the different layers of garment follow the covering relationship. For testing, we use BUFF [53] and Digital Wardrobe [1] datasets. The dataset preparation details are discussed in the supplementary document.

Implementation Details. The base architecture of our LGN is similar to that of PIFu [31] and PIFu-HD [32] since they also predict implicit fields aligned with image features. For SDF Subsystem, we first obtain garment masks from PGN segmentation of input image using Garment Mask Generators and obtain masked front and back normals using Normal Subnetworks, which are Pix2Pix-HD [50] networks. We use 4 stacks of Stacked Hourglass Network (HGN) [54] to encode the image features from the concatenation of normals and image. From spatial features from points and local encoded features by performing bi-linear interpolation of projected points on image feature space, different Multi-layer Perceptron (MLP) decoder layers predict SDF values for each layer of the garment, with layer 0 being the human body. Similarly, for GIF Subsystem, Indicator Mask Generators obtain GIF masks. 4 stacked-HGN encodes image features from the concatenation of PGN segmentation and indicator mask. Then, GIF is predicted using GIF Decoders. To optimize the network, we first pre-train Mask Generators. Then we individually train SDF Subsystem and GIF Subsystems. Thereafter, we use the covering loss to fine-tune SDF prediction to avoid the intersection, and GIF to ensure appropriate trimming of the open region on garment surfaces, and a consistent multi-layer covering relationship. We evaluate our methods and test with various state-of-the-art approaches on mainly two areas - 3D Clothed Human Reconstruction and Individual Garment Reconstruction. The quantitative comparison of 3D Clothed Human Reconstruction of our method with BCNet [2], PiFU-HD [32] and SMPLicit [5] are shown in the supplementary document. We omit comparison with MGN [1], Octopus [52] and PiFU [31] because of the availability of better reconstruction methods.

4.2 Quantitative Comparisons

We compare our methods with the state-of-the-art (Table 1) approaches on three publicly available datasets: Digital Wardrobe Dataset [1], SIZER Dataset [45] and BUFF Dataset [53]. We use Digital Wardrobe Dataset and SIZER Dataset to compare individual garment reconstructions, and BUFF Dataset [53] to compare full human body reconstruction. It is to be noted that since the datasets mentioned consist of only 2 layers of garments – upper and lower, we cannot make a comparison with them on multi-layer garment reconstruction. Please also note, we do not use BCNet [2] data set to have a fair comparison with BCNet. Also, we are unable to compare our results with DeepFashion3D [46] data set because the dataset only consists of garments and no human body.

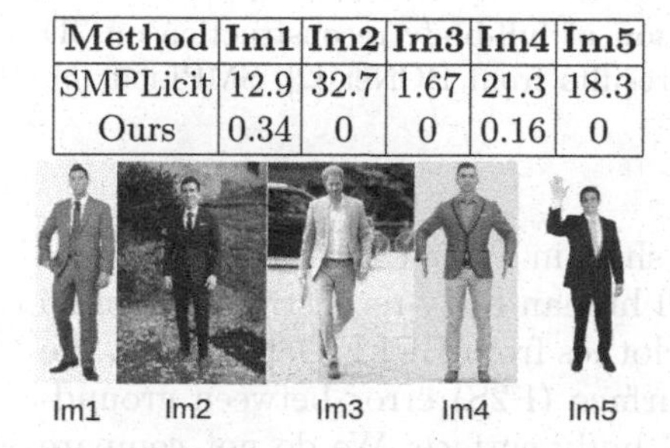

Method	Im1	Im2	Im3	Im4	Im5
SMPLicit	12.9	32.7	1.67	21.3	18.3
Ours	0.34	0	0	0.16	0

Fig. 4. Penetration depths (in cm).

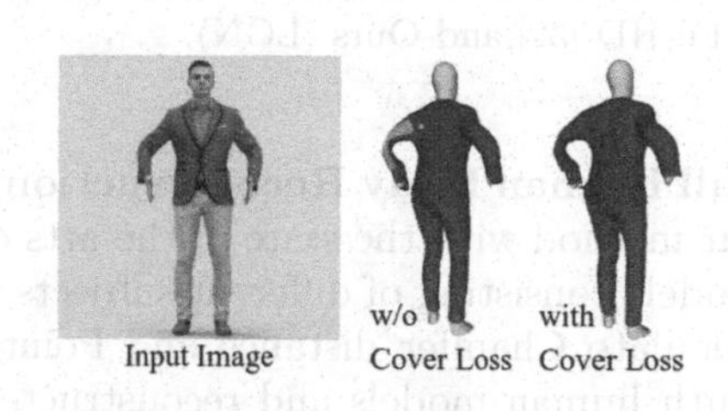

Fig. 5. Cover loss finetuning

Individual Garment Reconstruction. To compare our method with the state-of-the-art garment reconstruction approaches [2,5], we select 96 models from Digital Wardrobe Dataset [1] and 97 models from SIZER Dataset [45]. We use segmented Upper and Lower garments available with Dataset for comparison. We calculate the Mean P2S Error per garment between reconstructed garments and their ground-truth counterpart and observe the performance of our approach with other approaches in Table 1 A(i)&(ii). Our model outperforms state-of-the-arts on Digital Wardrobe Dataset [1]. For SIZER [45], BCNet performs better due to assuming reconstruction of 2 layer garments only, since segmentation for 3 layer of garments does not exist in the data set.

In Fig. 4, we calculate the Maximum Penetration Depth between different reconstructed garment layers and make a comparison with SMPLicit [5]. It can be seen that our work outperforms the state-of-the-art in this case.

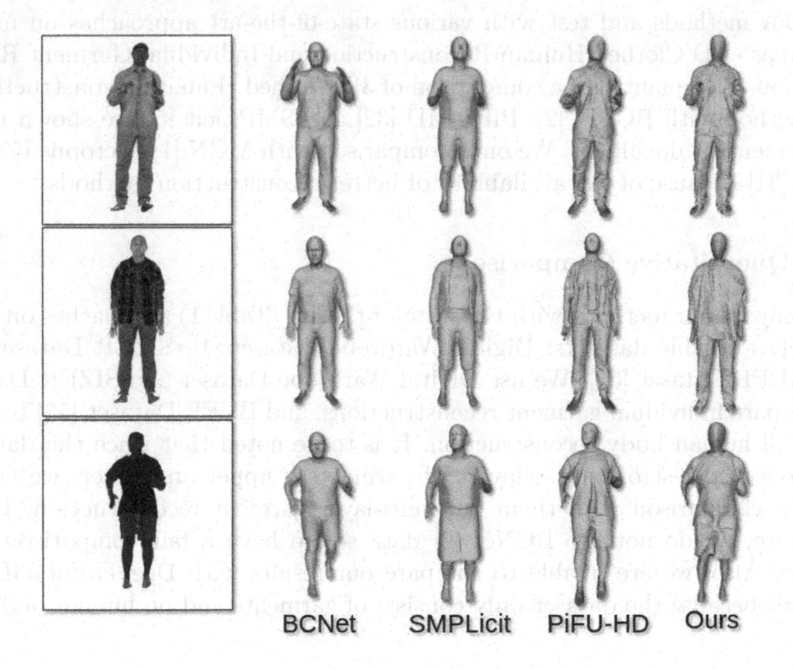

Fig. 6. From left to right, qualitative comparison of full-body reconstruction on 3D clothed human from ground truth (left), and results from BCNet [2], SMPLicit [5], PIFu-HD [32] and Ours (LGN).

Full Human Body Reconstruction. We show in Table 1 B the comparison of our method with the state-of-the-arts on full human body reconstruction, on 26 models consisting of different subjects and clothes from BUFF Dataset [53]. We calculate Chamfer distance and Point-to-surface (P2S) error between ground-truth human models and reconstructed full body surface. We do not compare with SMPLicit, because they have no method for full body reconstruction. Please

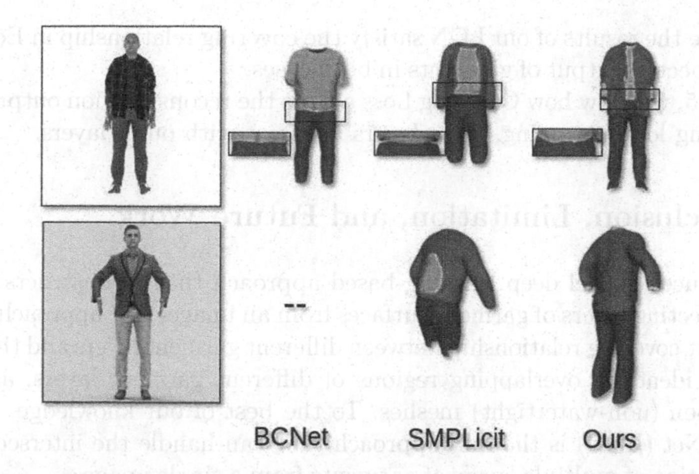

Fig. 7. Comparison between the reconstruction results of ours, BCNet [2] and SMPLicit [5]. Our reconstruction achieves intersection-free between different layers by satisfying the implicit covering relationship, while BCNet cannot reconstruct multi-layered garment structure, and the result from SMPLicit does not have such guarantee and has clear intersections between different layers.

note that we encounter lower results in this case than state-of-the-arts because we do not focus on accurate naked body (layer 0) reconstruction.

4.3 Qualitative Results

We compare the reconstruction quality of garment surfaces on the human body in Fig. 6. We can observe that our method (LGN) reconstructs a more detailed 3D human body than state-of-the-art explicit model reconstruction methods like BCNet, showing the effectiveness of implicit model reconstruction in comparison explicit approach. Also, SMPLicit generates a very coarse structure and loses many finer details for the clothes on the human body. Since we can generate individual implicit garment surfaces, we can retain finer details, especially between different layers. Since our networks are fine-tuned with IGR Loss, we reconstruct garments of similar quality to PIFu-HD without using fine-level networks.

In Fig. 7, we further show different challenges faced in the reconstruction of different garment surfaces. In the top row, we show the effect of covering relationship on multiple layers of garment reconstruction, specifically for the Shirt and Pant layers. From the given image, we expect the Pant layer to cover the Shirt layer without intersections. However, BCNet generates Shirt covering Pants, according to their pre-defined template. On the other hand, SMPLicit completely misses covering relation. In the bottom row, we show the reconstruction of the Coat layer above the Shirt layer as in the image. We expect two layers of garment reconstruction for the upper body. However, since BCNet is based on an explicit reconstruction of garments based on a displacement map on the SMPL body, it cannot reconstruct two-layer geometry for the upper body. Since SMPLicit does not guarantee intersection-free reconstruction, we find intersections between Shirt and Coat layer. Since the results of our LGN satisfy the covering relationship in Eq. (1), we get the expected output of garments in both cases.

In Fig. 5, we show how Covering Loss affects the reconstruction output. Without covering loss finetuning, inner layers intersect with outer layers.

5 Conclusion, Limitation, and Future Work

We introduce a novel deep learning-based approach that reconstructs multiple non-intersecting layers of garment surfaces from an image. Our approach enforces the implicit covering relationship between different garment layers and the human body and identifies overlapping regions of different garment layers, as well as extract open (non-watertight) meshes. To the best of our knowledge, Layered-Garment Net (LGN) is the first approach that can handle the intersection-free reconstruction of multiple layers of garments from a single image.

Our approach currently does not handle color information, since obtaining good texture for multiple reconstructed layers is difficult. Other neural implicit functions (e.g. Neural Radiance Fields) can address this issue. Our approach does not handle more challenging geometries consisting of manifold garment surfaces and details like pockets, hoodies, collars, etc., and some challenging poses, like limbs close to the body, etc. Also, since the naked human body model was not the focus of this work, the current approach does not handle the detailed full-body reconstruction. These issues can be a major improvement for future work.

References

1. Bhatnagar, B.L., Tiwari, G., Theobalt, C., Pons-Moll, G.: Multi-garment net: Learning to dress 3d people from images. In: Proceedings of the IEEE/CVF International Conference on Computer Vision, pp. 5420–5430 (2019)
2. Jiang, B., Zhang, J., Hong, Y., Luo, J., Liu, L., Bao, H.: BCNet: learning body and cloth shape from a single image. In: Vedaldi, A., Bischof, H., Brox, T., Frahm, J.-M. (eds.) ECCV 2020. LNCS, vol. 12365, pp. 18–35. Springer, Cham (2020). https://doi.org/10.1007/978-3-030-58565-5_2
3. Patel, C., Liao, Z., Pons-Moll, G.: TailorNet: predicting clothing in 3D as a function of human pose, shape and garment style. In: Proceedings of the IEEE/CVF Conference on Computer Vision and Pattern Recognition, pp. 7365–7375 (2020)
4. Su, Z., Yu, T., Wang, Y., Li, Y., Liu, Y.: Deepcloth: neural garment representation for shape and style editing. arXiv preprint arXiv:2011.14619 (2020)
5. Corona, E., Pumarola, A., Alenya, G., Pons-Moll, G., Moreno-Noguer, F.: Smplicit: topology-aware generative model for clothed people. In: Proceedings of the IEEE/CVF Conference on Computer Vision and Pattern Recognition, pp. 11875–11885 (2021)
6. Alldieck, T., Magnor, M., Xu, W., Theobalt, C., Pons-Moll, G.: Video based reconstruction of 3d people models. In: Proceedings of the IEEE Conference on Computer Vision and Pattern Recognition, pp. 8387–8397 (2018)
7. Loper, M., Mahmood, N., Romero, J., Pons-Moll, G., Black, M.J.: SMPL: a skinned multi-person linear model. ACM Trans. Graphics (TOG) **34**, 1–16 (2015)

8. Pavlakos, G., et al.: Expressive body capture: 3D hands, face, and body from a single image. In: Proceedings of the IEEE/CVF Conference on Computer Vision and Pattern Recognition, pp. 10975–10985 (2019)
9. Osman, A.A.A., Bolkart, T., Black, M.J.: STAR: sparse trained articulated human body regressor. In: Vedaldi, A., Bischof, H., Brox, T., Frahm, J.-M. (eds.) ECCV 2020. LNCS, vol. 12351, pp. 598–613. Springer, Cham (2020). https://doi.org/10.1007/978-3-030-58539-6_36
10. Bogo, F., Kanazawa, A., Lassner, C., Gehler, P., Romero, J., Black, M.J.: Keep it SMPL: automatic estimation of 3D human pose and shape from a single image. In: Leibe, B., Matas, J., Sebe, N., Welling, M. (eds.) ECCV 2016. LNCS, vol. 9909, pp. 561–578. Springer, Cham (2016). https://doi.org/10.1007/978-3-319-46454-1_34
11. Kanazawa, A., Black, M.J., Jacobs, D.W., Malik, J.: End-to-end recovery of human shape and pose. In: Proceedings of the IEEE Conference on Computer Vision and Pattern Recognition, pp. 7122–7131 (2018)
12. Pavlakos, G., Zhu, L., Zhou, X., Daniilidis, K.: Learning to estimate 3D human pose and shape from a single color image. In: Proceedings of the IEEE Conference on Computer Vision and Pattern recognition, pp. 459–468 (2018)
13. Smith, B.M., Chari, V., Agrawal, A., Rehg, J.M., Sever, R.: Towards accurate 3D human body reconstruction from silhouettes. In: 2019 International Conference on 3D Vision (3DV), pp. 279–288. IEEE (2019)
14. Omran, M., Lassner, C., Pons-Moll, G., Gehler, P., Schiele, B.: Neural body fitting: Unifying deep learning and model based human pose and shape estimation. In: 2018 international conference on 3D vision (3DV), pp. 484–494. IEEE (2018)
15. Lassner, C., Romero, J., Kiefel, M., Bogo, F., Black, M.J., Gehler, P.V.: Unite the people: closing the loop between 3D and 2D human representations. In: Proceedings of the IEEE Conference on Computer Vision and Pattern Recognition, pp. 6050–6059 (2017)
16. Alldieck, T., Magnor, M., Bhatnagar, B.L., Theobalt, C., Pons-Moll, G.: Learning to reconstruct people in clothing from a single RGB camera. In: Proceedings of the IEEE/CVF Conference on Computer Vision and Pattern Recognition, pp. 1175–1186 (2019)
17. Alldieck, T., Pons-Moll, G., Theobalt, C., Magnor, M.: Tex2shape: detailed full human body geometry from a single image. In: Proceedings of the IEEE/CVF International Conference on Computer Vision, pp. 2293–2303 (2019)
18. Su, Z., et al.: Mulaycap: multi-layer human performance capture using a monocular video camera. arXiv preprint arXiv:2004.05815 (2020)
19. Varol, G., et al.: BodyNet: volumetric inference of 3D human body shapes. In: Ferrari, V., Hebert, M., Sminchisescu, C., Weiss, Y. (eds.) ECCV 2018. LNCS, vol. 11211, pp. 20–38. Springer, Cham (2018). https://doi.org/10.1007/978-3-030-01234-2_2
20. Zheng, Z., Yu, T., Wei, Y., Dai, Q., Liu, Y.: Deephuman: 3D human reconstruction from a single image. In: Proceedings of the IEEE/CVF International Conference on Computer Vision, pp. 7739–7749 (2019)
21. Ma, Q., Saito, S., Yang, J., Tang, S., Black, M.J.: Scale: Modeling clothed humans with a surface codec of articulated local elements. In: Proceedings of the IEEE/CVF Conference on Computer Vision and Pattern Recognition, pp. 16082–16093 (2021)
22. Ma, Q., et al.: Learning to dress 3D people in generative clothing. In: Proceedings of the IEEE/CVF Conference on Computer Vision and Pattern Recognition, pp. 6469–6478 (2020)

23. Wang, L., Zhao, X., Yu, T., Wang, S., Liu, Y.: NormalGAN: learning detailed 3D human from a single RGB-D image. In: Vedaldi, A., Bischof, H., Brox, T., Frahm, J.-M. (eds.) ECCV 2020. LNCS, vol. 12365, pp. 430–446. Springer, Cham (2020). https://doi.org/10.1007/978-3-030-58565-5_26
24. Mescheder, L., Oechsle, M., Niemeyer, M., Nowozin, S., Geiger, A.: Occupancy networks: learning 3D reconstruction in function space. In: Proceedings of the IEEE/CVF Conference on Computer Vision and Pattern Recognition, pp. 4460–4470 (2019)
25. Chang, A.X., et al.: ShapeNet: an information-rich 3D model repository. arXiv preprint arXiv:1512.03012 (2015)
26. Lorensen, W.E., Cline, H.E.: Marching cubes: a high resolution 3D surface construction algorithm. ACM SIGGRAPH Comput. Graphics **21**, 163–169 (1987)
27. Chibane, J., Mir, A., Pons-Moll, G.: Neural unsigned distance fields for implicit function learning. arXiv preprint arXiv:2010.13938 (2020)
28. Park, J.J., Florence, P., Straub, J., Newcombe, R., Lovegrove, S.: DeepSDF: learning continuous signed distance functions for shape representation. In: Proceedings of the IEEE/CVF Conference on Computer Vision and Pattern Recognition, pp. 165–174 (2019)
29. Xu, Q., Wang, W., Ceylan, D., Mech, R., Neumann, U.: DISN: deep implicit surface network for high-quality single-view 3D reconstruction. arXiv preprint arXiv:1905.10711 (2019)
30. Sitzmann, V., Chan, E.R., Tucker, R., Snavely, N., Wetzstein, G.: MetaSDF: meta-learning signed distance functions. arXiv preprint arXiv:2006.09662 (2020)
31. Saito, S., Huang, Z., Natsume, R., Morishima, S., Kanazawa, A., Li, H.: PIFU: pixel-aligned implicit function for high-resolution clothed human digitization. In: Proceedings of the IEEE/CVF International Conference on Computer Vision, pp. 2304–2314 (2019)
32. Saito, S., Simon, T., Saragih, J., Joo, H.: Pifuhd: multi-level pixel-aligned implicit function for high-resolution 3D human digitization. In: Proceedings of the IEEE/CVF Conference on Computer Vision and Pattern Recognition, pp. 84–93 (2020)
33. Hong, Y., Zhang, J., Jiang, B., Guo, Y., Liu, L., Bao, H.: StereoPIFU: depth aware clothed human digitization via stereo vision. In: Proceedings of the IEEE/CVF Conference on Computer Vision and Pattern Recognition, pp. 535–545 (2021)
34. He, T., Collomosse, J., Jin, H., Soatto, S.: Geo-PIFU: geometry and pixel aligned implicit functions for single-view human reconstruction. arXiv preprint arXiv:2006.08072 (2020)
35. Zheng, Z., Yu, T., Liu, Y., Dai, Q.: Pamir: parametric model-conditioned implicit representation for image-based human reconstruction. IEEE Transactions on Pattern Analysis and Machine Intelligence (2021)
36. Wang, S., Mihajlovic, M., Ma, Q., Geiger, A., Tang, S.: MetaAvatar: learning animatable clothed human models from few depth images. In: Advances in Neural Information Processing Systems (NeurIPS) (2021)
37. Huang, Z., Xu, Y., Lassner, C., Li, H., Tung, T.: Arch: animatable reconstruction of clothed humans. In: Proceedings of the IEEE/CVF Conference on Computer Vision and Pattern Recognition, pp. 3093–3102 (2020)
38. Peng, S., et al.: Neural body: implicit neural representations with structured latent codes for novel view synthesis of dynamic humans. In: Proceedings of the IEEE/CVF Conference on Computer Vision and Pattern Recognition, pp. 9054–9063 (2021)

39. Peng, S., et al.: Animatable neural radiance fields for modeling dynamic human bodies. In: Proceedings of the IEEE/CVF International Conference on Computer Vision, pp. 14314–14323 (2021)
40. Liu, L., Habermann, M., Rudnev, V., Sarkar, K., Gu, J., Theobalt, C.: Neural actor: neural free-view synthesis of human actors with pose control. arXiv preprint arXiv:2106.02019 (2021)
41. Shao, R., Zhang, H., Zhang, H., Cao, Y., Yu, T., Liu, Y.: Doublefield: bridging the neural surface and radiance fields for high-fidelity human rendering. arXiv preprint arXiv:2106.03798 (2021)
42. Saito, S., Yang, J., Ma, Q., Black, M.J.: Scanimate: weakly supervised learning of skinned clothed avatar networks. In: Proceedings of the IEEE/CVF Conference on Computer Vision and Pattern Recognition, pp. 2886–2897 (2021)
43. Bhatnagar, B.L., Sminchisescu, C., Theobalt, C., Pons-Moll, G.: Combining implicit function learning and parametric models for 3D human reconstruction. In: Vedaldi, A., Bischof, H., Brox, T., Frahm, J.-M. (eds.) ECCV 2020. LNCS, vol. 12347, pp. 311–329. Springer, Cham (2020). https://doi.org/10.1007/978-3-030-58536-5_19
44. Gropp, A., Yariv, L., Haim, N., Atzmon, M., Lipman, Y.: Implicit geometric regularization for learning shapes. arXiv preprint arXiv:2002.10099 (2020)
45. Tiwari, G., Bhatnagar, B.L., Tung, T., Pons-Moll, G.: SIZER: a dataset and model for parsing 3D clothing and learning size sensitive 3D clothing. In: Vedaldi, A., Bischof, H., Brox, T., Frahm, J.-M. (eds.) ECCV 2020. LNCS, vol. 12348, pp. 1–18. Springer, Cham (2020). https://doi.org/10.1007/978-3-030-58580-8_1
46. Zhu, H., et al.: Deep Fashion3D: a dataset and benchmark for 3D garment reconstruction from single images. In: Vedaldi, A., Bischof, H., Brox, T., Frahm, J.-M. (eds.) ECCV 2020. LNCS, vol. 12346, pp. 512–530. Springer, Cham (2020). https://doi.org/10.1007/978-3-030-58452-8_30
47. Chi, C., Song, S.: Garmentnets: category-level pose estimation for garments via canonical space shape completion. arXiv preprint arXiv:2104.05177 (2021)
48. Jacobson, A., Kavan, L., Sorkine-Hornung, O.: Robust inside-outside segmentation using generalized winding numbers. ACM Trans. Graphics (TOG) **32**, 1–12 (2013)
49. Gong, K., Liang, X., Li, Y., Chen, Y., Yang, M., Lin, L.: Instance-level human parsing via part grouping network. In: Ferrari, V., Hebert, M., Sminchisescu, C., Weiss, Y. (eds.) ECCV 2018. LNCS, vol. 11208, pp. 805–822. Springer, Cham (2018). https://doi.org/10.1007/978-3-030-01225-0_47
50. Wang, T.C., Liu, M.Y., Zhu, J.Y., Tao, A., Kautz, J., Catanzaro, B.: High-resolution image synthesis and semantic manipulation with conditional GANs. In: Proceedings of the IEEE Conference on Computer Vision and Pattern Recognition, pp. 8798–8807 (2018)
51. : (Axyz design 3d people models and character animation software https://secure.axyz-design.com/)
52. Alldieck, T., Magnor, M., Xu, W., Theobalt, C., Pons-Moll, G.: Detailed human avatars from monocular video. In: 2018 International Conference on 3D Vision (3DV), pp. 98–109. IEEE (2018)
53. Zhang, C., Pujades, S., Black, M.J., Pons-Moll, G.: Detailed, accurate, human shape estimation from clothed 3D scan sequences. In: The IEEE Conference on Computer Vision and Pattern Recognition (CVPR) (2017)
54. Newell, A., Yang, K., Deng, J.: Stacked hourglass networks for human pose estimation. In: Leibe, B., Matas, J., Sebe, N., Welling, M. (eds.) ECCV 2016. LNCS, vol. 9912, pp. 483–499. Springer, Cham (2016). https://doi.org/10.1007/978-3-319-46484-8_29

SWPT: Spherical Window-Based Point Cloud Transformer

Xindong Guo[1,2], Yu Sun[2], Rong Zhao[1], Liqun Kuang[1], and Xie Han[1](✉)

[1] North University of China, Taiyuan, China
hanxie@nuc.edu.cn

[2] Shanxi Agricultural University, Jinzhong, China
gxd@sxau.edu.cn

Abstract. While the Transformer architecture has become the de-facto standard for natural language processing tasks and has shown promising prospects in image analysis domains, applying it to the 3D point cloud directly is still a challenge due to the irregularity and lack of order. Most current approaches adopt the farthest point searching as a downsampling method and construct local areas with the k-nearest neighbor strategy to extract features hierarchically. However, this scheme inevitably consumes lots of time and memory, which impedes its application to near-real-time systems and large-scale point cloud. This research designs a novel transformer-based network called Spherical Window-based Point Transformer (SWPT) for point cloud learning, which consists of a Spherical Projection module, a Spherical Window Transformer module and a crossing self-attention module. Specifically, we project the points on a spherical surface, then a window-based local self-attention is adopted to calculate the relationship between the points within a window. To obtain connections between different windows, the crossing self-attention is introduced, which rotates all the windows as a whole along the spherical surface and then aggregates the crossing features. It is inherently permutation invariant because of using simple and symmetric functions, making it suitable for point cloud processing. Extensive experiments demonstrate that SWPT can achieve the state-of-the-art performance with about 3-8 times faster than previous transformer-based methods on shape classification tasks, and achieve competitive results on part segmentation and the more difficult real-world classification tasks.

Keywords: Point cloud · Spherical projection · Transformer

1 Introduction

3D point clouds have been attracting more and more attention from both industry and academia due to their broad applications including autonomous driving, augmented reality, robotics, etc. Unlike images having regular pixel grids, 3D point clouds are irregular and unordered sets of points corresponding to object surfaces. This difference makes it challenging to apply traditional network architectures used widely in 2D computer vision directly into 3D point cloud.

L. Wang et al. (Eds.): ACCV 2022, LNCS 13841, pp. 396–412, 2023.
https://doi.org/10.1007/978-3-031-26319-4_24

Researchers propose a variety of models for deep learning on 3D point clouds which are classified into three categories according to data representations, i.e., projection-based, voxel-based and point-based models. The projection-based models [3,13,25,35] generally project 3D shapes into regular 2D representations, so that many classical convolutional models can be employed. However, these methods can only obtain limited receptive field from one or a few perspectives, which may induce losses of spatial relations between different parts. The voxel-based models [18,23,33] generally rasterize 3D point clouds onto regular grids and apply 3D convolutions for feature learning, which induces massive computational and memory costs due to the cubic growth in number of voxels related to the resolution. Although sparse convolutional models [2,23,29] alleviate this problem by performing only on voxels that are not empty, it may still not capture the spatial relations. The point-based models [20,22] operate directly on points and propagate features via pooling operators. Specifically, some works transform point cloud to a graph for message passing [14,24]. These models achieve more competitive results over previous methods. However, the neighborhood searching strategy which is a core component of these methods has a high computational complexity as the iterative number increasing or points scale getting larger.

In recent years, Transformer [1,28] has been dominating the natural language processing field, and has been applied to image vision tasks, also achieving encouraging performance [6,31]. As the core component of Transformer, the self-attention module computes the refined weighted features based on global context by considering the connections between any two words. So the output feature of each word is related to all input features, which make it capable of obtaining the global feature. Specifically, all operations of Transformer are parallelizable and order invariance, so it is naturally suitable for point cloud processing. Transformer-based model [8] is the pioneering work that introduces the self-attention in points processing, which employ the farthest point searching (FPS) as downsampling strategy and k-nearest neighbor (KNN) as local region searching strategy to perform local self-attention operation. It lacks spatial connections between different regions, and consumes a large amount of time when constructing local regions due to the high complexity of FPS and KNN.

To solve the problems mentioned above, we propose a noval Spherical Window based Point Cloud Transformer. Firstly, we project point cloud to a spherical surface to reduce the dimension of points, which makes the processing of points as simple as images. The Spherical Window (SW) module makes the points projected on the spherical surface more regular resembling pixels in an image, but with more spatial structure information. Then, we partition the points on the spherical surface into spherical windows and apply local self-attention hierarchically on each window. With this spherical window-based mechanism, local self-attention are performed in all windows parallelly, which would facilitate point cloud processing and is beneficial to its scalability for large point clouds. Finally, we introduce a Cross-Window operation to achieve connections between the neighboring windows, which is an efficient operation with computational complexity as low as $O(1)$. We use the SWPT as the backbone to a variety of point cloud recognition tasks, and extensive experiments demonstrate that it's an effective and efficient framework.

The main contributions of this paper are summarized as following:

- We propose a Spherical Projection (SP) Layer, which projects the points to a spherical surface, followed by a point-wise transformation to maintain the information between points. This layer is permutition-invariant, high-efficiency and thus is inherently suitable for point cloud processing.
- Based on the Spherical Projection Layer, we construct a Spherical Windows (SW) Transformer which partitions the spherical surface into windows comprising certain patches. Then we perform local self-attention on each window hierarchically.
- We introduce a cross window operation, which rotates the spherical surface as a whole by half of the window size to exploit the spatial connections between different windows. Extension experiments demonstrate that our network achieves the state-of-the-art performance on shape classification, part segmentation and semantic segmentation.

2 Related Works

2.1 Projection-Based Networks

As a pioneering work, MVCNN [25] simply use max-pooling to aggregate multi-view features into a global descriptor. But max-pooling only retains the top elements from a specific view, which result in loss of details. These methods [3,13] project 3D point clouds into various image planes, and then employ 2D CNNs to extract local features in these image planes followed by a multi-view features fusion module to obtain the global feature. This approach [35] propose a relation network to exploit the inter-relationships over a group of views which are aggregated to form a discriminative representation indicating the 3D object. Different from the previous methods, [30] construct a directed graph by regarding multiple views as graph nodes, and then design a GCNN over the view-graph to hierarchically achieve global shape descriptor.

2.2 Voxel-Based Networks

VoxNet [18] integrates volumetric Occupancy Grid representation with a supervised 3D CNN to utilize 3D information and deal with large amounts of point cloud. 3D ShapeNets [23,33] is another work using a Convolutional Deep Belief Network to represent a geometric 3D shape as a probability distribution of binary variables on a 3D voxel grid, which can learn the distribution of complex 3D shapes across different categories. However, these methods are unable to scale well to dense 3D data since the computation and memory consumption of such methods increases cubically with respect to the resolution of voxel.

2.3 Point-Based Networks

PointNet [20] as a pioneer work firstly utilizes permutation-invariant operators such as MLP and max-pooling to aggregate a global feature from a point cloud. As PointNet lacks local connections between points, the author propose PointNet++ [22] which applies tiny PointNet within a hierarchical spatial structure to extract local features with increasing contextual scales. Inspired by PointNet, many a recent works [9,11,32] are proposed with more sophisticated architecture achieving encouraging performance. Afterwards, a graph convolutional neural network(GCNN) is applied to extract features. DGCNN [24] is the first performing graph convolutions on KNN graphs. As the core component of EdgeConv in DGCNN, MLP is used to learn the features for each edge. Deepgcns [14] presents a new way to train a very deep GCNs to solve the vanishing gradient problem and shows a 56-layer GCN achieving positive effect. PointCNN [15] propose a X-transformation learnt from the point cloud, which weight the input features associated with the points and reorder them into latent potentially canonical order simultaneously. PointConv [32] extend the dynamic filter to a new convolution and take the local coordinates of 3D points as input to compute weight and density functions.

2.4 Transformer in NLP and Vision

[1] propose a neural machine translation with an attention mechanism allowing a model to automatically search for parts of a source sentence that are relevant to predicting a target word. [16] further propose a self-attention mechanism to extract an interpretable sentence embedding. Subsequent works employed self-attention layers to replace some or all of the spatial convolution layers. [28] propose Transformer based solely on self-attention mechanism without recurrence and convolutions entirely. [5] propose a new Bidirectional Transformers to pre-train deep bidirectional representations and it obtains competitive results.

Transformer and self-attention models have revolutionized natural language processing and inspired researchers to apply them to vision tasks. [31] proposed visual transformers that apply Transformer to token-based images from feature maps. [6] propose a pure transformer-based network partitioning a image to patches, and results show with sufficient training data, Transformer provides better performance than a traditional convolutional neural network. [17] propose a new vision Transformer, called Swin-Transformer, to incorporate inductive bias for spatial locality, as well as for hierarchy and translation invariance.

3 Spherical-Window Point Transformer

3.1 Overview

In this section we detail the design of SWPT. We first show how the spherical projection (SP) can provide an efficient representation for point clouds. Then we present the detail of the gridding module which splits the points on the spherical

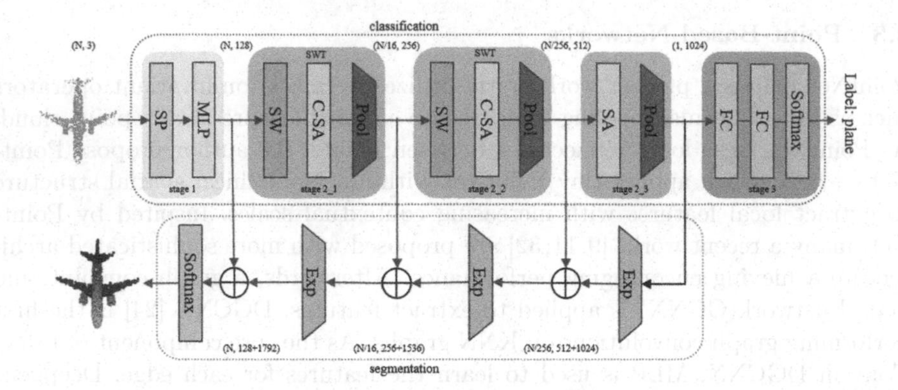

Fig. 1. Model architecture. The above is the classification network, and the below is the segmentation network. SP stands for the spherical projection. SW is the spherical window layer used to split points. C-SA is the crossing self-attention applied to calculate the cross window attentions. Exp means expanding high-dimension features to local windows.

surface into non-overlapping patches. Lastly, we introduce a high-performance self-attention module to extract local features, and with a rotation to enhance the cross-window connections.

The model architecture is illustrated in Fig. 1. The above of Fig. 1 illustrates the classification network which takes as input the raw point cloud with N points. Firstly, the network applies a Spherical Projection (SP) to all points, followed by a point-wise transformation, e.g. multi-layers perceptron (MLP). Then the outputs of the previous layer are fed into two stacked Spherical Window Transformer (SWT) which consists of a Spherical Window (SW) module, a Crossing self-attention (C-SA) module and a pooling module to learn hierarchical features. It is worth noting that the network achieves competitive results with only two iterations of the SWT module thanks to the high efficiency of the Spherical Window based neighborhood searching strategy. Finally, a self-attention (SA) and a pooling operation are employed to produce a global feature which is fed into a fully connected (FC) layer followed by a softmax function to predict the label. The below of Fig. 1 shows the segmentation network which expands the high-dimension features to the corresponding local windows in the previous layer and then concatenates them with features from the last layer. After three expanding and concatenation operations, the network produces the dense points features that a lightweight fully connected network use to generate labels.

3.2 Spherical Projection

Spherical Projection Layer plays a role of data pre-processing in the framework, which projects the points from the point cloud to a spherical surface, as illustrated in Fig. 2. For projecting the points on the spherical surface evenly, we normalize the points by the following formulation:

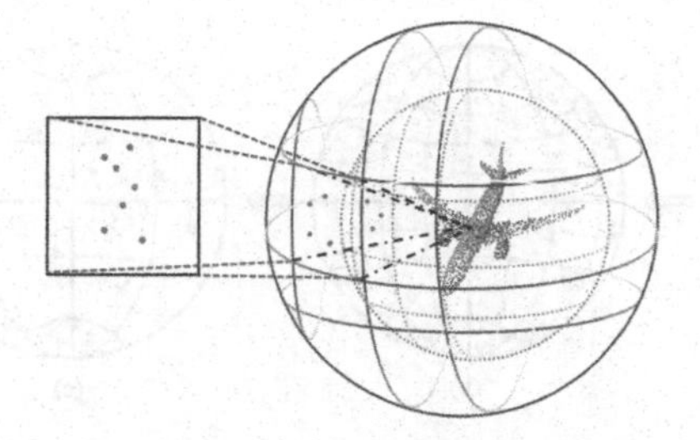

Fig. 2. Visualization of Spherical projection. For simplification, here we just show two spheres for projection. The points on different sections in the same frustum are gathered into a spherical window. The blue points and the green points in the window are from the blue spherical surface and green spherical surface respectively. (Color figure online)

$$p' = p - p_c \qquad p_c = \frac{1}{N}(\sum_{i=1}^{N} x_i, \sum_{i=1}^{N} y_i, \sum_{i=1}^{N} z_i) \tag{1}$$

where p_c is the centroid of point cloud and p' is the point translated from point $p = (x, y, z)$. N stands for the number of points. By doing so, we translate the centroid of point cloud to the origin of coordinates.

Then, we define a spherical projection method which can be formulated as follows:

$$\begin{aligned} r &= \sqrt{x^2 + y^2 + z^2} \\ \theta &= \arctan\frac{y}{x} \\ \varphi &= \arccos\frac{z}{r} \end{aligned} \tag{2}$$

Here (x, y, z) is the coordinate of point in the Cartesian system, θ and φ are the angle of azimuth and angle of pitch, respectively.

Although the spherical projection preserves most prominent features, the information of connection between points on the spherical surface may change compared to the original point cloud due to the reduction of dimension, causing structure information loss. To solve this problem, we apply a transformation to the points through a shared MLP, projecting the features of the points to a latent space, while maintaining the spatial relations between points:

$$f = T(p) \oplus P(p) \quad P(p) = r \oplus \theta \oplus \varphi, \tag{3}$$

where $p \in R^D$ is the feature of a point such as x-y-z coordinate, in our case D is 3. In other cases, D may be different when the point includes features like normal vector, etc. T is a pointwise transformation such as MLP. $\oplus$ is a vector concatenation operator. P indicates the spherical projection mentioned before.

Eventually, the output of the Spherical Projection module is a new set of features:

$$F = \{f_i \mid f_i = P(p_i) \oplus T(p_i),\ i \in N\}, \tag{4}$$

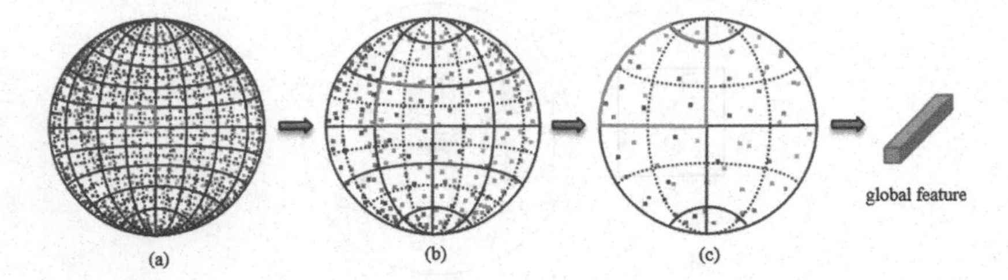

Fig. 3. The proposed Spherical Window Transformer obtains the global feature by merging features within a 2×2 spherical window hierarchically. The points features within a solid window are adopted to perform local self-attention and then aggregated to a point for the next stage denoted by the dotted box. The green box indicates a local spherical window to perform self-attention and pooling operation. Please note that here we simplify the figure structure to illustrate the principle of a module clearly. (Color figure online)

among which the angle of azimuth θ and angle of pitch φ are utilized to partition features into windows.

3.3 Spherical Window Transformer

The simplest way to apply Transformer on point clouds is to treat the entire points as a sequence and each point as an element, but it incurs massive computation and memory costs and local spatial relation loss. In addition, it lacks hierarchies which are significant in learning features proved by [22]. On the other hand, the farthest point sampling (FPS) and K-nearest neighbors (KNN) searching strategies adopted by most previous models introduce large computation and memory costs, making it unsuitable for large scale point clouds [7,8].

To this end, we propose the Spherical Window Transformer layer to compute local self-attention and to construct a hierarchical architecture, which has linear computational complexity to patch size. The Spherical Window Transformer layer aims to provide an efficient local area partitioner which splits the points on the spherical surface into lots of patches, a patch involve points from a neighborhood. To this end, we adopt a very simple and efficient approach formally defined as follow:

$$F' = R_{\varphi}(S_{\varphi}(R_{\theta}(S_{\theta}(F)))), \tag{5}$$

where $F \in R^{N \times C}$ is a set of point features from the last layer, S_θ is a sort function by θ, R_θ is capable of reshaping the set shape from $N \times C$ to $\Theta \times N_\theta \times C$ where $N = N_\theta \times \Theta$. S_φ and R_φ perform the similar thing. After then we obtain a new set with the shape $\Phi \times \Theta \times N_\phi \times N_\theta \times C$ meaning that the point cloud is partitioned into $\Phi \times \Theta$ spherical windows with $N_\theta \times N_\phi$ points in each window. Because of using sorting and reshaping only, this layer maintains the permutation-invariance of point cloud.

Then a local self-attention is employed in each window, followed by a patch merging block to gather the local features with low-dimension. Afterwards, local points are gathered to a high-dimension space with fewer points. The windows are arranged to evenly partition the points into non-overlapping areas which produce a higher layer local neighborhood, as shown in Fig. 3.

Given a window size of (θ, φ), which means it has a number of $\theta \times \varphi$ points, the computational complexity of global self-attention and window-based local self-attention on a spherical surface of $\Theta \times \Phi$ patches are:

$$\Omega(SA) = 4\Theta\Phi C^2 + 2(\Theta\Phi)^2 C, \tag{6}$$

$$\Omega(W-SA) = 4\Theta\Phi C^2 + 2(\Theta\Phi)(\theta\varphi)C, \tag{7}$$

where the former is quadratic to points number $\Theta \times \Phi$, and the latter is linear when the window size is fixed. The global self-attention computation cost is generally unaffordable for a large scale point cloud, while window-based self-attention is scalable. Another function of the Spherical Window module is to reduce the cardinality of patches as required, for example, from N points to N/W through a window of W size.

The strategy of choosing local neighborhoods of previous works is KNN with Euclidean distance, which has a linear computational complexity with respect to the number of neighborhoods, total points and the dimension of data. So it's still a challenge to extend a model using this to large scale point clouds. Our method using the Spherical Window to split points only needs to make a spherical surface grid by sorting points along the angle of azimuth and pitch, which only need to do once after the projection finished and is invariant to permutation of points. Extensive experiments demonstrates that the Spherical Window is a novel representation of point cloud on which local self-attention can be applied effectively and efficiently.

3.4 Crossing Windows Self-attention

Spherical window-based local self-attention performs well to obtain local features. However, it lacks the capacity of acquiring relationships across neighboring windows, which limits its modeling power. To capture cross window relationships while maintaining the efficiency of computation and memory cost, we propose a rotating spherical surface approach which rotates all the windows as a whole sphere along the angle of azimuth and the angle of pitch by half of window size respectively. As illustrated in Fig. 4, the module on the left calculates a regular self-attention within each window. Then the module on the right rotates windows along the angle of azimuth and pitch by $\lfloor \theta/2, \varphi/2 \rfloor$, followed by a self-attention computation same as before, where the θ and φ are window size. As the windows form a spherical surface and data points are distributed discretely over it, the rotation of windows does not need a masked operation resembling that in [17]. Results show that our Crossing Self-Attention approach introduces relationships between neighboring windows in previous layers and performs effectively and efficiently in several point cloud recognition tasks, as shown in Table 5.

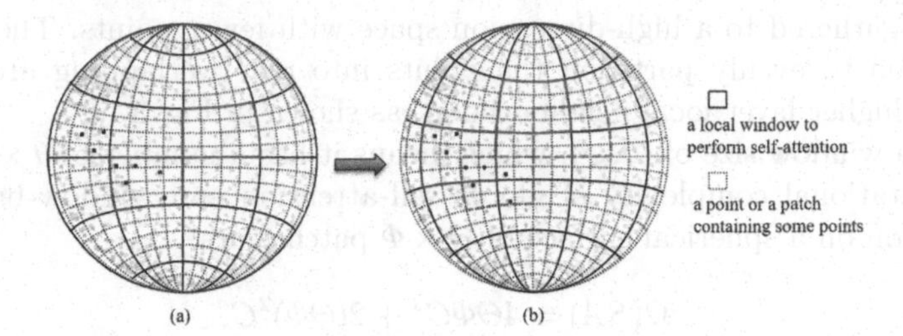

Fig. 4. Illustration of crossing windows self-attention considering the relationship between different windows. In this example, the window size is 4×4 and the rotating size is 2×2. After calculating the window-based self-attention showed by (a), the whole spherical window rotate along azimuth and pitch by 2×2 patches respectively, illustrated by (b). The green and blue dotted boxes in (b) indicate the spherical windows before rotation corresponding to windows in (a), and the solid boxes in (b) denote the spherical windows after rotation. (Color figure online)

4 Experiments

We evaluate the performance of SWPT on three recognization tasks: 3D shape classification, 3D part segmentation and real-world object classification, giving comprehensive analyses and comparing with other approaches. For 3D shape classification, we use the widely adopted ModelNet40 [33] dataset. For 3D part segmentation, we use the ShapeNet55 [36] dataset. For real-world object classification, we use the ScanObjectNN [27] dataset which is a recent point cloud object dataset constructed from the real-world indoor datasets such as SceneNN [10] and ScanNet [4].

We use PyTorch [19] as the framework to implement SWPT. We employ the stochastic gradient descent (SGD) optimizer with momentum 0.9 and weight decay 0.0001 for training. Other training parameters including number of patches, window size, learning rate and batch size are given later in each related section.

4.1 Shape Classification

Data and Metric. The ModelNet40 [33] dataset contains 12,311 CAD models in 40 categories, which is widely used in shape classification. For a fair comparison, we use the official split with 9,843 models for training and 2,468 models for testing, and use the same strategy as PointNet [20] to sample 1,024 points uniformly from each CAD model. For evaluation metrics, we adopt the mean accuracy within each category (mAcc) and the overall accuracy (OA) over all classes. In addition, we retrain the point-based methods to evaluate their efficiency.

Experiment Results. During training we used a random translation in [−0.2, 0.2], a random anisotropic scaling in [0.67, 1.5] and a random input dropout as the data augmentation strategy, while no data augmentation was used in testing.

Table 1. Comparison of state-of-the-art models on ModelNet40 dataset. The mAcc and OA results are quoted from the cited papers, while the latencies are obtained by training and testing the official code in the same environment.

Method	Input	Points	mAcc (%)	OA (%)	Latency (s)
3DShapeNets [33]	Voxel	–	77.3	84.7	–
VoxNet [18]	Voxel	–	83.0	85.9	–
Subvolumn [21]	Voxel	-	86.0	89.2	–
MVCNN [25]	Image	–	–	90.1	–
Kd-Net [12]	Point	-	–	91.8	–
PointNet [20]	Point	1k	86.2	89.2	11
PointNet++ [22]	Point	1k	–	91.9	21
PointCNN [15]	Point	1k	88.1	92.2	45
DGCNN [24]	Point	1k	**90.2**	92.2	10
PointConv [32]	Point	1k	–	92.5	58
KPConv [26]	Point	1k	–	92.9	25
PCT1 [7]	Point	1k	–	92.8	34
PCT2 [8]	Point	1k	–	93.2	18
SWPT (ours)	Point	1k	90.1	**93.5**	5

We set the batch size to 32, epochs to 200 and initial learning rate to 0.01 with a step schedule to adjust it at every 40 epochs. For a fair comparison of efficiency, we use the same hardware environments and only the official code from github. We experiment in the same test datasets and report the average latency time.

The results are presented in Table 1. The SWPT outperforms all prior methods in overall accuracy with 93.5%. As for Transformer-based methods, the SWPT achieves a better result with less time.

4.2 Part Segmentation

Data and Metric. Part Segmentation is a challenging task which aims to divide a 3D point cloud into multiple meaningful parts. We evaluate the models on the ShapeNet [36] dataset which contains 16,880 3D models consisting of 14,006 models for training and 2,874 models for testing. It has 16 categories and 50 parts, where each category has a number of parts between 2 to 6. Following PointNet [20], we downsample each model to 2,048 points with point-wise labels, which is widely used in prior works. For metrics, we evaluate the mean Intersection-over-Union (IoU) and IoU for each object category.

Experiment Results. During training we used a random translation in [−0.2, 0.2], a random anisotropic scaling in [0.5, 1.5] and a random input dropout as the data augmentation strategy, while no data augmentation was used in testing. We set the batch size to 32, epochs to 200 and initial learning rate to 0.001 with a step schedule to adjust it at every 40 epochs by 0.5. Table 2 shows the mean part IoU and category-wise IoU. The results show that our SWPT improves by 1.7% over PointNet and is competitive with most SOTA methods. The reason why SWPT does not achieve the best result in mIoU is that projecting overlapped and shaded points (of objects like motorbikes and cars) on a spherical surface may cause semantic ambiguity issues.

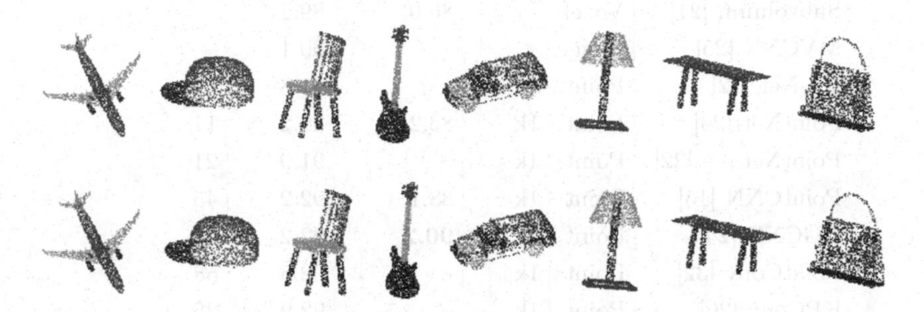

Fig. 5. Visualization of part segmentation results on the ShapeNet dataset. The top row shows the ground truth, results of SWPT are on the bottom row.

Visualization. Figure 5 shows the results of object part segmentation on a number of models in the ShapeNet dataset. The predictions of SWPT on the bottom row are precision and close to the ground truth, from which we can see some structural details captured by SWPT, such as the engines of the plane and wheels of the automobile.

4.3 Real-World Object Classification

Data and Metric. While ModelNet40 is the most popular benchmark for point cloud classification, it may lack a practical scenario due to its synthetic nature (i.e. complete, well-segmented and noisy-free). To evaluate the performance of our model on real-world objects, we conduct experiments on the ScanObjectNN benchmark which contains 15,000 objects categorized into 15 classes with 2,902 unique instances in read world. We adopt the following variants of ScanObjectNN in our experiments: (1) **OBJ_ONLY** which has only ground truth objects segmented from the scene datasets, and (2) **PB_T50_RS** which is the hardest variant for training and testing our model. For metrics, we report the overall accuracy (OA) over all classes. We use

Experiment Results. We use the same batch size, training epochs and other settings as in experiments on ModelNet40. As presented in Table 3, our SWPT achieves competitive results on both OBJ_ONLY and PB_T50_RS datasets compared with previous methods. There is still a small gap between SWPT and the SOTA method, mostly because of the dimensionality reduction caused by projection, however which significantly accelerates the speed of model inference.

Table 2. Comparison of part segmentation models on the ShapeNet. mIoU means part-average Intersection-over-Union. All results are quoted from the cited papers.

Method	mIoU	Aero	Bag	Cap	Car	Chair	Ear phone	Guitar	Knife	Lamp	Laptop	Motor	Mug	Pistol	Rocket	Skate board	Table
Yi [36]	81.4	81.0	78.4	77.7	75.7	87.6	61.9	92.0	85.4	82.5	95.7	70.6	91.9	**85.9**	53.1	69.8	75.3
Kd-Net [12]	82.3	80.1	74.6	74.3	70.3	88.6	73.5	90.2	87.2	81.0	94.9	57.4	86.7	78.1	51.8	69.9	80.3
PointNet [20]	83.7	83.4	78.7	82.5	74.9	89.6	73.0	91.5	85.9	80.8	95.3	65.2	93.0	81.2	57.9	72.8	80.6
PointNet++ [22]	85.1	82.4	79.0	87.7	77.3	90.8	71.8	91.0	85.9	83.7	95.3	71.6	94.1	81.3	58.7	76.4	82.6
PointCNN [15]	86.1	84.1	**86.5**	86.0	80.8	90.6	**79.7**	**92.3**	**88.4**	85.3	96.1	**77.2**	95.2	84.2	64.2	**80.0**	83.0
DGCNN [24]	85.2	84.0	83.4	86.7	77.8	90.6	74.7	91.2	87.5	82.8	95.7	66.3	94.9	81.1	**63.5**	74.5	82.6
PointConv [32]	85.7	–	–	–	–	–	–	–	–	–	–	–	–	–	–	–	–
PCT1 [7]	85.9	–	–	–	–	–	–	–	–	–	–	–	–	–	–	–	–
PCT2 [8]	**86.4**	**85.0**	82.4	**89.0**	**81.2**	**91.9**	71.5	91.3	88.1	**86.3**	95.8	64.6	95.8	83.6	62.2	77.6	83.7
SWPT (ours)	85.4	82.2	79.2	83.1	74.3	91.7	74.7	91.7	86.0	80.6	**97.7**	55.8	**96.6**	83.3	53.0	74.5	**83.9**

Visualization. Figure 6 illustrates the confusion matrices of the mainstream methods on the PB_T50_RS dataset which is the hardest variant. It can be seen that SWPT only has ambiguity issues between box and pillow, while PointNet and PointNet++ have ambiguity issues between table and desk besides. Even the PCT which obtains a good experimental result on the synthetic dataset has some ambiguity such as pillow vs bag and table vs desk. DGCNN has less ambiguity, which might benefit from the EdgeConv module recomputing graphs dynamically in each layer.

4.4 Ablation Study

We construct a number of ablation studies to analyze the performance of different components in SW-PCT. All studies are performed on the shape classification.

Spherical Window Size. We first evaluate the setting of window size (θ, φ), which determines the local areas to perform self-attention. As presented in Table 4, SWPT achieves the best performance when the window size is set to $(4, 4)$. When the windows size gets smaller, the model may not have sufficient context to compute self-attention, leading to some local detail loss. When the window size gets larger, the local area may have more data points with fewer relationships, which introduces extreme noise into self-attention computation, reducing the model's accuracy.

Cross-Attention. We conducted an ablation study on the Cross-Attention layer. As shown in Table 5, we investigate global self-attention, local self-attention and cross-attention with different times in each layer. The results indicate adopting cross-attention twice in each layer achieves the best performance, while continuously increasing cross-attention within a layer reduces the performance. This suggests that relationships between windows are critical in learning local features.

Table 3. Comparison of state-of-the-art methods on ScanObjectNN. All results are quoted from [27] except for the PCT.

Method	OBJ_ONLY (%)	PB_T50_RS (%)
3DmFV [2]	73.8	63
PointNet [20]	79.2	68.2
SpiderCNN [34]	79.5	73.7
PointNet++ [22]	84.3	77.9
DGCNN [24]	**86.2**	78.1
PointCNN [15]	85.5	78.5
BGA-DGCNN [27]	–	79.7
BGA-PN++ [27]	–	**80.2**
PCT [8]	80.7	71.4
SWPT (ours)	85.1	77.2

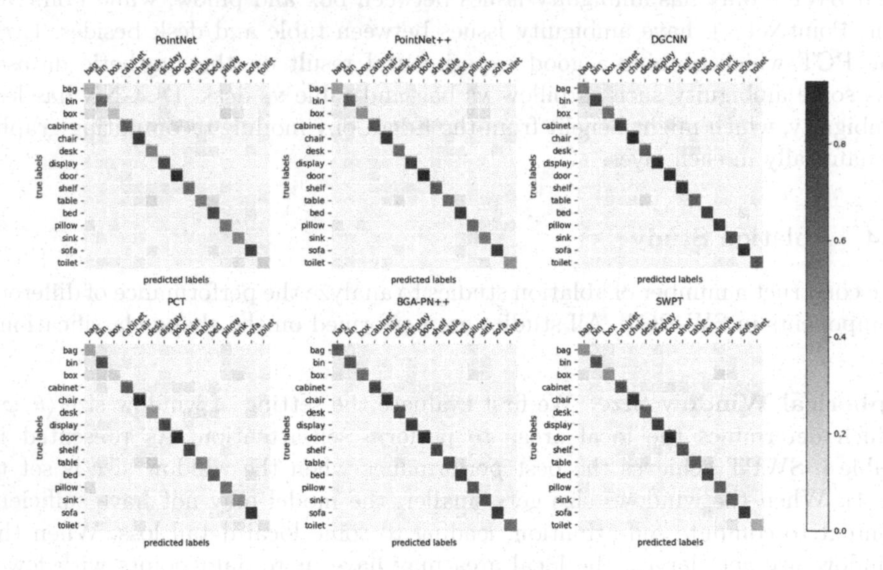

Fig. 6. Confusion matrices of some previous methods on the hardest real-world dataset PB_T50_RS.

Attention Type. We now study the attention type in self-attention layers. The results are shown in Table 6. It shows that scalar attention is more expressive than vector attention in our model. In addition, vector attention consumes more memory and time.

Pooling Type. Finally, we investigate the type of pooling used for gathering local features from a window. As shown in Table 7, Max-pooling outperforms other methods even the 'Max-Ave' pooling which is a concatenation of the results of Max-pooling and Average-pooling. 'Con-Pool' stands for the operation which concatenates all the features in a spherical window. This indicates that Max-pooling captures the most expressive features in the local area.

Table 4. Ablition study: window size to perform self-attention

Size	Layers	mAcc	OA
(2,2)	5	88.7	90.8
(3,3)	4	89.6	92.4
(4,4)	3	90.1	93.5
(8,8)	2	89.4	92.3

Table 5. Ablition study: cross-attention

Type	mAcc	OA
Local-AT	88.0	90.1
Cross-AT	89.5	92.2
Cross-AT × 2	90.1	93.5
Cross-AT × 3	89.7	92.7

Table 6. Ablition study: cross-attention

Type	mAcc	OA
Scalar attention	90.1	93.5
Vector attention	89.9	93.2

Table 7. Ablation study: type of pooling to gather local features, Con-Pool refers to the concatenation of features in a window.

Pooling	mAcc	OA
Max-Pool	90.1	93.5
Avg-Pool	87.9	90.2
Max-Avg	88.5	91.4
Con-Pool	88.6	92.3

5 Conclusion

In this paper, we present a Transformer-based architecture for 3D point cloud recognition which achieves competitive results with more efficiency compared with previous methods. To this end, we project the points on a spherical surface to reduce the dimension of each point, followed by a spherical window layer to perform local self-attention. We also introduce crossing window self-attention module to capture hierarchical features, which is proved that the relationships between different windows are effective to improve the accuracy of recognition tasks. In the future, we expect to study the semantic properties extracting, which would promote the network for other tasks such as 3D point cloud generation and 3D object detection.

Acknowledgements. This work was supported by the National Natural Science Foundation of China (NSFC, No. 62272426) and the Research Project by Shanxi Scholarship Council of China (No. 2020-113).

References

1. Bahdanau, D., Cho, K., Bengio, Y.: Neural machine translation by jointly learning to align and translate. arXiv preprint arXiv:1409.0473 (2014)
2. Ben-Shabat, Y., Lindenbaum, M., Fischer, A.: 3D point cloud classification and segmentation using 3D modified fisher vector representation for convolutional neural networks. arXiv preprint arXiv:1711.08241 (2017)
3. Chen, X., Ma, H., Wan, J., Li, B., Xia, T.: Multi-view 3D object detection network for autonomous driving. In: Proceedings of the IEEE Conference on Computer Vision and Pattern Recognition, pp. 1907–1915 (2017)
4. Dai, A.: ScanNet: richly-annotated 3D reconstructions of indoor scenes. In: 2017 IEEE Conference on Computer Vision and Pattern Recognition (CVPR) (2017)
5. Devlin, J., Chang, M.W., Lee, K., Toutanova, K.: BERT: pre-training of deep bidirectional transformers for language understanding. arXiv preprint arXiv:1810.04805 (2018)
6. Dosovitskiy, A., et al.: An image is worth 16 × 16 words: transformers for image recognition at scale. arXiv preprint arXiv:2010.11929 (2020)
7. Engel, N., Belagiannis, V., Dietmayer, K.: Point transformer. IEEE Access **9**, 134826–134840 (2021)
8. Guo, M.H., Cai, J.X., Liu, Z.N., Mu, T.J., Martin, R.R., Hu, S.M.: PCT: point cloud transformer. Comput. Visual Media **7**(2), 187–199 (2021)
9. Hu, Q., et al.: RandLA-Net: efficient semantic segmentation of large-scale point clouds. In: Proceedings of the IEEE/CVF Conference on Computer Vision and Pattern Recognition, pp. 11108–11117 (2020)
10. Hua, B.S., Pham, Q.H., Nguyen, D.T., Tran, M.K., Yeung, S.K.: SceneNN: a scene meshes dataset with annotations. In: Fourth International Conference on 3D Vision (2016)
11. Joseph-Rivlin, M., Zvirin, A., Kimmel, R.: Momen(e)t: flavor the moments in learning to classify shapes (2018)
12. Klokov, R., Lempitsky, V.: Escape from cells: deep KD-networks for the recognition of 3D point cloud models. In: Proceedings of the IEEE International Conference on Computer Vision, pp. 863–872 (2017)
13. Lang, A.H., Vora, S., Caesar, H., Zhou, L., Yang, J., Beijbom, O.: PointPillars: fast encoders for object detection from point clouds. In: Proceedings of the IEEE/CVF Conference on Computer Vision and Pattern Recognition, pp. 12697–12705 (2019)

14. Li, G., Muller, M., Thabet, A., Ghanem, B.: DeepGCNs: can GCNs go as deep as CNNs? In: Proceedings of the IEEE/CVF International Conference on Computer Vision, pp. 9267–9276 (2019)
15. Li, Y., Bu, R., Sun, M., Wu, W., Di, X., Chen, B.: PointCNN: convolution on X-transformed points. In: Advances in Neural Information Processing Systems, vol. 31 (2018)
16. Lin, Z., et al.: A structured self-attentive sentence embedding. arXiv preprint arXiv:1703.03130 (2017)
17. Liu, Z., et al.: Swin transformer: hierarchical vision transformer using shifted windows. In: Proceedings of the IEEE/CVF International Conference on Computer Vision, pp. 10012–10022 (2021)
18. Maturana, D., Scherer, S.: VoxNet: a 3D convolutional neural network for real-time object recognition. In: 2015 IEEE/RSJ International Conference on Intelligent Robots and Systems (IROS), pp. 922–928. IEEE (2015)
19. Paszke, A., Gross, S., Massa, F., Lerer, A., Chintala, S.: PyTorch: an imperative style, high-performance deep learning library (2019)
20. Qi, C.R., Su, H., Mo, K., Guibas, L.J.: PointNet: deep learning on point sets for 3D classification and segmentation. In: Proceedings of the IEEE Conference on Computer Vision and Pattern Recognition, pp. 652–660 (2017)
21. Qi, C.R., Su, H., Nießner, M., Dai, A., Yan, M., Guibas, L.J.: Volumetric and multi-view CNNs for object classification on 3D data. In: Proceedings of the IEEE Conference on Computer Vision and Pattern Recognition, pp. 5648–5656 (2016)
22. Qi, C.R., Yi, L., Su, H., Guibas, L.J.: PointNet++: deep hierarchical feature learning on point sets in a metric space. In: Advances in Neural Information Processing Systems, vol. 30 (2017)
23. Riegler, G., Osman Ulusoy, A., Geiger, A.: OctNet: learning deep 3D representations at high resolutions. In: Proceedings of the IEEE Conference on Computer Vision and Pattern Recognition, pp. 3577–3586 (2017)
24. Simonovsky, M., Komodakis, N.: Dynamic edge-conditioned filters in convolutional neural networks on graphs. In: Proceedings of the IEEE Conference on Computer Vision and Pattern Recognition, pp. 3693–3702 (2017)
25. Su, H., Maji, S., Kalogerakis, E., Learned-Miller, E.: Multi-view convolutional neural networks for 3D shape recognition. IEEE, December 2015
26. Thomas, H., Qi, C.R., Deschaud, J.E., Marcotegui, B., Goulette, F., Guibas, L.J.: KPConv: flexible and deformable convolution for point clouds. In: Proceedings of the IEEE/CVF International Conference on Computer Vision, pp. 6411–6420 (2019)
27. Uy, M.A., Pham, Q.H., Hua, B.S., Nguyen, T., Yeung, S.K.: Revisiting point cloud classification: a new benchmark dataset and classification model on real-world data. IEEE (2020)
28. Vaswani, A., et al.: Attention is all you need. In: Advances in Neural Information Processing Systems, vol. 30 (2017)
29. Wang, P.S., Liu, Y., Guo, Y.X., Sun, C.Y., Tong, X.: O-CNN: octree-based convolutional neural networks for 3D shape analysis. ACM Trans. Graph. (TOG) **36**(4), 1–11 (2017)
30. Wei, X., Yu, R., Sun, J.: View-GCN: view-based graph convolutional network for 3D shape analysis. In: 2020 IEEE/CVF Conference on Computer Vision and Pattern Recognition (CVPR), pp. 1847–1856. IEEE (2020)
31. Wu, B., et al.: Visual transformers: token-based image representation and processing for computer vision. arXiv preprint arXiv:2006.03677 (2020)

32. Wu, W., Qi, Z., Fuxin, L.: PointConv: deep convolutional networks on 3D point clouds. In: Proceedings of the IEEE/CVF Conference on Computer Vision and Pattern Recognition (CVPR), June 2019
33. Wu, Z., et al.: 3D ShapeNets: a deep representation for volumetric shapes. In: Proceedings of the IEEE Conference on Computer Vision and Pattern Recognition, pp. 1912–1920 (2015)
34. Xu, Y., Fan, T., Xu, M., Long, Z., Yu, Q.: SpiderCNN: deep learning on point sets with parameterized convolutional filters (2018)
35. Yang, Z., Wang, L.: Learning relationships for multi-view 3D object recognition. In: Proceedings of the IEEE/CVF International Conference on Computer Vision, pp. 7505–7514 (2019)
36. Yi, L., et al.: A scalable active framework for region annotation in 3D shape collections. ACM Trans. Graph. (ToG) **35**(6), 1–12 (2016)

Self-supervised Learning with Multi-view Rendering for 3D Point Cloud Analysis

Bach Tran[1(✉)], Binh-Son Hua[1], Anh Tuan Tran[1], and Minh Hoai[1,2]

[1] VinAI Research, Hanoi, Vietnam
{v.bachtx12,v.sonhb,v.anhtt152,v.hoainm}@vinai.io
[2] Stony Brook University, New York 11794, USA

Abstract. Recently, great progress has been made in 3D deep learning with the emergence of deep neural networks specifically designed for 3D point clouds. These networks are often trained from scratch or from pre-trained models learned purely from point cloud data. Inspired by the success of deep learning in the image domain, we devise a novel pre-training technique for better model initialization by utilizing the multi-view rendering of the 3D data. Our pre-training is self-supervised by a local pixel/point level correspondence loss computed from perspective projection and a global image/point cloud level loss based on knowledge distillation, thus effectively improving upon popular point cloud networks, including PointNet, DGCNN and SR-UNet. These improved models outperform existing state-of-the-art methods on various datasets and downstream tasks. We also analyze the benefits of synthetic and real data for pre-training, and observe that pre-training on synthetic data is also useful for high-level downstream tasks. Code and pre-trained models are available at https://github.com/VinAIResearch/selfsup_pcd.git.

Keywords: Self-supervised learning · Point cloud analysis · Multiple-view rendering · 3D deep learning

1 Introduction

Pixels and points are basic elements in computer vision for visual recognition. In the past decade, image collections have been successfully used to train neural networks for common visual recognition tasks, including object classification and semantic segmentation. Concurrently, advances in depth-sensing technologies, including RGB-D and LiDAR sensors, have enabled the acquisition of large-scale 3D data, facilitating the rapid development of visual recognition methods in 3D, notably neural networks for point cloud analysis in the last few years. Unlike images, annotation for point clouds are generally more expensive to acquire due to the laborious process of scene scanning, reconstruction, and annotation, and

---

Supplementary Information The online version contains supplementary material available at https://doi.org/10.1007/978-3-031-26319-4_25.

L. Wang et al. (Eds.): ACCV 2022, LNCS 13841, pp. 413–431, 2023.
https://doi.org/10.1007/978-3-031-26319-4_25

thus point cloud neural networks are often trained with datasets that are much smaller than image datasets.

A potential direction to improve the robustness for point cloud neural networks is self-supervised learning. By letting the point cloud network perform some pre-text tasks before supervised learning, a process commonly known as pre-training, the network can perform more effectively than that trained from scratch. With self-supervised learning, the pre-text tasks are designed so that the pre-training does not use additional labels. In 3D deep learning, some initial effort has been spent on investigating this direction [45,52,56]. However, most previous works solely considered self-supervised learning in the 3D domain; only a few works exploited images to support the learning of point cloud neural networks. In an early work, Pham et al. [39] trained autoencoders on both images and point clouds and applied constraints on the latent space of both domains, allowing feature transfers between 2D and 3D. Inspired by the recently growing literature on network pre-training, we explore how to use images to more effectively (self-)supervise point cloud neural networks.

Particularly, in this paper, we propose a method that utilizes multi-view rendering to generate pixel/point and image/point cloud pairs for self-supervising a point cloud neural network. We train two neural networks, one for image and one for point cloud, respectively, and direct both networks to agree upon their latent features in the 2D and 3D domains. To achieve this, we use the pixel and point correspondences to formulate a local loss function that encourages features of the correspondences to be similar. This is well-motivated by projective geometry in 3D computer vision that defines the coordinate mapping between the image and 3D space. To further regularize the self-supervision, we utilize knowledge distillation to formulate a global loss that encourages the feature distribution between images and point clouds to be similar as well. Our method works even when there is big domain gap between the pre-train data and test data, e.g., between synthetic and real data.

In summary, we make three technical contributions in this paper: (1) a pre-training technique built upon multi-view rendering that advocates the use of multi-view image features to self-supervise the point cloud neural network; (2) a local loss function that exploits pixel-point correspondence in the pre-training; (3) a global loss function that performs knowledge distillation from the images to the 3D point clouds.

2 Related Work

3D Deep Learning: Building a neural network to analyze 3D data is a non-trivial task. Early attempts involve extending neural networks for images to work with volumes [36], and projecting 3D data to 2D views that can be used with traditional neural networks [49]. Recent methods in deep learning with point clouds take a different approach by letting a network input point clouds directly. Two major directions can be taken to implement such idea: learning per-point features by pointwise MLP [42,43], and learning point features by examining points in a local neighborhood by custom convolutions tailored for point sets [26,30,32] and by graph neural networks [6,47,53]. Notable methods in such directions include PointNet [42] and DGCNN [53]. An inherent limitation of PointNet-based approaches is that they can only process a few thousands of points, limiting the ability to handle large-scale point clouds, where a sliding window is often used as a workaround [42]. More recent developments include the use of sparse tensor and sparse convolution [9,13,14] on large-scale point clouds for semantic segmentation and 3D object detection. We refer readers to [16] for a survey of deep learning methods for point clouds.

Self-supervised Learning: Unsupervised pre-training is a useful technique in deep learning with proven success in natural language processing [11] and representation learning [7,15,20,21,37,46,50,59]. For pre-training, one can use generative modeling techniques based on GANs [2,54] and autoencoders [19, 20,52,58], or other self-supervised learning techniques [1,4,8,24,40,45,56,57]. Pre-training is relevant to knowledge distillation [23], a class of techniques for transferring features learned from a large network (teacher network) to a small network (student network). Here we assume that the pre-text task is rather general and can be very different to the downstream tasks, and so a subset of the layers in the pre-trained can be transferred depending on the downstream task.

Self-supervised learning techniques for pre-training 3D point cloud networks have been recently explored from multiple perspectives. Early works use a pre-text task for self-supervised learning. The pre-text task can be solving a jigsaw puzzle [45], where a point cloud is subdivided into randomly arranged voxels, and the task is to predict for each point the voxel ID the point belongs to. Another pre-text task is point cloud completion [52] (OcCo), where a mechanism similar to mask-based pre-training in natural language processing is utilized. As an extension of autoencoder on 3D point clouds, Eckart et al. [12] apply soft segmentation on point clouds and enforces these partitions to comply a latent parametric model in an encoder-decoder network paradigm. Recent contrastive learning is also shown to be effective for pre-training 3D point clouds [56,60]. PointContrast [56] create positive pairs and negative pairs for contrastive learning by the correspondences between two camera views of a point cloud. Depth-Contrast [60] learn the representation with multiple 3D data formats including points and voxels.

Self-supervised learning with other 3D data modalities [3,17,18,25,33,34,39] has also been explored. Jing et al. [28,29] (CM) use 3D data with multi-modality and build cross-modal and cross-view invariant constraints, maximizing cross-modal agreement of the features of point cloud, mesh, and images, and maximizing cross-view agreement with the image features. Hou et al. [25] use contrastive learning on multi-view images constraints and image-geometry constraint. However, they only focus on 2D downstream tasks. Huang et al. [27] (STRL) proposed self-supervised learning for a sequence of point clouds which utilizes spatio-temporal cues. Pham et al. [39] (LCD) leverages a 2D-3D correspondence dataset and a triplet loss to transfer features from 2D to 3D only on *small cropped regions* of images and 3D point clouds. Compared with LCD [39], our method is largely different as we self-supervise 3D point features via multi-view projection in the *entire* image space. LCD [39] is suitable for image matching and point cloud registration tasks, while our method is suitable for point cloud analysis tasks such as classification and segmentation.

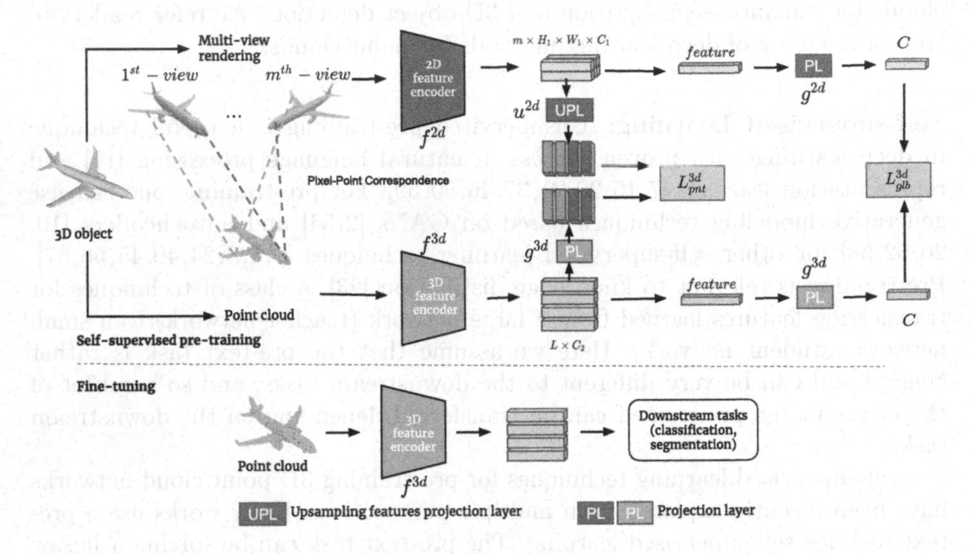

Fig. 1. Overview of our proposed method. The main proposal is pre-training steps that exploit multi-modal data, including multi-view images and point clouds, to learn a 3D feature encoder for effective point cloud representation. This model is then fine-tuned for different downstream tasks.

There are a few concurrent works [3,31]. In [3], the authors considered RGB rendering of the object surfaces but required the mesh textures in addition to the geometry for rendering. Our rendering is more practical in that we consider different rendering techniques and only require colorless point clouds. In [31], the authors focus on data from autonomous driving datasets including KITTI and nuScenes. Our method focuses on object datasets.

3 Self-supervised Learning for 3D Point Clouds

In this section, we describe the proposed self-supervised learning with multi-view rendering for point clouds. Our goal is to leverage multi-modal data of 3D objects, in which each object is associated with a 3D point cloud and multiple 2D images from various view points to pre-train the point cloud network. We propose to use multi-view rendering to generate images for input 3D objects that pair with the point clouds for pre-training. Using rendered images to pre-train point cloud networks implies an advantage that different 3D data representations, including triangle mesh and point cloud, can all be converted into a unified 2D representation to pre-train the networks. To ease this pre-training process, we do not require annotation for the 3D objects and rely on self-supervised learning techniques for the pre-training.

Our method consists of two steps. First, we learn feature representation for 2D images with self-supervised learning by ensuring the similarity between the representation features of two transformed images of the same original image. Second, we use the feature representation of 2D images to learn the 3D feature representation for 3D point clouds. We illustrate the overview of our method in Fig. 1, and we will describe the two steps in details in the next two subsections.

Let $\mathcal{D}$ denote the pre-training data, $\mathcal{D} = \{P_i, \{\mathbf{y}_{ij}, M_{ij}\}_{j=1}^m\}_{i=1}^n$, where n is the number of objects in the dataset and m the number of 2D views for each object. P_i is the 3D point cloud of the ith object, $\mathbf{y}_{ij}$ is the projected image of the ith object to the jth view using the projection matrix M_{ij}.

3.1 Learning Feature Representation for 2D Images

Let us start with learning the discriminative feature representation for multi-view images. In this step, we simply treat all object views $\{\{\mathbf{y}_{ij}\}_{j=1}^m\}_{i=1}^n$ as items of a set. Following SimCLR [7], we randomly sample a batch of k images from this image set in each training iteration. For each image in the batch, we randomly sample two augmentation operators (crop, color distortion, and Gaussian blur) to generate a pair of correlated images based on the original image. Given an image $\mathbf{x}_i$ in the batch, let $\mathbf{x}'_i$ and $\mathbf{x}''_i$ be its augmented images, respectively. Our objective is to learn a feature extraction network f^{2d} so that the resulting feature vectors for different augmentations of the same image $\mathbf{x}'_i$ and $\mathbf{x}''_i$ are similar, while both $\mathbf{x}'_i$ and $\mathbf{x}''_i$ should be different from the feature vectors of other image augmentations $\mathbf{x}'_j$ and $\mathbf{x}''_j$ for $j \neq i$. We therefore define the loss for image $\mathbf{x}_i$ as:

$$\mathcal{L}^{2d}(i) = -\log\left(\frac{\psi(\mathbf{x}'_i, \mathbf{x}''_i)}{\psi(\mathbf{x}'_i, \mathbf{x}''_i) + \sum_{j\neq i}(\psi(\mathbf{x}'_i, \mathbf{x}'_j) + \psi(\mathbf{x}'_i, \mathbf{x}''_j))}\right) \quad (1)$$
$$-\log\left(\frac{\psi(\mathbf{x}''_i, \mathbf{x}'_i)}{\psi(\mathbf{x}''_i, \mathbf{x}'_i) + \sum_{j\neq i}(\psi(\mathbf{x}''_i, \mathbf{x}'_j) + \psi(\mathbf{x}''_i, \mathbf{x}''_j))}\right).$$

Here, $\psi(\mathbf{x}_i, \mathbf{x}_j)$ is the function that measures the similarity between two images, and we use the exponential cosine similarity of the two feature vectors, i.e.,

$$\psi(\mathbf{x}_i, \mathbf{x}_j) = \exp\Big(\cos\left(g^{2d}(f^{2d}(\mathbf{x}_i)), g^{2d}(f^{2d}(\mathbf{x}_j))\right)/\tau\Big), \quad (2)$$

where τ is the temperature hyper-parameter, and g^{2d} is the projection layer (nonlinear projection layer).

The loss function for a batch of k images is: $\mathcal{L}^{2d} = \frac{1}{k}\sum_{i=1}^k \mathcal{L}^{2d}(i)$. In each optimization iteration, we calculate the gradient of this loss to optimize for the parameters of the feature extractor network f^{2d}, which is a fully convolutional neural network. The input to the network is an RGB image of dimensions $H \times W \times 3$ and the output is a 3D tensor of size $H_1 \times W_1 \times C_1$, where C_1 is the number of output channels and $H_1 = H/2^s, W_1 = W/2^s$, with s being the number of down-sampling layers in the network. The output tensor can be vectorized to form a global representation vector for the entire image. This output tensor can also be up-sampled to yield a feature map having the same width and height as those of the input image; in this case, there is a corresponding C_1-dim feature vector for each pixel of the input image.

3.2 Knowledge Transfer from 2D to 3D

Once the feature extraction function f^{2d} for 2D images has been learned, we will use it to learn a point-wise feature extraction function f^{3d} for 3D point clouds. The input to this feature extraction is a point cloud of L points, and the output is a 2D tensor of size $L \times C_2$, where C_2 is the number of feature dimensions. Each point of the point cloud has a corresponding C_2-dimensional feature vector. To learn the feature extraction f^{3d}, we develop a novel scheme that uses 2D-to-3D knowledge transfer. We use both global and point-wise knowledge transfer.

Global Knowledge Transfer. For global knowledge transfer, we minimize the distance between the aggregated 2D feature vector and the aggregated 3D feature vector by

$$\mathcal{L}_{glb}^{3d} = \frac{1}{n} \sum_{i=1}^{n} \left\| g^{2d}(\max_j f^{2d}(\mathbf{y}_{ij})) - g^{3d}(\max f^{3d}(P_i)) \right\|^2, \tag{3}$$

where P_i is the point cloud of the ith object and $\mathbf{y}_{ij}$ is the jth view of the ith object. f^{2d} is the feature extractor for 2D images, which was explained in Sect. 3.1. Function $\max_j f^{2d}(\mathbf{y}_{ij})$ is the pixel-wise max-pooling across different 2D views. Function f^{3d} is the feature extractor for 3D point cloud, which we seek to learn here. $\max f^{3d}(P_i)$ is element-wise max-pooling among all feature vectors of all points of point cloud P_i. Both g^{2d} and g^{3d} are nonlinear projection layers that transform 2D feature and 3D feature vectors to the feature space, respectively.

Point-Wise Knowledge Transfer: In addition to global knowledge transfer, we use contrastive learning that minimizes the distance between feature representation of a 3D point and its corresponding 2D pixel. To determine the point-to-pixel correspondences, we project each point of the point cloud P_i to each image view $\mathbf{y}_{ij}$ using the camera projection matrix M_{ij} to have $\mathbf{y}_{ij}^{2d} = M_{ij} P_i$, where $\mathbf{y}_{ij}^{2d}$ is a set of pixels from the rendered image $\mathbf{y}_{ij}$ corresponding to P_i. For point-wise knowledge transfer, in each optimization iteration, we sample a batch of k corresponding pixel-point pairs, and let $\{(\mathbf{z}_i^{2d}, \mathbf{z}_i^{3d})\}_{i=1}^{k}$ be the corresponding set of feature vector pairs. For the ith pixel-point pair, $\mathbf{z}_i^{2d}$ is obtained by:

(1) using the 2D feature extraction function f^{2d} on the image that contains the pixel; (2) passing the output to the upsampling feature projection module u^{2d}; and (3) extracting the feature vector at the location of the pixel in the projected feature map. $\mathbf{z}_i^{3d}$ is obtained by: (1) using the 3D feature extraction function f^{3d} on the point cloud containing that point; (2) passing the output through the projection function g^{3d}; and (3) extracting the corresponding feature vector of the point in the point cloud.

For point-wise knowledge transfer, we maximize the similarity between the pixel representation vector and the point representation vector, using a loss function inspired by SimCLR [7]. The loss term for the ith pixel-point pair is:

$$\begin{aligned}\mathcal{L}_{pnt}^{3d}(i) = &- \log\left(\frac{\psi(\mathbf{z}_i^{2d}, \mathbf{z}_i^{3d})}{\psi(\mathbf{z}_i^{2d}, \mathbf{z}_i^{3d}) + \sum_{j\neq i}(\psi(\mathbf{z}_i^{2d}, \mathbf{z}_j^{2d}) + \psi(\mathbf{z}_i^{2d}, \mathbf{z}_j^{3d}))}\right) \\ &- \log\left(\frac{\psi(\mathbf{z}_i^{3d}, \mathbf{z}_i^{2d})}{\psi(\mathbf{z}_i^{3d}, \mathbf{z}_i^{2d}) + \sum_{j\neq i}(\psi(\mathbf{z}_i^{3d}, \mathbf{z}_j^{2d}) + \psi(\mathbf{z}_i^{3d}, \mathbf{z}_j^{3d}))}\right),\end{aligned} \quad (4)$$

where $\psi(\cdot,\cdot)$ is the exponential cosine function defined in Eq. (2). Intuitively, both 2D and 3D features can be regarded as augmentations of a common latent feature, so they form a positive pair of which similarity can be maximized with the contrastive loss. The total loss function for a batch of k pixel-point pairs is: $\mathcal{L}_{pnt}^{3d} = \frac{1}{k}\sum_{i=1}^{k}\mathcal{L}_{pnt}^{3d}(i)$.

Combined Loss Function. To pre-train the point cloud network, we minimize a loss function that is the weighted combination of the global knowledge transfer loss and the point-wise knowledge transfer loss:

$$\mathcal{L}^{3d} = \lambda_{glb}\mathcal{L}_{glb}^{3d} + \lambda_{pnt}\mathcal{L}_{pnt}^{3d}. \quad (5)$$

In our experiments, we simply use $\lambda_{glb} = \lambda_{pnt} = 1$. After pre-training, we can now use the pre-trained weights to initialize the training of downstream tasks.

4 Experiments

4.1 Implementation Details

Dataset for Pre-training. Unless otherwise mentioned, we use ModelNet40 [55] for pre-training. ModelNet40 is a synthetic dataset that includes 9,480 training samples and 2,468 test samples in 40 categories. ModelNet40 represents each object by a 3D mesh, making it suitable for our multi-view rendering purpose. For each object in the training set of ModelNet40, we generate its point cloud using farthest-point sampling. We render the object into multi-view images by moving a camera around the object. Unless otherwise mentioned, each point cloud has 1024 points rendered into 12 views with 32×32 resolution. We use 12 views as they tend to cover all object directions in general. We keep the views in low resolution of 32×32 to avoid out of memory usage at training.

Our multi-view rendering is implemented as follows. First, each object is normalized into a unit cube. To generate multi-view images from a mesh object, we used Blender [44] with fixed camera parameters (focal length 35, sensor width 32, and sensor height 32) and a light source. The camera positions are chosen to cover the surrounding views of the object, and the distances from each camera to its neighbor positions are equal.

2D Feature Representation Learning. We use ResNet50 [22] as a 2D feature extractor f^{2d} in 2D self-supervised learning process. We use Adam optimizer with the initial learning rate 0.001 without learning decay. We then train the self-supervised model with 1000 epochs and batch size 512.

Table 1. Comparison among random, Jigsaw [45], OcCo [52], CM [29], STRL [27], and our initialization to the object classification downstream task. We reported the mean and standard deviation at the best epoch over three runs.

	DGCNN					
	Random	Jigsaw	OcCo	CM	STRL	Ours
MN40 [55]	92.7 ± 0.1	92.9 ± 0.1	92.9 ± 0.0	93.0 ± 0.1	93.1 ± 0.1	**93.2 ± 0.1**
SO [51]	82.8 ± 0.5	82.1 ± 0.2	83.2 ± 0.2	83.0 ± 0.2	83.2 ± 0.2	**84.3 ± 0.6**
SO BG [51]	81.4 ± 0.5	82.0 ± 0.4	82.9 ± 0.4	82.2 ± 0.2	83.2 ± 0.2	**84.5 ± 0.6**

3D Feature Representation Learning. We experiment with two common backbones PointNet [42] and DGCNN [53], which can be utilized for both classification and segmentation tasks. For PointNet [42], we use Adam optimizer with the initial learning rate 0.001, which decays 0.7 every 20 epochs. The momentum of batch normalization starts as 0.5, then divided by 2 every 20 epochs. For DGCNN [53], we use an SGD optimizer with the initial learning rate 0.1 and momentum 0.9. We use CosineAnnealingLR scheduler [35] for learning rate decay. For both backbones, we train the model with 250 epochs, 200 epochs, and 100 epochs for classification, part segmentation, and semantic segmentation task, respectively, with the same batch size as 32. After getting the pre-trained models, we test their effectiveness in training with different downstream tasks.

4.2 Object Classification

We first experiment with object classification for 3D point cloud analysis. Two standard benchmarks are used, namely ModelNet40 [55] and ScanObjectNN [51] dataset. We follow the previous paper [52] to use ModelNet40 in both pre-training and downstream tasks. ScanObjectNN is a real-world dataset that has 15 categories with 2,321 and 581 samples for training and testing, respectively. We use the default variant (OBJ_ONLY, denoted by ScanObjectNN) and the variant with background (OBJ_BG, denoted by ScanObjectNN BG). We follow the experimental setting in the original PointNet [42].

We compare the performance of DGCNN [53] with and without pre-training. The results are shown in Table 1. We also provide comparisons with the PointNet backbone [42] in the supplementary material. As can be seen, models with pre-training outperform their randomly initialized counterparts. We further compare our method to previous point cloud pre-training methods, including Jigsaw [45], OcCo [52], CM [29], and STRL [27]. Particularly, Jigsaw [45] learns to solve jigsaw puzzles as a pretext task for pre-training. OcCo [52] is based on mask-based pre-training from natural language processing to propose a point cloud completion task for pre-training. CM [29] considered self-supervision from cross-modality and cross-view feature learning, which shares some similarity to ours. Our method differs in that we use a contrastive loss to learn 2D features and an L2 loss to match 2D-3D features while CM [29] uses a triplet loss for 2D features and a cross-entropy loss for matching 2D-3D features. STRL [27] explored self-supervision with spatial-temporal representation learning. In Table 1, it can be seen that our proposed self-supervision with contrastive loss and multi-view rendering outperforms other initialization methods on both ModelNet40 and ScanObjectNN dataset.

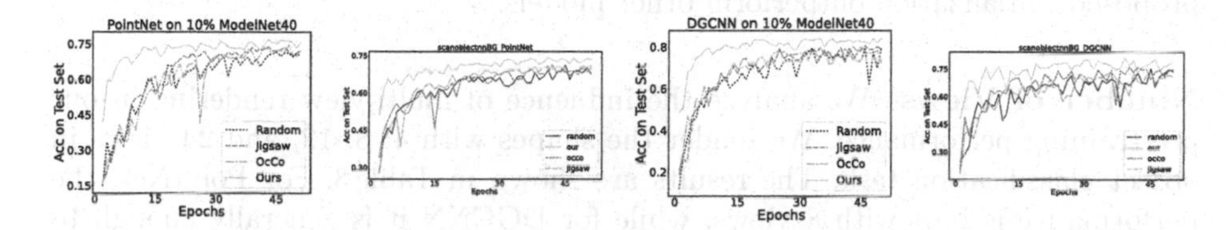

Fig. 2. Test-set accuracy over different training epochs in the object classification task. The plots show that previous pre-training methods are only marginally more effective than random initialization while our method shows a clear improvement.

Table 2. Performance of the object classification task trained with fewer data. Our method has significant gains compared to other initialization methods. We reported the mean at the best epoch over three runs.

	PointNet				DGCNN			
	Random	Jigsaw	OcCo	Ours	Random	Jigsaw	OcCo	Ours
5%	73.2	73.8	73.9	**77.9**	82.0	82.1	82.3	**84.9**
10%	75.2	77.3	75.6	**79.0**	84.7	84.1	84.9	**86.6**
20%	81.3	82.9	81.6	**84.6**	89.4	89.2	89.1	**90.2**
50%	86.6	86.5	87.1	**87.6**	91.6	91.8	91.7	**92.4**
70 %	88.3	88.4	88.4	**88.7**	92.3	92.4	92.5	**92.8**
90 %	88.5	88.8	88.8	**89.4**	92.6	92.9	92.9	**93.1**

4.3 Network Analysis

Accuracy Over Epochs. Figure 2 plots the accuracy on the test set over different training epochs. The proposed initialization helps both PointNet and DGCNN converge faster and obtain better accuracy than other initialization methods. For example, when we use ModelNet40 with 10% training dataset, the model with our initialization converges after around 15 epochs, while with other initialization methods, it takes around 40 epochs. For ScanObjectNN (OBJ_BG variant), the models converge after about 20 epochs with our initialization and about 45 epochs with other methods.

Training Size. Our pre-training allows the point cloud network to be trained with less data compared to initialization with random weights. To demonstrate this, we reduce the number of samples in the training set of ModelNet40 to 5%, 10%, 20%, 50%, and 70%. We then use these datasets to train the model for the object classification task. Finally, we evaluate these learned models on the test set of ModelNet40. Table 2 shows the results with random initialization, Jigsaw [45], OcCo [52], and our initialization, respectively. As can be seen, models using our proposed initialization outperform other models.

Number of Views. We analyze the influence of multi-view rendering in our pre-training performance. We render the shapes with 4, 8, 12, and 24 views in object classification task. The results are shown in Table 3. For PointNet, the performance is best with 8 views, while for DGCNN it is generally enough to use 4 views, but DGCNN for ScanObjectNN performs best with 24 views.

Table 3. Performance of object classification tested with different number of views in multi-view rendering.

	PointNet				DGCNN			
	4-views	8-views	12-views	24-views	4-views	8-views	12-views	24-views
MN40 [55]	88.9	**89.2**	88.9	88.9	**92.8**	92.3	92.5	92.3
SO [51]	79.0	**80.4**	79.3	79.1	82.7	82.6	82.8	**84.9**
SO BG [51]	74.2	**77.1**	75.7	76.6	**82.8**	81.9	82.6	81.4

Classification with SVM. To evaluate the generalization ability of our pre-trained model, we follow a similar test scenario in [45]. First, we freeze the weights of the pre-trained model and pass the 3D object through this model to obtain their embeddings. Then, we train a Support Vector Machine (SVM) on the embeddings of the train set and evaluate it on the test set on three datasets ModelNet40, ScanObjectNN, and ScanObjectNN (OBJ_BG variant). For SVM, we used grid search to find the best parameter of SVM with a Radial Basis Function kernel. The results are shown in Table 4. The proposed initialization outperforms the other initialization methods, Jigsaw and OcCo, sometimes by a wide margin as in the ScanObjectNN (OBJ_BG variant). We also provide additional comparisons to previous self-supervised methods on ModelNet40 in Table 5. As can be seen, our proposed method outperforms almost other methods in both PointNet and DGCNN, except in PointNet, our method is ranked second while ParAE [12] performs best.

Table 4. The result of SVM applied on the object embedding learned from different initializations. It shows that features learned by our method are more discriminative than other methods, resulting in more accurate classifications.

	PointNet			DGCNN		
	Jigsaw	OcCo	Ours	Jigsaw	OcCo	Ours
ModelNet40 [55]	82.5	87.2	**89.7**	83.1	89.5	**91.7**
ScanObjectNN [51]	49.7	62.1	**70.2**	57.8	69.0	**76.3**
ScanObjectNN BG [51]	48.9	61.7	**69.5**	51.1	67.5	**74.2**

Ablation Study of Loss Functions. Table 6 reports the performance of our method for the classification task when both global loss and pixel-point loss are used together or individually. The network achieves the best performance when trained with both losses. Using either global loss or pixel-point loss results in accuracy drop especially for the ScanObjectNN dataset [51]. This is because the global loss is only useful in distilling knowledge from an image view to a point cloud while the pixel-point loss encourages the model learn consistent features locally. Using both losses lets the model have the best of both worlds.

Table 5. More comparisons of SVM classification on ModelNet40.

	PointNet	DGCNN
Rotation [40]	88.6	99.8
STRL [27]	88.3	90.9
ParAE [12]	**90.3**	91.6
CrossPoint [3]	89.1	91.2
Ours	89.7	**91.7**

Table 6. Effect of loss function choice to our pre-training.

	PointNet			DGCNN		
	$\mathcal{L}^{3d}_{glb}$	$\mathcal{L}^{3d}_{pnt}$	$\mathcal{L}^{3d}$	$\mathcal{L}^{3d}_{glb}$	$\mathcal{L}^{3d}_{pnt}$	$\mathcal{L}^{3d}$
ModelNet40 [55]	88.5	88.5	**88.9**	92.4	92.1	**92.5**
ScanObjectNN [51]	77.6	78.8	**79.3**	81.8	81.1	**82.8**
ScanObjectNN BG [51]	74.5	74.2	**75.7**	81.6	81.6	**82.6**

Table 7. Effect of rendering techniques to the pre-training on PointNet [42].

	RGB	Silhouette	Shading
ModelNet40 [55]	88.3	**88.9**	**88.9**
ScanObjectNN [51]	**79.7**	78.8	79.3
ScanObjectNN BG [51]	75.1	75.6	**75.7**

Multi-view Rendering. Our pre-training requires multi-view image rendering, which can be implemented by a wide range of rendering techniques. We study the effect of images rendered from the object mesh, 3D position encoding, and object silhouette on the classification task (please refer to example rendering in Fig. 3). For the original object mesh, we use Blender [44] to render the object geometry with Phong shading, resulting in grayscale *shaded* images. For 3D position encoding, the images are rendered directly from point clouds. Specifically, we first assign a pseudo-color (RGB) to each point of a point cloud based on their 3D coordinates, then project the points to the image plane with preset camera projection matrices. For object silhouette, the process is generally similar except that we use black instead of the pseudo-color for each point in the point clouds. For pixel that has more than one corresponding point, we choose the point with minimum distance to the camera.

The performance of object classification with different rendering techniques is reported in Table 7, where there is no best technique overall. Using shaded images results in slightly higher accuracy than using position encoding and silhouette rendering in ModelNet40 and ScanObjectNN BG. This is because shaded images often have more details than position encoding and silhouette images since shaded images are rendered from meshes. Exploring more robust rendering techniques for self-supervised learning is left for future work.

Fig. 3. Object rendering with position encoding (RGB), silhouette, and shading.

Table 8. The result of four initialization in the part segmentation task on the ShapeNetPart dataset [53]. We reported the mean and standard error of mAcc and mIoU over three runs.

	PointNet				DGCNN			
	Random	Jigsaw	OcCo	Ours	Random	Jigsaw	OcCo	Ours
mAcc	93.3 ± 0.2	93.0 ± 0.0	93.3 ± 0.1	$\mathbf{93.4 \pm 0.0}$	94.2 ± 0.0	94.1 ± 0.0	$\mathbf{94.3 \pm 0.0}$	94.2 ± 0.1
mIoU	83.1 ± 0.3	83.2 ± 0.1	83.0 ± 0.2	$\mathbf{83.3 \pm 0.1}$	$\mathbf{84.7 \pm 0.0}$	84.5 ± 0.1	$\mathbf{84.7 \pm 0.1}$	$\mathbf{84.7 \pm 0.1}$

Table 9. Fold 1 of overall point accuracy (mAcc) and mean Intersection-over-Union (mIoU) on the S3DIS (Stanford Area 5 Test) [5].

	PointNet				DGCNN			
	Random	Jigsaw	OcCo	Ours	Random	Jigsaw	OcCo	Ours
mAcc	83.9	82.5	83.6	**85.0**	86.8	86.8	**87.0**	**87.0**
mIoU	43.6	43.6	44.5	**46.7**	49.3	48.2	49.5	**49.9**

4.4 Part Segmentation and Scene Segmentation

Beyond classification, we conduct experiments to validate our pre-training for semantic part segmentation and scene segmentation tasks. We first experiment with object part segmentation on the ShapeNetPart dataset [53] that includes 16,881 objects from 16 categories. Each object is represented by 2,048 points, and each point belongs to one of 50 part types. Most objects in the dataset are divided into two to five parts. As shown in Table 8, our initialization is slightly better than random initialization, Jigsaw, and OcCo in both overall point accuracy (mAcc) and mean Intersection-over-Union (mIoU) metric. We observed that the improvement is minor in the part segmentation task because the network architecture used for part segmentation is largely different from the pre-trained networks, and therefore only a few layers of the part segmentation networks can be initialized by the pre-trained model.

We also experiment with semantic scene segmentation on the Stanford Large-Scale 3D Indoor Spaces dataset (S3DIS) [5]. This dataset contains point clouds of 272 rooms from 6 areas and 13 categories. Each room is split into $1m\times1m$ blocks. Each point is represented by a 9D vector including XYZ, RGB, and normalized location in the room. During training, each block is sampled with 4096 points, but during testing, all points are used. The results are shown in Table 9. We see that models initialized by our method outperform others in both PointNet [42] and DGCNN [53].

Table 10. Comparison to PointContrast (PC) [56] on the semantic segmentation and 3D object detection task on S3DIS dataset [5], ScanNet dataset [10], and SUN RGB-D dataset [48]. Our method outperforms PointContrast when pre-trained on both datasets. The subscript indicates the performance difference compared to the Random case.

Dataset	Task (Metric)	Random	PC [56] ModelNet	Ours ModelNet	PC [56] ScanNet	Ours ScanNet
S3DIS (Area 5)	sem. seg. (Acc)	72.5	71.2 $_{-1.3}$	**73.2** $_{+0.7}$	73.0 $_{+0.5}$	73.0 $_{+0.5}$
S3DIS (Area 5)	sem. seg. (IoU)	64.5	64.1 $_{-0.4}$	66.0 $_{+1.5}$	66.1 $_{+1.6}$	**66.5** $_{+2.0}$
ScanNet	sem. seg. (Acc)	80.2	80.3 $_{+0.1}$	**81.1** $_{+0.9}$	80.8 $_{+0.6}$	81.0 $_{+0.8}$
ScanNet	sem. seg. (IoU)	72.4	72.5 $_{+0.1}$	73.3 $_{+0.9}$	73.1 $_{+0.7}$	**73.6** $_{+1.2}$
ScanNet	3D det. (AP_{50})	35.2	36.6 $_{+1.4}$	38.2 $_{+3.0}$	36.1 $_{+0.9}$	**39.2** $_{+4.0}$
ScanNet	3D det. (AP_{25})	56.5	58.2 $_{+1.7}$	58.4 $_{+1.9}$	59.5 $_{+3.0}$	**60.3** $_{+3.8}$
SUN RGB-D	3D det. (AP_{50})	32.3	34.8 $_{+2.5}$	34.9 $_{+2.6}$	34.8 $_{+2.5}$	**35.1** $_{+2.8}$
SUN RGB-D	3D det. (AP_{25})	55.5	57.8 $_{+2.3}$	**58.1** $_{+2.6}$	57.4 $_{+1.9}$	57.8 $_{+2.3}$

4.5 Pre-training with Synthetic vs. Real-World Data

Multi-view rendering can be easily used for self-supervised learning when working with synthetic data as we have shown with ModelNet40 [55]. Real-world 3D datasets, however, often do not provide such multi-view images, limiting our choices for pre-training. In this section, we investigate the role of synthetic and real-world data in pre-training by comparing to PointContrast [56] and DepthContrast [60] on the segmentation and detection task. We run different experiments using Sparse Residual U-Net (SR-UNet) [9] as the network backbone. Compared to PointNet and DGCNN backbone used in the previous sections, SR-UNet uses sparse convolutions to learn features on point clouds, discarding the need of a sliding window for processing large-scale point clouds. For pre-training data, we use ModelNet40 [55] as synthetic data and ScanNet [10] as real data.

Pre-training. As the original PointContrast [56] only supports ScanNet for pre-training, here we adapt ModelNet40 to PointContrast by using surface point cloud pairs, formed for every two continuous views, instead of the provided point cloud pairs from ScanNet. As for our model, we use two view images when their corresponding point cloud pairs have at least 30% overlapping. To define pixel-point pairs, we reconstruct a point cloud from the first depth image in an image pair, then project it to two color images to get pixel-point correspondences. During training, we follow original setting of PointContrast [56]. For our pre-trained model on ScanNet [10], we used the pre-trained ResNet50 [22] on ImageNet provided by Pytorch [38] as the 2D feature extractor. Besides, all images are resized to 240×320 and we only use the point-wise knowledge transfer loss for pre-training. We train the model with one GPU and four GPUs for ModelNet40 and ScanNet datasets, respectively.

Segmentation and Detection Results. We evaluate four pre-training configurations with the semantic segmentation task on two datasets S3DIS [5] and ScanNet [10]. We show the results in Table 10 (comparisons to PointContrast [56]) and Table 11 (comparisons to DepthContrast [60]). In Table 10, on both datasets, the performance gap between our models pre-trained on synthetic and real data is small. When testing on S3DIS, our pre-trained network on ModelNet even provides a slightly better performance compared to the pre-trained model on ScanNet on Acc. metric, and it offers 2% increase when compared with the PointContrast counterpart on both Acc. and IoU metric. In Table 11, we also compare with DepthContrast on semantic segmentation task. For S3DIS, our pre-trained models on both synthetic and real data achieve better performance approximately 2% on IoU and 4% on Acc. For ScanNet, our pre-trained model on synthetic data outperforms the random setting but is slightly less effective than DepthContrast. However, our pre-trained model on real data outperforms both random and DepthContrast initialization.

Table 11. Comparison to DepthContrast [60] on the semantic segmentation task on S3DIS dataset [5] and ScanNet dataset [10]. The subscript indicates the performance difference compared to the Random case.

Dataset	Task (Metric)	Random	DepthContrast [60]	Ours ModelNet	Ours ScanNet
S3DIS (Area 5)	sem. seg. (Acc)	70.9	72.1 $_{+1.2}$	**75.1** $_{+4.2}$	74.5 $_{+3.6}$
S3DIS (Area 5)	sem. seg. (IoU)	64.0	64.8 $_{+0.8}$	**66.8** $_{+2.8}$	66.5 $_{+2.5}$
ScanNet	sem. seg. (Acc)	77.2	77.6 $_{+0.4}$	77.4 $_{+0.2}$	**78.3** $_{+1.1}$
ScanNet	sem. seg. (IoU)	69.1	69.9 $_{+0.8}$	69.2 $_{+0.1}$	**70.7** $_{+1.6}$

We also perform comparison on the 3D object detection task on the ScanNet dataset [10] and SUN RGB-D dataset [48]. Following [56], we replace original PointNet++ [43] backbone of VoteNet [41] by SR-UNet [9]. The results are also shown in Table 10. As can be seen, our method outperforms PointContrast when pre-training on the same dataset. When using synthetic data, our model can obtain two points higher in mAP_{50} compared with the PointContrast counterpart. When using real data, the mAP scores increase slightly and achieve state-of-the-art performance.

5 Conclusion

We propose a self-supervised learning method based on multi-view rendering to pre-train 3D point cloud neural networks. Our self-supervision with multi-view rendering on global and local loss functions yield state-of-the-art performance on several downstream tasks including object classification, semantic segmentation and object detection. Our pre-training method works well on both synthetic and real-world data; it also proves the effectiveness of pre-training on synthetic data like ModelNet40 for downstream tasks with real data like semantic segmentation and 3D object detection.

References

1. Achituve, I., Maron, H., Chechik, G.: Self-supervised learning for domain adaptation on point clouds. In: Proceedings of the IEEE/CVF Winter Conference On Applications Of Computer Vision, pp. 123–133 (2021)
2. Achlioptas, P., Diamanti, O., Mitliagkas, I., Guibas, L.: Learning representations and generative models for 3D point clouds. In: International Conference on Machine Learning, pp. 40–49. PMLR (2018)
3. Afham, M., Dissanayake, I., Dissanayake, D., Dharmasiri, A., Thilakarathna, K., Rodrigo, R.: Crosspoint: Self-supervised cross-modal contrastive learning for 3D point cloud understanding. In: Proceedings of the IEEE/CVF Conference on Computer Vision and Pattern Recognition, pp. 9902–9912 (2022)
4. Alliegro, A., Boscaini, D., Tommasi, T.: Joint supervised and self-supervised learning for 3d real world challenges. In: 2020 25th International Conference on Pattern Recognition (ICPR), pp. 6718–6725. IEEE (2021)

5. Armeni, I., et al.: 3D semantic parsing of large-scale indoor spaces. In: Proceedings of the IEEE Conference on Computer Vision and Pattern Recognition, pp. 1534–1543 (2016)
6. Bruna, J., Zaremba, W., Szlam, A., LeCun, Y.: Spectral networks and locally connected networks on graphs. arXiv preprint arXiv:1312.6203 (2013)
7. Chen, T., Kornblith, S., Norouzi, M., Hinton, G.: A simple framework for contrastive learning of visual representations. In: International Conference on Machine Learning, pp. 1597–1607. PMLR (2020)
8. Chen, Y., et al.: Shape self-correction for unsupervised point cloud understanding. In: Proceedings of the IEEE/CVF International Conference on Computer Vision, pp. 8382–8391 (2021)
9. Choy, C., Gwak, J., Savarese, S.: 4D spatio-temporal convnets: Minkowski convolutional neural networks. In: Proceedings of the IEEE Conference on Computer Vision and Pattern Recognition, pp. 3075–3084 (2019)
10. Dai, A., Chang, A.X., Savva, M., Halber, M., Funkhouser, T., Nießner, M.: ScanNet: richly-annotated 3D reconstructions of indoor scenes. In: Proceedings of the Computer Vision and Pattern Recognition (CVPR). IEEE (2017)
11. Devlin, J., Chang, M.W., Lee, K., Toutanova, K.: BERT: pre-training of deep bidirectional transformers for language understanding. arXiv preprint arXiv:1810.04805 (2018)
12. Eckart, B., Yuan, W., Liu, C., Kautz, J.: Self-supervised learning on 3D point clouds by learning discrete generative models. In: Proceedings of the IEEE/CVF Conference on Computer Vision and Pattern Recognition, pp. 8248–8257 (2021)
13. Graham, B., Engelcke, M., Van Der Maaten, L.: 3D semantic segmentation with submanifold sparse convolutional networks. In: Proceedings of the IEEE Conference on Computer Vision and Pattern Recognition, pp. 9224–9232 (2018)
14. Graham, B., van der Maaten, L.: Submanifold sparse convolutional networks. arXiv preprint arXiv:1706.01307 (2017)
15. Grill, J.B., et al.: Bootstrap your own latent-a new approach to self-supervised learning. Adv. Neural. Inf. Process. Syst. **33**, 21271–21284 (2020)
16. Guo, Y., Wang, H., Hu, Q., Liu, H., Liu, L., Bennamoun, M.: Deep learning for 3D point clouds: a survey. IEEE Transactions on Pattern Analysis and Machine Intelligence (2020)
17. Gupta, S., Hoffman, J., Malik, J.: Cross modal distillation for supervision transfer. In: Proceedings of the IEEE Conference on Computer Vision and Pattern Recognition, pp. 2827–2836 (2016)
18. Hafner, F., Bhuiyan, A., Kooij, J.F., Granger, E.: A cross-modal distillation network for person re-identification in RGB-depth. arXiv preprint arXiv:1810.11641 (2018)
19. Han, Z., Wang, X., Liu, Y.S., Zwicker, M.: Multi-angle point cloud-VAE: unsupervised feature learning for 3D point clouds from multiple angles by joint self-reconstruction and half-to-half prediction. In: 2019 IEEE/CVF International Conference on Computer Vision (ICCV), pp. 10441–10450. IEEE (2019)
20. Hassani, K., Haley, M.: Unsupervised multi-task feature learning on point clouds. In: Proceedings of the IEEE/CVF International Conference on Computer Vision, pp. 8160–8171 (2019)
21. He, K., Fan, H., Wu, Y., Xie, S., Girshick, R.: Momentum contrast for unsupervised visual representation learning. In: Proceedings of the IEEE/CVF Conference on Computer Vision and Pattern Recognition, pp. 9729–9738 (2020)

22. He, K., Zhang, X., Ren, S., Sun, J.: Deep residual learning for image recognition. In: Proceedings of the IEEE Conference on Computer Vision and Pattern Recognition, pp. 770–778 (2016)
23. Hinton, G., Vinyals, O., Dean, J.: Distilling the knowledge in a neural network. arXiv preprint arXiv:1503.02531 (2015)
24. Hou, J., Graham, B., Nießner, M., Xie, S.: Exploring data-efficient 3D scene understanding with contrastive scene contexts. In: Proceedings of the IEEE/CVF Conference on Computer Vision and Pattern Recognition, pp. 15587–15597 (2021)
25. Hou, J., Xie, S., Graham, B., Dai, A., Nießner, M.: Pri3D: Can 3D priors help 2D representation learning? In: Proceedings of the IEEE/CVF International Conference on Computer Vision (ICCV), pp. 5693–5702 (2021)
26. Hua, B.S., Tran, M.K., Yeung, S.K.: Pointwise convolutional neural networks. In: Proceedings of the IEEE Conference on Computer Vision and Pattern Recognition, pp. 984–993 (2018)
27. Huang, S., Xie, Y., Zhu, S.C., Zhu, Y.: Spatio-temporal self-supervised representation learning for 3d point clouds. In: Proceedings of the IEEE/CVF International Conference on Computer Vision, pp. 6535–6545 (2021)
28. Jing, L., Chen, Y., Zhang, L., He, M., Tian, Y.: Self-supervised modal and view invariant feature learning. arXiv preprint arXiv:2005.14169 (2020)
29. Jing, L., Zhang, L., Tian, Y.: Self-supervised feature learning by cross-modality and cross-view correspondences. In: Proceedings of the IEEE/CVF Conference on Computer Vision and Pattern Recognition, pp. 1581–1591 (2021)
30. Li, Y., Bu, R., Sun, M., Wu, W., Di, X., Chen, B.: PointCNN: convolution on x-transformed points. Adv. Neural. Inf. Process. Syst. **31**, 820–830 (2018)
31. Li, Z., et al.: SimIPU: simple 2D image and 3D point cloud unsupervised pre-training for spatial-aware visual representations. In: Proceedings of the AAAI Conference on Artificial Intelligence, vol. 36, pp. 1500–1508 (2022)
32. Liu, Y., Fan, B., Xiang, S., Pan, C.: Relation-shape convolutional neural network for point cloud analysis. In: Proceedings of the IEEE/CVF Conference on Computer Vision and Pattern Recognition, pp. 8895–8904 (2019)
33. Liu, Y.C., et al.: Learning from 2D: pixel-to-point knowledge transfer for 3D pre-training. arXiv preprint arXiv:2104.04687 (2021)
34. Liu, Z., Qi, X., Fu, C.W.: 3D-to-2D distillation for indoor scene parsing. In: Proceedings of the IEEE/CVF Conference on Computer Vision and Pattern Recognition, pp. 4464–4474 (2021)
35. Loshchilov, I., Hutter, F.: SGDR: stochastic gradient descent with warm restarts. arXiv preprint arXiv:1608.03983 (2016)
36. Maturana, D., Scherer, S.: VoxNet: a 3D convolutional neural network for real-time object recognition. In: 2015 IEEE/RSJ International Conference on Intelligent Robots and Systems (IROS), pp. 922–928. IEEE (2015)
37. Misra, I., van der Maaten, L.: Self-supervised learning of pretext-invariant representations. In: Proceedings of the IEEE/CVF Conference on Computer Vision and Pattern Recognition, pp. 6707–6717 (2020)
38. Paszke, A., et al.: PyTorch: an imperative style, high-performance deep learning library. In: Advances in Neural Information Processing Systems (2019)
39. Pham, Q.H., Uy, M.A., Hua, B.S., Nguyen, D.T., Roig, G., Yeung, S.K.: LCD: learned cross-domain descriptors for 2D–3D matching. In: Proceedings of the AAAI Conference on Artificial Intelligence (2020)
40. Poursaeed, O., Jiang, T., Qiao, H., Xu, N., Kim, V.G.: Self-supervised learning of point clouds via orientation estimation. In: 2020 International Conference on 3D Vision (3DV), pp. 1018–1028. IEEE (2020)

41. Qi, C.R., Litany, O., He, K., Guibas, L.J.: Deep hough voting for 3D object detection in point clouds. In: Proceedings of the IEEE International Conference on Computer Vision (2019)
42. Qi, C.R., Su, H., Mo, K., Guibas, L.J.: PointNet: deep learning on point sets for 3D classification and segmentation. In: Proceedings of the IEEE Conference on Computer Vision and Pattern Recognition (CVPR) (2017)
43. Qi, C.R., Yi, L., Su, H., Guibas, L.J.: PointNet++: deep hierarchical feature learning on point sets in a metric space. In: Advances in Neural Information Processing Systems 30 (2017)
44. Roosendaal, T.: Blender - a 3D modelling and rendering package. Blender Foundation (2018). http://www.blender.org
45. Sauder, J., Sievers, B.: Self-supervised deep learning on point clouds by reconstructing space. In: Advances in Neural Information Processing Systems 32 (2019)
46. Sharma, C., Kaul, M.: Self-supervised few-shot learning on point clouds. Adv. Neural. Inf. Process. Syst. **33**, 7212–7221 (2020)
47. Simonovsky, M., Komodakis, N.: Dynamic edge-conditioned filters in convolutional neural networks on graphs. In: Proceedings of the IEEE Conference on Computer Vision and Pattern Recognition, pp. 3693–3702 (2017)
48. Song, S., Lichtenberg, S.P., Xiao, J.: Sun RGB-D: a RGB-D scene understanding benchmark suite. In: Proceedings of the IEEE Conference on Computer Vision and Pattern Recognition, pp. 567–576 (2015)
49. Su, H., Maji, S., Kalogerakis, E., Learned-Miller, E.: Multi-view convolutional neural networks for 3D shape recognition. In: Proceedings of the IEEE International Conference on Computer Vision, pp. 945–953 (2015)
50. Thabet, A., Alwassel, H., Ghanem, B.: Self-supervised learning of local features in 3D point clouds. In: Proceedings of the IEEE/CVF Conference on Computer Vision and Pattern Recognition Workshops, pp. 938–939 (2020)
51. Uy, M.A., Pham, Q.H., Hua, B.S., Nguyen, D.T., Yeung, S.K.: Revisiting point cloud classification: a new benchmark dataset and classification model on real-world data. In: International Conference on Computer Vision (ICCV) (2019)
52. Wang, H., Liu, Q., Yue, X., Lasenby, J., Kusner, M.J.: Unsupervised point cloud pre-training via occlusion completion. In: Proceedings of the IEEE/CVF International Conference on Computer Vision, pp. 9782–9792 (2021)
53. Wang, Y., Sun, Y., Liu, Z., Sarma, S.E., Bronstein, M.M., Solomon, J.M.: Dynamic graph CNN for learning on point clouds. ACM Trans. Graph. (TOG) **38**(5), 1–12 (2019)
54. Wu, J., Zhang, C., Xue, T., Freeman, W.T., Tenenbaum, J.B.: Learning a probabilistic latent space of object shapes via 3D generative-adversarial modeling. In: Proceedings of the 30th International Conference on Neural Information Processing Systems, pp. 82–90 (2016)
55. Wu, Z., et al.: 3D shapeNets: a deep representation for volumetric shapes. In: Proceedings of the IEEE Conference on Computer Vision and Pattern Recognition (CVPR) (2015)
56. Xie, S., Gu, J., Guo, D., Qi, C.R., Guibas, L., Litany, O.: PointContrast: unsupervised pre-training for 3d point cloud understanding. In: Vedaldi, A., Bischof, H., Brox, T., Frahm, J.-M. (eds.) ECCV 2020. LNCS, vol. 12348, pp. 574–591. Springer, Cham (2020). https://doi.org/10.1007/978-3-030-58580-8_34
57. Yamada, R., Kataoka, H., Chiba, N., Domae, Y., Ogata, T.: Point cloud pre-training with natural 3D structures. In: Proceedings of the IEEE/CVF Conference on Computer Vision and Pattern Recognition, pp. 21283–21293 (2022)

58. Yu, X., Tang, L., Rao, Y., Huang, T., Zhou, J., Lu, J.: Point-BERT: pre-training 3D point cloud transformers with masked point modeling. In: Proceedings of the IEEE/CVF Conference on Computer Vision and Pattern Recognition, pp. 19313–19322 (2022)
59. Zhang, L., Zhu, Z.: Unsupervised feature learning for point cloud understanding by contrasting and clustering using graph convolutional neural networks. In: 2019 International Conference on 3D Vision (3DV), pp. 395–404. IEEE (2019)
60. Zhang, Z., Girdhar, R., Joulin, A., Misra, I.: Self-supervised pretraining of 3D features on any point-cloud. In: Proceedings of the IEEE/CVF International Conference on Computer Vision, pp. 10252–10263 (2021)

PointFormer: A Dual Perception Attention-Based Network for Point Cloud Classification

Yijun Chen[1,2,3], Zhulun Yang[1,2,3], Xianwei Zheng[3](✉), Yadong Chang[1,2], and Xutao Li[1,2]

[1] Key Lab of Digital Signal and Image Processing of Guangdong Province, Shantou University, Shantou 515063, China
{21yjchen1,15zlyang3,21ydchang,lixt}@stu.edu.cn

[2] Department of Electronic Engineering, Shantou University, Shantou 515063, China

[3] School of Mathematics and Big Data, Foshan University, Foshan 52800, China
alex.w.zheng@hotmail.com

Abstract. Point cloud classification is a fundamental but still challenging task in 3-D computer vision. The main issue is that learning representational features from initial point cloud objects is always difficult for existing models. Inspired by the Transformer, which has achieved successful performance in the field of natural language processing, we propose a purely attention-based network, named PointFormer, for point cloud classification. Specifically, we design a novel simple point multiplicative attention mechanism. Based on that, we then construct both a local attention block and a global attention block to learn fine geometric features and overall representational features of the point cloud, respectively. Consequently, compared to the existing approaches, PointFormer has superior perception of local details and overall contours of the point cloud objects. In addition, we innovatively propose the Graph-Multiscale Perceptual Field (GMPF) testing strategy that can significantly improve the overall performance of the proposed PointFormer. We have conducted extensive experiments on the real-world dataset ScanObjectNN and the synthetic dataset ModelNet40. The results show that the PointFormer has stronger robustness and achieves highly competitive performance compared to other state-of-the-art approaches. The code is available at https://github.com/Yi-Jun-Chen/PointFormer.

Keywords: Point cloud classification · Attention mechanism · Feature extraction

1 Introduction

3-D vision is widely used in augmented reality, robotics, autonomous driving and many other fields. Voxels, meshes and point clouds can all be applied to effectively represent 3D data. In contrast to voxels and meshes, point clouds preserve

L. Wang et al. (Eds.): ACCV 2022, LNCS 13841, pp. 432–449, 2023.
https://doi.org/10.1007/978-3-031-26319-4_26

the original geometric information of 3D objects and are easy to collect. Therefore, it is more suitable for 3D object classification which is a fundamental and still challenging task in 3D vision. However, point clouds are irregular embeddings in continuous space which is different from 2D images. Therefore, existing methods are still challenging for learning representational features in point cloud classification task.

To address this challenge, much work [16,23] has been inspired by the successful application of convolutional neural networks to 2D computer vision, where 3D objects are projected onto different 2D planes and then the 2D convolution operation is performed. Other methods voxelize the point cloud and then apply 3D discrete convolution in 3D space. Unexpectedly, the performance of these approaches is largely limited by the computational and memory costs. In addition, the transformation of point clouds, such as voxelization and projection, leads to the loss of information, which is detrimental to the processing of point clouds. PointNet [3] is a pioneer in the direct manipulation of point cloud data. It extracts the features of point clouds through shared multilayer perceptron (MLP) and a max-pooling operation, achieving permutation invariance in point cloud processing and impressive classification results. However, PointNet ignores local structure information, which is proven to be important for feature learning in visual recognition. To better capture the local information of point cloud objects, some works [4,27] introduce attention mechanisms to enhance the local representation of features. Unfortunately, they usually have expensive overhead on memory and arithmetic power. Consequently, it is important to design networks with superior perception of the fine and overall geometric structure of point clouds while maintaining a relatively light weight.

To solve this problem, we propose a purely attention-based network, named PointFormer, for point clouds classification. In detail, we propose a novel and simple point multiplicative attention mechanism. Unlike traditional methods [3,17,29] that aggregate information from each point indiscriminately through pooling operations, the point multiplicative attention mechanism treats each point in point cloud differently to extract more discriminative features. Based on this, we then construct local attention block and global attention block that enables the network to have excellent perception of the local fine structure and overall shape of the point cloud. In addition, we propose the Graph-Multiscale Perceptual Field (GMPF) testing strategy for the first time to improve the overall performance of the PointFormer. In summary, our main contributions are displayed below.

- We design a point multiplicative attention mechanism with high expressiveness for point cloud objects. It applies shared multilayer perceptrons (MLP) to learn the encoding of points, which maintains low complexity while satisfying permutation invariance for point cloud processing.
- Based on the point multiplicative attention mechanism, we design both a local attention block and a global attention block for building PointFormer, which enables our model to have superior dual perception of local details and overall contours of point cloud.

- We propose a Graph-Multiscale Perceptual Field (GMPF) testing strategy, which greatly improves the performance of the model. Furthermore, as a general strategy, it can be easily transferred to other networks.
- Extensive experiments on real-world datasets ScanOjectNN and synthetic datasets ModelNet40 show that our proposed has strong robustness and superior classification performance.

2 Related Work

Unlike 2D images which have a regular grid structure, 3D point clouds have spatial continuity and irregularities that preclude direct applications of the existing deep learning models. To overcome this challenge, many approaches have been proposed, which can be broadly classified into three types: multi-view methods, voxel-based methods and point-based methods. We summarise the main features and limitations of these approaches as follows:

1) Multi-view Methods: Considering the success of CNNs in 2D images, these methods [6,9,23] project 3D point cloud objects onto multiple 2D planes. Then, convolutional neural networks are applied to perform feature extraction. Finally, the aggregation of multi-view features is performed to accurately classify the point cloud. For example, MVCNN [23] is a pioneering network that has achieved impressive results by aggregating multi-scale information through max-pooling operations. However, these methods may result in loss of information during the projection process. At the same time, these approaches usually lead to huge memory and arithmetic overheads because they do not make good use of the sparsity of point clouds.
2) Voxel-based Methods: voxel-based approaches regularize irregular point clouds by 3D voxelization [15,22] and then perform a 3D convolution operation. While these methods can achieve impressive results, the cubic increase in the number of grids can lead to an explosion in computation and memory overhead. To alleviate this problem, OctNet [20] uses shallow octrees to improve computational efficiency, but it is still a computationally intensive approach. In summary, these approaches not only consume computational and memory resources greatly, but also raise information loss during the voxelization process.
3) Point-based Methods: Instead of transforming point clouds into other data domains as the previous methods do, the point-based approaches directly use points as input for feature extraction. PointNet [3], as a pioneer model, realizes the real sense of a neural network acting directly on a point cloud. It is implemented by shared MLPs and symmetric operations (e.g. max-pooling) to satisfy the permutation invariance of point cloud processing. However, PointNet operates on each point individually, so it does not have the ability to extract local geometric features. To solve this problem, PointNet++ [17] designs a series of operations for furthest point sampling and grouping, hierarchically aggregating the features of neighbouring nodes. DGCNN [29] makes the local information spread to the global sufficiently by k-nearest neighbor

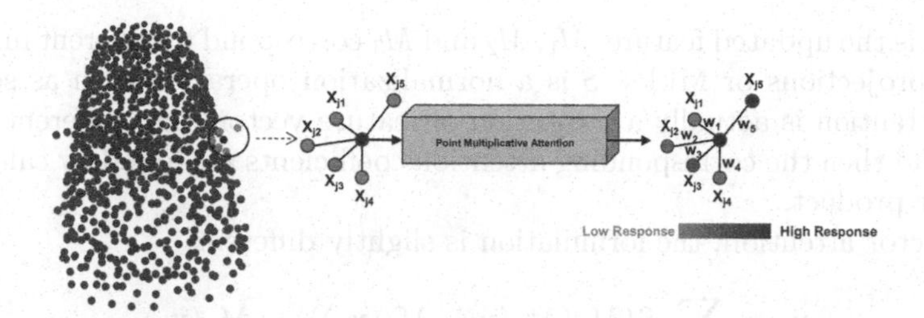

Fig. 1. The red point is selected as the central node, and its k-nearest neighbors (k=5 in the figure) are selected to form the k-NN graph. Then, neighboring nodes are weighted by the point multiplicative attention mechanism. The different depths of the colors respond to the magnitude of the weights. (Color figure Online)

(k-NN) composition and dynamic graph update between each layer, which learns the local geometric information of the point cloud very well. However, they uniformly use symmetric pooling operations such as max-pooling to retain the most significant elements, which leads to loss of information and is detrimental to the fine perception of point cloud objects. Other approaches [4,36] enhance the representation of local information by introducing attention. However, their large number of parameters and computational effort leads to a huge overhead on computational resources.

In contrast, as a point-based approach, our PointFormer has excellent dual perception of the local fine geometry and the overall contour of the point cloud. At the same time, our model is able to maintain a relatively light weight due to the proposed novel and simple point multiplicative attention mechanism. In addition, the proposed Graph-Multiscale Perceptual Field (GMPF) testing strategy gives PointFormer the ability to sense multi-scale information, thus greatly improving the performance of the model.

3 Our Approach

In this section, we first review the general formula for self-attention. Then, the point multiplicative attention mechanism is proposed for constructing our local attention block and global attention block. Finally, the complete framework of PointFormer and the Graph-Multiscale Perceptual Field(GMPF) testing strategy for model performance enhancement is presented.

3.1 Scalar and Vector Self-attention

Self-attention is a set operator [36] which can be classified into two types: scalar attention [27] and vector attention [35]. $\mathcal{X} = \{\mathbf{x}_i\}_i$ denotes the set of feature vectors. The scalar attention can be formulated as follows:

$$\mathbf{x}_i' = \sum_{\mathbf{x}_j \in \mathcal{X}} S\Big(M_1(\mathbf{x}_i)^\top M_2(\mathbf{x}_j)\Big) M_3(\mathbf{x}_j), \tag{1}$$

where $\mathbf{x}_i^{'}$ is the updated feature. M_1, M_2 and M_3 correspond to different mappings such as projections or MLPs. S is a normalization operation such as softmax. Scalar attention is actually a projection of feature vectors into different feature spaces and then the corresponding attention coefficients are found by calculating the inner product.

In vector attention, the formulation is slightly different:

$$\mathbf{x}_i^{'} = \sum_{\mathbf{x}_j \in \mathcal{X}} S\left(M_4\left(M_1\left(\mathbf{x}_i\right), M_2\left(\mathbf{x}_j\right)\right)\right) \odot M_3\left(\mathbf{x}_j\right), \tag{2}$$

M_4 represents a mapping (e.g. MLP) like M_1, M_2, M_3. Vector attention replaces the operation of inner product in scalar attention by introducing more learnable parameters in M_4. Both scalar and vector attention are set operator. Therefore, it satisfies the permutation invariance of point cloud processing and is well suited for processing point clouds.

3.2 Point Multiplicative Attention Mechanism

From the viewpoint of formulation and computation, scalar attention is simpler, but often less effective than vector attention. Conversely, vector attention is more effective but more complex. To reach a reliable balance between these mechanisms, we propose our point multiplicative attention mechanism:

$$\mathbf{x}_i^{'} = \sum_{\mathbf{x}_j \in \mathcal{X}} S\left(L_3^1\left(L_1\left(\mathbf{x}_i\right) + L_2\left(\mathbf{x}_j\right)\right)\right) L_2\left(\mathbf{x}_j\right), \tag{3}$$

where L_1,L_2 denote different single fully connected layers respectively. L_3^1 denotes a single fully connected layer with an output dimension of 1. As we can see, to reduce the complexity of vector attention, we replace all the MLPs in Eq. 3 with a linear layer. In particular, we replace the different MLPs M_2, M_3 in Eqs. 1 and 2 with the same single fully connected layer L_2. At the same time, unlike GAN [28], we use the vector addition instead of vector concatenation in Eq. 3 to reduce the dimension of features. Moreover, subject to scalar attention, we only output an attention coefficient for each neighbor feature $\mathbf{x}_j$ through the fully connected layer L_3^1 and then perform a weighted summation. This is the reason why our attention is called point multiplicative attention.

3.3 Local Attention Block

As point multiplicative attention follows a similar formulation as scalar and vector attention, it is still a set operator. We construct the k-nearest neighbor (k-NN) graph [29] for each node of the point cloud, as shown in Fig. 1. The features of the central node and the neighbor nodes constitute the set of feature vectors for the attention operation. We construct our local attention block by acting the point multiplicative attention mechanism on the k-NN graph of each point (as presented in Fig. 1).

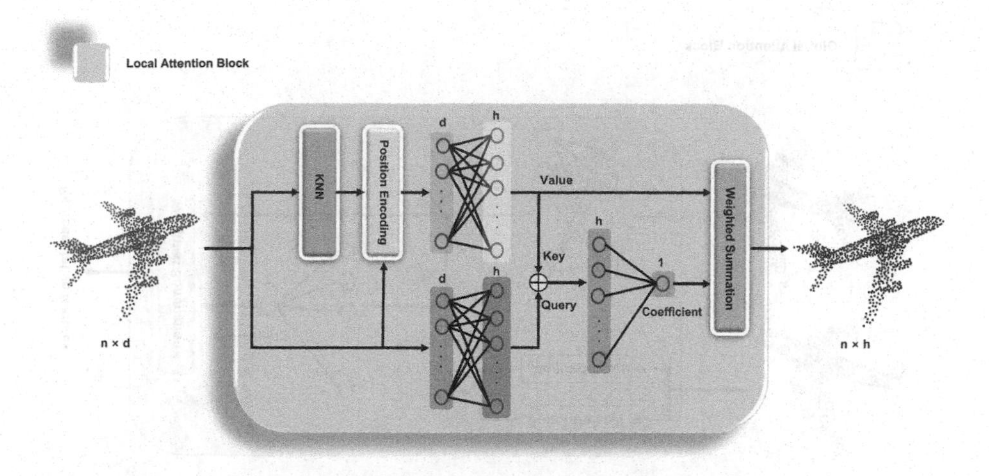

Fig. 2. Framework diagram of the local attention block. Input: Initial point cloud object features (blue). Output: Updated point cloud object features. The local attention block can be viewed as a d-dimensional to h-dimensional feature mapping. (Color figure online)

Our local attention block providing a better perception of local details of the point cloud objects. For example, in Fig. 1, the neighbor $\mathbf{x_{j5}}$ represents the sampled points from the rearview mirror of the car, while the other neighbors are the sampled points from the body. Point multiplicative attention mechanism can place on neighbor $\mathbf{x_{j5}}$ with a larger attention coefficient, thus paying more attention to this particular neighbor points while treating other neighbors differently.

Position Encoding: From the viewpoint of graph neural networks [7,34], attention is an information aggregation operator. Specifically, we learn the aggregation feature of neighboring points $\mathbf{x}_j$ to centroid $\mathbf{x}_i$ by L_2 in Eq. 3. Then, the aggregation of information is completed by weighted summation of the corresponding attention coefficients. Therefore, the linear layer L_2 can be viewed as a fit of the abstract function f(·) between $\mathbf{x}_i$ and $\mathbf{x}_j$. However, considering only the features $\mathbf{x}_j$ of neighbour nodes in L_2 may not make good use of the initial geometric information of the point cloud. In order to take the higher order features into considerations, we use $\mathbf{x}_i$, $\mathbf{x}_j$, $(\mathbf{x}_i - \mathbf{x}_j)$ and $\|\mathbf{x}_i - \mathbf{x}_j\|_2^2$ as inputs to L_2. This means that we use higher order difference terms $(\mathbf{x}_i - \mathbf{x}_j)$ and $\|\mathbf{x}_i - \mathbf{x}_j\|_2^2$ to retain high frequency detail information.

Finally, our local attention block can be formulated as:

$$\mathbf{x}_i' = \sum_{\mathbf{x}_j \in \mathcal{X}} S\left(L_3^1\left(L_1\left(\mathbf{x}_i\right) + L_2\left(PE\left(\mathbf{x}_j\right)\right)\right)\right) L_2\left(PE\left(\mathbf{x}_j\right)\right), \tag{4}$$

where PE denotes our position encoding strategy. The framework of the local attention block is illustrated in Fig. 2.

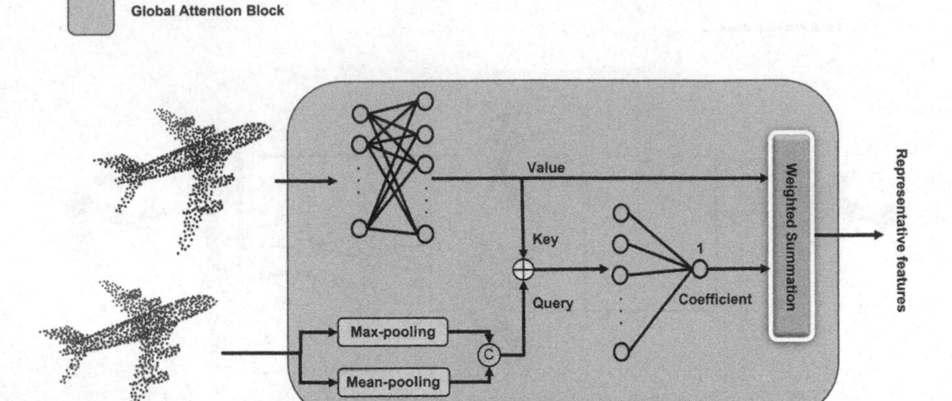

Fig. 3. Architecture of global attention block. The input is a point cloud carrying high-dimensional features [n × h], and the output is a representation vector of the point cloud [n × 1].

3.4 Global Attention Block

Local attention block provides excellent perception of local details, but lacks overall representation of the point cloud. To address this problem, global attention block is designed to extract highly representational features from point cloud objects. In detail, our global attention block is designed to apply point multiplicative attention to the entire point cloud object that carrying high-dimensional features, i.e., the red point cloud output from the local attention block in Fig. 2. Specifically, we concatenate the max-pooling and mean-pooling of the point cloud 3D coordinates to obtain the query vector. If $\mathcal{X} = \{\mathbf{x}_j\}_j$ denotes the set of features of all points of a point cloud object, the global attention block can be formulated as:

$$\mathbf{y} = \sum_{\mathbf{x}_j \in \mathcal{X}} S\left(L_3^1\left(L_1\left(query\right) + L_2\left(\mathbf{x}_j\right)\right)\right) L_2\left(\mathbf{x}_j\right), \tag{5}$$

where y denotes the representation vector of the point cloud. The complete structure of global attention block is presented in Fig. 3. The global attention block can be assigned an attention level for each point. This mechanism of treating each point differently allows our model to have superior perception of the overall contour of the point cloud. We will further visualize it in the experimental section.

3.5 Framework of PointFormer

In summary, we use four local attention blocks and one global attention block to build the PointFormer, as shown in Fig. 4. The output dimensions are set

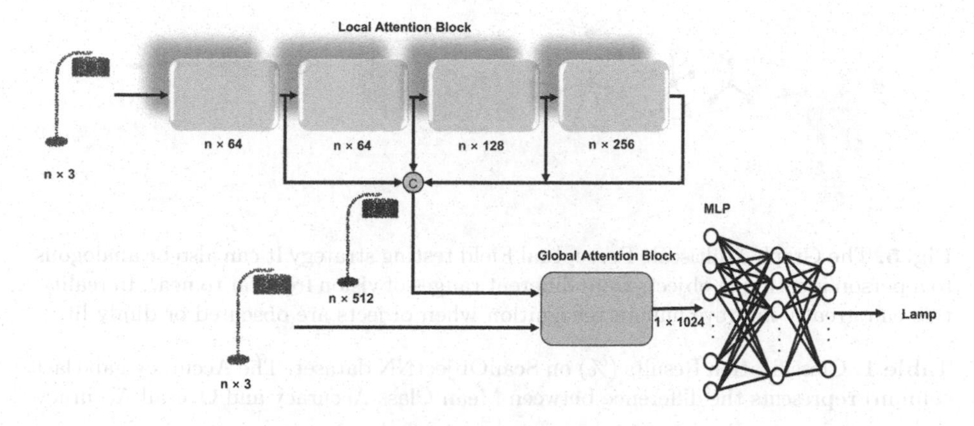

Fig. 4. Architecture of PointFormer.

to (64, 64, 128, 256) and (1024), respectively. Meanwhile, shortcut connections are applied to extract multi-scale features. Finally, the classification results are obtained by Multi-layer Perceptron (MLP).

Hierarchical Position Encoding (HPE): To maximize the initial geometric information of the point cloud while maintaining the lightweight of the model, our position encoding strategy in local attention block is further designed to use $\mathbf{x}_i$, $\mathbf{x}_j$, $(\mathbf{x}_i - \mathbf{x}_j)$ and $\|\mathbf{x}_i - \mathbf{x}_j\|_2^2$ as the input of L_2 only in the first local attention block. However, in the subsequent three blocks, we just use $\mathbf{x}_j$ and first-order difference term $(\mathbf{x}_i - \mathbf{x}_j)$ [29]. We refer to this location encoding strategy as Hierarchical Position Encoding (HPE).

3.6 Graph-Multiscale Perceptual Field (GMPF) Testing Strategy

In order to improve the performance while without introducing any burden (e.g., parameters, inference time), we design a novel Graph-Multiscale Perceptual Field (GMPF) testing strategy. Briefly, in the training phase, we construct the k-nearest neighbor (k-NN) graph using a specific number of neighbors (e.g., k = 5 in Fig. 1), while in the testing phase, multi-scale k-NN graphs are used for testing (e.g., k=3, 5, 7). Equivalently, the GMPF strategy is applied to learn a model on a fixed-scale k-NN graph while observing the information on the multi-scale graph during testing as shown in Fig. 5. In the experimental section, we will further show that this strategy can greatly improve the robustness and overall performance of the model. Moreover, as a general strategy, it can be naturally transferred to other graph-based networks.

4 Experiments

In this section, we first provide the experimental setup for evaluation. We then comprehensively evaluate PoinrFormer on ScanObjectNN [26] and ModelNet40

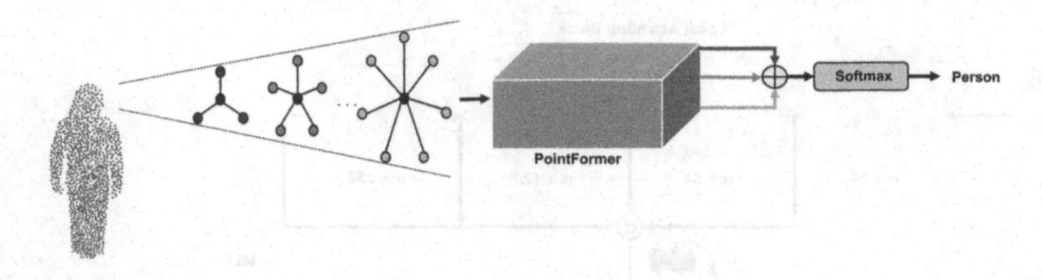

Fig. 5. The Graph-Multiscale Perceptual Field testing strategy.It can also be analogous to a person identifying objects from different ranges of vision from far to near. In reality, this can greatly improve human recognition when objects are obscured or dimly lit.

Table 1. Classification Results (%) on ScanObjectNN dataset. The Accuracy gap (last column) represents the difference between Mean Class Accuracy and Overall Accuracy (%).

Method	Mean Class Accuracy	Overall Accuracy	Accuracy gap
3DmFV [2]	58.1	63.0	4.9
PointNet [3]	63.4	68.2	4.8
SpiderCNN [32]	69.8	73.7	3.9
PointNet++ [17]	75.4	77.9	2.5
DGCNN [29]	73.6	78.1	4.5
PointCNN [11]	75.1	78.5	3.4
BGA-DGCNN [26]	75.7	79.7	4.0
BGA-PN++ [26]	77.5	80.2	2.7
DRNet [18]	78.0	80.3	2.3
GBNet [19]	77.8	80.5	2.7
PointFormer	**78.9**	**81.1**	**2.2**

[30] benchmarks. Finally, we conducted a large number of careful and reasonable ablation experiments to prove the rationality and validity of each component of the proposed network.

4.1 Experimental Setup

1) Training: We implement PointFormer using PyTorch and employ SGD optimizer with an initial learning rate of 0.1 and the momentum of 0.9 for training. We use the cosine annealing [14] to reduce the learning rate until 0.001. The batch size is 36, number of neighbors is k=20 and we do not use batch normalization decay. The epochs on the ScanObjectNN dataset and the ModelNet40 dataset are 350 and 250, respectively.
2) Testing: In our testing, we use the GMPF testing strategy. Specifically, we set the multi-scale k-NN graph to k=(15,16,20,25,28) on the ScanObjectNN dataset and k=(15,20,28) on the ModelNet40 dataset respectively. The selection will be further explained in the experimental section.

Table 2. Classification Results (%) on ModelNet40 dataset. (coords: 3-dimensional coordinates, norm: point normal vector, voting: multivotes test method, k: ×1024, -: unknown)

Method	Input	#point	Mean Class Accuracy	Overall Accuracy
ECC [21]	coords	1k	83.2	87.4
PointNet [3]	coords	1 k	86.0	89.2
SCN [31]	coords	1 k	87.6	90.0
Kd-Net [8]	coords	1 k	–	90.6
PointCNN [11]	coords	1 k	88.1	92.2
PCNN [1]	coords	1 k	–	92.3
DensePoint [12]	coords	1 k	–	92.8
RS-CNN [13]	coords	1 k	–	92.9
DGCNN [29]	coords	1 k	90.2	92.9
DGCNN [29]	coords	2 k	90.7	93.5
KP-Conv [25]	coords	1 k	–	92.9
PointASNL [33]	coords	1 k	–	92.9
SO-Net [10]	coords	2 k	87.3	90.9
SO-Net [10]	coords + norm	5 k	90.8	93.4
RGCNN [24]	coords + norm	1 k	87.3	90.5
PointNet++ [17]	coords + norm	5 k	-	91.9
SpiderCNN [32]	coords + norm	5 k	–	92.4
DensePoint [12]	coords + voting	1 k	–	93.2
RS-CNN [13]	coords + voting	1 k	–	93.6
PCT [5]	coords	1 k	–	93.2
Point Transformer [36]	coords	1 k	90.6	93.7
GB-Net [19]	coords	1 k	91.0	93.8
PointFormer	coords	1 k	**90.7**	**93.7**

4.2 Classification Results

1) Classification on ScanObjectNN: ScanObjectNN consists of approximately 15,000 objects, which belong to 15 categories. Table 1 shows the classification results comparison between PointFormer and each competitive benchmark on ScanObjectNN. It can be seen that our model achieves 78.9% mean class accuracy and 81.1% overall class accuracy, which is 0.9% and 0.6% higher than the previous best results, respectively. Note that the number of parameters of PointFormer is 3.99 M and FLOPs is 3.48 G, while for GBNet [19] is of 8.78 M and 11.57 G, respectively, which is much larger than our model. However, we achieve a superior performance than it, which means we have earned a better balance between model complexity and performance. Besides, our PointFormer attains the smallest gap between mean class accuracy and overall accuracy. This displays that our network has superior performance for each categorie, exhibiting better inter-class robustness.

The objects in ScanObjectNN contain many distortion points, background points and missing regions [26], which poses a serious challenge to the robustness of the model. However, the local attention block and global attention block in PointFormer bring an effective solution to this problem. For example, in the local attention block, the attention weights the features of each neighbor node, which allows the model to learn to discard more useless nodes during training. Moreover, when the features are finally aggregated, our global attention block can even abandon unwanted nodes directly by setting the coefficient to low response. *More importantly, our GMPF testing strategy can fully extract useful information from the multi-scale k-NN graph.* This can greatly improve the noise immunity of the model when the dataset is not pure. This means that our model is able to eliminate the detrimental effects of more noise points. So PointFormer is able to demonstrate booming robustness in unclean real-world datasets.

Table 3. Graph-Multiscale Perceptual Field testing strategy (GMPF) effectiveness analysis on the ScanObjectNN and ModelNet40 datasets. We are comparing mean class and overall class accuracy (%)

	ScanObjectNN		ModelNet40	
	Mean Class Acc.	Overall Acc.	Mean Class Acc.	Overall Acc.
PointFormer	77.2	80.1	90.5	93.6
PointFormer (**GMPF**)	**78.9**	**81.1**	**90.7**	**93.7**

2) Classification on ModelNet40: This dataset, specifically, contains 12,311 synthetic pure CAD models in which the point cloud data is divided into 40 categories. Typically, 9843 models are used for training and 2468 models are reserved for testing in our experiment. Table 2 presents the highly competitive classification result for our network (mean class accuracy: 90.7% and overall accuracy: 93.7%). Note that our model uses only the 1k 3D coordinates of the point cloud as input, but carries out even better results than some approaches that use more information. As can be observed in Table 2, DGCNN [29] takes 2k points as input and achieves 90.7% average class accuracy and 93.5% overall accuracy. RGCNN [24] additionally used the normal vector of points as input to attain 90.5% overall accuracy. SO-Net [10] even used 5k points and normal vectors, achieving an average classification accuracy of 90.8% and an overall accuracy of 93.4%. Moreover, our model also outperforms the models that employ the voting strategy such as RSCNN [13] and DensePoint [12].

It is worth noting that although the classification performance of PointFormer on ModelNet40 is slightly worse than GB-Net [19], our model is far lighter and more robust than GB-Net as analyzed earlier, which implies that our PointFormer is more practical in real-world applications.

4.3 PointFormer Design Analysis

To demonstrate the soundness and validity of our model design, we conduct comparative and ablation experiments in this section, as well as conducting the visualization and analysing the complexity of the model.

1) Effectiveness of GMPF testing strategy: GMPF is our innovative work, which does not bring any parametric quantity to the model, but greatly improves the robustness and performance of our model. We verify the effectiveness of this strategy by comparing the performance of the model with GMPF and the model without GMPF on the ScanObjectNN and ModelNet40 dataset. The results are shown in Table 3. As can be seen, on the real-world ScanObjectNN dataset, the GMPF strategy *improved the average accuracy and overall accuracy by 1.7% and 1%,respectively,* which is a stunning result with significant relevance! Besides, the GMPF strategy on clean synthetic ModelNet40 dataset also earns good gains. This strategy is like learning at a specific angle while recognizing point cloud objects at multiscale related angles. The effect may not be as dramatic when the data is very pure, but when faced with realistic unclean datasets, GMPF can greatly improve the robustness and overall performance of the network.

Table 4. The effects of different choices of multiscale k-NN graph on the ScanObjectNN datasets. We are comparing mean class accuracy and overall accuracy (%).

	k = (20)	k = (20, 28)	k = (15, 20, 25)	k = (15, 20, 28)	k = (15, 16, 20, 25, 28)
Mean Class Acc	77.2	77.9	78.8	78.7	**78.9**
Overall Acc	80.1	80.2	80.9	80.9	**81.1**

Table 5. The results (%) of different HPE strategies. We are comparing the overall class accuracy on the ModelNet40 dataset. L_2: the l_2 paradigm (3D Euclidean distance) of x_i-x_j, x_i: center vertex, x_j: neighborhood nodes, Feature Dimension: the input channel of the four local attention block respectively.

Model	Hierarchical Position Encoding	Feature Dimension	Overall Accuracy
A	4*(x_i, x_i-x_j)	(6, 128, 128, 256)	93.3
B	(x_i, x_j, L_2^2) and 3*(x_i, x_j)	(7, 128, 128, 256)	93.1
C	(x_i, x_j, x_i-x_j, L_2) and 3*(x_i, x_j, x_i-x_j)	(10, 192, 192, 384)	92.9
D	(x_i, x_j, x_i-x_j, L_2^2) and 3*(x_i, x_i-x_j)	(10, 128, 128, 256)	**93.7**

On the ScanObjectNN and ModelNet40 datasets, the multiscale k-nearnest neighbor graphs were selected as k=(15,16,20,25,28) and k=(15,20,28), respectively. In Table 4, we compare the effects of different choices of k-NN graph on the performance of the PointFormer on the ScanObjectNN dataset. As we can see, our selection based on which makes the testing result best. Similarly, on the ModelNet40 dataset, we do the same.

2) Effectiveness of Hierarchical Position Encoding (HPE): We verified the effectiveness of HPE and compared different encoding strategies on ModelNet40 dataset. The results are presented in Table 5. Compared to models A and B, our model D retains and utilises more point cloud geometry information. At the same time, it does not increase the size of the model and introduce too many parameters. The model C, on the other hand, exponentially increases the FLOPs and the size of the model when the features are high-dimensional. This greatly boost the risk of overfitting. Thus, it can be seen that our Hierarchical Position Relation strikes a good balance between model complexity and maximizing the use of the original geometric information in point cloud object.
3) Network Complexity: We analyze the network complexity of PointFormer and other advanced approaches by comparing the number of parameters and FLOPs on ModelNet40 dataset. For the sake of fairness, we standardize the testing conditions (GeForce RTX 3090, batch size = 1). The results are displayed in Table 6. As we can see, the complexity of our model is slightly higher than that of the traditional DGCNN approach, but compared with current Attention-based methods such as (2) and (4), our model is much lighter and achieves better results. It is worth noting that, for simplicity, we perform the comparison on a clean synthetic dataset ModelNet40, but the greater strength of our network is the strong robustness demonstrated on the impure real-world dataset ScanObjectNN. However, the more complex approaches (2) and (4) do not necessarily stand up to the test in this case. At the same time, compared to GBNet [19], PointFormer is much lighter and has better robustness, as mentioned in the previous analysis.

Table 6. Complexity analysis of classification network on the ModelNet40 dataset.

Method	Param	FLOPs	Overall acc. (%)
(1)DGCNN [29]	1.81 M	2.72 G	92.9
(2)GBNet [19]	8.78 M	11.57 G	93.8
(3)Point Transformer [36]	9.58 M	18.41 G	93.7
(4)Point Transformer [4]	21.67 M	4.79 G	92.8
PointFormer	**3.99 M**	**3.48 G**	**93.7**

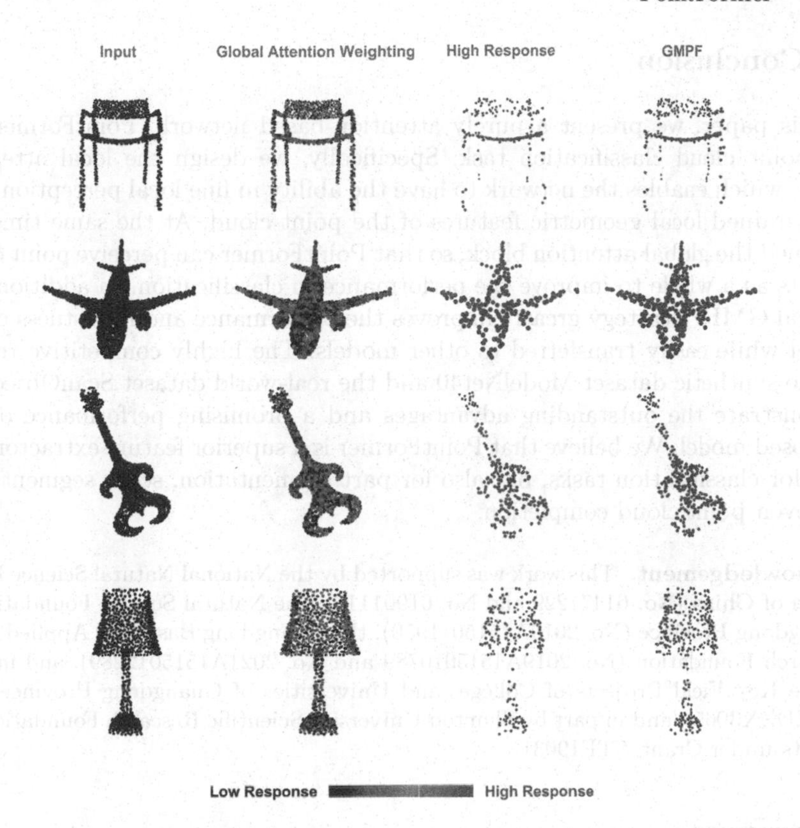

Fig. 6. Visualisation of the importance of each point in the point cloud. Each point is coloured by the attention factor. Blue represents the lowest attention and red the highest. (Color figure online)

4) Visualization: To illustrate PointFormer's excellent perception of point cloud objects, we visualize the attention factor of the global attention block at each point. As shown in Fig. 6, PointFormer focuses more on recognizable contour points in the point cloud, while treats other unimportant points differently. In the case of airplane, the points that PointFormer pays the most attention to are distributed in various parts of the airplane, such as wings, tail, fuselage, etc., which well form a recognizable outline of an airplane. However, PointFormer can selectively ignore some optional points, as shown in the blue points in Fig. 6. In addition, our testing strategy (GMPF) allows the model to be more perceptive of the overall contours of the point cloud. As presented in the last column of Fig. 6, the highly responsive points provide a more complete and accurate representation of the point cloud shape.

5 Conclusion

In this paper, we present a purely attention-based network, PointFormer, for the point cloud classification task. Specifically, we design the local attention block, which enables the network to have the ability in fine local perception with finer-grained local geometric features of the point cloud. At the same time, we designed the global attention block, so that PointFormer can perceive point cloud objects as a whole to improve the performance in classification. In addition, the general GMPF strategy greatly improves the performance and robustness of the model while easily transferred to other models. The highly competitive results on the synthetic dataset ModelNet40 and the real-world dataset ScanObjectNN demonstrate the outstanding advantages and a promising performance of the proposed model. We believe that PointFormer is a superior feature extractor, not only for classification tasks, but also for part segmentation, scene segmentation and even point cloud completion.

Acknowledgement. This work was supported by the National Natural Science Foundation of China (No. 61471229 and No. 61901116), the Natural Science Foundation of Guangdong Province (No. 2019A1515011950), the Guangdong Basic and Applied Basic Research Foundation (No. 2019A1515010789 and No. 2021A1515012289), and in part by the Key Field Projects of Colleges and Universities of Guangdong Province (No. 2020ZDZX3065), and in part by Shantou University Scientific Research Foundation for Talents under Grant NTF19031.

References

1. Atzmon, M., Maron, H., Lipman, Y.: Point convolutional neural networks by extension operators. CoRR abs/1803.10091 (2018). http://arxiv.org/abs/1803.10091
2. Ben-Shabat, Y., Lindenbaum, M., Fischer, A.: 3DmFV: three-dimensional point cloud classification in real-time using convolutional neural networks. IEEE Robotics and Automation Letters **3**(4), 3145–3152 (2018). https://doi.org/10.1109/LRA.2018.2850061
3. Charles, R.Q., Su, H., Kaichun, M., Guibas, L.J.: PointNet: deep learning on point sets for 3D classification and segmentation. In: 2017 IEEE Conference on Computer Vision and Pattern Recognition (CVPR), pp. 77–85 (2017). https://doi.org/10.1109/CVPR.2017.16
4. Engel, N., Belagiannis, V., Dietmayer, K.: Point transformer. IEEE Access **9**, 134826–134840 (2021). https://doi.org/10.1109/ACCESS.2021.3116304
5. Guo, M., Cai, J., Liu, Z., Mu, T., Martin, R.R., Hu, S.: PCT: point cloud transformer. Comput. Vis. Media **7**(2), 187–199 (2021)
6. Kanezaki, A., Matsushita, Y., Nishida, Y.: RotationNet: joint object categorization and pose estimation using multiviews from unsupervised viewpoints. In: 2018 IEEE/CVF Conference on Computer Vision and Pattern Recognition(CVPR), pp. 5010–5019 (2018). https://doi.org/10.1109/CVPR.2018.00526

7. Kipf, T.N., Welling, M.: Semi-supervised classification with graph convolutional networks. In: 5th International Conference on Learning Representations, ICLR 2017, Toulon, France, 24–26 April 2017, Conference Track Proceedings (2017). https://openreview.net/forum?id=SJU4ayYgl
8. Klokov, R., Lempitsky, V.: Escape from cells: Deep KD-networks for the recognition of 3D point cloud models. In: 2017 IEEE International Conference on Computer Vision (ICCV), pp. 863–872 (2017). https://doi.org/10.1109/ICCV.2017.99
9. Lang, A.H., Vora, S., Caesar, H., Zhou, L., Yang, J., Beijbom, O.: PointPillars: fast encoders for object detection from point clouds. In: 2019 IEEE/CVF Conference on Computer Vision and Pattern Recognition (CVPR), pp. 12689–12697 (2019). https://doi.org/10.1109/CVPR.2019.01298
10. Li, J., Chen, B.M., Lee, G.H.: SO-Net: self-organizing network for point cloud analysis. In: 2018 IEEE Conference on Computer Vision and Pattern Recognition(CVPR), pp. 9397–9406 (2018). https://doi.org/10.1109/CVPR.2018.00979
11. Li, Y., Bu, R., Sun, M., Wu, W., Di, X., Chen, B.: PointCNN: convolution on X-transformed points. In: Advances in Neural Information Processing Systems 31: Annual Conference on Neural Information Processing Systems 2018, NeurIPS 2018, 3–8 December 2018, Montréal, Canada, pp. 828–838 (2018)
12. Liu, Y., Fan, B., Meng, G., Lu, J., Xiang, S., Pan, C.: DensePoint: learning densely contextual representation for efficient point cloud processing. In: 2019 IEEE/CVF International Conference on Computer Vision (ICCV), pp. 5238–5247 (2019). https://doi.org/10.1109/ICCV.2019.00534
13. Liu, Y., Fan, B., Xiang, S., Pan, C.: Relation-shape convolutional neural network for point cloud analysis. In: 2019 IEEE/CVF Conference on Computer Vision and Pattern Recognition (CVPR), pp. 8887–8896 (2019). https://doi.org/10.1109/CVPR.2019.00910
14. Loshchilov, I., Hutter, F.: SGDR: stochastic gradient descent with warm restarts. In: 5th International Conference on Learning Representations, ICLR 2017, Toulon, France, 24–26 April 2017, Conference Track Proceedings (2017)
15. Maturana, D., Scherer, S.: VoxNet: a 3D convolutional neural network for real-time object recognition. In: 2015 IEEE/RSJ International Conference on Intelligent Robots and Systems (IROS), pp. 922–928 (2015). https://doi.org/10.1109/IROS.2015.7353481
16. Qi, C.R., Su, H., Nießner, M., Dai, A., Yan, M., Guibas, L.J.: Volumetric and multi-view cnns for object classification on 3D data. In: 2016 IEEE Conference on Computer Vision and Pattern Recognition (CVPR), pp. 5648–5656 (2016). https://doi.org/10.1109/CVPR.2016.609
17. Qi, C.R., Yi, L., Su, H., Guibas, L.J.: PointNet++: deep hierarchical feature learning on point sets in a metric space. In: Advances in Neural Information Processing Systems, vol. 30 (2017)
18. Qiu, S., Anwar, S., Barnes, N.: Dense-resolution network for point cloud classification and segmentation. In: 2021 IEEE Winter Conference on Applications of Computer Vision (WACV), pp. 3812–3821 (2021). https://doi.org/10.1109/WACV48630.2021.00386
19. Qiu, S., Anwar, S., Barnes, N.: Geometric back-projection network for point cloud classification. IEEE Trans. Multimedia **24**, 1943–1955 (2022). https://doi.org/10.1109/TMM.2021.3074240
20. Riegler, G., Ulusoy, A.O., Geiger, A.: OctNet: learning deep 3D representations at high resolutions. In: 2017 IEEE Conference on Computer Vision and Pattern Recognition (CVPR), pp. 6620–6629 (2017). https://doi.org/10.1109/CVPR.2017.701

21. Simonovsky, M., Komodakis, N.: Dynamic edge-conditioned filters in convolutional neural networks on graphs. In: 2017 IEEE Conference on Computer Vision and Pattern Recognition (CVPR), pp. 29–38 (2017). https://doi.org/10.1109/CVPR.2017.11
22. Song, S., Yu, F., Zeng, A., Chang, A.X., Savva, M., Funkhouser, T.: Semantic scene completion from a single depth image. In: 2017 IEEE Conference on Computer Vision and Pattern Recognition (CVPR), pp. 190–198 (2017). https://doi.org/10.1109/CVPR.2017.28
23. Su, H., Maji, S., Kalogerakis, E., Learned-Miller, E.: Multi-view convolutional neural networks for 3d shape recognition. In: 2015 IEEE International Conference on Computer Vision (ICCV), pp. 945–953 (2015). https://doi.org/10.1109/ICCV.2015.114
24. Te, G., Hu, W., Zheng, A., Guo, Z.: RGCNN: regularized graph CNN for point cloud segmentation. In: Boll, S., et al. (eds.) 2018 ACM Multimedia Conference on Multimedia Conference, MM 2018, Seoul, Republic of Korea, 22–26 October 2018, pp. 746–754. ACM (2018). https://doi.org/10.1145/3240508.3240621
25. Thomas, H., Qi, C.R., Deschaud, J.E., Marcotegui, B., Goulette, F., Guibas, L.: KPConv: flexible and deformable convolution for point clouds. In: 2019 IEEE/CVF International Conference on Computer Vision (ICCV), pp. 6410–6419 (2019). https://doi.org/10.1109/ICCV.2019.00651
26. Uy, M.A., Pham, Q.H., Hua, B.S., Nguyen, T., Yeung, S.K.: Revisiting point cloud classification: a new benchmark dataset and classification model on real-world data. In: 2019 IEEE/CVF International Conference on Computer Vision (ICCV), pp. 1588–1597 (2019). https://doi.org/10.1109/ICCV.2019.00167
27. Vaswani, A., et al.: Attention is all you need. In: Advances in Neural Information Processing Systems 30: Annual Conference on Neural Information Processing Systems 2017, 4–9 December 2017, Long Beach, CA, USA, pp. 5998–6008 (2017)
28. Velickovic, P., Cucurull, G., Casanova, A., Romero, A., Liò, P., Bengio, Y.: Graph attention networks. CoRR abs/1710.10903 (2017)
29. Wang, Y., Sun, Y., Liu, Z., Sarma, S.E., Bronstein, M.M., Solomon, J.M.: Dynamic graph CNN for learning on point clouds. ACM Trans. Graph. **38**(5), 3326362 (2019). https://doi.org/10.1145/3326362
30. Wu, Z., et al.: 3d shapeNets: a deep representation for volumetric shapes. In: 2015 IEEE Conference on Computer Vision and Pattern Recognition (CVPR), pp. 1912–1920 (2015). https://doi.org/10.1109/CVPR.2015.7298801
31. Xie, S., Liu, S., Chen, Z., Tu, Z.: Attentional shapeContextNet for point cloud recognition. In: 2018 IEEE/CVF Conference on Computer Vision and Pattern Recognition, pp. 4606–4615 (2018). https://doi.org/10.1109/CVPR.2018.00484
32. Xu, Y., Fan, T., Xu, M., Zeng, L., Qiao, Yu.: SpiderCNN: deep learning on point sets with parameterized convolutional filters. In: Ferrari, V., Hebert, M., Sminchisescu, C., Weiss, Y. (eds.) ECCV 2018. LNCS, vol. 11212, pp. 90–105. Springer, Cham (2018). https://doi.org/10.1007/978-3-030-01237-3_6
33. Yan, X., Zheng, C., Li, Z., Wang, S., Cui, S.: PointASNL: robust point clouds processing using nonlocal neural networks with adaptive sampling. In: 2020 IEEE/CVF Conference on Computer Vision and Pattern Recognition (CVPR), pp. 5588–5597 (2020). https://doi.org/10.1109/CVPR42600.2020.00563
34. Yu, Z., Zheng, X., Yang, Z., Lu, B., Li, X., Fu, M.: Interaction-temporal GCN: a hybrid deep framework for covid-19 pandemic analysis. IEEE Open J. Eng. Med. Biol. **2**, 97–103 (2021). https://doi.org/10.1109/ojemb.2021.3063890

35. Zhao, H., Jia, J., Koltun, V.: Exploring self-attention for image recognition. In: 2020 IEEE/CVF Conference on Computer Vision and Pattern Recognition (CVPR), pp. 10073–10082 (2020). https://doi.org/10.1109/CVPR42600.2020.01009
36. Zhao, H., Jiang, L., Jia, J., Torr, P.H., Koltun, V.: Point transformer. In: Proceedings of the IEEE/CVF International Conference on Computer Vision(ICCV), pp. 16259–16268 (2021)

Neural Deformable Voxel Grid for Fast Optimization of Dynamic View Synthesis

Xiang Guo[1], Guanying Chen[2], Yuchao Dai[1(✉)], Xiaoqing Ye[3], Jiadai Sun[1], Xiao Tan[3], and Errui Ding[3]

[1] Northwestern Polytechnical University, Xi'an, China
{guoxiang,sunjiadai}@mail.nwpu.edu.cn, daiyuchao@nwpu.edu.cn
[2] FNii and SSE, CUHK-Shenzhen, Shenzhen, China
chenguanying@cuhk.edu.cn
[3] Baidu Inc., Beijing, China
{yexiaoqing,dingerrui}@baidu.com, tanxchong@gmail.com

Abstract. Recently, Neural Radiance Fields (NeRF) is revolutionizing the task of novel view synthesis (NVS) for its superior performance. In this paper, we propose to synthesize dynamic scenes. Extending the methods for static scenes to dynamic scenes is not straightforward as both the scene geometry and appearance change over time, especially under monocular setup. Also, the existing dynamic NeRF methods generally require a lengthy per-scene training procedure, where multi-layer perceptrons (MLP) are fitted to model both motions and radiance. In this paper, built on top of the recent advances in voxel-grid optimization, we propose a fast deformable radiance field method to handle dynamic scenes. Our method consists of two modules. The first module adopts a deformation grid to store 3D dynamic features, and a light-weight MLP for decoding the deformation that maps a 3D point in the observation space to the canonical space using the interpolated features. The second module contains a density and a color grid to model the geometry and density of the scene. The occlusion is explicitly modeled to further improve the rendering quality. Experimental results show that our method achieves comparable performance to D-NeRF using only 20 minutes for training, which is more than 70× faster than D-NeRF, clearly demonstrating the efficiency of our proposed method.

Keywords: Dynamic view synthesis · Neural radiance fields · Voxel-grid representation · Fast optimization

1 Introduction

Novel view synthesis (NVS) is a long-standing problem in computer vision and graphics, and has many applications in augmented reality, virtual reality,

X. Guo and G. Chen—Authors contributed equally to this work.

Supplementary Information The online version contains supplementary material available at https://doi.org/10.1007/978-3-031-26319-4_27.

L. Wang et al. (Eds.): ACCV 2022, LNCS 13841, pp. 450–468, 2023.
https://doi.org/10.1007/978-3-031-26319-4_27

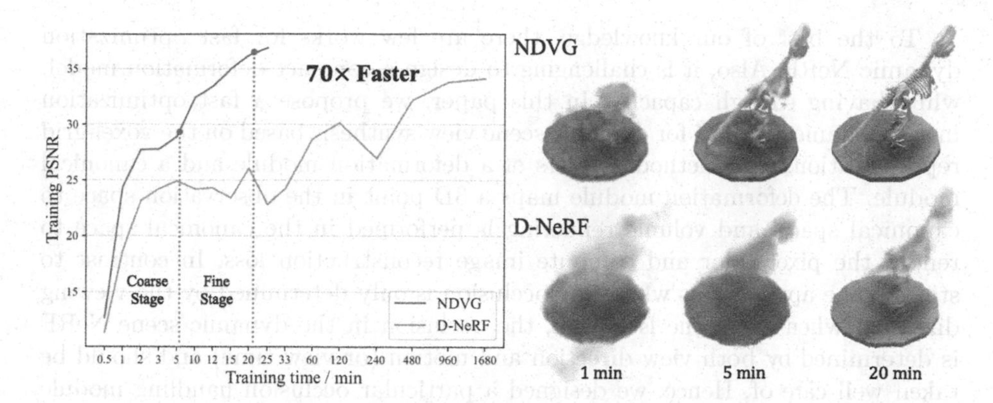

Fig. 1. Neural Deformable Voxel Grid (NDVG) for fast optimization of dynamic view synthesis. Left side of the figure shows that our method achieves a super fast convergence in 20 min, which is 70× faster than the D-NeRF method. Right side of the figure visualizes the results after training with 1, 5 and 20 min.

content creation, *etc.* Recently, neural rendering methods have achieved significant progress in this problem [31,34,60]. In particular, the neural radiance fields (NeRF) [31] produces photorealistic rendering by representing a static scene with a multi-layer perception (MLP), which maps a 5D input (3D coordinate and 2D view direction) to its density and color. Recently, a series of works extend NeRF based framework from static scenes to dynamic scenes [8,13,22,35,39,52,55,58].

Novel view synthesis of a dynamic scene from a monocular video is still a very challenging problem. Besides the difficulties to recover motions and geometries with only one observation at each time step, the training process usually takes days which hinders applications in practice. The NeRF based methods, at each iteration, require millions of network queries to obtain colors and densities of the sampled points for the sampled rays, based on which volume rendering computes the pixel colors [18]. In dynamic condition, the methods are even more complex with deformation model, *e.g.*, D-NeRF [39] optimizes a large deformation network and a canonical network to fit a dynamic scene, requiring more than 27 hours to converge. How to develop an efficient and accurate dynamic view synthesis method remains an open problem.

In a static scenario, to reduce the training time for a scene, some methods propose first to train the model on a dataset consisting of multiple scenes [5,53, 56,65], and then finetune it on the target scene, reducing the optimization time to several minutes. However, these methods rely on a large training dataset and a lengthy pre-training time.

Very recently, the voxel-grid representation has been exploited to speed up the optimization of radiance fields [32,48,64]. These methods are able to optimize a scene representation from scratch within just a few minutes, significantly accelerating the training speed without any pre-training. The key idea is to replace the time-consuming deep network query with the fast trilinear voxel-grid interpolation. However, these methods are tailored for static scenes and cannot be directly applied to handle dynamic scenes.

To the best of our knowledge, there are few works for fast optimization dynamic NeRF. Also, it is challenging to design a compact deformation model, while having enough capacity. In this paper, we propose a fast optimization method, named NDVG for dynamic scene view synthesis based on the voxel-grid representation. Our method consists of a deformation module and a canonical module. The deformation module maps a 3D point in the observation space to canonical space, and volume rendering is performed in the canonical space to render the pixel color and compute image reconstruction loss. In contrast to static scene applications where the occlusion is only determined by the viewing direction when the scene is known, the occlusion in the dynamic scene NeRF is determined by both view direction and motion (or view time) and should be taken well care of. Hence, we designed a particular occlusion handling module to explicitly model the occlusion to further improve the rendering quality.

In summary, the key contributions of this paper are as follows:

- We propose a fast deformable radiance field method based on the voxel-grid representation to enable space-time view synthesis for dynamic scenes. To the best of our knowledge, this is the first method that integrates the voxel-grid optimization with deformable radiance field.
- We introduce a deformation grid to store the 3D dynamic features and adopt a light-weight MLP to decode the feature to deformation. Our method explicitly models occlusion to improve the results.
- Our method produces rendering results comparable to D-NeRF [39] within only 20 minutes, which is more than 70× faster (see Fig. 1).

2 Related Work

Novel View Synthesis. Rendering a scene from arbitrary views has a long history in both vision and graphics [4,6,14,21], and surveys of recent methods can be found in [44,49,50]. Traditional methods need to explicitly build a 3D model for the scene, such as point clouds [1] or meshes [16,42,43,51], and then render a novel view from this geometry. Another category of methods explicitly estimates depth and then uses it to warp pixels or learned feature to a novel view, such as [7,11,19,37,42,43,59]. Numerous other works using multi-plane images (MPIs) to represent scenes [10,15,17,30,46,47,54,69], but the MPIs representation can only support relatively limited viewpoint changes during inference.

Neural Scene Representation. Recently, neural scene representations dominate novel view synthesis. In particular, Mildenhall *et al.* propose NeRF [31] to use MLPs to model a 5D radiance field, which can render impressive view synthesis for static scenes captured. Since then, many follow-up methods have extended the capabilities of NeRF, including relighting [3,45,68], handling in-the-wild scenarios [29], extending to large unbounded 360° scenes [66], removing the requirement for pose estimation [23,57], incorporating anti-aliasing for multi-scale rendering [2], and estimating the 6-DoF camera poses [61].

Fast NeRF Rendering and Optimization. Rendering and optimization in NeRF-like schemes are very time-consuming, as they require multiple samples along each ray for color accumulation. To speed up the rendering procedure, some methods predict the depth or sampling near the surface to guide more efficient sampling [33,38]. Other methods use the octree or sparse voxel grid to avoid sampling points in empty spaces [26,28,64]. In addition, some methods subdivided the 3D volume into multiple cells that can be processed more efficiently, such as DeRF [40] and KiloNeRF [41]. AutoInt [24] reduces the number of evaluations along a ray by learning partial integrals. However, these methods still need to optimize a deep implicit model, leading to a lengthy training time. To accelerate the optimization time on a test scene, some methods first train the model on a large dataset to learn a scene prior, and then finetune it on the target scene [5,53,56,65]. However, these methods require time-consuming pretraining.

More recently, the voxel-grid representation has been exploited to speed up the optimization of radiance field [32,48,63]. Plenoxels [63] represents a scene as a 3D grid with spherical harmonics, which can be optimized from calibrated images via gradient methods and regularization without any neural components. Similarly, DVGO [48] optimizes a hybrid explicit-implicit representation that consists of a dense grid and a light-weight MLP. Although these methods achieve a fast optimization speed, they are only applicable to static scenes and thus cannot be used to render dynamic scenes.

Dynamic Scene Modeling. Recently, several concurrent methods have extended NeRF to deal with dynamic scenes [8,13,22,35,39,52,55,58].

NeRFlow [8] learns a 4D spatial-temporal representation of a dynamic scene from a set of RGB images. Yoon *et al.* [62] propose to use an underlying 4D reconstruction, combining single-view depth and depth from multi-view stereo to render virtual views with 3D warping. Gao *et al.* [13] jointly train a time-invariant model (static) and a time-varying model (dynamic), and regularize the dynamic NeRF by scene flow estimation, finally blending the results in an unsupervised manner. NSFF [22] models the dynamic scene as a time-variant continuous function of appearance, geometry, and 3D scene motion. DCT-NeRF [55] uses the Discrete Cosine Transform (DCT) to capture the dynamic motion and learn smooth and stable trajectories over time for each point in space.

D-NeRF [39], NR-NeRF [52] and Nerfies [35] first learn a static canonical radiance field for capturing geometry and appearance, and then learn the deformation/displacement field of the scene at each time instant w.r.t. the canonical space. Xian *et al.* [58] represent a 4D space-time irradiance field as a function that maps a spatial-temporal location to the emitted color and volume density.

Although promising results have been shown for dynamic view synthesis, these methods all require a long optimization time to fit a dynamic scene, limiting their wider applications.

As the concurrent research of our work, there are few works which aim to speed up training of dynamic NeRF [9,12,25]. TiNeuVox [9] uses a small MLP to model the deformation and uses the multi-distance interpolation to get the feature for radiance network which estimates the density and color. Compared to TiNeuVox [9], we propose a deformation feature grid to enhance the capability of the small deformation network and not effect training speed. V4D [12] uses the 3D feature voxel to model the 4D radiance field with additional time dimension concatenated and proposes look-up tables for pixel-level refinement. While V4D [12] focuses mainly on improving image quality, the speed up of training is not significant compared with TiNeuVox [9] and ours. DeVRF [25] also builds on voxel-grid representation, which proposes to use multi-view data to overcome the nontrivial problem of the monocular setup. Multi-view data eases the learning of motion and geometry compared with ours which uses monocular images.

3 Neural Deformable Voxel Grid

Given an image sequence $\{I_t\}_{t=1}^T$ with camera poses $\{\mathbf{T}_t\}_{t=1}^T$ of a dynamic scene captured by a monocular camera, our goal is to develop a fast optimization method based on the radiance fields to represent this dynamic scene and support novel view synthesis at different times.

3.1 Overview

Mathematically, given a query 3D point $\mathbf{p}$, the view direction $\mathbf{d}$, and a time instance t, we need to estimate the corresponding density σ and color $\mathbf{c}$.

A straightforward way is to directly use an MLP to learn the mapping from $(\mathbf{p}, \mathbf{d}, t)$ to $(\sigma, \mathbf{c})$. However, existing dynamic methods show that such a high-dimensional mapping is difficult to learn, and they propose a framework based on the canonical scene to ease learning difficulty [36,39,52]. These methods adopt a deformation MLP to map a 3D point in the observation space to a static canonical space as $\Psi_t : (\mathbf{p}, t) \rightarrow \Delta\mathbf{p}$. The density and color are then estimated in the canonical space as $\Psi_p : (\sigma, \mathbf{c}) = f(\mathbf{p}+\Delta\mathbf{p}, \mathbf{d})$. However, to render the network with the capability of handling complex motion, large MLPs are inevitable in existing methods and therefore result in a long optimization time (*i.e.*, from hours to days).

Motivated by the recent successes of voxel-grid optimization in accelerating the training of static radiance field [48,63], we introduce the voxel-grid representation into the canonical scene representation based framework to enable fast dynamic scene optimization. Our method consists of a *deformation module* and a *canonical module* (see Fig. 2). The key idea is to replace most of the heavy MLP computations with the fast voxel-grid feature interpolation.

3.2 Deformation Module for Motion Modeling

Assuming the canonical space is at time t_{can}, the deformation module estimates the offset $\Delta\mathbf{p}$ of a point $\mathbf{p}$ at any time t to the canonical space. As the input has 4 dimensions (*i.e.*, 3 for $\mathbf{p}$ and 1 for t), it is inefficient to directly store the offsets in a 4D feature grid. Therefore, we adopt a hybrid explicit-implicit representation that consists of a 3D feature grid and a light-weight MLP to decode the interpolated feature.

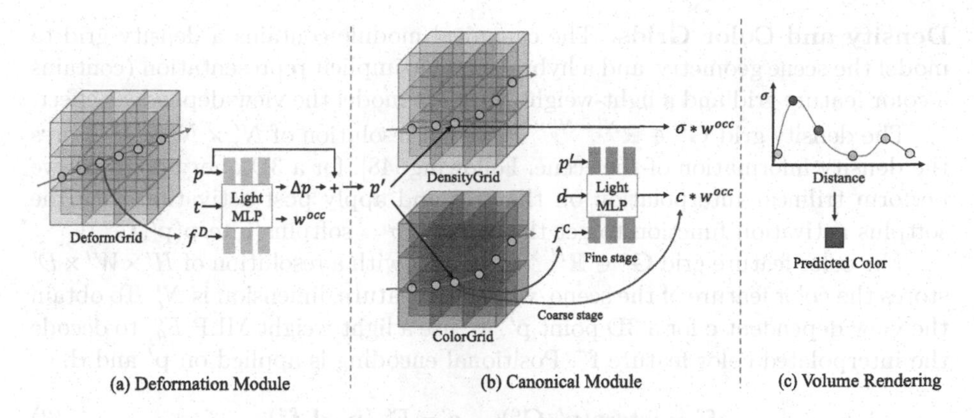

Fig. 2. Overview of our proposed method. Our method consists of (a) a *deformation module* to model the motion of the space points and (b) a *canonical module* to model the radiance field of the static scene at the canonical time. To render a ray shooting from the camera center, we compute the deformation of all sampled points and transform the sampled points to the canonical space, where the density and color are computed. The pixel color can then be rendered by (c) volume rendering.

Neural Deformable Voxel Grid (NDVG). We first build a 3D deformation feature grid $\mathbf{G}^d \in \mathbb{R}^{N_x \times N_y \times N_z \times N_c}$ with a resolution of $N_x \times N_y \times N_z$, and the feature vector in each voxel has a dimension of N_c. For a continuous 3D coordinate $\mathbf{p}$, its deformation feature can be quickly queried from the deformation feature grid using trilinear interpolation:

$$\text{interp}(\mathbf{p}, \mathbf{G}^d) : (\mathbb{R}^3, \mathbb{R}^{N_x \times N_y \times N_z \times N_c}) \to \mathbb{R}^{N_c}. \tag{1}$$

To obtain the offset from the observation space to the canonical space for a query point, we introduce a light-weight MLP $F^d_{\theta_1}$. It takes the coordinate $\mathbf{p}$, time t, and the interpolated feature as input, and regresses the offset $\Delta\mathbf{p}$.

$$\mathbf{f}^d = \text{interp}(\mathbf{p}, \mathbf{G}^d), \quad \Delta\mathbf{p} = \begin{cases} F^d_{\theta_1}(\mathbf{p}, t, \mathbf{f}^d) & \text{if } t \neq t_{can}, \\ 0 & \text{otherwise.} \end{cases} \tag{2}$$

Following [31], we apply positional encoding for $\mathbf{p}$ and t. Finally, we get the position of $\mathbf{p}$ in the canonical space $\mathbf{p}'$ as: $\mathbf{p}' = \mathbf{p} + \Delta\mathbf{p}$.

The deformation feature grid can provide learnable features that encode the deformation information of the points in the 3D space, so that a light-weight MLP is capable to model the deformation of the 3D space.

3.3 Canonical Module for View Synthesis

To render the pixel color of a camera ray in the observation space, we first sample K points on the ray and transform them to the canonical space via our deformation module. The canonical module then computes the corresponding densities and colors for volume rendering to composite the pixel color.

Density and Color Grids. The canonical module contains a density grid to model the scene geometry, and a hybrid explicit-implicit representation (contains a color feature grid and a light-weight MLP) to model the view-dependent effect.

The density grid $\mathbf{G}^\sigma \in \mathbb{R}^{N'_x \times N'_y \times N'_z}$ with a resolution of $N'_x \times N'_y \times N'_z$ stores the density information of the scene. Following [48], for a 3D query point $\mathbf{p}'$, we perform trilinear interpolation on the grid and apply post-activation with the softplus activation function to get the density: $\sigma = \text{softplus}(\text{interp}(\mathbf{p}', \mathbf{G}^\sigma))$

The color feature grid $\mathbf{G}^c \in \mathbb{R}^{H' \times W' \times D' \times N'_c}$ with a resolution of $H' \times W' \times D'$ stores the color feature of the scene, where the feature dimension is N'_c. To obtain the view dependent $\mathbf{c}$ for a 3D point $\mathbf{p}'$, we use a light-weight MLP $F^c_{\theta_2}$ to decode the interpolated color feature $\mathbf{f}^c$. Positional encoding is applied on $\mathbf{p}'$ and $\mathbf{d}$.

$$\mathbf{f}^c = \text{interp}(\mathbf{p}', \mathbf{G}^c), \quad \mathbf{c} = F^c_{\theta_2}(\mathbf{p}', \mathbf{d}, \mathbf{f}^c). \tag{3}$$

3.4 Occlusion-Aware Volume Rendering

Volume rendering is the key in radiance fields based methods to render pixel color in a differentiable manner. For a ray $\mathbf{r}(w) = \mathbf{o} + w\mathbf{d}$ emitted from the camera center $\mathbf{o}$ with view direction $\mathbf{d}$ through a given pixel on the image plane, the estimated color $\hat{\mathbf{C}}(\mathbf{r})$ of this ray is computed as

$$\hat{\mathbf{C}}(\mathbf{r}) = \sum_{k=1}^{K} T(w_k)\, \alpha\left(\sigma(w_k)\delta_k\right) \mathbf{c}(w_k), \quad T(w_k) = \exp\left(-\sum_{j=1}^{k-1} \sigma(t_j)\delta_j\right), \tag{4}$$

where K is the sampled points in the ray, δ_k is the distance between adjacent samples on the ray, and $\alpha\left(\sigma(w_k)\delta_k\right) = 1 - \exp(-\sigma(w_k)\delta_k)$.

Occlusion Problem. Note that for dynamic scenes, the occlusion will cause problems in the canonical scene-based methods. We assume the "empty" points (*i.e.*, non-object points) in the space are static. If a 3D point is the empty point at time t but occupied by an object point in the canonical space, the occlusion happens. This is because the ideal deformation for an empty point in the observation space is zero, then this point will be mapped to the same location (but occupied by an object point) in the canonical space (see Fig. 3 for illustration). As a result, the obtained density and color value for this empty point will be non-zero, leading to incorrect rendering color.

Occlusion Reasoning. To tackle the occlusion problem, we additionally estimate an occlusion mask for a query point using the deformation MLP $F^d_{\theta_1}$. Then Eq. (2) is revised as:

$$(\Delta\mathbf{p}, w^{occ}) = F^d_{\theta_1}(\mathbf{p}, t, \mathbf{f}^D). \tag{5}$$

If a query point is an empty point at time t but is occupied by an object point in the canonical space, the estimated w^{occ} should be 0. Based on the estimated

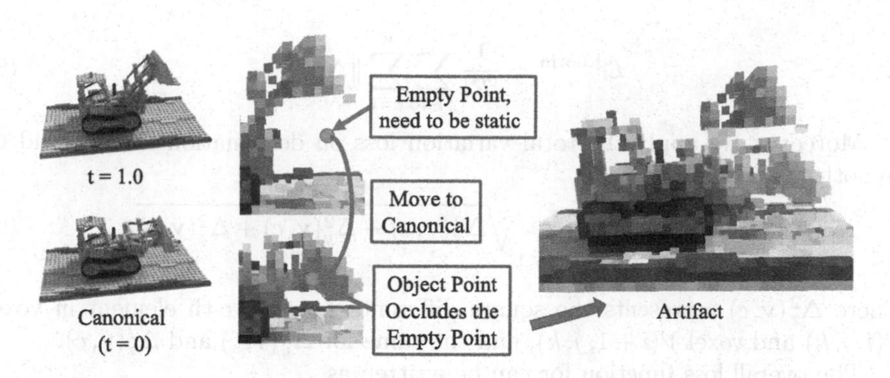

Fig. 3. Effect of Occlusion. If the static empty point in current time is occluded by the object point in canonical time, this empty point will bring the object point back to current time, which creates artifacts.

occlusion mask w^{occ}, we filter the estimated density and color before sending them to volume rendering to remove the influence of the occluded points:

$$\sigma' = \sigma \times w^{occ}, \quad \mathbf{c}' = \mathbf{c} \times w^{occ}. \tag{6}$$

The occlusion estimation is optimized by the image reconstruction loss in an end-to-end manner. By taking occlusion into account, our method is able to achieve better rendering quality.

4 Model Optimization

There is a trade-off between grid resolution and time consumption. Using a higher grid resolution generally leads to better image quality but at the cost of a higher computational time. Meanwhile, the space is dominated by empty space, which takes no effect on the synthesized images.

Loss Function. We utilize a series of losses to optimize the model. The first is the photometric loss, which is the mean square error (MSE) between the estimated ray color $\hat{\mathbf{C}}(\mathbf{r})$ and the ground-truth color $\mathbf{C}(\mathbf{r})$ as

$$\mathcal{L}^{\text{photo}} = \frac{1}{|\mathcal{R}|} \sum_{r \in \mathcal{R}} \left\| \hat{\mathbf{C}}(\mathbf{r}) - \mathbf{C}(\mathbf{r}) \right\|_2^2, \tag{7}$$

where $\mathcal{R}$ is the set of rays sampled in one batch.

Also, similarly to [48], we use the sampled point color supervision $\mathcal{L}^{\text{ptc}}$, which optimizes the color of sampled points with top N weights that contribute most to the rendered color of the ray. In addition, we use the background entropy loss $\mathcal{L}^{\text{bg}}$ to encourage the points to concentrate on either foreground or background.

For the deformation module, we apply the L_1 norm regularization to the estimated deformation based on the prior that most points are static:

$$\mathcal{L}^{\text{d\_norm}} = \frac{1}{|\mathcal{R}|} \sum_{r \in \mathcal{R}} \sum_{i=1}^{K} \|\Delta \mathbf{p}_i\|_1 . \tag{8}$$

Moreover, we apply the total variation loss on deformation feature grid to smooth the voxel features as

$$\mathcal{L}^{\text{d\_tv}} = \frac{1}{|\mathcal{V}|} \sum_{\mathbf{v} \in \mathcal{V} c \in [N_c]} \sqrt{\Delta_x^2(\mathbf{v}, c) + \Delta_y^2(\mathbf{v}, c) + \Delta_z^2(\mathbf{v}, c)}, \tag{9}$$

where $\Delta_x^2(\mathbf{v}, c)$ represents the square difference between c-th element in voxel $\mathcal{V}(i, j, k)$ and voxel $\mathcal{V}(i+1, j, k)$, and the same for $\Delta_y^2(\mathbf{v}, c)$ and $\Delta_z^2(\mathbf{v}, c)$.

The overall loss function for can be written as

$$\mathcal{L} = \mathcal{L}^{\text{photo}} + w^{\text{ptc}} \cdot \mathcal{L}^{\text{ptc}} + w^{\text{bg}} \cdot \mathcal{L}^{\text{bg}} + w^{\text{d\_norm}} \cdot \mathcal{L}^{\text{d\_norm}} + w^{\text{d\_tv}} \cdot \mathcal{L}^{\text{d\_tv}}, \tag{10}$$

where w^{ptc}, w^{bg}, $w^{\text{df\_norm}}$, and $w^{\text{df\_tv}}$ are weights to balance each components in the final coarse loss.

Following [48], we use the same strategy for voxel allocation, model point sampling, and low-density initialization. Considering that the deformation is more sensitive to resolution compared with radiance (density and color), we set a higher grid resolution to the deformation module than the canonical module.

Progressive Training. Typically, when a time step goes further away from the canonical time, the deformation between this time step and the canonical time step get larger. To reduce the learning difficulty, we use a progressive training strategy. The optimization starts with training images close to canonical time, and progressively adds images with further time steps.

Coarse-to-Fine Optimization. To speed up the training while maintaining the model capacity, we adopt a coarse-to-fine optimization procedure following [31,48]. We first optimize a coarse model to roughly recover the deformation and the canonical space geometry. Then, we use the coarse model to locate the object region and filter out a large portion of empty space, after which a fine model is optimized to recover a more accurate and detailed deformation and geometry. In the coarse model, we do not model the view-dependent effect, and the feature dimension of color grid K' is set to 3, which directly corresponds to the RGB color. Based on the optimized coarse module, we apply empty space filtering strategies, including finding fine-stage bounding box and empty point filtering to speed up training. Also, we initialize the fine-stage model with model weight trained during coarse stage. We present details of our empty space filtering strategies and fine model design in the supplementary material.

5 Experiments

5.1 Dataset and Metrics

We evaluate our method on the D-NeRF dataset [39], which contains eight dynamic scenes with 360° viewpoint settings. Beside synthetic dataset, we also

Table 1. Quantitative comparison. We report LPIPS (lower is better) and PSNR/SSIM (higher is better) on eight dynamic scenes of the D-NeRF dataset.

Methods	Hell warrior			Mutant			Hook			Bouncing balls		
	PSNR↑	SSIM↑	LPIPS↓	PSNR↑	SSIM↑	LPIPS↓	PSNR↑	SSIM↑	LPIPS↓	PSNR↑	SSIM↑	LPIPS↓
D-NeRF (half[1]) [39]	25.03	**0.951**	**0.069**	31.29	0.974	0.027	29.26	**0.965**	0.117	**38.93**	**0.990**	**0.103**
NDVG (half[1])	**25.53**	0.949	0.073	**35.53**	**0.988**	**0.014**	**29.80**	**0.965**	**0.037**	34.58	0.972	0.114
NDVG (w/o occ[2])	25.16	0.956	0.067	34.14	**0.980**	0.026	29.88	**0.963**	0.047	37.14	0.986	0.080
NDVG (w/o grid[3])	26.45	0.959	**0.065**	**34.42**	**0.980**	**0.025**	29.08	0.956	0.050	**37.78**	**0.988**	0.063
NDVG (w/o refine[4])	24.30	0.946	0.092	28.59	0.940	0.070	26.85	0.935	0.081	28.17	0.954	0.178
NDVG (w/o filter[5])	19.55	0.927	0.104	31.75	0.961	0.051	27.71	0.948	0.068	35.47	**0.988**	**0.054**
NDVG (full[6])	**26.49**	**0.960**	0.067	34.41	**0.980**	0.027	**30.00**	**0.963**	**0.046**	37.52	0.987	0.075
Methods	Lego			T-Rex			Stand up			Jumping jacks		
	PSNR↑	SSIM↑	LPIPS↓	PSNR↑	SSIM↑	LPIPS↓	PSNR↑	SSIM↑	LPIPS↓	PSNR↑	SSIM↑	LPIPS↓
D-NeRF (half) [39]	21.64	0.839	0.165	**31.76**	**0.977**	**0.040**	32.80	0.982	**0.021**	**32.80**	**0.981**	**0.037**
NDVG (half)	**25.23**	**0.931**	**0.049**	30.15	0.967	0.047	**34.05**	**0.983**	0.022	29.45	0.960	0.078
NDVG (w/o occ)	24.77	0.935	0.059	32.57	**0.979**	**0.031**	18.78	0.899	0.112	30.87	0.973	0.044
NDVG (w/o grid)	24.18	0.916	0.078	31.64	0.976	0.034	32.99	0.980	**0.027**	30.64	0.971	0.044
NDVG (w/o refine)	23.30	0.851	0.167	27.35	0.940	0.079	29.80	0.965	0.051	26.13	0.920	0.150
NDVG (w/o filter)	22.75	0.887	0.140	28.58	0.952	0.067	32.36	0.976	0.035	28.19	0.957	0.077
NDVG (full)	**25.04**	**0.940**	**0.053**	**32.62**	0.978	0.033	**33.22**	**0.979**	0.030	**31.25**	**0.974**	**0.040**

[1] half: using half resolution of the original dataset images.
[2] w/o occ: not using occlusion reasoning.
[3] w/o grid: not using deformation feature grid, only deform MLP.
[4] w/o refine: only using coarse training stage.
[5] w/o filter: not using coarse training, direct optimize fine module.
[6] full: using full resolution of the original dataset images.

conduct experiments on real scenes, proposed by HyperNeRF [36]. This dataset captures images of real dynamic scenes with a multi-view camera rig consisting of 2 phones with around a 16cm baseline. The dynamic motion consists of both rigid and non-rigid deformation with unbounded scenes. We use several metrics for the evaluation: Peak Signal-to-Noise Ratio (PSNR), Structural Similarity (SSIM), and Learned Perceptual Image Patch Similarity (LPIPS) [67].

5.2 Implementation Details

We set the expected voxel number to 1,664k and 190^3 for the grid in Deformation Module in the coarse and fine stages, respectively. For Canonical Module, we set the expected voxel number to 1,024k and 160^3. The light-weight MLP in the deformation module has 4 layers, each with a width of 64. The light-weight MLP in the canonical module has 3 layers, each with a width of 128.

When training with full resolution images, we train 10k and 20k iterations for the coarse and fine stages for all scenes. When training with half-resolution images, we reduce the iteration to 5k and 10k for coarse and fine stages. In terms of positional encoding, we set the frequency to 5 for position and time, and 4 for direction. We use the Adam optimizer [20] and sample 8,192 rays per iteration. More details of settings can be found in our supplementary material.

5.3 Comparisons

Quantitative Evaluation on the Dataset. We first quantitatively compare the results in Table 1. To compare with D-NeRF [39], we set the same 400×400 image resolution, and present average results on each scene assessed by metrics which are mentioned above. According to Table 1, we could see that our method NDVG achieves comparable results with D-NeRF [39] for all three metrics. For real scenes, we test on four scenes, namely the *Peel Banana*, *Chicken*, *Broom* and *3D Printer* following HyperNeRF [36]. We follow the same settings of the experiments of TiNueVox [9] and report the metrics of PSNR and MS-SSIM in Table 2 (results of other methods are taken from the TiNuxVox paper). As shown in Table 2, our method could achieve comparable or even better results and at least 10x faster, which clearly demonstrates the effectiveness of our method.

Table 2. Quantitative comparison on real scenes.

Methods	Time	3D printer		Broom		Chicken		Peel banana		Mean	
		PSNR↑	MS-SSIM↑	PSNR↑	MS-SSIM↑	PSNR↑	MS-SSIM↑	PSNR↑	MS-SSIM↑	PSNR↑	MS-SSIM↑
NeRF [31]	~hours	20.7	0.780	19.9	0.653	19.9	0.777	20.0	0.769	20.1	0.745
NV [27]	~hours	16.2	0.665	17.7	0.623	17.6	0.615	15.9	0.380	16.9	0.571
NSFF [22]	~hours	27.7	0.947	26.1	0.871	26.9	0.944	24.6	0.902	26.3	0.916
Nerfies [35]	~hours	20.6	0.830	19.2	0.567	26.7	0.943	22.4	0.872	22.2	0.803
HyperNeRF [36]	~hours	20.0	0.821	19.3	0.591	26.9	0.948	23.3	0.896	22.4	0.814
TiNeuVox [9]	30 mins	22.8	0.841	21.5	0.686	28.3	0.947	24.4	0.873	24.3	0.837
NDVG (Ours)	35 mins	22.4	0.839	21.5	0.703	27.1	0.939	22.8	0.828	23.3	0.823

Table 3. Training time and rendering speed comparison. We report these using the public code of D-NeRF [39] on the same device (RTX 3090 GPU) with our method. We include the mean PSNR across eight scenes in the D-NeRF dataset [39] for comparison of synthesized image quality. Our method could achieve good PSNR, while spend much less optimization time and have faster rendering speed.

Methods	PSNR↑	Training time (s/scene)↓	Rendering speed (s/img)↓
NeRF(half)† [31]	19.00	60185	4.5
D-NeRF(half) [39]	30.02	99034	8.7
NDVG(half)	**30.32**	**1380**	**0.4**
NDVG(w/o refine)	26.73	**708**	2.6
NDVG(w/o filter)	27.85	2487	3.5
NDVG(full)	**31.08**	2087	**1.7**

† We use implementation of D-NeRF [39] to train NeRF on dynamic dataset.

Training Time and Rendering Speed Comparison. The key contribution of our work is to accelerate the optimization speed of novel view synthesis models on dynamic scenes. In Table 3, our method with the same half resolution setting with D-NeRF [39], achieves 70× faster convergence with an even higher average PSNR. Though our main purpose is to speed up training, the proposed method also has a reasonably fast rendering speed, compared to complete neural network based model. According to Table 3, our method has a 20× faster rendering speed compared with D-NeRF [39].

For real scenes in Table 2, compared with previous methods without acceleration which takes hours or even days to train, our method could finish training in 35 min which is at least 10x faster. Compared with the concurrent research TiNueVox [9], which also aims to speed up training, we could achieve comparable results with the same training time, without using cuda acceleration for ray points sampling and total variation computation.

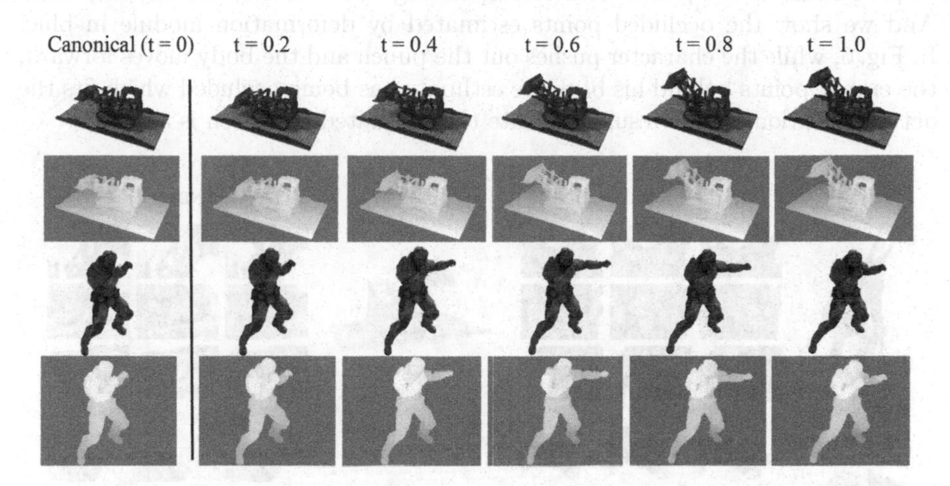

Fig. 4. Learned Geometry. We show examples of geometries learned by our model. For each, we show rendered images and corresponding disparity under two novel views and six time steps.

Qualitative Comparison. We provide some visualization of the learned scene representation in Fig. 4. We can see that our method can successfully recover the canonical geometry, and render high-quality dynamic sequences. The results indicate that our method can faithfully model the motion of the dynamic scenes.

In Fig. 5, we show the rendering results of more difficult situations and compare them with the results of D-NeRF [39]. Our method achieves comparable or even better image results using only 1/70 of the training time. If zoom in for more details, we could see that our method actually recovers more high-frequency details, taking the armour and the cloth of the worker as examples.

5.4 Method Analysis

In this section, we aim to study and prove the effectiveness of three designs in our proposed method: the occlusion reasoning, the coarse-to-fine optimization strategy, and the deformation feature grid.

Occlusion Reasoning. We show that occlusion reasoning is critical in Sect. 3.4 for the canonical-based pipeline under the assumption that the empty point is static. Results of NDVG (full) and NDVG (w/o occ) in Table 1 compare the quantitative results of models with and without occlusion reasoning. We can see that the model with occlusion reasoning achieves more accurate results, verifying the effectiveness of our method.

We also visualize the estimated occlusion at different time steps in Fig. 6. We warp the canonical grid (density grid and color grid) into the different time steps, and show the points with corresponding colors if they are object points. And we show the occluded points estimated by deformation module in blue. In Fig. 6, while the character pushes out the punch and the body moves forward, the empty points behind his back are estimated as being occluded which fits the actual situation. These results indicate the estimated occlusion is accurate.

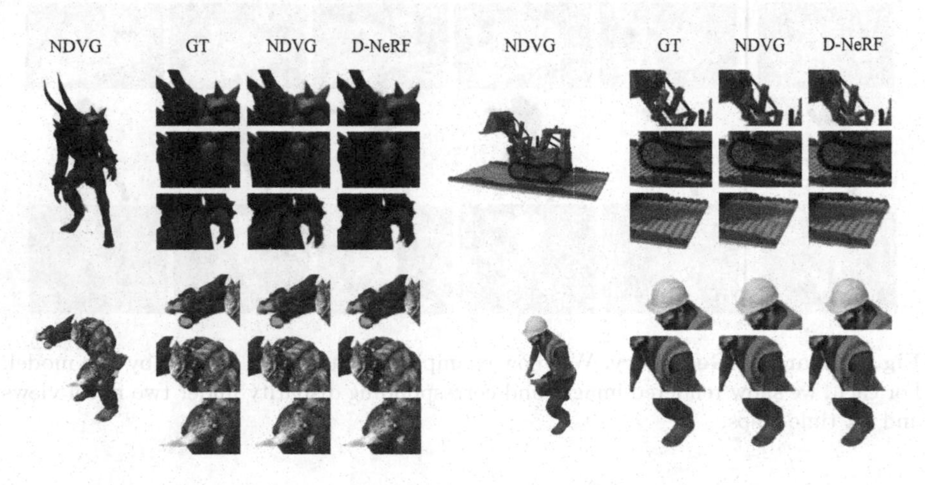

Fig. 5. Qualitative Comparison. Synthesized images on test set of the dataset. For each scene, we show an image rendered at novel view, and followed by zoom in of ground truth, our NDVG, and D-NeRF [39].

Effectiveness of Coarse-to-Fine Optimization. The coarse-to-fine optimization could not only improve the image quality rendered by our model, but also help speed up the training and rendering process significantly. To prove the efficiency of the coarse-to-fine optimization, we conduct two extra experiments (see Table 1). The first one is NDVG (w/o refine), which only contains the coarse training stage. As NDVG (w/o refine) has limited grid resolution and does not have view-independent color representation, the performance is expected to be low. The second one is NDVG (w/o filter), which is not initialized by coarse training and directly begins from scratch for fine stage. As NDVG (w/o filter) does not have a trained coarse model for object region location, the fine model could not shrink the bounding box of the scene which means most of the grid space is wasted, which lead to obvious worse results. Also, the empty points cannot be filtered, which could increase training time and decrease rendering speed significantly, which is evidenced in Table 3.

Deformation Feature Gird. In our deformation module, we use a deformation feature grid to encode dynamic information of 3D points, which will be decoded by a light-weight MLP to regress the deformation. A light-weight MLP is sufficient to model the complex motion of scenes by integrating with this feature grid. Table 1 shows that the method integrated with a deformation feature grid, NDVG (full), achieves better novel-view synthesis results than the one without, NDVG (w/o grid). This result clearly demonstrates the effectiveness of the proposed deformation feature grid. Please refer to our supplementary material for further study of the deformation feature grid design.

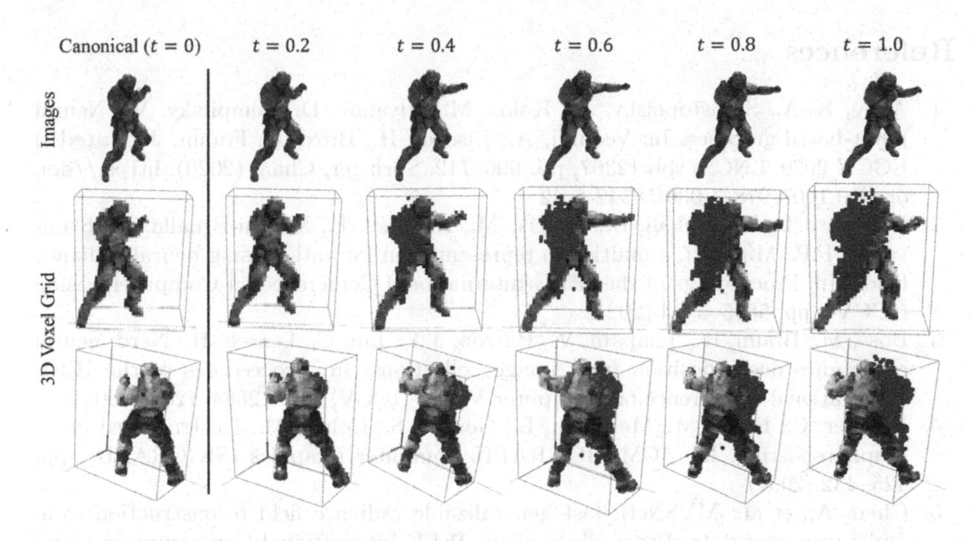

Fig. 6. Occlusion Estimation Visualization. We visualize the estimated occlusion points at different times in blue color. The first row shows the images at each time step which gives an insight of the motion. We warp the canonical grids into corresponding time step by the deformation estimated by deformation module. We visualize points with their rgb colors whose density is over 0.8. (Color figure online)

6 Conclusions

In this paper, we have presented a fast optimization method for dynamic view synthesis based on a hybrid implicit-explicit representation. Our method consists of a deformation module to map a 3D point in observation space to canonical space, and a canonical module to represent the scene geometry and appearance. In each module, explicit dense grids and the light-weight MLP are used for fast feature interpolation and decoding, which significantly accelerates the optimization time compared with methods that rely on heavy MLP queries. Moreover, occlusion is explicitly modeled to improve the rendering quality. Experiments show that our method only requires 30 min to converge on a dynamic scene. Compared with the existing D-NeRF method, our method achieves a 70× acceleration with comparable rendering quality.

Limitation. Despite our method greatly speeds up the training of radiance field methods for dynamic view synthesis, it is mainly designed for bounded scenes. In the future, we will extend our method to deal with real-world unbounded scenes (*e.g.*, unbounded 360 and face-forward scenes).

Acknowledgements. This work was supported in part by the National Natural Science Foundation of China (Nos. 61871325, 61901387, 62001394, 62202409) and the National Key Research and Development Program of China (No. 2018AAA0102803).

References

1. Aliev, K.-A., Sevastopolsky, A., Kolos, M., Ulyanov, D., Lempitsky, V.: Neural point-based graphics. In: Vedaldi, A., Bischof, H., Brox, T., Frahm, J.-M. (eds.) ECCV 2020. LNCS, vol. 12367, pp. 696–712. Springer, Cham (2020). https://doi.org/10.1007/978-3-030-58542-6_42
2. Barron, J.T., Mildenhall, B., Tancik, M., Hedman, P., Martin-Brualla, R., Srinivasan, P.P.: Mip-nerf: a multiscale representation for anti-aliasing neural radiance fields. In: Proceedings of the IEEE International Conference on Computer Vision (ICCV), pp. 5855–5864 (2021)
3. Boss, M., Braun, R., Jampani, V., Barron, J.T., Liu, C., Lensch, H.: Nerd: neural reflectance decomposition from image collections. In: Proceedings of the IEEE International Conference on Computer Vision (ICCV), pp. 12684–12694 (2021)
4. Buehler, C., Bosse, M., McMillan, L., Gortler, S., Cohen, M.: Unstructured lumigraph rendering. In: ACM SIGGRAPH Computer Graphics (SIGGRAPH), pp. 425–432 (2001)
5. Chen, A., et al.: MVSNerf: fast generalizable radiance field reconstruction from multi-view stereo. In: Proceedings of the IEEE International Conference on Computer Vision (ICCV), pp. 14124–14133 (2021)
6. Chen, S.E., Williams, L.: View interpolation for image synthesis. In: ACM SIGGRAPH Computer Graphics (SIGGRAPH), pp. 279–288 (1993)
7. Choi, I., Gallo, O., Troccoli, A., Kim, M.H., Kautz, J.: Extreme view synthesis. In: Proceedings of the IEEE International Conference on Computer Vision (ICCV), pp. 7781–7790 (2019)
8. Du, Y., Zhang, Y., Yu, H.X., Tenenbaum, J.B., Wu, J.: Neural radiance flow for 4D view synthesis and video processing. In: Proceedings of the IEEE International Conference on Computer Vision (ICCV), pp. 14304–14314 (2021)
9. Fang, J., et al.: Fast dynamic radiance fields with time-aware neural voxels. arXiv preprint arXiv:2205.15285 (2022)
10. Flynn, J., et al.: Deepview: view synthesis with learned gradient descent. In: Proceedings of the IEEE Conference on Computer Vision and Pattern Recognition (CVPR), pp. 2367–2376 (2019)
11. Flynn, J., Neulander, I., Philbin, J., Snavely, N.: Deepstereo: learning to predict new views from the world's imagery. In: Proceedings of the IEEE Conference on Computer Vision and Pattern Recognition (CVPR), pp. 5515–5524 (2016)

12. Gan, W., Xu, H., Huang, Y., Chen, S., Yokoya, N.: V4d: Voxel for 4D novel view synthesis. arXiv preprint arXiv:2205.14332 (2022)
13. Gao, C., Saraf, A., Kopf, J., Huang, J.B.: Dynamic view synthesis from dynamic monocular video. In: Proceedings of the IEEE International Conference on Computer Vision (ICCV), pp. 5712–5721 (2021)
14. Greene, N.: Environment mapping and other applications of world projections. IEEE Comput. Graph. Appl. (CG&A) **6**(11), 21–29 (1986)
15. Habtegebrial, T., Jampani, V., Gallo, O., Stricker, D.: Generative view synthesis: from single-view semantics to novel-view images. In: Proceedings of the Advances in Neural Information Processing Systems (NeurIPS) (2020)
16. Hedman, P., Philip, J., Price, T., Frahm, J.M., Drettakis, G., Brostow, G.: Deep blending for free-viewpoint image-based rendering. ACM Trans. Graph. (TOG) **37**(6), 1–15 (2018)
17. Huang, H.-P., Tseng, H.-Y., Lee, H.-Y., Huang, J.-B.: Semantic view synthesis. In: Vedaldi, A., Bischof, H., Brox, T., Frahm, J.-M. (eds.) ECCV 2020. LNCS, vol. 12357, pp. 592–608. Springer, Cham (2020). https://doi.org/10.1007/978-3-030-58610-2_35
18. Kajiya, J.T., Von Herzen, B.P.: Ray tracing volume densities. ACM SIGGRAPH Comput. Graph. (SIGGRAPH) **18**(3), 165–174 (1984)
19. Kalantari, N.K., Wang, T.C., Ramamoorthi, R.: Learning-based view synthesis for light field cameras. ACM Trans. Graph. (TOG) **35**(6), 1–10 (2016)
20. Kingma, D.P., Ba, J.: Adam: a method for stochastic optimization. In: Proceedings of the The International Conference on Learning Representations (ICLR) (2015)
21. Levoy, M., Hanrahan, P.: Light field rendering. In: ACM SIGGRAPH Computer Graphics (SIGGRAPH), pp. 31–42 (1996)
22. Li, Z., Niklaus, S., Snavely, N., Wang, O.: Neural scene flow fields for space-time view synthesis of dynamic scenes. In: Proceedings of the IEEE Conference on Computer Vision and Pattern Recognition (CVPR), pp. 6498–6508 (2021)
23. Lin, C.H., Ma, W.C., Torralba, A., Lucey, S.: Barf: bundle-adjusting neural radiance fields. In: Proceedings of the IEEE International Conference on Computer Vision (ICCV), pp. 5741–5751 (2021)
24. Lindell, D.B., Martel, J.N., Wetzstein, G.: Autoint: automatic integration for fast neural volume rendering. In: Proceedings of the IEEE Conference on Computer Vision and Pattern Recognition (CVPR), pp. 14556–14565 (2021)
25. Liu, J.W., et al.: Devrf: fast deformable voxel radiance fields for dynamic scenes. arXiv preprint arXiv:2205.15723 (2022)
26. Liu, L., Gu, J., Lin, K.Z., Chua, T.S., Theobalt, C.: Neural sparse voxel fields. In: Proceedings of the Advances in Neural Information Processing Systems (NeurIPS), vol. 33, pp. 15651–15663 (2020)
27. Lombardi, S., Simon, T., Saragih, J., Schwartz, G., Lehrmann, A., Sheikh, Y.: Neural volumes: learning dynamic renderable volumes from images. ACM Trans. Graph. (TOG) **38**(4), 65:1–65:14 (2019)
28. Lombardi, S., Simon, T., Schwartz, G., Zollhoefer, M., Sheikh, Y., Saragih, J.: Mixture of volumetric primitives for efficient neural rendering. ACM Trans. Graph. (TOG) **40**(4), 1–13 (2021)
29. Martin-Brualla, R., et al.: Nerf in the wild: neural radiance fields for unconstrained photo collections. In: Proceedings of the IEEE Conference on Computer Vision and Pattern Recognition (CVPR), pp. 7210–7219 (2021)
30. Mildenhall, B., et al.: Local light field fusion: practical view synthesis with prescriptive sampling guidelines. ACM Trans. Graph. (TOG) **38**(4), 1–14 (2019)

31. Mildenhall, B., Srinivasan, P.P., Tancik, M., Barron, J.T., Ramamoorthi, R., Ng, R.: Nerf: representing scenes as neural radiance fields for view synthesis. In: Proceedings of the European Conference on Computer Vision (ECCV), pp. 405–421 (2020)
32. Müller, T., Evans, A., Schied, C., Keller, A.: Instant neural graphics primitives with a multiresolution hash encoding. arXiv preprint arXiv:2201.05989 (2022)
33. Neff, T., et al.: DONeRF: towards real-time rendering of compact neural radiance fields using depth oracle networks. Comput. Graph. Forum **40**(4) (2021)
34. Niemeyer, M., Mescheder, L., Oechsle, M., Geiger, A.: Differentiable volumetric rendering: learning implicit 3D representations without 3D supervision. In: Proceedings of the IEEE Conference on Computer Vision and Pattern Recognition (CVPR), pp. 3504–3515 (2020)
35. Park, K., et al.: Nerfies: deformable neural radiance fields. In: Proceedings of the IEEE International Conference on Computer Vision (ICCV) (2021)
36. Park, K., et al.: HyperNeRF: a higher-dimensional representation for topologically varying neural radiance fields. ACM Trans. Graph. (TOG) **40**(6), 1–12 (2021)
37. Penner, E., Zhang, L.: Soft 3D reconstruction for view synthesis. ACM Trans. Graph. (TOG) **36**(6), 1–11 (2017)
38. Piala, M., Clark, R.: Terminerf: ray termination prediction for efficient neural rendering. In: Proceedings of the International Conference on 3D Vision (3DV), pp. 1106–1114 (2021)
39. Pumarola, A., Corona, E., Pons-Moll, G., Moreno-Noguer, F.: D-nerf: neural radiance fields for dynamic scenes. In: Proceedings of the IEEE Conference on Computer Vision and Pattern Recognition (CVPR), pp. 10318–10327 (2021)
40. Rebain, D., Jiang, W., Yazdani, S., Li, K., Yi, K.M., Tagliasacchi, A.: Derf: decomposed radiance fields. In: Proceedings of the IEEE Conference on Computer Vision and Pattern Recognition (CVPR), pp. 14153–14161 (2021)
41. Reiser, C., Peng, S., Liao, Y., Geiger, A.: Kilonerf: speeding up neural radiance fields with thousands of tiny mlps. In: Proceedings of the IEEE International Conference on Computer Vision (ICCV), pp. 14335–14345 (2021)
42. Riegler, G., Koltun, V.: Free view synthesis. In: Vedaldi, A., Bischof, H., Brox, T., Frahm, J.-M. (eds.) ECCV 2020. LNCS, vol. 12364, pp. 623–640. Springer, Cham (2020). https://doi.org/10.1007/978-3-030-58529-7_37
43. Riegler, G., Koltun, V.: Stable view synthesis. In: Proceedings of the IEEE Conference on Computer Vision and Pattern Recognition (CVPR), pp. 12216–12225 (2021)
44. Shum, H., Kang, S.B.: Review of image-based rendering techniques. In: Visual Communications and Image Processing (VCIP), vol. 4067, pp. 2–13 (2000)
45. Srinivasan, P.P., Deng, B., Zhang, X., Tancik, M., Mildenhall, B., Barron, J.T.: Nerv: neural reflectance and visibility fields for relighting and view synthesis. In: Proceedings of the IEEE Conference on Computer Vision and Pattern Recognition (CVPR), pp. 7495–7504 (2021)
46. Srinivasan, P.P., Mildenhall, B., Tancik, M., Barron, J.T., Tucker, R., Snavely, N.: Lighthouse: predicting lighting volumes for spatially-coherent illumination. In: Proceedings of the IEEE Conference on Computer Vision and Pattern Recognition (CVPR), pp. 8080–8089 (2020)
47. Srinivasan, P.P., Tucker, R., Barron, J.T., Ramamoorthi, R., Ng, R., Snavely, N.: Pushing the boundaries of view extrapolation with multiplane images. In: Proceedings of the IEEE Conference on Computer Vision and Pattern Recognition (CVPR), pp. 175–184 (2019)

48. Sun, C., Sun, M., Chen, H.T.: Direct voxel grid optimization: super-fast convergence for radiance fields reconstruction. arXiv preprint arXiv:2111.11215 (2021)
49. Tewari, A., et al.: Advances in neural rendering. In: ACM SIGGRAPH Computer Graphics (SIGGRAPH), pp. 1–320 (2021)
50. Tewari, A., et al.: State of the art on neural rendering. In: Computer Graphics Forum, pp. 701–727 (2020)
51. Thies, J., Zollhöfer, M., Nießner, M.: Deferred neural rendering: image synthesis using neural textures. ACM Trans. Graph. (TOG) **38**(4), 1–12 (2019)
52. Tretschk, E., Tewari, A., Golyanik, V., Zollhöfer, M., Lassner, C., Theobalt, C.: Non-rigid neural radiance fields: Reconstruction and novel view synthesis of a dynamic scene from monocular video. In: Proceedings of the IEEE International Conference on Computer Vision (ICCV), pp. 12959–12970 (2021)
53. Trevithick, A., Yang, B.: Grf: learning a general radiance field for 3D representation and rendering. In: Proceedings of the IEEE International Conference on Computer Vision (ICCV), pp. 15182–15192 (2021)
54. Tucker, R., Snavely, N.: Single-view view synthesis with multiplane images. In: Proceedings of the IEEE Conference on Computer Vision and Pattern Recognition (CVPR), pp. 551–560 (2020)
55. Wang, C., Eckart, B., Lucey, S., Gallo, O.: Neural trajectory fields for dynamic novel view synthesis. arXiv preprint arXiv:2105.05994 (2021)
56. Wang, Q., et al.: Ibrnet: learning multi-view image-based rendering. In: Proceedings of the IEEE Conference on Computer Vision and Pattern Recognition (CVPR), pp. 4690–4699 (2021)
57. Wang, Z., Wu, S., Xie, W., Chen, M., Prisacariu, V.A.: Nerf-: neural radiance fields without known camera parameters. arXiv preprint arXiv:2102.07064 (2021)
58. Xian, W., Huang, J.B., Kopf, J., Kim, C.: Space-time neural irradiance fields for free-viewpoint video. In: Proceedings of the IEEE Conference on Computer Vision and Pattern Recognition (CVPR), pp. 9421–9431 (2021)
59. Xu, Z., Bi, S., Sunkavalli, K., Hadap, S., Su, H., Ramamoorthi, R.: Deep view synthesis from sparse photometric images. ACM Trans. Graph. (TOG) **38**(4), 1–13 (2019)
60. Yariv, L., et al.: Multiview neural surface reconstruction by disentangling geometry and appearance. In: Proceedings of the Advances in Neural Information Processing Systems (NeurIPS), vol. 33, pp. 2492–2502 (2020)
61. Yen-Chen, L., Florence, P., Barron, J.T., Rodriguez, A., Isola, P., Lin, T.Y.: inerf: inverting neural radiance fields for pose estimation. In: Proceedings of the International Conference on Intelligent Robots and Systems (IROS), pp. 1323–1330 (2021)
62. Yoon, J.S., Kim, K., Gallo, O., Park, H.S., Kautz, J.: Novel view synthesis of dynamic scenes with globally coherent depths from a monocular camera. In: Proceedings of the IEEE Conference on Computer Vision and Pattern Recognition (CVPR), pp. 5336–5345 (2020)
63. Yu, A., Fridovich-Keil, S., Tancik, M., Chen, Q., Recht, B., Kanazawa, A.: Plenoxels: radiance fields without neural networks. In: Proceedings of the IEEE Conference on Computer Vision and Pattern Recognition (CVPR) (2021)
64. Yu, A., Li, R., Tancik, M., Li, H., Ng, R., Kanazawa, A.: Plenoctrees for real-time rendering of neural radiance fields. In: Proceedings of the IEEE International Conference on Computer Vision (ICCV), pp. 5752–5761 (2021)
65. Yu, A., Ye, V., Tancik, M., Kanazawa, A.: pixelnerf: neural radiance fields from one or few images. In: Proceedings of the IEEE Conference on Computer Vision and Pattern Recognition (CVPR), pp. 4578–4587 (2021)

66. Zhang, K., Riegler, G., Snavely, N., Koltun, V.: NERF++: analyzing and improving neural radiance fields. arXiv preprint arXiv:2010.07492 (2020)
67. Zhang, R., Isola, P., Efros, A.A., Shechtman, E., Wang, O.: The unreasonable effectiveness of deep features as a perceptual metric. In: Proceedings of the IEEE Conference on Computer Vision and Pattern Recognition (CVPR), pp. 586–595 (2018)
68. Zhang, X., Srinivasan, P.P., Deng, B., Debevec, P., Freeman, W.T., Barron, J.T.: Nerfactor: neural factorization of shape and reflectance under an unknown illumination. ACM Trans. Graph. (TOG) **40**(6), 1–18 (2021)
69. Zhou, T., Tucker, R., Flynn, J., Fyffe, G., Snavely, N.: Stereo magnification: learning view synthesis using multiplane images. ACM Trans. Graph. (TOG) **37**(4), 1–12 (2018)

Spotlights: Probing Shapes from Spherical Viewpoints

Jiaxin Wei[1], Lige Liu[2], Ran Cheng[2], Wenqing Jiang[1], Minghao Xu[2], Xinyu Jiang[2], Tao Sun[2(✉)], Sören Schwertfeger[1], and Laurent Kneip[1]

[1] ShanghaiTech University, Shanghai, China
{weijx,jiangwq,soerensch,lkneip}@shanghaitech.edu.cn
[2] RoboZone, Midea Inc., Foshan, China
{liulg12,chengran1,xumh33,jxy77,tsun}@midea.com

Abstract. Recent years have witnessed the surge of learned representations that directly build upon point clouds. Inspired by spherical multiview scanners, we propose a novel sampling model called *Spotlights* to represent a 3D shape as a compact 1D array of depth values. It simulates the configuration of cameras evenly distributed on a sphere, where each virtual camera casts light rays from its principal point to probe for possible intersections with the object surrounded by the sphere. The structured point cloud is hence given implicitly as a function of depths. We provide a detailed geometric analysis of this new sampling scheme and prove its effectiveness in the context of the point cloud completion task. Experimental results on both synthetic and real dataset demonstrate that our method achieves competitive accuracy and consistency while at a lower computational cost. The code and dataset will be released at https://github.com/goldoak/Spotlights.

Keywords: Shape representation · Point cloud completion

1 Introduction

Over the past decade, the community has put major efforts into the recovery of explicit 3D object models, which is of eminent importance in many computer vision and robotics applications [2,10,13,21–23,33,36]. Traditionally, the recovery of complete object geometries from sensor readings relies on scanning systems that employ spherical high-resolution camera arrangements [11,26]. If additionally equipped with light sources, such scanners are turned into light stages. However, without such expensive equipment, shape reconstruction often turns out to be erroneous and incomplete owing to occlusions, limited viewpoints, and measurement imperfections.

Previous work has investigated many different representations of objects to deal with those problems, such as volumetric occupancy grids [41], distance

---

Supplementary Information The online version contains supplementary material available at https://doi.org/10.1007/978-3-031-26319-4_28.

L. Wang et al. (Eds.): ACCV 2022, LNCS 13841, pp. 469–485, 2023.
https://doi.org/10.1007/978-3-031-26319-4_28

fields [24], and more recently, point cloud-based representations [30,31]. The latter are preferred in terms of efficiency and capacity of maintaining fine-grained details. Yet, each point in a point cloud needs to be represented by a 3D coordinate, and the large number of points generated in reality can strain computational resources, bandwidth, and storage.

To this end, we propose a novel structured point cloud representation inspired by spherical multi-view 3D object scanners [11,26]. Sparse point clouds are generated by a simulated configuration of multiple views evenly distributed on a sphere around the object where each view casts a preset bundle of rays from its center towards the object. Due to its similarity to stage lighting instruments, we name it as *Spotlights*. It represents point clouds in a more compact way since it only needs to store a 1D array of scalars, i.e. the depths along the rays, and thus serves as a basis for efficient point-based networks. Moreover, Spotlights is a structured representation that produces ordered point clouds. This order-preserving property is of great benefit to spatial-temporal estimation problems such as Simultaneous Localization And Mapping (SLAM) [36] and 3D object tracking [17].

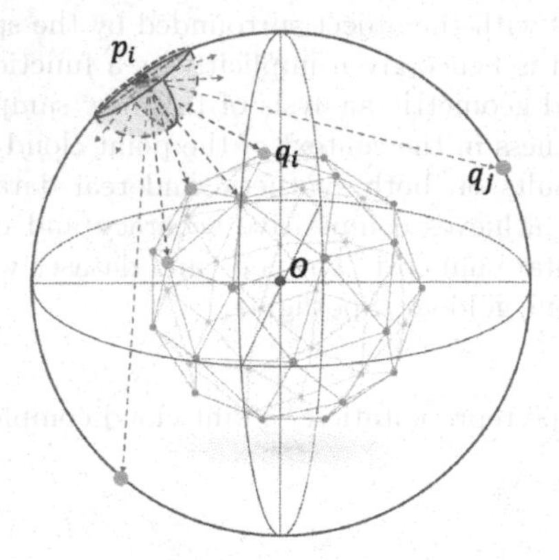

Fig. 1. Illustration of Spotlights with only one view. Dashed arrows indicate the ray bundle cast from p_i. Some rays intersect with the bounded object (purple) while others spread in void space (orange). (Color figure online)

To demonstrate the superiority of Spotlights, we focus on the 3D shape completion task which aims to predict a complete point cloud with respect to a partial observation. Previous efforts are devoted to inferring fine-grained details [5,14,44] and designing delicate point-based decoders [43] [37]. In this work, we present a lightweight neural network for fast point cloud prediction based on our proposed shape representation. It achieves competitive results in accuracy without substantially sabotaging completeness. Furthermore, our completed point clouds can easily find correspondences across temporal states, therefore, are useful for some subsequent tasks. Our main contributions are as follows:

- We introduce a novel point representation denoted *Spotlights*. It is a unified and compact representation. The effectiveness of the proposed representation is shown by the theoretical analysis of local point density.
- We apply our proposed representation to the task of 3D shape completion and achieve remarkable results on par with existing supervised state-of-the-art approaches on both synthetic and real data. Notably, our network consumes less memory and is fast at inference time.
- We build a new synthetic dataset for 3D shape completion, comprising 128000 training samples and 30000 testing samples over 79 unique car models (i.e. 2000 different observations per object). The ground truth point clouds are extracted by our Spotlights.

2 Related Work

Nowadays, shape prediction is mostly studied in the context of shape completion and reconstruction from 3D data (e.g. depth images or point clouds). Though this problem can be solved using lower-level geometric principles (e.g. volumetric diffusion [6], symmetry [38]) or shape retrieval methods [19,25], the focus of our work lies on learning-based 3D point cloud completion.

In recent years, the community has demonstrated the feasibility of learning highly efficient representations for geometric problems by relying directly on 3D point clouds [30,31,34]. Fan et al. [9] and Achlioptas et al. [1] introduce auto-encoder CNNs for direct point cloud completion, showing that point clouds encode only points on the surface, naturally permit direct manipulation and put fewer restrictions on the topology or fine-grainedness of a shape. Yuan et al. [44] extend the idea by permutation invariant and convolution-free feature extraction, as well as a coarse-to-fine feature generation process. Xie et al. [42] internally revert to a volumetric representation by adding differentiable layers to transform back and forth between point clouds and grids. Huang et al. [14] furthermore propose to limit the prediction to the missing part of a point cloud, thus preserving details in the original input measurement. Wen et al. [40] mimic an earth mover and complete point clouds by moving points along the surface.

While direct 3D point-based representations have lead to significant advances in terms of efficiency, they suffer from disorder and the inability to focus on inherent local structures. Yang et al. [43] propose *FoldingNet*, a deep auto-encoder that generates ordered point clouds by a folding-based decoder that deforms a regular 2D grid of points onto the underlying 3D object surface of a point cloud. Although they claim that their representation can learn "cuts" on the 2D grid and represent surfaces that are topologically different from a single plane, they at the same time admit that the representation fails on more complex geometries. Li et al. [18] propose to use GAN to generate realistic shapes from unit spheres, establishing dense correspondences between spheres and generated shapes. Wen et al. [39] propose an extension with hierarchical folding by adding skip-attention connections. Groueix et al. [12] introduce *AtlasNet*. Similar to the structure of FoldingNet, this work deforms multiple 2D grids. Nonetheless, the modelling as a finite collection of manifolds still imposes constraints on the overall topology of an object's shape. Tchapmi et al. [37] propose *TopNet*. Their decoder generates structured point clouds using a hierarchical rooted tree structure. The root node embeds the entire point cloud, child nodes subsets of the points, and leaf nodes individual points. Rather than sampling a finite set of manifolds, this architecture generates structured point clouds without any assumptions on the topology of an object's shape.

3 Method

In this section, we first review relevant background knowledge on Fibonacci spheres and solid angles (see Sect. 3.1). Then, we describe the detailed construction of our newly proposed Spotlights representation (see Sect. 3.2), and further prove its theoretical validity (see Sect. 3.3).

3.1 Background

Spherical Fibonacci Point Set. One of the most common approximation algorithms to generate uniformly distributed points on a unit sphere $\mathcal{S}^2 \subset \mathbb{R}^3$ is mapping a 2D Fibonacci lattice onto the spherical surface. The Fibonacci lattice is defined as a unit square $[0,1)^2$ with an arbitrary number of n points evenly distributed inside it. The 2D point with index i is expressed as

$$\mathbf{p}_i = (x_i, y_i) = \left(\left[\frac{i}{\Phi} \right], \frac{i}{n} \right), 0 \leq i < n.$$

Note that $[\cdot]$ takes the fractional part of the argument and $\Phi = \frac{1+\sqrt{5}}{2}$ is the golden ratio. The 2D points are then mapped to 3D points on the unit sphere using cylindrical equal-area projection. The 3D points are easily parameterized using the spherical coordinates

$$\mathbf{P}_i = (\varphi_i, \theta_i) = (2\pi x_i, \arccos(1 - 2y_i)), \varphi_i \in [0, 2\pi], \theta_i \in [0, \pi].$$

We refer the readers to [15] for more details.

Solid Angle. Similar to the planar angle in 2D, the solid angle for a spherical cap is a 3D angle subtended by the surface at its center. Suppose the area of the cap is A and the radius is r, then the solid angle is given as $\Omega = \frac{A}{r^2}$. Thus, the solid angle of a sphere is 4π. Extended to arbitrary surfaces, the solid angle of a differential area is

$$d\Omega = \frac{dA}{r^2} = \frac{r\text{sin}\theta d\varphi \cdot rd\theta}{r^2} = \text{sin}\theta d\theta d\varphi,$$

where $\theta \in [0, \pi]$ and $\varphi \in [0, 2\pi]$. By calculating the surface integral, we can obtain

$$\Omega = \iint \text{sin}\theta d\theta d\varphi.$$

3.2 Spotlights Model Construction

In this subsection, we show how to construct the Spotlights model (see Fig. 1 for an illustration). There are three consecutive steps described in the following.

Bounding Sphere Extraction. To find the bounding sphere of an object of interest, we first compute its bounding box by determining the difference between the extrema of the surface point coordinates along each dimension. Practically, we can also easily get the bounding box parameters (i.e. center, size and orientation) for real-world objects using state-of-the-art 3D object detection methods [7,16,27–29]. The bounding sphere is then given as the circumscribed sphere of the bounding box. The radius of the sphere is used as a scaling factor to normalize the real-world object into a unit sphere.

Ray Bundle Organization. We leverage the Fibonacci lattice mapping method presented in Sect. 3.1 to evenly sample the unit sphere obtained from step 1 and place a bundle of virtual viewing rays at each position to mimic the behavior of a camera. The motivation of using a sphere instead of a cuboid arrangement of views is given by the fact that a spherical arrangement has more diversity in the distribution of principal viewing directions, thereby alleviating self-occlusion in the data. Also, spherical multi-view configurations are commonly adopted for 2.5D multi-surface rendering [4,35].

To define structured ray bundles emitted from each viewpoint, we can imagine placing a smaller sphere at each primary sampling point on the unit sphere. The intersection plane between the unit sphere and each small sphere forms an internal spherical cap, on which we again use Fibonacci sampling to define a fixed arrangement of sampling points. Thus, the rays originate from the primary sampling points and traverse the secondary sampling points on the spherical cap. It is easy to adjust the opening angle of the ray bundle. Suppose the radii of the outer sphere and the small sphere are R and r, respectively. The maximum polar angle of the spherical cap is then given by $\omega = \arccos\frac{r}{2R}$. The opening angle ω as well as the number of rays serve as design hyper-parameters. We refer to Sect. 5.3 for a detailed discussion.

Ray Casting. We finally cast the rays onto the object and retrieve the distance between the view center and the first intersection point with the object's surface. For rays that miss the surface, we manually set the distances to a sentinel value of zero. With known origins and viewing directions, the 3D point cloud can be easily recovered from the depths along the rays. Compared to 3D point coordinates, 1D distances are more efficient to store and generate. To validate this property, we apply our representation to the 3D point cloud completion task. While obtaining competitive results in terms of accuracy, our network is fast at inference time and has fewer parameters than previous methods (see Sect. 5.1). Besides, the fixed ray arrangement enables the reproduction of ordered point clouds, thus providing one-to-one correspondences across temporal states.

3.3 Analysis of Spotlights

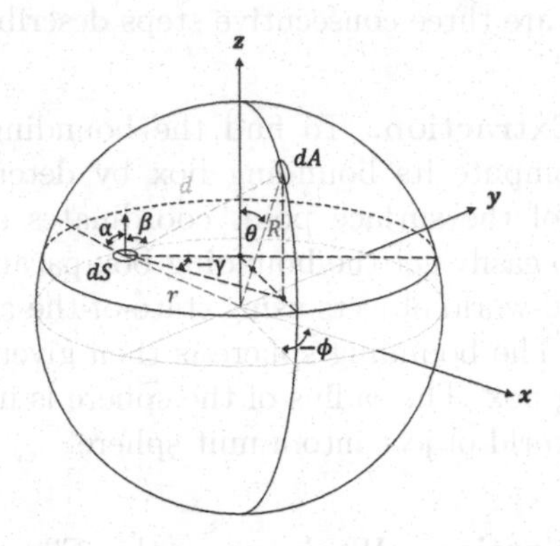

Fig. 2. Mathematical model of Spotlights.

The quality of the sampling is measured from two aspects, namely the uniformity of the ray density for surfels at different positions and orientations within the bounding sphere when 1) there is no occlusion, or 2) there is occlusion. The first measure analyses how much the sampling density of the object surface depends on its location. The second measure provides a means to analyse the impact of occlusion that typically occurs in practical scenarios. To simplify the theoretical analysis, we make the following assumptions:

1. The object is bounded by a unit sphere, that is, $R = 1$.
2. An arbitrary surfel dS on the object can be parameterized by the distance of the surfel to the center of the sphere r and the orientation of the surfel α, where α is defined as the angle between the surface normal and the vector connecting the centers of the sphere and the surfel (see Fig. 2). The two parameters are uniformly distributed in their respective ranges, i.e. $r \sim U(0, 1)$ and $\alpha \sim U(0, \pi)$.
3. There are N ray sources uniformly distributed on the sphere and each of them emits M rays within a certain solid angle Ω oriented towards the center of the sphere. For convenience, we set $\Omega = 2\pi$. Note that we also assume $N, M \gg 1$. The source density ρ_s and the ray density ρ_r are then given by

$$\rho_s = \frac{N}{4\pi R^2} = \frac{N}{4\pi}, \quad \text{and} \quad \rho_r = \frac{M}{2\pi}.$$

In order to calculate the ray density of dS, we take another surfel dA on the sphere, calculate the ray density of dS caused by dA, and then do the integration over dA. The area of surfel dA with polar angle θ and azimuth angle ϕ is

$dA = \sin\theta d\theta d\phi$. Using trigonometry function, we can calculate the distance between dS and dA:

$$d^2 = 1 + r^2 - 2r(\cos\alpha\cos\theta - \sin\alpha\sin\theta\cos\phi).$$

The rays that can be received by the surfel dS from dA are the solid angle occupied by the view of dS at dA multiplied by the ray density ρ_r. Then, the ray density ρ of surfel dS is the integration over dA, i.e.

$$\begin{aligned}\rho(r,\alpha) &= \frac{1}{dS}\int dA\cdot\rho_s\cdot\rho_r\cdot\frac{dS\cos\beta}{d^2}\\ &= \frac{NM}{8\pi^2}\int d\theta\int d\phi\frac{\sin\theta(\cos\theta - r\cos\alpha)}{(r^2 - 2r(\cos\alpha\cos\theta - \sin\alpha\sin\theta\cos\phi)+1)^{3/2}}.\end{aligned}$$

The ray density ρ as a function of r and α is shown in Fig. 3. We can observe that ρ has no divergence and is a fairly smooth function of r and α when the surfel dS is not too close to the sphere surface. Quantitatively, when $0 \le r \le 0.8$, ρ takes values between 0.039 and 0.057. The proposed sampling strategy therefore ensures that the object surfaces at arbitrary positions and orientations in the sphere have comparable sampling density.

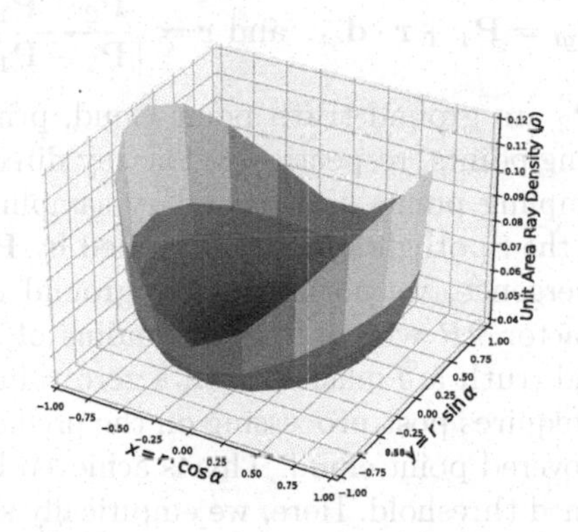

Fig. 3. Ray density of surfel dS.

Next, we consider the case where certain directions seen from dS are occluded, therefore no rays can be received from the related sources on the surface of the sphere. The blocked direction is defined by the polar and azimuth angles (β, γ) (see Fig. 2). We assume that the blocking is random and uniform, i.e., $\beta \sim U(0, \pi)$ and $\gamma \sim U(0, 2\pi)$. The unit area ray density (ρ) as a function of $(r, \alpha, \beta, \gamma)$ is then given by

$$\begin{aligned}\rho(r,\alpha,\beta,\gamma) &= \frac{1}{dS}\cdot dA\cdot\rho_s\cdot\rho_r\cdot\frac{dS\cos\beta}{d^2}\\ &= \frac{NM}{8\pi^2}\frac{\cos\beta}{(\xi r^2+1)^{1/2}}, \quad \xi = (\sin\alpha\sin\beta\cos\gamma - \cos\alpha\cos\beta)^2 - 1.\end{aligned}$$

As can be observed, ρ is a continuous function of β and γ, which ensures that when part of the directions are occluded, there are still rays that can hit dS from other angles, resulting in better sampling coverage. A simplified version of our point cloud sampling strategy only consists of casting a single ray towards the center of the sphere from each viewpoint. However, it is intuitively clear that this would limit the local inclination of rays sampling dS to a single direction, and thereby lead to an increased likelihood of occlusions.

Note that in practice the solid angle covered by each bundle of viewing rays should be chosen smaller such that more rays intersect with the object surface, and fewer rays are wasted. We experiment with different solid angles and find a balance between completeness and hit ratio in Sect. 5.3.

4 Shape Completion with Spotlights

Next we introduce our 3D shape completion network based on the Spotlights representation, named *Spotlights Array Network (SA-Net)*.

By leveraging the Spotlights model, we reformulate the 3D shape generation into a 1D array regression problem. The ground truth depth values $\mathbf{d}_{gt}$ satisfies that

$$\mathbf{P}_{gt} = \mathbf{P}_1 + \mathbf{r} \cdot \mathbf{d}_{gt}, \text{ and } \mathbf{r} = \frac{\mathbf{P}_2 - \mathbf{P}_1}{\|\mathbf{P}_2 - \mathbf{P}_1\|}$$

where $\mathbf{P}_{gt}$, $\mathbf{P}_1$ and $\mathbf{P}_2$ are ground truth point cloud, primary sampling points and secondary sampling points, respectively. The ray directions $\mathbf{r}$ are computed using the primary sampling points and secondary sampling points as discussed in Sect. 3.2. Actually, the Spotlights model is formed by $\mathbf{P}_1$, $\mathbf{P}_2$ and $\mathbf{r}$.

To speed up convergence, we normalize the ground truth depths into the range $[0, 1]$ using a factor $2R$ where R is the radius of the bounding sphere. Missing rays in ground truth are masked with a zero value. Note that the existence of missing rays requires post-processing on the predicted array to filter out the outliers in the recovered point cloud. This is achieved by simply clipping the array using a pre-defined threshold. Here, we empirically set the threshold value to be 0.2.

The pipeline of SA-Net is shown in Fig. 4. It consists of a point-based encoder and a decoder purely made up of multi-layer perceptrons (MLPs). Similar to previous work, we feed the partial observation into the encoder to obtain a 1024-dim latent vector. The difference is that we regress an array of scalars instead of 3D coordinates, which significantly reduces the computational cost. We calculate a L1 training loss between the predicted values and ground truth. At inference time, a complete 3D point cloud of the object can be recovered from the predicted 1D scalars using the Spotlights sampling model, that is,

$$\mathbf{P}_{pred} = \mathbf{P}_1 + \mathbf{r} \cdot \mathbf{d}_{pred}.$$

In addition, 3D points in $\mathbf{P}_{pred}$ are ordered due to the structure in the Spotlights model.

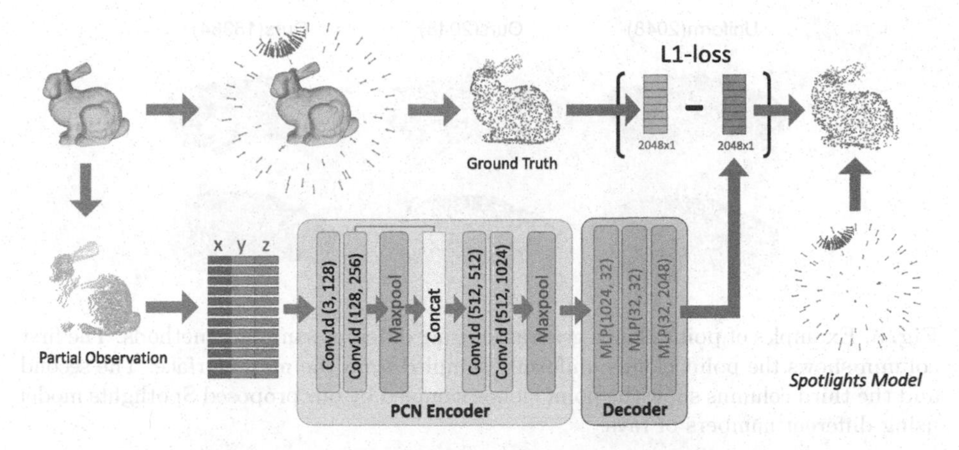

Fig. 4. Architecture of SA-Net. The network takes incomplete point cloud observations as input and predicts a 1D array, from which the complete point cloud is recovered by an inverse application of the Spotlights Model. Note that the PCN backbone can be replaced with other point-based encoders.

5 Experiments

We first describe how to build a synthetic dataset using Spotlights. Next, we perform 3D shape completion on both the newly made dataset and KITTI-360 [20], and compare our results against existing state-of-the-art methods. Moreover, we provide ablation studies on ShapeNet [3] to figure out the importance of the hyper-parameters involved in the Spotlights model.

5.1 Shape Completion on Synthetic Dataset

Data Generation. We export 79 different vehicle assets of 3 categories from CARLA [8] and randomly select 64 vehicles for training, 8 vehicles for validation and the rest for testing. We use Blender to simulate a HDL-64 lidar sensor to further generate 2000 partial observations for each object and downsample those point clouds to 256 points. As for the ground truth, we uniformly sample 2048 points from the mesh surface for other methods while using Spotlights to get a 1D array containing 2048 values for SA-Net. Specifically, we evenly sample 32 primary points and 64 secondary points with 60° opening angle to construct the Spotlights model. For a fair comparison, we also densely sample 16384 points for each object to calculate the evaluation metrics introduced in the following. Examples of point clouds generated using different sampling methods are shown in Fig. 5. Despite the sparsity of the point cloud sampled by Spotlights, it still captures the overall shape of the target object. Besides, increasing the number of rays can notably improve the point density.

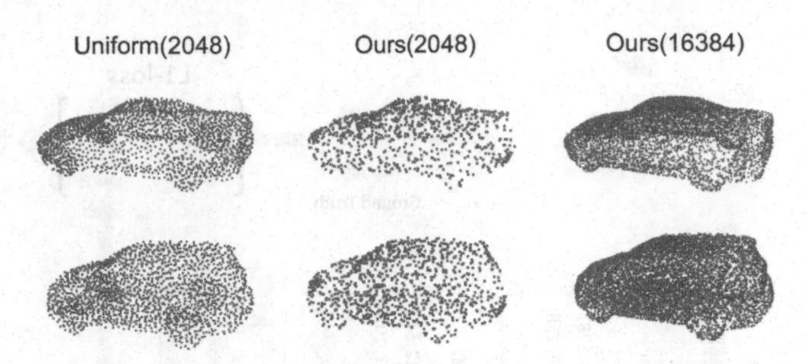

Fig. 5. Examples of point clouds generated using different sampling methods. The first column shows the point clouds uniformly sampled from the mesh surface. The second and the third columns show the point clouds sampled by our proposed Spotlights model using different numbers of rays.

Comparison Methods. We compare all methods on the Completion3D benchmark [37], an evaluation platform designed for 3D point cloud completion, and replace the original dataset with our generated synthetic one outlined above. In our setting, TopNet [37], PCN [44], FoldingNet [43] and AtlasNet [12] directly predict a complete point cloud containing 2048 points, whereas our SA-Net outputs a 1D array with 2048 values from which we further decode the sparse point cloud. According to the analysis in [44], we use the PCN encoder for all methods.

Evaluation Metrics. The Chamfer Distance (CD) introduced by [9] is one of the most widely used metrics to measure the discrepancy between a predicted and the ground truth point cloud. Suppose the two point clouds are denoted as $P_1, P_2 \subseteq \mathbb{R}^3$. The CD is given by

$$d_{CD}(P_1, P_2) = \frac{1}{|P_1|} \sum_{x \in P_1} \min_{y \in P_2} \|x - y\| + \frac{1}{|P_2|} \sum_{y \in P_2} \min_{x \in P_1} \|x - y\|, \qquad (1)$$

and it computes the average nearest neighbor distance between P_1 and P_2. Here we use CD in a single direction (i.e. we only employ one of the terms in (1)) to calculate the distance from the prediction to ground truth. As in [32], this metric indicates the accuracy of each of the predicted points with respect to the ground truth point cloud. To prevent large distances caused by too sparse reference point clouds, we furthermore choose dense ground truth point clouds with 16384 points. In addition, the L1-norm is used to improve robustness against outliers.

Performance. The quantitative results of SA-Net compared with alternative methods [12,37,43,44] is shown in Table 1. As can be observed, our SA-Net achieves competitive accuracy, while being faster and using less memory. Note that SA-Net performs better after applying an additional clip operation for statistical outlier removal. The qualitative results are shown in Fig. 6. The sparsity of our predicted point clouds is caused by missing rays.

Table 1. Accuracy and efficiency of comparisons on the synthetic dataset. Accuracy indicates the quality of completed point clouds, which is measured by the single direction CD. Efficiency is evaluated in terms of inference time and the number of network parameters.

Method	Accuracy				Network	
	Hatchback	Sedan	SUV	Average	#Params	Runtime
AtalsNet	0.06586	0.07325	0.06051	0.06462	8.3M	14.1 ms
PCN	0.02528	0.02842	0.02221	0.02428	7.6M	34.1 ms
FoldingNet	0.02437	0.02701	0.02166	0.02345	8.5M	34.6 ms
TopNet	0.02396	**0.02631**	0.02051	0.02252	7.5M	37.9 ms
Ours (w/o clip)	0.02356	0.03095	0.02171	0.02442	**0.9M**	**0.80** ms
Ours (w/ clip)	**0.02124**	0.02775	**0.01850**	**0.02133**	**0.9M**	1.05 ms

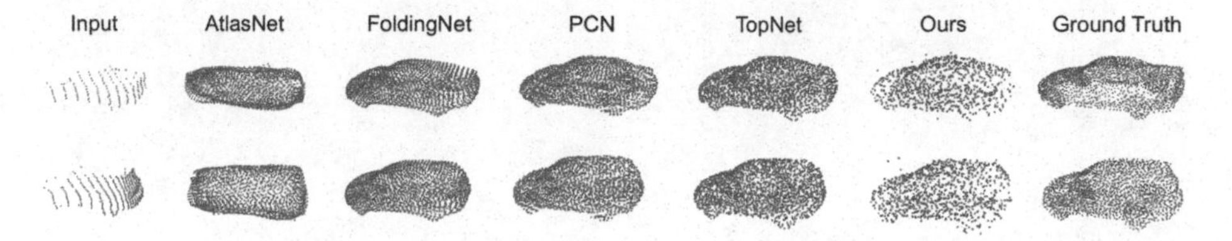

Fig. 6. Qualitative Results of shape completion on the synthetic dataset.

5.2 Shape Completion on Real Dataset

We transfer all models trained on our synthetic dataset to KITTI-360 [20], a large-scale outdoor dataset, to further evaluate generalization ability. Specifically, we take real-world LiDAR scans and segment out points within the bounding boxes of each frame. This preprocessing generates a total of 4382 partial observations spreading over the classes of *car*, *truck*, *caravan*, and *trailer*. All point clouds are represented in object coordinates and normalized into a unit sphere using the bounding box parameters. Since there is no ground truth point clouds on KITTI-360, we adopt the consistency metric introduced in [44]. It averages the CD results over multiple predictions of the same object in different frames. The results are listed in Table 2. It is obvious that our SA-Net generates more consistent point clouds, indicating that it can better deal with large variations in the input point clouds. The qualitative results on real scans are shown in Fig. 7.

5.3 Ablation Study

In this subsection, we further evaluate the influence of the number of rays, polar angles used in Spotlights. The controlled experiments are conducted on a subset of the ShapeNet dataset [3] containing 6 different categories. Here, we uniformly sample 16384 points on the mesh surface of each object as our ground truth. In addition, the impact of different sampling resolutions is measured through shape completion on our synthetic dataset.

Table 2. Consistency of comparisons on KITTI-360. We report the average CD of completed point clouds of the same target to evaluate the consistency of predictions.

Model	Consistency				
	Car	Truck	Caravan	Trailer	Average
AtalsNet	0.0649	0.0674	0.0651	0.0710	0.0671
PCN	0.0493	0.0612	0.0644	0.0580	0.0582
FoldingNet	0.0514	0.0611	0.0629	0.0595	0.0587
TopNet	0.0513	0.0644	0.0648	0.0593	0.0599
Ours	**0.0410**	**0.0576**	**0.0613**	**0.0498**	**0.0524**

Fig. 7. Qualitative results of point cloud completion on real scans in KITTI-360. The gray points are from raw scans. The purple points and the green points are partial inputs and completed results, respectively. (Color figure online)

Completeness of Sampled 3D Point Clouds. Following [32], we measure the completeness of the 3D point cloud sampled from Spotlights, which indicates how much of the object surface is covered by sample points. The metric again relies on the single-direction CD, however this time iterating through the ground truth point cloud finding the nearest point in the Spotlights representation.

For Spotlights, completeness is closely related to both the number of rays and the polar angle. Hence, we vary the number of rays from 2048 to 8192 and test 3 different polar angles under each setting. Note that the maximum polar angle in our case is about 83 degrees ($r = \frac{1}{4}R$). The average score on each shape category is reported in Table 3. The results coincide with the intuition that more rays usually produce more points on the surface, whereas polar angles that are too large or too small can lead to degraded point cloud quality since most of the rays will either scatter into the void or concentrate on small local regions.

Table 3. Completeness of point clouds sampled from Spotlights. We report the average distance from the dense ground truth to the sampled point cloud with varying numbers of rays and polar angles.

#Rays	Polar angle	Completeness						
		Bathtub	Bottle	Cabinet	Chair	Lamp	Sofa	Average
2048	30°	0.0474	0.0547	0.0678	0.0444	0.0426	0.0619	0.0506
	60°	0.0349	0.0380	0.0450	0.0339	0.0352	0.0447	0.0376
	83°	0.0382	0.0397	0.0457	0.0389	0.0407	0.0469	0.0415
4096	30°	0.0326	0.0367	0.0476	0.0302	0.0298	0.0450	0.0353
	60°	0.0260	0.0278	0.0349	0.0253	0.0263	0.0346	0.0285
	83°	0.0290	0.0297	0.0361	0.0293	0.0312	0.0367	0.0318
8192	30°	0.0233	0.0259	0.0345	0.0216	0.0212	0.0331	0.0255
	60°	**0.0197**	**0.0208**	**0.0279**	**0.0192**	**0.0200**	**0.0274**	**0.0220**
	83°	0.0222	0.0225	0.0294	0.0223	0.0244	0.0295	0.0248

Table 4. Hit ratio of rays with different polar angles. It evaluates how many rays actually hit the object.

Polar angle	Hit ratio (%)						
	Bathtub	Bottle	Cabinet	Chair	Lamp	Sofa	Average
30°	**93.8**	**81.6**	**93.9**	**79.1**	**47.2**	**85.0**	**78.6**
60°	63.4	45.2	67.0	49.1	31.5	57.1	50.9
83°	43.2	29.9	47.7	32.6	22.6	39.8	34.8

Hit Ratio of Virtual Rays. In most cases, objects cannot fill the entire space bounded by the outer sphere of Spotlights, which results in wasted rays that have no intersection with the object. To measure the utilization of rays, we compute the hit ratio defined as the number of surface intersecting rays over the total number of casted rays. As shown in Table 4, the hit ratio is inversely proportional to the polar angle and much lower for thin objects such as lamps. Although small polar angles waste fewer rays, we strongly suggest not to use those values as completeness also needs to be considered.

Sampling Resolution. As discussed in Sect. 3.2, we sample twice in Spotlights model to better control the ray directions. The total number of rays is then the multiplication of the number of primary and secondary sampling points. To evaluate the performance of Spotlights model with different sampling resolutions, we try several resolution combinations on the task of shape completion and report the accuracy as in Sect. 5.1 in Table 5. By comparing the first and the second row, we can find that changing the sampling resolutions has no big difference on performance as long as the total number of rays are the same. Also, the third row shows that adding more rays can increase the accuracy, which confirms our intuition.

Table 5. Accuracy of completed point clouds with different sampling resolutions on the synthetic dataset. For A*B, A and B are the number of primary and secondary sampling points, respectively. Accuracy measures the single direction CD from the prediction to ground truth.

Resolution	Accuracy			
	Hatchback	Sedan	SUV	Average
32*64	0.02124	0.02775	0.01850	0.02133
64*32	0.02219	0.02771	0.01907	0.02179
256*64	**0.02058**	**0.02700**	**0.01839**	**0.02098**

5.4 Limitations

Although Spotlights representation is compact compared to 3D coordinate-based point cloud, it still has several limitations.

First, the ray bundle needs to be carefully organized to reach the balance between efficiency and accuracy. Since the rays spread across the space bounded by the sphere, around 50% to 60% of the rays fail to actually hit the object surface, which is a big waste of model capacity and also results in the sparsity of sampled point clouds.

Second, adjacent rays do not assure intersections in a local surface, which makes it hard to learn depth variation along rays, especially for complex concave objects. To address these issues, in future extensions, we will try a differentiable variant of our Spotlights model to actively learn the orientation of each ray and dynamically adjust it based on the observations.

6 Conclusions

In this paper, we propose a novel compact point cloud representation denoted *Spotlights*, which is useful for applications demanding for high efficiency. In contrast to prior art, the structure in our representation is not imposed in an implicit way using complex network architectures, but in an explicit way through a handcrafted object surface sampling model, which is highly inspired by the spherical view arrangement in 3D object scanners. Also, no strong assumptions about the enclosed objects are being made, and a relatively homogeneous sampling distribution is achieved.

We demonstrate its potential in the context of shape completion where experiments on both synthetic and real dataset show that it can achieve competitive accuracy and consistency while being orders of magnitude faster. The fact that we directly predict depths along rays suggests that our representation has the ability to predict any shape of similar complexity. In future work, we will try to minimize the invalid rays in Spotlights model and figure out a better way to learn the depth variation along the ray.

References

1. Achlioptas, P., Diamanti, O., Mitliagkas, I., Guibas, L.J.: Learning representations and generative models for 3D point clouds. In: Proceedings of the IEEE International Conference on Machine Learning (ICML), pp. 40–49 (2018)
2. Campos, C., Elvira, R., Rodríguez, J.J.G., Montiel, J.M., Tardós, J.D.: ORB-SLAM3: an accurate open-source library for visual, visual-inertial and multi-map slam. IEEE Trans. Rob. (T-RO) **37**(6), 1874–1890 (2021)
3. Chang, A.X., et al.: Shapenet: an information-rich 3D model repository. arXiv preprint arXiv:1512.03012 (2015)
4. Chen, D.Y., Tian, X.P., Shen, Y.T., Ouhyoung, M.: On visual similarity based 3d model retrieval. In: Computer Graphics Forum, vol. 22, pp. 223–232. Wiley Online Library (2003)
5. Dai, A., Qi, C.R., Nießner, M.: Shape completion using 3D-encoder-predictor cnns and shape synthesis. In: Proceedings of the IEEE Conference on Computer Vision and Pattern Recognition (CVPR), pp. 5868–5877 (2017)
6. Davis, J., Marschner, S.R., Garr, M., Levoy, M.: Filling holes in complex surfaces using volumetric diffusion. In: In Proceedings of FIrst International Symposium on 3D Data Processing Visualization and Transmission (2002)
7. Deng, J., Shi, S., Li, P., Zhou, W., Zhang, Y., Li, H.: Voxel r-cnn: towards high performance voxel-based 3d object detection. arXiv preprint arXiv:2012.15712 1(2), 4 (2020)
8. Dosovitskiy, A., Ros, G., Codevilla, F., Lopez, A., Koltun, V.: CARLA: an open urban driving simulator. In: Proceedings of the 1st Annual Conference on Robot Learning, pp. 1–16 (2017)
9. Fan, H., Su, H., Guibas, L.J.: A point set generation network for 3D object reconstruction from a single image. In: Proceedings of the IEEE Conference on Computer Vision and Pattern Recognition (CVPR), pp. 605–613 (2017)
10. Furukawa, Y., Ponce, J.: Accurate, dense, and robust multi-view stereopsis. IEEE Trans. Pattern Anal. Mach. Intell. (PAMI) **32**, 1362–1376 (2010)
11. Garsthagen, R.: An open source, low-cost, multi camera full-body 3D scanner. In: Proceedings of 5th International Conference on 3D Body Scanning Technologies, pp. 174–183 (2014)
12. Groueix, T., Fisher, M., Kim, V., Russell, B., Aubry, M.: AtlasNet: a papier-mache approach to learning 3d surface generation. In: Proceedings of the IEEE Conference on Computer Vision and Pattern Recognition (CVPR) (2018)
13. Hartley, R., Zisserman, A.: Multiple View Geometry in Computer Vision, 2nd edn. Cambridge University Press, Cambridge (2003)
14. Huang, Z., Yu, Y., Xu, J., Ni, F., Le, X.: PF-Net: point fractal network for 3D point cloud completion. In: Proceedings of the IEEE Conference on Computer Vision and Pattern Recognition (CVPR), pp. 7662–7670 (2020)
15. Keinert, B., Innmann, M., Sänger, M., Stamminger, M.: Spherical fibonacci mapping. ACM Trans. Graph. (TOG) **34**(6), 1–7 (2015)
16. Lang, A.H., Vora, S., Caesar, H., Zhou, L., Yang, J., Beijbom, O.: Pointpillars: fast encoders for object detection from point clouds. In: Proceedings of the IEEE/CVF Conference on Computer Vision and Pattern Recognition, pp. 12697–12705 (2019)
17. Li, P., Qin, T., Shen, S.: Stereo vision-based semantic 3D object and ego-motion tracking for autonomous driving. In: Proceedings of the European Conference on Computer Vision (ECCV), pp. 646–661 (2018)
18. Li, R., Li, X., Hui, K.H., Fu, C.W.: Sp-gan: sphere-guided 3D shape generation and manipulation. ACM Trans. Graph. **40**(4) (2021)

19. Li, Y., Dai, A., Guibas, L., Nießner, M.: Database-assisted object retrieval for real-time 3d reconstruction. Comput. Graph. Forum **34**, 435–446 (2015)
20. Liao, Y., Xie, J., Geiger, A.: KITTI-360: a novel dataset and benchmarks for urban scene understanding in 2D and 3D. arXiv preprint arXiv:2109.13410 (2021)
21. McCormac, J., Clark, R., Bloesch, M., Davison, A., Leutenegger, S.: Fusion++: volumetric object-level slam. In: Proceedings of the International Conference on 3D Vision (3DV), pp. 32–41 (2018)
22. Newcombe, R., et al.: KinectFusion: real-time dense surface mapping and tracking. In: Proceedings of the International Symposium on Mixed and Augmented Reality (ISMAR) (2011)
23. Newcombe, R., Lovegrove, S., Davison, A.: DTAM: dense tracking and mapping in real-time. In: Proceedings of the International Conference on Computer Vision (ICCV), pp. 2320–2327 (2011)
24. Park, J.J., Florence, P., Straub, J., Newcombe, R., Lovegrove, S.: DeepSDF: learning continuous signed distance functions for shape representation. In: Proceedings of the IEEE Conference on Computer Vision and Pattern Recognition (CVPR) (2019)
25. Pauly, M., Mitra, N.J., Giesen, J., Gross, M.H., Guibas, L.J.: Example-based 3D scan completion (EPFL-CONF-149337), pp. 23–32 (2005)
26. Pesce, M., Galantucci, L., Percoco, G., Lavecchia, F.: A low-cost multi camera 3D scanning system for quality measurement of non-static subjects. Procedia CIRP **28**, 88–93 (2015)
27. Qi, C.R., Chen, X., Litany, O., Guibas, L.J.: Imvotenet: boosting 3D object detection in point clouds with image votes. In: Proceedings of the IEEE/CVF Conference on Computer Vision and Pattern Recognition, pp. 4404–4413 (2020)
28. Qi, C.R., Litany, O., He, K., Guibas, L.J.: Deep hough voting for 3D object detection in point clouds. In: Proceedings of the IEEE/CVF International Conference on Computer Vision, pp. 9277–9286 (2019)
29. Qi, C.R., Liu, W., Wu, C., Su, H., Guibas, L.J.: Frustum pointnets for 3D object detection from rgb-d data. In: Proceedings of the IEEE Conference on Computer Vision and Pattern Recognition, pp. 918–927 (2018)
30. Qi, C.R., Su, H., Mo, K., Guibas, L.J.: Pointnet: deep learning on point sets for 3D classification and segmentation. In: Proceedings of the IEEE Conference on Computer Vision and Pattern Recognition (CVPR), pp. 652–660 (2017)
31. Qi, C.R., Yi, L., Su, H., Guibas, L.J.: Pointnet++: deep hierarchical feature learning on point sets in a metric space. Adv. Neural Inf. Process. Syst. **30**, 1–10 (2017)
32. Seitz, S.M., Curless, B., Diebel, J., Scharstein, D., Szeliski, R.: A comparison and evaluation of multi-view stereo reconstruction algorithms. In: 2006 IEEE Computer Society Conference on Computer Vision and Pattern Recognition (CVPR 2006), vol. 1, pp. 519–528. IEEE (2006)
33. Shan, T., Englot, B., Meyers, D.: LIO-SAM: tightly-coupled lidar inertial odometry via smoothing and mapping. In: Proceedings of the IEEE/RSJ Conference on Intelligent Robots and Systems (IROS) (2020)
34. Shin, D., Fowlkes, C.C., Hoiem, D.: Pixels, voxels, and views: a study of shape representations for single view 3D object shape prediction. In: Proceedings of the IEEE Conference on Computer Vision and Pattern Recognition (CVPR) (2018)
35. Shin, D., Fowlkes, C.C., Hoiem, D.: Pixels, voxels, and views: a study of shape representations for single view 3D object shape prediction. In: Proceedings of the IEEE Conference on Computer Vision and Pattern Recognition, pp. 3061–3069 (2018)

36. Sucar, E., Wada, K., Davison, A.: NodeSLAM: neural object descriptors for multi-view shape reconstruction. In: Proceedings of the International Conference on 3D Vision (3DV) (2020)
37. Tchapmi, L.P., Kosaraju, V., Rezatofighi, H., Reid, I., Savarese, S.: Topnet: structural point cloud decoder. In: Proceedings of the IEEE Conference on Computer Vision and Pattern Recognition (CVPR), pp. 383–392 (2019)
38. Thrun, S., Wegbreit, B.: Shape from symmetry. In: Proceedings of the International Conference on Computer Vision (ICCV) (2005)
39. Wen, X., Li, T., Han, Z., Liu, Y.S.: Point cloud completion by skip-attention network with hierarchical folding. In: Proceedings of the IEEE Conference on Computer Vision and Pattern Recognition (CVPR) (2020)
40. Wen, X., et al.: PMP-Net: point cloud completion by learning multi-step point moving paths. In: Proceedings of the IEEE Conference on Computer Vision and Pattern Recognition (CVPR) (2021)
41. Wu, Z., et al.: 3D ShapeNets: a deep representation for volumetric shapes. In: Proceedings of the IEEE Conference on Computer Vision and Pattern Recognition (CVPR) (2015)
42. Xie, H., Yao, H., Zhou, S., Mao, J., Zhang, S., Sun, W.: GRNet: gridding residual network for dense point cloud completion. In: Proceedings of the European Conference on Computer Vision (ECCV) (2020)
43. Yang, Y., Feng, C., Shen, Y., Tian, D.: Foldingnet: point cloud auto-encoder via deep grid deformation. In: Proceedings of the IEEE Conference on Computer Vision and Pattern Recognition (CVPR), pp. 206–215 (2018)
44. Yuan, W., Khot, T., Held, D., Mertz, C., Hebert, M.: PCN: point completion network. In: Proceedings of the International Conference on 3D Vision (3DV), pp. 728–737. IEEE (2018)

OVPT: Optimal Viewset Pooling Transformer for 3D Object Recognition

Wenju Wang, Gang Chen(✉), Haoran Zhou, and Xiaolin Wang

University of Shanghai for Science and Technology, Shanghai 200093, China
203592861@st.usst.edu.cn

Abstract. The current methods for multi-view-based 3D object recognition have the problem of losing the correlation between views and rendering 3D objects with multi-view redundancy. This makes it difficult to improve recognition performance and unnecessarily increases the computational cost and running time of the network. Especially in the case of limited computing resources, the recognition performance is further affected. Our study developed an optimal viewset pooling transformer (OVPT) method for efficient and accurate 3D object recognition. The OVPT method constructs the optimal viewset based on information entropy to reduce the redundancy of the multi-view scheme. We used convolutional neural network (CNN) to extract the multi-view low-level local features of the optimal viewset. Embedding class token into the headers of multi-view low-level local features and splicing with position encoding generates local-view token sequences. This sequence was trained parallel with a pooling transformer to generate a local view information token sequence. At the same time, the global class token captured the global feature information of the local view token sequence. The two were aggregated next into a single compact 3D global feature descriptor. On two public benchmarks, ModelNet10 and ModelNet40, for each 3D object we only need a smaller number of optimal viewsets, achieving an overall recognition accuracy (OA) of 99.33% and 97.48%, respectively. Compared with other deep learning methods, our method still achieves state-of-the-art performance with limited computational resources. Our source code is available at https://github.com/shepherds001/OVPT.

1 Introduction

With the rapid development of 3D acquisition technology, 3D scanners, depth scanners, and 3D cameras have become popular and inexpensive. The acquisition of 3D data such as point clouds and meshes has become more convenient and accurate [1]. These factors have promoted the widespread application of 3D data-based object recognition techniques in the fields of environment perception for autonomous driving [2], object recognition for robots [3], and scene understanding for augmented reality [4]. Therefore, 3D object recognition has become a hotspot for current research (Fig. 1).

Supplementary Information The online version contains supplementary material available at https://doi.org/10.1007/978-3-031-26319-4_29.

L. Wang et al. (Eds.): ACCV 2022, LNCS 13841, pp. 486–503, 2023.
https://doi.org/10.1007/978-3-031-26319-4_29

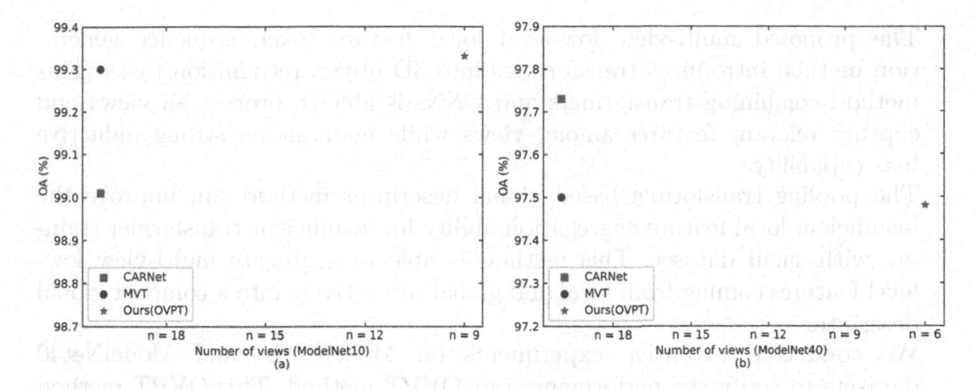

Fig. 1. Comparison with the state-of-the-art approaches in terms of recognition performance, and the number of views required per 3D object.

Deep learning-based methods have become mainstream research techniques for 3D object recognition tasks. In general, these methods can be divided according to the type of data input to the deep neural network: voxel-based methods [5–12], point-cloud-based methods [13–20], and multi-view-based methods [21–30].

Multi-view-based methods render 3D data objects into multiple 2D views, so they no longer need to rely on complex 3D features. Instead, the rendered multi-views are fed into multiple well-established image classification networks to extract multi-view low-level local features. Finally, the multi-view low-level local features are aggregated into global descriptors to complete the 3D object recognition task. Especially when 3D objects are occluded, such methods can complement each other's detailed features of 3D objects according to views from different perspectives. Compared with voxel-based and point cloud-based methods, the multi-view-based method achieves the best 3D object recognition performance.

However, this type of method still has the shortcomings that feature information cannot be extracted for all views simultaneously during training and that related feature information between multiple views cannot be efficiently captured. In addition, there is redundancy when rendering 3D objects into multiple views. The relevant feature information between multiple views is indispensable for aggregating the multi-view local features into a compact global descriptor. The omission of these relevant features is why it is difficult to improve the recognition accuracy of this type of method. The view redundancy problem increases unnecessary network running time and affects final recognition accuracy. This paper researches this and proposes the optimal viewset pooling transformer (OVPT) method. Our main contributions are summarized as follows:

- The method to construct the optimal viewset based on information entropy solves the view redundancy problem of the multi-viewpoint rendering method [24], which reduces computational cost of the network.

- The proposed multi-view low-level local feature token sequence generation method introduces transformers into 3D object recognition tasks. This method combining transformers and CNNs is able to process all views and capture relevant features among views while maintaining strong inductive bias capability.
- The pooling transformer-based global descriptor method can improve the insufficient local feature aggregation ability for insufficient transformer training with small dataset. This method is able to aggregate multi-view low-level features coming from local and global respectively into a compact global descriptor.
- We conducted extensive experiments on ModelNet10 and ModelNet40 datasets to verify the performance our OVPT method. This OVPT method can achieve respectively 99.33% and 97.48% of overall recognition accuracy (OA) in two datasets, only requiring a smaller number of optimal viewsets. Compared with other state-of-the-art methods, our OVPT network achieves the best performance.

2 Related Work

Voxel-Based Methods. VoxNet [5] uses 3D CNN to extract voxelized 3D object features and processes non-overlapping voxels through max pooling. However, it cannot compactly represent the structure of 3D objects. Therefore, Kd-network [9] was proposed. It creates a structural graph of 3D objects based on a Kd-tree structure, and computes a sequence of hierarchical representations in a feed-forward bottom-up fashion. Its network structure occupies less memory and is more computationally efficient, but loses information about local geometry. These voxel-based methods solve the problems of large memory footprint and long training time of point cloud voxelization, but still suffer from the problems of lost information and high computational cost.

Point Cloud-Based Methods. Point cloud voxelization inevitably loses information that may be essential. Some methods consider processing point clouds directly. PointNet [13] was the earliest method to process point clouds directly. It uses T-Net to perform an affine transformation on the input point matrix, and extract each point feature through a multi-layer perceptron. But it could not capture the local neighborhood information between points. Thus, PointNet++ [14] was developed, which constructs local neighborhood subsets by introducing a hierarchical neural network, and then extracted local neighborhood features based on PointNet. PointNet++ solved the local neighborhood information extraction of PointNet to a certain extent. Dynamic Graph CNN (DGCNN) [19] uses EdgeConv to build a dynamic graph convolutional neural network for object recognition. EdgeConv could extract local domain feature information, and the local shape features of the extracted point cloud could keep the arrangement invariance. Although the point cloud-based method can directly process the point cloud to reduce the loss of information, its network is often complex, the training time is long, and the final recognition accuracy is not high enough.

Multi-view-Based Methods. Multi-view-based approaches render 3D data objects into multiple 2D views, so they no longer need to rely on complex 3D features. This class of methods achieves the best 3D object recognition performance. Multi-view Convolutional Neural Networks (MVCNN) [21] employs a 2D CNN network to process rendered multiple views individually. MVCNN then uses view pooling to combine information from multiple views into a compact shape descriptor. However, it lost the position information of the views when pooling multiple views. Multi-view convolutional neural network (GVCNN) [23] mines intra-group similarity and inter-group distinguishability between views by grouping multi-view features to enhance the capture of location information. Hierarchical Multi-View Context Modeling (HMVCM) [30] adopted adaptive calculation of feature weights to aggregate features into compact 3D object descriptors. This type of hierarchical multi-view context modeling used a module that combined a CNN and a Bidirectional Long Short-Term Memory (Bi-LSTM) network to learn the visual context features of a single view and its neighborhood. This network had an overall recognition accuracy (OA) on the ModelNet40 dataset of 94.57%. However, it did not consider the local features of all views in parallel during training, and lost relevant information between views, so the aggregated global descriptors were not sufficiently compact. Its 3D object recognition accuracy has room for improvement.

3 Methods

The OVPT network proposed in this paper is shown in Fig. 2. OVPT has three parts: (a) Optimal viewset construction based on information entropy; (b) Multi-view low-level local feature token sequence generation; (c) Global descriptor generation based on the pooling transformer.

3.1 Optimal Viewset Construction Based on Information Entropy

The input stage of the OVPT network renders 3D objects (represented by point cloud or meshes) into multiple 2D views. This study selected a mesh representation with higher accuracy for 3D object recognition. Of course, 3D objects in the form of point clouds can also be reconstructed into mesh forms [31–33]. The specific process is as follows:

Multi-view Acquisition. For a 3D object O, different 2D rendering views $V = \{v_1, ...v_i..., v_N\}$ can be obtained by setting the camera in different positions, where v_i represents the view taken from the i-th viewpoint, $v_i \in \mathbb{R}^{C\times H\times W}$.

$$[v_1, ...v_i..., v_N] = Render(O) \tag{1}$$

We use the dodecahedron camera viewpoint setting [24]. It places the 3D object in the center of the dodecahedron then sets the camera viewpoint at the

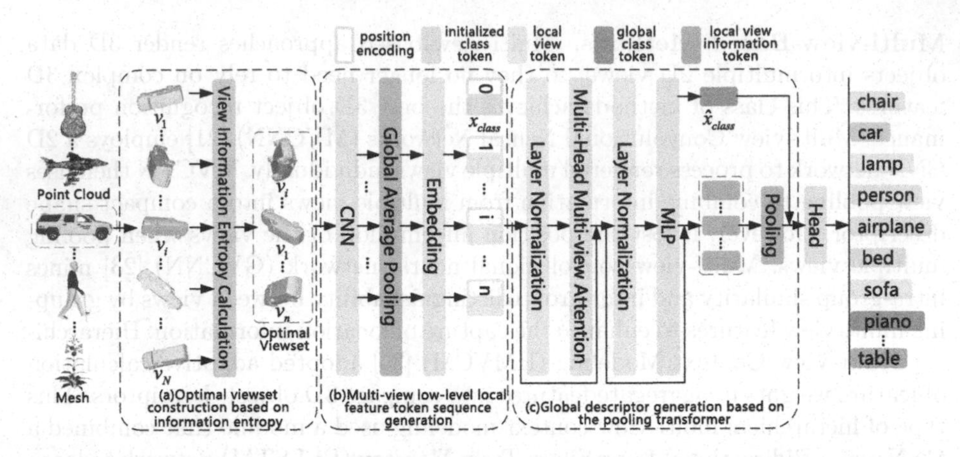

Fig. 2. The architecture of the proposed optimal viewset pooling transformer (OVPT).

vertices of the dodecahedron. This setup evenly distributes the camera viewpoints in 3D space, capturing as much global spatial information as possible and reducing information loss. However, among the 20 views rendered in this way, there are always duplicates. This may lead to redundant features being extracted by the deep neural network and increase the running time of the network, eventually decreasing the accuracy of the recognition.

Optimal Viewset Construction. Information entropy can highlight the grayscale information for the pixel position in the view and the comprehensive characteristics of the grayscale distribution in the pixel neighborhood. Assuming a given amount of information contained in the view, information entropy can be an effective means to evaluate view quality. Aiming at the problem of repetitive in the viewpoint settings of the dodecahedron camera, we introduce the information entropy [34] of 2D views as an evaluation criterion to construct the optimal viewset to reduce redundant views. Different from the previous viewpoint selection methods based on information entropy [35,36], our method does not require human intervention and only considers the quality of the view itself without calculating the projected area of the viewpoint, which is more reliable. Dodecahedron camera viewpoint settings and optimal viewset are shown in Fig. 3. The optimal viewset is constructed as follows:

(1) Information entropy calculation of N views ($N = 20$): H_i represents the information entropy of the i-th view v_i. The specific calculation is shown in formulas (2) and (3):

$$P_{a,b} = f(a,b)/W \cdot H \tag{2}$$

$$H_i = -\sum_{a=0}^{255} P_{a,b} log P_{a,b} \tag{3}$$

where (a, b) is a binary group, a represents the gray value of the center in a sliding window, and b is the average gray value of the center pixel in the window. $P_{a,b}$

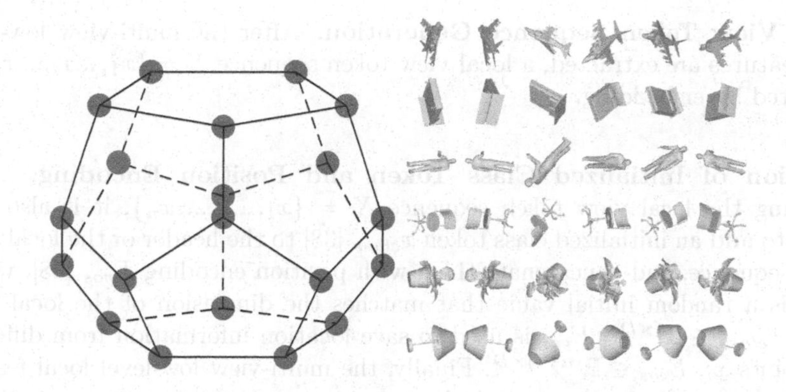

Fig. 3. Dodecahedron camera viewpoint settings [24] and optimal viewset.

is the probability that (a, b) appears in the full view v_i. $f(a, b)$ is the number of times the binary group (a, b) appears in the full view v_i. W and H represent the width and height of the view.

(2) Rank the information entropy values $H_i (i = 1, ..., N, N = 20)$. (3) Construct the optimal viewset: The views with the best information entropy ranking n ($n < N, n = 6, 9$ in this paper) are regarded as the optimal viewset. When $n = 1$ in the optimal viewset, the single view with the highest information entropy value is selected, which is called the optimal view.

3.2 Multi-view Low-level Local Feature Token Sequence Generation

Because the transformer was first proposed for natural language processing, its input requirements are two-dimensional matrix sequences. For the optimal viewset $V = \{v_1, ...v_i..., v_n\}$, $v_i \in \mathbb{R}^{C \times H \times W}$, where n is the number of views, C is the number of channels, H is the height of the image, and W is the width of the image. Therefore, the obtained view cannot be directly input to the transformer, and it needs to be flattened into a local view token sequence $X = \{x_1, ...x_i..., x_n\}$. x_i represents the local view token generated by the i-th view, $x_i \in \mathbb{R}^{1 \times D}$, where D is the constant latent vector size used in all transformer layers. To this end, this paper proposes a multi-view low-level local feature token sequence generation method. The specific process is as follows:

Low-level Local Feature Extraction. We used multiple CNNs pretrained on ImageNet [37] to extract low-level local features of multi-view $V = \{v_1, ...v_i, ...v_n\}$. Any well-established 2D image classification network can be used as a multi-view low-level feature extractor.

Local View Token Sequence Generation. After the multi-view low-level local features are extracted, a local view token sequence $X = \{x_1, ...x_i..., x_n\}$ is generated by embedding.

Addition of Initialized Class Token and Position Encoding. After obtaining the local view token sequence $X = \{x_1, ...x_i..., x_n\}$, it is also necessary to add an initialized class token x_{class} [38] to the header of the local view token sequence, and concatenate them with position encoding E_{pos} [38]. where x_{class} is a random initial value that matches the dimension of the local view token, $x_{class} \in \mathbb{R}^{1\times D}$, E_{pos} is used to save location information from different viewpoints x_i, $E_{pos} \in \mathbb{R}^{(n+1)\times D}$. Finally, the multi-view low-level local feature token sequence X_0 can be generated.

$$X_0 = [x_{class}; x_1, ...x_i..., x_n] \oplus E_{pos}, X_0 \in \mathbb{R}^{(n+1)\times D} \tag{4}$$

3.3 Global Descriptor Generation Based on the Pooling Transformer

The method uses a pooling transformer to aggregate X_0 into one compact 3D global descriptor in two steps:

Global Feature Information Generation Based on Transformer. The generation of the global feature information has three steps: layer normalization [39] processing; multi-head multi-view attention calculation; and residual connection and the use of multi-layer perceptron:

(1) Layer Normalization Processing: X_0 is input to the pooling transformer as a sequence of multi-view low-level local feature tokens. Before the calculating Multi-Head Multi-View Attention (MHMVA), X_0 undergoes Layer Normalization (LN), see formula (5):

$$\hat{X}_0 = LN(X_0), \hat{X}_0 \in \mathbb{R}^{(n+1)\times D} \tag{5}$$

(2) Multi-Head Multi-View Attention Calculation: We use the normalized $\hat{X}_0$ to generate $Query$, Key, and $Value$ through linear transformations. MHMVA performs multiple parallel Multi-View Attention (MVA). The inputs q_i, k_i and v_i of each MVA can be obtained by equally dividing the $Query$, Key and $Value$ vectors, where $q_i \in \mathbb{R}^{(n+1)\times \frac{D_Q}{N}}$, $k_i \in \mathbb{R}^{(n+1)\times \frac{D_K}{N}}$, $v_i \in \mathbb{R}^{(n+1)\times \frac{D_V}{N}}$, $D_Q = D_K = D_V$ represents the vector dimension of Query, Key and Value respectively. We obtained multiple MVA according to the number of heads N, and multiple subspaces can be formed. Therefore, MHMVA can pay to take the information of various parts of the input features into account. The calculation result X_i^{MVA} of each MVA is obtained from formula (6):

$$X_i^{MVA} = Softmax(\frac{q_i k_i^T}{\sqrt{D_K/N}})v_i, X_i^{MVA} \in \mathbb{R}^{(n+1)\times \frac{D_K}{N}} \tag{6}$$

Concat is performed on each X_i^{MVA} after calculation, and the MHMVA calculation is finally completed after a linear transformation, as shown in formula (7):

$$X_{MHMVA} = h_{\Theta}(\sum_{i=1}^{N} X_i^{MVA}), X_{MHMVA} \in \mathbb{R}^{(n+1)\times D} \tag{7}$$

where h_{Θ} represents a linear function with dropout.

(3) Residual Connection and the Use of Multi-Layer Perceptrons: The X_{MHMVA} obtained after the MHMVA calculation uses residual connections [40] to avoid vanishing gradients.

$$X_1 = X_{MHMVA} + X_0, X_1 \in \mathbb{R}^{(n+1)\times D} \tag{8}$$

After X_1 is obtained, it is also processed by layer normalization and input to a multi-layer perceptron (MLP). Because MHMVA does not fit the complex process sufficiently, MLP is added after them to enhance the model's generalization. The MLP consists of linear layers that use the GELU [41] activation function shown in formula (9):

$$MLP(x) = GELU(W_1 x + b_1)W_2 + b_2 \tag{9}$$

where W_1 and b_1 are the weights of the first fully connected layer, W_2 and b_2 are the weights of the second fully connected layer, and x represents the input feature information.

There is also a residual connection between the output of MLP and X_1, and the calculation is given by formula (10):

$$\hat{X_1} = MLP(LN(X_1)) + X_1 \tag{10}$$

The final $\hat{X_1}$ is the output of global feature information generation based on the transformer method, where $\hat{X_1} \in \mathbb{R}^{(n+1)\times D}$. It consists of a global class token $\hat{x}_{class}$ and the local view information token sequence $\{\hat{x}_1, ...\hat{x}_i..., \hat{x}_n\}$, where $\hat{x}_{class} \in \mathbb{R}^{1\times D}, \hat{x}_i \in \mathbb{R}^{1\times D}$. The global class token $\hat{x}_{class}$ stores the global feature information of the local view token sequence.

Local View Information Token Sequence Aggregation Based on Pooling. After parallel transformer training, the global class token $\hat{x}_{class}$ saves the global feature information of the local view token sequence, but the single best local view information token may be lost. It is very efficient to aggregate this part of the information into a 3D global descriptor. Our local view information token sequence aggregation based on the pooling method can solve this problem. It can simultaneously capture the single best local view information token while preserving the global feature information of the local view token sequence. This method pools the local view information token sequence $\{\hat{x}_1, ...\hat{x}_i..., \hat{x}_n\}$ to obtain the best local view information token then splices the best local view information token and the global class token $\hat{x}_{class}$. After these processes, we can aggregate multi-view low-level local feature token sequences locally and globally, then generate a more compact 3D global descriptor $Y, Y \in \mathbb{R}^{1\times D}$. The 3D global descriptor is input to the head layer to complete the object recognition task.

$$Y = max[\hat{x}_1, ...\hat{x}_i..., \hat{x}_n] + \hat{x}_{class} \quad (11)$$

4 Experimental Results and Discussion

4.1 Dataset

ModelNet [6] is a widely used 3D object recognition dataset, popular for its diverse categories, clear shapes, and well-built advantages. The two benchmark datasets ModelNet40 and ModelNet10 are its subsets. Among them, ModelNet40 consists of 40 categories (such as airplanes, cars, plants, and lights), with 12,311 CAD models, including 9,843 training samples and 2,468 test samples. ModelNet10 consists of 10 categories, with 4,899 CAD models, including 3,991 training samples and 908 test samples.

4.2 Implementation Details

We conducted extensive comparative experiments using PyCharm on a computer with the Windows10 operating system. The relevant configuration of this computer is as follows: (1) Central Processing Unit (CPU) was an Intel(R) Xeon CPU @2.80 GHz, (2) Graphic Processing Unit (GPU) was RTX2080, (3) Random Access Memory (RAM) was 64.0 GB, and (4) Pytorch 1.6 was used. For all our experiments, the learning rate was initialized to 0.0001, the epoch was set to 20, the batch size is set to 8 by default. On ModelNet10 and ModelNet40, the CNNs are resnet34 [40] and densenet121 [42], respectively. We used the Adam [43] algorithm to optimize the network structure based on the learning rate decay and the L2 regularization weight decay strategy to avoid overfitting in our network.

4.3 The Influence of the Number of Views

When 3D objects are rendered into multiple 2D views, the number of views has different effects on the object recognition performance of the network. We selected eight different view numbers to quantitatively analyze the recognition accuracy of the OVPT method on ModelNet40 (including 1, 3, 6, 9, 12 and 15 views selected using the optimal viewset construction method based on information entropy, and the batch size is uniformly set to 8).

98.0
97.5
97.0
96.5
96.0
95.5
95.0
94.5
94.0
Accuracy (%)
OA
AA
97.48
97.28
97.19
97.03
96.92
96.74
96.16
95.82
95.91
95.9
95.66
94.3
n=1
n=3
n=6
n=9
n=12
n=15
Number of views (ModelNet40)

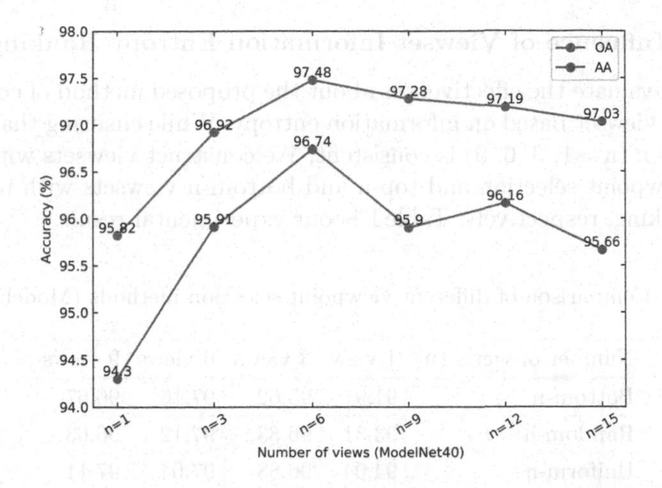

Fig. 4. Recognition performance with different numbers of views.

The object recognition performance of the OVPT method with different number of views is shown in Fig. 4. We can find that more views are not necessarily conducive to 3D object recognition. That is, the object recognition performance can't always improve with the increase of the number of views. On the ModelNet40 dataset, when the number of views $n <= 6$, the recognition accuracy is proportional to the number of views. When $6 < n <= 15$, the recognition performance generally appears a downward trend with the increase of the number of views. This experimental result verifies our proposal: the 20 views of the 3D object obtained by the dodecahedron viewpoint method have redundancy, that is, the features of these views have repeated parts. Redundant view features will lead to the degradation of the object recognition performance of the network.

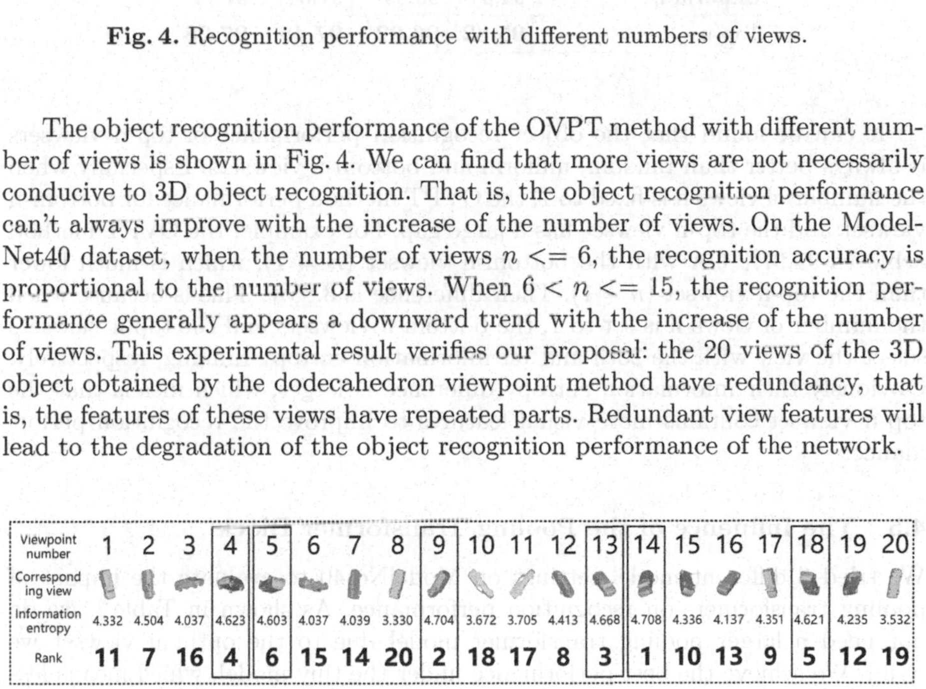

Viewpoint number
1 2 3 4 5 6 7 8 9 10 11 12 13 14 15 16 17 18 19 20
Corresponding view
Information entropy
4.332 4.504 4.037 4.623 4.603 4.037 4.039 3.330 4.704 3.672 3.705 4.413 4.668 4.708 4.336 4.137 4.351 4.621 4.235 3.532
Rank
11 7 16 4 6 15 14 20 2 18 17 8 3 1 10 13 9 5 12 19

Fig. 5. Optimal viewset construction method based on information entropy.

The proposed optimal viewset construction method solves this problem (see Fig. 5). On the ModelNet40 dataset, selecting the top six views according to the information entropy value as the best viewset, the OVPT method achieves the best OA of 97.48% and AA of 96.74%. Compared with 97.03% OA and 95.66% AA obtained by 15 views, this result is improved by 0.45% and 1.08%.

4.4 The Influence of Viewset Information Entropy Ranking

We further evaluate the effectiveness about the proposed method of constructing the optimal viewset based on information entropy. While ensuring that the number of views n (n = 1, 3, 6, 9) is consistent, We construct viewsets with random, uniform viewpoint selection and top-n and bottom-n viewsets with information entropy ranking, respectively. Table 1 is our experimental result.

Table 1. Comparison of different viewpoint selection methods (ModelNet40).

Number of views (n)	1 view	3 views	6 views	9 views
Bottom-n	91.97	95.62	97.16	96.67
Random-n	93.31	96.83	97.12	96.63
Uniform-n	94.04	96.88	97.04	97.11
Top-n	**95.82**	**96.92**	**97.48**	**97.28**

It can be found that the object recognition performance of top-n viewsets is always better than random, uniform and bottom-n viewsets. Especially when the number of views n is fixed to 1, the OVPT method performance for bottom-n viewset and the top-n viewset has a large gap. For example, the OVPT method achieves 91.97% OA with the bottom-n viewset ($n = 1$), which is much lower than the top-n viewset ($n = 1$). Their difference is 3.85%. This is because when the number of views n is set to 1, the bottom n viewsets and the top n viewsets select the view with the 20th and 1st information entropy ranking, respectively. Obviously their information entropy difference is larger, which means that the top-n viewset contains more visual features to improve the recognition performance.

4.5 The Influence of the Pooling Transformer Block

We tried 3 different model settings on ModelNet40 to evaluate the impact of pooling transformers on recognition performance. As shown in Table 2, we do not need a larger pooling transformer model due to the optimal viewset we build. We achieve the best performance under the tiny model, which means less computational cost. The size of the tiny model is only 29.7MB, which is more lightweight than other model settings.

Table 2. The influence of the pooling transformer block.

Model	Hidden size	Heads	OA (%)	AA (%)	Model size
Tiny	192	3	**97.48**	**96.74**	**29.7 MB**
Small	384	6	97.20	96.16	35.6 MB
Base	768	12	97.28	96.13	57.4 MB

Table 3. Ablation study.

Optimal viewset	Transformer	Pooling transformer	OA (%)	AA (%)
	✓		98.45	98.26
		✓	98.78	98.66
✓		✓	**99.33**	**99.21**

4.6 Ablation Study

We performed an ablation study on the OVPT network on ModelNet10. The results experiment are shown in Table 3. It can be found that the best recognition performance (99.33% for OA and 99.21% for AA) is achieved with our optimal viewset and pooling transformer method. The main reason is that the optimal viewset solves the redundancy problem of current viewpoint rendering methods. At the same time, the pooling transformer method solves the problem of insufficient local feature aggregation ability of the transformer, which can obtain the feature information of all local view token sequences from local and global aggregation respectively.

4.7 Model Complexity

The results of our experiments comparing model complexity with other methods are shown in Table 4. It can be found that our OVPT method outperforms these enumerated methods in both time and space complexity. In terms of space complexity, the model size of our OVPT method is 29.7MB, and the model sizes of MVCNN and CARNet are 44.8MB and 45.7MB, respectively. It can be seen that the model size of OVPT is reduced by 35% compared to the previous SOTA method CARNet. Even though CARNet uses ResNet18 as the multi-view feature extractor, we use DenseNet121 which is more complex than ResNet18. This is because in the subsequent part of OVPT, which requires only one Tiny Pooling Transformer block (see Fig. 2) to aggregate multi-view features, CARNet also contains more smaller hand-designed components. In terms of time complexity, the running time of our OVPT method is 0.10 s, and the running times of MVCNN and CARNet are 0.12 s and 0.20 s, respectively. Obviously, the running time of our OVPT method is the shortest among these methods, and only needs half of the running time of the previous SOTA method CARNet.

Table 4. Model size and running time comparison (ModelNet40).

Model	Model size	Running time	Relative time cost
MVCNN [21]	44.8 MB	0.12 s	0.6×
CARNet [44]	45.7 MB	0.20 s	1×
OVPT (Ours)	**29.7 MB**	**0.10 s**	**0.5×**

4.8 Visual Analysis of Confusion Matrix

We use confusion matrix to visually analyze the recognition performance of the OVPT method on the ModelNet40 dataset. It can help us understand which categories are easier or harder to identify. The values on the diagonal of the confusion matrix represent the number of correct identifications, and the values outside the diagonal indicate the number of incorrect identifications. When the value outside the diagonal of the category is lower, it means that the OVPT method is more accurate in identifying the category. As shown in Fig. 6, OVPT achieves 100% recognition accuracy on categories such as airplane, bed, and car. Of course, there are also some mis-judgments like cup, night_stand and vase. For the vase object containing 100 samples, 4 samples were misjudged as lamps. We attribute this to the fact that some objects have similar visual features, leading to in model recognition errors. Notably, OVPT has excellent recognition performance in most categories with complex visual features.

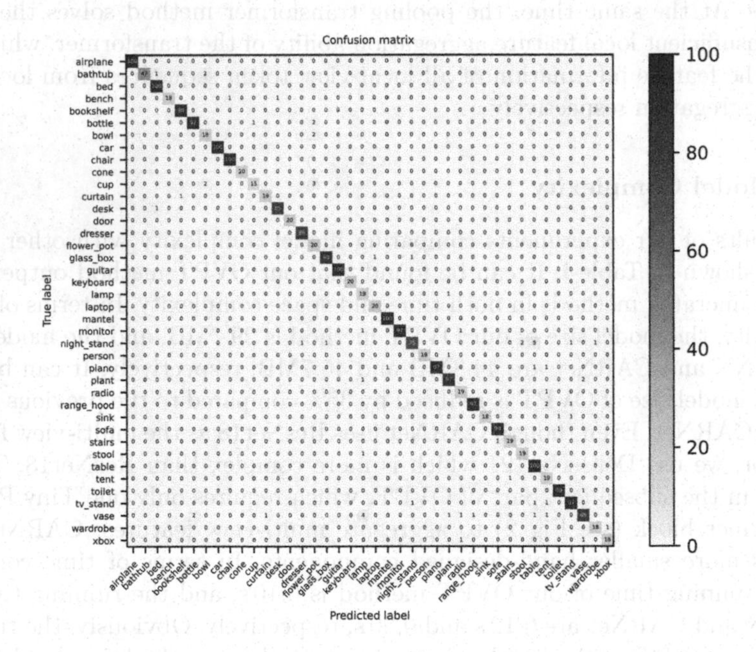

Fig. 6. Confusion matrix.

4.9 Comparsions with State-of-the-Art Methods

As seen in the Table 5, the OVPT method achieves the state-of-the-art recognition accuracy on the ModelNet10 dataset with OA of 99.33% and AA of 99.21%. Compared with the CARNet [44] method, OVPT improves OA by 0.32%. However, we only need nine views of each 3D object to accomplish the object recognition task. Compared with other multi-view-based methods, the view number is also minimal, which helps reduce computational cost and running time. We can also find that our OVPT method outperforms other point cloud-based and voxel-based methods.

On ModelNet40, the Transformer architecture of MVT [45] follows DeiT's setting [46] to solve the problem of insufficient local feature aggregation capability of the transformer on smaller-scale datasets, and requires stacking multiple local-global Transformer blocks (12 blocks in total) for 3D objects identify. When 20 views of each 3D object are input, the OA reaches 97.50%. Our OVPT method also can achieve close to its recognition accuracy (OA reaches 97.48%). However, one our proposed pooling transformer block and 6 views of each 3D object are only employed, which are much less than MVT. CARNet [44] combines the dodecahedron view rendering method and the knn search method to exploit the latent correspondence of views and viewpoints to aggregate shape features in communication. In contrast, our OVPT method is simpler and easier to use.

Table 5. Recognition performance comparison with present state-of-the-art methods.

Methods	Input modality	ModelNet40		ModelNet10	
		OA (%)	AA (%)	OA (%)	AA (%)
3D ShapeNets [6]	Voxel	–	77.32	–	83.54
VoxNet [5]	Voxel	–	83.00	–	92.00
O-CNN [8]	Voxel	90.60	–	–	–
PointNet [13]	Point cloud	89.20	86.20	–	–
PointNet++ [14]	Point cloud	91.90	–	–	–
DGCNN [19]	Point cloud	93.50	90.70	–	–
MVCNN [21]	12 views	92.10	89.90	–	–
HMVCM [30]	12 views	94.57	–	95.70	–
MHBN [47]	6 views	94.70	93.10	95.00	95.00
RotationNet [24]	20 views	97.37	95.84	98.46	95.99
MVT [45]	20 Views	97.50	–	99.30	–
CARNet [44]	20 views	**97.73**	–	99.01	–
OVPT (Ours)	**1 view**	95.82	94.30	98.45	97.98
OVPT (Ours)	6 views	97.48	**96.74**	98.89	98.88
OVPT (Ours)	9 views	97.28	95.90	**99.33**	**99.21**

5 Conclusion

This paper proposes an OVPT network for efficient, accurate 3D object recognition tasks. Compared with other deep learning methods, OVPT introduces information entropy to solve the problem of redundancy when rendering 3D objects into multiple views. The pooling transformer can efficiently capture the relevant feature information between multiple views and realize the global and local aggregation of multi-view low-level local feature token sequences into compact 3D global descriptors. We conducted a series of experiments on two popular ModelNet datasets, and the results show that OVPT achieves state-of-the-art performance while using the least number of views. Our method significantly improves the accuracy and efficiency of 3D object recognition tasks and reduces the computational cost, and is especially suitable for computationally resource-constrained environments. This method can be widely used in areas such as autonomous driving, augmented reality, interior design, and robotics.

Acknowledgments. The fnancial support for this work was sponsored by Natural Science Foundation of Shanghai under Grant No. 19ZR1435900.

References

1. Liang, Z., et al.: Stereo matching using multi-level cost volume and multi-scale feature constancy. IEEE Trans. Pattern Anal. Mach. Intell. **43**(1), 300–315 (2021). https://doi.org/10.1109/TPAMI.2019.2928550
2. Li, Y., et al.: Deep learning for lidar point clouds in autonomous driving: a review. IEEE Trans. Neural Netw. Learn. Syst. **32**(8), 3412–3432 (2021). https://doi.org/10.1109/TNNLS.2020.3015992
3. Kästner, L., Frasineanu, V.C., Lambrecht, J.: A 3D-deep-learning-based augmented reality calibration method for robotic environments using depth sensor data. In: 2020 IEEE International Conference on Robotics and Automation (ICRA), pp. 1135–1141 (2020). https://doi.org/10.1109/ICRA40945.2020.9197155
4. Lee, D., et al.: Large-scale localization datasets in crowded indoor spaces. In: 2021 IEEE/CVF Conference on Computer Vision and Pattern Recognition (CVPR), pp. 3226–3235 (2021). https://doi.org/10.1109/CVPR46437.2021.00324
5. Maturana, D., Scherer, S.: Voxnet: a 3D convolutional neural network for real-time object recognition. In: 2015 IEEE/RSJ International Conference on Intelligent Robots and Systems (IROS), pp. 922–928 (2015). https://doi.org/10.1109/IROS.2015.7353481
6. Wu, Z., et al.: 3D shapenets: a deep representation for volumetric shapes. In: 2015 IEEE Conference on Computer Vision and Pattern Recognition (CVPR), pp. 1912–1920 (2015). https://doi.org/10.1109/CVPR.2015.7298801
7. Riegler, G., Ulusoy, A.O., Geiger, A.: Octnet: learning deep 3D representations at high resolutions. In: 2017 IEEE Conference on Computer Vision and Pattern Recognition (CVPR), pp. 6620–6629 (2017). https://doi.org/10.1109/CVPR.2017.701
8. Wang, P.S., Liu, Y., Guo, Y.X., Sun, C.Y., Tong, X.: O-cnn: octree-based convolutional neural networks for 3d shape analysis. ACM Trans. Graph. **36**(4) (2017). https://doi.org/10.1145/3072959.3073608
9. Klokov, R., Lempitsky, V.: Escape from cells: Deep kd-networks for the recognition of 3D point cloud models. In: 2017 IEEE International Conference on Computer Vision (ICCV), pp. 863–872 (2017). https://doi.org/10.1109/ICCV.2017.99
10. Zeng, W., Gevers, T.: 3DContextNet: K-d tree guided hierarchical learning of point clouds using local and global contextual cues. In: Leal-Taixé, L., Roth, S. (eds.) ECCV 2018. LNCS, vol. 11131, pp. 314–330. Springer, Cham (2019). https://doi.org/10.1007/978-3-030-11015-4_24

11. Le, T., Duan, Y.: Pointgrid: a deep network for 3D shape understanding. In: 2018 IEEE/CVF Conference on Computer Vision and Pattern Recognition, pp. 9204–9214 (2018). https://doi.org/10.1109/CVPR.2018.00959
12. Meng, H.Y., Gao, L., Lai, Y.K., Manocha, D.: Vv-net: voxel vae net with group convolutions for point cloud segmentation. In: 2019 IEEE/CVF International Conference on Computer Vision (ICCV), pp. 8499–8507 (2019). https://doi.org/10.1109/ICCV.2019.00859
13. Charles, R.Q., Su, H., Kaichun, M., Guibas, L.J.: Pointnet: Deep learning on point sets for 3D classification and segmentation. In: 2017 IEEE Conference on Computer Vision and Pattern Recognition (CVPR), pp. 77–85 (2017). https://doi.org/10.1109/CVPR.2017.16
14. Qi, C.R., Yi, L., Su, H., Guibas, L.J.: Pointnet++: deep hierarchical feature learning on point sets in a metric space. In: Guyon, I., et al. (eds.) Advances in Neural Information Processing Systems, vol. 30. Curran Associates, Inc. (2017). https://proceedings.neurips.cc/paper/2017/file/d8bf84be3800d12f74d8b05e9b89836f-Paper.pdf
15. Jiang, M., Wu, Y., Zhao, T., Zhao, Z., Lu, C.: Pointsift: a sift-like network module for 3D point cloud semantic segmentation (2018). https://doi.org/10.48550/ARXIV.1807.00652
16. Zhao, H., Jiang, L., Fu, C.W., Jia, J.: Pointweb: enhancing local neighborhood features for point cloud processing. In: 2019 IEEE/CVF Conference on Computer Vision and Pattern Recognition (CVPR), pp. 5560–5568 (2019). https://doi.org/10.1109/CVPR.2019.00571
17. Fu, J., Liu, J., Tian, H., Li, Y., Bao, Y., Fang, Z., Lu, H.: Dual attention network for scene segmentation. In: 2019 IEEE/CVF Conference on Computer Vision and Pattern Recognition (CVPR), pp. 3141–3149 (2019). https://doi.org/10.1109/CVPR.2019.00326
18. Feng, M., Zhang, L., Lin, X., Gilani, S.Z., Mian, A.: Point attention network for semantic segmentation of 3D point clouds. Pattern Recogn. **107**, 107446 (2020). https://doi.org/10.1016/j.patcog.2020.107446, https://www.sciencedirect.com/science/article/pii/S0031320320302491
19. Wang, Y., Sun, Y., Liu, Z., Sarma, S.E., Bronstein, M.M., Solomon, J.M.: Dynamic graph cnn for learning on point clouds. ACM Trans. Graph. **38**(5) (2019). https://doi.org/10.1145/3326362
20. Zhang, K., Hao, M., Wang, J., de Silva, C.W., Fu, C.: Linked dynamic graph cnn: learning on point cloud via linking hierarchical features (2019). https://doi.org/10.48550/ARXIV.1904.10014, https://arxiv.org/abs/1904.10014
21. Su, H., Maji, S., Kalogerakis, E., Learned-Miller, E.: Multi-view convolutional neural networks for 3D shape recognition. In: 2015 IEEE International Conference on Computer Vision (ICCV), pp. 945–953 (2015). https://doi.org/10.1109/ICCV.2015.114
22. Wang, C., Pelillo, M., Siddiqi, K.: Dominant set clustering and pooling for multi-view 3d object recognition (2019). https://doi.org/10.48550/ARXIV.1906.01592, https://arxiv.org/abs/1906.01592
23. Feng, Y., Zhang, Z., Zhao, X., Ji, R., Gao, Y.: Gvcnn: group-view convolutional neural networks for 3D shape recognition. In: 2018 IEEE/CVF Conference on Computer Vision and Pattern Recognition, pp. 264–272 (2018). https://doi.org/10.1109/CVPR.2018.00035

24. Kanezaki, A., Matsushita, Y., Nishida, Y.: Rotationnet: joint object categorization and pose estimation using multiviews from unsupervised viewpoints. In: 2018 IEEE/CVF Conference on Computer Vision and Pattern Recognition, pp. 5010–5019 (2018). https://doi.org/10.1109/CVPR.2018.00526
25. Esteves, C., Xu, Y., Allec-Blanchette, C., Daniilidis, K.: Equivariant multi-view networks. In: 2019 IEEE/CVF International Conference on Computer Vision (ICCV), pp. 1568–1577 (2019). https://doi.org/10.1109/ICCV.2019.00165
26. Han, Z.: 3d2seqviews: aggregating sequential views for 3D global feature learning by CNN with hierarchical attention aggregation. IEEE Trans. Image Process. **28**(8), 3986–3999 (2019). https://doi.org/10.1109/TIP.2019.2904460
27. He, X., Huang, T., Bai, S., Bai, X.: View n-gram network for 3D object retrieval. In: 2019 IEEE/CVF International Conference on Computer Vision (ICCV), pp. 7514–7523 (2019). https://doi.org/10.1109/ICCV.2019.00761
28. Yang, Z., Wang, L.: Learning relationships for multi-view 3D object recognition. In: 2019 IEEE/CVF International Conference on Computer Vision (ICCV), pp. 7504–7513 (2019). https://doi.org/10.1109/ICCV.2019.00760
29. Chen, S., Zheng, L., Zhang, Y., Sun, Z., Xu, K.: Veram: view-enhanced recurrent attention model for 3D shape classification. IEEE Trans. Visualization Comput. Graph. **25**(12), 3244–3257 (2019). https://doi.org/10.1109/TVCG.2018.2866793
30. Liu, A.A., et al.: Hierarchical multi-view context modelling for 3D object classification and retrieval. Inf. Sci. **547**, 984–995 (2021). https://doi.org/10.1016/j.ins.2020.09.057, https://www.sciencedirect.com/science/article/pii/S0020025520309671
31. Lin, C., Li, C., Liu, Y., Chen, N., Choi, Y.K., Wang, W.: Point2skeleton: learning skeletal representations from point clouds. In: 2021 IEEE/CVF Conference on Computer Vision and Pattern Recognition (CVPR). pp. 4275–4284 (2021). https://doi.org/10.1109/CVPR46437.2021.00426
32. Liu, M., Zhang, X., Su, H.: Meshing point clouds with predicted intrinsic-extrinsic ratio guidance. In: Vedaldi, A., Bischof, H., Brox, T., Frahm, J.-M. (eds.) ECCV 2020. LNCS, vol. 12353, pp. 68–84. Springer, Cham (2020). https://doi.org/10.1007/978-3-030-58598-3_5
33. Rakotosaona, M.J., Guerrero, P., Aigerman, N., Mitra, N., Ovsjanikov, M.: Learning delaunay surface elements for mesh reconstruction. In: 2021 IEEE/CVF Conference on Computer Vision and Pattern Recognition (CVPR), pp. 22–31 (2021). https://doi.org/10.1109/CVPR46437.2021.00009
34. Shannon, C.E.: A mathematical theory of communication. Bell Syst. Tech. J. **27**(3), 379–423 (1948). https://doi.org/10.1002/j.1538-7305.1948.tb01338.x
35. Vázquez, P.P., Feixas, M., Sbert, M., Heidrich, W.: Viewpoint selection using viewpoint entropy. In: VMV, vol. 1, pp. 273–280. Citeseer (2001)
36. Vázquez, P.P., Feixas, M., Sbert, M., Heidrich, W.: Automatic view selection using viewpoint entropy and its application to image-based modelling. Comput. Graph. Forum **22**(4), 689–700. https://doi.org/10.1111/j.1467-8659.2003.00717.x, https://onlinelibrary.wiley.com/doi/abs/10.1111/j.1467-8659.2003.00717.x
37. Deng, J., Dong, W., Socher, R., Li, L.J., Li, K., Fei-Fei, L.: Imagenet: a large-scale hierarchical image database. In: 2009 IEEE Conference on Computer Vision and Pattern Recognition, pp. 248–255 (2009). https://doi.org/10.1109/CVPR.2009.5206848
38. Dosovitskiy, A., et al.: An image is worth 16×16 words: transformers for image recognition at scale (2020). https://doi.org/10.48550/ARXIV.2010.11929, https://arxiv.org/abs/2010.11929

39. Ba, J.L., Kiros, J.R., Hinton, G.E.: Layer normalization (2016). https://doi.org/10.48550/ARXIV.1607.06450, https://arxiv.org/abs/1607.06450
40. He, K., Zhang, X., Ren, S., Sun, J.: Deep residual learning for image recognition. In: 2016 IEEE Conference on Computer Vision and Pattern Recognition (CVPR), pp. 770–778 (2016). https://doi.org/10.1109/CVPR.2016.90
41. Hendrycks, D., Gimpel, K.: Gaussian error linear units (gelus) (2016). https://doi.org/10.48550/ARXIV.1606.08415, https://arxiv.org/abs/1606.08415
42. Huang, G., Liu, Z., Van Der Maaten, L., Weinberger, K.Q.: Densely connected convolutional networks. In: 2017 IEEE Conference on Computer Vision and Pattern Recognition (CVPR), pp. 2261–2269 (2017). https://doi.org/10.1109/CVPR.2017.243
43. Kingma, D.P., Ba, J.: Adam: a method for stochastic optimization (2014). https://doi.org/10.48550/ARXIV.1412.6980, https://arxiv.org/abs/1412.6980
44. Xu, Y., Zheng, C., Xu, R., Quan, Y., Ling, H.: Multi-view 3D shape recognition via correspondence-aware deep learning. IEEE Trans. Image Process. **30**, 5299–5312 (2021). https://doi.org/10.1109/TIP.2021.3082310
45. Chen, S., Yu, T., Li, P.: Mvt: multi-view vision transformer for 3D object recognition (2021). https://doi.org/10.48550/ARXIV.2110.13083, https://arxiv.org/abs/2110.13083
46. Touvron, H., Cord, M., Douze, M., Massa, F., Sablayrolles, A., Jegou, H.: Training data-efficient image transformers & distillation through attention. In: Meila, M., Zhang, T. (eds.) Proceedings of the 38th International Conference on Machine Learning. Proceedings of Machine Learning Research, vol. 139, pp. 10347–10357. PMLR (2021). https://proceedings.mlr.press/v139/touvron21a.html
47. Yu, T., Meng, J., Yuan, J.: Multi-view harmonized bilinear network for 3D object recognition. In: 2018 IEEE/CVF Conference on Computer Vision and Pattern Recognition, pp. 186–194 (2018). https://doi.org/10.1109/CVPR.2018.00027

Structure Guided Proposal Completion for 3D Object Detection

Chao Shi[1], Chongyang Zhang[1,2](✉), and Yan Luo[1]

[1] School of Electronic Information and Electrical Engineering, Shanghai Jiao Tong University, Shanghai 200240, China
{shichaostone,sunny_zhang,luoyan_bb}@sjtu.edu.cn

[2] Shanghai Key Laboratory of Digital Media Processing and Transmission, Shanghai 200240, China

Abstract. 3D object detection from point clouds is one of the key components in autonomous driving. Current two-stage detectors generate a small number of proposals, and then refine them in the second RCNN procedure. However, due to the inherent sparsity of point clouds, the first stage may predict some low quality proposals with incomplete structure and inaccurate localization. These low quality proposals fail to obtain adequate and precise proposal features which are essential for the following refinement, inevitably degrading the overall detection performance. To alleviate this problem, we propose Structure guided Proposal Completion (SPC) for 3D object detection from point clouds. Specifically, two completion strategies are developed to obtain high quality proposals: one is Structure Completion, in which a group of structural proposals are obtained by traversing most structures, and thus at least one proposal with ground truth similar structure can be guaranteed. The other is RoI Feature Completion, which is used to fill the empty area of proposals with virtual points under structure-aware manner. With the proposed SPC, high quality proposals with clearer structure and more precise localization can be obtained, and further promote the RCNN to perceive adequate proposal features. Extensive experiments on KITTI benchmark demonstrate the effectiveness of our proposed method, especially for hard setting objects with fewer LiDAR points.

Keywords: 3D object detection · Point cloud · Proposal completion

1 Introduction

3D object detection is one of the core computer vision tasks, since it benefits wide applications in various fields, such as autonomous driving, virtual reality and robot perception [1,5]. Point clouds from LiDAR sensors are often adopted to detect objects from 3D space for its less sensitivity to weather and time of the day. Hence, point clouds based 3D detectors [3,6,20,21,29,30] are one of the main research topics in both academic and industrial area.

L. Wang et al. (Eds.): ACCV 2022, LNCS 13841, pp. 504–520, 2023.
https://doi.org/10.1007/978-3-031-26319-4_30

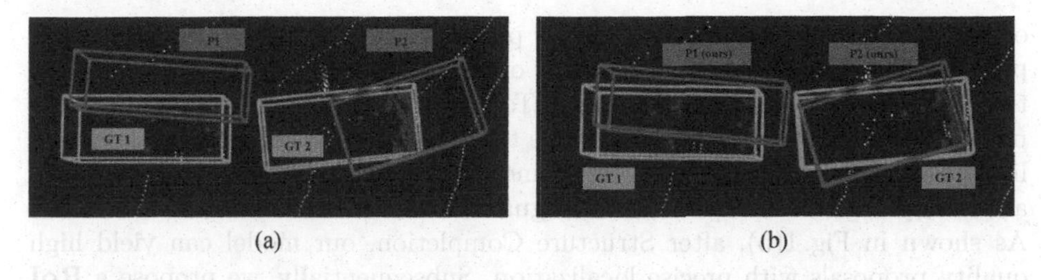

Fig. 1. Examples of RPN predicted proposals from sparse point clouds: (a) proposals from the first stage of PointRCNN [22], denoted as P1 and P2; and (b) augmented proposals by our Structure Completion module, denoted as P1(ours) and P2(ours). The object points, ground-truth boxes, and the predicted proposals are shown in blue, green and red colors, respectively. After applying Structure Completion, we can obtain a set of augmented proposals and we only show the most precisely localized proposals for better view. (Color figure online)

Despite the great success of the above methods, their performance is still limited by the sparsity of point clouds. It means that the LiDAR sensors can solely capture small portion of object point clouds, especially in distant or occluded regions, making it difficult to generate precise 3D bounding boxes. Concretely, we analyzed the statistical results of point clouds in KITTI dataset [4], and found that: more than 20.2% objects have less than 30 points, including 10.8% objects have less than 10 points, and even worse for pedestrian and cyclist in KITTI dataset. Most of them are incomplete with some important parts missing, such as tyre of a bicycle or arm of a man.

To alleviate the sparsity of point clouds, some methods [2,24,26] combine the data of multi sensors with the usage of complementary information. MV3D [2] directly merges the image features and birds-eye-view (BEV) features. PointPainting [24] first paints points with semantic segmentation score, then uses the painted points to point-based 3D detectors. However, these approaches are complex with multiple fusion modules. Pyramid R-CNN [15] designs RoI-grid Pyramid to gather more points of interest outside proposals for accurate object recognition, but the background points may also be gathered. ImpDet [19] uses evenly distributed virtual points to learn richer context features, but ignores the structural information of each object. In addition to the widely studied issues above, we have further discovered that the sparsity also brings the difficulty to localize object accurately. As shown in Fig. 1(a), region proposal network (RPN) may yield low quality proposals with inaccurate localization due to the sparsity of points, which brings difficulties for further regression. Moreover, constrained by sparse points and features, it is hard for the second refinement stage [22,30] to generate adequate proposal features.

To overcome the above limitations, we propose a novel two-stage 3D detection framework, namely **S**tructure guided **P**roposal **C**ompletion (SPC). To mitigate the problem of low quality proposals, we innovatively add a **Structure Completion** module after RPN, to augment the proposals containing insufficient points and features in structure guided manner. This module is inspired by the observation that for objects with fewer points, it is difficult for the network to predict where to regress and which part of the object these points belong to due to the insufficient context information. To this end, based on the predicted proposals and object structure, we traverse the missing structure of proposals with limited points and obtain a set of augmented proposals, in which at least one accurately localized proposal can be guaranteed for second refinement stage. As shown in Fig. 1(b), after Structure Completion, our model can yield high quality proposals with precise localization. Subsequentially, we propose a **RoI Feature Completion** module, aiming to obtain more adequate proposal features by virtual points under structure-aware manner. Firstly, we use 2D CNN to extract BEV feature from pseudo BEV map. Then we fill the empty area in each proposal with virtual points which are evenly distributed in the proposal. The proposal features are obtained by adaptively aggregating the virtual point features and raw point features with attention mechanism. Combining all the above components, our approach can accurately detect objects with a few points and effectively improve the performance of 3D object detection. Eventually, for the 3D detection under challenging settings, our method presents outstanding performance compared with the state-of-the-art methods.

We summarize our contributions as follows: (1) We propose a Structure guided Proposal Completion 3D object detector (SPC), which mitigates the low quality proposal and inadequate proposal feature problems caused by the sparsity issue. (2) We propose two simple yet effective completion strategies. One is Structure Completion, by traversing the missing structures of object to obtain accurately localized proposal, which can be served as a plug-in to enhance point-based 3D detection models. The other is RoI Feature Completion, by filling the empty area of proposals with virtual points to generate adequate proposal features. (3) We conduct extensive experiments on KITTI dateset and demonstrate the effectiveness of our approach, especially for hard setting objects.

2 Related Work

In generally, LiDAR-based 3D object detectors can be categorized into two streams: (i) single-stage detectors predict object bounding boxes and scores directly in one stage, which usually run effectively due to simpler network structures; and (ii) two-stage detectors usually generate some coarse proposals in the first stage, then these proposals and corresponding features are fed into second stage for refinement, which help detectors attain higher precision.

2.1 Single-Stage Object Detectors

VoxelNet [35] first encodes point clouds as voxels, then proposes voxel feature encoding layer to extract voxel-wise feature. But it is computationally expensive due to the 3D convolution operation. SECOND [28] adopts the sparse convolution to accelerate voxel feature extraction process. PointPillar [8] collapses the points in vertical columns (pillars) instead of voxels for effective encoding process, then uses pseudo-image for feature learning and object detection. 3D-SSD [29] introduces a novel sampling strategy named Feature-FPS for better classification by combining feature-based and point-based sampling distances. SA-SSD [6] proposes an auxiliary network and losses on the basis of 3D voxel CNN to preserve structure information. SE-SSD [34] utilizes teacher SSD and student SSD to get more training data, meanwhile it also consumes more time to train the model.

2.2 Two-Stage Object Detectors

PointRCNN [22] first uses PointNet [17] to segment foreground objects and generate 3D proposals, then refines them with semantic features. Based on PointRCNN, Part-A^2 [23] introduces an intra-object part supervision to improve the feature representation. STD [30] proposes PointsPool operation for RoI refinement, which converts sparse feature to dense feature representation. PV-RCNN [21] combines point-based and voxel-based network to extract features from keypoints and voxels, then aggregates them by RoI-grid pooling. Voxel R-CNN [3] proposes Voxel RoI Pooling to extract RoI feature from voxels for refinement. Pyramid R-CNN [15] alleviates the sparsity and imbalanced distribution problems of points by RoI-grid pyramid and density-aware radius prediction. CT3D [20] uses channel-wise attention to reweight the proposal features for refinement. SFD [25] is multi-modality method with image and LiDAR as inputs for 3D detection. Different from the previous methods, we add a Structure Completion after RPN to generate more accurately localized proposals.

2.3 Point Cloud Augmentation

Point cloud augmentation aims to generate denser points representation from sparse LiDAR points or RGB images. LiDAR-based methods like PUNet [32] reconstructs multiple upsampled points from high level feature vectors. Image-based methods [11,31] first perform depth completion and then convert to point clouds. MVP [31] predicts object depth in image space and further generates dense points. ImpDet [19] uniformly places virtual points around candidate point, then randomly chooses some points for boundary generation. In contrast, we generate virtual points in structure-aware manner and obtain their features from BEV feature, which is more accurate for refinement stage.

3 Methodology

The overall framework of Structure guided Proposal Completion (SPC) is shown in Fig. 2. Given the input raw points, the first RPN predicts a number of proposals, and the following Structure Completion augments the sampled proposals by traversing most structures. We further introduce the RoI Feature Completion module for more adequate proposal features generation, which will be fed into the final detection head. Details are shown in the following sections.

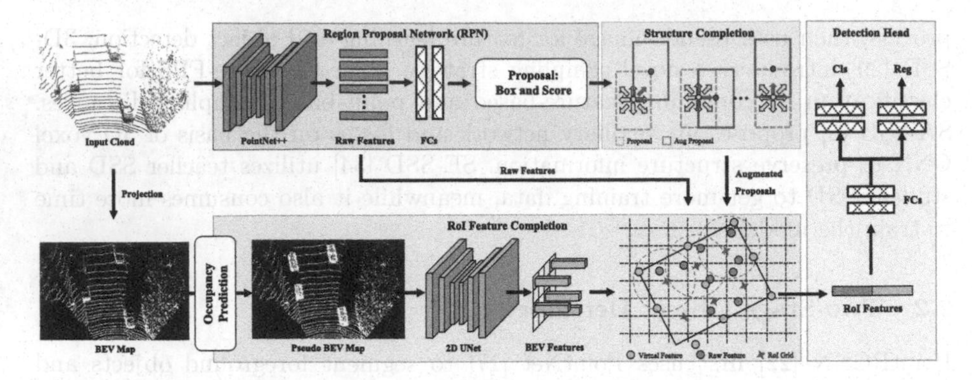

Fig. 2. Framework overview. We first generate 3D proposals from raw points with RPN, then a set of more accurately localized proposals will be generated by Structure Completion module. Subsequently, RoI Feature Completion module is introduced to generate more adequate proposal features by attentively aggregating raw point features and virtual point features. The final one is detection head for box refinement.

3.1 RPN for Proposals Generation

For each 3D scene, let $\mathcal{P} = \{(x, y, z, r)_n, n = 1, \dots, N\}$ be a set of raw point clouds, where (x_n, y_n, z_n) means 3D location in LiDAR coordinate system and r_n means the reflectance. RPN takes $\mathcal{P}$ as input and generates a set of 3D bounding boxes $\mathcal{B} = \{B_1, B_2, \cdots, B_K\} \in \mathbb{R}^{K\times 8}$ that represent the detected objects, where K denotes the number of objects in each scene. Each 3D bounding box B_k is represented as $(x_k, y_k, z_k, h_k, w_k, l_k, \theta_k, c_k)$, where (x_k, y_k, z_k) is object center, (h_k, w_k, l_k) is object size, θ_k is object orientation, and c_k is classification score. In this paper, we choose the first stage of PointRCNN [22] as our default RPN to generate 3D proposals due to its effectiveness and accuracy. It is worth noting that PointRCNN can be replaced by other high quality RPN. During the first stage, RPN outputs a set of 3D proposals and extracts corresponding point features fed to the next stage.

3.2 Structure Completion Module

Due to distance and different forms of occlusion, the number of point clouds in certain objects may be quite limited. When further integrated with the indispensable point sampling strategy, some essential foreground points are discarded, which makes the sparsity issue worse. Hence, it is difficult for RPN to predict where to regress and which portion these points belong to, resulting in low quality proposal with inaccurate localization. This motivates us to generate precisely localized proposals. Accordingly, we propose **Structure Completion** module to augment the existing predicted proposals.

Preliminary. For each object, we divide the captured points into four parts (front-left, front-right, behind-left and behind-right, respectively) in 3D space.

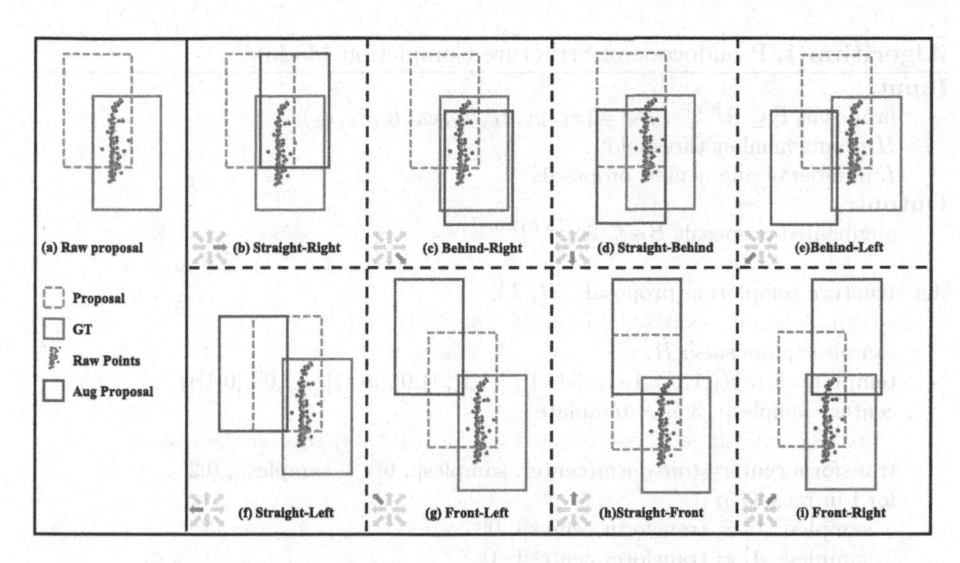

Fig. 3. The illustration of Structure Completion module. We show the captured points, ground truth, predicted proposal and augmented proposals in red, green, blue and purple, respectively. (a) shows the captured points, ground truth box and raw proposal. (b)-(i) represent the Structure Completion process from raw proposal to augmented proposals according to the missing structures. Specifically, we take proposal center as origin and shift the proposal by $l/2$ or $w/2$ in eight directions. For example, we shift the raw proposal by $w/2$ in straight-right direction and obtain the augmented proposal as shown in the (b). Particularly, we can obtain at least one more precisely localized proposal after Structure Completion, as shown in the (c). (Color figure online)

The origin is located at the bottom center of the object, and the front direction is same with the heading direction. Inspired by [14], we find that the low quality proposals are mainly caused by localization error in 3D space, so our Structure Completion module focuses on improving the accuracy of localization prediction.

Structure Completion. Since proposals with fewer points suffer from structure missing and inaccurate localization, our mechanism pays more attention on these proposals, denoted as $\mathcal{B}_s = \{B_1, B_2, \cdots, B_s\}$. In details, given the full-set proposals generated by RPN, we first count the number of points in each proposal, and then sample the proposals with small number of points for subsequent completion. Considering that large objects like cars naturally have more points than small objects like pedestrians and cyclists, it is unfair to directly select proposals with the smaller number of points among all proposals. So we sample the proposals with fewer points in the same class objects instead.

As shown in the Fig. 3(a), LiDAR captures points belonging to front-left part of the object, denoted by red points, and ground truth is drawn in green box. We show them in BEV map for better visualization. The subsequent downsampling operation will worsen this problem and make their structure incomplete, then RPN may make a wrong prediction with inaccurate localization.

Algorithm 1. Pseudocode of Structure Completion Module

Input:

proposals $\mathcal{B} \subseteq \mathbb{R}^{K\times 8}$, $B_k = (x_k, y_k, z_k, h_k, w_k, l_k, \theta_k, c_k)$
H points number threshold
L number of augmented proposals

Output:

augmented proposals $\mathcal{B}_H \subseteq \mathbb{R}^{(H\times L)\times 8}$

```
def structure_completion(proposals, H, L):
    # sample proposals based on points number threshold
    samples=proposals[: H, :]
    template=array(([1,1], [1,-1], [-1,-1], [-1,1], [1,0], [0,-1], [-1,0], [0,1]))
    center=samples[:, 3:5] × template
    # transform from local coordinate system to LiDAR coordinate system
    transform_center=transform(center, samples[:, 6]) + samples[:, 0:2]
    for i in range(L):
      samples[:, 0] = transform_center[i, 0]
      samples[:, 1] = transform_center[i, 1]
    return samples
```

It means the predicted proposal may be distributed around the object with relative small IoU. For example, RPN incorrectly predicts these points as behind-right structure and generates a proposal on the front-left direction, as shown in the Fig. 3(a), which is hard to refine. In order to obtain accurately localized proposals, we propose to augment these predictions by traversing object structure. Concretely, for each selected proposal B_s, we take the predicted center (x, y, z) as the origin, then shift the proposal by $w/2$, $l/2$, $(w/2, l/2)$ in the eight different directions, including straight-right, behind-right, straight-behind, behind-left, straight-left, front-left, straight-front and front-right directions, respectively, as shown in the Fig. 3 (b)–(i). All structures of the object, *i.e.*, the front-right, behind-left or behind-right structure, can be included through this method. Finally, for each selected proposal, we will get eight new proposals. Since we have traversed possible missing structures, at least one accurately localized proposal can be obtained for refinement stage, as shown in the Fig. 3(c).

The pseudocode of Structure Completion module is presented in Algorithm 1. Given $\mathcal{B} = \{B_1, B_2, \cdots, B_K\}$ proposals generated from the first stage, we first sample $\mathcal{B}_s$ proposals based on the points number threshold H. For each sampled proposal, we will generate L new augmented proposals. Finally, we will obtain $\mathcal{B}_{aug} = \{K + L \times H\}$ proposals. By default, we set $L = 8$. Despite its simplicity, this module makes a significant contribution and we will conduct ablation experiments in the Experiments Section to analyze the effects.

3.3 RoI Feature Completion Module

On top of high quality proposals generated by our Structure Completion, we further attempt to refine these proposals including localization and heading. However, previous methods [10,22] may fail to obtain complete proposal features because certain proposals have fewer points and features. This motivates us to generate adequate proposal feature for each proposal. Specifically, we propose RoI Feature Completion module, which consists of BEV feature learning, virtual points generation, and feature aggregation.

BEV Feature Learning. BEV image has several advantages compared to front view image. Such as, objects preserve origin physical size and no occlusion, which make BEV image more feasible in 3D object detection, especially for localization prediction. So we utilize BEV features for further virtual point features generation. Concretely, by projecting the points, we can obtain BEV image $H \times W \times C$, in which H and W are height and width of BEV image. However, using 2D CNN to extract raw BEV features directly may be inappropriate due to the sparsity, which makes objects on BEV image incomplete. So we first use the occupancy prediction [27] to generate more complete pseudo BEV map, then use 2D CNN to extract BEV features f^{bev}.

Virtual Points Generation. For each generated 3D proposal B_k, we enlarge its width and length to contain more points and features. Nevertheless, as distance increases, points become sparser and fewer points can be captured. To this end, we propose structure-aware virtual points generation module, which generates uniformly distributed virtual points to enrich original proposal points and features. This module is inspired by the observation that 3D objects have relatively fixed structures and resemble shape prototypes.

Concretely, we first generate one type of template from original dataset for each category, and then uniformly place M virtual points in each template, denoted as $\mathcal{V} = \{(x, y, z, r)_m, m = 1, \dots, M\}$. These points encode the structural information of the object. Then we select corresponding template and place them in the canonical coordinate system. The heading and the center of the template remain the same as the original proposal. We project $\mathcal{V}$ to BEV map and apply bilinear interpolation to get virtual points feature f_k^v from BEV feature f^{bev} due to its larger receptive fields. After that, for each proposal B_k, we have raw points p_k and corresponding features f_k^r, and we also have virtual points v_k and corresponding features f_k^v. Hence, the original proposal features f_k^p are enriched by concatenating virtual point features.

$$f_k^p = [f_k^r; f_k^v], \text{for } k = 1, 2, \cdots, K \quad (1)$$

where $[*; *]$ denotes the concatenation operation.

Feature Aggregation. After virtual points generation, original proposal feature is enriched by virtual points. However, the proposal region may contain background points due to inaccurate localization. Intuitively, it is inappropriate to utilize foreground and background points equally, because foreground points and virtual points indicated object boundary should make more contributions to the refinement stage, while background points should contribute less. To adaptively aggregate raw point features and virtual point features, we adopt the attention-based aggregation module to re-weight above features. In practice, we apply point-wise attention and channel-wise attention to strengthen the features, respectively. The point-wise attention can be represented as:

$$R_k = W_2 \delta \left(W_1 e_k \right), e_k = \text{POOL} \left(f_k^p \right) \tag{2}$$

where e_k represents the pooled features across the channel-wise dimensions, POOL represents the average-pooling, W_1, W_2 are the weight parameters of two fully-connected layers, and δ is the ReLU activation function. Similar to the point-wise attention, the channel-wise attention can also be computed and we will get $S_k = W_2' \delta \left(W_1' \left(g^k \right)^T \right)$, where g^k represents the pooled features across the point-wise dimensions.

By element-wise multiply, we obtain the attention matrix $A_k = \sigma \left(R_k \bigodot S_k \right)$, where σ is sigmoid function and $\bigodot$ is element-wise multiply operation. After that, the re-weighting features can be formulated as:

$$\widetilde{f_k^p} = f_k^p A_k \tag{3}$$

Eventually, the re-weighting features are fed into the network [18] to obtain proposal features for the final confidence classification and box refinement.

3.4 Detection Head and Loss Function

Detection Head. The detection head takes proposal features as input for refinement stage. Specifically, we first transform the proposal features into feature vectors by a shared two layers MLP. Then, the feature vectors are fed into two branches for bounding box regression and confidence prediction. In practice, our detection head adopts the same architecture as that in [22] with several fully connected layers.

Loss Function. We use multi-task loss to train our model. Specifically, the total loss is composed of region proposal loss $\mathcal{L}_{rpn}$ and refinement loss $\mathcal{L}_{rcnn}$ as

$$\mathcal{L}_{total} = \mathcal{L}_{rpn} + \mathcal{L}_{rcnn} \tag{4}$$

For region proposal loss $\mathcal{L}_{rpn}$, we adopt the same loss function with [22] as

$$\mathcal{L}_{rpn} = \mathcal{L}_{focal} + \mathcal{L}_{reg} \tag{5}$$

where classification loss $\mathcal{L}_{focal}$ are focal loss and we keep the default parameters $\alpha_t = 0.25$ and $\gamma = 2$, and smooth-L1 loss is utilized for regression loss $\mathcal{L}_{reg}$.

The refinement loss $\mathcal{L}_{rcnn}$ consists of two parts, including the smooth L1 loss for box refinement $\mathcal{L}_{box}$, and the multi-class cross entropy loss for box class prediction $\mathcal{L}_{ce}$,

$$\mathcal{L}_{rcnn} = \mathcal{L}_{box} + \mathcal{L}_{ce} \tag{6}$$

Table 1. Comparison with state-of-the-art methods on KITTI *val* set, the results are evaluated by the 3D Average Precision with 40 recall positions. The 3D APs with 11 recall positions are also reported under moderate setting. "-" means that results are not reported in the published version. Best in bold.

Type	Method	Modality	Car 3D AP_{R40}			Ped. 3D AP_{R40}			Cyc. 3D AP_{R40}			3D AP_{R11}		
			Easy	Mod.	Hard	Easy	Mod.	Hard	Easy	Mod.	Hard	Car	Ped.	Cyc.
1-stage	SECOND [28]	L	90.97	79.94	77.09	58.01	51.88	47.05	78.50	56.74	52.83	76.48	52.98	67.15
	PointPillars [8]	L	87.75	78.35	75.18	57.30	51.41	46.87	81.57	62.94	58.98	77.28	52.29	62.68
	SA-SSD [6]	L	92.23	84.30	81.36	-	-	-	-	-	-	79.91	-	-
2-stage	PointRCNN [22]	L	89.47	80.32	77.92	62.82	53.66	47.01	91.62	71.34	66.77	78.61	53.85	71.62
	PV-RCNN [21]	L	92.57	84.43	82.69	64.26	56.67	**51.91**	88.88	71.95	66.79	83.24	57.37	69.48
	Voxel R-CNN [3]	L	92.38	85.29	82.86	-	-	-	-	-	-	84.52	-	-
	Ours(SPC)	L	**92.71**	**85.75**	**83.35**	**66.18**	**58.92**	51.38	**93.66**	**75.55**	**70.97**	**85.85**	**59.63**	**74.72**
	Improvement	-	*+0.14*	*+0.46*	*+0.49*	*+1.92*	*+2.25*	*-0.53*	*+2.04*	*+3.60*	*+4.18*	*+1.33*	*+2.26*	*+3.10*

4 Experiments

We evaluate the effectiveness of our proposed method on the KITTI dataset [4], which consists of 7481 training samples and 7518 testing samples in 3D object detection task. The training data is divided into a *train* set with 3712 samples and a *val* set with 3769 samples [2]. We report our results on *val* set and *test* set for *Car*, *Pedestrian* and *Cyclist* categories. The 3D Average Precision (3D AP) is utilized as the evaluation metric. We adopt the official evaluation protocol for fair comparisons. Specifically, the IoU threshold is set to 0.7 for *Car* and 0.5 for *Pedestrian* and *Cyclist*. APs are computed by recalling 11 and 40 positions on the *val* and *test* splits respectively.

4.1 Experimental Setup

Network Architecture. We randomly choose 16384 points from the entire point clouds per scene, and the detection range is limited to [0,70.4]m for the x axis, [-40,40]m for the y axis and [-3,1]m for the z axis. We adopt PointRCNN [22] as our default RPN due to its efficiency. And we use 2D UNet [9] to extract pseudo BEV map features. For the virtual points generation strategy, we empirically generate 27 virtual points for each proposals, and the corresponding virtual point feature dimension is 128. For the box refinement module in SPC, we follow the usual RoI pooling strategies in PointRCNN [22].

Training and Inference Details. The SPC model is end-to-end optimized for 80 epochs with the ADAM [7] optimizer with the batch size 4. The learning rate is initialized as 0.01 and is decayed 10x at 50 and 70 epochs. We conduct data augmentation at training stage following strategies in [3], including gt-sampling, random rotation, random flipping and random scaling.

4.2 Comparison with State-of-the-Arts

We compare our model with the state-of-the-art competitors on *Car*, *Pedestrian*, *Cyclist* using AP under 40 recall positions (AP_{R40}), as shown in Table 1.

Table 2. Comparison with state-of-the-art methods on the KITTI *test* set, with average precision of 40 recall positions evaluated on the KITTI server. "-" means that the results are not reported in the published version. "L" means LiDAR-only methods, and "L+R" means the detector makes the use of both LiDAR and RGB modality. Best in bold.

Type	Method	Modality	Car 3D AP_{R40}			Ped. 3D AP_{R40}			Cyc. 3D AP_{R40}		
			Easy	Mod.	Hard	Easy	Mod.	Hard	Easy	Mod.	Hard
1-stage	VoxelNet [35]	L	77.47	65.11	57.73	39.48	33.69	31.51	61.22	48.36	44.37
	ContFuse [12]	L+R	83.68	68.78	61.67	-	-	-	-	-	-
	SECOND [28]	L	84.65	75.96	68.71	45.31	35.52	33.14	75.83	60.82	53.67
	PointPillars [8]	L	82.58	74.31	68.99	51.45	41.92	38.89	77.10	58.65	51.92
	TANet [13]	L	84.39	75.94	68.82	**53.72**	**44.34**	**40.49**	75.70	59.44	52.53
	SA-SSD [6]	L	88.75	79.79	74.16	-	-	-	-	-	-
	CIA-SSD [33]	L	89.59	80.28	72.87	-	-	-	-	-	-
2-stage	MV3D [2]	L+R	74.97	63.63	54.00	-	-	-	-	-	-
	F-PointNet [16]	L+R	82.19	69.79	60.59	50.53	42.15	38.08	72.27	56.12	49.01
	PointRCNN [22]	L	86.96	75.64	70.70	47.98	39.37	36.01	74.96	58.82	52.53
	STD [30]	L	87.95	79.71	75.09	53.29	42.47	38.35	78.69	61.59	55.30
	MMF [11]	L+R	88.40	77.43	70.22	-	-	-	-	-	-
	PV-RCNN [21]	L	**90.25**	81.43	76.82	52.17	43.29	40.29	78.60	**63.71**	**57.65**
	PI-RCNN [26]	L+R	84.37	74.82	70.03	-	-	-	-	-	-
	CT3D [20]	L	87.83	81.77	77.16	-	-	-	-	-	-
	Ours	L	87.69	**81.80**	**77.22**	48.37	39.88	37.21	**80.66**	63.04	56.88

The AP_{R40} of SA-SSD [6], PV-RCNN [21] and Voxel R-CNN [3] come from published papers, and the AP_{R40} of SECOND [28], PointPillars [8] and PointRCNN [22] come from the results of the officially released code. In addition, we also report the 3D APs under 11 recall positions (AP_{R11}) under moderate setting. Our method achieves the best performance among all competitors. Specifically, our SPC outperforms other methods by 1.33%, 2.26% and 3.10% 3D AP_{R11} on car, pedestrian, cyclist class under moderate setting. Figure 4 shows some predicted results and we project them onto color images for better visualization. As observed, our SPC can produce high quality 3D bounding box in different kinds of scenes.

We also compare our model with state-of-the-art methods on the KITTI *test* set by submitting our results to KITTI online test server. As Table 2 displays, compared with all the models, SPC surpasses them on moderate and hard setting of car class. SPC also shows competitive results on the pedestrians and cyclists class. Those methods include the models that take LiDAR and RGB images as inputs and the ones taking LiDAR input only.

4.3 Ablation Study

Here we provide extensive experiments to analyze the effectiveness of our method. All of the experiments in this section are conducted on the KITTI *val* set.

Component Analysis. We individually evaluate the contributions of Structure Completion and RoI Feature Completion module in our model. We choose

Table 3. Component analysis of our SPC model on KITTI *val* set. **Structure** means the Structure Completion, while the **Feature** means the RoI Feature Completion module. Performance comparisons of different components with 3D AP_{R40}.

Method	Structure	Feature	Car			Ped.			Cyc.		
			Easy	Mod.	Hard	Easy	Mod.	Hard	Easy	Mod.	Hard
Baseline			89.47	80.32	77.92	62.82	53.66	47.01	91.62	71.34	66.77
SPC	✓		91.87	82.68	80.38	65.74	58.21	51.18	93.52	75.48	70.78
SPC		✓	92.62	83.69	81.27	64.17	55.24	48.21	91.67	73.99	69.15
SPC	✓	✓	**92.71**	**85.75**	**83.35**	**66.18**	**58.92**	**51.38**	**93.66**	**75.55**	**70.97**

Fig. 4. Qualitative results on KITTI *val* set. The predicted bounding boxes and ground-truth bounding boxes are shown in red and green, respectively, and we project them back onto the color images for better visualization. (Color figure online)

PointRCNN [22] as our baseline and all hyperparameters and training procedures are the same, whose mAP for car, pedestrian and cyclist are 81.70%, 56.31% and 74.43%, respectively. Table 3 shows the importance of each component of our SPC model. Simply adding Structure Completion after RPN, the process boosts the performance beyond the baseline, improving the AP_{R40} by 2.41%, 3.88% and 3.34% for three categories, respectively. It clearly demonstrates the effectiveness of Structure Completion module, which helps to alleviate the low quality proposal generation problem of RPN, especially for pedestrian and cyclist under moderate and hard setting. The main reason that leads to this phenomenon, is the inaccurate localization caused by structure missing has bigger impact for pedestrian and cyclist. Compared with car class, pedestrian and cyclist are relative smaller, then the same localization error has a more severe influence on the IoU between the predicted bounding boxes and the ground truth of the small object. Our RoI Feature Completion module contributes an improvement of 3.39%, 1.37% and 1.69% on the three categories, respectively. Finally, we combine the two module and obtain the full model of SPC.

Table 4. Effects of Structure Completion module in different LiDAR-based detection paradigms on KITTI *val* set with 3D AP_{R40}. **Structure** means the Structure Completion module.

Method	Car			Ped.			Cyc.		
	Easy	Mod.	Hard	Easy	Mod.	Hard	Easy	Mod.	Hard
PointRCNN	89.47	80.32	77.92	62.82	53.66	47.01	91.62	71.34	66.77
PointRCNN+**Structure**	91.87	82.68	80.38	65.74	58.21	51.18	93.52	75.48	70.78
Improvement	*+2.40*	*+2.36*	*+2.36*	*+2.92*	*+4.55*	*+4.17*	*+1.90*	*+4.14*	*+4.01*
PV-RCNN	92.57	84.43	82.69	64.26	56.67	51.91	88.88	71.95	66.79
PV-RCNN+**Structure**	92.65	85.57	83.12	67.43	58.46	51.30	90.42	72.68	67.12
Improvement	*+0.08*	*+1.14*	*+0.43*	*+3.17*	*+1.79*	*-0.61*	*+1.54*	*+0.73*	*+0.33*

Table 5. Ablation on Structure Completion module. Performance comparisons of different Structure Completion strategies with 3D AP_{R40}. H means the points number threshold to sample proposals. L means the number of augmented proposals.

Method	Parameter	Car			Ped.			Cyc.		
		Easy	Mod.	Hard	Easy	Mod.	Hard	Easy	Mod.	Hard
Baseline		89.47	80.32	77.92	62.82	53.66	47.01	91.62	71.34	66.77
H	20	91.42	50.54	78.10	62.65	55.15	48.54	90.90	73.62	69.16
	40	**91.87**	**82.68**	**80.38**	65.74	**58.21**	**51.18**	**93.52**	**75.48**	**70.78**
	60	91.01	82.12	79.70	65.32	56.52	49.74	92.95	74.42	69.89
L	4	89.67	80.63	78.21	**66.09**	58.03	50.71	92.20	71.65	67.04
	6	89.01	78.87	77.90	64.83	55.93	48.70	92.07	73.15	68.82
	8	**91.87**	**82.68**	**80.38**	65.74	**58.21**	**51.18**	**93.52**	**75.48**	**70.78**

Effects of Structure Completion. The Structure Completion is easily extended to LiDAR-based 3D detectors. To verify it can play a plug-in to other models, we select PointRCNN [22] and PV-RCNN [21] to test on KITTI *val* set. As shown in Table 4, the Structure Completion module can bring +0.08% $\sim$ +4.55% AP_{R40} to the original 3D detector. We also test our model with different Structure Completion strategies. We first analyze the effect of hyperparameter H, as shown in Table 5. When we augment the proposals with fewer points, SPC shows better capacity than the baseline, especially at moderate and hard setting. When proposals with enough points ($H = 60$) are selected to augment, the performance slightly decrease. The reason is that for objects with sufficient points, they are conducive to network's inference of the localization and size of the objects. Augmenting these relatively accurate localization proposals may yield some proposals that are easily misclassified as positive, resulting in performance degradation. The best result comes from the $H = 40$ setting and we select as our default setting. Besides, we further analyze the effect of hyperparameter L, as displayed in Table 5. We get the best result by setting $L = 8$, that means each augmented proposal may make contributions to the finally result, because it is hard to identify which structure is missing.

Table 6. Ablation on RoI Feature Completion module. Performance comparisons of different number of virtual points with 3D AP_{R40}.

Number	Car			Ped.			Cyc.		
	Easy	Mod.	Hard	Easy	Mod.	Hard	Easy	Mod.	Hard
8	92.65	85.57	83.11	64.95	56.07	49.44	91.29	72.46	68.66
27	**92.71**	**85.75**	**83.35**	**66.18**	**58.92**	**51.38**	**93.66**	**75.55**	**70.97**
64	92.54	85.52	83.11	66.00	56.39	48.60	93.23	75.30	70.60

Table 7. Speed and Accuracy comparisons on KITTI *val* set. The inference speed is tested under single RTX 3090 GPU with batch size 1. PR means PointRCNN [22], Structure means Structure Completion and Feature means RoI Feature Completion.

Inference speed			AP_{R40} for car detection		
PR	PR + Structure	PR + Feature	PR	PR + Structure	PR + Feature
71.1 ms	72.0 ms	90.4 ms	80.32	82.68	83.69

Effects of RoI Feature Completion. We further validate the effectiveness of RoI Feature Completion module. As Table 6 displays, if we place too many virtual points in each proposal, the number of virtual points may be greater than the raw points and degrade the overall performance, because it cannot help to fit a bounding box well. Moreover, the more virtual points we generate, the higher computation costs. We choose the optimal value when the model achieves the best performance, *i.e.*, M =27.

Efficiency Analysis. We analyze the efficiency of our proposed methods from inference speed and accuracy, as shown in Table 7. The proposed method only adds little latency, including 0.9ms for Structure Completion and 19.3ms for RoI Feature Completion module. And we achieve 2.36% and 3.37% accuracy improvements on moderate setting for car detection than the baseline.

5 Conclusion

We present a two-stage 3D object detection framework SPC for outdoor point clouds. It aims to mitigate the problems of low quality proposal generation in RPN and insufficient proposal features in refinement stage. To this end, we introduce a Structure Completion strategy that generates at least one proposal structure similar to the ground truth boxes by traversing most structures. Moreover, we propose a RoI Feature Completion module that helps to obtain adequate proposal features by attentively aggregating raw point features and virtual point features. Our approach achieves competitive performance compared with state-of-the-art methods on KITTI dataset, especially on the hard setting objects.

Acknowledgments. This work was partly funded by NSFC(No.61971281), Shanghai Municipal Science and Technology Major Project (2021SHZDZX0102), and STCSM(18DZ2270700).

References

1. Bansal, M., Krizhevsky, A., Ogale, A.: ChauffeurNet: learning to drive by imitating the best and synthesizing the worst. arXiv preprint arXiv:1812.03079 (2018)
2. Chen, X., Ma, H., Wan, J., Li, B., Xia, T.: Multi-view 3D object detection network for autonomous driving. In: Proceedings of the IEEE Conference on Computer Vision and Pattern Recognition, pp. 1907–1915 (2017)
3. Deng, J., Shi, S., Li, P., Zhou, W., Zhang, Y., Li, H.: Voxel r-CNN: towards high performance voxel-based 3D object detection. arXiv preprint arXiv:2012.15712 1(2), 4 (2020)
4. Geiger, A., Lenz, P., Stiller, C., Urtasun, R.: Vision meets robotics: the kitti dataset. The Int. J. Robot. Res. **32**(11), 1231–1237 (2013)
5. Grigorescu, S., Trasnea, B., Cocias, T., Macesanu, G.: A survey of deep learning techniques for autonomous driving. Int. J. Robot. Res. **37**(3), 362–386 (2020)
6. He, C., Zeng, H., Huang, J., Hua, X.S., Zhang, L.: Structure aware single-stage 3D object detection from point cloud. In: Proceedings of the IEEE/CVF Conference on Computer Vision and Pattern Recognition, pp. 11873–11882 (2020)
7. Kingma, D.P., Ba, J.: Adam: a method for stochastic optimization. arXiv preprint arXiv:1412.6980 (2014)
8. Lang, A.H., Vora, S., Caesar, H., Zhou, L., Yang, J., Beijbom, O.: Pointpillars: fast encoders for object detection from point clouds. In: Proceedings of the IEEE/CVF Conference on Computer Vision and Pattern Recognition, pp. 12697–12705 (2019)
9. Li, X., Chen, H., Qi, X., Dou, Q., Fu, C.W., Heng, P.A.: H-DenseUNet: hybrid densely connected UNet for liver and tumor segmentation from CT volumes. IEEE Trans. Med. Imaging **37**(12), 2663–2674 (2018)
10. Li, Z., Wang, F., Wang, N.: Lidar r-CNN: an efficient and universal 3d object detector. In: Proceedings of the IEEE/CVF Conference on Computer Vision and Pattern Recognition, pp. 7546–7555 (2021)
11. Liang, M., Yang, B., Chen, Y., Hu, R., Urtasun, R.: Multi-task multi-sensor fusion for 3d object detection. In: Proceedings of the IEEE/CVF Conference on Computer Vision and Pattern Recognition, pp. 7345–7353 (2019)
12. Liang, M., Yang, B., Wang, S., Urtasun, R.: Deep continuous fusion for multi-sensor 3d object detection. In: Ferrari, V., Hebert, M., Sminchisescu, C., Weiss, Y. (eds.) ECCV 2018. LNCS, vol. 11220, pp. 663–678. Springer, Cham (2018). https://doi.org/10.1007/978-3-030-01270-0_39
13. Liu, Z., Zhao, X., Huang, T., Hu, R., Zhou, Y., Bai, X.: TaNet: robust 3D object detection from point clouds with triple attention. In: Proceedings of the AAAI Conference on Artificial Intelligence, pp. 11677–11684 (2020)
14. Ma, X., et al.: Delving into localization errors for monocular 3D object detection. In: Proceedings of the IEEE/CVF Conference on Computer Vision and Pattern Recognition, pp. 4721–4730 (2021)
15. Mao, J., Niu, M., Bai, H., Liang, X., Xu, H., Xu, C.: Pyramid r-CNN: towards better performance and adaptability for 3D object detection. In: Proceedings of the IEEE/CVF International Conference on Computer Vision, pp. 2723–2732 (2021)

16. Qi, C.R., Liu, W., Wu, C., Su, H., Guibas, L.J.: Frustum pointnets for 3D object detection from RGB-d data. In: Proceedings of the IEEE Conference on Computer Vision and Pattern Recognition, pp. 918–927 (2018)
17. Qi, C.R., Su, H., Mo, K., Guibas, L.J.: PointNet: deep learning on point sets for 3D classification and segmentation. In: Proceedings of the IEEE Conference on Computer Vision and Pattern Recognition, pp. 652–660 (2017)
18. Qi, C.R., Yi, L., Su, H., Guibas, L.J.: Pointnet++: deep hierarchical feature learning on point sets in a metric space. In: 30th Proceedings of Conference on Advances in Neural Information Processing Systems (2017)
19. Qian, X., Wang, L., Zhu, Y., Zhang, L., Fu, Y., Xue, X.: Impdet: exploring implicit fields for 3D object detection. arXiv preprint arXiv:2203.17240 (2022)
20. Sheng, H., et al.: Improving 3d object detection with channel-wise transformer. In: Proceedings of the IEEE/CVF International Conference on Computer Vision, pp. 2743–2752 (2021)
21. Shi, S., et al.: PV-RCNN: point-voxel feature set abstraction for 3D object detection. In: Proceedings of the IEEE/CVF Conference on Computer Vision and Pattern Recognition, pp. 10529–10538 (2020)
22. Shi, S., Wang, X., Li, H.: Point RCNN: 3D object proposal generation and detection from point cloud. In: Proceedings of the IEEE/CVF Conference On Computer Vision and Pattern Recognition, pp. 770–779 (2019)
23. Shi, S., Wang, Z., Wang, X., Li, H.: Part-a^2 Net: 3D part-aware and aggregation neural network for object detection from point cloud. arXiv preprint arXiv:1907.03670 2(3) (2019)
24. Vora, S., Lang, A.H., Helou, B., Beijbom, O.: Pointpainting: sequential fusion for 3d object detection. In: Proceedings of the IEEE/CVF Conference on Computer Vision and Pattern Recognition, pp. 4604–4612 (2020)
25. Wu, X., et al.: Sparse fuse dense: towards high quality 3D detection with depth completion. In: Proceedings of the IEEE/CVF Conference on Computer Vision and Pattern Recognition, pp. 5418–5427 (2022)
26. Xie, L., et al.: Pi-RCNN: an efficient multi-sensor 3D object detector with point-based attentive CONT-conv fusion module. In: Proceedings of the AAAI Conference on Artificial Intelligence, vol. 34, pp. 12460–12467 (2020)
27. Xu, Q., Zhong, Y., Neumann, U.: Behind the curtain: learning occluded shapes for 3D object detection. In: Proceedings of the AAAI Conference on Artificial Intelligence, vol. 36, pp. 2893–2901 (2022)
28. Yan, Y., Mao, Y., Li, B.: Second: sparsely embedded convolutional detection. Sensors **18**(10), 3337 (2018)
29. Yang, Z., Sun, Y., Liu, S., Jia, J.: 3DSSD: point-based 3D single stage object detector. In: Proceedings of the IEEE/CVF Conference On Computer Vision and Pattern Recognition, pp. 11040–11048 (2020)
30. Yang, Z., Sun, Y., Liu, S., Shen, X., Jia, J.: STD: sparse-to-dense 3D object detector for point cloud. In: Proceedings of the IEEE/CVF International Conference on Computer Vision, pp. 1951–1960 (2019)
31. Yin, T., Zhou, X., Krähenbühl, P.: Multimodal virtual point 3D detection. In: 34th Proceedings of Advances in Neural Information Processing Systems (2021)
32. Yu, L., Li, X., Fu, C.W., Cohen-Or, D., Heng, P.A.: Pu-Net: point cloud upsampling network. In: Proceedings of the IEEE Conference on Computer Vision and Pattern Recognition, pp. 2790–2799 (2018)
33. Zheng, W., Tang, W., Chen, S., Jiang, L., Fu, C.W.: CIA-SSD: confident IOU-aware single-stage object detector from point cloud. arXiv preprint arXiv:2012.03015 (2020)

34. Zheng, W., Tang, W., Jiang, L., Fu, C.W.: SE-SSD: self-ensembling single-stage object detector from point cloud. In: Proceedings of the IEEE/CVF Conference on Computer Vision and Pattern Recognition, pp. 14494–14503 (2021)
35. Zhou, Y., Tuzel, O.: VoxelNet: end-to-end learning for point cloud based 3D object detection. In: Proceedings of the IEEE Conference on Computer Vision and Pattern Recognition, pp. 4490–4499 (2018)

Optimization Methods

MaxGNR: A Dynamic Weight Strategy via Maximizing Gradient-to-Noise Ratio for Multi-task Learning

Caoyun Fan[1], Wenqing Chen[2], Jidong Tian[1], Yitian Li[1], Hao He[1(✉)], and Yaohui Jin[1]

[1] Shanghai Jiao Tong University, Shanghai, China
{fcy3649,frank92,yitian_li,hehao,jinyh}@sjtu.edu.cn
[2] Sun Yat-sen University, Guangzhou, China
chenwq95@mail.sysu.edu.cn

Abstract. When modeling related tasks in computer vision, Multi-Task Learning (MTL) can outperform Single-Task Learning (STL) due to its ability to capture intrinsic relatedness among tasks. However, MTL may encounter the insufficient training problem, i.e., some tasks in MTL may encounter non-optimal situation compared with STL. A series of studies point out that too much gradient noise would lead to performance degradation in STL, however, in the MTL scenario, Inter-Task Gradient Noise (ITGN) is an additional source of gradient noise for each task, which can also affect the optimization process. In this paper, we point out ITGN as a key factor leading to the insufficient training problem. We define the Gradient-to-Noise Ratio (GNR) to measure the relative magnitude of gradient noise and design the MaxGNR algorithm to alleviate the ITGN interference of each task by maximizing the GNR of each task. We carefully evaluate our MaxGNR algorithm on two standard image MTL datasets: NYUv2 and Cityscapes. The results show that our algorithm outperforms the baselines under identical experimental conditions.

Keywords: Multi-task learning · Gradient noise · Weight strategy

1 Introduction

Deep learning [10] has achieved significant success in Multi-Task Learning (MTL) in the field of computer vision. The multi-task model captures the intrinsic correlation among tasks and also allows multiple inferences in one single forward pass, so MTL could achieve the unity of high performance and efficiency in the optimization process [24]. However, MTL may suffer from the insufficient training problem

C. Fan and W. Chen—These authors contributed equally.

Supplementary Information The online version contains supplementary material available at https://doi.org/10.1007/978-3-031-26319-4_31.

L. Wang et al. (Eds.): ACCV 2022, LNCS 13841, pp. 523–538, 2023.
https://doi.org/10.1007/978-3-031-26319-4_31

Table 1. Percentage change in performance (Compared STL and MTL at the same settings. Details in Sect. 5.) of MTL relative to STL. NYUv2 and CityScapes are two one-to-many image datasets, where **Seg.**, **Dep.**, and **SN.** denote Segmentation task, Depth task, and Surface Normal task, respectively.

Task	NYUv2			CityScapes	
	Seg.	Dep.	SN.	Seg.	Dep.
Performance	8.3% ↑	6.9% ↑	9.7% ↓	1.1% ↑	13.2% ↓

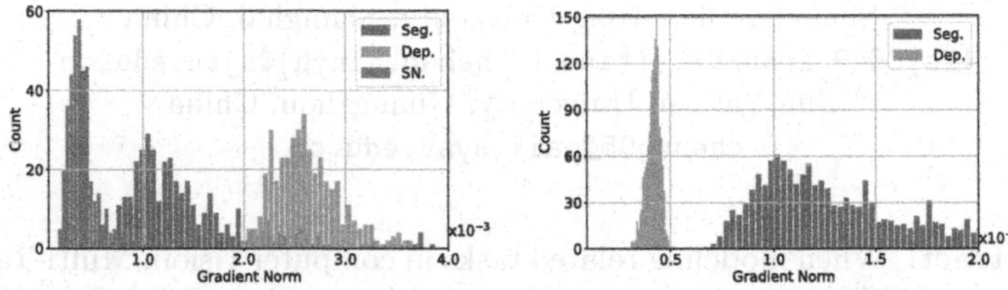

(a) NYUv2

(b) CityScapes

Fig. 1. Gradient Norm distribution on NYUv2 and CityScapes.

[16], which means some tasks in MTL are not optimal compared with the single-task solution.

Most deep learning models are trained to be optimum by the Stochastic Gradient Descent (SGD) technique [1]: models are optimized by the loss gradient based on a mini-batch randomly selected from all data. The gradient obtained by the SGD algorithm has noise [2,26], and many studies [13,14] point out that the gradient noise magnitude could affect the SGD generalization in Single-Task Learning (STL). This viewpoint gives us a key insight: in the MTL scenario, due to the joint learning of multiple tasks, the Inter-Task Gradient Noise (ITGN) could also interfere with the optimization of specific tasks, which may be the cause of the insufficient training problem.

Following this insight, we analyze the performance of each task (Table 1) as well as gradient norm distribution (Fig. 1) on two image MTL datasets. We find that the variance of the gradient norm varies widely across tasks, which implies that different tasks typically have widely varying gradient noise. At the same time, small gradient tasks tend to suffer from performance deterioration in MTL, which shows the relative disadvantage of small gradient tasks in MTL. In this paper, we demonstrate that Inter-Task Gradient Noise (ITGN) is one of the key optimization challenges in MTL (Sect. 3). In fact, the gradient in MTL optimization contains the noises of all tasks, so ITGN is an additional source of gradient noise compared to STL. As a result, some tasks (especially those with small gradients) may suffer from performance deterioration due to the effect of ITGN because the model has difficulty converging to the optimal position based on the gradient when the gradient noise is too high.

To alleviate the ITGN effect in MTL, we design a dynamic weight strategy called MaxGNR. Specifically, we realize that the magnitude of gradients and noise may vary greatly from task to task, and to quantitatively describe the effect of noise on SGD optimization, we define the Gradient-to-Noise Ratio (GNR) to measure the relative magnitude of gradient noise. Further, we explain that maximizing the GNR of each task in the MTL scenario is a reasonable method to mitigate ITGN interference (Sect. 4.1). Then, we propose the momentum method to approximate the theoretical GNR to a computable expression (Sect. 4.2), therefore, we can dynamically select the weights that maximize GNR at each iteration (Sect. 4.3). As a result, the ITGN interference is reduced through the MaxGNR algorithm, and the insufficient training problem could be mitigated.

Furthermore, we extensively examine our MaxGNR algorithm on two standard image MTL datasets: NYUv2 [22] and CityScapes [7], and the experimental results reveal that the MaxGNR algorithm obtain superior performance than other baselines under the same premise of other settings. Our contributions are:

- We analyze the effect of gradient noise on MTL and connect the insufficient training problem with ITGN interference.
- We define the Gradient-to-Noise Ratio (GNR) to measure the relative magnitude of gradient noise in MTL, and quantitatively describe the ITGN interference by GNR.
- We propose the MaxGNR algorithm, a dynamic weight strategy to alleviate ITGN interference and experiments demonstrate the algorithm's validity.

2 Related Work

2.1 Multi-task Learning

There are mainly two aspects in the field of MTL: network architecture improvement [25,31] and optimization strategy development, aiming at feature extraction and balance of performance in MTL, respectively. There have been lots of studies on multi-task architecture improvements, some works focus on the encoder structure [9,17,19,20] and others focus on the decoder part [25,27,31]. In this paper, we focus on optimization strategy development. A major challenge of MTL is how to balance the performance of tasks in joint learning [24]. To solve this challenge, researchers have proposed many algorithms, here we divide these algorithms into two categories: *coarse-grained* and *fine-grained.* The coarse-grained algorithm is to design optimization strategy according to the metrics of prediction, Uncertainty [6] utilizes homeostatic uncertainty to balance each loss; Dynamic Weight Averaging [17] tries to balance the decline rate of different task losses; Dynamic Task Prioritization [11] focuses on the balance of key performance indicators for multiple tasks; GLS [5] uses the geometric mean of task-specific losses as the target loss. The fine-grained algorithm is based on the gradient of model parameters, which can more accurately reflect the updating direction in each iteration. In recent years, such algorithms have been developed rapidly. GradNorm [4] stimulates the task-specific gradients to be of similar magnitude; MGDA [21] regards MTL as multi-objective optimization; PCGrad [28] cut the conflict gradient of different tasks. However, although gradient noise is an important problem in STL, the effect of gradient noise in MTL has never been considered.

2.2 Stochastic Gradient Descent

Stochastic Gradient Descent (SGD) is one of the standard methods [1] to optimize machine learning models. It is originally proposed to make up for the computational bottleneck of gradient descent (GD). Some studies focus on the generation capability of SGD. [32] reports SGD outperforms GD, [13,14] deduce theoretically that small-batch SGD generalizes better than large-batch SGD. The gradient obtained by the SGD algorithm has un-biased gradient noise [18]. To suppress the optimization oscillation caused by gradient noise, [23] proposes to introduce the momentum method in the optimization process, and many optimizers absorb the idea of momentum method, such as Adagrad [8], Adadelta [29], Adam [15].

3 Effect of Gradient Noise

In this section, we analyze the source of gradient noise from the perspective of SGD (Sect. 3.1), and analyze the effect of gradient noise on STL (Sect. 3.2) and MTL (Sect. 3.3).

3.1 Preliminaries of SGD

Formally, the design of the machine learning algorithm is based on data $D = \{(X_i, Y_i)\}$, the hypothesis function F, and loss function l. Algorithms are designed to search the optimal parameter $\{F_\theta | \theta \in \Theta \subset \mathbb{R}^d\}$ to get the lowest expected risk $\mathcal{R}(\theta)$ under the loss function l, where θ is the parameter of the hypothesis function and d is the dimension of the parameter. The expected risk $\mathcal{R}(\theta)$ cannot be obtained, so stochastic Gradient Descent algorithm (SGD) [1] selects a mini-batch $S = \{(X_i, Y_i)\}_{|S|}$ independent and identically distributed (i.i.d.) from the data D, and the expected risk $\mathcal{R}(\theta)$ can be estimated by the empirical risk $\widehat{\mathcal{R}}(\theta)$ calculated by mini-batch. In this paper, $\nabla_\theta \mathcal{R}(\theta)$ and $\nabla_\theta \widehat{\mathcal{R}}(\theta)$ are denoted as $g(\theta)$ and $\widehat{g_S}(\theta)$, and $l(F_\theta(X_i), Y_i)$ is expressed as $l_i(\theta)$.

The gradient $\widehat{g_S}(\theta)$ obtained by SGD is stochastic, so the gradient can be decomposed into the expected gradient and gradient noise. Many works [12,18] assume that the gradient noise belongs to the Gaussian class due to the classical central limit theorem [18]. In this paper, we follow this assumption in order to describe the effect of gradient noise on MTL quantitatively.

Assumption 1. An individual data (X_i, Y_i) is selected independent and identically distributed (i.i.d.) from data D, and its gradient $\nabla_\theta l_i(\theta)$ obeys the Gaussian distribution:

$$\nabla_\theta l_i(\theta) \sim \mathcal{N}(g(\theta), C) \tag{1}$$

where $g(\theta)$ is the expected gradient, C is the covariance matrix, and is approximately constant for θ, which is determined by data D and loss function l.

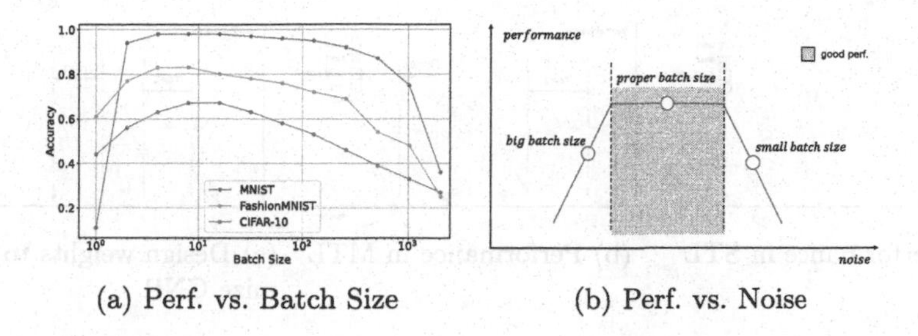

(a) Perf. vs. Batch Size (b) Perf. vs. Noise

Fig. 2. The effect of gradient noise magnitude on the performance of STL. The noise magnitude is controlled by the batch size. We conduct experiments on MNIST, FashionMNIST, and CIFAR10. As the batch size increases, performance first increases, then stabilizes, and finally decreases on all three datasets.

3.2 Gradient Noise in STL

Many studies [13,14] emphasize the magnitude of gradient noise could effect the generalization of SGD. In practice, noise magnitude can be easily adjusted by batch size $|S|$ [12]. According to Assumption 1, $\widehat{g_S}(\theta)$ can be simplified as:

$$\begin{aligned} \widehat{g_S}(\theta) =& \frac{1}{|S|}\sum_{i=1}^{|S|} \nabla_\theta l_i(\theta) = g(\theta) + n_{g(\theta)} \\ \text{where} \quad & n_{g(\theta)} \sim \mathcal{N}(0, \frac{C}{|S|}) \end{aligned} \tag{2}$$

where $n_{g(\theta)}$ is the gradient noise. Because of the Positive Semi-definite of C, the noise magnitude can be measured as:

$$\mathbb{E}[\|n_{g(\theta)}\|^2] = tr(C)/|S| \tag{3}$$

By adjusting the batch size $|S|$, we evaluated the impact of noise magnitude on model performance. The result in Fig. 2(a) showed that the performance can be maintained only in the appropriate noise magnitude range. In fact, large noise would interfere with the convergence of the model, while the model is unable to escape the local optimum with small noise, so the effect of gradient noise on performance can be expressed as Fig. 2(b).

3.3 Inter-task Gradient Noise

In the multi-task SGD process, multiple tasks have their own loss functions $\{l^k\}$, and Weighted Average Method (WAM) is a mainstream method to combine multiple loss functions. Specifically, WAM can be expressed as:

$$L = \sum_{k=1}^{n} \omega_k l^k(\theta) \quad s.t. \sum_{k=1}^{n} \omega_k = 1 \tag{4}$$

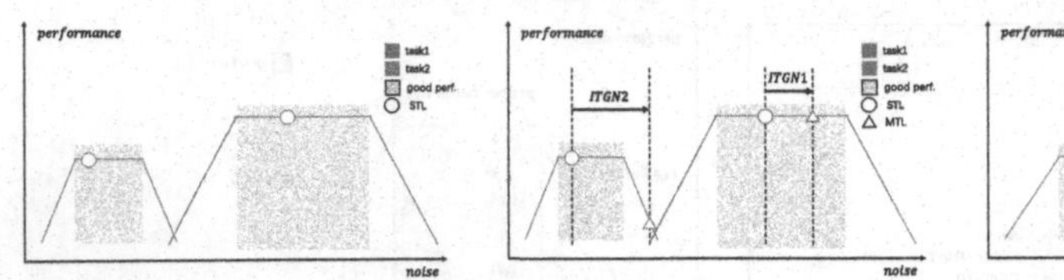

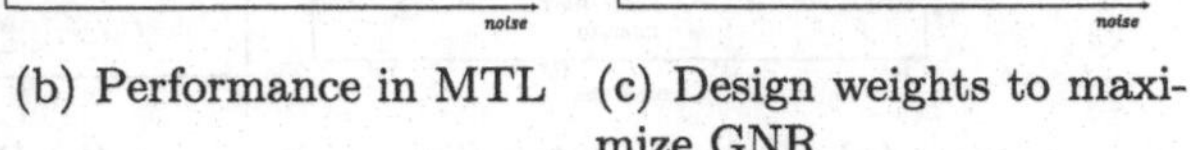

(a) Performance in STL (b) Performance in MTL (c) Design weights to maximize GNR

Fig. 3. Illustration of MaxGNR algorithm. In the STL scenario, reasonable settings (model, batch size, learning rate, etc.) can be selected to ensure that the gradient noise is in the appropriate range, as in Fig. 3(a). In the MTL scenario, ITGN is also a source of gradient noise that can affect the optimization of specific tasks. Therefore, ITGN may cause task-specific performance deterioration due to the large differences in ITGN, as is illustrated in Fig. 3(b). Our MaxGNR algorithm attempts to alleviate the ITGN interference as Fig. 3(c). According to Eq. 12, We can design appropriate weights to maximize GNR for the purpose of mitigating ITGN interference.

According to Assumption 1, the SGD gradient in MTL is expressed as follows:

$$\widehat{g_S}(\theta) = \sum_{k=1}^{n} \omega_k g^k(\theta) + \sum_{k=1}^{n} \omega_k n_{g^k(\theta)} \tag{5}$$

Based on Eq. 5, ITGN also contributes to task i's optimization in the MTL scenario, where ITGN_i is as follows:

$$\text{ITGN}_i = \sum_{k \neq i} \omega_k n_{g^k(\theta)} \sim \mathcal{N}(0, \frac{\sum_{k \neq i} \omega_k^2 C^k}{|S|}) \tag{6}$$

From Eq. 3, we can infer that $\{C^k\}$ and $|S|$ control the magnitude of $\{n_{g^k(\theta)}\}$, although multiple tasks share $|S|$, $\{C^k\}$ are task-specific and determined by the intrinsic characteristics of tasks, so $\{n_{g^k(\theta)}\}$ usually exist huge discrepancies, as shown in Fig. 1. Therefore, ITGN can cause performance deterioration in MTL (similar to the performance deterioration due to large noise in STL).

4 Method

We describe the necessity of controlling the noise magnitude range to maintain excellent model performance in Sect. 3.2 as Fig. 3(a) and explain that ITGN could also interfere with the specific task's optimization in MTL in Sect. 3.3. In summary, the range of ITGN may vary greatly, so it can be inferred that ITGN is likely to cause performance deterioration on some tasks as Fig. 3(b). In this section, we propose the MaxGNR algorithm, a dynamic weight strategy to minimize the ITGN interference in MTL as Fig. 3(c).

Gradient-to-Noise Ratio (GNR) is the core concept of this paper, which is defined to measure the relative magnitude of gradient noise (Sect. 4.1), and it can

Algorithm 1: MaxGNR algorithm

Input: data $\{X_i, Y_i^1, \ldots, Y_i^n\}_{|D|}$, learning rate μ.

Output: optimal network parameters θ.

Initialize network parameters θ_0 ;

for $t = 0$ *to* T **do**

select mini-batch data $S = \{X_i, Y_i^1, \ldots, Y_i^n\}_{|S|}$;

compute each empirical risk $\{\widehat{\mathcal{R}}^k(\theta_t)\}$;

compute the gradient of empirical risk $\{\nabla\widehat{\mathcal{R}}^k(\theta_t)\}$;

compute the momentum gradient $\{m_t^k\}$ and noise $\{n_t^k\}$;

select $\{\omega_t^k\}$ by $\arg\max_{\{\omega_k\}}\{\min\{\mathrm{GNR}_k(m_t^k, \{n_t^k\})\}\}$;

update network parameters $\theta_{t+1} = \theta_t - \mu\nabla_{\theta_t}\sum_{k=1}^{n}\omega_t^k\widehat{\mathcal{R}}^k(\theta_t)$;

end

be found that ITGN would reduce GNR compared to STL because ITGN is an additional gradient noise source. So it is a viable method to mitigate ITGN interference by maximizing GNR. To solve the dilemma that GNR is not computable, we propose to estimate $\{g^k(\theta)\}$ and $\{n_{g^k(\theta)}\}$ using the momentum method, and thus estimate a computable GNR (Sect. 4.2). Eventually, we can dynamically select the weights that maximize GNR in each iteration, thus achieving the purpose of alleviating ITGN interference (Sect. 4.3). The whole process of the MaxGNR algorithm is shown in Alg. 1.

4.1 Gradient-to-Noise Ratio

In order to measure the noise interference on stochastic gradient optimization, it is necessary to define an appropriate metric to measure the relative magnitude of gradient noise to the gradient.

Definition 1. *In the process of SGD, for a specific task, the expected gradient is* $g(\theta)$*, the random variable that could make* $\widehat{g_S}(\theta)$ *deviate from* $g(\theta)$ *is the gradient noise* $n_{g(\theta)}$*, we define* ***Gradient-to-Noise Ratio (GNR)*** *as:*

$$GNR = \frac{\|g(\theta)\|^2}{\mathbb{E}[\|n_{g(\theta)}\|^2]} \tag{7}$$

GNR reflects the relative magnitude of gradient noise to the gradient. The homogeneity ensures that GNR is independent of weight in STL, and according to Eq. 3, $\mathbb{E}[\|n_{g(\theta)}\|^2]$ is an approximate constant and easy to represent.

There is only one source of gradient noise in STL. According to definition 1, GNR_S is represented as:

$$\text{GNR}_S = \frac{\|g(\theta)\|^2}{\mathbb{E}[\|n_{g(\theta)}\|^2]} = \frac{\|g(\theta)\|^2}{tr(C)/|S|} \tag{8}$$

In the MTL scenario, due to the ITGN interference, task k's GNR is represented as:

$$\text{GNR}_M^k = \frac{\|\omega_k g^k(\theta)\|^2}{\mathbb{E}[\|\omega_k n_{g^k(\theta)} + \text{ITGN}_k\|^2]} \tag{9}$$

According to Eq. 3 and Eq. 5, $\mathbb{E}[\|\text{ITGN}_k\|^2]$ could be represented by $\{C^k\}$ and $|S|$, and due to the linear property of matrix's trace:

$$tr(m \cdot A + n \cdot B) = m \cdot tr(A) + n \cdot tr(B) \tag{10}$$

GNR_M^k is eventually simplified as:

$$\text{GNR}_M^k = \frac{\|\omega_k g^k(\theta)\|^2}{\sum_{k=1}^n \omega_k^2 tr(C^k)/|S|} \leq \text{GNR}_S^k \tag{11}$$

In general, hyperparameters with superior performance in STL would be selected for MTL, so we assume that GNR_S^k can be controlled as an appropriate GNR. However, ITGN interference reduces the GNR of each task, if $\text{ITGN}_k \gg n_{g^k(\theta)}$, GNR_M^k could be much smaller than GNR_S^k, which is likely to cause performance deterioration of task k. Therefore, selecting weights $\{\omega_k\}$ to maximize GNR_M^k is a feasible method to alleviate ITGN interference for task k. In the MTL scenario, to ensure that all tasks could alleviate ITGN interference as much as possible, we should maximize $\min\{\text{GNR}_M^k\}$ as:

$$\{\omega_k\} = \arg\max_{\{\omega_k\}} \left\{ \frac{\min\left\{\|\omega_k g^k(\theta)\|^2\right\}}{\sum_{k=1}^n \omega_k^2 tr(C^k)/|S|} \right\} \tag{12}$$

4.2 Estimation of Expected Gradient

However, GNR is a theoretical concept and it is not realistic to calculate $\{g^k(\theta_t)\}$ and $\{C^k\}$, therefore, we should design a method to estimate them. During the training process, we can only get the empirical gradient $\{\widehat{g_S}^k(\theta_t)\}$. In this paper, we propose that the momentum method can be used to estimate the expected gradient and gradient noise.

Momentum Method. The momentum method is widely applied in various optimizers. AdaGrad, AdaDelta, Adam, and so on adopt the momentum method as a technique to stabilize gradient. Momentum is expressed as m, and we set $m_0 = \widehat{g_S}(\theta_0)$. The basic process of the momentum method is as follows:

$$\begin{aligned} m_t &= \gamma \cdot m_{t-1} + (1-\gamma) \cdot \widehat{g_S}(\theta_t) \\ \theta_t &= \theta_{t-1} - \mu \cdot m_t \end{aligned} \tag{13}$$

where γ is the decay rate, μ is the learning rate, both are hyperparameters. It is an option to estimate the expected gradient by momentum gradient, because the momentum gradient could reduce noise and stabilize gradient. According to Eq. 2 and Eq. 13, the momentum gradient can be derived as:

$$m_t^k \approx (1-\gamma)\sum_{i=1}^{t}\gamma^{t-i}\cdot g^k(\theta_i) + n_{m_t^k}$$
$$\text{where} \quad n_{m_t^k} \sim \sqrt{\frac{1-\gamma}{1+\gamma}}\mathcal{N}(0,\frac{C}{|S|}) \tag{14}$$

By comparing Eq. 2 and Eq. 14, it can be found that the momentum method produces a contraction coefficient of $\sqrt{(1-\gamma)/(1+\gamma)}$ on noise magnitude, the larger γ is, the smaller the estimated noise magnitude is. At the same time, the cost is that the momentum gradient m_t^k is no longer the un-biased estimation of $g^k(\theta_t)$, the larger γ is, the greater the estimation error is. So γ becomes a trade-off factor of the expected gradient estimation, if the estimation error of $g^k(\theta_t)$ is too large, the momentum method is also invalid. Fortunately, parameter optimization is a long process, and the change of parameters in each iteration is small compared to the parameters themselves. Based on this fact, we make the following assumption:

Assumption 2. the expected gradient changes slowly and the expected gradients in adjacent iterations are similar:

$$g^k(\theta_n) \approx g^k(\theta_{n+\Delta n}), \text{ when } \Delta n \text{ is small} \tag{15}$$

On the premise of Assumption 2, the estimation error of $g^k(\theta_t)$ in Eq. 14 can be reduced to a large extent, so it is reasonable to estimate $g^k(\theta_t)$ with $m^k(\theta_t)$.

4.3 Weight Selection

According to the estimation method proposed by Eq. 14, the empirical gradient of each task can be decomposed into momentum gradient $m^k(\theta)$ and gradient noise $n_{g^k}(\theta)$. Therefore, we can estimate $n_{g^k}(\theta)$:

$$n_{g^k}(\theta) = \widehat{g_S}^k(\theta) - m^k(\theta) \tag{16}$$

To calculate the GNR of each task, we can replace $g^k(\theta)$ with $m^k(\theta)$ according to Eq. 14 and Assumption 2, and although $\mathbb{E}[\|n_{g^k(\theta)}\|^2]$ is approximately determined, each gradient noise is a random variable, so it is reasonable to use $\|\sum_{k=1}^{n}\omega_k n_{g^k(\theta)}\|^2$ to replace $\sum_{k=1}^{n}\omega_k^2 tr(C^k)/|S|$. Therefore, the MaxGNR algorithm can ultimately expressed as:

$$\{\omega_k\} = \arg\max_{\{\omega_k\}}\left\{\frac{\min\{\|\omega_k m^k(\theta)\|^2\}}{\|\sum_{k=1}^{n}\omega_k n_{g^k(\theta)}\|^2}\right\} \tag{17}$$

Table 2. Experiment results of different algorithms for NYUv2, and we split the table into coarse-grained and fine-grained algorithms. The **bold** represents the top2 scores.

Algorithm	Segmentation		Depth		Surface Normal	
	mIoU ↑	Pix Acc ↑	Abs Err ↓	Rel Err ↓	Mean ↓	Median ↓
Single task	26.15	53.27	0.6655	27.89	30.57	25.12
Equal weights	28.32	55.60	0.6196	27.61	31.80	26.82
coarse-grained						
DWA	28.43	56.08	0.6075	25.08	31.98	26.89
Uncertainty	28.06	54.46	0.6059	25.69	31.49	26.32
fine-grained						
GradNorm	28.92	56.26	0.6057	**24.06**	31.66	26.49
MGDA	22.38	49.59	0.7414	27.51	30.50	**23.85**
PCGrad	28.72	56.17	0.6131	27.33	32.19	27.00
MaxGNR ($\gamma = 0.5$)	**29.76**	**57.64**	**0.5943**	**24.35**	**29.89**	**24.38**
MaxGNR ($\gamma = 0.9$)	**29.48**	**56.81**	**0.5904**	24.91	**30.43**	24.85

Because of the non-linearity of Eq. 17, a general analytical solution cannot be found. We use the steepest descent method [3] to obtain the optimal $\{\omega_k\}$ at each iteration. Compared with the number of parameters in the neural network, the number of parameters in the steepest descent method is the number of tasks, so the computation cost can be ignored. At the same time, the computation cost of the steepest descent method is not sensitive to the number of tasks, which is different from grid search.

5 Experiments

5.1 Experimental Settings

Datasets. We focused on one-to-many predictions datasets $\{X_i, Y_i^1, \cdots, Y_i^n\}_{|D|}$ [30] in this paper. We evaluated the proposed MaxGNR algorithm on two image datasets: NYUv2 [22] and CityScapes [7]. NYUv2 is a challenging indoor scene dataset in various room types (bathrooms, living rooms, studies, etc.), and this dataset has three tasks: 13-class semantic segmentation, depth estimation, and surface normal prediction. Although NYUv2 is a relatively small dataset (795 training, 654 test images), it contains both regression and classification tasks, making it a good candidate for testing the robustness of different types of loss functions. CityScapes is a high-resolution street-view images dataset, we chose two sub-tasks for our experiment: semantic segmentation and depth estimation. Compared to NYUv2, the street-view image contained in CityScapes is relatively simpler, because the viewpoints and lighting conditions are relatively fixed, and the appearance of each object class changes little in shape.

Table 3. Experiment results of different algorithms for CityScapes.

Algorithm	Segmentation		Depth	
	mIoU ↑	Pix Acc ↑ ↑	Abs Err ↓	Rel Err ↓
Single task	66.91	90.03	0.0153	35.85
Equal weights	67.63	90.09	0.0163	57.13
GradNorm	67.20	90.76	0.0169	57.21
MGDA	66.83	90.88	0.0170	47.14
PGGrad	**67.75**	**91.01**	0.0164	54.95
MaxGNR ($\gamma = 0.5$)	**68.01**	**91.07**	**0.0162**	**44.11**
MaxGNR ($\gamma = 0.7$)	67.33	90.76	**0.0160**	**42.48**

Implementation Details. We implemented our experiments based on the network architecture in [17]. The network architecture had an encoder-decoder structure that was homogeneous across all tasks, where the encoder extracted the representation and the decoder matched individual tasks. For the NYUv2 and CityScapes datasets, we set the batch size to 2 and 128, respectively, and the learning rate of the Adam optimizer to 1e-4. After training each model for 100/200 epochs, we selected the best model on the training set for testing. In our experiment, we set the learning rate of the steepest descent method to 1e-3, and $\{\omega_k\}$ updated 100 steps in each iteration.

Evaluation Metrics. To compare the baselines and our method, we chose two metrics for each task. Following the settings in [6,17,28], we chose mIoU and Pixel Accuracy for Segmentation task, Absolute and Relative Error for Depth task, and Mean and Median Angle Distance for Surface Normal task.

5.2 Baselines

In addition to equal weights and single-task models, the baselines can be divided into two types: coarse-grained algorithms and fine-grained algorithms. Coarse-grained algorithms included DWA [17] and Uncertainty [6], fine-grained algorithms included GradNorm [4], MGDA [21] and PCGrad [28].

5.3 Main Results

Results on NYUv2 was shown in Table 2. Our MaxGNR algorithm outperformed the baselines in almost all metrics. When specific tasks were analyzed, most approaches outperformed the single-task model in the Depth task, and most baselines performed marginally better than the single-task model in the Segmentation task, while, in the Surface Normal task, almost all baselines suffered from performance deterioration. It was important to note that MGDA,

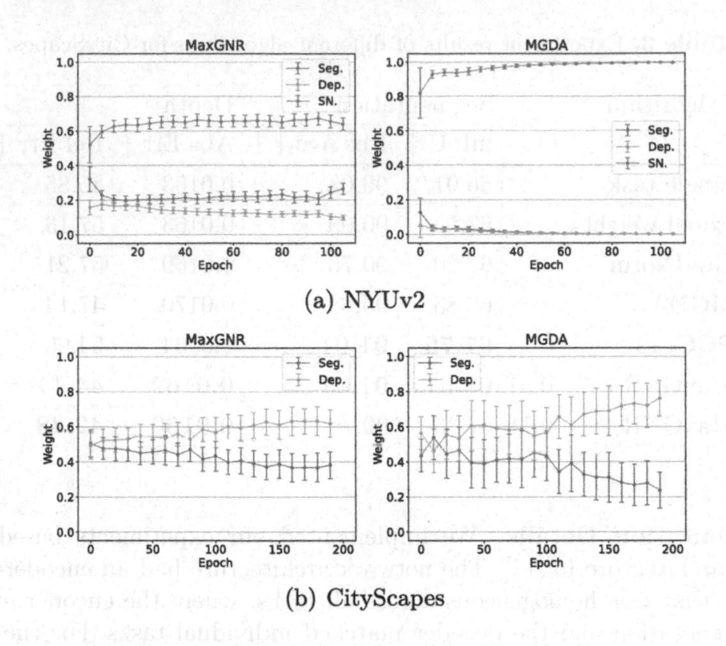

Fig. 4. Dynamic change of weights and the training process.

which outperformed the single-task model in the Surface Normal task, cannot balance the optimization performance on all tasks, resulting in MGDA's bad performance in the other two tasks. Our MaxGNR algorithm not only outperformed the single-task model and baselines in the Segmentation and Depth tasks, but it also performed well in the Surface Normal task and overcame the insufficient training problem.

Results on CityScapes was shown in Table 3. In CityScapes, the performance of the single-task model was similar to all MTL algorithms on the Segmentation task (about 91%), while there was a serious performance deterioration in the Depth task. We analyzed the performance of baselines and our algorithm in two tasks. In the Segmentation task, our algorithm was not significantly better than other baselines, but in the Depth task, our algorithm improved the performance of the model and greatly narrowed the gap with the single-task model.

5.4 Dynamic Change of Weights

The dynamic weight strategy can assign appropriate weights to each task at different training stages. We compared the weights of the different stages obtained by MaxGNR and MGDA in Fig. 4. We found that the weight trends obtained by the two algorithms were similar, but MGDA's weights were more extreme compared to MaxGNR's weights. Specifically, in NYUv2, the weight assignment of MGDA led the model to focus only on the Surface Normal task, while MaxGNR's assignment was intuitively more balanced. The experimental results (MGDA in Table 2)

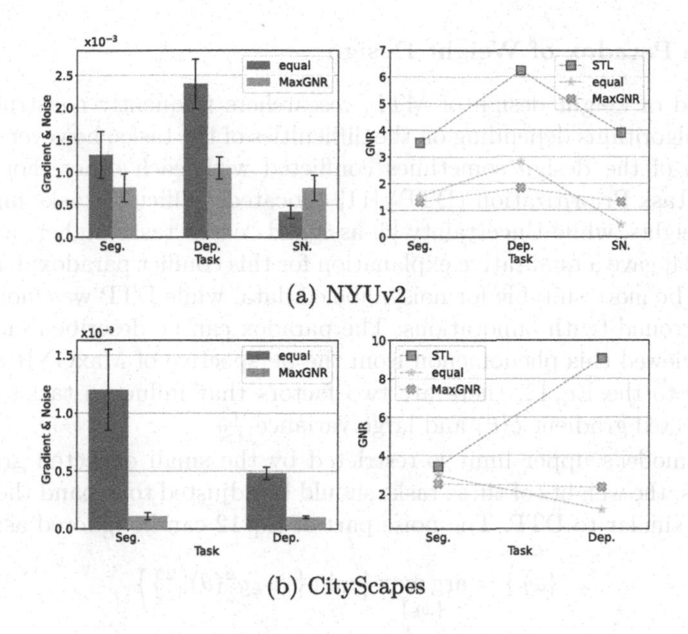

(a) NYUv2

(b) CityScapes

Fig. 5. Gradient Norm Distribution and GNR of each task in NYUv2 and CityScapes.

showed that extreme weight assignments may lead to insufficient training problems. The situation was similar in CityScapes. We believed that the weight assignment difference came from the object being balanced: MGDA attempted to balance each task's gradient, while MaxGNR attempted to balance each task's GNR.

6 Discussion

6.1 Gradient Noise and Performance

We fixed the best models under equal weights and MaxGNR settings and counted the distribution of the gradient norm for all data and calculated the current GNR of the models based on the gradient norm distribution, as shown in Fig. 5.

We found that in the equal weights model, the gradients and noises of each task differed significantly, and our algorithm clearly balanced the gradients and noises of each task (Fig. 5 Left). Compared with GNR_S, the GNR_M of Surface Normal task in NYUv2 and Depth task in CityScapes decreased most significantly in the equal weights model, and the performances of these tasks also exhibited a huge decrease. Our MaxGNR algorithm balanced each GNR_M (Fig. 5 Right), by slightly reducing the GNR_M of some tasks (usually harmless), the tasks suffering from performance deterioration would mitigate the interference of ITGN, and gained performance improvements. The experimental results also showed that the model in the MaxGNR setting had significant performance improvements in the Surface Normal task in NYUv2 and the Depth task in CityScapes. At the same time, the performance improvement of some tasks can help the model to get better representation, which can improve the performance of the model in general.

6.2 The Paradox of Weight Design

In the field of weight design of MTL, researchers frequently constructed optimization algorithms depending on the difficulties of the tasks, however, the starting points of the design sometimes conflicted with each other. For example, Dynamic Task Prioritization (DTP) [11] allocated "difficult" tasks higher task-specific weights, while Uncertainty [6] assigned "easy" tasks higher task-specific weights. [24] gave a qualitative explanation for this conflict paradox: Uncertainty seemed to be more suitable for noisy labeled data, while DTP was more suitable for clean ground-truth annotations. The paradox can be described more clearly when we viewed this phenomenon from the perspective of MaxGNR algorithm. According to the Eq. 12, there are two factors that influence task's difficulty: small expected gradient $g(\theta)$ and large variance $\frac{C}{|S|}$.

If the model's upper limit is restricted by the small expected gradients of some tasks, the weights of these tasks should be adjusted to expand the expected gradients, similar to DTP. The noise part of Eq. 12 can be ignored as:

$$\{\omega_k\} = \arg\max_{\{\omega_k\}} \left\{ \min \left\{ \|\omega_k g^k(\theta)\|^2 \right\} \right\} \tag{18}$$

On the contrary, if the model's upper limit is restricted by the large variances of specific tasks, the weights of these tasks should be reduced, similar to Uncertainty, to reduce the impact on other small gradient tasks. So the gradient part of Eq. 12 can be ignored as:

$$\{\omega_k\} = \arg\max_{\{\omega_k\}} \left\{ \frac{1}{\sum_{k=1}^{n} \omega_k^2 tr(C^k)/|S|} \right\} \tag{19}$$

In conclusion, the MaxGNR algorithm not only considers the magnitude of the gradient but also considers the noise interference, which to some extent unifies these two kinds of algorithms with distinct starting points.

7 Conclusion

In this work, we attributed the insufficient training problem in MTL to the ITGN interference, and we proposed MaxGNR algorithm, a novel dynamic weight strategy to alleviate this interference. Experiments verified the effectiveness of our algorithm. Looking ahead, gradient noise in MTL is a new field, and we hope to explore the influence of gradient noise on more complex tasks. Besides, how to choose appropriate tasks for joint learning is an open question, and the GNR framework may be a possible research direction.

Acknowledgements. This work was supported by the Shanghai Municipal Science and Technology Major Project (2021SHZDZX0102), and the Shanghai Science and Technology Innovation Action Plan (20511102600).

References

1. Bottou, L., et al.: Stochastic gradient learning in neural networks. In: NeurIPS (1991)
2. Bottou, L., Curtis, F.E., Nocedal, J.: Optimization methods for large-scale machine learning. Siam Review (2018)
3. Boyd, S.P., Vandenberghe, L.: Convex Optimization. Cambridge University Press, Cambridge (2014)
4. Chen, Z., Badrinarayanan, V., Lee, C.Y., Rabinovich, A.: Gradnorm: Gradient normalization for adaptive loss balancing in deep multitask networks. In: ICML (2018)
5. Chennupati, S., Sistu, G., Yogamani, S., Rawashdeh, S.A.: Multinet++: multi-stream feature aggregation and geometric loss strategy for multi-task learning. In: CVPR Workshops (2019)
6. Cipolla, R., Gal, Y., Kendall, A.: Multi-task learning using uncertainty to weigh losses for scene geometry and semantics. In: CVPR (2018)
7. Cordts, M., et al.: The cityscapes dataset for semantic urban scene understanding. In: CVPR (2016)
8. Duchi, J., Hazan, E., Singer, Y.: Adaptive subgradient methods for online learning and stochastic optimization. JMLR **12**, 2121–2159 (2011)
9. Gao, Y., Ma, J., Zhao, M., Liu, W., Yuille, A.L.: NDDR-CNN: layerwise feature fusing in multi-task CNNs by neural discriminative dimensionality reduction. In: CVPR (2019)
10. Goodfellow, I.J., Bengio, Y., Courville, A.C.: Deep Learning. MIT Press, Adaptive computation and machine learning (2016)
11. Guo, M., Haque, A., Huang, D.A., Yeung, S., Fei-Fei, L.: Dynamic task prioritization for multitask learning. In: ECCV (2018)
12. He, F., Liu, T., Tao, D.: Control batch size and learning rate to generalize well: theoretical and empirical evidence. In: NeurIPS (2019)
13. Hoffer, E., Hubara, I., Soudry, D.: Train longer, generalize better: closing the generalization gap in large batch training of neural networks. In: NeurIPS (2017)
14. Keskar, N.S., Mudigere, D., Nocedal, J., Smelyanskiy, M., Tang, P.T.P.: On large-batch training for deep learning: generalization gap and sharp minima. In: ICLR (2017)
15. Kingma, D.P., Ba, J.L.: Adam: a method for stochastic optimization. In: ICLR (2015)
16. Liu, L., et al.: Towards impartial multi-task learning. In: ICLR (2021)
17. Liu, S., Johns, E., Davison, A.J.: End-to-end multi-task learning with attention. In: CVPR (2019)
18. Mandt, S., Hoffman, M.D., Blei, D.M.: Stochastic gradient descent as approximate bayesian inference. JMLR **18**, 1–35 (2017)
19. Misra, I., Shrivastava, A., Gupta, A., Hebert, M.: Cross-stitch networks for multi-task learning. In: CVPR (2016)
20. Ruder, S., Bingel, J., Augenstein, I., Søgaard, A.: Latent multi-task architecture learning. In: AAAI (2019)
21. Sener, O., Koltun, V.: Multi-task learning as multi-objective optimization. In: NeurIPS (2018)
22. Silberman, N., Hoiem, D., Kohli, P., Fergus, R.: Indoor segmentation and support inference from RGBD images. In: ECCV (2012)

23. Sutskever, I., Martens, J., Dahl, G., Hinton, G.: On the importance of initialization and momentum in deep learning. In: ICML (2013)
24. Vandenhende, S., Georgoulis, S., Gansbeke, W.V., Proesmans, M., Dai, D., Gool, L.V.: Multi-task learning for dense prediction tasks: a survey. TPAMI (2021)
25. Vandenhende, S., Georgoulis, S., Gool, L.V.: MTI-Net: multi-scale task interaction networks for multi-task learning. In: ECCV (2020)
26. Wu, J., Hu, W., Xiong, H., Huan, J., Braverman, V., Zhu, Z.: On the noisy gradient descent that generalizes as SGD. In: ICML (2020)
27. Xu, D., Ouyang, W., Wang, X., Sebe, N.: Pad-net: Multi-tasks guided prediction-and-distillation network for simultaneous depth estimation and scene parsing. In: CVPR (2018)
28. Yu, T., Kumar, S., Gupta, A., Levine, S., Hausman, K., Finn, C.: Gradient surgery for multi-task learning. In: NeurIPS (2020)
29. Zeiler, M.D.: Adadelta: an adaptive learning rate method. arXiv preprint (2012)
30. Zhang, H., Xiao, L., Wang, Y., Jin, Y.: A generalized recurrent neural architecture for text classification with multi-task learning. In: IJCAI (2017)
31. Zhang, Z., Cui, Z., Xu, C., Jie, Z., Li, X., Yang, J.: Joint task-recursive learning for semantic segmentation and depth estimation. In: ECCV (2018)
32. Zhu, Z., Wu, J., Yu, B., Wu, L., Ma, J.: The anisotropic noise in stochastic gradient descent: its behavior of escaping from sharp minima and regularization effects. In: ICML (2019)

Adaptive FSP: Adaptive Architecture Search with Filter Shape Pruning

Aeri Kim, Seungju Lee, Eunji Kwon, and Seokhyeong Kang(✉)

Department of Electrical Engineering, Pohang University of Science and Technology, Pohang, South Korea
{arkim17,seungju825,eunjikwon,shkang}@postech.ac.kr

Abstract. Deep Convolutional Neural Networks (CNNs) have high memory footprint and computing power requirements, making their deployment in embedded devices difficult. Network pruning has received attention in reducing those requirements of CNNs. Among the pruning methods, Stripe-Wise Pruning (SWP) achieved a further network compression than conventional filter pruning methods and can obtain optimal kernel shapes of filters. However, the model pruned by SWP has filter redundancy because some filters have the same kernel shape. In this paper, we propose the Filter Shape Pruning (FSP) method, which prunes the networks using the kernel shape while maintaining the receptive fields. To obtain an architecture that satisfies the target FLOPs with the FSP method, we propose the Adaptive Architecture Search (AAS) framework. The AAS framework adaptively searches for the architecture that satisfies the target FLOPs with the layer-wise threshold. The layer-wise threshold is calculated at each iteration using the metric that reflects the filters influence on accuracy and FLOPs together. Comprehensive experimental results demonstrate that the FSP can achieve a higher compression ratio with an acceptable reduction in accuracy.

Keywords: Deep learning optimization · Structured pruning · Convolution neural networks

1 Introduction

Deep Convolutional Neural Networks (CNNs) have been widely used in computer vision applications, such as image classification [6,31], object detection [21,29,30], and segmentation [1,24]. These successes rely on the tremendous number of parameters and computations of the networks. However, their high requirements in storage and computing resource make the networks hard to deploy in edge devices. To address this problem, numerous studies have proposed different approaches to compress the networks. Popular approaches include quantization [2], compact network design [10,39], and network pruning [5,23].

Supplementary Information The online version contains supplementary material available at https://doi.org/10.1007/978-3-031-26319-4_32.

L. Wang et al. (Eds.): ACCV 2022, LNCS 13841, pp. 539–555, 2023.
https://doi.org/10.1007/978-3-031-26319-4_32

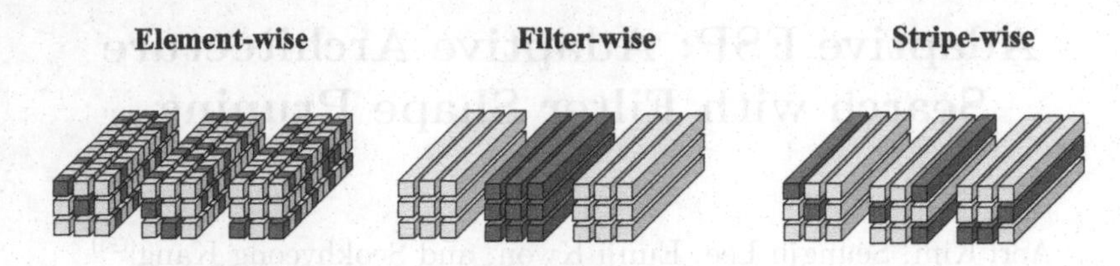

Fig. 1. Different pruning types of convolution layers.

Network pruning methods can be classified into several categories: element-wise pruning, filter pruning, and stripe-wise pruning. Element-wise pruning [5] is the most fine-grained non-structured pruning method that removes the individual weights to obtain a high compression rate without sacrificing accuracy. However, because the positions of non-zero weights are irregular and random, element-wise pruning has irregular memory access and thus cannot achieve practical performance improvement without dedicated hardware. In contrast, filter pruning [9,15,23] is a structured method that removes the unimportant filters, which can directly achieve real performance improvement in general processors. However, filter pruning is less fine-grained than element-wise pruning; therefore, the further compression is limited. To overcome this weakness of filter pruning, Stripe-Wise Pruning (SWP) [27] has been introduced. SWP combines the strengths of element-wise pruning and filter pruning methods. Thus, SWP achieves finer granularity than traditional filter pruning and can still be accelerated in general processors. Figure 1 shows the different pruning types of convolution layers.

SWP can obtain the optimal kernel shapes of filters by pruning unimportant stripes after training the importance of stripes in the filter using a learnable matrix called a FilterSkeleton (FS). The follow-up studies of SWP [12,20,22] focused on obtaining further optimized kernel shapes; therefore, the improvement of compression rate is insignificant or rather diminished. Moreover, when we visualize the filters of each layer in the model after SWP, several filters are found to have the same kernel shape as shown in Fig. 2. The kernel shapes significantly influence receptive fields, which are crucial for feature extraction. This indicates that the features extracted by filters with the same kernel shape have higher similarity than the others. Furthermore, this property causes filter redundancy, a phenomenon in which filters extract similar features that exist in duplicate. Thus, by pruning these filters, filter redundancy can be effectively reduced with little impact on accuracy. Motivated by these observations, we propose the Filter Shape Pruning (FSP) method, which prunes the model after SWP by kernel shape while preserving the receptive fields of kernel shapes. Moreover, we use "the FSP rule", a crucial concept in FSP, that maintains receptive fields. The FSP method can highly compress networks with a slight loss in accuracy.

In this study, we propose the Adaptive Architecture Search (AAS) framework that efficiently applies FSP to the networks. The AAS framework adaptively searches for an architecture, that meets the target FLOPs and fine-tunes the pruned model from scratch. Previous adaptive pruning studies [33,40] proposed a metric that only considers the accuracy or computational intensity of the

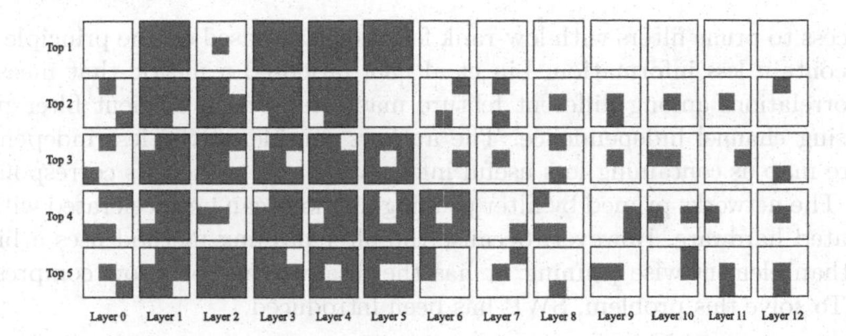

Fig. 2. Visualization of the VGG-16 filters pruned by SWP [27]. We illustrate the Top-5 filters according to their frequency in each layer, and the blue color indicates the remaining stripe after SWP.(Color figure online)

pruned model. In contrast, we propose a metric that considers the effect of the filters on accuracy and FLOPs. This metric can find filters that generate many FLOPs while less sensitive in accuracy. Furthermore, using the proposed metric during the AAS framework, we use the adaptive layer-wise threshold that reflects the architecture, which varies at each iteration. Using our proposed framework, we can obtain an architecture consisting of filters that have good scores in both accuracy and FLOPs.

The main contributions of this study can be summarized as follows:

- We propose the FSP method that efficiently reduces filter redundancy while maintaining receptive fields of kernel shapes. Furthermore, we prune the network using "the FSP rule" to preserve the receptive fields. Using FSP, we can achieve a higher compression ratio than other pruning methods.
- We propose the AAS framework using a layer-wise threshold that adaptively changes. The layer-wise threshold is calculated by a metric reflecting the influence of the filter on accuracy and FLOPs together. The framework obtains the architecture that satisfies the target FLOPs using the FSP method.

The rest of the paper is organized as follows. Section 2 covers related works. Section 3 presents preliminaries. Section 4 describes the FSP method and the AAS framework. Section 5 provides the experimental results and analysis. Section 6 covers additional experiments. Finally, we summarize and conclude this study in Sect. 7.

2 Related Work

Filter Pruning. Filter pruning is a structured pruning method that prunes unimportant filters. Therefore, it does not require dedicated hardware/libraries. Li et al. [15] pruned unimportant filters based on the l_1 norm of the filter weights. Liu et al. [23] used the scaling factor γ, a trainable variable of batch normalization, to remove unimportant channels from the output channels of each layer. He et al. [17] discovered that the average rank of several feature maps generated by a single filter is always the same, regardless of batch size. The authors formulated a process to prune filters with low-rank feature maps based on the principle that they contain less information. Lin et al. [35] proposed a metric that measures the correlations among different feature maps to perform efficient filter pruning using channel independence. The authors considered the less independent feature map as containing less useful information and pruned its corresponding filter. The networks pruned by filter pruning methods can be accelerated without dedicated hardware. However, because the filter pruning method uses a bigger unit than element-wise pruning, it has the disadvantage of a low compression rate. To solve this problem, SWP has been introduced.

Stripe-Wise Pruning. Meng et al. [27] proposed Stripe-Wise Pruning (SWP) to prune more fine-grained networks than conventional filter pruning methods and can still be accelerated without dedicated hardware. SWP obtains the optimal kernel shape of the filter by learning the importance of the filter stripes. Liu et al. [22] developed the pruning framework called Squeezing More Out of Filters (SMOF) that reduces the kernel size with the "peeling" method and the number of filters of CNNs with Filter Mask, which learns the importance of filters. Huo et al. [12] proposed Balanced-Stripe-Wise Pruning (BSWP). BSWP balances stripes by adding a regulation term to the loss, reflecting the frequency of kernel stripes and the number of filter stripes within each layer. Liu et al. [20] proposed a two-stage approach that automatically finds the optimal kernel shape. The authors add three regulation terms to the loss: sparse regulation, direction-wise regulation, and group-wise regulation. Furthermore, they proposed a binary search algorithm to find the pruning threshold to meet the constraint. Because these studies focused on obtaining further optimized kernel shapes, the improvement in compression ratio is small or rather decreased. Therefore, the benefits of the fine granularity of SWP are underused. To address this problem, we propose the Filter Shape Pruning (FSP) method, which prunes the networks using the kernel shape while fully using the fine granularity of SWP.

Adaptive Pruning Method. The adaptive pruning method determines the pruning rate by considering the trade-off between accuracy and reduction in computations at each iteration. Therefore, this method has a smaller accuracy drop than the non-adaptive pruning strategy, which prunes a fixed percentage of filters. Singh et al. [33] proposed a framework called Play and Prune (PP), which jointly prunes and fine-tunes the networks with an adaptive weight threshold. The initial weight threshold is obtained using an optimization formula, and the weight threshold is adaptively determined by considering the accuracy of the remaining filters. Zhao et al. [40] presented an adaptive and activation-based pruning approach to generate models that automatically satisfy each of the three target tasks: accuracy-critical, memory-constructed, and latency-sensitive. This work uses the average of activation-based attention maps to determine the importance of filters and prunes the networks by adaptively adjusting the global threshold. However, these studies used a metric that reflects either the accuracy or FLOPs when adaptively obtaining pruning rates. To solve this problem, we propose a metric that considers both accuracy and FLOPs, which allows us to find unimportant filters in both aspects.

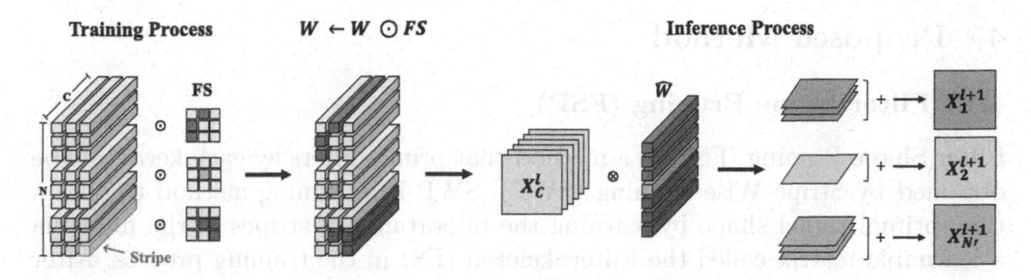

Fig. 3. Overall process of SWP. During the training process, FilterSkeleton (FS) learns the importance of stripes. After the training process, valid stripes are multiplied by filters and we obtain the optimal kernel shapes. Only valid stripes create an output feature map in the inference process.

3 Preliminaries

In this section, we review the key concepts of Stripe-Wise Pruning (SWP) [27] as shown in Fig. 3. SWP has been introduced to learn the optimal kernel shape with pruning and achieves $K^2\times$ finer granularity than filter pruning (assuming the kernel size is $K \times K$). Meng et al. [27] introduced a learnable matrix I, namely FilterSkeleton (FS), to learn the optimal kernel shape. In the convolution process, I is multiplied by filter W, which can be expressed as

$$X_{n,h,w}^{l+1} = \sum_{c}^{C}\sum_{i}^{K}\sum_{j}^{K} I_{n,i,j}^{l} \times W_{n,c,i,j}^{l} \times X_{c,h+i-1,w+j-1}^{l}, \tag{1}$$

where $X_{c,h,w}^{l}$ denotes a l-th convolutional layer feature map. N and C are the number of filters and channels of the input feature map in layer l, respectively; and n and c are the n-th filter and c-th input channel, respectively. K denotes kernel size, and i and j are indexes of the width and height of the kernel. I is first initialized with an all-one matrix, and the l_1 norm penalty of I is added to the original loss term.

For efficient pruning, stripes with $I_{n,i,j}^{l}$ less than threshold T are no longer updated in the training process, and these stripes are pruned after the training process. We define the pruned FS as $\hat{I}_{n',i,j}^{l}$, where n' denotes the n'-th filter in N', and N' is the number of filters with at least one stripe remaining in the kernel after pruning. Finally, $\hat{I}_{n',i,j}^{l}$ is merged to W (*i.e.*, $\hat{W} \leftarrow W \odot \hat{I}$), and only $\hat{W}$ is used during inference. In the inference process shown in Fig. 3, $\hat{W}$ has only valid stripes; therefore, the calculation order of Equ. (1) can be modified so that the channel-direction is calculated first. In other words, Equ. (1) can be reformulated as follows:

$$X_{n,h,w}^{l+1} = \sum_{i}^{K}\sum_{j}^{K}(\sum_{c}^{C} \hat{W}_{n,c,i,j}^{l} \times X_{c,h+i-1,w+j-1}^{l}) . \tag{2}$$

The channel-direction calculation of Equ. (2) can be implemented by 'im2col'; therefore, the model after SWP is available in the general processor.

4 Proposed Method

4.1 Filter Shape Pruning (FSP)

Filter Shape Pruning (FSP) is a method that prunes filters by each kernel shape obtained by Stripe-Wise Pruning (SWP). SWP is a pruning method to obtain the optimal kernel shape by learning the importance of stripes in the filter via a learnable matrix called the FilterSkeleton (FS) in the training process. After training, stripes with a FS value less than the global threshold are pruned, and then we can gain the optimal kernel shapes consisting of valid stripes in each filter. As shown in Fig. 2, filters with the same kernel shape exist at each layer. As the kernel shape of the filter is a significant factor affecting receptive fields, filters with the same kernel shape have similar receptive fields, which causes filter redundancy. FSP can efficiently reduce this filter redundancy to obtain a highly compressed model.

The pruning targets of FSP are filters with overlapping kernel shapes in each layer, and filters with the same kernel shape can be grouped together. Thus, N' filters in layer l can be expressed as follows:

$$W^l_{N'} = \{S^l_0, S^l_1, \cdots, S^l_a, \cdots, S^l_A\}, \tag{3}$$

where A is the number of types of kernel shape in layer l, and a is the a-th kernel shape type. S^l_a is a set of filters with the same kernel shape type. In other words, N' filters in layer l can be classified by kernel shape; therefore, there can be multiple filters with kernel shape of a-th type, and we define a set of those filters as S^l_a. If more than two filters exist within S^l_a, the corresponding S^l_a is the pruning target set. Here, the most important point is to retain at least one filter within each S^l_a to preserve the receptive field of the kernel shape. In other words, the maximum compression rate of FSP is defined when all S^l_a have only one filter. Therefore, it is not necessary to search for an architecture adaptively to gain the maximum compression rate. We observed that small N networks such as ResNet56/110 are less likely to overlap kernel shapes than others; therefore, we propose the weight inheritance technique which uses the weights after SWP. The weight inheritance technique can reduce the overhead of searching and fine-tuning for small N networks while obtaining the maximum compression rate; it describes how to combine all filters in S^l_a as one new filter, and only the new filter is used for fine-tuning. To create the new filter, the weights $W^l_{n'}$ of all filters in S^l_a are multiplied with their FilterSkeleton $I^l_{n'}$, and the multiplication values are added together. Also, a new FilterSkeleton is initialized with the matrix in which only the kernel shape position (i, j) is filled with '1'; otherwise filled with '0'. This process can be formulated as

$$W^l_m = \sum^{num\_f} W^l_{n'} \odot I^l_{n'}, \quad I^l_{m,i,j} = \begin{cases} 1, & \text{if (i, j)} \subset \text{validstripes} \\ 0, & \text{otherwise} \end{cases}, \tag{4}$$

where num_f denotes the number of filters in S^l_a. We applied FSP to ResNet56/110 using the weight inheritance technique in our experiments.

The purpose of FSP is to remove redundant filters with the same kernel shape until reaching the target FLOPs while preserving the minimum diversity of kernel shapes. Therefore, FSP can be expressed as a problem of determining the number of filters to remove in each pruning target set S_a^l. To solve this problem, we propose the Adaptive Architecture Search (AAS) framework. In FSP, removing redundant filters with the same kernel shape is more important if there are multiple filters in S_a^l. However, if there is only one filter in S_a^l, FSP stops the removal of filters in S_a^l to preserve the receptive field of the kernel shape, and we define this"the FSP rule". We describe the AAS framework and "the FSP rule" in Sect. 4.2.

4.2 Adaptive Architecture Search (AAS)

We propose the AAS framework that determines the number of filters to be removed from each S_a^l to obtain the architecture that meets the target FLOPs. For selecting filters to remove, we introduce a metric that considers the influence of the filter on accuracy and FLOPs and use the layer-wise threshold T^l obtained by this metric. In other words, the framework removes filters that have smaller importance than T^l. Then, T^l adaptively changes according to the proposed metric updated at each iteration. In the framework, if only one filter remains in an S_a^l, the S_a^l is excluded from the pruning target sets to preserve the receptive field of the kernel shape.

Layer-Wise Threshold. We introduce a metric that determines how much to increase the layer-wise threshold, and this metric simultaneously considers the influence of the filter on accuracy and FLOPs. To calculate the metric, we define the filter importance as $F_{n'}^l$ and FLOPs importance as $M_{n'}^l$. $F_{n'}^l$ represents the filter's influence on the accuracy, and the larger the $F_{n'}^l$, the greater the impact of the filter on accuracy. $F_{n'}^l$ is defined follows:

$$F_{n'}^l = \frac{\sum_i^K \sum_j^K \hat{I}_{n',i,j}^l}{n_s}, \tag{5}$$

where n_s is the number of valid stripes in the filter, which is the number of nonzero elements in $\hat{I}_{n'}^l$. $M_{n'}^l$ represents the FLOPs generated by the filter and is defined as follows:

$$M_{n'}^l = H' \times W' \times C \times n_s, \tag{6}$$

where H' and W' denote the output feature map height and width, respectively.

To determine how much to increase the layer-wise threshold, we first need to determine the effect of each layer on accuracy and FLOPs. Therefore, we calculate the layer score L_s^l and layer FLOPs L_m^l using Equ. (5) and Equ. (6), and they are expressed as

$$L_s^l = \frac{\sum_{n'}^{N'} F_{n'}^l}{N'}, \quad L_m^l = \frac{\sum_{n'}^{N'} M_{n'}^l}{N'}. \tag{7}$$

Using Equ. (7), the relative layer's accuracy importance AL^l and FLOPs importance FL^l compared to all convolution layers are as follows:

$$AL^l = \frac{L_s^l}{\sum_l^L L_s^l}, \quad FL^l = \frac{L_m^l}{\sum_l^L L_m^l}. \tag{8}$$

Finally, the formulation of the metric with AL^l and FL^l is as follows:

$$\Delta T^l = \frac{\alpha \cdot (1 - AL^l) + \beta \cdot FL^l}{metric\_norm}, \tag{9}$$

where α and β are the weight parameters of the metric. We will discuss the variation of parameters and FLOPs according to the values of α and β in Sect. 6.2. *metric_norm* regulates the variation of the threshold. Since the threshold is too large if the model's FLOPs are smaller than the target FLOPs, *metric_norm* is multiplied by ϵ ($\epsilon > 1$) to make ΔT^l small. We used $\epsilon = 2$ in our experiments. The larger the AL^l, the greater the effect of the filter on accuracy; thus, we have to prune the smaller AL^l filters. Therefore, we use $(1 - AL^l)$ for the metric.
Using the metric (Equ. (9)), the layer-wise threshold T^l is updated as follows:

$$T^{l'} = T^l \cdot (1 + \Delta T^l). \tag{10}$$

The FSP Rule. The key concept of FSP is to retain at least one filter in S_a^l to preserve the receptive fields of kernel shapes. To maintain the key concept of FSP, the AAS framework follows the FSP rule, which deals with maintaining receptive fields throughout the framework. After pruning, if there is only one filter in S_a^l, S_a^l is eliminated from the pruning target sets in the next iteration. In the case that all filters in S_a^l are less than T^l, only the filter with the most significant $F_{n'}^l$ remains, and S_a^l is eliminated from the pruning target in the next iteration.

The Procedure of Adaptive Architecture Search (AAS). The AAS procedure is illustrated in Fig. 4. First, the framework classifies the filters of the model after SWP (Equ. (3)), and sets S_a^l, which has more than two filters, as the pruning target set. Second, our framework saves or rewinds the state. The framework evaluates the FLOPs of the pruned model in the end, and if the FLOPs of the pruned model are smaller than the target FLOPs in iteration i, *metric_norm* increases to make smaller ΔT^L of iteration $i+1$. To obtain a model close to the target FLOPs by applying the adjusted ΔT^L, the model before pruning at iteration i is required, so the framework has to save the pruned model of iteration $i-1$. Therefore, the framework checks whether *metric_norm* is updated and if not updated, the framework saves the latest model. If *metric_norm* is updated, the framework rewinds the current state to the saved state. Next, the framework calculates the metric (Equ. (9)) and update the layer-wise threshold T^L (Equ. (10)). Then, the framework compares the filter importance $F_{n'}^l$ with T^l for each layer and prunes the model according to "the FSP rule". Finally, the framework evaluates the FLOPs of the pruned model, which are described as

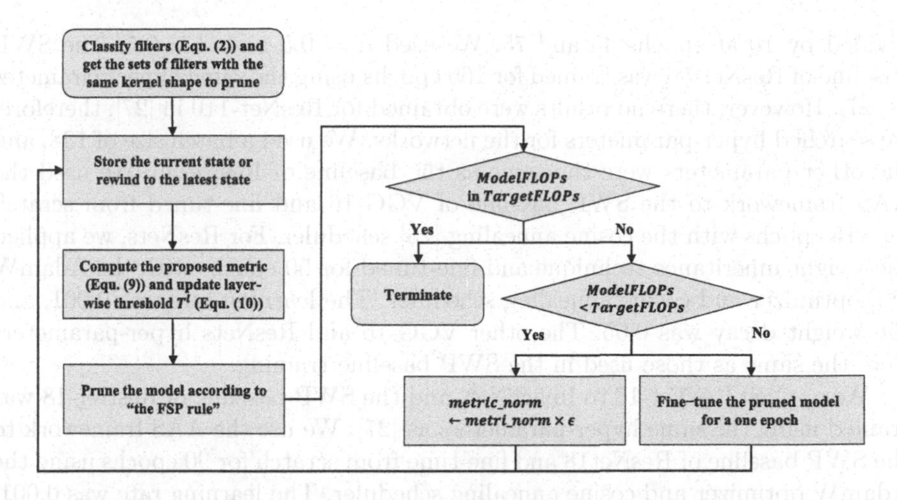

Fig. 4. AAS framework procedure. $ModelFLOPs$ represents the FLOPs of the pruned model and $TargetFLOPs$ represents the target FLOPs boundary. When the pruned model meets $TargetFLOPs$, it is fine-tuned from scratch.

$ModelFLOPs$. If $ModelFLOPs$ is within the target FLOPs, the framework is terminated. If it is smaller than the target FLOPs, ΔT^L is adjusted by increasing *metric_norm*. The adjusted ΔT^L makes $ModelFLOPs$ at the next iteration close to the target FLOPs. If $ModelFLOPs$ does not reach the target FLOPs, the framework fine-tunes the pruned model for one epoch. This process adjusts the FS values of the pruned model in iteration i so that the FS values suitable for the current architecture can be reflected in the next iteration. The proposed metric reflects the changed architecture for each iteration; therefore, the framework can find the optimal architecture that satisfies the target FLOPs. After obtaining the pruned model that satisfies the target FLOPs, the pruned model is fine-tuned from scratch.

5 Experiments

5.1 Experimental Settings

We conducted experiments on two popular datasets (Cifar-10 [13] and ImageNet [3]) and two different architectures (VGG-16 [32] and ResNet [7]). We applied VGG-16 and ResNet-56/110 to Cifar-10. Our method was based on a model after Stripe-Wise Pruning (SWP); thus, we obtain the baseline of SWP[1] mainly using hyper-parameters similar to those used in [27]. The SWP baseline of VGG-16 was trained using the same hyper-parameter as [27], except the training epoch. We used 115 epochs for training, and the initial learning rate was 0.1 and was

[1] The baseline obtained by SWP influences the performance of our method. The better the SWP baseline, the better the performance compared with our results. We report the SWP re-implementation results used as the baseline.

divided by 10 at epochs 45 and 75. We used $\alpha = 0.5$ and $\beta = 0.5$. The SWP baseline of ResNet-56 was trained for 160 epochs using the same hyper-parameter as [27]. However, there no results were obtained for ResNet-110 in [27]; therefore, we searched hyper-parameters for the networks. We used a batch size of 128, and the other parameters were the same as the baseline of ResNet56. We used the AAS framework to the SWP baseline of VGG-16 and fine-tuned from scratch for 115 epochs with the cosine annealing [25] scheduler. For ResNets, we applied the weight inheritance technique and fine-tuned for 50 epochs with the AdamW [26] optimizer and cosine annealing scheduler. The learning rate was 0.001, and the weight decay was 0.05. The other VGG-16 and ResNets hyper-parameters were the same as those used in the SWP baseline training.

We applied ResNet-18 to ImageNet, and the SWP baseline of ResNet-18 was trained using the same hyper-parameters as [27]. We use the AAS framework to the SWP baseline of ResNet18 and fine-tune from scratch for 90 epochs using the AdamW optimizer and cosine annealing scheduler. The learning rate was 0.001, and the weight decay was 0.01. We used $\alpha = 1$ and $\beta = 2$. The other ResNet-18 parameters were the same as those used for the SWP baseline training.

5.2 Results on Cifar-10

VGG-16. Table 1 shows the performance of our method and other pruning methods. AAS-15% indicates that the target FLOPs were 15% less than that of the SWP baseline. Furthermore, our SWP baseline is the re-implementation result of SWP in Table 1. Compared to L1, Hinge, GAL, SSS, SMOF, and AAB, AAS-15% achieved a significantly large reduction in both FLOPs and parameters with higher accuracy. AAS-30% was advantageous in all aspects compared to ABCPruner and HRank. AAS-50% significantly reduced the FLOPs (83.6%) and parameters (96.73%) and was the only one that reduced the FLOPs by more than 80%. Figure 5 shows the comparison of our method with other pruning methods, and the higher the dot in the upper right, the better the performance of the method. In Fig. 5a, the advantages of the FSP were clearer. Our results showed better performance considering both the FLOPs reduction and accuracy. These results indicate that FSP reduces filter redundancy while maintaining the receptive field of the kernel so that it can further compress the networks with a small accuracy loss.

ResNet-56/110. Table 1 shows the experimental results of ResNet-56/110 on Cifar-10. Max indicates that we apply the maximum compression rate of FSP to the networks using the weight inheritance technique. Our SWP baselines are re-implementation results of SWP in Table 1. The Max of ResNet-56 reduced FLOPs by 80.51% and deleted parameters by 78.22%. In addition, our results had the best performance in all aspects among the studies in Table 1. Compared to our SWP baseline, our results achieved higher accuracy and improved reduction of parameters and FLOPs. Figure 5b shows a graph comparing Max to other studies in terms of FLOPs reduction and accuracy. Max is at the furthest upper right than the other works, which confirms that our method performed better in terms of accuracy and FLOPs.

Table 1. Comparing the FSP method with state-of-the-art pruning methods for VGG-16 and ResNet-56/110 on Cifar-10. We have sorted in the order of small reduction in FLOPs, and our methods show the most significant reduction in FLOPs and parameters.

Model	Method	Accuracy (%)	FLOPs ↓ (%)	Param ↓ (%)
VGG-16	Baseline	93.25	0	0
	L1 [15]	93.40	34.2	64
	Hinge [16]	93.59	39.07	80.05
	GAL [19]	92.03	39.6	77.6
	SSS [11]	93.02	41.6	73.8
	SMOF [22]	93.50	58	80.9
	AAB [40]	93.41	61.17	72.85
	SWP (reimp.)	93.52	67.02	93.1
	SWP [27]	93.65	71.16	92.66
	Ours (AAS-15%)	**93.61**	**72.01**	**94.43**
	ABCPruner [18]	93.08	73.68	88.68
	HRank [17]	91.23	76.5	92
	Ours (AAS-30%)	**93.40**	**76.99**	**95.47**
	Ours (AAS-50%)	**92.71**	**83.6**	**96.73**
ResNet-56	Baseline	93.10	0	0
	CP [9]	91.8	50	–
	DSA [28]	92.91	52.2	–
	SOKS [20]	93.08	51.73	54.12
	IR [36]	92.70	67.7	–
	AAB [40]	92.54	71.44	–
	CHIP [34]	92.05	72.3	71.8
	HRank [17]	90.72	74.1	68.1
	SWP [27]	92.98	77.7	75.6
	SWP (reimp.)	92.82	77.95	73.82
	TRP [38]	91.62	77.83	–
	FP [14]	91.54	79.50	70.59
	Ours (Max)	**92.88**	**80.51**	**78.22**
ResNet-110	Baseline	93.50	0	0
	L1 [15]	93.30	38.6	32.4
	GAL [19]	92.55	48.5	44.8
	FP [14]	93.73	48.52	44.77
	ABCPruner [18]	93.58	65.04	67.41
	HRank [17]	92.65	68.6	68.7
	SWP(reimp.)	93.63	70	71.43
	CHIP [34]	93.63	71.6	68.3
	Ours (Max)	**93.42**	**74.8**	**77.11**

The results of ResNet-110 is also reported in Table 1. Compared with L1, GAL, and HRank, our results had better performance on all fronts. Particularly, our method removed 2.38× parameters and 1.94× FLOPs than L1. In compari-

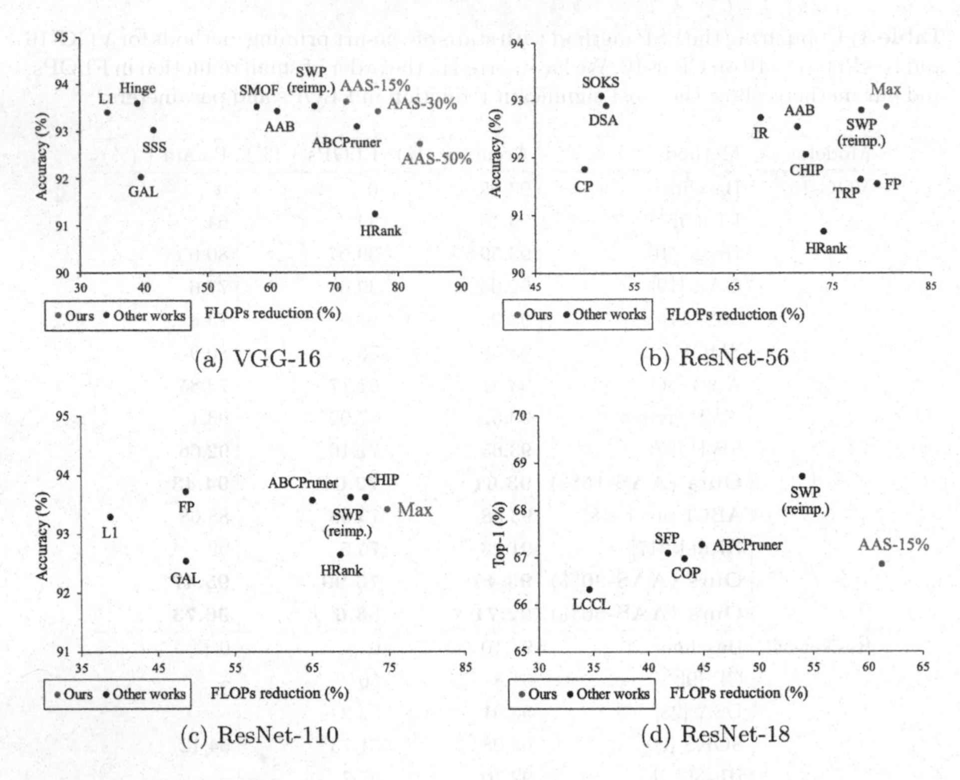

(a) VGG-16

(b) ResNet-56

(c) ResNet-110

(d) ResNet-18

Fig. 5. FLOPs reduction and accuracy comparison with other pruning methods on Cifar-10 and ImageNet. (a)-(c) were the results of Cifar-10, and (d) was the result of ImageNet.

son with FP and CHIP, Max reduced FLOPs and parameters with little drop in accuracy. Figure 5c demonstrates that our method achieved the highest reduction in FLOPs with only a slight drop in accuracy. These results demonstrate that FSP can obtain a high compression rate with little accuracy loss by maintaining the receptive fields of kernel shapes.

5.3 Results on ImageNet

The results for ResNet-18 on ImageNet are shown in Table 2. Our SWP baseline is the re-implementation result of SWP in Table 2. AAS-15% reduced the FLOPs by 15% and deleted parameters by 36.74% based on the SWP baseline. Compared with LCCL, our results reduced more FLOPs with higher accuracy. Furthermore, our results achieved a 61.2% FLOPs and 57.74% parameters reduction with small accuracy loss compared to SFP, COP, and ABCPruner. Figure 5d shows that AAS-15% had the highest compression rate among the other pruning methods with similar accuracy.

Table 2. Comparing the AAS framework with state-of-the-art pruning methods on ImageNet. We applied our framework to ResNet-18. We sorted in the order of smallest reduction in FLOPs, and our method obtained the highest compression ratio in FLOPs and parameters.

Model	Method	Top-1 (%)	Top-5 (%)	FLOPs ↓ (%)	Param ↓ (%)
ResNet-18	Baseline	69.76	89.08	0	0
	LCCL [4]	66.33	86.94	34.6	–
	SFP [8]	67.1	87.78	41.8	–
	COP [37]	66.98	–	43.3	45.1
	ABCPruner [18]	67.28	87.67	44.88	43.55
	SWP (reimp.)	68.72	88.63	53.98	41.83
	SWP [27]	69.59	89.04	54.58	–
	Ours (AAS-15%)	**66.86**	**87.05**	**61.2**	**57.74**

Table 3. Effectiveness of pruning filters while preserving the receptive field of the kernel shape. We applied VGG-16 on Cifar-10 and ResNet-18 on ImageNet.

Model	Method	Param (M)	FLOPs (M)	Accuracy (%)
VGG-16	**FSP**	**0.68**	**144.33**	**93.40**
	Simple baseline	0.78	156.14	92.79
ResNet-18	**FSP**	**4.94**	**1415.95**	**66.86**
	Simple baseline	6.01	1424.57	65.64

6 Ablation Study

6.1 Effect of Preserving the Receptive Field

We experimentally proved the effect of preserving the receptive field of the kernel shape, which is the core concept of FSP. We compared FSP with a 'Simple baseline,' which removes filters with the same kernel shape after SWP in the order of the smallest filter importance (Equ. (5)). We experimented VGG-16 on Cifar-10 and ResNet-18 on ImageNet, and used the same SWP baseline for both FSP and the simple baseline. We fine-tuned both FSP and the simple baseline to the same conditions as Sect. 5.1. The result is shown in Table 3.

In Table 3, although FSP has fewer parameters and FLOPs than the simple baseline, it showed higher accuracy in both VGG-16 and ResNet-18. In the simple baseline case, we observed that some kernel shape sets had no filters. That is, the receptive fields corresponding to those sets were erased, which resulted in an additional accuracy drop. The results demonstrate that the core concept of FSP, which is to preserve the receptive field of the kernel shape, plays a key role in reducing the loss in accuracy.

Table 4. Variance of parameters, FLOPs, and accuracy according to α and β. We set $\alpha = 1 - \beta$.

α	β	Param ↓ (%)	FLOPs ↓ (%)	FLOPs/Param	Accuracy (%)
0.1	0.9	20.19(0.83M)	30.18(144.46M)	1.49	93.07
0.3	0.7	27.88(0.75M)	30.35(144.10M)	1.09	93.21
0.5	0.5	34.62(0.68M)	30.24(144.33M)	0.87	93.40
0.7	0.3	38.46(0.64M)	30.22(144.37M)	0.79	93.18
0.9	0.1	40.38(0.62M)	30.03(144.76M)	0.74	92.96

6.2 Weight Parameters of the Metric: α and β

α and β are weight parameters used for the proposed metric and determine whether to assign weight to accuracy importance or FLOPs importance. We investigated the difference between the parameters and FLOPs according to α and β for the general case. We experimented with VGG-16 on Cifar-10 under the same conditions except for α and β and used AAS-30% as a pruning method. We reported parameters and FLOPs reduction compared to the SWP baseline. Table 4 shows the results. The larger the β compared to the α, the larger the FLOPs reduction compared to the parameters reduction. In other words, the larger the β, the more the framework preferentially prunes filters that generate many FLOPs. Considering the accuracy, the overall performance was good when $\alpha = 0.5$ and β=0.5.

7 Conclusion

In this study, we propose the Filter Shape Pruning (FSP) method, which prunes networks using the kernel shape of the filter while preserving the receptive field of the kernel shape. In addition, we proposed the Adaptive Architecture Search (AAS) framework to search for the architecture that satisfies the target FLOPs with the FSP method. The AAS framework adaptively searches the architecture that meets the target FLOPs with the layer-wise threshold. The layer-wise threshold is updated at each iteration by the proposed metric that considers the effect of the filter on both accuracy and FLOPs. The experimental results demonstrated that the FSP method could obtain a higher compression rate than other pruning methods with an acceptable accuracy loss by preserving the receptive fields of kernel shapes.

Acknowledgements. This work was supported by Institute of Information & communications Technology Planning & Evaluation (IITP) grant funded by the Korea government (MSIT) (No.2022-0-01172, DRAM PIM Design Base Technology Development) and by the National Research Foundation of Korea(NRF) grant funded by the Korea government(MSIT) (No.2019R1A5A1027055).

References

1. Chen, L.C., Papandreou, G., Kokkinos, I., Murphy, K., Yuille, A.L.: DeepLab: semantic image segmentation with deep convolutional nets, atrous convolution, and fully connected CRFs. IEEE Trans. Pattern Anal. Mach. Intell. **40**(4), 834–848 (2017)
2. Chen, W., Wilson, J., Tyree, S., Weinberger, K., Chen, Y.: Compressing neural networks with the hashing trick. In: International Conference on Machine Learning, pp. 2285–2294. PMLR (2015)
3. Deng, J., Dong, W., Socher, R., Li, L.J., Li, K., Fei-Fei, L.: ImageNet: a large-scale hierarchical image database. In: 2009 IEEE Conference on Computer Vision and Pattern Recognition, pp. 248–255. IEEE (2009)
4. Dong, X., Huang, J., Yang, Y., Yan, S.: More is less: a more complicated network with less inference complexity. In: Proceedings of the IEEE Conference on Computer Vision and Pattern Recognition, pp. 5840–5848 (2017)
5. Han, S., Pool, J., Tran, J., Dally, W.: Learning both weights and connections for efficient neural network. In: Advances in Neural Information Processing Systems, vol. 28 (2015)
6. He, K., Zhang, X., Ren, S., Sun, J.: Deep residual learning for image recognition. In: Proceedings of the IEEE Conference on Computer Vision and Pattern Recognition, pp. 770–778 (2016)
7. He, K., Zhang, X., Ren, S., Sun, J.: Deep residual learning for image recognition. In: Proceedings of the IEEE Conference on Computer Vision and Pattern Recognition, pp. 770–778 (2016)
8. He, Y., Kang, G., Dong, X., Fu, Y., Yang, Y.: Soft filter pruning for accelerating deep convolutional neural networks. In: IJCAI International Joint Conference on Artificial Intelligence (2018)
9. He, Y., Zhang, X., Sun, J.: Channel pruning for accelerating very deep neural networks. In: Proceedings of the IEEE International Conference on Computer vision, pp. 1389–1397 (2017)
10. Howard, A.G., et al.: MobileNets: efficient convolutional neural networks for mobile vision applications. arXiv preprint arXiv:1704.04861 (2017)
11. Huang, Z., Wang, N.: Data-driven sparse structure selection for deep neural networks. In: Proceedings of the European Conference on Computer Vision (ECCV), pp. 304–320 (2018)
12. Huo, Z., Wang, C., Chen, W., Li, Y., Wang, J., Wu, J.: Balanced stripe-wise pruning in the filter. In: ICASSP 2022–2022 IEEE International Conference on Acoustics, Speech and Signal Processing (ICASSP), pp. 4408–4412. IEEE (2022)
13. Krizhevsky, A., Hinton, G., et al.: Learning multiple layers of features from tiny images (2009)
14. Li, D., Chen, S., Liu, X., Sun, Y., Zhang, L.: Towards optimal filter pruning with balanced performance and pruning speed. In: Proceedings of the Asian Conference on Computer Vision (2020)
15. Li, H., Kadav, A., Durdanovic, I., Samet, H., Graf, H.P.: Pruning filters for efficient convnets. arXiv preprint arXiv:1608.08710 (2016)
16. Li, Y., Gu, S., Mayer, C., Gool, L.V., Timofte, R.: Group sparsity: the hinge between filter pruning and decomposition for network compression. In: Proceedings of the IEEE/CVF Conference on Computer Vision and Pattern Recognition, pp. 8018–8027 (2020)

17. Lin, M., Ji, R., Wang, Y., Zhang, Y., Zhang, B., Tian, Y., Shao, L.: HRank: filter pruning using high-rank feature map. In: Proceedings of the IEEE/CVF Conference on Computer Vision and Pattern Recognition, pp. 1529–1538 (2020)
18. Lin, M., Ji, R., Zhang, Y., Zhang, B., Wu, Y., Tian, Y.: Channel pruning via automatic structure search. In: Proceedings of the Twenty-Ninth International Conference on International Joint Conferences on Artificial Intelligence, pp. 673–679 (2021)
19. Lin, S., Ji, R., Yan, C., Zhang, B., Cao, L., Ye, Q., Huang, F., Doermann, D.: Towards optimal structured CNN pruning via generative adversarial learning. In: Proceedings of the IEEE/CVF Conference on Computer Vision and Pattern Recognition, pp. 2790–2799 (2019)
20. Liu, G., Zhang, K., Lv, M.: SOKs: automatic searching of the optimal kernel shapes for stripe-wise network pruning. IEEE Trans. Neural Netw. Learn. Syst. (2022)
21. Liu, W., et al.: SSD: single shot multibox detector. In: Leibe, B., Matas, J., Sebe, N., Welling, M. (eds.) ECCV 2016. LNCS, vol. 9905, pp. 21–37. Springer, Cham (2016). https://doi.org/10.1007/978-3-319-46448-0_2
22. Liu, Y., Guan, B., Xu, Q., Li, W., Quan, S.: SMOF: squeezing more out of filters yields hardware-friendly CNN pruning. arXiv preprint arXiv:2110.10842 (2021)
23. Liu, Z., Li, J., Shen, Z., Huang, G., Yan, S., Zhang, C.: Learning efficient convolutional networks through network slimming. In: Proceedings of the IEEE International Conference on Computer Vision, pp. 2736–2744 (2017)
24. Long, J., Shelhamer, E., Darrell, T.: Fully convolutional networks for semantic segmentation. In: Proceedings of the IEEE Conference on Computer Vision and Pattern Recognition, pp. 3431–3440 (2015)
25. Loshchilov, I., Hutter, F.: SGDR: stochastic gradient descent with warm restarts. arXiv preprint arXiv:1608.03983 (2016)
26. Loshchilov, I., Hutter, F.: Decoupled weight decay regularization. arXiv preprint arXiv:1711.05101 (2017)
27. Meng, F., et al.: Pruning filter in filter. Adv. Neural. Inf. Process. Syst. **33**, 17629–17640 (2020)
28. Ning, X., Zhao, T., Li, W., Lei, P., Wang, Yu., Yang, H.: DSA: more efficient budgeted pruning via differentiable sparsity allocation. In: Vedaldi, A., Bischof, H., Brox, T., Frahm, J.-M. (eds.) ECCV 2020. LNCS, vol. 12348, pp. 592–607. Springer, Cham (2020). https://doi.org/10.1007/978-3-030-58580-8_35
29. Redmon, J., Divvala, S., Girshick, R., Farhadi, A.: You only look once: unified, real-time object detection. In: Proceedings of the IEEE Conference on Computer Vision and Pattern Recognition, pp. 779–788 (2016)
30. Ren, S., He, K., Girshick, R., Sun, J.: Faster R-CNN: towards real-time object detection with region proposal networks. In: Advances in Neural Information Processing Systems, vol. 28 (2015)
31. Simonyan, K., Zisserman, A.: Very deep convolutional networks for large-scale image recognition. arXiv preprint arXiv:1409.1556 (2014)
32. Simonyan, K., Zisserman, A.: Very deep convolutional networks for large-scale image recognition. arXiv preprint arXiv:1409.1556 (2014)
33. Singh, P., Verma, V.K., Rai, P., Namboodiri, V.P.: Play and Prune: adaptive filter pruning for deep model compression. arXiv preprint arXiv:1905.04446 (2019)
34. Sui, Y., Yin, M., Xie, Y., Phan, H., Aliari Zonouz, S., Yuan, B.: Adv. Neural. Inf. Process. Syst. **34**, 24604–24616 (2021)
35. Sui, Y., Yin, M., Xie, Y., Phan, H., Aliari Zonouz, S., Yuan, B.: CHIP: channel independence-based pruning for compact neural networks. In: Advances in Neural Information Processing Systems, vol. 34 (2021)

36. Wang, H., Zhang, Q., Wang, Y., Yu, L., Hu, H.: Structured pruning for efficient convnets via incremental regularization. In: 2019 International Joint Conference on Neural Networks (IJCNN), pp. 1–8. IEEE (2019)
37. Wang, W., Fu, C., Guo, J., Cai, D., He, X.: COP: customized deep model compression via regularized correlation-based filter-level pruning. In: Proceedings of the 28th International Joint Conference on Artificial Intelligence, pp. 3785–3791 (2019)
38. Xu, Y., et al.: TRP: trained rank pruning for efficient deep neural networks. In: Proceedings of the Twenty-Ninth International Conference on International Joint Conferences on Artificial Intelligence, pp. 977–983 (2021)
39. Zhang, X., Zhou, X., Lin, M., Sun, J.: ShuffleNet: an extremely efficient convolutional neural network for mobile devices. In: Proceedings of the IEEE Conference on Computer Vision and Pattern Recognition, pp. 6848–6856 (2018)
40. Zhao, K., Jain, A., Zhao, M.: Adaptive activation-based structured pruning. arXiv preprint arXiv:2201.10520 (2022)

DecisioNet: A Binary-Tree Structured Neural Network

Noam Gottlieb(✉) and Michael Werman

The Hebrew University of Jerusalem, Jerusalem, Israel
{noam.gottlieb2,michael.werman}@mail.huji.ac.il

Abstract. Deep neural networks (DNNs) and decision trees (DTs) are both state-of-the-art classifiers. DNNs perform well due to their representational learning capabilities, while DTs are computationally efficient as they perform inference along one route (root-to-leaf) that is dependent on the input data. In this paper, we present DecisioNet (DN), a binary-tree structured neural network. We propose a systematic way to convert an existing DNN into a DN to create a lightweight version of the original model. DecisioNet takes the best of both worlds - it uses neural modules to perform representational learning and utilizes its tree structure to perform only a portion of the computations. We evaluate various DN architectures, along with their corresponding baseline models on the FashionMNIST, CIFAR10, and CIFAR100 datasets. We show that the DN variants achieve similar accuracy while significantly reducing the computational cost of the original network.

Keywords: Neural network optimization · Decision trees

1 Introduction

Deep neural networks (DNNs) have achieved exceptional performance in various visual recognition tasks in recent years, such as image classification, object detection, and semantic segmentation. That is mostly due to their representational learning capabilities. However, deploying DNN models in an industrial environment is challenging - especially when the computational resources are low (which is the case for many mobile device applications) or where the model's inference time has to be fast enough (*e.g.*, real-time applications). In addition, a DNN is seen in many cases as a "black box" - one cannot easily figure out *why* a final prediction is made.

Another powerful machine learning model is the Decision Tree (DT). A DT model learns a routing function, where each node of the tree routes the data to one of its children until it reaches a leaf with the final output. This conditional computation means that only part of the DT is visited for each input thus achieving high efficiency. Moreover, DTs provide an interpretable structure, allowing the user to understand why a decision was made. On the downside, DTs usually require hand-engineered data features, and they cannot be trained with gradient-based optimization methods, which limits their expressiveness.

L. Wang et al. (Eds.): ACCV 2022, LNCS 13841, pp. 556–570, 2023.
https://doi.org/10.1007/978-3-031-26319-4_33

In this paper, we propose a novel general model with the benefits of both DNNs and DTs - the DecisioNet (DN). This is a binary-tree structured DNN, derived from any other DNN that we wish to reduce its computational cost. It can be trained end-to-end using backpropagation [18] just like any other DNN. In addition, the DN has routing modules which play the role of the DT within this model, routing the input through the tree. The outcome is a lighter model - in terms of parameters and even more in terms of computational cost - whose performance is at par with the baseline model. We evaluate and compare full baseline models against their DN variants on the FashionMNIST [22], CIFAR10, and CIFAR100 [14] datasets. In this paper, we examine DN only for image classification tasks but this method can be used for other types of tasks as well.

Contributions. The contributions of this paper are: i) we propose a systematic way of transforming an existing DNN into a tree-structured version of it (its DecisioNet), yielding a lightweight model with comparable performance. ii) we propose a way of training the DN end-to-end, despite the explicit data routing functions within the DN (which involves non-differentiable operations) using Improved Semantic Hashing. iii) we provide an open-source PyTorch [16] implementation of DecisioNet, available at https://github.com/noamgot/DecisioNet.

2 Related Work

Soft Decision Trees. The Soft Decision Tree (SDT) mechanism has been studied in various works, including [5,10,17,20]. SDT is a DT with neural routing nodes sending the data to a sub-tree, multiplied by a probabilistic factor in the range $[0, 1]$. The final prediction of the tree is the weighted sum of all leaves, where the weight of each leaf is the probability of arriving at it.

This method differs from traditional DTs, whose decisions are binary and deterministic. Another SDT-like method is the Neural Decision Forest (NDF), proposed in [13]. This method achieved great performance on the Imagenet dataset [4], replacing the last fully-connect layer of a neural network with decision forest nodes. These nodes yield a prediction which is a weighted sum of all the trees' predictions. The major drawback of SDTs is that all paths of the tree must be executed to get a prediction. These methods lack the advantage of traditional decision trees using only a single computational path, and therefore their efficiency is limited. In this work, we focus on hard decision trees only, which perform only one path of the tree.

Neural Trees with Hard Routing. Another type of neural DT is proposed in [6,8], where the forward pass utilizes the routing nodes to make a hard decision, and in this way indeed only the relevant nodes of the tree are visited. [8] introduce a routing function which outputs binary values at test time only, such that the tree still performs soft decisions during training, allowing it to train end-to-end. [6] on the other hand introduce the Tree Ensemble Layer (TEL), which is capable of performing hard decisions at test time and even during training, by simply skipping unreachable nodes (*i.e.*, nodes whose ancestors reach probability is 0). Notice that the number of reachable leaves is not necessarily limited to 1, hence a soft routing is still performed. In contrast, our method applies fully hard routing (with a single output leaf) both at training time (where we also use soft decision behavior) and at test time.

Tree-Structured Neural Networks. Instead of replacing DTs decision nodes with neural modules, the other direction is also possible: creating a DNN with a structure of a DT, such that the routing behavior of a DT is present. This network consists of multiple sub-networks, that are traversed conditionally based on the routing nodes' decisions. [23] proposed HD-CNN, where a small CNN is first activated to classify inputs into coarse categories. This network has a hierarchical architecture which is essentially an SDT-structured DNN with a single decision node with k branches. HD-CNN is not trained end-to-end but in a modular way: there is a common (shared) network for predicting coarse categories and coarse-category expert networks. Each of these networks is trained alone (the expert modules are trained only with their corresponding classes; The coarse-category labels are achieved by clustering similar classes based on an existing network), and all of them are fined tuned together in the end. [9] extended the tree-structured net of HD-CNN to conditional networks. These are DAG-based CNN architectures with data routing. They distinguish between two types of routing: i) explicit routing - where data is conditionally passed to a node's children (one or more) based on a routing function. This method is similar to DTs. ii) implicit routing - where the data is split into portions that are sent to the node's children unconditionally. In the paper, this method is done using filter groups (*i.e.*, splitting outputs feature maps into groups, where each group goes to a separate route). Another tree-structured NN is Adaptive Neural Trees (ANT) [21]. In ANT, the tree structure is learned together with the model's weights; This approach is different from tree-structured NNs with a static architecture that is commonly used (in this work as well).

Differences and Similarities. DecisioNet is, to some extent, a combination of HD-CNN [23] and conditional networks [9]. We create labels for training the routing modules with a method that's based on the one proposed in HD-CNN, but unlike [23], we train deeper, tree-structured DNNs with explicit routing, and we do the training end-to-end. We use explicit routing, similar to the one that was suggested for conditional networks. However, we use a different method for training end-to-end (improved semantic hashing [11]), and more importantly, we show actual results for explicitly-routed DNN trees (in [9] the main results were achieved using implicit routing; The only result that contained explicit routing was a CNN ensemble with a single decision for choosing which models to use for the final prediction). Finally, the approach used by ANT [21] is building the tree-structured net from scratch; We aim to optimize the computational cost of an *existing network* by transforming its structure into a tree.

3 DecisioNet

The proposed model is based on any existing deep neural network (DNN) such as VGG [19], ResNet [7], *etc.*We'll denote such a DNN as the "baseline model".

3.1 Architecture

The DN architecture is based directly on the baseline model's architecture. It is a transformation of the baseline model into a binary decision tree, whose primary goal is to achieve similar performance with fewer computational operations. We begin by choosing *split points* - these are the places the new DN model will decide to route the data into one of two routes. In our method, after the i-th split (Starting with $i = 1$), we split the relevant layers to 2^i equal portions, in terms of the number of filters in convolutional layers, *etc.*A toy example of this idea is displayed in Fig. 1. This way we get a balanced tree whose number of parameters is fewer than that of the baseline model. Moreover, since at test time our model chooses a single path from the root to a leaf, traversing only through a portion of the nodes (just like a DT), the number of multiply-accumulate (MAC) operations is also smaller. The routing at each split is done using a trainable *routing module* (RM). The RM is a small, lightweight network that makes a binary decision (a detailed explanation is found in Sect. 3.3).

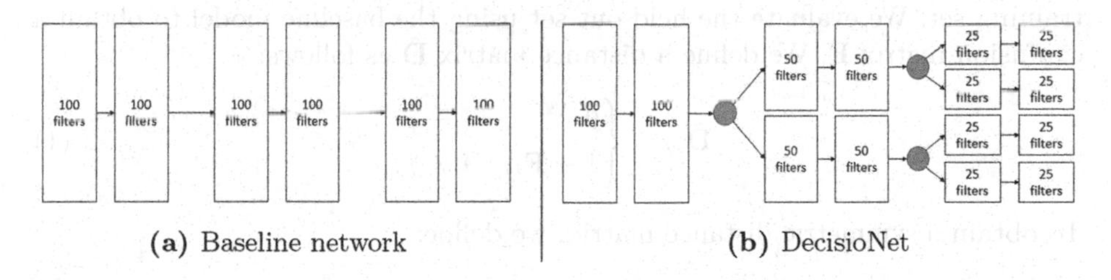

Fig. 1. A toy example of transforming a baseline model (a) into its respective DecisioNet (b). For the sake of clarity, the baseline model consists of 6 convolutional layers, each with 100 filters. The blue circles represent routing modules. In the DN model, the layers after the first split (leftmost blue circle) have half the number of filters compared to their corresponding layers in the baseline model. Similarly, the layers after the second split(s) have a quarter the number of filters. This idea can be easily generalized to an arbitrary number of splits. At test time, each RM picks one of two routes and the data is passed to the next module through this route only. (Color figure online)

3.2 Classes Hierarchical Clustering

After deciding the depth of the DecisioNet (*i.e.*, the number of splits), we perform hierarchical clustering of the dataset's classes to obtain intermediate labels for the routing modules. The goal of this phase is to extract the hierarchical structure of the data, grouping similar classes into nested clusters. The routing labels are derived directly from the clustering. The general idea is demonstrated in Fig. 2.

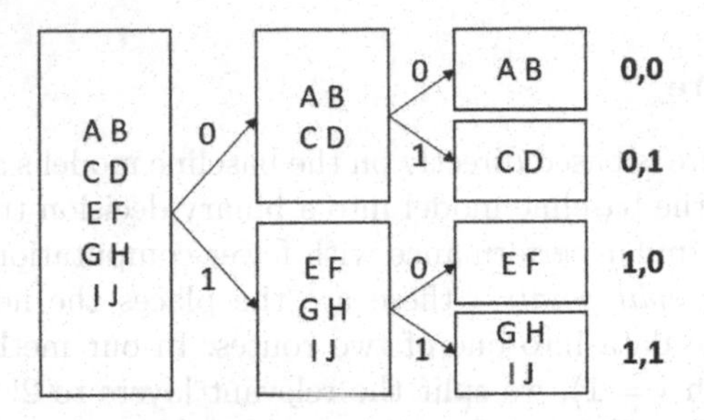

Fig. 2. Hierarchical clustering example - 10 classes (A–J) are clustered into a clustering tree of depth 2. The deepest (rightmost) clusters share the same routing labels, stated to their right. For example, classes E and F labels are 1 for the first routing, and 0 for the second.

We use a method similar to the one used by [23], extending it to handle more than just 2 hierarchical levels (coarse and fine), using a different clustering method. We begin by randomly sampling a balanced subset of images out of the training set. We evaluate the held-out set using the baseline model to obtain a confusion matrix $\mathbf{F}$. We define a distance matrix $\hat{\mathbf{D}}$ as follows:

$$\hat{\mathbf{D}}_{ij} = \begin{cases} 0 & i = j \\ 1 - \mathbf{F}_{ij} & i \neq j \end{cases} \tag{1}$$

To obtain a symmetric distance matrix, we define:

$$\mathbf{D} = \frac{1}{2}\left(\hat{\mathbf{D}} + \hat{\mathbf{D}}^T\right) \tag{2}$$

At this point, $\mathbf{D}_{ij}$ measures the similarity between classes i and j. Having a distance matrix, we use it to perform hierarchical agglomerative clustering. This phase gives us a division of the dataset's classes into 2 clusters, a division of each of these clusters into 2 sub-clusters, and so on. This can be seen as a mapping of a class to its set of clusters $F_c : \{i\}_{i=1}^{C} \rightarrow \{0,1\}^k$, where C is the number of classes and k is the depth of the DecisioNet tree.

Unlike [23], we do not allow overlap between same-level clusters, *i.e.*, at each clustering depth, each class can be found in exactly one sub-cluster. The main problem with this choice is that images that are routed to a wrong branch (at any level, even in the deepest routing module) end up in the wrong leaf and are miss-classified. We overcome this problem by allowing each leaf of the DN to predict all classes - even classes that shouldn't have been routed to this leaf. Allowing cluster overlap is possible though, and we leave this for future research.

3.3 Routing

One of the key questions when dealing with tree-structured neural networks is which routing method to use. There are 2 main types of routing found in the literature:

1. *Soft routing* - in this method, the output of each branch is multiplied by a real value (usually between 0 and 1) and the final output of some routing point is the sum of its branches' outputs.
2. *Hard routing* - in this method, each branch either passes its input or not.

The main advantage of the hard-routing method (and the disadvantage of the soft-routing method), is that it allows us to save computational cost by performing only the chosen path of a given input. To decrease the computational cost, DN uses the hard-routing approach.[1]

The Routing Module. To choose a route for input, we use a routing module (RM) - this module is a small efficient neural network which outputs a single binary value. Inspired by [2,3], we used a similar routing module. These papers used this module to choose filters of a convolutional layer, hence they needed multiple outputs; However, we use this module to choose a *single* branch, so we use a slightly modified version of it.

Let $x \in \mathbb{R}^{C \times H \times W}$ be the output of the last layer before a split in the tree. Our routing module can be defined as follows:

$$RM(x) = B\left(FC\left(GAP\left(x\right)\right)\right) \tag{3}$$

where GAP is a Global Average Pooling operation, defined as:

$$GAP(x) = \frac{1}{HW} \sum_{i=0}^{H-1} \sum_{j=0}^{W-1} x_{ij} \tag{4}$$

FC is a linear projection layer (*i.e.*, a fully-connected layer) with C inputs and a single output and B is a binarization function which will be introduced in the next paragraph. A schematic diagram of the proposed routing module can be found in Fig. 3.

Improved Semantic Hashing. We want our module to output a binary value - either 0 or 1 - to choose the next computational path. Performing simple binarization using a threshold is not an option, as this operation is almost always with zero derivative and therefore we will not be able to perform backpropagation properly. Like [2,3], we adopt the Improved Semantic Hashing method, proposed by [11].

[1] To the best of our knowledge, popular deep learning frameworks, such as PyTorch [16] and TensorFlow [1], do not support tree-structured neural nets when using batches of inputs. It means that in practice, during training and evaluating the data passes through all the nodes of the tree where we zero out the "untraversed" ones.

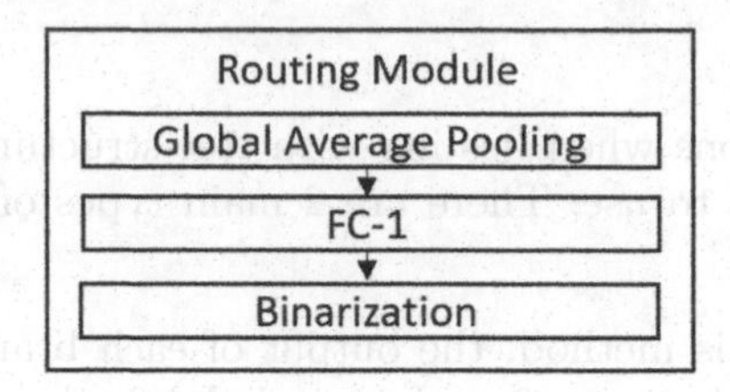

Fig. 3. A schematic diagram of the proposed routing module. FC-1 is a fully-connected layer with a single output neuron. Full details are in Sect. 3.3

During training, we draw a random noise ϵ with mean 0 and standard deviation 1. We define the following values:

$$\begin{aligned} g_\epsilon(z) &= z + \epsilon \\ g_r(z) &= \sigma'\left(g_\epsilon\left(z\right)\right) \\ g_b(z) &= \mathbb{1}\left(g_\epsilon\left(z\right) > 0\right) \end{aligned} \tag{5}$$

$\mathbb{1}(z)$ is an indicator function (evaluated to 1 when z is true and 0 otherwise); σ' is the saturating sigmoid function from [12]:

$$\sigma'(z) = \max\left(0, \min\left(1, 1.2\sigma\left(z\right) - 0.1\right)\right) \tag{6}$$

where σ is the well-known sigmoid function. Notice that g_r is a real-valued number in the range $[0,1]$ while g_b is a binary value; Moreover, the gradient of g_r is well defined w.r.t g_ϵ, while the gradient of g_b w.r.t g_ϵ is 0.

Finally, we can define our binarization function B as follows:

$$B(z) = g_{b/r}(z) \tag{7}$$

The term $g_{b/r}$ denotes that in the forward pass we use either g_r or g_b, and that is eventually the output of the RM. The choice is done at random with a 50% chance for each. In the backward pass, we *always* use the gradients of g_r to backpropagate meaningful gradients[2]. To allow this dual behavior, we sum the branches' outputs in the following way:

$$f(x) = \left(1 - \left(RM\left(x\right)\right)\right) f_L\left(x\right) + RM\left(x\right) f_R\left(x\right) \tag{8}$$

where f_L, f_R are the functions applied to x by the left and right branches. Notice that if RM outputs a binary value, only one of the branches is used, while in the real-valued case the branches' outputs are mixed (*i.e.*, soft routing is applied).

During evaluation and inference, we perform the same procedure with 2 changes: first, there is no additive noise ($\epsilon = 0$); second, the routing module only outputs binary values, so in this phase, we only use g_b as an output of RM.

To conclude, the Improved Semantic Hashing method allows us to train the DecisioNet model in an end-to-end manner and to apply hard decisions at inference time.

[2] In PyTorch, this operation can be performed like this:
```
out = g_b + g_r - g_r.detach().
```

3.4 Loss Function

DecisioNet is trained with 2 sets of labels - classification labels and routing labels (extracted in the clustering phase). During training, we want our model to classify inputs correctly while passing classes to their desired routes, based on the routing labels. Therefore, we design a loss function that balances these 2 demands. This loss is described as follows:

$$\mathcal{L} = \mathcal{L}_{cls} + \beta\mathcal{L}_{\sigma} \tag{9}$$

where $\mathcal{L}_{cls}$ is a classification loss (*e.g.*, cross-entropy loss) and $\mathcal{L}_{\sigma}$ is a MSE loss for the routing labels. β is a hyper-parameter for balancing the classification accuracy and the routing accuracy. We also tried to use a MSE variant which penalizes routing mistakes with different weights depending on the RM depth. The assumption was that it would make sense to give a higher weight to routing mistakes that happen earlier. However, we did not find evidence that this method gives a significant improvement, hence we use the vanilla MSE instead.

4 Experiments

4.1 FashionMNIST and CIFAR10

Fashion-MNIST [22] consists of a training set of 60K examples and a test set of 10K examples. Each example is a 28×28 grayscale image, associated with a label from 10 classes. The CIFAR10 dataset [14] consists of 60K 32×32 color images in 10 classes, with 6000 images per class. There are 5000 training images and 1000 test images per class. Both datasets were trained using their training images and evaluated on the test images.

Model. We used Network-In-Network (NiN) [15] as our baseline model. The architecture we used is displayed in Fig. 4a. We treat it as a stack of 3 small networks (which we refer to as "blocks"). The DecisioNet variants of this model are generated using the method described in Sect. 3. We experimented with 4 varieties of DecisioNets:

1. *DN2* - A basic version with 2 splits, located in-between the baseline NiN model blocks. An example of a single DN2 computational path is displayed in Fig. 4b. Since there are 2 splits in the DN2 tree, there are 4 different computational paths (from root to leaf).
2. *DN1-early* - this version contains a single split point, between the first and second NiN blocks. It is equivalent to the DN2 without the deeper routing modules.
3. *DN1-late* - this version contains a single split point, between the second and third NiN blocks. It is equivalent to the DN2 without the first routing module and replacing the 2 deeper RMs with a single one.

4. *DN2-slim* - A slimmer version of DN2, where we "push" the routing modules to earlier stages compared to DN2. It results in an even lighter version of DN2 (fewer parameters and MAC operations). An example of a single DN2-slim computational path is displayed in Fig. 4c.

In all cases, the last layer of the DN model outputs 10 values. It means that classes that reached the wrong leaf can still be predicted correctly. This way we overcome the aforementioned choice of not allowing overlapping between clusters.

Preprocessing. For each dataset we calculate its training set mean and standard deviation values (per channel) and use them to normalize the images before feeding them into the model. We also created the routing labels using the method described in Sect. 3.2. For CIFAR10, we conduct 2 sets of experiments - one with data augmentation (which includes random flips and crops, zero-padded by 4 pixels from each side) and one without it. The data-augmented experiments are marked in the tables with a "+" sign.

Training. We apply a similar training method to the one used in [15]: we used stochastic gradient descent with mini-batches of 128 samples, momentum of 0.9, and weight decay of 0.0005. The initial learning rates of FashionMNIST and CIFAR10 were 0.01 and 0.05, respectively. We decrease the learning rate by a factor of 10 when there are 10 epochs without training accuracy improvement. We repeat this process twice during the training period. The training stops if either we reach 300 epochs, or the test loss stops improving for 30 epochs (the latter is to prevent overfitting).

The DecisioNet training process is identical and has the additional hyperparameter β for the loss function (Eq. 9). The value of β varies between datasets and models. The specific value that was used for every experiment appears next to the results (Table 2).

We initialize the weights of the convolutional layers from a zero-centered Gaussian distribution with a standard deviation of 0.05.

Clustering Interpretability. Aside from the performance of the different models, we show that the clustering method partitions the data in an intuitive way. For example, the CIFAR10 first partition is into 2 clusters containing animals and vehicles. The vehicle cluster is then separated into land vehicles (car, truck) and non-land vehicles (plane, ship). In FashionMNIST the first division is into footwear (sandal, snicker, ankle-boot) and non-footwear. The latter is then partitioned into legwear (a singleton cluster with only pants) and non-legwear (T-shirt, dress, bag, etc.). The full hierarchical clustering (for 2 decision levels) is displayed in Table 1.

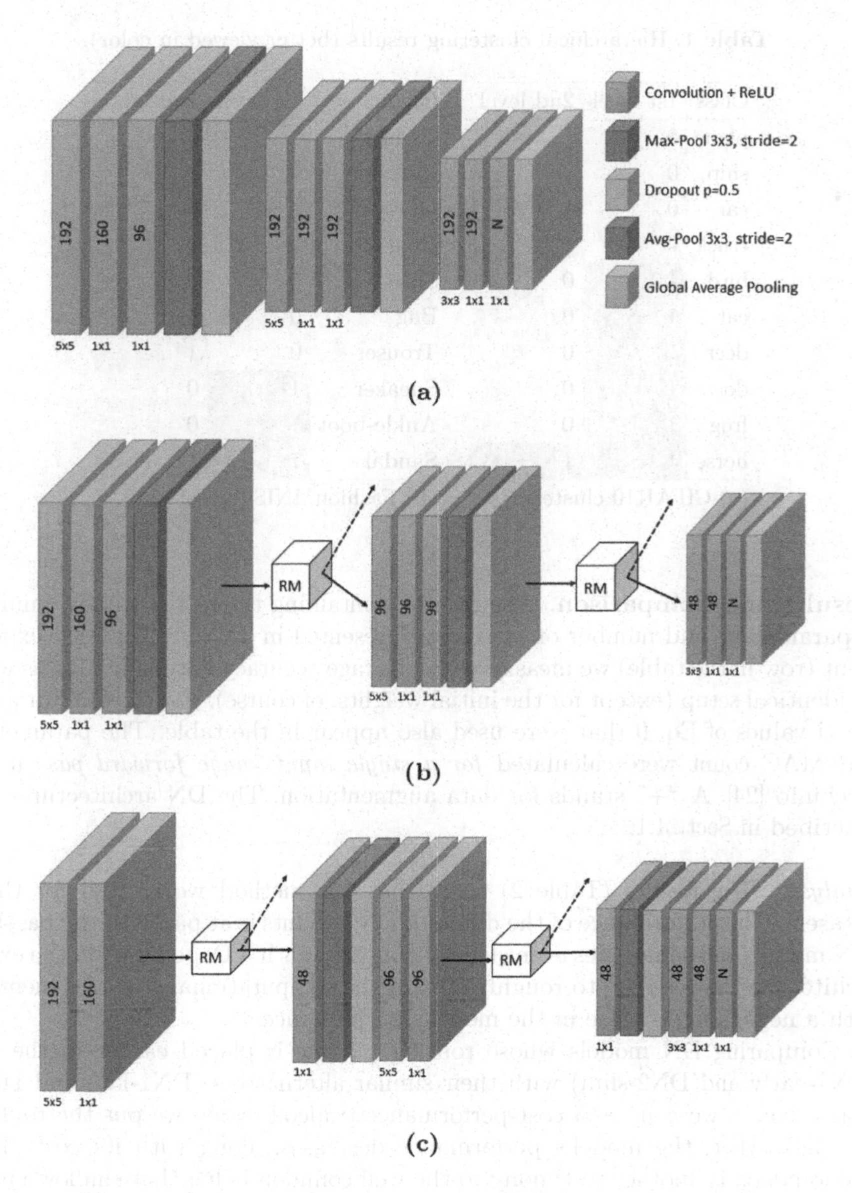

Fig. 4. In (a) we display the Network-in-Network (NiN) baseline model architecture. For convolutional layer blocks, the kernel size is stated below each block, and the number of filters is written on the side. We always add zero-padding such that the output and input spatial dimensions are the same (only pooling layers reduce the spatial dimensions). N is the number of output classes (in our case $N = 10$). In (b) and (c) we display a computational path of 2 DecisioNet variants of the NiN model - DN2 and DN2-slim, respectively. The routing module (RM) is the one that is described in Sect. 3.3. Notice that the first block after the RM has convolutional layers with half the amount of filters compared to the baseline model; After the second RM, the corresponding block contains a quarter of the filters in the convolutional layers.

Table 1. Hierarchical clustering results (better viewed in color).

Class	1st level	2nd level
plane	0	0
ship	0	0
car	0	1
truck	0	1
bird	1	0
cat	1	0
deer	1	0
dog	1	0
frog	1	0
horse	1	1

(a) CIFAR10 clustering

Class	1st level	2nd level
T-shirt	0	0
Pullover	0	0
Dress	0	0
Coat	0	0
Shirt	0	0
Bag	0	0
Trouser	0	1
Sneaker	1	0
Ankle-boot	1	0
Sandal	1	1

(b) FashionMNIST clustering

Results and Comparison. The results, containing the test accuracy, number of parameters, and number of MACs are presented in Table 2. For each experiment (row in the table) we measured the average accuracy based on 10 runs with an identical setup (except for the initial weights, of course). For the DN variants, the β values of Eq. 9 that were used also appear in the table. The parameters and MAC count were calculated *for a single input image forward pass* using torchinfo [24]. A "+" stands for data augmentation. The DN architectures are described in Sect. 4.1.

Analysis: The results (Table 2) show that our method works well for these datasets. The performance of the different DN variants is at par with the baseline NiN model (and sometimes even slightly outperforms it). Depending on the exact architecture, we save up to roughly 60% of the computational cost and memory with a negligible decrease in the model's performance.

Comparing DN models whose routing module is placed earlier in the net (DN1-early and DN2-slim) with their similar alternatives (DN1-late and DN2, respectively), we can see a cost-performance tradeoff: when we put the routing module earlier, the model's performance decreases, along with its cost. This phenomenon is another testimony to the well common belief that shallow layers extract features that are important for all classes.

We also would like to emphasize the significance of choosing an appropriate β: when setting $\beta = 0$, we essentially encourage the DN to only optimize the classification loss, ignoring the routing loss. We found that the performance in this case, is similar to a random choice (*i.e.*, DN outputs a constant prediction). This is surprising because in some cases the DN ignored the routing labels at some point, while still improving its classification accuracy.

Table 2. Accuracy results, along with parameters and MACs count. For experiments with augmented data (marked with "+") we omit the parameters and MACs count, as they are identical to those of the non-augmented experiments. More details in Sect. 4.1.

Dataset	Architecture	β	Acc. (%)	Acc. change (%)	Params (K)	Params change (%)	MACs (M)	MACs change (%)
FashionMNIST	Baseline	–	92.7	–	957.386	–	163.3	
	DN1-early	3.0	92.9	↑ 0.2	736.309	↓23.1	93.64	↓42.7
	DN1-late	3.0	93.4	↑ 0.7	939.157	↓1.9	153.76	↓5.8
	DN2	3.0	92.8	↑ 0.1	727.307	↓24.0	91.24	↓44.1
	DN2-slim	3.0	92.7	0.0	414.027	↓56.8	60.68	↓62.8
CIFAR10	Baseline	–	87.0	–	966.986	–	223.12	–
	DN1-early	3.0	86.6	↓ 0.4	745.909	↓22.9	132.14	↓40.8
	DN1-late	3.0	88.2	↑1.2	948.757	↓1.9	210.66	↓5.6
	DN2	0.5	86.8	↓ 0.2	736.907	↓23.8	129.00	↓42.2
	DN2-slim	1.0	86.2	↓ 0.8	423.627	↓56.2	89.08	↓60.1
CIFAR10+	Baseline	–	88.4	–	–	–	–	–
	DN1-early	3.0	87.8	↓ 0.6	–	–	–	–
	DN1-late	3.0	89.4	↑ 1.0	–	–	–	–
	DN2	0.5	87.6	↓ 0.8	–	–	–	–
	DN2-slim	1.0	86.3	↓ 2.1	–	–	–	–

4.2 CIFAR100

The CIFAR100 dataset [14] is like the CIFAR10, except that it has 100 classes containing 600 images each (500 training images and 100 test images per class).

For this dataset, we experimented with Wide ResNet (WRN) [25] as our baseline network. We used the version with 28 layers and a depth factor of $k = 10$ (denoted as WRN-28-10 in the original paper). We created 3 DN variants out of this network: DN2, whose splits are in between the residual blocks (2 splits), along with DN1-early and DN1-late - each of them with a single split point, positioned at the first or last DN2 split point position. A diagram of the baseline model and the DN2 variant are displayed in Fig. 5, and the results of our experiments are in Table 3.

Training. We train the network using the same method as in [25]. That is, we use SGD with Nesterov momentum of 0.9 and a weight decay of 0.0005. The initial learning rate is set to 0.1 and is decreased by a factor of 5 when we reach 60, 120 and 160 epochs. The training is stopped after 200 epochs. The batch size used for training is 128. We augmented the dataset with the same method as we did with CIFAR10, except padding crops with reflections instead of zeros (that is done to follow the training procedure from the original paper).

Analysis: We see similar behavior to the former experiments. The top1 accuracy drop is larger than in the former experiments. One possible explanation to this is the fact that the parameter ratio between the DNs and the baseline model is significantly higher. Different routing module positioning might yield better performance (at some computational cost).

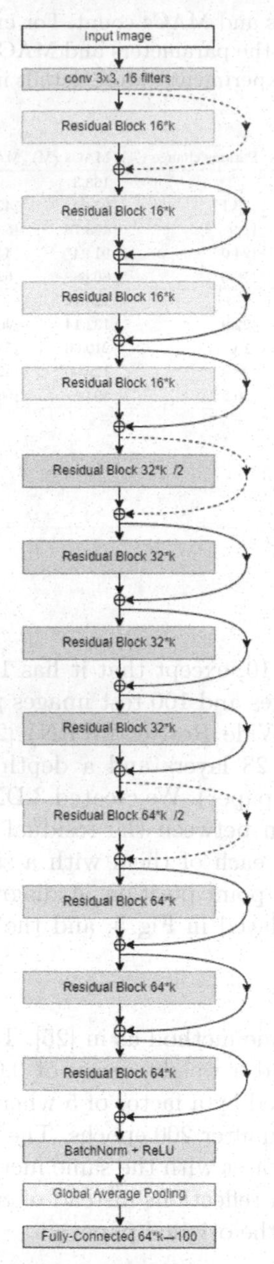

(a) Baseline WRN-n-k architecture with $n = 28$. The dashed arrows are 1x1 convolutions for dimensions matching

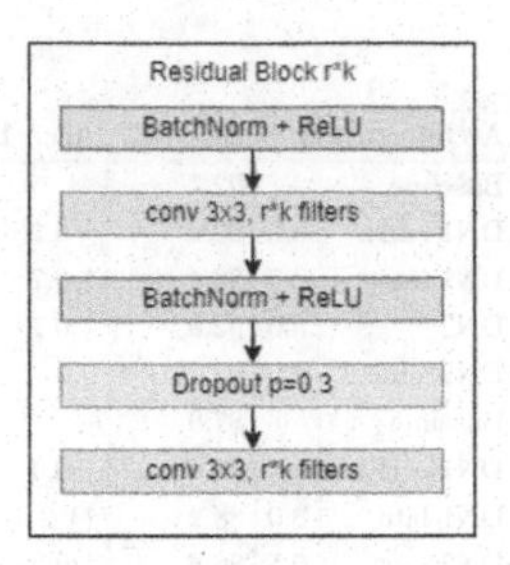

(b) Residual block architecture. In some cases, the first convolution has a stride of size 2 (In (a) these blocks has the additional "/2" text)

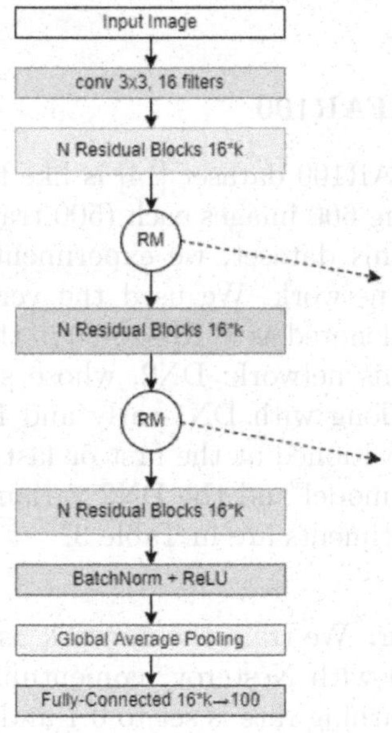

(c) WRN-DN2 single computational branch (we group the residual blocks from (a) into a single block for clarity; Generally, $N = (n-4)/6$, so in our case $N = 4$). RM is the routing module.

Fig. 5. Wide-ResNet-n-k architectures - in this paper we only use $n = 28$ and $k = 10$.

Table 3. CIFAR100 results

Architecture	β	Top-1 Acc. (%)	Top-5 Acc. (%)	Params (M)	MACs (G)
WRN baseline	–	80.7	95.4	36.537	5.24
WRN-DN1-early	3.0	77.6 (↓ 3.1)	93.9 (↓ 1.5)	19.385 (↓ 46.9%)	2.6 (↓ 50.4%)
WRN-DN1-late	3.0	78.7 (↓ 2.0)	94.5 (↓ 0.9)	23.635 (↓ 35.3%)	3.94 (↓ 24.8%)
WRN-DN2	5.0	75.7 (↓ 5.0)	92.7 (↓ 2.7)	12.935 (↓ 64.6%)	2.28 (↓ 56.5%)

5 Conclusion

We introduced DecisioNet (DN), a binary-tree structured network with conditional routing. We proposed a systematic way of building it based on an existing baseline network. DN utilizes Improved Semantic Hashing for training end-to-end, applying conditional routing both during training and evaluation. DN takes advantage of its tree structure, passing inputs only through a portion of the net's neural modules saving a lot of the computational cost. We evaluated multiple DN variants along with their baseline models on multiple image classification datasets and showed that DN is capable of achieving similar performance (w.r.t.its baseline model) while significantly decreasing the computational cost.

Future Research. One clear drawback of our method is the use of additional labels. In addition, these labels are generated using a pre-trained model. Learning hard routing without such auxiliary labels is challenging (as we saw when we set $\beta = 0$), and it might be an interesting direction for future research. Another direction is creating DN with unbalanced trees: in this paper, we only examined balanced DNs, while our clustering trees were not necessarily balanced. A possible idea is to divide the computational power of branches proportionally to the size of the data they should process (*e.g.*, a dataset with 100 classes which is split into clusters of size 70 and 30, would have one branch with 70% of the computational power and the other would have 30%).

Acknowledgements. Thanks to the ISF (1439/22) and the DFG for funding.

References

1. Abadi, M., et al.: TensorFlow: large-scale machine learning on heterogeneous systems (2015). https://www.tensorflow.org/, Software available from tensorflow.org
2. Chen, J., Zhu, Z., Li, C., Zhao, Y.: Self-adaptive network pruning (2019)
3. Chen, Z., Li, Y., Bengio, S., Si, S.: You look twice: GaterNet for dynamic filter selection in CNNs (2019)
4. Deng, J., Dong, W., Socher, R., Li, L.J., Li, K., Fei-Fei, L.: ImageNet: a large-scale hierarchical image database. In: 2009 IEEE Conference on Computer Vision and Pattern Recognition, pp. 248–255. IEEE (2009)
5. Frosst, N., Hinton, G.: Distilling a neural network into a soft decision tree. arXiv preprint arXiv:1711.09784 (2017)

6. Hazimeh, H., Ponomareva, N., Mol, P., Tan, Z., Mazumder, R.: The tree ensemble layer: differentiability meets conditional computation. In: International Conference on Machine Learning, pp. 4138–4148. PMLR (2020)
7. He, K., Zhang, X., Ren, S., Sun, J.: Deep residual learning for image recognition (2015)
8. Hehn, T.M., Kooij, J.F.P., Hamprecht, F.A.: End-to-end learning of decision trees and forests. Int. J. Comput. Vision **128**(4), 997–1011 (2020). https://doi.org/10.1007/s11263-019-01237-6
9. Ioannou, Y., et al.: Decision forests, convolutional networks and the models in-between. arXiv preprint arXiv:1603.01250 (2016)
10. Jordan, M.I., Jacobs, R.A.: Hierarchical mixtures of experts and the EM algorithm. Neural Comput. **6**(2), 181–214 (1994)
11. Kaiser, Ł., Bengio, S.: Discrete autoencoders for sequence models (2018)
12. Kaiser, Ł., Sutskever, I.: Neural GPUs learn algorithms (2016)
13. Kontschieder, P., Fiterau, M., Criminisi, A., Bulo, S.R.: Deep neural decision forests. In: Proceedings of the IEEE International Conference on Computer Vision, pp. 1467–1475 (2015)
14. Krizhevsky, A., Hinton, G., et al.: Learning multiple layers of features from tiny images (2009)
15. Lin, M., Chen, Q., Yan, S.: Network in network. arXiv preprint arXiv:1312.4400 (2013)
16. Paszke, A., et al.: PyTorch: an imperative style, high-performance deep learning library. In: Wallach, H., Larochelle, H., Beygelzimer, A., d'Alché-Buc, F., Fox, E., Garnett, R. (eds.) Advances in Neural Information Processing Systems, vol. 32, pp. 8024–8035. Curran Associates, Inc. (2019). http://papers.neurips.cc/paper/9015-pytorch-an-imperative-style-high-performance-deep-learning-library.pdf
17. Rota Bulo, S., Kontschieder, P.: Neural decision forests for semantic image labelling. In: Proceedings of the IEEE Conference on Computer Vision and Pattern Recognition, pp. 81–88 (2014)
18. Rumelhart, D.E., Hinton, G.E., Williams, R.J.: Learning representations by back-propagating errors. Nature **323**(6088), 533–536 (1986)
19. Simonyan, K., Zisserman, A.: Very deep convolutional networks for large-scale image recognition (2015)
20. Suárez, A., Lutsko, J.F.: Globally optimal fuzzy decision trees for classification and regression. IEEE Trans. Pattern Anal. Mach. Intell. **21**(12), 1297–1311 (1999)
21. Tanno, R., Arulkumaran, K., Alexander, D., Criminisi, A., Nori, A.: Adaptive neural trees. In: International Conference on Machine Learning, pp. 6166–6175. PMLR (2019)
22. Xiao, H., Rasul, K., Vollgraf, R.: Fashion-MNIST: a novel image dataset for benchmarking machine learning algorithms. arXiv preprint arXiv:1708.07747 (2017)
23. Yan, Z., et al.: HD-CNN: hierarchical deep convolutional neural network for large scale visual recognition (2015)
24. Yep, T.: torchinfo, March 2020. https://github.com/TylerYep/torchinfo
25. Zagoruyko, S., Komodakis, N.: Wide residual networks. arXiv preprint arXiv:1605.07146 (2016)

An RNN-Based Framework for the MILP Problem in Robustness Verification of Neural Networks

Hao Xue[1], Xia Zeng[1,2], Wang Lin[3], Zhengfeng Yang[1(✉)], Chao Peng[1], and Zhenbing Zeng[4]

[1] Shanghai Key Lab of Trustworthy Computing, East China Normal University, Shanghai, China
hxue@stu.ecnu.edu.cn, {zfyang,cpeng}@sei.ecnu.edu.cn

[2] School of Computer and Information Science, Southwest University, Chongqing, China
xzeng0712@swu.edu.cn

[3] School of Information Science and Technology, Zhejiang Sci-Tech University, Hangzhou, China
linwang@zstu.edu.cn

[4] Department of Mathematics, Shanghai University, Shanghai, China
zbzeng@shu.edu.cn

Abstract. Robustness verification of 's is becoming increasingly crucial for their potential use in many safety-critical applications. Essentially, the problem of robustness verification can be encoded as a typical Mixed-Integer Linear Programming (MILP) problem, which can be solved via branch-and-bound strategies. However, these methods can only afford limited scalability and remain challenging for verifying large-scale neural networks. In this paper, we present a novel framework to speed up the solving of the MILP problems generated from the robustness verification of deep neural networks. It employs a semi-planet relaxation to abstract ReLU activation functions, via an RNN-based strategy for selecting the relaxed ReLU neurons to be tightened. We have developed a prototype tool L2T and conducted comparison experiments with state-of-the-art verifiers on a set of large-scale benchmarks. The experiments show that our framework is both efficient and scalable even when applied to verify the robustness of large-scale neural networks.

Keywords: Robustness verification · Learning methods · Semi-planet relaxation · Neural networks

1 Introduction

While the deep neural network has been widely applied to various tasks and achieved outstanding performance in recent years, it also encounters robustness problems from many adversarial examples [14,30], which could raise serious consequences in safety-critical systems such as autonomous driving, aircraft flight control, and nuclear systems. For example, a slightly perturbed image could make an autonomous car incorrectly classify a white truck to cloud and cause a crash accident [13]. Consequently, robustness

L. Wang et al. (Eds.): ACCV 2022, LNCS 13841, pp. 571–586, 2023.
https://doi.org/10.1007/978-3-031-26319-4_34

verification of neural networks becomes increasingly crucial, which aims to check that the outputs for all inputs in a given domain have the correct corresponding label.

Several methods [3,16,28] have been proposed to use abstraction for achieving the robustness. With the wide use of the ReLU activation function in neural networks, it is more useful in practice to investigate the robustness verification problem [20].

Due to the piece-wise linear structure of ReLU networks, the problem of robustness verification can be rewritten into a typical Mixed-Integer Linear Programming (MILP) problem, which can be solved via branch-and-bound strategies [7]. However, for large-scale ReLU networks, solving the MILP problems stemming from robustness verification still remains a challenge. To this end, several convex relaxation approaches have been studied for approximating ReLU activation functions [5,9,22,24,26]. Nevertheless, as the error accumulation arising from the over-relaxation is often hard to estimate, these approaches may fail to prove the robustness.

Meanwhile, machine learning methods have already been used in solving MILP problems. In [1] and [15], regression and ranking approaches are used to assign a branching score to each potential branching choice. Bayesian optimization is also applied in [11] to learn verification policies. The scheme in [21] captures the structure of a neural network and makes tightening decisions through a graph neural network. Recently, symbolic interval propagation and linear programming are used to improve the verification performance [17]. However, the above schemes are either limited to specific neural networks or relied heavily on hand-designed features.

To address the issues mentioned above, in this paper, we propose a novel framework to speed up the solving of MILP problems generated from the robustness verification of neural networks. Since the relaxation of ReLU activation functions may result in failure of the original robustness verification, we propose a semi-planet relaxation, which adopts an RNN-based strategy for selecting the relaxed ReLU neurons to be tightened. Our framework entails an iterative process to gradually repair failures caused by over-relaxation, hence it can perform effective verification while ensuring the efficiency of the MILP problem solving.

To summarize, the main contributions of this paper are:

- We present a general scheme, which augments the original neural network by additional layers with one output neuron, and transforms the verification problem into the output positiveness checking of the augmented network. Thus we transform the robustness verification problem into a single MILP problem, making our RNN-based solving method much easier to implement and scale.
- We build an iterative framework that enhances the efficiency of robustness verification through an intelligent neuron selection strategy by an RNN to gradually tighten the relaxed MILP solution.
- Our RNN embraces scalability by utilizing the network's structure information, and the trained RNN is generalizable for robustness verification on different network structures. We also provide an approach to generate the training dataset, which makes RNN much easier to train and implement.
- We conduct comparison experiments with state-of-the-art verifiers on large-scale benchmarks, the results show that our framework has excellent applicability: it outperforms most other verifiers in terms of average solving time and has lower timeout rate in most cases.

2 Preliminaries

In this section, we provide the background on Deep Neural Networks (DNNs) and describe the robustness verification of DNNs.

2.1 Deep Neural Networks

A deep neural network $\mathcal{N}$ is a tuple $\langle \mathcal{X}, \mathcal{H}, \Phi \rangle$, where $\mathcal{X} = \{\mathbf{x}^{[0]}, \dots, \mathbf{x}^{[n]}\}$ is a set of layers, $\mathcal{H} = \{H_1, \dots, H_n\}$ consists of affine functions between layers, and $\Phi = \{\phi_1, \dots, \phi_n\}$ is a set of activation functions. More specifically, $\mathbf{x}^{[0]}$ is the input layer, $\mathbf{x}^{[n]}$ is the output layer, and $\mathbf{x}^{[1]}, \dots, \mathbf{x}^{[n-1]}$ are called hidden layers. Each layer $\mathbf{x}^{[i]}, 0 \le i \le n$ is associated with an s_i-dimensional vector space, in which each dimension corresponds to a neuron. For each $1 \le i \le n$, the affine function is written as $H_i(\mathbf{x}^{[i-1]}) = W^{[i]}\,\mathbf{x}^{[i-1]} + \mathbf{b}^{[i]}$, where $W^{[i]}$ and $\mathbf{b}^{[i]}$ are called the weight matrix and the bias vector, respectively. Furthermore, for each layer $\mathbf{x}^{[i]}$, the value of each neuron in the hidden layer is assigned by the affine function H_i for the values of neurons in the previous layer, and then applying the activation function ϕ_i, i.e., $\mathbf{x}^{[i]} = \phi_i(W^{[i]}\,\mathbf{x}^{[i-1]} + \mathbf{b}^{[i]})$, with the activation function ϕ_i being applied element-wise.

In this paper, we focus on DNNs with Rectified Linear Unit (ReLU) activation functions [12], defined as $\mathrm{ReLU}(x) = \max(0, x)$. From the mapping point of view, the i-th layer can be seen as the mapping image of the $(i-1)$-th layer. For each non-input layer $\mathbf{x}^{[i]}, 1 \le i \le n$, we can define a function $f^{[i]} : \mathbb{R}^{s_{i-1}} \longrightarrow \mathbb{R}^{s_i}$, with

$$f^{[i]}(\mathbf{x}) = \phi_i(W^{[i]}\mathbf{x} + \mathbf{b}^{[i]}), \quad 1 \le i \le n. \tag{1}$$

A DNN $\mathcal{N}$ is expressed as the composition function $f : \mathbb{R}^{s_0} \longrightarrow \mathbb{R}^{s_n}$, i.e.,

$$f(\mathbf{x}) = f^{[n]}(f^{[n-1]}(\cdots(f^{[1]}(\mathbf{x})))). \tag{2}$$

Given an input $\mathbf{x}^{[0]}$, the DNN $\mathcal{N}$ assigns the label selected with the largest logit of the output layer $\mathbf{x}^{[n]}$, that is, $k = \arg\max(f(\mathbf{x}^{[0]})) = \arg\max_{1 \le j \le s_n}(\mathbf{x}_j^{[n]})$.

2.2 Robustness Verification of DNNs

We begin with formally defining the notion of robustness, and introducing the robustness verification of DNNs.

Given a DNN $\mathcal{N}$ with its associated function f and an input point $\mathbf{x}_c$, we now define ϵ-robustness for a DNN $\mathcal{N}$ on $\mathbf{x}_c$. Let $B(\mathbf{x}_c, \epsilon)$ be the ℓ_∞-ball of radius $\epsilon \in \mathbb{R}_{>0}$ around the point $\mathbf{x}_c$, i.e., $B(\mathbf{x}_c, \epsilon) = \{\mathbf{x}^{[0]} \in \mathbb{R}^{s_0} \mid \|\mathbf{x}^{[0]} - \mathbf{x}_c\|_\infty \le \epsilon\}$.

We say that an input domain $B(\mathbf{x}_c, \epsilon)$ has the same label if any input $\mathbf{x}^{[0]}$ chosen from $B(\mathbf{x}_c, \epsilon)$ has the same label as $\mathbf{x}_c$ has. This property that the ϵ-ball around the point having the same label is called as *ϵ-robustness*, i.e.,

$$\arg\max(f(\mathbf{x}^{[0]})) = \arg\max(f(\mathbf{x}_c)), \ \forall \mathbf{x}^{[0]} \in B(\mathbf{x}_c, \epsilon). \tag{3}$$

Determining the ϵ-robustness of $\mathbf{x}_c$ with the label k, can be transformed into the following equivalent optimization problem

$$\left.\begin{array}{ll} p^* = \min_{\mathbf{x},\hat{\mathbf{x}}}\{\ \mathbf{x}_k^{[n]} - \mathbf{x}_{k'}^{[n]},\ \forall k' \neq k\ \} & \\ \text{s.t.} \quad \mathbf{x}^{[0]} \in B(\mathbf{x}_c, \epsilon), & \\ \qquad \hat{\mathbf{x}}^{[i]} = W^{[i]}\mathbf{x}^{[i-1]} + \mathbf{b}^{[i]}, & i = 1, \ldots, n, \\ \qquad \mathbf{x}^{[i]} = \phi_i(\hat{\mathbf{x}}^{[i]}), & i = 1, \ldots, n. \end{array}\right\} \quad (4)$$

Remark that $\mathcal{N}$ with respect to $B(\mathbf{x}_c, \epsilon)$ is robust if and only if the optimum of (4) is positive, i.e., $p^* > 0$.

The main challenge of optimization solving (4) is to handle ReLU activation functions, which bring non-linearities to the problem of robustness verification. Seeking the optimal solution of (4) is an arduous task, as the optimization problem is generally NP-hard [18]. An effective technique to eliminate the non-linearities is to encode them with the help of binary variables. Let $\mathbf{l}^{[i]}$ and $\mathbf{u}^{[i]}$ be lower bounds and upper bounds on $\hat{\mathbf{x}}^{[i]}$, respectively. The ReLU activations $\mathbf{x}^{[i]} = \phi^{[i]}(\hat{\mathbf{x}}^{[i]})$ can be encoded by the following constraints:

$$\mathcal{I}(\mathbf{x}^{[i]}) \triangleq \begin{cases} \mathbf{x}^{[i]} \succcurlyeq \mathbf{0}, \quad \mathbf{x}^{[i]} \succcurlyeq \hat{\mathbf{x}}^{[i]}, \\ \mathbf{u}^{[i]} \odot \mathbf{z}^{[i]} \succcurlyeq \mathbf{x}^{[i]}, \\ \hat{\mathbf{x}}^{[i]} - \mathbf{l}^{[i]} \odot (\mathbf{1} - \mathbf{z}^{[i]}) \succcurlyeq \mathbf{x}^{[i]}, \\ \mathbf{z}^{[i]} \in \{0, 1\}^{s_i}, \end{cases} \quad (5)$$

where s is the binary vector with length s_i, $\odot$ denotes the Hadamard product of matrices, and the generalized inequality $\succcurlyeq$ denotes the componentwise inequality between vectors. Therefore, the robustness verification problem (4) can be transformed into an equivalent Mixed-Integer Linear Programming (MILP) problem [6]:

$$\left.\begin{array}{ll} p^* = \min_{\mathbf{x},\hat{\mathbf{x}},\mathbf{z}}\{\ \mathbf{x}_k^{[n]} - \mathbf{x}_{k'}^{[n]},\ \forall k' \neq k\ \} & \\ \text{s.t.} \quad \mathbf{x}^{[0]} \in B(\mathbf{x}_c, \epsilon), & \\ \qquad \hat{\mathbf{x}}^{[i]} = W^{[i]}\mathbf{x}^{[i-1]} + \mathbf{b}^{[i]}, & i = 1, \ldots, n, \\ \qquad \mathcal{I}(\mathbf{x}^{[i]}), & i = 1, \ldots, n. \end{array}\right\} \quad (6)$$

It is easy to verify that $\mathbf{z}_j^{[i]} = 0 \Leftrightarrow \mathbf{x}_j^{[i]} = 0$ and $\mathbf{z}_j^{[i]} = 1 \Leftrightarrow \mathbf{x}_j^{[i]} = \hat{\mathbf{x}}_j^{[i]}$.

3 Semi-Planet Relaxation for MILP Problems

3.1 Formulating the Minimum Function in MILP Model

The objective function of the optimization problem (6) is piecewise-linear [7]. There are some approaches [8,10,25] making it amenable to mathematical programming solvers, but either discard the network structure or sacrifice the scalability. In the following, we propose an approach for rewriting the objective function with multiple neurons into an equivalent single-neuron setting by adding a few layers at the end of the network $\mathcal{N}$. This allows us to preserve network structure and ensure the scalability of solving the problem. Note that only linear layers and ReLU activation functions are added for the sake of simplicity.

Before doing this, we first add a new layer to represent the difference between $\mathbf{x}_k^{[n]}$ and all other output neurons:

$$\begin{aligned}\mathbf{x}^{[n+1]} &= \phi_{n+1}\left(W^{[n+1]}\mathbf{x}^{[n]} + \mathbf{b}^{[n+1]}\right),\\ W^{[n+1]} &= \mathbf{1}^{\mathrm{T}}\mathbf{e}_k - I, \quad \mathbf{b}^{[n+1]} = \mathbf{0},\end{aligned} \tag{7}$$

where $\mathbf{e}_k$ is one-hot encoding vector for label k, I is an identity matrix. Now, the remaining task is to compute the minimum of $\mathbf{x}^{[n+1]}$. For simplicity, we illustrate how to use the neural network structure to represent *Min Function* of two elements,

$$\min(a,b) = \frac{a+b-|a-b|}{2} = \frac{\phi(a+b)-\phi(-a-b)-\phi(a-b)-\phi(b-a)}{2}, \tag{8}$$

that is, $\min(a,b) = W_2 \cdot \phi(W_1 \cdot \begin{bmatrix} a \\ b \end{bmatrix})$, where

$$W_1 = \begin{bmatrix} 1 & -1 & 1 & -1 \\ 1 & -1 & -1 & 1 \end{bmatrix}^{\mathrm{T}}, \quad W_2 = \left[\tfrac{1}{2}, -\tfrac{1}{2}, -\tfrac{1}{2}, -\tfrac{1}{2}\right]. \tag{9}$$

By calling the above process recursively, it is easy to build a neural network to express *Min Function* for greater than 2 elements. Therefore, computing the minimum of s_{n+1} neurons, can be represented as a neural network $\mathcal{N}_g$,

$$g(\mathbf{x}) = W_2 g_{\lceil \log_2 s_{n+1} \rceil}\left(\cdots\left(g_2\left(g_1(\mathbf{x})\right)\right)\right), \tag{10}$$

where $g_j(\mathbf{x}) = \phi_{n+1+j}\left(W^{[n+1+j]}\mathbf{x} + \mathbf{b}^{[n+1+j]}\right)$, $j = 1,\ldots,\lceil \log_2 s_{n+1} \rceil$, which halves the input vector space by finding the minimum in pairs.

Combining $\mathcal{N}$ and $\mathcal{N}_g$ can produce an augmented network, denoted by $\widehat{\mathcal{N}}$, which has $m = n+1+\lceil \log_2 s_{n+1} \rceil$ layers. Now, the original robustness verification problem can be transformed into that of determining the positiveness of the output of the augmented network $\widehat{\mathcal{N}}$. Therefore, the problem (6) can be rewritten as an equivalent MILP problem:

$$\left.\begin{aligned} J_{\mathcal{I}}^* = \min_{\mathbf{x},\hat{\mathbf{x}},\mathbf{z}} \ & W_2\mathbf{x}^{[m]} \\ \text{s.t.} \ & \mathbf{x}^{[0]} \in B(\mathbf{x}_c, \epsilon), \\ & \hat{\mathbf{x}}^{[i]} = W^{[i]}\mathbf{x}^{[i-1]} + \mathbf{b}^{[i]}, \quad & i = 1,\ldots,m, \\ & \mathcal{I}(\mathbf{x}^{[i]}), & i = 1,\ldots,m. \end{aligned}\right\} \tag{11}$$

Notably, the optimum $J_{\mathcal{I}}^*$ is positive if and only if the original network $\mathcal{N}$ with respect to $B(\mathbf{x}_c, \epsilon)$ is robust. Hereafter, we abbreviate the optimization problem (11) as following

$$J_{\mathcal{I}}^* = \min J\left(\widehat{\mathcal{N}}, \mathcal{I}(X)\right), \tag{12}$$

where $\mathcal{I}(X)$ denotes the binary constraints for the set of ReLU neurons X of $\widehat{\mathcal{N}}$.

3.2 Relaxation and Tightening for MILP Constraints

When encountering practical neural networks, due to binary constraints on overwhelming $\mathcal{I}(\mathbf{x}_j^{[i]})$, how to solve MILP problem (12) is the key challenge. Planet relaxation [9]

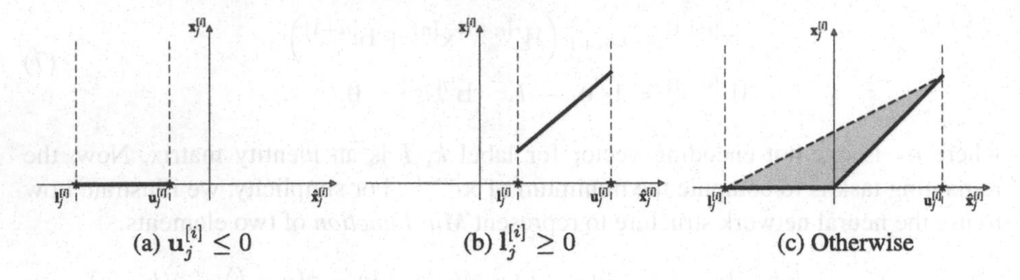

Fig. 1. The form of $\mathcal{P}(\mathbf{x}_j^{[i]})$ is contingent on the value of $\mathbf{u}_j^{[i]}$ and $\mathbf{l}_j^{[i]}$.

has been widely used for ReLU neurons relaxation [5,6,21] due to its tightness of formulation. Applying planet relaxation to ReLU activation functions $\mathbf{x}_j^{[i]} = \phi(\hat{\mathbf{x}}_j^{[i]})$ may yield the following over-approximation with linear inequalities,

$$\mathcal{P}(\mathbf{x}_j^{[i]}) \triangleq \begin{cases} \mathbf{x}_j^{[i]} = 0, & \text{if } \mathbf{u}_j^{[i]} \leq 0 \\ \mathbf{x}_j^{[i]} = \hat{\mathbf{x}}_j^{[i]}, & \text{if } \mathbf{l}_j^{[i]} \geq 0 \\ \mathbf{x}_j^{[i]} \geq 0, \mathbf{x}_j^{[i]} \geq \hat{\mathbf{x}}_j^{[i]}, \mathbf{x}_j^{[i]} \leq \dfrac{\mathbf{u}_j^{[i]}(\hat{\mathbf{x}}_j^{[i]} - \mathbf{l}_j^{[i]})}{\mathbf{u}_j^{[i]} - \mathbf{l}_j^{[i]}}. & \text{otherwise} \end{cases} \tag{13}$$

Figure 1 is depicted to show the planet relaxation of a ReLU unit.

By introducing $\mathcal{P}(\mathbf{x}_j^{[i]})$ to relax $\mathcal{I}(\mathbf{x}_j^{[i]})$, the MILP problem (12) is transformed into the following LP problem:

$$J_{\mathcal{P}}^* = \min J\left(\widehat{\mathcal{N}}, \mathcal{P}(X)\right), \tag{14}$$

where $\mathcal{P}(X)$ denotes the planet relaxation for the set of ReLU neurons X of $\widehat{\mathcal{N}}$.

Remark 1. In comparison with the MILP problem (12), the feasible set of the LP problem (14) is a superset of that of (12). Consequently, the optimum of (14) is a lower bound of the one of (12), i.e., $J_{\mathcal{P}}^* \leq J_{\mathcal{I}}^*$. Therefore, the relaxation transformation between the optimization problems derives sufficient conditions for robustness verification of the neural networks.

The gap between $J_{\mathcal{P}}^*$ and $J_{\mathcal{I}}^*$ depends on the tightness of planet relaxation. Unfortunately, the relaxation may yield too conservative optimum $J_{\mathcal{P}}^*$, such that $J_{\mathcal{P}}^* < 0 < J_{\mathcal{I}}^*$ holds, which fails to prove the robustness property for the given neural network. Therefore, how to provide adequate precision and scalability to satisfy the requirement of robustness verification is crucial. In this work, we suggest a semi-planet relaxation scheme where relaxation and tightening processes cooperate to approximate the ReLU neurons. Concretely, when the optimum $J_{\mathcal{P}}^* < 0$, a tightening procedure is performed to tighten the relaxations of some ReLU neurons and recover their original binary constraints. Let $T \subseteq X$ be a subset of ReLU neurons that are chosen to tighten the relaxation, based on the above semi-planet relaxation, (14) can be transformed into

$$J_T^* = \min J\left(\widehat{\mathcal{N}}, \mathcal{I}(T) \wedge \mathcal{P}(X - T)\right). \tag{15}$$

Obviously, the optimum J_T^* of (15) satisfies the inequalities, i.e., $J_{\mathcal{P}}^* \leq J_T^* \leq J_{\mathcal{I}}^*$, which derives that the augmented network $\widehat{\mathcal{N}}$ with respect to $B(\mathbf{x}_c, \epsilon)$ is robust when $J_T^* > 0$. As the number of tightened neurons in T increases, J_T^* will be closer to $J_{\mathcal{I}}^*$. As a result, the complexity of solving the problem (15) will also grow exponentially. So the challenge is how to balance the relaxation and tightening, that is, how to choose as few neurons as possible to tighten for proving the robustness of $\widehat{\mathcal{N}}$.

4 Learning to Tighten

The MILP problem (12), derived from robustness verification of neural networks, transfers a more tractable linear programming (LP) problem when all ReLU neurons are relaxed by the planet relaxation technique. However, this relaxation of non-linearity may yield too conservative constraints so that the optimum of the resulted LP is negative, which fails to prove the original robustness verification problem.

A natural idea to avoid this situation would be to tight only a part of neurons that have more influence on the verification problem. The key is how to determine those key neurons. In this work, we will propose a framework for solving the MILP problem which incorporates an RNN-based strategy for selecting neurons to be tightened.

4.1 RNN-Based Framework

In this section, we describe an RNN-based framework for solving the MILP problem in robustness verification of neural networks. To build the framework we adopt an iterative process to gradually repair failures caused by over-relaxation, and pursue effective verification while ensuring the efficiency of the MILP problem solving. Specifically, we begin the iteration with a complete relaxed problem in (14) from the original problem, by applying the planet relaxation (12) for all ReLU neurons. If the solution satisfies $J_{\mathcal{P}}^* > 0$, then the robustness of the original problem has been verified. Otherwise, we apply the one-neuron-tighten-test (15) one by one for the set X consisting of all current relaxed neurons, and further solve each updated semi-relaxed problem.

Once the optimal solution satisfies $J_T^* > 0$ after testing on certain set T of the relaxed neurons, we derive that the neural network has been verified to be robust. Otherwise one more neuron is chosen to be tightened. To guarantee the completeness, suppose $\mathbf{x}^*$ is the minimizer of J_T^*, then assign $\mathbf{x}^*$ to (12) can arrive at the upper bound J_T', and the constraint $J_T' > 0$ is incorporated to ensure that our tightening scheme is always available for robustness verification. Meanwhile, if it occurs that $J_T' < 0$ at one stage, it is easy to show that the neural network $\widehat{\mathcal{N}}$ is unrobust.

As shown in Fig. 2, the framework consists of two main functional modules, the MILP-solving module and the neuron-selection module. Given a feed-forward neural network $\mathcal{N}$ with label k, we first augment $\mathcal{N}$ into $\widehat{\mathcal{N}}$ for generating a generic MILP problem (Sect. 3.1), and then relax all or some of the neurons (Sect. 3.2). Next, we check whether $J_T^* > 0$ or $J_T' < 0$ by the MILP-solving module, and terminate the iteration if the condition is satisfied at certain steps. Otherwise, the process enters the neuron-selection module, i.e., the Recurrent Tightening Module (RTM), as shown inside the dashed-round-box in Fig. 2. The RTM module is designed to implement the neuron selection strategy for solving the MILP problem. The trained RTM works by grading

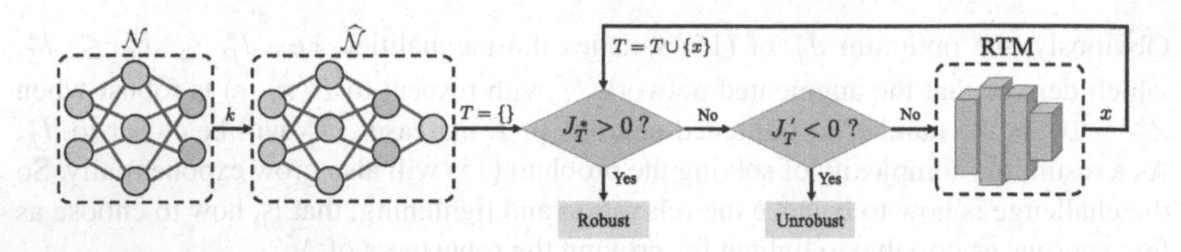

Fig. 2. The architecture of RNN-based framework.

Algorithm 1. The RNN-based Framework for Robustness Verification

Require: Network $\mathcal{N}$, ℓ_∞-ball $B(\mathbf{x}_c, \epsilon)$, label k
1: $\widehat{\mathcal{N}}, X \leftarrow$ Augment$(\mathcal{N}, B(\mathbf{x}_c, \epsilon), k)$ ▷ Augment network $\mathcal{N}$ and initialize ReLU neurons
2: $T \leftarrow \{\}$ ▷ Initialize an empty tighten set T
3: **while** $\neg(J_T^* > 0) \wedge \neg(J_T' < 0)$ **do**
4: $\quad \Theta \leftarrow \left\{\widehat{\mathcal{N}}, \mathcal{I}(T), \mathcal{P}(X - T)\right\}$ ▷ Gather information of current MILP problem
5: $\quad x \leftarrow$ RTM(Θ) ▷ Select a relaxed neuron to tighten
6: $\quad T \leftarrow T \cup \{x\}$
7: $\quad$ Update J_T^* and J_T' with $J\left(\widehat{\mathcal{N}}, \mathcal{I}(T) \wedge \mathcal{P}(X - T)\right)$
8: **if** $J_T^* > 0$ **then return** Robust
9: **else if** $J_T' < 0$ **then return** Unrobust

each neuron from the currently relaxed collection X for optimal tightening choice. A more detailed explanation on how to train such RTM will be given in Sect. 4.2.

The main data flow of the RNN-based framework is given in Algorithm 1. The algorithm takes a neural network $\mathcal{N}$ with its input domain $B(\mathbf{x}_c, \epsilon)$ and labels it by k. In line 1, the algorithm augments $\mathcal{N}$ to $\widehat{\mathcal{N}}$ and extracts all the ReLU neurons X. In line 2, the set T for collecting neurons to tighten is initialized to the empty set. The main loop from Line 3 to Line 8 is to check the robustness condition and do the neuron selection, where line 5 is implemented by the neuron-selection module (cf. RTM in Fig. 2), and line 7 carries out semi-relaxed MILP problem by the MILP-solving module. The algorithm either returns 'Robust' when $J_T^* > 0$ is satisfied in line 8, or reaches 'Unrobust' if $J_T' < 0$ (Line 9).

4.2 Recurrent Tightening Module

The Recurrent Tightening Module (RTM) implements a learned tightening strategy, the key module of the RNN-based framework introduced in Sect. 4.1. It takes the network presented by a numerical tensor as input. The final output of the RTM is a selected neuron which has the highest score based on the result of the RNN whose training process has been illustrated in Sect. 4.3.

The neuron selection process of RTM is shown in Fig. 3. Actually, the input of the module at the beginning is a semi-relaxed MILP problem to be verified, for simplicity we shall express it in an associate network form. Without causing ambiguity, we denote the input as Θ which carries the parameter information of the current MILP

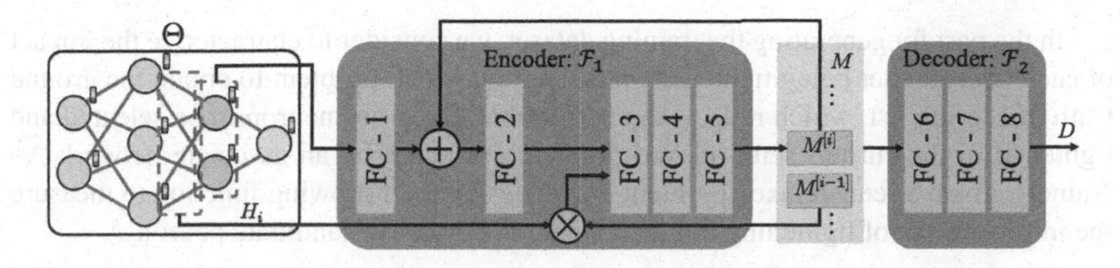

Fig. 3. The flow-chart of RTM.

problem, including the structure parameters of the augmented network $\widehat{\mathcal{N}}$ to be verified, collection of all ReLU neurons X, and the tightened ReLU neuron collection T. The green rectangle submodule in Fig. 3, an RNN consists of the Encoder and the Decoder, is designed to score the tighten. The current neurons in X of the network Θ will be graded and recorded layer-by-layer through this RNN-based scorer. The scoring results of all neurons will be recorded in a list denoted by D whose dimension is the same as the number of neurons. In the final stage, the RTM selects the neuron with the highest score as the output.

In RTM, the neuron selecting is processing in the following way. In the initial, a trained RNN equipped with parameter $\boldsymbol{\theta} = \{\boldsymbol{\theta}_1, \boldsymbol{\theta}_2\}$ encodes $\widehat{\mathcal{N}}$ in layer-by-layer wise according to the structure of the input network, and produces a list of feature matrices $M = \{M^{[1]}, \dots, M^{[m]}\}$ which has the same layer with $\widehat{\mathcal{N}}$, where each $M^{[i]}$ denotes the extracted feature matrix on the i-th layer, and each row of $M^{[i]}$ is the feature vector for a neuron in the layer. To improve the efficiency of the overall algorithm, we have also stored M in order to reuse it in the subsequent encoding process. In addition, we denote the fully-connected Encoder and Decoder by functions $\mathcal{F}_1$ and $\mathcal{F}_2$, respectively. Mathematically, the RTM can be expressed as follows,

$$\begin{aligned} M^{[i]} &= \mathcal{F}_1(\Theta, H_i(M^{[i-1]}), M^{[i]}; \boldsymbol{\theta}_1), \quad i = 1, \dots, m, \\ D &= \mathcal{F}_2(M; \boldsymbol{\theta}_2), \quad d = \arg\max(D), \quad x = \text{Index}(X, d), \end{aligned} \tag{16}$$

where x is the selected neuron to be added into T for the next tightening.

Remark 2. In addition to update $M^{[i]}$ by Θ, we also allow the updating process reusing $M^{[i]}$ to collect previous information and capturing the local information in $M^{[i-1]}$ by utilizing the feed-forward function H_i of the network $\widehat{\mathcal{N}}$. Therefore, (16) is a recurrent function for tightening the neuron successively.

Remark 3. To increase the scalability of the method for different network structures, the RTM module also encode the structure of network and decode all the neurons in the network layer-by-layer processing.

4.3 Training the RNN in RTM

In this subsection, we explain the two important parts of the RNN training process that is based on supervised learning for making the neuron selection policy.

In the part for generating the training dataset, we consider to characterize the impact of each neuron after being tightened on the current MILP problem to obtain the ground truth of scoring list, which reflects the performance of every neuron to be selected and tightened in the current semi-relaxed problem presented by an associate network $\widehat{\mathcal{N}}$. Namely, given a semi-relaxed problem Θ, we design the following function to measure the improvement of tightening for each tightening decision candidate neuron x,

$$y_T(x) = \left(J_T^* - \min(J_{T\cup\{x\}}^*, 0)\right) / J_T^*. \quad (17)$$

Here notice that for each neuron $x \in X$, if $x \in T$ then $y_T(x) = 0$, else if $x \notin T$ and $J_{T\cup\{x\}}^* > 0$ then $y_T(x) = 1$, which means the current MILP has been verified after tightening the neuron x and x is the one we prefer to select. Otherwise $0 < y_T(x) < 1$ for all $x \notin T$ and $J_{T\cup\{x\}}^* < 0$, and the value $y_T(x)$ measures the relative improvement for the result of the current MILP after tightening the neuron x; briefly, $y_T(x)$ is the target score of selecting and tightening x. Hereafter, we use Y_T to denote the ground truth of score list of all neurons ordered layer-by-layer.

The second important part for constructing the neuron selection policy is to design a loss function that is best fitting the relax-and-tighten of the MILP related to robustness verification. For a given RNN parameterized by $\boldsymbol{\theta}$, The following two principles are adopted in the design of the loss function.

- For the current MILP which is presented by Θ, the score list D_T as the output of the training RNN should fit the ground truth of score list Y_T well, i.e.,

$$\mathcal{L}_1(\boldsymbol{\theta}) = \|Y_T \odot (D_T - Y_T)\|_1, \quad (18)$$

  where D_T denotes the predicted tightening scores at Θ, i.e. $D_T = \text{RNN}(\Theta; \boldsymbol{\theta})$. And the Hadamard product is introduced in $\mathcal{L}_1$ for assigning higher weights to those with higher scores according to Y_T.
- It is worth noting that tightening the majority of neurons may give similar improvements, which leads to the inconvenience of ranking the predicted scores by RNN. So we introduce a small constant $s > 0$ and construct the following pairwise ranking loss function $\mathcal{L}_2$ to regularize the scores are at least s-apart for each pair of neurons, i.e.,

$$\mathcal{L}_2(\boldsymbol{\theta}) = \frac{1}{r^2} \sum_{i=1}^{r-1} \sum_{j=i+1}^{r} \max(s - d_T^{[i]} + d_T^{[j]}, 0), \quad (19)$$

  where $d_T^{[i]}$ and $d_T^{[j]}$ are i-th and j-th highest scores predicted in D_T respectively. For the learning efficiency, the pairwise ranking loss $\mathcal{L}_2$ only involves the top r scores in D_T.

In summary, the total loss function is constructed as follows

$$\mathcal{L}(\boldsymbol{\theta}) = \lambda \mathcal{L}_1(\boldsymbol{\theta}) + (1 - \lambda)\mathcal{L}_2(\boldsymbol{\theta}), \quad (20)$$

where $0 < \lambda < 1$ is the parameter to control the weights between $\mathcal{L}_1$ and $\mathcal{L}_2$. In fact, the experiment performance in Sect. 5 reflects that the design of the loss function is reasonable and helps to improve training efficiency, which validates the above analysis.

Algorithm 2. Generating Training Dataset

Require: $\mathcal{N}, \{B(\mathbf{x}_i, \epsilon_i), k_i\}_{i=1,\dots,P}$
1: $S \leftarrow \{\}$ ▷ Initialize training dataset S
2: **for** $i = 1, \dots, P$ **do** ▷ Generate dataset using P images
3: $\widehat{\mathcal{N}}, X \leftarrow$ Augment$(\mathcal{N}, B(\mathbf{x}_i, \epsilon_i), k_i)$; $T \leftarrow \{\}$
4: **for** $j = 1, \dots, t$ **do** ▷ Tighten neurons at most t times
5: $\Theta \leftarrow \left\{\widehat{\mathcal{N}}, \mathcal{I}(T), \mathcal{P}(X - T)\right\}$
6: Construct ground truth score list Y_T with current Θ
7: $S \leftarrow S \cup \{(\Theta, Y_T)\}$ ▷ Add sample (Θ, Y_T) to dataset S
8: Index the optimal neuron x with highest tightening score in Y_T
9: $T \leftarrow T \cup \{x\}$
10: **if** Robustness is verified **then Break**
11: **return** S

Training Dataset. We build a subset of CIFAR-10 images to generate dataset S by the following steps. At first, we randomly pick $P = 400$ images and the corresponding perturbation ϵ with adversarial labels determined from [21], which can cover most robustness verification problems on CIFAR-10. An adversarially trained network called Base model is also chosen according to [21]. In the second step, we initialize an empty set T for each image and measure the improvements for all the relaxed ReLU neurons to construct the ground truth of score list Y_T. At last, we refer to the score list Y_T to select the optimal neuron with the highest improvement and put it into T, and then update the current MILP problem to repeat the above process until the recursion depth reaches the threshold t or the robustness is verified. The procedure is given in Algorithm 2.

Training Details. In the training setting, we have employed `Adam Optimizer` [19] with $\beta_1 = 0.9$ and $\beta_2 = 0.999$ for backward propagation of the RNN. The RNN is trained for 100 epochs with a learning rate of 0.001 and a weight decay of 0.0001. We have also set $r = 35, s = 0.04$, and $\lambda = 0.5$. For generating datasets, we have set $t = 20$ and divided the training set and validation set randomly according to the ratio of 7:3. In terms of training accuracy, we rank tightening scores predicted by RNN and consider the top 5 scores that are also in the top 5 of ground truth scores as correct choices since we are mainly interested in tightening neurons with great improvement using the metric defined by (17). Finally, our RNN reaches 71.6% accuracy on training dataset and 63.5% accuracy on validation dataset. We also run an ablation study by removing $\mathcal{L}_2$, which reduced the accuracy of the validation dataset from 63.5% to 59.7%.

5 Experiments

We have implemented a tool in Pytorch [23] called L2T for the neural network robustness verification, which is based on the policy of learning to tighten (L2T for short) in our proposed framework. The code is available at https://github.com/Vampire689/L2T. For demonstrating the advantages of L2T, we first make a comparison with different tighten-based strategies on the dataset provided in [21] (c.f. Sect. 5.1). And L2T has been compared with the current mainstream robustness verification tools on two large-scale robustness verification benchmarks, i.e., the OVAL benchmark [22] and COLT

Table 1. The result of strategies on Easy, Medium and Hard dataset.

Dataset	Average number of neurons tightened			
	Rand	BaBSR	GNN	L2T
CIFAR - Easy	31.65	24.10	22.82	22.53
CIFAR - Medium	60.71	38.67	27.24	24.90
CIFAR - Hard	122.90	48.74	35.74	31.98

benchmark [4] to show the performance of two aspects: (1) the efficiency of L2T on OVAL benchmark for verifying adversarial properties (c.f. Sect. 5.2); (2) the good performance on scalability to verify larger models on the COLT benchmark (c.f. Sect. 5.3).

5.1 Effectiveness of RNN-Based Tightening Strategy

We analyse L2T by comparing with three different tightening strategies: 1) Random selection (Rand), 2) hand-designed heuristic (BaBSR) [6], and 3) GNN learning-based strategy (GNN) [21]. We use the dataset of three different difficulty levels (Easy, Medium and Hard) with the network of 3172 neurons provided in [21] and compute the average number of neurons tightened on solved properties of all strategies. Results of different tightening strategies are listed in Table 1. L2T achieves the least average tightening number of neurons for solving the properties compared to other strategies. In particular, L2T outperforms GNN in all levels, showing that our RNN-based framework has learned the empirical information from continuous tightening strategy.

5.2 Performance on the OVAL Benchmark

We evaluate the performance of L2T and five baselines on OVAL benchmark [22] used in [31]. The benchmark consists of sets of adversarial robustness properties on three adversarially trained CIFAR-10 CNNs, which are Base, Wide and Deep models with 3172, 6244 and 6756 ReLUs respectively [22]. All three models are robustly trained using the method from [33]. There are 100 properties for each of the three models, and each property is assigned a non-correct class and associated with a specifically designed perturbation radius $\epsilon \in [0, 16/255]$ on correctly classified CIFAR-10 images. The benchmark is to verify that for the given ϵ, the trained network will not misclassify by labelling the image as the non-correct class. A timeout of one hour per property is suggested. We conduct experimental comparisons with five state of the art verifiers: 1) OVAL [5], a strong verification framework based on Lagrangian decomposition on GPUs. 2) GNN [21], a state-of-the-art tightening-based verifier using a learned graph neural network to imitate optimal ReLU selecting strategy. 3) ERAN [26–29], a scalable verification toolbox based on abstract interpretation. 4) A.set [22], a latest dual-space verifier with a tighter linear relaxation than Planet relaxations. 5) $\alpha\beta$-crown [32,34,35], a most recent fast and scalable verifier with efficient bound propagation.

We report the average solving time and the percentage of timeout over all properties in Table 2. It can be shown that L2T ranks the top two in average time efficiency

Table 2. The performance of verifiers on the Base, Wide and Deep model.

Network	Metric	OVAL	GNN	ERAN	A.set	$\alpha\beta$-crown	L2T
Base-3172-[0,16/255]	Time (s)	835.2	662.7	805.9	377.3	711.5	442.8
	Timeout (%)	20	15	5	7	15	9
Wide-6244-[0,16/255]	Time (s)	539.4	268.5	607.1	162.7	354.7	228.8
	Timeout (%)	12	6	7	3	5	5
Deep-6756-[0,16/255]	Time (s)	258.6	80.8	574.6	190.8	55.9	26.1
	Timeout (%)	4	1	1	2	1	0

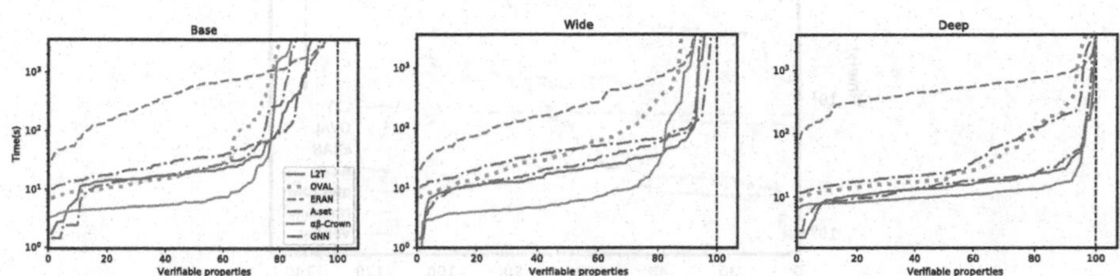

Fig. 4. Cactus plots for Base, Wide and Deep model on OVAL benchmark.

among the current popular tools, and L2T performs the best on the Deep model with no timeout. L2T provides an additional average time saving by 37.8%, 35.5% and 53.3% on three models respectively over the very recent $\alpha\beta$-crown. Although A.set achieves a slightly lower time of verifying properties on the Base and Wide model, L2T leads to 7x faster than A.set on Deep model. Taken together, L2T costs the least total average time on all three models. The cactus plots in Fig. 4 show the increasing solving time with the accumulative verified properties for each verifier. L2T consistently provides great performance on the majority of properties.

5.3 Performance on the COLT Benchmark

Furthermore, we evaluate L2T and six baselines on COLT benchmark [4], which is quite challenging that includes the two largest CIFAR-10 CNNs with tens of thousands ReLUs in VNN-COMP 2021 [2]. The goal is to verify that the classification is robust within an adversarial region defined by l_∞-ball of radius ϵ around an image. The COLT benchmark considers the first 100 images of the test datasets and discards those that are incorrectly classified. The values of ϵ for two CNNs are 2/255 and 8/255 respectively, and the verification time is limited to five minutes per property. To evaluate the performance on COLT benchmark, we introduce another two well-performed verifiers instead the underperforming GNN: 1) Nnenum [3], a recent tool using multi-level abstraction to achieve high-performance verification of ReLU networks. 2) VeriNet [16], a sound and complete symbolic interval propagation-based toolkit for robustness verification.

As shown in Table 3, L2T significantly outperforms the majority of verifiers in terms of average solving time and achieves lower timeout rate for two large-scale CIFAR-10

Table 3. The performance of verifiers on two large-scale CIFAR-10 networks.

Network	Metric	OVAL	Nnenum	VeriNet	ERAN	A.set	$\alpha\beta$-crown	L2T
CIFAR-49402-2/255	Time (s)	85.3	115.2	80.9	87.8	104.5	35.0	48.6
	Timeout (%)	26.3	30	23.8	26.3	32.5	3.7	11.2
CIFAR-16634-8/255	Time (s)	124.8	135.0	105.6	111.5	166.1	43.8	88.0
	Timeout (%)	27.5	27.5	23.8	16.3	35	6.3	13.8

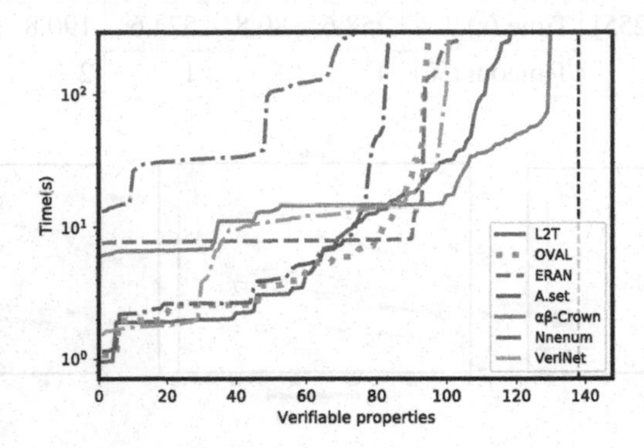

Fig. 5. COLT benchmark cactus plot for L2T and six baselines.

networks. In comparison with A.set, which performed excellent in previous experiments, the average time of L2T on these two networks was reduced by 53.5% and 47.0%, respectively. And the percentage of timeout also drops quite a lot. Although L2T is the second only to $\alpha\beta$-crown in terms of time efficiency, the comprehensive performance of our tool is still competitive as shown in the cactus plot in Fig. 5, which shows the increasing solving time required with accumulative number of verified properties of two networks for each verifier. Within the limited solving time of 10s, L2T verified around 35 more properties than $\alpha\beta$-crown. As verifying robustness properties becomes increasingly complex, $\alpha\beta$-crown verified more properties than L2T, while L2T is more efficient than other verifiers.

6 Conclusion

We have proposed an RNN-based framework for solving MILP problems generated from the robustness verification of deep neural networks. This framework adopts an iterative process to gradually repair failures caused by over-relaxation. Boosted by the intelligent neuron selection strategy through a well-trained RNN in its key recurrent tightening module, our framework can achieve effective verification while ensuring the efficiency of MILP solving. A prototype tool named L2T has been implemented and compared with state-of-the-art verifiers on some large-scale benchmarks. The experimental results demonstrate the efficiency and scalability of our approach for robustness verification on large-scale neural networks.

Acknowledgements. This work is supported by the National Natural Science Foundation of China under Grant 12171159, 61902325, and Shanghai Trusted Industry Internet Software Collaborative Innovation Center.

References

1. Alvarez, A.M., Louveaux, Q., Wehenkel, L.: A machine learning-based approximation of strong branching. INFORMS J. Comput. **29**, 185–195 (2017)
2. Bak, S., Liu, C., Johnson, T.: The second international verification of neural networks competition (VNN-COMP 2021): summary and results. arXiv preprint arXiv:2109.00498 (2021)
3. Bak, S., Tran, H.D., Hobbs, K., Johnson, T.T.: Improved geometric path enumeration for verifying relu neural networks. In: International Conference on Computer Aided Verification, pp. 66–96 (2020)
4. Balunovic, M., Vechev, M.: Adversarial training and provable defenses: bridging the gap. In: International Conference on Learning Representations (2020)
5. Bunel, R., et al.: Lagrangian decomposition for neural network verification. In: Conference on Uncertainty in Artificial Intelligence, pp. 370–379 (2020)
6. Bunel, R., Mudigonda, P., Turkaslan, I., Torr, P., Lu, J., Kohli, P.: Branch and bound for piecewise linear neural network verification. J. Mach. Learn. Res. 42:1–42:39 (2020)
7. Bunel, R., Turkaslan, I., Torr, P.H.S., Kohli, P., Mudigonda, P.K.: A unified view of piecewise linear neural network verification. In: Neural Information Processing Systems (2018)
8. Dvijotham, K., Stanforth, R., Gowal, S., Mann, T.A., Kohli, P.: A dual approach to scalable verification of deep networks. In: Conference on Uncertainty in Artificial Intelligence, p. 3 (2018)
9. Ehlers, R.: Formal verification of piece-wise linear feed-forward neural networks. In: International Symposium on Automated Technology for Verification and Analysis, pp. 269–286 (2017)
10. Elboher, Y.Y., Gottschlich, J., Katz, G.: An abstraction-based framework for neural network verification. In: International Conference on Computer Aided Verification, pp. 43–65 (2020)
11. Gasse, M., Chetelat, D., Ferroni, N., Charlin, L., Lodi, A.: Exact combinatorial optimization with graph convolutional neural networks. In: Advances in Neural Information Processing Systems. Curran Associates, Inc. (2019)
12. Glorot, X., Bordes, A., Bengio, Y.: Deep sparse rectifier neural networks. In: Proceedings of the Fourteenth International Conference on Artificial Intelligence and Statistics, pp. 315–323 (2011)
13. Golson, J.: Tesla driver killed in crash with autopilot active, NHTSA investigating. The Verge (2016)
14. Goodfellow, I.J., Shlens, J., Szegedy, C.: Explaining and harnessing adversarial examples. arXiv preprint arXiv:1412.6572 (2014)
15. Hansknecht, C., Joormann, I., Stiller, S.: Cuts, primal heuristics, and learning to branch for the time-dependent traveling salesman problem. arXiv preprint arXiv:1805.01415 (2018)
16. Henriksen, P., Lomuscio, A.: Efficient neural network verification via adaptive refinement and adversarial search. In: European Conference on Artificial Intelligence, pp. 2513–2520 (2020)
17. Henriksen, P., Lomuscio, A.: Deepsplit: an efficient splitting method for neural network verification via indirect effect analysis. In: International Joint Conference on Artificial Intelligence (2021)
18. Katz, G., Barrett, C., Dill, D.L., Julian, K., Kochenderfer, M.J.: Reluplex: an efficient SMT solver for verifying deep neural networks. In: International Conference on Computer Aided Verification, pp. 97–117 (2017)

19. Kingma, D.P., Ba, J.: Adam: a method for stochastic optimization. arXiv preprint arXiv:1412.6980 (2014)
20. Lin, W., et al.: Robustness verification of classification deep neural networks via linear programming. In: Conference on Computer Vision and Pattern Recognition, pp. 11418–11427 (2019)
21. Lu, J., Kumar, M.P.: Neural network branching for neural network verification. In: International Conference on Learning Representations (2020)
22. Palma, A.D., Behl, H., Bunel, R.R., Torr, P., Kumar, M.P.: Scaling the convex barrier with active sets. In: International Conference on Learning Representations (2021)
23. Paszke, A., et al.: Pytorch: an imperative style, high-performance deep learning library. In: Conference on Neural Information Processing Systems, pp. 8024–8035 (2019)
24. Raghunathan, A., Steinhardt, J., Liang, P.: Semidefinite relaxations for certifying robustness to adversarial examples. In: Conference on Neural Information Processing Systems, pp. 10900–10910 (2018)
25. Ruan, W., Huang, X., Kwiatkowska, M.: Reachability analysis of deep neural networks with provable guarantees. In: International Joint Conference on Artificial Intelligence (2018)
26. Singh, G., Ganvir, R., Puschel, M., Vechev, M.: Beyond the single neuron convex barrier for neural network certification. In: Conference on Neural Information Processing Systems (2019)
27. Singh, G., Gehr, T., Mirman, M., Püschel, M., Vechev, M.T.: Fast and effective robustness certification. In: Conference on Neural Information Processing Systems (2018)
28. Singh, G., Gehr, T., Püschel, M., Vechev, M.: An abstract domain for certifying neural networks. In: Proceedings of the ACM on Programming Languages, pp. 1–30 (2019)
29. Singh, G., Gehr, T., Püschel, M., Vechev, M.: Robustness certification with refinement. In: International Conference on Learning Representations (2019)
30. Szegedy, C., et al.: Intriguing properties of neural networks. In: International Conference on Learning Representations (2014)
31. VNN-COMP: International verification of neural networks competition (VNN-COMP). In: International Conference on Computer Aided Verification (2020)
32. Wang, S., et al.: Beta-CROWN: efficient bound propagation with per-neuron split constraints for complete and incomplete neural network verification. In: Advances in Neural Information Processing Systems (2021)
33. Wong, E., Kolter, Z.: Provable defenses against adversarial examples via the convex outer adversarial polytope. In: International Conference on Machine Learning, pp. 5286–5295 (2018)
34. Xu, K., et al.: Fast and complete: enabling complete neural network verification with rapid and massively parallel incomplete verifiers. In: International Conference on Learning Representations (2021)
35. Zhang, H., Weng, T.W., Chen, P.Y., Hsieh, C.J., Daniel, L.: Efficient neural network robustness certification with general activation functions. In: Advances in Neural Information Processing Systems (2018)

Image Denoising Using Convolutional Sparse Coding Network with Dry Friction

Yali Zhang[1], Xiaofan Wang[1], Fengpin Wang[1], and Jinjia Wang[1,2](✉)

[1] School of Information Science and Engineering, Yanshan University, Qinhuangdao 066004, China
wjj@ysu.edu.cn

[2] Hebei Key Laboratory of Information Transmission and Signal Processing, Yanshan University, Qinhuangdao 066004, China

Abstract. Convolutional sparse coding model has been successfully used in some tasks such as signal or image processing and classification. The recently proposed supervised convolutional sparse coding network (CSCNet) model based on the Minimum Mean Square Error (MMSE) approximation shows the similar PSNR value for image denoising problem with state of the art methods while using much fewer parameters. The CSCNet uses the learning convolutional iterative shrinkage-thresholding algorithms (LISTA) based on the convolutional dictionary setting. However, LISTA methods are known to converge to local minima. In this paper we proposed one novel algorithm based on LISTA with dry friction, named LISTDFA. The dry friction enters the LISTDFA algorithm through proximal mapping. Due to the nature of dry friction, the LISTDFA algorithm is proven to converge in a finite time. The corresponding iterative neural network preserves the computational simplicity of the original CSCNet, and can reach a better local minima practically.

Keywords: Image denoising · Convolutional sparse coding · Iterative shrinkage thresholding algorithms · Dry friction

1 Introduction

Noise is the pollution of image in the process of acquisition, compression and transmission, which is easy to cause the loss of image information and bring adverse effects on image processing [13,27]. Image denoising is the process of removing noise from the image polluted by noise and restoring the original image. In recent years, it has been paid attention as the basis of other image processing and is one of the key issues in the field of image processing. In order to achieve image denoising, we must know the prior information of the original signal. Image priors, also known as image models, involve the mathematical description of the

This work was supported in part by Basic Research Cooperation Projects of Beijing, Tianjin and Hebei (F2019203583, 19JCZDJC65600), and the Central Funds Guiding the Local Science and Technology Development (Basic Research Projects) (206Z5001G).

L. Wang et al. (Eds.): ACCV 2022, LNCS 13841, pp. 587–601, 2023.
https://doi.org/10.1007/978-3-031-26319-4_35

true distribution of images. Convolutional sparse coding (CSC) is a popular and important prior model in the field of signal processing and machine learning [29]. CSC with the banded convolutional structure constrained forms has a solid theoretical foundation and an uniqueness of the solution. CSC has many applications in the field of image processing, such as inverse problems [21], image reconstruction [20,31,36,37], image denoising [28,30], image inpainting [16,20,37] and so on. Along with the application of deep learning, some advanced image reconstruction methods based on convolutional neural networks have achieved excellent performance in image denoising [10,33–35]. In state of the art studies, the improvement of neural network module makes the reconstruction model more interactive in industrial and commercial scenarios [17,18].

Iterative neural network [7] is a kind of unfold network based on the iterative algorithm. Iterative neural network combines a forward neural network and an iterative algorithm, leading to good generalization capability over an iterative algorithm. For example, an iterative threshold shrink algorithm (ISTA) is a simple algorithm for the sparse representation problem. The learning iterative threshold shrink algorithm (LISTA) [15] is an iterative neural network that can be trained through supervised learning. Moreover, a convolutional iterative threshold shrink algorithm is a simple algorithm for the CSC problem. The learning convolutional iterative threshold shrink algorithm, named learning convolutional sparse coding, is an corresponding iterative neural network [24]. Similarly, a deep coupled ISTA network is proposed for multi-modal image super-resolution problem [9]. A deep convolutional sparse coding network is proposed for jpeg artifacts reduction [14]. A deep coupled convolutional sparse coding network is proposed for pan-sharpening CSC problem [32]. Inspired by multilayer neural network, a multilayer ISTA is proposed and a multilayer ISTA network is proposed for classification [26].

Recently, a supervised convolution sparse coding network (CSCNet) is proposed for image denoising problem [22]. CSCNet is based on the ISTA algorithm and LISTA network [15,24]. It is trained via stochastic gradient-descent using self-supervised form. Thus, naturally CSCNet can learn the convolutional dictionary over very large datasets. Although it first answers a problem why the CSC model denoise natural images poorly, there are some problems in CSCNet. The ISTA iterative algorithm tends to jump into a local minima, leading to CSCNet producing the worse results for the application in the noise measurements.

In order to jump out of a local minimal of the ISTA algorithm, an iterative shrinkage-thresholding with dry friction algorithms (ISTDFA) is proposed. It is an improved proximal gradient algorithm with forward-backward splitting based on the a heavy ball system with dry friction [1]. The nature of dry friction can make the system reach a stable state in a finite time. Dry friction enters into the algorithm based on the proximal mapping. The proposed ISTDFA algorithm can effectively reduce the value of objective function of the ISTA algorithm. However, dry friction has a certain effect on the update of the convolution dictionary. The back propagation algorithms commonly used in deep learning do not take into account the effect of dry friction. The ISTDFA algorithm solves the problem throughout the recently proposed variance regularization [12]. It makes the update of the convolution dictionary independent of dry friction. Moreover, ISTDFA is extended to a learned iterative shrinkage-thresholding with dry friction algorithms (LISTDFA) based on iterative neural network. The novel iterative neural network, named CSCNet-DF, is proposed. CSCNet-DF includes an encoder and a decoder. The encoder uses LISTDFA to produce the coding from input images. The decoder is used to reconstruct the images. The minimum mean square error between the input images and output images is used to update the convolution dictionary by the back propagation algorithms.

Finally, the experimental results show that the proposed CSCNet-DF network is superior to CSCNet.

2 Related Work

2.1 Convolutional Sparse Coding Model

This sparse coding model assumes that a signal $\mathbf{y} = \mathbf{D}\boldsymbol{\Gamma}$ is a linear combination of atoms, where $\mathbf{D} \in \mathbb{R}^{N\times M}$ is a dictionary and $\boldsymbol{\Gamma} \in \mathbb{R}^M$ is a sparse vector. For a given signal $\mathbf{y}$ and dictionary $\mathbf{D}$, the sparse representation of $\mathbf{y}$ is solved by the following optimization problem:

$$\min_{\boldsymbol{\Gamma}} F(\boldsymbol{\Gamma}) = \min_{\boldsymbol{\Gamma}} \frac{1}{2}\|\mathbf{y} - \mathbf{D}\boldsymbol{\Gamma}\|_2^2 + \lambda\|\boldsymbol{\Gamma}\|_1 \,. \tag{1}$$

The solution of this sparse coding problem (1) is unique and can be obtained by many classical algorithms, such as the Orthogonal Matching Pursuit (OMP) [5], and Basis Pursuit (BP) [6]. The corresponding classical dictionary learning methods include MOD [11], trainlets [25], online dictionary learning [19], K-SVD [2], and so on. If D is a shift-invariant convolutional dictionary, this problem (1) is changed as the convolutional sparse coding problem. The solution is obtained by the Fourier-based fast methods [4,21]. But the Fourier-based signal representation loses signal localization.

Assume that a signal $\mathbf{y}$ can be expressed as $\mathbf{y} = \mathbf{D}\boldsymbol{\Gamma} = \sum_{i=1}^{N} \mathbf{P}_i^T \mathbf{D}_L \boldsymbol{\alpha}_i$. All shifted version of the local dictionary $\mathbf{D}_L \in \mathbb{R}^{n\times m}$ compose the convolutional dictionary $\mathbf{D}$. And all sparse vectors $\boldsymbol{\alpha}_i$ compose the sparse vector $\boldsymbol{\Gamma}$. The slice $\mathbf{s}_i$ is defined as $\mathbf{s}_i = \mathbf{D}_L\boldsymbol{\alpha}_i$, represents the i-th slice. The $\mathbf{P}_i^T$ represents the operation that put $\mathbf{D}_L\boldsymbol{\alpha}_i$ in the i-th position of the signal. The slice-based convolution sparse coding is proposed [20], which can be reformulated as

$$\min_{\boldsymbol{\alpha}_i} \frac{1}{2}\|\mathbf{y} - \sum_{i=1}^{n} \mathbf{P}_i^T \mathbf{s}_i\|_2^2 + \lambda \sum_{i=1}^{n} \|\boldsymbol{\alpha}_i\|_1 \,. \tag{2}$$

As far as we know, K-SVD [2] is one of the method to resolve the CSC. Both SBDL [20] and LoBCoD [37] are effective algorithms to solve the CSC problem (2). All of them have great dependence on alternating optimization algorithm, which makes the algorithms have more iterations.

2.2 Convolutional Sparse Coding Network

We believe that the CSC model is suitable for processing texture signals, leading to Gabor-like non-smooth filters. It is used in some problems, such as cartoon texture separation, image fusion or single image super-resolution to model the texture part of an image. However, the CSC model cannot cope with image denoising or other inverse problems involving noisy signals. In other words, the CSC model can only be used to model noise-free images. Simon et al. [23] argue that a convolutional dictionary with high coherence over-fits the noisy signal, so the CSC model is not suitable for the natural image denoising problem. This is because the filters in the CSC model cannot simultaneously meet the conditions of low global cross-correlation and low auto-correlation. To solve this problem, a new idea has been proposed [23]. To suppress the dictionary coherence and obtain the optimal solution, a larger stride size in the convolutional dictionary should be chosen. Unlike the basic CSC model with $q = 1$, the filters can be partially overlap when the stride q is chosen in the range of $1 \leq q \leq n$ and q is large enough. In this case, the convolutional dictionary is guaranteed to be mutually consistent even for the smooth filters. A CSCNet network is proposed for the image denoising problem [22]. CSCNet is an iterate neural network that unfolds the ISAT algorithm [24], which is a supervised denoising model. CSCNet first replicates the input noisy images q^2 times to obtain the possible q^2 shifted versions. Then the sparse coefficients is computed by CSCNet. Finally, the denoised reconstructed image is obtained by calculating the average image of all the shifted reconstructed images. The convolutional dictionary and network parameters are updated by back-propagation algorithm based on the minimization of the mean square error between the clean and denoised images. Experimental results show that CSCNet not only obtains similar denoising performance to SOTA supervised methods, but also uses only fewer neural network parameters.

2.3 Variance Regularization

The goal of convolution sparse coding is to find the corresponding sparse vector $\mathbf{\Gamma}$ and convolution dictionary $\mathbf{D}$ on the premise of a given signal $\mathbf{y}$. For any positive constant c, we can obtain the same reconstruction from the re-scaled convolutional dictionary $D' = c\mathbf{D}$ and sparse vector $\mathbf{\Gamma}' = \mathbf{\Gamma}/c$. If $c > 1$ holds, then we can get the same reconstruction from a smaller ℓ_1 norm of $\mathbf{\Gamma}'$ than $\mathbf{\Gamma}$. If there is no upper limit on the values of the convolution dictionary $\mathbf{D}$, which means the values of the needles can be arbitrarily small, leading to the collapse of the ℓ_1 norm of the sparse vector. In order to avoid the collapse of ℓ_1 regularization of sparse vector, it is often necessary to restrict convolution dictionary $\mathbf{D}$ in optimization problems. In practice, the column of the convolution dictionary $\mathbf{D}$ is often limited to a constant ℓ_2 norm. However, this is a very challenging thing to expand to a network.

In recent work, the variance principle [3] is presented, which uses a regularization term on the variance of the embeddings on each dimension individually to avoid the collapse of the sparse vector. Similar to this method, Yann et al.

[12] add a regularization term to the minimized objection function to neutralize the effect of $\mathbf{D}$'s weights, which ensures the latent sparse vector components have greater variances than a fixed threshold over the sparse representations for given inputs. This strategy also plays a role in one iterative neural network because the variance regularization term is independent from the dictionary $\mathbf{D}$'s architecture.

3 Proposed CSCNet-DF Network

3.1 Iterative Shrinkage-Thresholding with Dry Friction Algorithm

Throughout this section $\mathbb{H}$ is a real Hilbert space, and the associated norm $\|\cdot\|$. The minimized objective function has the following form $f+g$, where the non-convex smooth function $f:\mathbb{H}\to\mathbb{R}$ represents a L-Lipschitz continuous function, and the convex non-smooth function $g:\mathbb{H}\to\mathbb{R}\cup\{+\infty\}$ denotes a proper lower semi-continuous function.

For the non-smooth non-convex potential function $f+g$, we can get the corresponding heavy ball with dry friction system [1]

$$\mathbf{x}''(t)+\gamma\mathbf{x}'(t)+\partial\phi(\mathbf{x}'(t))+\nabla f(\mathbf{x}(t))+\partial g(\mathbf{x}(t))\ni 0\,. \tag{3}$$

The system contains two different kinds of damping:

(1) Viscous damping: $\gamma\mathbf{x}'(t)$ is viscous damping, and γ is a viscous damping coefficient satisfying $\gamma>0$.
(2) Dry friction damping: $\partial\phi(\mathbf{x}'(t))$ represents dry friction. ϕ is the dry friction potential function, which has the following properties: ϕ is a convex and lower semi-continuous function, and ϕ has a sharp minimum at the origin, that is, $\min_{\xi\in H}\phi(\xi)=\phi(0)=0$.

According to the properties of dry friction ϕ, we assume that $\phi(\mathbf{x})=r\|\mathbf{x}\|_1$, and r is the dry friction coefficient satisfying $r>0$. From the definition of the function ϕ, the dry friction properties are satisfied with $B(0,r)\subset\partial\phi(0)$.

The time discretization of Eq. (3) with a fixed time step size $h>0$ is given as follows

$$\frac{(\mathbf{x}^{k+1}-\mathbf{x}^k)-(\mathbf{x}^k-\mathbf{x}^{k-1})}{h^2}+\frac{\gamma(\mathbf{x}^{k+1}-\mathbf{x}^k)}{h}+\partial\phi\left(\frac{\mathbf{x}^{k+1}-\mathbf{x}^k}{h}\right)+\nabla f(\mathbf{x}^k)+\partial g(\mathbf{x}^{k+1})\ni 0\,. \tag{4}$$

Equation (4) relates the classical dynamics method to the proximal gradient methods: the smooth function f is implicit in Eq. (4), corresponding to the gradient step of the proximal gradient method; similarly, non-smooth functions g and ϕ are explict in Eq. (4), corresponding to the proximal step of the proximal gradient method.

For each $k\in\mathbb{N}$, let's introduce the auxiliary convex function defined by

$$\phi_k(\mathbf{x}^k+hx):=h\phi(x)\,. \tag{5}$$

Then, we have $\partial\phi_k\left(\mathbf{x}^{k+1}\right)=\partial\phi_k\left(\mathbf{x}^k+h\frac{\mathbf{x}^{k+1}-\mathbf{x}^k}{h}\right)=\partial\phi\left(\frac{\mathbf{x}^{k+1}-\mathbf{x}^k}{h}\right)$. And then we can induce that from (4)

$$\begin{aligned}&\tfrac{1}{h^2}\left(\mathbf{x}^{k+1}-2\mathbf{x}^k+\mathbf{x}^{k-1}\right)+\tfrac{\gamma}{h}\left(\mathbf{x}^{k+1}-\mathbf{x}^k\right)\\&+\partial\phi_k\left(\mathbf{x}^{k+1}\right)+\nabla f\left(\mathbf{x}^k\right)+\partial g\left(\mathbf{x}^{k+1}\right)\ni 0\end{aligned}\quad (6)$$

Since ϕ is continuous, we have $\partial\phi_k+\partial g=\partial\left(\phi_k+g\right)$, which implies

$$\tfrac{1+\gamma h}{h^2}\mathbf{x}^{k+1}+\partial\left(g+\phi_k\right)\left(\mathbf{x}^{k+1}\right)\ni\tfrac{2+\gamma h}{h^2}\mathbf{x}^k-\tfrac{1}{h^2}\mathbf{x}^{k-1}-\nabla f\left(\mathbf{x}^k\right). \quad (7)$$

We finally get the iterative shrinkage-thresholding with dry friction algorithm (ISTDFA),

$$\mathbf{x}^{k+1}=\operatorname{prox}_{\frac{h^2}{1+\gamma h}(g+\phi_k)}\left(\mathbf{x}^{k+1/2}-\frac{h^2}{1+\gamma h}\nabla f\left(\mathbf{x}^k\right)\right), \quad (8)$$

where $\mathbf{x}^{k+1/2}=\mathbf{x}^k+\frac{1}{1+\gamma h}\left(\mathbf{x}^k-\mathbf{x}^{k-1}\right)$, and the proximal map is defined as follows:

$$\operatorname{prox}_{\eta p}(\mathbf{x}):=\arg\min_{\boldsymbol{\xi}\in\mathbb{H}}\left\{\eta p(\boldsymbol{\xi})+\frac{1}{2}\|\mathbf{x}-\boldsymbol{\xi}\|^2\right\}. \quad (9)$$

Theorem 1. *Suppose that the parameters $h>0$, $\gamma>0$ satisfy $h<\frac{2\gamma}{L}$. Then for the sequence $\left\{\mathbf{x}^k\right\}$ genereted by the algorithm (ISTDFA) we have:*

(1) $\sum_k\left\|\boldsymbol{x}^{k+1}-\boldsymbol{x}^k\right\|<+\infty$, and therefore $\lim_k\boldsymbol{x}^k=\boldsymbol{x}^*$ exists for the strong topology of $\mathbb{H}$.
(2) The vector $\boldsymbol{x}^*$ satisfies $0\in\partial\phi(0)+\nabla f(\boldsymbol{x}^*)+\partial g(\boldsymbol{x}^*)$.
(3) Suppose that $\mathbb{H}$ is finite dimensional, and suppose that $-\left(\nabla f(\boldsymbol{x}^*)+\partial g(\boldsymbol{x}^*)\right)\in int(\partial(0))$. Then, the sequence $\left\{\mathbf{x}^k\right\}$ is finitely convergent.

3.2 Application ISTDFA to CSC Model

Consider the novel minimization problem for CSC model via local processing and variance regularization as follows:

$$\arg\min_{\{\boldsymbol{\alpha}_{l,i}\}_{i=1}^N}\frac{1}{2}\sum_{l=1}^{I}\left\|\sum_{i=1}^{N}\mathbf{P}_i^T\mathbf{D}_L\boldsymbol{\alpha}_{l,i}-\mathbf{y}_l\right\|_2^2+\lambda\sum_{l=1}^{I}\sum_{i=1}^{N}\|\boldsymbol{\alpha}_{l,i}\|_1+\beta\sum_{i=1}^{N}\left[\left(T-\sqrt{\operatorname{Var}(\boldsymbol{\alpha}_{\cdot i})}\right)_+\right]^2, \quad (10)$$

where $\mathbf{D}_L$ is the local convolutional dictionary which has n rows and m columns; $\boldsymbol{\alpha}_{l,i}$ is the sparse coding of each component i of each sample l; $\mathbf{P}_i^T$ which has N rows and n columns is the operator that puts $\mathbf{D}_L\boldsymbol{\alpha}_{l,i}$ in the i-th position and pads the rest of the entries with zero; $\mathbf{y}_l$ is the nosiy signal; $\|\cdot\|_2$ stands for vector of ℓ_2 norm or Frobenius norm of matrix; λ is the super parameter. When a noisy signal $\mathbf{y}_l=\mathbf{x}_l+\mathbf{v}_l\in\mathbb{R}^N$ is at hand, seeking for its sparse representation $\hat{\boldsymbol{\alpha}}_{l,i}$, leads to an estimation of the original signal via $\hat{\mathbf{x}}_l=\mathbf{P}_i^T\mathbf{D}_L\hat{\boldsymbol{\alpha}}_{l,i}$.

In order to keep the variance of each potential code component remains above a preset threshold, a regularization term is added in (10). For the non-smooth convex optimization model, we define f and g as follows:

$$f\left(\boldsymbol{\alpha}_{l,i}\right)=\frac{1}{2}\sum_{l=1}^{I}\left\|\sum_{i=1}^{N}\mathbf{P}_i^T\mathbf{D}_L\boldsymbol{\alpha}_{l,i}-\mathbf{y}_l\right\|_2^2+\beta\sum_{i=1}^{N}\left[\left(T-\sqrt{\operatorname{Var}\left(\boldsymbol{\alpha}_{\cdot i}\right)}\right)_+\right]^2, \quad (11)$$

$$g\left(\boldsymbol{\alpha}_{l,i}\right)=\lambda\sum_{l=1}^{I}\sum_{i=1}^{N}\left\|\boldsymbol{\alpha}_{l,i}\right\|_1. \quad (12)$$

The first item in (11) is the reconstruction term. The second item in (11) is over squared hinge terms involving the variance of each latent component $\boldsymbol{\alpha}_{\cdot i}\in\mathbb{R}^n$ across the batch where $\operatorname{Var}\left(\boldsymbol{\alpha}_{\cdot i}\right)=\frac{1}{I-1}\sum_{l=1}^{I}\left(\boldsymbol{\alpha}_{l,i}-\mu_i\right)^2$ and μ_i is the mean across the i-th latent component, namely $\mu_i=\frac{1}{I}\sum_{l=1}^{I}\boldsymbol{\alpha}_{l,i}$. The hinge terms are non-zero for any latent dimension whose variance is below the fixed threshold of $\sqrt{T}$.

For solving the CSC problem, we extended the LISTA to the learning ISTDFA. We proposed convolutional learning ISTDFA algorithm to approximate the convolutional sparse coding model, which is presented in Algorithm 1. The input of the proposed convolutional learning ISTDFA algorithm are the noise signal $\mathbf{y}_l$, the dictionary $\mathbf{D}_L$. Some parameters are firstly initialized. Going into the algorithm, we first compute the gradient of f and the local Lipschitz constant respectively, then $\boldsymbol{\alpha}_{l,i}^{k+1}$ can be updated by the proximal mapping.

Next we can induce the gradient of f,

$$\nabla f=\begin{cases}\mathbf{D}_L^T\mathbf{P}_b\left(\sum_{i=1}^{N}\mathbf{P}_i^T\mathbf{D}_L\boldsymbol{\alpha}_{a,i}-\mathbf{y}_a\right)-\frac{2\beta}{I-1}\frac{\left(T-\sqrt{\operatorname{Var}(\boldsymbol{\alpha}_{\cdot b})}\right)}{\sqrt{\operatorname{Var}(\boldsymbol{\alpha}_{\cdot b})}}\left(\boldsymbol{\alpha}_{a,b}-\mu_b\right), & \sqrt{\operatorname{Var}(\boldsymbol{\alpha}_{\cdot b})}<T\\ \mathbf{D}_L^T\mathbf{P}_b\left(\sum_{i=1}^{N}\mathbf{P}_i^T\mathbf{D}_L\boldsymbol{\alpha}_{a,i}-\mathbf{y}_a\right), & otherwise\end{cases}. \quad (13)$$

Now, let us analyze the computation of proximal function. From the definition of ϕ and the relationship of ϕ and ϕ_k, we induce that $\phi_k\left(\mathbf{x}\right)=r\left\|\mathbf{x}-\mathbf{a}\right\|_1$, where $\mathbf{a}=\mathbf{x}_k$.

Setting $\lambda=\frac{h^2}{1+\gamma h}$, from the definition of proximal mapping, we can induce that

$$\operatorname{prox}_{\lambda(\phi_k+g)}\left(\mathbf{x}\right)=\arg\min_{\mathbf{u}}\frac{1}{2}\left\|\mathbf{x}-\mathbf{u}\right\|^2+\lambda\left\|\mathbf{u}\right\|_1+\lambda r\left\|\mathbf{u}-\mathbf{a}\right\|_1. \quad (14)$$

By noticing that $\operatorname{prox}_{\lambda(\phi_k+g)}$ is a separable optimization problem, it can be reduced to the computation component-wise of a one dimensional optimization problem. For each $a\in\mathbb{R}$, set

$$T_a\left(x\right)=\arg\min_{u}\frac{1}{2}\left(u-x\right)^2+\lambda r\left|u-a\right|_1+\lambda\left|u\right|_1. \quad (15)$$

Observe that $T_a\left(x\right)=-T_{-a}\left(-x\right)$. So, we just need to consider the case $a\geq 0$.

Using the discontinuous but differentiable of ℓ_1 regularization, we can get that $\lambda r\partial\left|u-a\right|_1+\lambda\partial\left|u\right|_1\ni x-u$.

(1) When $u > a$, $\partial |u-a|_1 = 1$, $\partial |u|_1 = 1$, then $u = \text{prox}_{\lambda(\phi_k+g)}(x) = x - \lambda(1+r) > a$;
(2) When $a < u < 0$, $\partial |u-a|_1 = -1$, $\partial |u|_1 = 1$, then $a < u = \text{prox}_{\lambda(\phi_k+g)}(x) = x - \lambda(1-r) < 0$;
(3) When $u < 0$, $\partial |u-a|_1 = -1$, $\partial |u|_1 = -1$, then $u = \text{prox}_{\lambda(\phi_k+g)}(x) = x + \lambda(1+r) < 0$;
(4) When $u = a$, $\partial |u-a|_1 = \{-1,1\}$, $\partial |u|_1 = 1$, then $u = \text{prox}_{\lambda(\phi_k+g)}(x) = x - \lambda(1 + r\{-1,1\}) = a$;
(5) When $u = 0$, $\partial |u-a|_1 = 1$, $\partial |u|_1 = \{-1,1\}$, then $u = \text{prox}_{\lambda(\phi_k+g)}(x) = x - \lambda(\{-1,1\} + r) = 0$;

Through the above analysis, we can get that the unique solution of $T_a(x)$ is

$$T_a(x) = \begin{cases} x - \lambda(1+r) & x > \lambda(1+r) + a \\ a & \lambda(1-r) + a < x < \lambda(1+r) + a \\ x - \lambda(1-r) & \lambda(1-r) < x < \lambda(1-r) + a \\ 0 & -\lambda(1+r) < x < \lambda(1-r) \\ x + \lambda(1+r) & x < -\lambda(1+r) \end{cases}. \tag{16}$$

This is a threshold operator with two critical values a and 0. Consequently, for each $i = 1, 2, ..., n$, we have $\left(\text{prox}_{\lambda(\phi_k+g)}(x)\right)_i$ is

$$\left(\text{prox}_{\lambda(\phi_k+g)}(x)\right)_i = \begin{cases} T_{a_i}(x_i) & a_i \geq 0 \\ -T_{-a_i}(-x_i) & a_i \leq 0 \end{cases}, \tag{17}$$

with a_i the i-th component of the vector $\mathbf{a} = \mathbf{x}^k$ and $\lambda = \frac{h^2}{1+\gamma h}$.

The iterative shrinkage-thresholding with dry friction algorithm for the CSC model is proposed in Algorithm 1.

Algorithm 1. Iterative shrinkage-thresholding with dry friction algorithm for CSC.

Input: noised signal $\mathbf{y}_l$, local convolutional dictionary $\mathbf{D}_L$.
Output: Estimated coding $\boldsymbol{\alpha}_{l,i}^k$.
Initialization: $\boldsymbol{\alpha}_{l,i}^0 = \boldsymbol{\alpha}_{l,i}^1, \gamma > 0, h < \frac{2\gamma}{L}$
For iteration k= 0:$K-1$
compute $\nabla f\left(\boldsymbol{\alpha}_{l,i}^k\right)$ using 13,
$\boldsymbol{\alpha}_{l,i}^{k+1/2} = \boldsymbol{\alpha}_{l,i}^k + \frac{1}{1+\gamma h}\left(\boldsymbol{\alpha}_{l,i}^k - \boldsymbol{\alpha}_{l,i}^{k-1}\right)$,
$\boldsymbol{\alpha}_{l,i}^{k+1} = \text{prox}_{\frac{h^2}{1+\gamma h}(\phi_k+g)}\left(\boldsymbol{\alpha}_{l,i}^{k+1/2} - \frac{h^2}{1+\gamma h}\nabla f\left(\boldsymbol{\alpha}_{l,i}^k\right)\right)$,
end for

3.3 Convolutional Sparse Coding Network with Dry Friction

Through Algorithm 1, the update formula of needles can be written as:

$$\alpha_{l,i}^{k+1} = \operatorname{prox}_{\frac{h^2}{1+\gamma h}(\phi_k+g)}\left(\alpha_{l,i}^{k} + \frac{1}{1+\gamma h}\left(\alpha_{l,i}^{k} - \alpha_{l,i}^{k-1}\right) - \frac{h^2}{1+\gamma h}\nabla f\left(\alpha_{l,i}^{k}\right)\right), \tag{18}$$

where $\frac{1}{1+\gamma h}\left(\alpha_{l,i}^{k} - \alpha_{l,i}^{k-1}\right)$ represents the inertia item.

The CSCNet network with dry friction (CSCNet-DF) is proposed based on (18) in the Algorithm 1. The network diagram of CSCNet-DF is presented in Fig. 1. Generally speaking, convergence requires a lot of times to achieve, which will bring great computational burden. In order to overcome this burden, we adopt the calculation idea of LISTA algorithm by learning the parameters $\mathbf{A}$ and $\mathbf{B}$ of the nonlinear recursive encoder strictly following (18), where $\mathbf{A}$ is a convolution operator and $\mathbf{B}$ is a transposed-convolution operation. Once the needles are at hand, the estimated clean image is then obtained by a linear transposed-convolutional decoder, $\hat{\mathbf{x}} = \mathbf{C}\boldsymbol{\Gamma}$, where $\boldsymbol{\Gamma}$ is a matrix composed of needles $\boldsymbol{\alpha}_{l,i}$. In this paper, matrices $\mathbf{A}, \mathbf{B}$ and $\mathbf{C}$ are constructed as a set of bounded shift invariant filters, and self-supervised learning is carried out together with thresholds $\frac{h^2}{1+\gamma h}$.

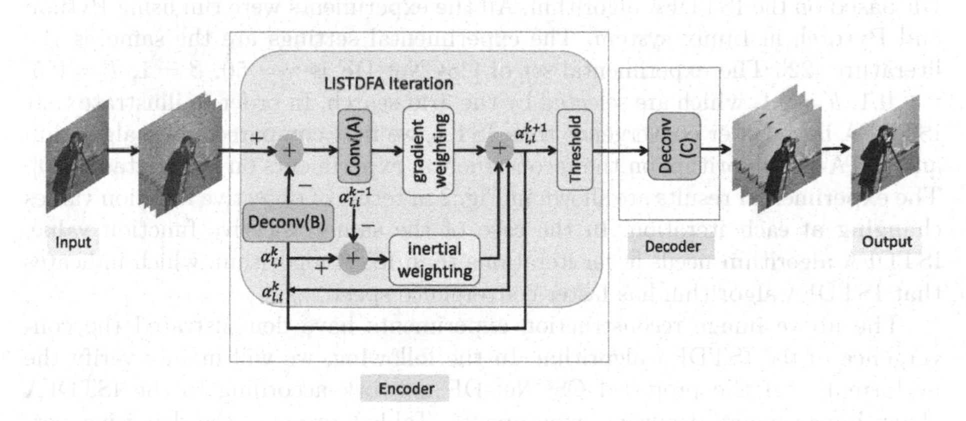

Fig. 1. The CSCNet-DF network. The encoder is LISTDFA Iteration. The decoder is Deconv(C) module.

Our proposed network structure is based on the CSCNet structure. The input is the noised image, which is duplicated many times using a shifted version. The output is denoised image, which is a simple average of the estimates of all the shifts. The main body of the network structure is the encoder and decoder. The encoder is the iterative network unfolded by the LISTDFA algorithm. The decoder is the reconstruction process of an image. Compared with CSCNet, CSCNet-DF has the same decoder, while they have the different encoder unfolded by the LISTA and LISTDFA, respectively. In detail, our proposed network has some different unit, such as, gradient weighting, inertia weighting and threshold modules. Compared with CSCNet, our proposed network adds a gradient weight module and an inertia weight module. In a way, CSCNet is a special case of our proposed network without dry friction.

In order to improve the efficiency of the network, we use the back propagation algorithm to update the convolution dictionary by minimizing the loss function. The loss function is the Mean Square Error (MSE) between the reconstructed signal $\hat{\mathbf{y}}_l^K = \mathbf{P}_i^T \mathbf{D}_L \hat{\alpha}_{l,i}^K$ and the original signal $\mathbf{y}_l$:

$$L_{MSE} = \frac{1}{I} \sum_{l=1}^{I} \left\| \mathbf{y}_l - \hat{\mathbf{y}}_l^K \right\|_2^2 , \tag{19}$$

where $\hat{\mathbf{y}}_l^K$ represents the reconstruction result of K-th iteration, and $\mathbf{y}_l$ represents the l-th of the original signal $\mathbf{y}$.

4 Experiments and Results

This section provides numerical results for image denoising on the Set12, BSD68 and color-BSD68 datasets to analyse the performances of the proposed CSCNet-DF based on the ISTDFA algorithm. All the experiments were run using Python and Pytorch in Linux system. The experimental settings are the same as the literature [22]. The experimental set of CSCNet-DF is $\gamma = 50$, $\beta = 1$, $T = 0.5$, $r = 0.1$, $h = 0.1$, which are selected by the grid search. In order to illustrate that ISTDFA have faster convergence than ISTA, we first compared ISTA algorithm and ISTA-DF algorithm on the reconstruction experiments on City dataset [20]. The experimental results are shown in Fig. 2 in terms of objective function values changing at each iteration. In the case of the same objective function value, ISTDFA algorithm needs fewer iterations than ISTA algorithm, which indicates that ISTDFA algorithm has faster convergence speed.

The above image reconstruction experiments have demonstrated the convergence of the ISTDFA algorithm. In the following, we will mainly verify the performance of the proposed CSCNet-DF network according to the ISTDFA algorithm in image denoising experiments. Table 1 presents the denoising performance (PSNR) results of the CSCNet [22] and CSCNet-DF networks on the Set12 datasets. Due to limited space, we do not show the learned filters. We observe Table 1 and know that the proposed CSCNet-DF network outperforms CSCNet. To further validate the proposed iterative neural networks, Table 2 presents the PSNR results of CSCNet, CSCNet-DF, the well-known BM3D [8] and DnCNN [34] models on the BSD68 dataset. Table 3 presents the PSNR results of two CSCNet networks and the well-known BM3D [8], DnCNN [34], AdmFM-Net [18], SwinIR [17] models on color-BSD68 dataset. Experimental results on BSD68 and color-BSD68 datasets show that our proposed CSCNet-DF network outperforms the BM3D, CSCNet and DnCNN. Experimental results on color-BSD68 dataset show that our proposed CSCNet-DF network outperforms

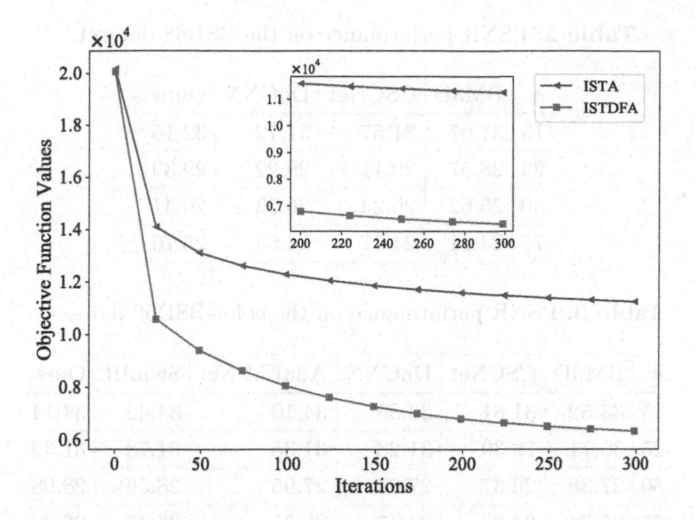

Fig. 2. Performance comparison of ISTDFA and ISTA in terms of objective function values changing at each iterations.

the AdaFM-Net. The performance of our proposed network is slightly worse than that of SwinIR using Swin transformer, but relatively speaking, our network is simpler, with less parameters and less computation. Through the comparative analysis of the PSNR of the denoised images using different models, it can be found that the performance of CSCNet-DF is better than BM3D, CSCNet and DnCNN in practice. In addition, in order to make the experiment more convincing, we conducted a denoising experiment on the Set12 dataset, and the images are shown in Fig. 3. It can be seen that compared with CSCNet, our proposed network texture is clearer and achieves better denoising performance.

Table 1. Denoising performance (PSNR) on the Set12 dataset.

σ	CSCNet	Ours
15	31.23	32.66
25	28.73	30.24
50	25.32	27.14
75	23.28	25.35
Time	1.67 s	0.79 s

Finally, we design an ablation experiment to illustrate the role of dry friction in the network. We compare the image denoising performance with different dry friction coefficients on color datasets which is shown in Table 4. The results show that dry friction can improve the network performance to a certain extent.

Table 2. PSNR performance on the BSD68 dataset.

σ	BM3D	CSCNet	DnCNN	Ours
15	31.07	31.57	31.72	32.15
25	28.57	29.11	29.22	29.39
50	25.62	26.24	26.23	26.41
75	24.21	24.77	24.64	25.76

Table 3. PSNR performance on the color-BSD68 dataset.

σ	BM3D	CSCNet	DnCNN	AdaFM-Net	SwinIR	Ours
15	33.52	31.81	33.89	34.10	34.42	34.14
25	30.71	29.30	31.23	31.35	31.78	31.39
50	27.38	26.37	27.92	27.95	28.56	28.08
75	25.74	24.87	24.47	26.35	26.45	26.38

(a) original image

(b) noised image

(c) CSCNet

(d) CSCNet-DF

Fig. 3. Illustration of images denosing by CSCNet and CSCNet-DF. (a) The original image, (b) the noising image, (c) the denoising image using CSCNet, (d) the denoising image using CSCNet-DF.

Table 4. Comparison of PSNR performance under different dry friction coefficients on the color-BSD68 dataset ($\sigma = 15$).

r	0	0.01	0.05	0.1
PSNR(dB)	31.81	32.85	33.45	34.14

5 Conclusion

In this paper the iterative neural network based on the convolutional sparse coding model using the learned ISTDFA algorithm is proposed. Introduce of the dry friction, achieves fast convergence and low values of objection function for image denoising problem. The forward process of the proposed CSCNet-DF network includes encoder and decoder, giving the sparse coding and reconstructed signal. The backward process uses back propagation algorithms to update the convolutional dictionary. The experimental results on three dataset show that the CSCNet-DF network is superior to the CSCNet. In the future, the Nesterov accelerated method with dry friction will be studied.

Acknowledgements. We thank referee for comments that greatly improved the manuscript. Thanks to Jin Li and Ze Li of Yanshan University for their help in programming and experiment.

References

1. Adly, S., Attouch, H.: First-order inertial algorithms involving dry friction damping. Math. Program. **193**(1), 405–445 (2022)
2. Aharon, M., Elad, M., Bruckstein, A.: K-SVD: an algorithm for designing overcomplete dictionaries for sparse representation. IEEE Trans. Signal Process. **54**(11), 4311–4322 (2006). https://doi.org/10.1109/TSP.2006.881199
3. Bardes, A., Ponce, J., LeCun, Y.: VICReg: variance-invariance-covariance regularization for self-supervised learning. In: International Conference on Learning Representations (2022). https://openreview.net/forum?id=xm6YD62D1Ub
4. Chalasani, R., Principe, J.C., Ramakrishnan, N.: A fast proximal method for convolutional sparse coding. In: 2013 The International Joint Conference on Neural Networks (IJCNN), pp. 1–5 (2013). https://doi.org/10.1109/IJCNN.2013.6706854
5. Chen, S., Billings, S.A., Luo, W.: Orthogonal least squares methods and their application to non-linear system identification. Int. J. Control **50**(5), 1873–1896 (1989)
6. Chen, S.S., Saunders, D.M.A.: Atomic decomposition by basis pursuit. SIAM Rev. **43**(1), 129–159 (2001)
7. Chun, I.Y., Huang, Z., Lim, H., Fessler, J.: Momentum-net: fast and convergent iterative neural network for inverse problems. IEEE Trans. Pattern Anal. Mach. Intell. (2020). https://doi.org/10.1109/TPAMI.2020.3012955
8. Dabov, K., Foi, A., Katkovnik, V., Egiazarian, K.: Image denoising by sparse 3-D transform-domain collaborative filtering. IEEE Trans. Image Process. **16**(8), 2080–2095 (2007). https://doi.org/10.1109/TIP.2007.901238
9. Deng, X., Dragotti, P.L.: Deep coupled ISTA network for multi-modal image super-resolution. IEEE Trans. Image Process. **29**, 1683–1698 (2019)
10. Dong, C., Deng, Y., Loy, C.C., Tang, X.: Compression artifacts reduction by a deep convolutional network. In: 2015 IEEE International Conference on Computer Vision (ICCV), pp. 576–584 (2015). https://doi.org/10.1109/ICCV.2015.73
11. Engan, K., Aase, S., Hakon Husoy, J.: Method of optimal directions for frame design. In: 1999 IEEE International Conference on Acoustics, Speech, and Signal Processing (ICASSP), vol. 5, pp. 2443–2446 (1999). https://doi.org/10.1109/ICASSP.1999.760624
12. Evtimova, K., LeCun, Y.: Sparse coding with multi-layer decoders using variance regularization. Trans. Mach. Learn. Res. (2022). https://openreview.net/forum?id=4GuIi1jJ74
13. Fan, L., Zhang, F., Fan, H., Zhang, C.: Brief review of image denoising techniques. Vis. Comput. Ind. Biomed. Art **2**(1), 1–12 (2019). https://doi.org/10.1186/s42492-019-0016-7
14. Fu, X., Zha, Z.J., Wu, F., Ding, X., Paisley, J.: JPEG artifacts reduction via deep convolutional sparse coding. In: 2019 IEEE/CVF International Conference on Computer Vision (CVPR), pp. 2501–2510 (2019)

15. Gregor, K., LeCun, Y.: Learning fast approximations of sparse coding. In: The 27th International Conference on Machine Learning (ICML), pp. 399–406 (2010)
16. Guillemot, C., Le Meur, O.: Image inpainting: overview and recent advances. IEEE Signal Process. Mag. **31**(1), 127–144 (2014). https://doi.org/10.1109/MSP.2013.2273004
17. He, J., Dong, C., Qiao, Y.: Modulating image restoration with continual levels via adaptive feature modification layers. In: 2019 IEEE/CVF Conference on Computer Vision and Pattern Recognition (CVPR), pp. 11048–11056 (2019). https://doi.org/10.1109/CVPR.2019.01131
18. Liang, J., Cao, J., Sun, G., Zhang, K., Van Gool, L., Timofte, R.: SwinIR: image restoration using swin transformer. In: 2021 IEEE/CVF International Conference on Computer Vision Workshops (ICCVW), pp. 1833–1844 (2021). https://doi.org/10.1109/ICCVW54120.2021.00210
19. Mairal, J., Bach, F., Ponce, J., Sapiro, G.: Online dictionary learning for sparse coding. In: The 26th Annual International Conference on Machine Learning (ICML), pp. 689–696 (2009). https://doi.org/10.1145/1553374.1553463
20. Papyan, V., Romano, Y., Elad, M., Sulam, J.: Convolutional dictionary learning via local processing. In: 2017 IEEE International Conference on Computer Vision (ICCV), pp. 5306–5314 (2017). https://doi.org/10.1109/ICCV.2017.566
21. Papyan, V., Sulam, J., Elad, M.: Working locally thinking globally: theoretical guarantees for convolutional sparse coding. IEEE Trans. Signal Process. **65**(21), 5687–5701 (2017). https://doi.org/10.1109/TSP.2017.2733447
22. Simon, D., Elad, M.: Rethinking the CSC model for natural images. In: 2019 Neural Information Processing Systems (NeurIPS), pp. 2274–2284 (2019). https://doi.org/10.5555/3454287.3454491
23. Simon, D., Sulam, J., Romano, Y., Lu, Y.M., Elad, M.: MMSE approximation for sparse coding algorithms using stochastic resonance. IEEE Trans. Signal Process. **67**(17), 4597–4610 (2019). https://doi.org/10.1109/TSP.2019.2929464
24. Sreter, H., Giryes, R.: Learned convolutional sparse coding. In: 2018 IEEE International Conference on Acoustics, Speech and Signal Processing (ICASSP), pp. 2191–2195 (2018). https://doi.org/10.1109/ICASSP.2018.8462313
25. Sulam, J., Ophir, B., Zibulevsky, M., Elad, M.: Trainlets: dictionary learning in high dimensions. IEEE Trans. Signal Process. **64**(12), 3180–3193 (2016)
26. Sulam, J., Aberdam, A., Beck, A., Elad, M.: On multi-layer basis pursuit, efficient algorithms and convolutional neural networks. IEEE Trans. Pattern Anal. Mach. Intell. **42**(8), 1968–1980 (2019)
27. Tian, C., Fei, L., Zheng, W., Xu, Y., Zuo, W., Lin, C.W.: Deep learning on image denoising: an overview. Neural Netw. **131**, 251–275 (2020)
28. Wang, W., Xia, X.G., He, C., Ren, Z., Wang, T., Lei, B.: A noise-robust online convolutional coding model and its applications to poisson denoising and image fusion. Appl. Math. Model. **95**(1), 644–666 (2021). https://doi.org/10.1016/j.apm.2021.02.023
29. Wohlberg, B.: Efficient algorithms for convolutional sparse representations. IEEE Trans. Image Process. **25**(1), 301–315 (2016). https://doi.org/10.1109/TIP.2015.2495260
30. Xu, J., Deng, X., Xu, M.: Revisiting convolutional sparse coding for image denoising: from a multi-scale perspective. IEEE Signal Process. Lett. **29**, 1202–1206 (2022). https://doi.org/10.1109/LSP.2022.3175096
31. Yang, L., et al.: Image reconstruction via manifold constrained convolutional sparse coding for image sets. IEEE J. Sel. Top. Sign. Process. **11**(7), 1072–1081 (2017). https://doi.org/10.1109/JSTSP.2017.2743683

32. Yin, H.: PSCSC-Net: a deep coupled convolutional sparse coding network for pansharpening. IEEE Trans. Geosci. Remote Sens. **60**, 1–16 (2022). https://doi.org/10.1109/TGRS.2021.3088313
33. Zhang, K., Li, Y., Zuo, W., Zhang, L., Van Gool, L., Timofte, R.: Plug-and-play image restoration with deep denoiser prior. IEEE Trans. Pattern Anal. Mach. Intell. **44**(10), 6360–6376 (2022). https://doi.org/10.1109/TPAMI.2021.3088914
34. Zhang, K., Zuo, W., Chen, Y., Meng, D., Zhang, L.: Beyond a gaussian denoiser: residual learning of deep CNN for image denoising. IEEE Trans. Image Process. **26**(7), 3142–3155 (2017). https://doi.org/10.1109/TIP.2017.2662206
35. Zhang, K., Zuo, W., Zhang, L.: FFDNet: toward a fast and flexible solution for CNN-based image denoising. IEEE Trans. Image Process. **27**(9), 4608–4622 (2018). https://doi.org/10.1109/TIP.2018.2839891
36. Zheng, J., Zhang, X., Wang, W., Jiang, X.: Handling slice permutations variability in tensor recovery. In: 2022 the AAAI Conference on Artificial Intelligence, pp. 3499–3507 (2022). https://doi.org/10.1609/aaai.v36i3.20261
37. Zisselman, E., Sulam, J., Elad, M.: A local block coordinate descent algorithm for the CSC model. In: 2019 IEEE/CVF Conference on Computer Vision and Pattern Recognition (CVPR), pp. 8200–8209 (2019). https://doi.org/10.1109/CVPR.2019.00840

Neural Network Panning: Screening the Optimal Sparse Network Before Training

Xiatao Kang[1], Ping Li[1,3](✉), Jiayi Yao[1], and Chengxi Li[2]

[1] School of Computer and Communication Engineering, Changsha University of Science and Technology, Changsha, China
lping9188@163.com
[2] School of Computer, Xidian University, Xi'an, China
[3] Hunan Provincial Key Laboratory of Intelligent Processing of Big Data on Transp, Changsha, China

Abstract. Pruning on neural networks before training not only compresses the original models, but also accelerates the network training phase, which has substantial application value. The current work focuses on fine-grained pruning, which uses metrics to calculate weight scores for weight screening, and extends from the initial single-order pruning to iterative pruning. Through these works, we argue that network pruning can be summarized as an expressive force transfer process of weights, where the reserved weights will take on the expressive force from the removed ones for the purpose of maintaining the performance of original networks. In order to achieve optimal expressive force scheduling, we propose a pruning scheme before training called Neural Network Panning which guides expressive force transfer through multi-index and multi-process steps, and designs a kind of panning agent based on reinforcement learning to automate processes. Experimental results show that Panning performs better than various available pruning before training methods. Our code is made public at: https://github.com/kangxiatao/RLPanning.

Keywords: Deep learning · Reinforcement learning · Network pruning · Pruning before training

1 Introduction

In recent years, neural networks have achieved breakthrough results in computer vision [15,40] and natural language processing [16]. More complex architectural designs improve model performance but significantly increase the number of parameters of the neural network. Research on neural network compression [10] is devoted to reducing the memory overhead and computational cost of neural networks and maintaining the performance of the network architecture. Neural network pruning [11] is the most direct method of neural network compression, and the purpose of compression is achieved by filtering out unimportant weights.

L. Wang et al. (Eds.): ACCV 2022, LNCS 13841, pp. 602–617, 2023.
https://doi.org/10.1007/978-3-031-26319-4_36

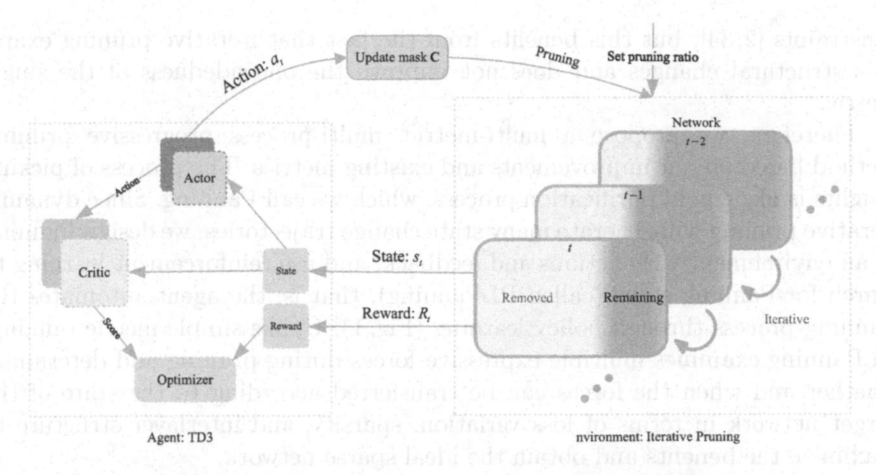

Fig. 1. Overview of Panning based on reinforcement learning. On the right is the Panning environment, which iteratively prunes the network after setting the pruning rate, and the feedback includes the current state of the sparse network and the set reward. On the left is the TD3 agent, which acts through the feedback given by the environment, that is, it selects the optimal pruning strategy based on the network state during iterative pruning.

The latest research focuses on network pruning before training [5,12,17], with progress in theory and application. A series of studies of the lottery ticket hypothesis [4,6,25] demonstrates the effectiveness of pruning before training. Many methods of pruning before training achieve desirable performance [14,33,34]. However, some work argues that pruning before training should be considered a search of the network architecture [24] and that random selection of weights with equal layer pruning quotas [5,32] can even achieve better generalization. Drop weights randomly or by structure from left to right during training using sparse constraints [1], and both are very effective. We believe that the idea of these works is that weights are equal, and the network structure is directly examined when pruning the network rather than the weight ordering under a particular index.

The characterization of weight expressiveness has always been the research focus of neural network pruning algorithms. Using sensitivity [18] as a metric is currently the mainstream method for pruning before training. In our research, we found that the process of pruning can be regarded as the process of transferring the expressive force of the removed weight to the remaining weight so that any weight has an equal opportunity to express. Assuming that the weight sensitivity is a representation of the weight expressive force, Fig. 2 shows the expressive force transfer through iterative pruning. Different measurement methods indicate different affordability of weights, and a single measurement index is always flawed. Loss-sensitive correlation metrics [18,35] are prone to inter-layer disconnection at high compression rates, while metrics that only consider inter-layer equilibrium [33] do not generalize well at low compression rates. Some methods improve performance using iterative pruning with a single metric and adding constraints [2,34], but this benefits from the fact that iterative pruning examines structural changes and does not improve the one-sidedness of the single metric.

Therefore, we propose a multi-metric, multi-process progressive pruning method based on our improvements and existing metrics. The process of picking weights is like a gold purification process, which we call Panning. Since dynamic iterative pruning will generate many state change trajectories, we design Panning as an environment with actions and feedback, and use reinforcement learning to search for Panning steps (called RLPanning), that is, the agent automates the Panning process through policy learning (Fig. 1). Unlike simple metric ranking, RLPanning examines multiple expressive forces during pruning and determines whether and when the forces can be transferred according to the state of the target network in terms of loss variation, sparsity, and interlayer structure to maximize the benefits and obtain the ideal sparse network.

The contributions of this paper are as follows:

- We generalize the pruning process as the expressive force transfer process, analyze various weight metrics, and make improvements.
- We propose a multi-metric, multi-process iterative pruning method before training, called Panning. Panning performs well by pruning the network before training, maintaining generalization well under extremely high compression rates.
- We design a Panning environment and use reinforcement learning to learn a Panning strategy by sampling spatial actions, called RLPanning. RLPanning obtains the optimal pruning strategy with a comprehensive measure of multiple indicators through dynamic expressive force transfer.

2 Related Work

Neural network pruning has been a hot research topic along with the development of artificial intelligence so far. The purpose of pruning is to reduce the parameters and computation of neural networks to achieve compression of neural networks [10], which can be divided into two major categories of unstructured and structured pruning according to the way of pruning networks. Unstructured pruning performs fine-grained pruning intending to minimize the number of parameters, and the classical process is training, pruning, and fine-tuning [11]. Sparse constraints in training to obtain sparse structures are also more widely used [27,37]. Structured pruning focuses on the structure of the network, and the pruning results in a lean, compact model. In convolutional networks, structured pruning is usually considered the pruning of the filter [19], and most of the work sets structured constraints [8,23,36] in training to prune the network.

In order to obtain better compression structure and performance, many works are devoted to automatic neural network architecture search [31,38,39]. Although automatic search network architecture is prevalent, it also suffers from substantial computational overhead due to the infinite search space. Subsequent work has made new attempts to use adversarial learning [21] or reinforcement learning [13] to search for pruning strategies under a fixed framework to achieve network simplification.

Pruning Before Training. Pruning before training has substantial theoretical and applied value. The lottery ticket hypothesis [4] demonstrates the possibility of obtaining superior sparse networks before training through extensive experiments. Since the lottery hypothesis was put forward, much work has been done on expansion [3,6,28] and theoretical research [25,29].

In the current application, the pruning before training focuses on fine-grain pruning, and the weight is usually measured by examining the effect of removing a certain weight on a certain state of the network. Specifically, SNIP [18] examines the impact of pruning weights on loss, that is, the sensitivity of weights to model losses. GraSP [35] examines the degree to which the removal weight decreases the loss, that is, the sensitivity of the weight to the change in loss. SynFlow [33] starts from the trainability of the maximum compression network and examines the balance of weights between layers. FORCE [14] progressively prunes the network based on the SNIP method and allows for the resurrection of removed weights, taking into account changes in the network structure. Among them, SNIP and GraSP methods can obtain sparse networks with only a single network pruning. Methods that consider inter-layer balance and network structure require iterative network pruning. In addition, there are iterative pruning methods to make dynamic expansion [2,22] or set constraints [34] to obtain good performance.

For weight measurement, SynFlow summarizes the mathematical expression of sensitivity, called synaptic saliency [33] $\mathcal{S}(\theta) = \frac{\partial \mathcal{R}}{\partial \theta} \odot \theta$, where $\mathcal{R}$ is a scalar loss function, and θ is a parameter to be examined. Hence $\mathcal{S}_{SNIP}(\theta) = \left|\frac{\partial \mathcal{L}}{\partial \theta} \odot \theta\right|, \mathcal{S}_{GraSP}(\theta) = \frac{\partial g^T g}{\partial \theta} \odot \theta$, $g^T g$ is the first-order Taylor approximation of loss drop.

3 Methodology

In this section, we introduce Panning in detail and then automate Panning using reinforcement learning. Our goal is to have the agent schedule the expressive force of the weights according to the state of the network, cross-validate the weights with a variety of meaningful metrics, and find the optimal sparse sub-network before training. Finally, we discuss and improve the existing metrics and apply them in Panning.

3.1 Problem Definition

We mainly consider fine-grained pruning of neural networks, that is, removing parts of weights that are considered unimportant, regardless of the structure of the network. In order to describe this irregular sparse network conveniently, it is usually represented by the Hadamard product of the mask $\mathbf{c}$ and the weight $\mathbf{w}$, then the optimization problem of the network is:

$$\begin{aligned} &\min_{\mathbf{c},\mathbf{w}} \mathcal{L}(\mathbf{c} \odot \mathbf{w}; \mathcal{D}) \\ &\text{s.t. } \mathbf{c} \in \{0,1\}^m, \|\mathbf{c}\|_0 \leq m(1-\rho) \end{aligned} \tag{1}$$

where $\mathcal{L}(\cdot)$ represents the loss function, $\mathcal{D}$ is the dataset, m is the total number of weights in the network, and ρ is the target pruning ratio (sparse degree). When the mask $\mathbf{c}$ corresponding to the weight is set to zero, it means that the weight is removed.

3.2 Transfer of Expressive Force

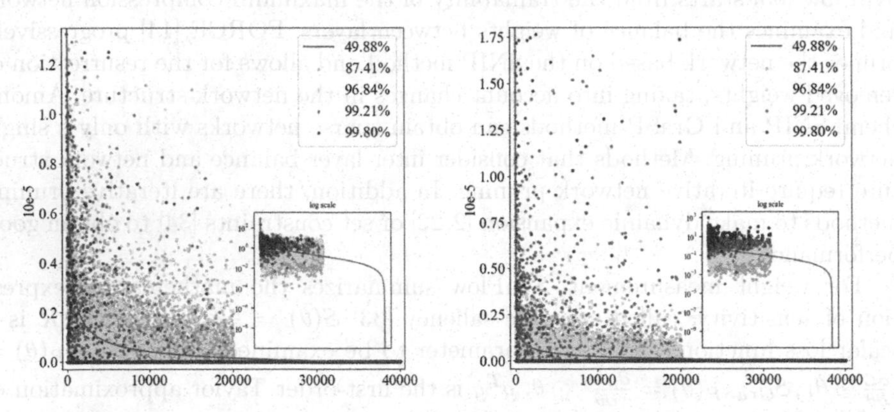

Fig. 2. Comparison of metric scores at different pruning ratios during the iterative pruning process of SNIP (left) and GraSP (right). Experiments were performed under VGG19/CIFAR10. The vertical axis is the metric score $\mathcal{S}$ (where $\mathcal{S}_{GraSP}$ is the absolute value), and the horizontal axis is the number of weights (in order to reduce the amount of data, after flattening all weights into vectors, take a value every 500 weights). After sorting the weight scores for the first time (line), the score changes (dots) under other pruning ratios are displayed at the corresponding weight positions. The subplots are the logarithmic changes in the vertical axis scale.

We use SNIP and GraSP for iterative pruning, respectively, and show in Fig. 2 the changes in the metric scores of all weights in the network under different compression ratios (the two adjacent pruning rates can represent the change in scores before and after pruning), and more obvious hierarchical changes can be observed after taking the logarithm of the vertical axis (as shown in the subplot). We find that the weight expressive force is sensitive to the compression ratio. As the compression ratio increases, some of the less expressive weight metrics improve significantly (above the red line), and vice versa.

Iterative dynamic pruning is widely considered to perform better, especially before training. Through continuous compression and gradient iterations, weights that were evaluated as strong in performance in the most recent iteration are retained. The invariant properties of flow transported between layers are given in the SynFLow [33]. Naturally, we view model compression as a process of transferring expressive force between different weights based on flow sensitivity. Take $\mathcal{S}(\theta) = |\frac{\partial \mathcal{R}}{\partial \theta} \odot \theta|$ as the weight expressive force, and only observe the magnitude of the force.

Specifically, the expressive force of the removed weight will be transferred to the remaining weight, and the force of the remaining part will also be adjusted due to structural changes. Therefore, weights with low importance at the beginning may become critical weights in the future, and a single pruning can easily remove this part of the weights, resulting in the unreasonable distribution of weight expressive force. It is challenging to maintain layer balance even if layer collapse does not occur. The advantage of iterative pruning is that it gradually transfers the expressive force and examines the changes in the structure so that the connection structure between the remaining weights is better maintained.

3.3 Artificial Panning

Although iterative pruning solves the layer collapse problem, existing iterations use a single metric pruning throughout, ignoring the expressive force of weights under other metrics. Simply summing up multiple metrics for measurement will easily cover up specific properties, resulting in an indistinct degree of weight discrimination. We design the Panning algorithm based on dynamic iterations to establish multiple mappings between sparse network structure changes and weights. The characteristic of Panning is to observe the state of the network in the iterative process, and select the most favorable indicator to measure the weight according to the state change. Based on this, the network is progressively pruned in a loop, and the removed weights are allowed to re-live until the desired sparse network is finally obtained.

Specifically, the pruning rate changes exponentially during the iteration process, and the pruning ratio ρ_i of round i is:

$$\rho_i = 1 - (1 - \rho_{target})^{(\frac{i}{T})} \tag{2}$$

where ρ_{target} is the target pruning ratio, and T is the total the number of iterations. Then calculate the various metrics in the current state of the model, and obtain the score of the weight $\mathbf{w}$ through a specific fusion:

$$\mathcal{S}_{\mathbf{w}} = \sum_{i=1}^{k} p_i \mathcal{N}(\mathcal{S}_i) \tag{3}$$

Algorithm 1. Panning

Require: Weight $\mathbf{w}$, batch data $\mathcal{D}^b$, target pruning rate ρ_{target}.
1: Initialize $\mathbf{w}$, $\mathbf{c}$
2: **for** $i = 0$ **to** T **do**
3: Remove weight $\mathbf{w}^* = \mathbf{c} \odot \mathbf{w}$
4: Calculate loss $\mathcal{R}$
5: Reset weight $\mathbf{w}^* = \mathbf{w}$
6: Calculate score $\mathcal{S}_{\mathbf{w}} = \sum_{i=1}^{k} p_i \mathcal{N}(\mathcal{S}_i)$
7: Update pruning rate $\rho_i = 1 - (1 - \rho_{target})^{(\frac{i}{T})}$
8: Update mask $\mathbf{c}$
9: **end for**
10: return $\mathbf{c}$

where p_i is a hyperparameter that adjusts the proportion of the metric, and $\mathcal{N}(\mathcal{S}_i)$ represents the normalization of the metric score. Finally, the network is gradually pruned according to the score $\mathcal{S}_{\mathbf{w}}$ and the pruning ratio ρ_i:

$$\mathbf{c} = \mathcal{S}_{\mathbf{w}}(\boldsymbol{\theta}) > \mathcal{S}_{top(m(1-\rho_i))} \tag{4}$$

where $\mathcal{S}_{top(\kappa)}$ is the κ-th value after the descending order, and m is the total number of weights. When the weight $\mathbf{w}_i$ does not satisfy the conditions of the above formula, the corresponding mask $\mathbf{c}_i$ is set to zero, the weight is removed and exists as a zero value. The value of $\mathbf{w}_i$ is reset the next time the weight sensitivity $\mathcal{S}$ is calculated, so that the removed weight can be restored. Algorithm 1 shows the complete gold panning steps.

3.4 Panning Based on Reinforcement Learning

According to our observation, an ideal sparse network can be obtained by setting the hyperparameters of Panning based only on the compression of the network. Naturally, if more network states are examined and dynamic expressive force transfer is achieved based on state changes, this can theoretically lead to a sparse network with better performance. Therefore, we design Panning as an environment containing action and state space, so that the intelligent body can automate Panning through policy learning. Figure 1 gives an overview of RLPanning, which is described in detail in this section.

Panning Environment. Each Panning iteration results in a new sparse network, that is, different sparsity and performance. The number of iterations is taken as the time t, so the state space s_t of the Panning environment is:

$$s_t = (\mathcal{L}, \varDelta\mathcal{L}, \mathcal{L}_s, \varDelta\mathcal{L}_s, \rho_t, \rho_e, t) \tag{5}$$

where $\mathcal{L}_s, \varDelta\mathcal{L}_s$ are the loss and loss reduction of the sparse sub-network in the current state, respectively, ρ_e represents the effective compression rate of the network [34] (Invalid retention due to disconnection between layers is prone to occur at high compression rates). Note that all states are normalized to facilitate agent training. The action space a of Panning is the hyperparameter p_i that regulates the proportion of metrics, and assuming that there are κ metrics, the action space dimension is κ and is continuous for the action:

$$a \in [-1,1], p_i = \frac{a_i + 1}{2} \tag{6}$$

Our target is to affect $\mathcal{L}, \Delta\mathcal{L}$ as little as possible during pruning and to ensure that the effective compression rate is close to the target compression rate. So we design a reward term R_t consisting of four components:

$$R_t = -\left|\mathcal{N}(\mathcal{L}) - \mathcal{N}(\mathcal{L}_s)\right| - \left|\mathcal{N}(\Delta\mathcal{L}) - \mathcal{N}(\Delta\mathcal{L}_s)\right| - \alpha\left|\rho_e - \rho_t\right| - r_{\text{done}}$$
$$r_{\text{done}} = \begin{cases} T - t, & \text{if } \rho_e = 1 \\ 0, & \text{otherwise} \end{cases} \tag{7}$$

where $\mathcal{N}(\cdot)$ represents normalization. T is the maximum number of iterations. r_{done} is the reward in the final state, $done = True$ when $t = T$ or effective compression ratio $\rho_e = 1$, $\rho_e = 1$ means that all weights are removed, and the iteration will end early. α is the reward strength that adjusts the quality of network compression. Usually the ineffective compression weight is less than one percentage point, and we set α to 100.

Panning Agent. The continuous action of Panning is a deterministic strategy, and we adopt TD3 [7], which performs well in the field of continuous control, as the Panning agent. TD3 is an actor-critic network architecture. As an optimized version of DDPG [20], the delay and smoothing processing of TD3 largely mitigates the effects of sudden changes in environmental states due to different batches of samples, which is well suited to our Panning environment.

The workflow of the TD3 agent is shown on the left in Fig. 1. The actor network gives an action, and the critic network estimates the reward for the action. The critic network is equivalent to the Q network in DQN [26], and the goal is to solve the action a that maximizes the Q value. The actions and states of the Panning environment are all standardized, the structure and optimization of the TD3 agent basically do not need to be changed, and the network scale and all hyperparameters are shown in the experimental part.

In addition, after resetting the environment, in order to better sample in the space, the final target compression ratio ρ_T is randomly selected between {80%,99.99%}. In addition, we have set a curriculum learning [30] from easy to complex, that is, the probability of the compression target ρ_T being a low compression ratio in the early stage of the agent training is high, and the probability of $\rho_T > 99\%$ in the later stage is higher, thereby speeding up the convergence speed of the agent training.

3.5 Selection of Metrics

The currently widely accepted singe-shot pruning metrics are SNIP and GraSP, which are very sample-dependent and perform poorly at high compression ratios.

SynFlow is more prominent in iterative pruning and achieves a good balance between layers. The most significant advantage of SynFlow is that it does not require samples, but the lack of consideration of sample characteristics also makes it generalize poorly at low compression rates. Actually, SNIP, GraSP, and SynFlow are complementary and computationally very similar. Therefore, we choose these three metrics to apply to Panning, with action space $\kappa = 3$ and weight score $\mathcal{S}_w = p_1\mathcal{N}(\mathcal{S}_{SynFlow}) + p_2\mathcal{N}(\mathcal{S}_{SNIP}) + p_3\mathcal{N}(\mathcal{S}_{GraSP})$.

Due to gradient decay in the deep network, when the SNIP pruning ratio is large, it is easy to remove too many subsequent convolutional layers and cause faults. The idea of GraSP is to maintain the gradient flow and prune the weights to maximize the loss drop, which improves compressibility. However, there is a significant error in the calculation of the first-order approximation $g^T g$, especially in the ill-conditioned problem [9] that the first-order term of the loss function is smaller than the second-order term during training. Based on this, we believe that maximizing the gradient norm is not equivalent to maintaining the gradient flow, and it is more accurate to remove weights that have little influence on the gradient flow. Therefore, we changed $\mathcal{S}_{GraSP}$ to $|\frac{\partial g^T g}{\partial \theta} \odot \theta|$, that is, to examine the impact of weights on model trainability. If it is analyzed from the perspective of acting force, after taking the absolute value, it means that only the magnitude of the force is considered, and the direction of action of the force is ignored. Moreover, the experiment proves that the modified $\mathcal{S}_{GraSP}$ has better performance.

In addition, in the classification task, when $\mathcal{R}$ is the fitting loss $\mathcal{L}$ of the network, the gradient information brought by different samples is more abundant. Therefore, when calculating the loss $\mathcal{L}$, we set a specific sampling to ensure that $\mathcal{D}^b$ contains all categories and an equal number of samples for each category. Suppose the dataset has l categories, k samples are taken from each category, then $\mathcal{D}^b = \{(x^b, y^b)\}$, $x^b = \left(x^1_{1,2,\ldots,k}, x^2_{1,2,\ldots,k}, \ldots, x^l_{1,2,\ldots,k}\right)$.

4 Experiments

4.1 Experimental Setup

In the image classification task, we prune LeNet5, VGG19, and ResNet18 convolutional networks with panning before training and select MNISIT, Fashion-MNISIT, CIFAR10/100, and TinyImageNet datasets to evaluate our method. In this experiment, images are enhanced by random flipping and cropping. Details and hyperparameters of sparse network training after pruning are shown in Table 1.

4.2 Performance of the Improved Metric

The performance comparison between the improved SNIP and GraSP metrics and the original method on VGG19/CIFAR10 is shown in Table 2. The results were obtained by taking the average of three experiments.

Table 1. Model optimizer and hyperparameters.

	MNIST	CIFAR	ImageNet	
	LeNet	VGG/ResNet	VGG	ResNet
Optimizer	Momentum (0.9)			
Learning rate	Cosine annealing (0.1)			
Training epochs	80	180	200	300
Batch size	256	128	128	128
Weight decay	1e-4	5e-4	5e-4	1e-4

Table 2. Performance comparison of improved SNIP and GraSP.

VGG19/CIFAR10	Acc: 94.20%				
Pruning ratio	85%	90%	95%	98%	99%
SNIP	93.91	93.82	93.72	91.16	10.00
Ours SNIP	**94.05**	**93.98**	**93.86**	**91.83**	10.00
GraSP	93.59	93.50	92.90	92.39	91.04
Ours GraSP	**93.77**	**93.69**	**93.48**	**92.78**	**91.84**

Note that the loss is computed using a batch of samples containing all labels, although the overall improved SNIP has only a slight improvement in accuracy, which is essential on datasets with more complex labels. Under the 99% pruning rate, the test set accuracy is still 10%, which is caused by the fault of SNIP itself. For the improvement of GraSP, the improvement of accuracy is more prominent, especially after the compression rate is higher than 90%. This suggests that sensitivity to the gradient norm is better than maximizing the gradient norm, validating our previous analysis.

4.3 Experimental Results of Artificial Panning

In Sect. 3.5, we introduce the metrics selected by Panning. In the process of Panning, when the compression rate is low, the balance between layers is better, and the weight sensitive to loss should have the highest priority. When the compression rate is high, the number of weights is minimal, and it is easy to have connectivity matters [34], so the sensitivity between layers is more important. Combined with the influence of gradient norm on trainability, the hyperparameter settings under different compression stages in the Panning process are shown in Table 3.

We compare the compression ratio of Panning, SynFlow, SNIP iteration, and GraSP iteration between $\{10^{-1}, 10^{-4}\}$. In order to highlight the effectiveness of Panning, the SNIP and GraSP iterations also use the FORCE [14] method to prune the network dynamically, and the number T of iterative pruning is unified as 100 times. Taking the L2 regularized complete network as the baseline, Fig. 3

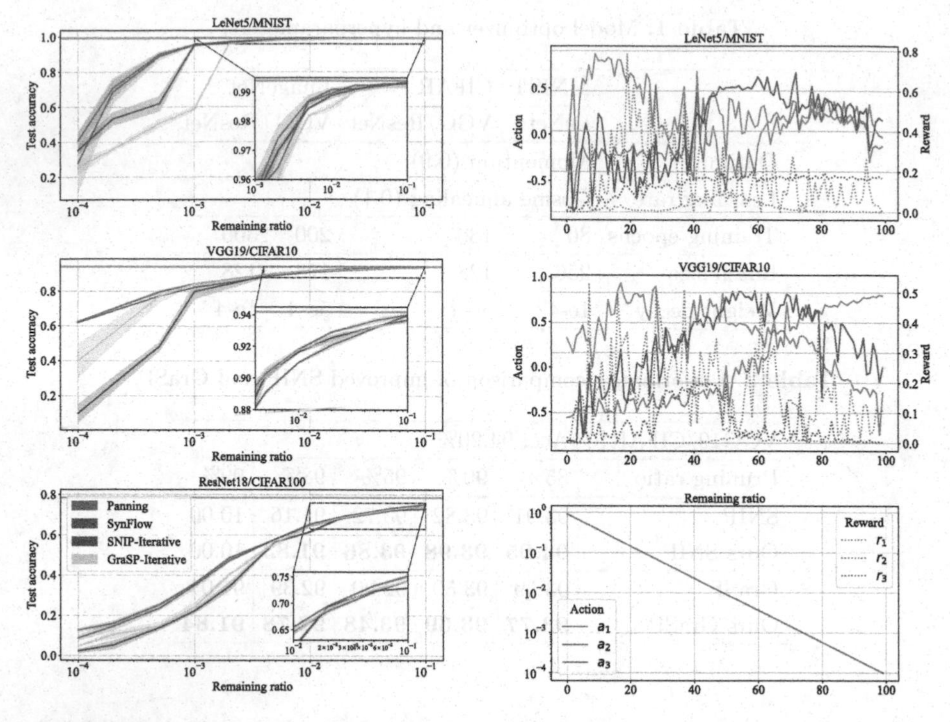

Fig. 3. Performance comparison of Panning and iterative pruning. The shaded area indicates the error of repeated experiments. The gray horizontal line is the baseline.

Fig. 4. Action and reward changes generated by RLPanning during pruning. The left axis is the output action a of the agent. The right axis is the reward r of environmental feedback. The bottom graph shows the change in the remaining ratio over iterations.

shows the experimental results on VGG19/CIFAR10, ResNet18/CIFAR100, and LeNet5/MNIST.

Note that the curve corresponding to Panning in the figure is always above the other methods, whether with a high or low compression ratio. The MNIST dataset is relatively simple, and the differences between the various methods are minor. The advantage of Panning on VGG19/CIFAR10 is mainly manifested in the low compression rate. On ResNet18/CIFAR100, the generalization ability through Panning is also maintained well at extremely high compression rates. In addition, the hyperparameters that adjust the proportion of metrics in Panning are set based on our estimates. In fact, if the parameters in Table 3 are further tuned, Panning will achieve better performance, but our estimated settings have demonstrated the effectiveness of Panning for multiple screening.

Table 3. Hyperparameter setting of artificial Panning.

Pruning ratio	p_1(SynFlow)	p_2(SNIP)	p_3(GraSP)
(0, 0.8]	0.2	0.5	0.3
(0.8, 0.9]	0.2	0.4	0.4
(0.9, 0.98]	0.2	0.3	0.5
(0.98, 0.99]	0.4	0.2	0.4
(0.99, 1)	0.5	0	0.5

Table 4. TD3 training details and hyperparameters.

Training		TD3	
Optimizer	Adam	Exploration noise	0.1
Learning rate	3e-4	Discount factor	0.99
Start timesteps	2e3	Network update rate	0.01
Max timesteps	2e5	Policy noise	0.2
Batch size	256		

4.4 Experimental Results of RLPanning

The actor and critic networks of the TD3 agent in RLPanning are both three-layer linear networks, using the Tanh activation function, and the dimension of the hidden layer is 256. The training details and key hyperparameters are shown in Table 4. When training the agent, we only use LeNet5/MNIST as the network and dataset for the Panning environment and do not need other network structures and datasets. The resulting Panning agent can be adapted to other convolutional network pruning tasks.

We apply the trained Panning agent to prune LeNet5 and VGG19. The feedback of the Panning environment and the actions of the Panning agent are shown in Fig. 4. Where $\{r_1, r_2, r_3\}$ correspond to the first three unsigned items in Eq. (7), respectively. The reward r comes from the state of the sparse network, and the r_2 change is more drastic than other reward items due to different data batches. It is not difficult to find from the figure that the action a_2 plays a leading role in the early stage of the iteration. As the remaining weights decrease, the magnitudes of actions a_1, a_3 begin to increase. By the late iteration, the reward r gradually stabilizes, with comparable action magnitude in LeNet5/MNIST and even much higher action a_1 in VGG19/CIFAR10, due to the deeper network layers of VGG and the more significant challenge of inter-layer equalization. Note that the actions produced by the agent are roughly consistent with our preset changes, and the same pattern occurs on untrained networks and datasets.

The performance of RLPanning pruned LeNet5 and VGG19 is shown in Table 5. We take the average of three experiments to get the results. It is observed that there is less room for improvement at low compression ratios, and RLPan-

Table 5. Performance comparison of RLPanning and Panning.

LeNet5/MNIST	Acc: 99.40%					
Pruning ratio	90%	95%	98%	99%	99.9%	99.99%
Panning	99.37	**99.21**	99.11	99.09	95.51	40.85
RLPanning	**99.40**	99.18	**99.23**	**99.15**	**97.89**	**45.43**
LeNet5/FashionMNIST	Acc: 91.98%					
Panning	90.67	90.14	88.74	86.92	68.97	26.67
RLPanning	**90.84**	**90.53**	**89.35**	**87.41**	**69.02**	**30.26**
VGG19/CIFAR10	Acc: 94.20%					
Panning	94.02	93.66	92.81	91.75	83.87	62.32
RLPanning	**94.09**	**93.80**	**92.98**	**92.12**	**84.52**	**64.51**

Table 6. Performance comparison of pruned VGG19 and ResNet18 on Tiny-ImageNet.

Network	VGG19: 63.29%			ResNet18: 63.92%		
Pruning ratio	90%	95%	98%	90%	95%	98%
SNIP	61.15	59.32	49.04	60.42	58.56	50.66
GraSP	60.26	59.53	56.54	60.18	58.84	55.79
SynFlow	59.54	58.06	45.29	59.03	56.77	46.34
Panning	61.67	**60.25**	**58.33**	60.74	**58.92**	**56.48**
RLPanning	**61.84**	59.83	58.17	**61.07**	58.73	55.82

ning is very close to Panning in terms of test accuracy. However, RLPanning still shows the advantage of dynamic pruning according to the environment at a high compression rate. Table 6 shows the performance comparison on TinyImageNet. It is observed that Panning has very good performance on large datasets, or rather more obvious improvement in the underfitting state. Note that RLPanning is analogous to Panning on TinyImageNet and significantly outperforms other methods. RLPanning is a little more prominent at low compression ratios. Note that the TD3 model we trained based on LeNet5/MNIST, and it is impossible to accurately capture features for large datasets. But even so, the experimental data verifies the effectiveness of the agent trained on the small model can still be applied to the large model and large dataset. This means that agents trained with reinforcement learning can suppress the interference caused by changes in dataset size and network scale, and can more profoundly reflect the inherent nature of weight screening.

5 Conclusion

In the experiments of this paper, we generalize the pruning process as the process of expressive force transfer, and analyze and improve the existing weight metrics before training. We then propose a pruning before training approach for Panning and automate the Panning process through reinforcement learning. Our experimental results show that Panning further improves the performance of pruned neural networks before training. The expressive force transfer is a fascinating phenomenon, and in the follow-up work, we will consider the higher-order terms of the Taylor approximation to design a new metric. In addition, we will make further explorations in structured sparsity and constraint-guided sparsity based on expressive force transfer.

Acknowledgements. This research was partially supported by Hunan Provincial Key Laboratory of Intelligent Processing of Big Data on Transp. Thanks to Google Cloud and Huawei Cloud for cloud computing services.

References

1. Bu, J., Daw, A., Maruf, M., Karpatne, A.: Learning compact representations of neural networks using discriminative masking (DAM). In: NeurIPS, pp. 3491–3503 (2021)
2. Cho, M., Joshi, A., Hegde, C.: ESPN: extremely sparse pruned networks. CoRR abs/2006.15741 (2020). https://arxiv.org/abs/2006.15741
3. Desai, S., Zhan, H., Aly, A.: Evaluating lottery tickets under distributional shifts. In: EMNLP-IJCNLP 2019, p. 153 (2019). https://doi.org/10.18653/v1/D19-6117
4. Frankle, J., Carbin, M.: The lottery ticket hypothesis: finding sparse, trainable neural networks. In: ICLR (2019). https://openreview.net/forum?id=rJl-b3RcF7
5. Frankle, J., Dziugaite, G.K., Roy, D., Carbin, M.: Pruning neural networks at initialization: why are we missing the mark? In: ICLR (2021). https://openreview.net/forum?id=Ig-VyQc-MLK
6. Frankle, J., Dziugaite, G.K., Roy, D.M., Carbin, M.: Stabilizing the lottery ticket hypothesis. arXiv preprint arXiv:1903.01611 (2019). https://doi.org/10.48550/arXiv.1903.01611
7. Fujimoto, S., van Hoof, H., Meger, D.: Addressing function approximation error in actor-critic methods. In: ICML. Proceedings of Machine Learning Research, vol. 80, pp. 1582–1591. PMLR (2018)
8. Gao, S., Liu, X., Chien, L., Zhang, W., Alvarez, J.M.: VACL: variance-aware cross-layer regularization for pruning deep residual networks. In: ICCV Workshops, pp. 2980–2988. IEEE (2019). https://doi.org/10.1109/ICCVW.2019.00360
9. Goodfellow, I., Bengio, Y., Courville, A.: Deep Learning. MIT Press, Cambridge (2016)
10. Han, S., Mao, H., Dally, W.J.: Deep compression: compressing deep neural network with pruning, trained quantization and Huffman coding. In: ICLR (2016)
11. Han, S., Pool, J., Tran, J., Dally, W.J.: Learning both weights and connections for efficient neural network. In: NIPS, pp. 1135–1143 (2015)
12. Hayou, S., Ton, J., Doucet, A., Teh, Y.W.: Robust pruning at initialization. In: ICLR (2021). https://openreview.net/forum?id=vXj_ucZQ4hA
13. He, Y., Lin, J., Liu, Z., Wang, H., Li, L.-J., Han, S.: AMC: AutoML for model compression and acceleration on mobile devices. In: Ferrari, V., Hebert, M., Sminchisescu, C., Weiss, Y. (eds.) ECCV 2018. LNCS, vol. 11211, pp. 815–832. Springer, Cham (2018). https://doi.org/10.1007/978-3-030-01234-2_48

14. de Jorge, P., Sanyal, A., Behl, H.S., Torr, P.H.S., Rogez, G., Dokania, P.K.: Progressive skeletonization: trimming more fat from a network at initialization. In: ICLR (2021). https://openreview.net/forum?id=9GsFOUyUPi
15. Krizhevsky, A., Sutskever, I., Hinton, G.E.: Imagenet classification with deep convolutional neural networks. In: NIPS, pp. 1106–1114 (2012)
16. Lan, Z., Chen, M., Goodman, S., Gimpel, K., Sharma, P., Soricut, R.: ALBERT: a lite BERT for self-supervised learning of language representations. In: ICLR. OpenReview.net (2020)
17. Lee, N., Ajanthan, T., Gould, S., Torr, P.H.S.: A signal propagation perspective for pruning neural networks at initialization. In: ICLR (2020). https://openreview.net/forum?id=HJeTo2VFwH
18. Lee, N., Ajanthan, T., Torr, P.H.S.: SNIP: single-shot network pruning based on connection sensitivity. In: ICLR (Poster) (2019). https://openreview.net/forum?id=B1VZqjAcYX
19. Li, H., Kadav, A., Durdanovic, I., Samet, H., Graf, H.P.: Pruning filters for efficient convnets. In: ICLR (Poster). OpenReview.net (2017)
20. Lillicrap, T.P., et al.: Continuous control with deep reinforcement learning. In: ICLR (Poster) (2016)
21. Lin, S., et al.: Towards optimal structured CNN pruning via generative adversarial learning. In: CVPR, pp. 2790–2799. Computer Vision Foundation/IEEE (2019)
22. Liu, S., Yin, L., Mocanu, D.C., Pechenizkiy, M.: Do we actually need dense overparameterization? In-time over-parameterization in sparse training. In: ICML. Proceedings of Machine Learning Research, vol. 139, pp. 6989–7000. PMLR (2021)
23. Liu, Z., Li, J., Shen, Z., Huang, G., Yan, S., Zhang, C.: Learning efficient convolutional networks through network slimming. In: ICCV, pp. 2755–2763. IEEE Computer Society (2017)
24. Liu, Z., Sun, M., Zhou, T., Huang, G., Darrell, T.: Rethinking the value of network pruning. In: ICLR (Poster) (2019). https://openreview.net/forum?id=rJlnB3C5Ym
25. Malach, E., Yehudai, G., Shalev-Shwartz, S., Shamir, O.: Proving the lottery ticket hypothesis: pruning is all you need. In: ICML. Proceedings of Machine Learning Research, vol. 119, pp. 6682–6691. PMLR (2020). http://proceedings.mlr.press/v119/
26. Mnih, V., et al.: Human-level control through deep reinforcement learning. Nature **518**(7540), 529–533 (2015)
27. Mocanu, D.C., Mocanu, E., Stone, P., Nguyen, P.H., Gibescu, M.: Scalable training of artificial neural networks with adaptive sparse connectivity inspired by network science. Nat. Commun. (2018). https://doi.org/10.1038/s41467-018-04316-3
28. Morcos, A.S., Yu, H., Paganini, M., Tian, Y.: One ticket to win them all: generalizing lottery ticket initializations across datasets and optimizers. In: NeurIPS, pp. 4933–4943 (2019). https://proceedings.neurips.cc/paper/2019/hash/a4613e8d72a61b3b69b32d040f89ad81-Abstract.html
29. Orseau, L., Hutter, M., Rivasplata, O.: Logarithmic pruning is all you need. In: NeurIPS (2020). https://proceedings.neurips.cc/paper/2020/hash/1e9491470749d5b0e361ce4f0b24d037-Abstract.html
30. Qu, M., Tang, J., Han, J.: Curriculum learning for heterogeneous star network embedding via deep reinforcement learning. In: WSDM, pp. 468–476. ACM (2018)
31. Real, E., et al.: Large-scale evolution of image classifiers. In: ICML. Proceedings of Machine Learning Research, vol. 70, pp. 2902–2911. PMLR (2017)
32. Su, J., et al.: Sanity-checking pruning methods: random tickets can win the jackpot. In: NeurIPS (2020)

33. Tanaka, H., Kunin, D., Yamins, D.L., Ganguli, S.: Pruning neural networks without any data by iteratively conserving synaptic flow. In: NeurIPS (2020). https://proceedings.neurips.cc/paper/2020/hash/46a4378f835dc8040c8057beb6a2da52-Abstract.html
34. Vysogorets, A., Kempe, J.: Connectivity matters: neural network pruning through the lens of effective sparsity (2021). https://doi.org/10.48550/ARXIV.2107.02306
35. Wang, C., Zhang, G., Grosse, R.B.: Picking winning tickets before training by preserving gradient flow. In: ICLR (2020). https://openreview.net/forum?id=SkgsACVKPH
36. Wen, W., Wu, C., Wang, Y., Chen, Y., Li, H.: Learning structured sparsity in deep neural networks. In: NIPS, pp. 2074–2082 (2016)
37. Zhu, M., Gupta, S.: To prune, or not to prune: exploring the efficacy of pruning for model compression. In: ICLR (Workshop). OpenReview.net (2018)
38. Zoph, B., Le, Q.V.: Neural architecture search with reinforcement learning. In: ICLR. OpenReview.net (2017)
39. Zoph, B., Vasudevan, V., Shlens, J., Le, Q.V.: Learning transferable architectures for scalable image recognition. In: CVPR, pp. 8697–8710. Computer Vision Foundation/IEEE Computer Society (2018)
40. Zou, Z., Shi, Z., Guo, Y., Ye, J.: Object detection in 20 years: a survey. CoRR abs/1905.05055 (2019)

Network Pruning via Feature Shift Minimization

Yuanzhi Duan[1], Yue Zhou[1], Peng He[1], Qiang Liu[2], Shukai Duan[1], and Xiaofang Hu[1](✉)

[1] College of Artificial Intelligence, Southwest University, Chongqing, China
{swuyzhid,hepeng5}@email.swu.edu.cn, {duansk,huxf}@swu.edu.cn

[2] Harbin Institute of Technology, Harbin, China
18b933041@stu.hit.edu.cn

Abstract. Channel pruning is widely used to reduce the complexity of deep network models. Recent pruning methods usually identify which parts of the network to discard by proposing a channel importance criterion. However, recent studies have shown that these criteria do not work well in all conditions. In this paper, we propose a novel Feature Shift Minimization (FSM) method to compress CNN models, which evaluates the feature shift by converging the information of both features and filters. Specifically, we first investigate the compression efficiency with some prevalent methods in different layer-depths and then propose the feature shift concept. Then, we introduce an approximation method to estimate the magnitude of the feature shift, since it is difficult to compute it directly. Besides, we present a distribution-optimization algorithm to compensate for the accuracy loss and improve the network compression efficiency. The proposed method yields state-of-the-art performance on various benchmark networks and datasets, verified by extensive experiments. Our codes are available at: https://github.com/lscgx/FSM.

1 Introduction

The rapid development of convolutional neural networks (CNN) has obtained remarkable success in a wide range of computer vision applications, such as image classification [11,18,24], video analysis [5,21,30], object detection [6,8,39], semantic segmentation [1,2,41], etc. The combination of CNN models and IoT devices yields significant economic and social benefits in the real world. However, better performance for a CNN model usually means higher computational complexity and a greater number of parameters, limiting its application in resource-constrained devices. Therefore, model compression techniques are required.

To this end, compressing the existing network is a popular strategy, including tensor decomposition [38], parameter quantification [32], weight pruning [9,25], structural pruning [28], etc. Moreover, another strategy is to create a new small network directly, including knowledge distillation [16] and compact network design [17]. Among these compression techniques, structural pruning has significant performance and is usable for various network architectures. In

L. Wang et al. (Eds.): ACCV 2022, LNCS 13841, pp. 618–634, 2023.
https://doi.org/10.1007/978-3-031-26319-4_37

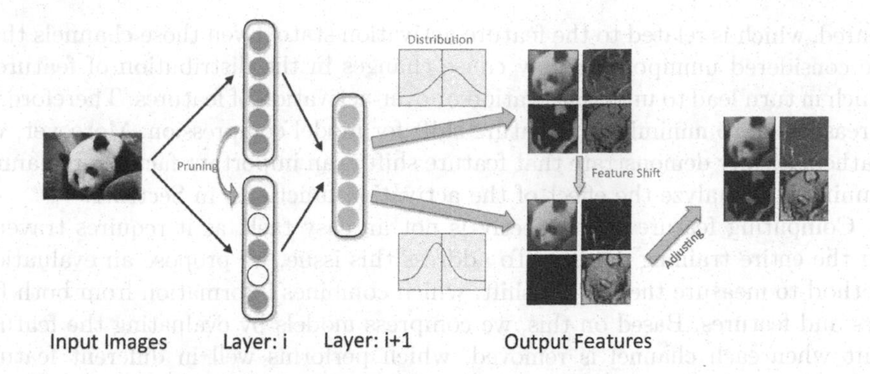

Fig. 1. Diagram of the proposed Feature Shift Minimization method (FSM).

this paper, we present a channel pruning method, belonging to structural pruning, for model compression.

In the process of channel pruning, some channels that are considered unimportant or redundant are discarded and get a sub-network of the original network. The core of channel pruning lies in the design of filter or feature selection criteria. One idea is to directly use the inherent properties of filters or features, such as statistical properties, matrix properties, etc. For example, Li *et al.* [14,25] sorting the filters by l_1 norm of filters, and Lin *et al.* [28] prune channels by calculating the rank of features. Another idea is to evaluate the impact of removing a channel on the next layer or the accuracy loss [34]. These criteria, which are based on different properties, succeeded in model compression, but the performance faded in some conditions. Because it is difficult for a single criterion to take into account all factors such as feature size or feature dimension at the same time. So, many pruning criteria are suboptimal because they only work well locally, not globally. As shown in Fig. 2, in some cases, random pruning can even outperform well-designed methods.

To compensate for the limitations of a single criterion, some multi-criteria approaches have been proposed recently. For example, in [27], a collaborative compression method was proposed for channel pruning, which uses the information of compression sensitivity for each layer. Similarly, in [46], the structural redundancy information of each layer is used to guide pruning. Although these methods take into account the influence of other factors on pruning efficiency, they make no improvements to the criteria and require additional computational consumption.

In this paper, we propose a new channel pruning method by minimizing feature shift. We define feature shift as the changes in the distribution of features in a layer due to pruning. As shown in Fig. 1, the feature details changed or disappeared due to pruning, but when the distribution of the features is properly adjusted, the disappeared feature details emerge again. This experiment was done on ImageNet, using the ResNet-50 model. This suggests that the loss of feature detail, caused by pruning some channels, may not have really disappeared, which is related to the feature activation state. Even those channels that are considered unimportant may cause changes in the distribution of features, which in turn lead to under-activation or over-activation of features. Therefore, it is reasonable to minimize the feature shift for model compression. Moreover, we mathematically demonstrate that feature shift is an important factor for channel pruning and analyze the effect of the activation functions, in Sect. 3.1.

Computing feature shift directly is not an easy task, as it requires traversing the entire training dataset. To address this issue, we propose an evaluation method to measure the feature shift, which combines information from both filters and features. Based on this, we compress models by evaluating the feature shift when each channel is removed, which performs well in different feature dimensions. In particular, the proposed pruning method does not require sampling the features and uses only the parameters of the pre-trained models. In addition, to prove that the feature shift occurs in the pruning process, we design a distribution-optimization algorithm to adjust the pruned feature distribution, which significantly recovers the accuracy loss, as will be discussed in Sect. 3.4.

To summarize, our main contributions are as follows:

1) We investigate the relationship between compression rate, layer depth, and model accuracy with different pruning methods. It reveals that the feature shift is an important factor affecting pruning efficiency.
2) We propose a novel channel pruning method, Feature Shift Minimization (FSM), which combines information from both features and filters. Moreover, a distribution-optimization algorithm is designed to accelerate network compression.
3) Extensive experiments on CIFAR-10 and ImageNet, using VGGNet, MobileNet, GoogLeNet, and ResNet, demonstrate the effectiveness of the proposed FSM.

2 Related Work

Channel Pruning. Channel pruning has been successfully applied to the compression of various network architectures, which lies in the selection of filters or features. Some previous methods use the intrinsic properties of filters to compress models. For example, [14,25] prune filters according to their l_i norms. In [28], the rank of the features is used to measure the information richness of the features and retain the features with high information content. Some other approaches go beyond the limits of a single layer. For example, Chin *et al.* [3] removes filters by calculating the global ranking. Liu *et al.* [33] analyzes the relationship between the filters and BN layers and designs a scale factor to represent the importance. In addition, computing the statistical information of the next layer and minimizing the feature reconstruction error is also a popular idea [15,34]. Different from methods that are based on only a single filter or feature, [43,46,48] remove redundant parts of a model by measuring the difference between the filters or features.

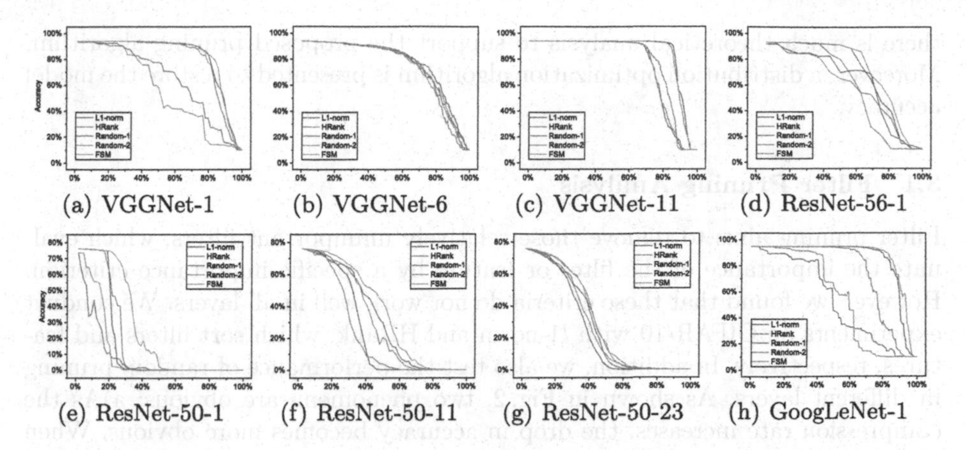

Fig. 2. Analysis of the relationship between compression rate, layer depth, accuracy, and methods. Names such as "ResNet-50-1" mean experiment on the 1st layer of ResNet-50. For each subfigure, the horizontal axis is the compression rate, and the vertical axis is the accuracy.

Effect of Layer-Depth. For different layers, the feature size, dimensions, and redundancy are different. Recent works have found that different layers of a model have different sensitivities to the compression rate. Wang *et al.* [46] proposes that the structural redundancy of each layer are different, which plays a key role in the pruning process. In layers with high structural redundancy, more filters can be safely discarded with little loss of accuracy. Li *et al.* [27] represents that the compression sensitivity of each layer is distinct, and it is used to guide pruning. In addition, He *et al.* [12] propose that a pruning criterion does not always work well in all layers, and they achieve better performance by using different pruning criteria on different layer-depths. Earlier methods [13,19,51] required iterative tuning of the compression ratio to achieve better model accuracy. It is tedious and time-consuming. Considering the difference between different layers can greatly improve the pruning efficiency.

We investigate the trend of accuracy when adopting different compression rates in different layer-depths. Two methods (L1 and HRank) are investigated, which sort filters and features, respectively. Based on this, we propose a new perspective to explore the redundancy of neural networks.

3 Methodology

In this section, we analyze the process of filter pruning and point out that the distribution of features is changed after pruning. Then we propose a novel pruning criterion that minimizes the feature shift. Unlike some heuristic pruning criteria, there is much theoretical analysis to support the proposed pruning algorithm. Moreover, a distribution optimization algorithm is presented to restore the model accuracy.

3.1 Filter Pruning Analysis

Filter pruning aims to remove those relatively unimportant filters, which evaluate the importance of the filter or feature by a specific importance criterion. However, we found that these criteria do not work well in all layers. We conduct experiments on CIFAR-10 with $l1$-norm and HRank, which sort filters and features, respectively. In addition, we also test the performance of random pruning in different layers. As shown in Fig. 2, two phenomena are obvious: a)As the compression rate increases, the drop in accuracy becomes more obvious. When the compression rate exceeds a critical value, the accuracy declines dramatically. b) For high-dimensional features, sorting filters or feature maps by these criteria isn't substantially better than random pruning.

The phenomena reveal that: a) Exist other factors damage the accuracy, and their impact grows with the compression rate. b) For low-dimensional features, pruning criteria are effective. However, for high-dimensional features, random pruning also performs well. This indicates that the gap between the importance scores of filters/features becomes smaller as the dimensionality increases.

To clarify why these phenomena occur, we mathematically analyze the changes that occur in one feature map before and after pruning. A typical inference process in one layer is Input-Conv-BN [20]-ReLU [7]-Output.

Many benchmark network architectures, such as ResNet, DenseNet, and GoogLeNet, are based on it. In particular, we let $X_i = (x_i^{(1)} \cdots x_i^{(d_i)})$ represent the input for i-th layer of one CNN model, where d_i stand for the number of the dimensions. Firstly, we normalize each dimension on BN layer

$$\widehat{x}_i^{(k)} = \frac{x_i^{(k)} - E[x_i^{(k)}]}{\sqrt{Var[x_i^{(k)}]}}, \; y_i^{(k)} = \gamma_i^{(k)} \widehat{x}_i^{(k)} + \beta_i^{(k)}, \tag{1}$$

where E is the expectation, Var is the variance, and k stand for the k-th dimension. We let $Y_i = (y_i^{(1)} \cdots y_i^{(d_i)})$ denotes the output of the BN layer. The output of the k-th dimension has a mean of $\beta_i^{(k)}$ and a standard deviation of $\gamma_i^{(k)}$, which can be directly obtained by the pre-trained models. Then, Y_i is passed on to the ReLU layer. For each of the values in Y_i, the ReLU activation function is applied to it to get thresholded values. The output can be formulated as:

$$\widehat{y}_i^{(k)} = ReLU(y_i^{(k)}) = max(0, y_i^{(k)}). \tag{2}$$

Obviously, the distribution of $\widehat{y}_i^{(k)}$ changes after passing through the ReLU layer. The output expectation

$$E[\widehat{y}_i^{(k)}] = E[max(0, y_i^{(k)})], \tag{3}$$

and the input of the $(i+1)$-th layer

$$x_{i+1}^{(k)} = (\widehat{y}_i^{(1)} \cdots \widehat{y}_i^{(d_i)}) \cdot w_{i+1} \neq (\widehat{y}_i^{(1)} \cdots \widehat{y}_i^{(\widehat{d}_i)}) \cdot w_{i+1}, \ s.t. \ \widehat{d}_i \leq d_i, \tag{4}$$

where w represents the weights and $\widehat{d}_i$ stands for the i-th dimension after pruning. As can be seen, the distribution of input x of one channel will shift when pruning some channels in the previous layer. We assume that the shifted expectation of $x_i^{(k)}$ as $\widehat{E}[x_i^{(k)}]$, and the shifted variance as $\widehat{Var}[x_i^{(k)}]$. More, we let $\Delta_E = E[x_i^{(k)}] - \widehat{E}[x_i^{(k)}]$. So we can reformulate the Eq. (1) as:

$$y_i^{(k)} = \gamma_i^{(k)} \widehat{x}_i^{(k)} + \beta_i^{(k)} = \gamma_i^{(k)} \frac{x_i^{(k)} - E[x_i^{(k)}]}{\sqrt{Var[x_i^{(k)}]}} + \beta_i^{(k)} \tag{5}$$

$$= \gamma_i^{(k)} \frac{x_i^{(k)} - (\widehat{E}[x_i^{(k)}] + \Delta_E)}{\sqrt{Var[x_i^{(k)}]}} + \beta_i^{(k)} \tag{6}$$

$$= \frac{\sqrt{\widehat{Var}[x_i^{(k)}]}}{\sqrt{Var[x_i^{(k)}]}} \gamma_i^{(k)} \frac{x_i^{(k)} - \widehat{E}[x_i^{(k)}]}{\sqrt{\widehat{Var}[x_i^{(k)}]}} + \beta_i^{(k)} - \gamma_i^{(k)} \frac{\Delta_E}{\sqrt{Var[x_i^{(k)}]}} \tag{7}$$

The mean of $y_i^{(k)}$ is $\beta_i^{(k)} - \gamma_i^{(k)} \frac{\Delta_E}{\sqrt{Var[x_i^{(k)}]}}$, and the standard deviation shifted to $\frac{\sqrt{\widehat{Var}[x_i^{(k)}]}}{\sqrt{Var[x_i^{(k)}]}} \gamma_i^{(k)}$.

As a result, the expectation $E[\widehat{y}_i^{(k)}]$ is shifted after pruning, which directly affects the input to the next layer. At the same time, due to the ReLU activation function, some values that were not activated before being activated, or the activated values are deactivated. We refer to the change in the distributions after the ReLU layer in the process of pruning as *feature shift*.

Predictably, at high compression rates, the magnitude of the feature shift may be greater compared to low rates, and the drop in accuracy may also be greater. To prove it, we consider the feature shift of each channel as the selection criterion in pruning. We prune channels by their expectation of the feature shift. The channels with the least shift are discarded first. As shown in Fig. 2, many experiments show that the decreasing trend of the accuracy of the proposed method is less pronounced than other prevalent methods. Our method can maintain higher accuracy compared to others at the same compression rate (e.g., 11.3%/L1 vs. 8.79%/HRank vs. 66.52%/FSM with a ratio 40% on ResNet-50-11). It reveals that the feature shift is one of the critical factors for performance degradation. In particular, when we prune the model in reverse order, the accuracy catastrophically decreases, as shown in Table 1. It demonstrates that the greater the magnitude of the feature

Table 1. Experiments on ResNet-50-23. FSM-R means pruning in reverse order.

Rate	20%	30%	40%
FSM (%)	70.42	56.50	32.20
FSM-R (%)	32.89	8.128	1.62

shift, the greater the loss of accuracy. The decrease in accuracy has a positive correlation with the feature shift. So, it is reasonable to use the feature shift to guide pruning. Features that exhibit less feature shift are relatively unimportant.

3.2 Evaluating the Feature Shift

As demonstrated in Eq. (7), we need to calculate $\widehat{E}[x_i^{(k)}]$, which stands for the expectation of the k-th dimension in the i-th layer. It is difficult and time-consuming to calculate it directly over the entire training dataset. We present an approximation method to estimate the expectation of features and prove the feasibility in Sect. 5.2. Specifically, we first expand the $\widehat{E}[x_i^{(k)}]$ as:

$$\widehat{E}[x_i^{(k)}] = \widehat{E}[(\widehat{y}_{i-1}^{(1)} \cdots \widehat{y}_{i-1}^{(\widehat{d}_i-1)}) \cdot w_i^{(k)}] \tag{8}$$

$$= \widehat{E}[\widehat{y}_{i-1}^{(1)}] * w_i^{(k,1)} + \cdots + \widehat{E}[\widehat{y}_{i-1}^{(\widehat{d}_i-1)}] * w_i^{(k,\widehat{d}_i-1)}, \tag{9}$$

where $\widehat{d}_{i-1}$ stands for the $(i-1)$-th dimension after pruning, and $\widehat{d}_{i-1} \leq d_{i-1}$. Moreover, according to Eq. (1), we get

$$\widehat{E}[\widehat{y}_{i-1}^{(k)}] = E[max(0, y_{i-1}^{(k)})] \approx \int_0^{\infty} x \frac{1}{\gamma_{i-1}^{(k)}\sqrt{(2\pi)}} e^{-\frac{(x-\beta_{i-1}^{(k)})^2}{2(\gamma_{i-1}^{(k)})^2}} dx. \tag{10}$$

Without computing the expectation over the entire train dataset, the approximation of $\widehat{E}[x_i^{(k)}]$ can be obtained by combining Eq. (8) and Eq. (10). Then, according to Eq. (7), we get the shifted expectation and the standard deviation of the features.

3.3 Filter Selection Strategy

The pruning strategy of a layer can be regarded as minimizing the sum of the feature shift of all channels of the pruned model. For a CNN model with N layers, the optimal pruning strategy can be expressed as an optimization problem:

$$arg\,min \sum_{i=1}^{N} \sum_{k=1}^{\widehat{d}_i} \left| \Delta_{E_i^{(k)}} \right| = arg\,min \sum_{i=1}^{N} \sum_{k=1}^{\widehat{d}_i} \left| E[x_i^{(k)}] - \widehat{E}[x_i^{(k)}] \right|, \tag{11}$$

where $\widehat{d}_i$ is the dimension of each layer after pruning. During the pruning of a layer, we evaluate the effect of each feature for the feature shift of the next layer, as follows:

$$\delta(x_i^{(k)}) = \sum_{j=1}^{d_{i+1}} \left| w_{i+1}^{(j,k)} * E[y_i^{(k)}] \right|, \tag{12}$$

where $w_{i+1}^{(j,k)}$ denotes the k-th vector of the $(i+1)$-th layer's j-th dimension. $\delta(x_i^{(k)})$ represents the sum of feature shift of one output feature to all channels in the $(i+1)$-th layer. Then, we sort all features according to the values $\delta(\cdot)$. Features with low values are considered unimportant and will be discarded in preference.

3.4 Distribution Optimization

For a pruned model, we present a simple distribution optimization method to recover accuracy. As presented in Eq. (7), the distribution of $y_i^{(k)}$ was changed due to the shift of $E[x_i^{(k)}]$ and $Var[x_i^{(k)}]$ after pruning. Hence, we can adjust them to $\widehat{E}[x_i^{(k)}]$ and $\widehat{Var}[x_i^{(k)}]$ reduce the impact of the shift.

In Eq. (8), we present a evaluation method for $\widehat{E}[x_i^{(k)}]$. Let λ represent the evaluation error, $\lambda = \widehat{E}[x_i^{(k)}] \,/\, E[x_i^{(k)}]$, where the λ is computed on unpruned model. After pruning, we set

$$E[x_i^{(k)}] = \widehat{E}[x_i^{(k)}] \,/\, \lambda \tag{13}$$

For $\widehat{Var}[x_i^{(k)}]$, it is difficult to evaluate, or the error is large. We experimentally show that set

$$Var[x_i^{(k)}] = \frac{\widehat{d_i}}{d_i} \times Var[x_i^{(k)}], \tag{14}$$

performs well in most layers, especially in the deeper layers.

One possible explanation is that the differences between high-dimensional features are not obvious, and the statistical properties are approximate. Random pruning in deep layers produces great results, as shown in Fig. 2, supporting this explanation. Furthermorc, as shown in Fig. 3, extensive experiments on different architectures reveal that the proposed distribution optimization strategy is effective and proves that feature shift occurs during network pruning. Obviously, the algorithm is plug-and-play and may also be combined with other pruning methods.

3.5 Pruning Procedure

The pruning procedure is summarized as follows:

1) For each channel $x_i^{(k)}$ of one layer, we first calculate its output expectation $E[y_i^{(k)}]$.
2) Then, we calculate the feature shift $\delta(x_i^{(k)})$ caused by channel $x_i^{(k)}$, according to Eq. (12).
3) Sort all channels through $\delta(\cdot)$ and discard those channels with small values.
4) After pruning, using Eq. (13) and Eq. (14) to recover accuracy.
5) Fine-tuning the model one epoch, and back to the first step to prune the next layer.

After all layers have been pruned, we train the pruned model for some epochs.

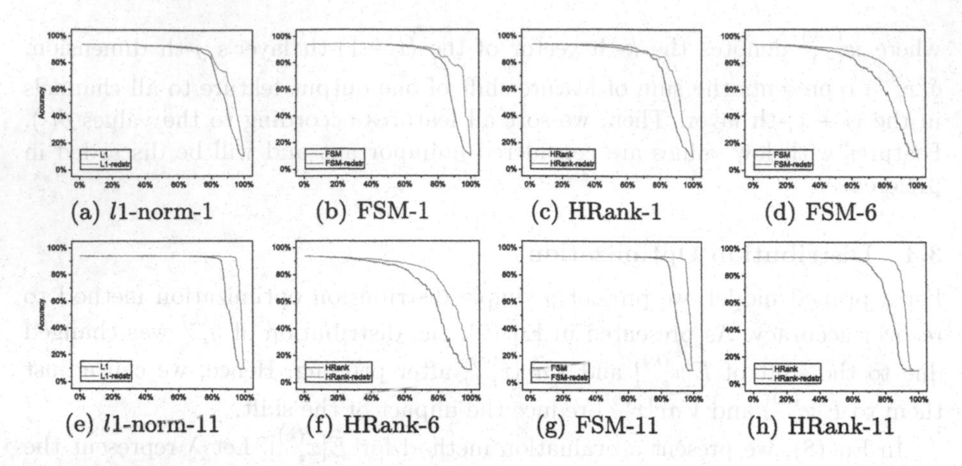

Fig. 3. Comparison of the accuracy before and after using distribution optimization on VGGNet. Similarly, the horizontal axis represents the compression rate, and the vertical axis is the accuracy. Names such as "HRank-1" mean to adopt the HRank method on the 1st layer of VGGNet. The results suggest that the feature shift is a critical factor that damages model accuracy.

4 Experiments

In this section, in order to verify the effectiveness of the proposed FSM, we conduct extensive experiments on CIFAR-10 [23] and ImageNet [24]. Prevalent models, such as ResNet [11], GoogLeNet [44], MobileNet-V2 [40], and VGGNet [42], are adopted to test the performance. All experiments are running on Pytorch 1.8 [36] under Intel i7-8700K CPU @3.70GHz and NVIDIA GTX 1080Ti GPU.

4.1 Implementation Details

For all models, the pruning process is performed in two steps:1) pruning the filters layer by layer and fine-tune 1 epoch for each layer. 2) After pruning, we train the model for some epochs using the stochastic gradient descent algorithm (SGD) with momentum 0.9. For VGGNet and GoogLeNet on CIFAR-10, we train the model for 200 epochs, in which the initial learning rate, batch size, and weight decay are set to 0.01, 128, and 0, respectively. The learning rate is divided by 10 at epochs 100 and 150. For ResNet-56 on CIFAR-10, we train 300 epochs and set the weight decay to 0.0005. The learning rate is divided by 10 at epochs 150 and 225. For ResNet-50, we train the model for 120 epochs with an initial learning rate of 0.01, and the learning rate is divided by 10 at epochs 30, 60, and 90. The batch size is set to 64. For MobileNet-V2, we fine-tune the pruned model for 150 epochs, and the learning rate is decayed by the cosine annealing scheduler with an initial learning rate of 0.01. The weight decay is set to 4×10^{-5}, following the original MobileNet-V2 paper setup. We train all the models three times and report the mean.

Table 2. Comparison with other prevalent pruning methods on CIFAR-10

Model	Method	Baseline Acc	Top-1 Acc	FLOPs ↓	Param. ↓
ResNet-56	SFP [13]	93.59%	93.35%	52.6%	–
	CCP [37]	93.50%	93.42%	52.6%	–
	HRank [28]	93.26%	92.17%	50.0%	42.4%
	NPPM [4]	93.04%	93.40%	50.0%	–
	DHP [26]	–	93.58%	49.0%	41.6%
	SCP [22]	93.69	93.23%	51.5%	–
	FSM (ours)	**93.26%**	**93.63%**	**51.2%**	**43.6%**
	FSM (ours)	**93.26%**	**92.76%**	**68.2%**	**68.5%**
VGGNet	L1 [25]	93.25%	93.40%	34.2%	63.3%
	GAL [29]	93.96%	92.03%	39.6%	77.2%
	HRank [28]	93.96%	93.42%	53.7%	82.9%
	EEMC [50]	93.36%	93.63%	56.6%	–
	FSM (ours)	**93.96%**	**93.73%**	**66.0%**	**86.3%**
	FSM (ours)	**93.96%**	**92.86%**	**81.0%**	**90.6%**
GoogLeNet	L1 [25]	–	94.54%	32.9%	42.9%
	GAL [29]	95.05%	93.93%	38.2%	49.3%
	HRank [28]	95.05%	94.53%	54.6%	55.4%
	FSM (ours)	**95.05%**	**94.72%**	**62.9%**	**55.5%**
	FSM (ours)	**95.05%**	**94.29%**	**75.4%**	**64.6%**

We set the compression rate of each layer according to its accuracy drop curve, which maintains the model accuracy to the greatest extent possible. The details are provided in the supplementary.

4.2 Results and Analysis

CIFAR-10. In Table 2, we compare the proposed FSM method with other prevalent algorithms on VGGNet, GoogLeNet, and ResNet-56. Our method achieves a significantly compression efficiency on reductions of FLOPs and parameters, but with higher accuracy. For example, compared with HRank, FSM yields a better accuracy (93.73% vs. 93.42%) under a greater FLOPs reduction (66.0% vs. 53.7%) and parameters reduction (86.3% vs. 82.9%). For ResNet-56 on CIFAR-10, FSM prunes 51.2% of FLOPs and 43.6% of parameters, but the accuracy improved by 0.37%. NPPM and DHP have been proposed recently and they have a great performance in network compression. Compared with NPPM, our method shows a better performance, and with a higher FLOPs reduction. Compared with DHP, under similar accuracy, FSM performs better at the reduction of FLOPs (51.2% vs. 49.0%) and parameters (43.6% vs. 41.6%). For GoogLeNet, FSM outperforms HRank and GAL in all aspects. With over 62% FLOPs reduction, FSM still maintains 94.72% accuracy.

Furthermore, we verified the performance of FSM at high compression rates. For instance, with more than 90% of the parameters discarded, FSM reduces

Table 3. Comparison with other prevalent pruning methods on ImageNet

Model	Method	Top-1 Acc	Top-5 Acc	Δ Top-1	Δ Top-5	FLOPs ↓
ResNet-50	DCP [51]	74.95%	92.32%	−1.06%	−0.61%	55.6%
	CCP [37]	75.21%	92.42%	−0.94%	−0.45%	54.1%
	Meta [31]	75.40%	–	−1.20%	–	51.2%
	GBN [49]	75.18%	92.41%	−0.67%	−0.26%	55.1%
	BNFI [35]	75.02%	–	−1.29%	–	52.8%
	HRank [28]	74.98%	92.44%	−1.17%	−0.54%	43.8%
	SCP [22]	75.27%	92.30%	−0.62%	−0.68%	54.3%
	SRR-GR [46]	75.11%	92.35%	−1.02%	−0.51%	55.1%
	GReg [45]	75.16%	–	−0.97%	–	56.7%
	FSM (ours)	**75.43%**	**92.45%**	**−0.66%**	**−0.53%**	**57.2%**
MobileNet-V2	CC [47]	70.91%	–	−0.89%	–	30.7%
	BNFI [35]	70.97%	–	−1.22%	–	30.0%
	FSM (ours)	**71.18%**	**89.81%**	**−0.70%**	**−0.48%**	**30.6%**

the accuracy by only 1.1% on VGGNet. The same results can be observed on ResNet-56 and GoogLeNet.

ImageNet 2012. ImageNet 2012 has over 1.28 million training images and 50,000 validation images divided into 1,000 categories. ResNet-50, compared to VGGNet and ResNet-56, has a larger feature size. To verify the applicability of the proposed FSM, on ResNet-50, we test the performance at different compression rates and different layer-depth, by comparing it with L1 and HRank, as shown in Fig. 2. The results show that FSM successfully picks out the more important channels, even for large-size features. In contrast, L1 and HRank performed poorly, even less well than the random method. In most cases, the proposed FSM shows the best performance.

In Table 3, we compare the proposed FSM method with other prevalent methods on ResNet-50. Our method achieves significant performance, reducing the FLOPs by more than 57% with only 0.66% loss in Top-1 accuracy. Compared with HRank, our method shows a better performance (75.43% vs. 74.98%), and with a higher FLOPs reduction (57.2% vs. 43.8%). Compared with GBN and DCP, under similar FLOPs reduction of about 55%, FSM obtains better accuracy. Our method reduces FLOPs by 6% more than MetaPruning under similar Top-1 accuracy. More comparison details can be found in Table 3.

MobileNet-V2 is a memory-efficient model that is suitable for mobile devices. However, to further compress it is challenging while maintaining the model accuracy. When pruning around 30% of FLOPs, FSM achieves better performance (71.18%) than CC (70.91%) and BNFI (70.97%) on Top-1 accuracy. The results demonstrate that FSM has excellent performance on the large-scale ImageNet dataset and shows higher applicability.

The Speedup on GPU. To show the practical speedup of our method in real scenarios, we measure the inference time by time/images on the NVIDIA 1080Ti

Table 4. The practical speedup for the pruned model over the unpruned model. "T": inference time. "S": practical speedup on GPU (NVIDIA 1080Ti).

Model	FLOPs (%) ↓	Param. (%) ↓	T (ms)	S (×)
GoogLeNet	62.9	55.5	5.03 (±0.01)	1.47×
	75.4	64.7	4.58 (±0.02)	1.62×
ResNet-50	57.2	42.8	13.81 (±0.04)	1.30×

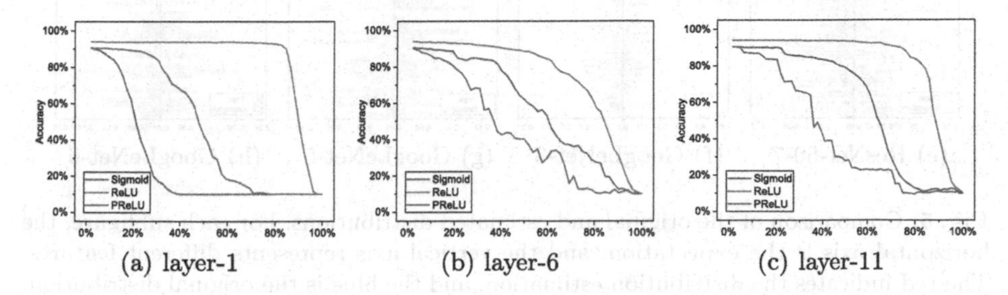

Fig. 4. The effect of different activation functions on channel pruning for VGG. For each subfigure, the horizontal axis is the compression rate, and the vertical axis is the accuracy.

GPU. As shown in Table 4, the proposed FSM achieves 1.62× and 1.30× speedup with batch size 64 on GoogLeNet and ResNet-50, respectively.

5 Ablation Study

5.1 Effect of the Activation Function

In Sect. 3.1, we illustrate that the activation function affects the distribution of the features. In channel pruning, the threshold property of the ReLU activation function results in changes in the feature activation state. In this section, we experiment with the performance of three popular activation functions (Sigmoid, ReLU, and PReLU [10]) on different layers in VGGNet. As shown in Fig. 4, the trend of accuracy with Sigmoid and PReLU changes smoother than ReLU. Although they do not have the problem of gradient plunges, there is a large loss of accuracy at small compression rates. In contrast, ReLU achieves better performance at most rates. In particular, $PReLU(x) = max(0, x) + a \times min(0, x)$, and $Sigmoid(x) = \sigma(x) = \frac{1}{1+e^{-x}}$. Different from the ReLU, their gradient is greater than 0 when $x < 0$. Some values should reach 0 under ReLU, but return a negative value under PReLU, which leads to more changes in features. This explains why PReLU loses more accuracy when the feature shift occurs compared to Sigmoid and ReLU. In general, models with ReLU activation functions are more tolerable for channel pruning than Sigmoid and PReLU.

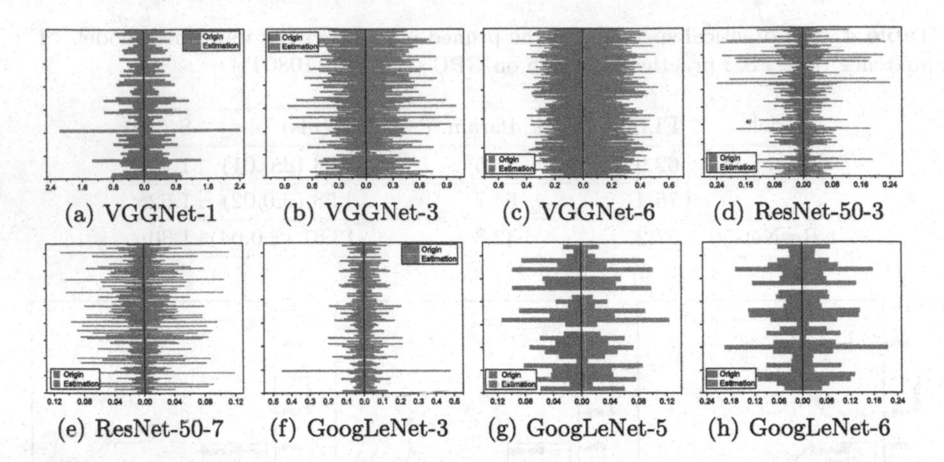

(a) VGGNet-1 (b) VGGNet-3 (c) VGGNet-6 (d) ResNet-50-3

(e) ResNet-50-7 (f) GoogLeNet-3 (g) GoogLeNet-5 (h) GoogLeNet-6

Fig. 5. Comparison of the original and estimated distributions. For each subfigure, the horizontal axis is the expectation, and the vertical axis represents different features. The red indicates the distribution estimation, and the blue is the original distribution. More comparisons are provided in the supplementary. (Color figure online)

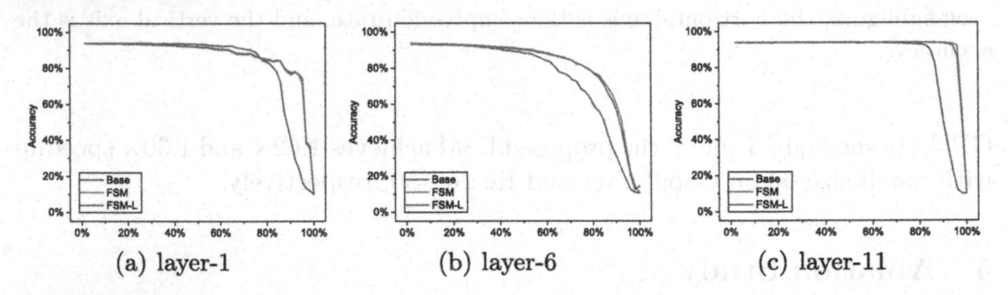

(a) layer-1 (b) layer-6 (c) layer-11

Fig. 6. The effect of evaluation error λ for channel pruning on VGGNet. For each subfigure, the FSM-L means that the expectation evaluation values are amended with λ.

5.2 Error of the Feature Shift Evaluation

The error of the feature distribution evaluation is shown in Fig. 5. The differences between the proposed evaluation method and the true distribution are compared at different layer depths. It can be seen that the evaluation error is in an acceptable range. In addition, our experiments show that expectation and layer depth are negatively correlated. For very small expectation values, our method can still estimate them accurately.

5.3 Effect of the Evaluation Error λ

In Sect. 3.4, we present the estimation error λ to rectify the evaluated distribution. As shown in Fig. 6, the method using λ generates a better performance at low compression rates. Also, no additional accuracy loss is caused at high compression rates.

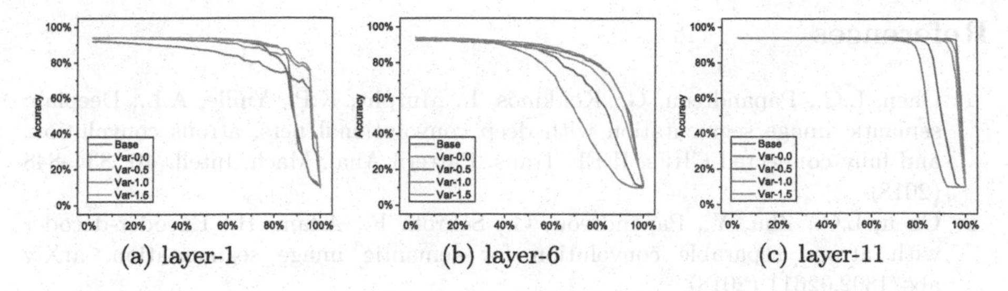

Fig. 7. Comparisons of the variance estimation with different magnitude coefficients on VGGNet. Names such as "Var-1.5" stand for reduce the variance $Var[x_i^{(k)}]$ by $(1.5 \times \frac{\hat{d}_i}{d_i})$%. Similarly, the horizontal axis represents the compression rate, and the vertical axis is the accuracy.

5.4 Effect of the Variance Adjustment Coefficients

In Sect. 3.4, we empirically reduce the variance $Var[x_i^{(k)}]$ by $\frac{\hat{d}_i}{d_i}$% because it is difficult to calculate it directly. As shown in Fig. 7, we adopt different magnitude coefficients to adjust the variances of pruned models. The results show that the VAR-1.0, adopted in our work, achieves the best performance in almost all cases.

6 Conclusions

In this paper, we propose a novel channel pruning method by minimizing feature shift. We first prove the existence of feature shift mathematically, inspired by the investigation of accuracy curves in channel pruning. The feature shift is used to explain why the accuracy plummets when the compression rate exceeds a critical value. Based on this, an efficient channel pruning method (FSM) is proposed, which performs well, especially at high compression rates. Then, we present a feature shift evaluation method that does not traverse the training data set. In addition, a distribution optimization method is designed to improve the efficiency of model compression and is plug-and-play. Extensive experiments and rigorous ablation studies demonstrate the effectiveness of the proposed FSM for channel pruning. In the future, we will further research the impact of multi-branch architecture on the FSM. Also, the combination with other pruning methods is worth trying.

Acknowledgements. The work was supported by National Natural Science Foundation of China (Grant Nos. 61976246 and U20A20227), Natural Science Foundation of Chongqing (Grant No. cstc2020jcyj-msxm X0385).

References

1. Chen, L.C., Papandreou, G., Kokkinos, I., Murphy, K.P., Yuille, A.L.: Deeplab: semantic image segmentation with deep convolutional nets, atrous convolution, and fully connected CRFs. IEEE Trans. Pattern Anal. Mach. Intell. **40**, 834–848 (2018)
2. Chen, L.C., Zhu, Y., Papandreou, G., Schroff, F., Adam, H.: Encoder-decoder with atrous separable convolution for semantic image segmentation. arXiv abs/1802.02611 (2018)
3. Chin, T.W., Ding, R., Zhang, C., Marculescu, D.: Towards efficient model compression via learned global ranking. In: 2020 IEEE/CVF Conference on Computer Vision and Pattern Recognition (CVPR), pp. 1515–1525 (2020)
4. Gao, S., Huang, F., Cai, W.T., Huang, H.: Network pruning via performance maximization. In: 2021 IEEE/CVF Conference on Computer Vision and Pattern Recognition (CVPR), pp. 9266–9276 (2021)
5. Girdhar, R., Tran, D., Torresani, L., Ramanan, D.: Distinit: learning video representations without a single labeled video. In: 2019 IEEE/CVF International Conference on Computer Vision (ICCV), pp. 852–861 (2019)
6. Girshick, R.B., Donahue, J., Darrell, T., Malik, J.: Rich feature hierarchies for accurate object detection and semantic segmentation. In: 2014 IEEE Conference on Computer Vision and Pattern Recognition, pp. 580–587 (2014)
7. Glorot, X., Bordes, A., Bengio, Y.: Deep sparse rectifier neural networks. In: AISTATS (2011)
8. Guo, J., et al.: Hit-detector: hierarchical trinity architecture search for object detection. In: 2020 IEEE/CVF Conference on Computer Vision and Pattern Recognition (CVPR), pp. 11402–11411 (2020)
9. Han, S., Pool, J., Tran, J., Dally, W.J.: Learning both weights and connections for efficient neural network. arXiv abs/1506.02626 (2015)
10. He, K., Zhang, X., Ren, S., Sun, J.: Delving deep into rectifiers: surpassing human-level performance on imagenet classification. In: 2015 IEEE International Conference on Computer Vision (ICCV), pp. 1026–1034 (2015)
11. He, K., Zhang, X., Ren, S., Sun, J.: Deep residual learning for image recognition. In: 2016 IEEE Conference on Computer Vision and Pattern Recognition (CVPR), pp. 770–778 (2016)
12. He, Y., Ding, Y., Liu, P., Zhu, L., Zhang, H., Yang, Y.: Learning filter pruning criteria for deep convolutional neural networks acceleration. In: 2020 IEEE/CVF Conference on Computer Vision and Pattern Recognition (CVPR), pp. 2006–2015 (2020)
13. He, Y., Kang, G., Dong, X., Fu, Y., Yang, Y.: Soft filter pruning for accelerating deep convolutional neural networks. arXiv abs/1808.06866 (2018)
14. He, Y., Liu, P., Wang, Z., Yang, Y.: Pruning filter via geometric median for deep convolutional neural networks acceleration. arXiv abs/1811.00250 (2018)
15. He, Y., Zhang, X., Sun, J.: Channel pruning for accelerating very deep neural networks. In: 2017 IEEE International Conference on Computer Vision (ICCV), pp. 1398–1406 (2017)
16. Hinton, G.E., Vinyals, O., Dean, J.: Distilling the knowledge in a neural network. arXiv abs/1503.02531 (2015)
17. Howard, A.G., et al.: Mobilenets: efficient convolutional neural networks for mobile vision applications. arXiv abs/1704.04861 (2017)

18. Huang, G., Liu, Z., Weinberger, K.Q.: Densely connected convolutional networks. In: 2017 IEEE Conference on Computer Vision and Pattern Recognition (CVPR), pp. 2261–2269 (2017)
19. Huang, Z., Wang, N.: Data-driven sparse structure selection for deep neural networks. arXiv abs/1707.01213 (2018)
20. Ioffe, S., Szegedy, C.: Batch normalization: accelerating deep network training by reducing internal covariate shift. arXiv abs/1502.03167 (2015)
21. Jiang, L., Xu, M., Liu, T., Qiao, M., Wang, Z.: DeepVS: a deep learning based video saliency prediction approach. In: ECCV (2018)
22. Kang, M., Han, B.: Operation-aware soft channel pruning using differentiable masks. In: ICML (2020)
23. Krizhevsky, A.: Learning multiple layers of features from tiny images (2009)
24. Krizhevsky, A., Sutskever, I., Hinton, G.E.: Imagenet classification with deep convolutional neural networks. Commun. ACM **60**, 84–90 (2012)
25. Li, H., Kadav, A., Durdanovic, I., Samet, H., Graf, H.P.: Pruning filters for efficient convnets. arXiv abs/1608.08710 (2017)
26. Li, Y., Gu, S., Zhang, K., Gool, L.V., Timofte, R.: DHP: differentiable meta pruning via hypernetworks. arXiv abs/2003.13683 (2020)
27. Li, Y., et al.: Towards compact CNNs via collaborative compression. In: 2021 IEEE/CVF Conference on Computer Vision and Pattern Recognition (CVPR), pp. 6434–6443 (2021)
28. Lin, M., et al.: HRank: filter pruning using high-rank feature map. In: 2020 IEEE/CVF Conference on Computer Vision and Pattern Recognition (CVPR), pp. 1526–1535 (2020)
29. Lin, S., et al.: Towards optimal structured CNN pruning via generative adversarial learning. In: 2019 IEEE/CVF Conference on Computer Vision and Pattern Recognition (CVPR), pp. 2785–2794 (2019)
30. Lin, T., Liu, X., Li, X., Ding, E., Wen, S.: BMN: boundary-matching network for temporal action proposal generation. In: 2019 IEEE/CVF International Conference on Computer Vision (ICCV), pp. 3888–3897 (2019)
31. Liu, Z., et al.: Metapruning: meta learning for automatic neural network channel pruning. In: 2019 IEEE/CVF International Conference on Computer Vision (ICCV), pp. 3295–3304 (2019)
32. Liu, Z., Shen, Z., Savvides, M., Cheng, K.T.: Reactnet: towards precise binary neural network with generalized activation functions. arXiv abs/2003.03488 (2020)
33. Liu, Z., Li, J., Shen, Z., Huang, G., Yan, S., Zhang, C.: Learning efficient convolutional networks through network slimming. In: 2017 IEEE International Conference on Computer Vision (ICCV), pp. 2755–2763 (2017)
34. Luo, J.H., Wu, J., Lin, W.: ThiNet: a filter level pruning method for deep neural network compression. In: 2017 IEEE International Conference on Computer Vision (ICCV), pp. 5068–5076 (2017)
35. Oh, J., Kim, H., Baik, S., Hong, C., Lee, K.M.: Batch normalization tells you which filter is important. In: 2022 IEEE/CVF Winter Conference on Applications of Computer Vision (WACV), pp. 3351–3360 (2022)
36. Paszke, A., et al.: Automatic differentiation in pytorch (2017)
37. Peng, H., Wu, J., Chen, S., Huang, J.: Collaborative channel pruning for deep networks. In: ICML (2019)
38. Raja, K.B., Raghavendra, R., Busch, C.: Obtaining stable iris codes exploiting low-rank tensor space and spatial structure aware refinement for better iris recognition. In: 2019 International Conference on Biometrics (ICB), pp. 1–8 (2019)

39. Ren, S., He, K., Girshick, R.B., Sun, J.: Faster R-CNN: towards real-time object detection with region proposal networks. IEEE Trans. Pattern Anal. Mach. Intell. **39**, 1137–1149 (2015)
40. Sandler, M., Howard, A.G., Zhu, M., Zhmoginov, A., Chen, L.C.: Mobilenetv 2: inverted residuals and linear bottlenecks. In: 2018 IEEE/CVF Conference on Computer Vision and Pattern Recognition, pp. 4510–4520 (2018)
41. Shelhamer, E., Long, J., Darrell, T.: Fully convolutional networks for semantic segmentation. IEEE Trans. Pattern Anal. Mach. Intell. **39**, 640–651 (2017)
42. Simonyan, K., Zisserman, A.: Very deep convolutional networks for large-scale image recognition. CoRR abs/1409.1556 (2015)
43. Singh, P., Verma, V.K., Rai, P., Namboodiri, V.P.: Leveraging filter correlations for deep model compression. In: 2020 IEEE Winter Conference on Applications of Computer Vision (WACV), pp. 824–833 (2020)
44. Szegedy, C., et al.: Going deeper with convolutions. In: 2015 IEEE Conference on Computer Vision and Pattern Recognition (CVPR), pp. 1–9 (2015)
45. Wang, H., Qin, C., Zhang, Y., Fu, Y.: Neural pruning via growing regularization. In: International Conference on Learning Representations (2020)
46. Wang, Z., Li, C., Wang, X.: Convolutional neural network pruning with structural redundancy reduction. In: 2021 IEEE/CVF Conference on Computer Vision and Pattern Recognition (CVPR), pp. 14908–14917 (2021)
47. Wang, Z., Li, C., Wang, X.: Convolutional neural network pruning with structural redundancy reduction. In: Proceedings of the IEEE/CVF Conference on Computer Vision and Pattern Recognition, pp. 14913–14922 (2021)
48. Wang, Z., et al.: Model pruning based on quantified similarity of feature maps. arXiv abs/2105.06052 (2021)
49. You, Z., Yan, K., Ye, J., Ma, M., Wang, P.: Gate decorator: global filter pruning method for accelerating deep convolutional neural networks. arXiv abs/1909.08174 (2019)
50. Zhang, Y., Gao, S., Huang, H.: Exploration and estimation for model compression. In: 2021 IEEE/CVF International Conference on Computer Vision (ICCV), pp. 477–486 (2021)
51. Zhuang, Z., et al.: Discrimination-aware channel pruning for deep neural networks. In: NeurIPS (2018)

Training Dynamics Aware Neural Network Optimization with Stabilization

Zilin Fang[1,3], Mohamad Shahbazi[1], Thomas Probst[1(✉)], Danda Pani Paudel[1], and Luc Van Gool[1,2]

[1] Computer Vision Laboratory, ETH Zurich, Zürich, Switzerland
{mshahbazi,probstt,paudel,vangool}@vision.ee.ethz.ch
[2] VISICS, ESAT/PSI, KU Leuven, Leuven, Belgium
[3] School of Computing, NUS, Singapore, Singapore
zilin.fang@u.nus.edu

Abstract. We investigate the process of neural network training using gradient descent-based optimizers from a dynamic system point of view. To this end, we model the iterative parameter updates as a time-discrete switched linear system and analyze its stability behavior over the course of training. Accordingly, we develop a regularization scheme to encourage stable training dynamics by penalizing divergent parameter updates. Our experiments show promising stabilization and convergence effects on regression tasks, density-based crowd counting, and generative adversarial networks (GAN). Our results indicate that stable network training minimizes the variance of performance across different parameter initializations, and increases robustness to the choice of learning rate. Particularly in the GAN setup, the stability regularization enables faster convergence and lower FID with more consistency across runs. Our source code is available at: https://github.com/fangzl123/stableTrain.git.

1 Introduction

Ever since Augustin-Louis Cauchy first described the method of gradient descent in 1847 [1], its efficiency and flexibility has inspired countless solvers and optimization algorithms until this day [2–9]. Gradient descent is, in fact, the backbone of the recent advancements in deep learning, in conjunction with the error back-propagation technique [10]. In particular, auto-differentiation offers gradient computation with negligible overhead on the function evaluation, making possible the optimization of large-scale non-linear functions with millions of parameters – such as deep neural networks – using gradient descent [11,12].

Gradient descent (GD) and its derivatives have been extensively studied with regards to their convergence properties on various problems [13,14]. For instance, the choice of the learning rate is crucial for (fast) convergence, and depends on the curvature around a local optimum. Choosing a high learning rate may cause oscillating and divergent behavior, whereas a low learning rate may cause the optimizer to never reach a good solution. Moreover, when using stochastic gradients in the case of typical neural network training, gradient noise can cause

L. Wang et al. (Eds.): ACCV 2022, LNCS 13841, pp. 635–651, 2023.
https://doi.org/10.1007/978-3-031-26319-4_38

undesired updates. Other sources of noise can be multi-task training [15–17], data augmentation [18–20], or re-sampling operations [21–23].

To this end, numerous GD-based optimization algorithms have been proposed to deal with the aforementioned issues. It has shown to be beneficial to modify the gradients before the update step to enforce a desired behavior [24, 25]. For instance, gradient clipping [24] offers an intuitive solution to the problem of diverging gradients. Introducing momentum can reduce noise and accelerate convergence in the presence of high curvature regions and is used by many modern optimizers [8,26]. Furthermore, adaptive learning rates for each parameter (e.g. RMSprop, Adam) often offer robust learning behavior. With AdaDelta [27] there is even a hyperparameter-free optimizer with adaptive learning rates.

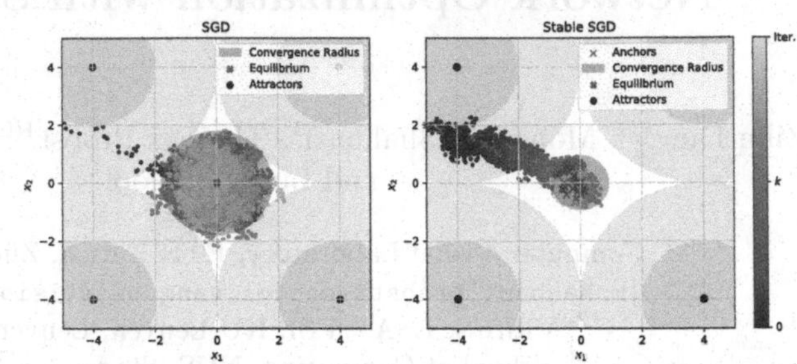

Fig. 1. Trajectories with and without stabilization. The loss function randomly switches between 4 quadratic attractors, and the convergence to the equilibrium at origin is desired. Switching causes SGD to oscillate, which is dampened by stabilization.

In general, the behavior of GD optimization can be understood from the viewpoint of a corresponding dynamic system [14]. The system intuitively models the evolution of the parameters (as the state variable) under the update given by the optimizer. For instance, in the case of a quadratic cost function, the dynamic system is given by a linear system (as discussed in Sect. 2.1). Based on the spectrum of the update matrix with respect to the learning rate, convergence criteria can be developed [13,14]. In particular, we are interested in the stability of the weight updates, while stability analysis has been previously applied to the input-output dynamics of neural networks [28,29].

In the general case of neural networks, however, mini-batch training, nonlinearities, and non-convex loss functions complicate such analysis. In this paper, we attempt a stability analysis of network training with the tools of switched linear systems (SLS) [30–34]. The key assumption is that evolution of parameters in a specific layer follows a time-variant linear dynamic over the course of iterations. This is motivated by two observations. First, each mini-batch contains different random samples. Second, the input activation to the layer change after each update iteration. In both cases, the update dynamics change accordingly, which we model by switching between corresponding linear systems.

The paper is structured as follows. We begin by presenting the theoretical concept of SLS in the context of gradient descent training in Sect. 2. Based on this model, we analyse the stability of the optimization procedure in Sect. 3 and develop a mechanism to control the parameter updates, such that the update dynamics remain stable. For our experiments, we approximate this mechanism by introducing a stability regularizer to be jointly optimized with the loss function. In Sect. 4, we empirically demonstrate the effect of stabilization in terms of variability and performance on various computer vision problems, including generative adversarial networks [35,36] and crowd counting [37,38].

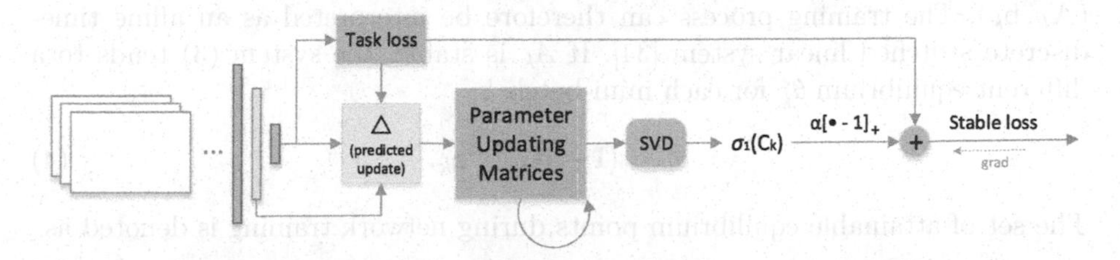

Fig. 2. Stability regularization pipeline. For training, the features of the last linear/convolutional layer or the weights and bias themselves are taken out to construct a time-variant state matrix which represents the parameters updating process at the current step. After left multiplying with the accumulated history matrix, singular value decomposition will be applied to obtain the matrix norm. By scaling with a hyperparameter α and adding to the original loss, the training loss now is a combination of the task-specific loss and the regularization loss.

2 Neural Network Training Dynamics

We introduce the theoretical part by analyzing the update process of mini-batch gradient descent in Sect. 2.1, and by formulating a dynamic system to model the corresponding dynamics. Then we discuss how several aspects of GD can be modeled using linear systems in Sect. 2.2. Finally, we present a model for generic updates, and formulate the joint dynamic system evolving over training iterations. The joint dynamics will be analysed with regards to stability in Sect. 3.

2.1 Mini-Batch Training as a Dynamic System

Let us consider one update iteration for a parameter vector $\theta \subseteq \Theta$ of one layer during training of a network f_Θ using stochastic gradient descent with mini-batches. At iteration k, the update for batch $(\mathbf{x}_k, \mathbf{y}_k)$ is

$$\theta_{k+1} = \theta_k - \eta \frac{\partial}{\partial \theta} \underbrace{\mathcal{L}(f_\Theta, \mathbf{x}_k, \mathbf{y}_k)}_{:=\mathcal{L}^k}, \tag{1}$$

with learning rate η and loss functional $\mathcal{L}$.

We start with the example of a linear last layer with ℓ_2 regression loss,

$$\mathcal{L}(\mathsf{a}(x), y) = (\theta^T \mathsf{a}(x) + b - y)^2, \tag{2}$$

where a is the input activation, and y the label. Since the gradient $\frac{\partial}{\partial\theta}\mathcal{L}$ is a linear function of the parameters θ, the update (1) can be formulated as an affine system $(\mathbf{A}_k, \mathsf{b}_k)$,

$$\theta_{k+1} = \mathbf{A}_k\theta_k + \mathsf{b}_k. \tag{3}$$

For each iteration k, the parameter update is given by a corresponding system $(\mathbf{A}_k, \mathsf{b}_k)$. The training process can therefore be interpreted as an affine time-discrete switched linear system [34]. If A_k is stable, the system (3) tends to a different equilibrium θ_k^e for each mini-batch k,

$$\theta_k^e = (\mathbf{I} - \mathbf{A}_k)^{-1}\mathsf{b}_k. \tag{4}$$

The set of attainable equilibrium points during network training is denoted as,

$$\vartheta^e = \{\theta_k^e \in \mathbb{R}^m : \theta_k^e = (\mathbf{I} - \mathbf{A}_\mu)^{-1}\mathsf{b}_\mu\}, \tag{5}$$

formed by convex combinations of stable linear systems

$$\begin{gathered}\mathbf{A}_\mu = \sum_{i=0\ldots k} \mu_i \mathbf{A}_i, \quad \mathsf{b}_\mu = \sum_{i=0\ldots k} \mu_i \mathsf{b}_i, \\ \mu \in \{\mu \in \mathbb{R}^k : \sum_i \mu_i = 1, \mu_i \geq 0 \forall i\}.\end{gathered} \tag{6}$$

Any solution θ^* of the neural network training is therefore found in $\theta^* \in \vartheta^e$.

However, not only can any number of systems in $\Sigma = \{(\mathbf{A}_i, \mathsf{b}_i)\}$ be unstable, but even switching between stable systems can create instability. This leads to oscillating behavior, especially with high learning rates η. On the other hand, clever switching of systems can stabilize the overall dynamic. In this paper, we are interested in studying the stability of deep learning optimization problems and finding ways for stabilising the network training.

2.2 Dynamics of Gradient Descent Optimizers

We can model the update dynamics of GD-based optimizers with learning rate η using a first order system as follows:

$$\begin{bmatrix}\theta_{k+1} \\ \Delta\theta_{k+1}\end{bmatrix} = \begin{bmatrix}\mathbf{D}_k & -\eta\mathbf{H}_k \\ \mathbf{L}_k & \mathbf{M}_k\end{bmatrix}\begin{bmatrix}\theta_k \\ \Delta\theta_k\end{bmatrix} + \mathbf{B}_k\mathsf{u}_k, \tag{7}$$

with time-variant subsystems $\mathbf{D}_k$, $\mathbf{H}_k$, $\mathbf{L}_k$, and $\mathbf{M}_k$. $\mathbf{D}_k$ performs parameter updates proportional to the parameters θ, and can be used to represent ℓ_2 weight decay for $\mathbf{D}_k = \mathbf{I} - \eta\tau$, with regularization strength τ. The additive update is performed with subsystem $\mathbf{H}_k$, and is typically chosen as $\mathbf{H}_k = \mathbf{I}$, in case the learning rate is the same for all parameters. System $\mathbf{L}_k$ handles linear gradients, such as in the case of (3). The choice of $\mathbf{M}_k$ allows for introduction of gradient momentum, e.g. with $\mathbf{M}_k = \beta\mathbf{I}$. The affine bias terms u_k act on the update through the control matrix $\mathbf{B}_k$.

For many optimizers and loss functions, the linear assumption as in (3) does not apply. In this case, we omit the linear dynamics $\mathbf{L}_k = \mathbf{0}$ and introduce the additive update through the affine term as,

$$\begin{bmatrix} \theta_{k+1} \\ \Delta\theta_{k+1} \end{bmatrix} = \begin{bmatrix} \mathbf{D} & -\eta\mathbf{I} \\ \mathbf{0} & \mathbf{0} \end{bmatrix} \begin{bmatrix} \theta_k \\ \Delta\theta_k \end{bmatrix} + \begin{bmatrix} \mathbf{0} \\ \mathbf{I} \end{bmatrix} \mathbf{O}(\frac{\partial}{\partial\theta}\mathcal{L}^k), \tag{8}$$

where $\mathbf{O}(\frac{\partial}{\partial\theta}\mathcal{L}^k)$ is the update from the optimizer (e.g. Adam [8]). Since $\mathbf{O}$ already includes momentum terms, we set $\mathbf{M} = \mathbf{0}$. To simplify further analysis, we rewrite (8) into a homogeneous form with state $\tilde{\theta}_k$ and state transition matrix $\mathbf{A}_k$ as,

$$\tilde{\theta}_{k+1} = \begin{bmatrix} \theta_{k+1} \\ 1 \end{bmatrix} = \begin{bmatrix} \mathbf{D} & -\eta\mathbf{O}(\frac{\partial}{\partial\theta}\mathcal{L}^k) \\ 0 & I \end{bmatrix} \begin{bmatrix} \theta_k \\ 1 \end{bmatrix} = \mathbf{A}_k\tilde{\theta}_k. \tag{9}$$

The state of this affine switched linear system is therefore directly adapted by the optimizer algorithm. We can write the finite matrix left-multiplication chain up to iteration k as,

$$\begin{aligned} \tilde{\theta}_{k+1} &= \mathbf{A}_k \dots \mathbf{A}_0 x_0 = \prod_{i=0}^{k} \mathbf{A}_i x_0 \\ &= \underbrace{\begin{bmatrix} \mathbf{D}^k & -\eta\sum_{i=1}^{k} \mathbf{D}^{k-i}\mathbf{O}(\frac{\partial}{\partial\theta}\mathcal{L}^i) \\ 0 & I \end{bmatrix}}_{\mathbf{C}_k} \tilde{\theta}_0. \end{aligned} \tag{10}$$

$\mathbf{C}_k$ represents the joint dynamic evolving over iterations.

3 Stable Network Optimization

Motivated by stabilization through switching, in the following, we develop a stability criterion for the system at the next time step in order to stabilize the update dynamic. With the tools of Liapunov functions, we derive a constraint on the joint dynamic $\mathbf{C}_k$, to avoid unstable updates in Sect. 3.1. In practice, we relax this constraint in to ways: by introducing temporary anchors in Sect. 3.3, and by approximation through regularization in Sect. 3.4, to obtain an efficient algorithm.

3.1 Liapunov Stability

To analyse the stability of the system (10), we define a Liapunov energy function,

$$V(\xi) = \xi^T\mathbf{P}\xi - 1, \tag{11}$$

with $\xi = \left[(\theta - \theta_0)^T \; 1\right]^T$ around the initial parameter θ_0. For time-discrete systems, we consider the energy difference between two steps,

$$\Delta V = V(\xi_{k+1}) - V(\xi_k). \tag{12}$$

According to Liapunov's theorem, for $V(0) = 0$, and $\Delta V(x) \leq 0$, the system is stable around the origin [39]. Note that we do no desire asymptomatic stability, since we do not know the solution of the learning problem beforehand. Thus, we merely aim for stable behavior in the vicinity of the initial parameter θ_0. For our dynamic (10) we obtain,

$$\begin{aligned} \Delta V &= V(\mathbf{C}_k\xi) - V(\xi) \\ &= \xi^T(\mathbf{C}_k^T\mathbf{P}\mathbf{C}_k - \mathbf{P})\xi. \end{aligned} \tag{13}$$

Therefore, we achieve $\Delta V \leq 0$ for,

$$\mathbf{C}_k^T\mathbf{P}\mathbf{C}_k - \mathbf{P} = -\mathbf{Q}. \tag{14}$$

with positive definite matrices $\mathbf{P}, \mathbf{Q} \succeq 0$. Choosing $\mathbf{P} = \mathbf{I}_{3x3}$, we obtain $\mathbf{Q} \succeq 0$, if

$$\begin{aligned} &\mathbf{I} - \mathbf{C}_k^T\mathbf{C}_k \succeq \mathbf{0} \\ \Rightarrow &\max_i \lambda_i(\mathbf{C}_k^T\mathbf{C}_k) \leq \min_i \lambda_i(\mathbf{I}) \\ \Rightarrow &\sigma_1(\mathbf{C}_k) \leq 1, \end{aligned} \tag{15}$$

where $\sigma_1(\mathbf{C}_k)$ is the largest singular value of $\mathbf{C}_k$.

3.2 Stable Network Training

To facilitate stable training, the goal is to find network parameters Θ, that minimize the loss $\mathcal{L}$ under the stability constraints given by (15). At iteration k, we therefore wish to control the network parameters Θ_k by solving,

$$\begin{aligned} \Theta_k = \arg\min_{\Theta} \quad & \mathcal{L}(f_\Theta, \mathbf{x}_k, \mathbf{y}_k) \\ s.t. \quad & \mathbf{C}_k(\Theta) = \mathbf{A}_k(\Theta)\mathbf{C}_{k-1}, \\ & \sigma_1(\mathbf{C}_k) \leq 1. \end{aligned} \tag{16}$$

This can be seen as a model predictive controller with a single step time horizon. Here we predict how θ_k would be updated according to $\mathcal{L}$, as represented by $\mathbf{A}_k(\Theta_k)$ (9). The actual update of $\theta_k \subseteq \Theta_k$ however is given by the solution Θ_k of (16), taking the constraints (15) into consideration.

In practice, we approximate the constrained problem of (16) by introducing a regularizer to the original loss as,

$$\mathcal{L}_{stable} = \mathcal{L}^k + \alpha\left[\sigma_1(\mathbf{C}_k) - 1\right]_+ \tag{17}$$

where $\mathcal{L}^k$ is the task-specific mini-batch loss, $[\bullet]_+$ is a rectifier, and $\sigma_1(\mathbf{C}_k)$ is the stability-based regularizer with strength α.

Algorithm 1: Stability-regularized training process

```
#theta: layer weights
#eta: learning rate
#alpha: regularizer strength
def train_epoch(batch_data, theta, eta, alpha):
  # reset history (optional)
  C = eye(len(theta)+1)

  for (x,y) in batch_data:

    #predict update O(∂/∂θ ℒ), e.g. with GD:
    update = -eta * compute_grad(theta, x, y)

    #construct joint system Eq.(9) and Eq.(10)
    A_tilde = affine_system(update)
    C_tilde = A.matmul(C_tilde)

    #compute total loss Eq.(17)
    loss = ℒ(x,y)+alpha*relu(svd(C_tilde)[0]-1)

    #backprop and parameter update
    loss.backwards()
    update_final = optimizer.step()

    #update history
    A = affine_system(update_final)
    C = A.matmul(C_tilde)
```

3.3 Anchoring

When encouraging Liapunov stability according to (17), we effectively search only in the vicinity of the initial solution. This is too restrictive in some cases. We therefore introduce a series of anchors, where the history gets re-initialized as $\mathbf{C}_k \leftarrow \mathbf{I}$. In our experiments, we investigate epoch-wise resetting, trading of stability and exploration. Figure 1 shows the behavior of stabilization with anchors, in a 2D toy example. In every iteration, one of 4 noisy attractors is randomly chosen to compute the gradient (mimicking mini-batch noise), and we desire to reach the equilibrium point at the origin. Stabilization (15) can mitigate the oscillations, while moving the anchor enables asymptotic convergence.

3.4 Training Algorithm

To optimize (17) using SGD, we first perform a forward pass to construct the current $\mathbf{A}_k(\Theta)$ and update the history matrix $\mathbf{C}_{k-1}$. Note that $\mathbf{C}_{k-1}$ contains updates from preceding mini-batches and is treated as a constant, while the updated $\mathbf{C}_k$ involves taking the derivative of $\mathcal{L}$ w.r.t. the parameters Θ. Since the gradients of (17) are a function of the update $\mathbf{O}$, we express the update analytically as a function of θ, if possible. For complex optimizers and loss functions, we assuming a locally constant gradient, and approximate the affine part as $\mathbf{O}(\theta) \approx \theta + \mathbf{O}(\frac{\partial}{\partial\theta}\mathcal{L}) - \theta_{k-1}$. We summarize the training process in Algorithm 1. An illustration of the loss computation is given in Fig. 2.

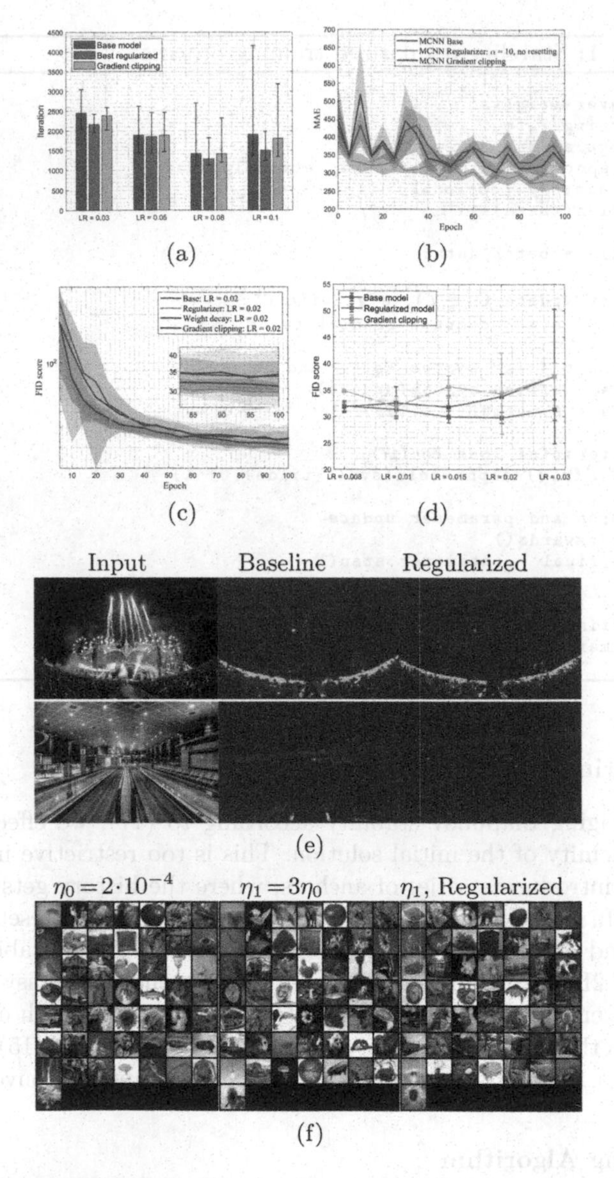

(a) (b) (c) (d) (e) (f)

Fig. 3. Effect of Stabilizing Regularizer. (a) Average iterations needed to reach 90% accuracy on MNIST, with maximal and minimal iterations. Regularization reduces fluctuation across runs and can speed up convergence. (b) MAE (mean absolute error) for MCNN crowd counting network. Better and more stable results are obtained by regularizing. (c) SNGAN training with different regularizing techniques. (d) FID statistics of SNGAN achieved at different learning rates. Regularization reduces variance and can achieve better results. (e) Predicted crowd density maps from MCNN. Stabilization yields more distinct results for crowds (1st row) and sparser dots for negative samples (2nd row). (f) Generated images from fixed noise by BigGAN trained on CIFAR-100. The results of the base (2nd column) and regularized (3rd column) models are obtained with higher learning rates and half of the iterations. See Table 2 for more details. Besides, our method compares favourably also in terms of the effective training time.

4 Experiments

We now are going to empirically investigate our theoretical considerations. In particular, we evaluate Algorithm 1 on different tasks, network architectures, and datasets. The following presentation of our experiments is focused on regression-type problems, since we do not observe any significant beneficial nor detrimental effects of stability-aware training in the case of classification. Possible reasons are discussed in Sect. 5. Additionally, we experiment on the notoriously unstable training of GANs to show the potential of our approach in such highly dynamic environments. Our guiding hypothesis is that stability regularization can improve performance, and achieves higher consistency across multiple independent trials with random initialization. We start by providing an overview of the datasets and the tasks investigated for stability-aware training.

MNIST Digit Classification. We start with the popular MNIST [40] dataset which is designed for 10 digit classification. We convert the task into a scalar regression problem, and add a linear layer with one output unit to the end of the LeNet-5 [40] network. For evaluation, we bin the continuous output back into 10 classes. In particular, we are interested in the convergence with and without the regularizer, as measured by crossing the 90% accuracy threshold.

NWPU Crowd Counting. Next, we evaluate our regularizer on crowd counting. The task involves counting the number of people in an image of large crowds. NWPU-Crowd dataset [41] contains 5,109 images with various illumination and density range. With 351 negative samples and large appearance variations within the data, it is a challenging dataset. State-of-the-art methods typically approach crowd counting through regression of a pixel-wise density map [37,38,42]. The final count is obtained by summing over the density map.

CIFAR10/100 Image Generation. Furthermore, we evaluate stability regularization in the GAN setup, where gradients are naturally unstable, making it hard to train. This is due to the dynamics of the two-player game between the discriminator network (judging an image to be real or fake), and a generator, trying to fool the discriminator by generating realistic images. As training of a GAN aims to find the balance (i.e. the Nash equilibrium), unstable gradients will cause problems or even failure in training the model. We focus on CIFAR-10 [43] first to demonstrate the ability of our regularizer and compare to several other regularization methods. Test on CIFAR-100 [43] is to study the stability behavior further since there are fewer samples per class, which implicitly means the model suffers mode collapse easier.

Implementation Details. In all experiments, we stabilize the last layer only. As optimizers, we use SGD and Adam. As discussed in Sect. 3.4 and Algorithm 1, the generic training procedure includes two backpropagation steps. Because of the stateful nature of Adam, we copy weights and optimizer state for the last layer. Then the forward pass is performed, and we back-propagate to obtain the dynamics for the task-specific loss $\mathcal{L}^k$. Next, A_k is constructed by subtracting the old parameters from the updated ones. Finally, we update the original weights based on the gradients with respect to the complete loss $\mathcal{L}_{stable}$.

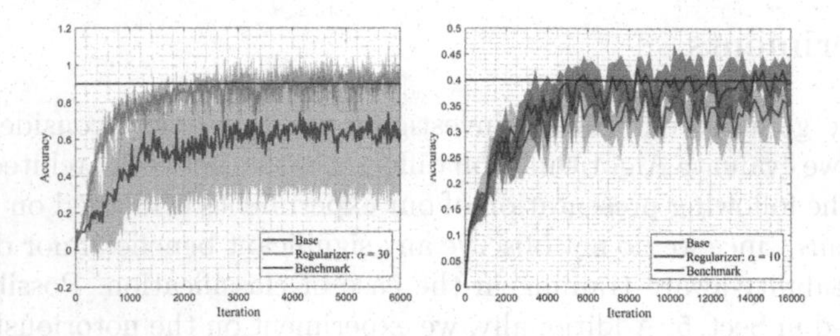

Fig. 4. High Learning Rate behavior. Accuracy curve across iterations for MNIST and CIFAR-10 when the learning rate is higher than the optimum setting. Top is the result for MNIST (LR = 0.12) and bottom is the result for CIFAR-10 (LR = 0.2). In both cases, the regularize improves convergence and reduces variance across runs.

4.1 Results

We summarize our main results for all tasks in Fig. 3. To indicate the spread in some figures, we either plot the min/max bar, or one standard deviation as a shaded area. As a general trend, we observe that across tasks, the variance in the evaluation metric is significantly reduced when training with stabilization. In many cases, we can further see an accuracy improvement on average. The MNIST experiments in (a) show that the regularizer enables a significantly quicker convergence on average. In figure (b), we observe that a lower MAE with less variance is achieved in the case of crowd counting. The bottom row (c, d) summarizes our GAN experiments with the SNGAN [44] architecture, also exhibiting reduced variance across runs, as well as a significant improvement in FID across a range of learning rates. In the following, we present each of the experiments in more detail.

MNIST. We train the networks with different values for the learning rate η and the hyper-parameter α, that is, $\eta \in \{0.03, 0.05, 0.08, 0.1\}$ and $\alpha \in \{1, 3, 10, 30, 100\}$. Figure 3 (a) compares the iteration numbers needed for the base and the best regularized model to reach the benchmark for the first time under three different learning rate settings. The bar indicates the average speed over ten independent tests. The two models are comparable when the learning rate is below 0.1, and the best learning rate for the base model is around 0.08. As the learning rate increases, the base model takes more time to achieve the same accuracy, and the training process is more unstable, as indicated by the min-max distance in the figure. In contrast, our regularized model behaves better when using higher learning rates, resulting in both less average iterations and smaller variation range. From this experiment we can conclude that the regularization is orthogonal to a reduction in learning rate.

To further demonstrate the effectiveness of stabilization in the high learning rate regime, we experimented on both MNIST ($\eta = 0.12$) and CIFAR-10 ($\eta = 0.2$). The accuracy curves for the base and regularized models with error regions defined by one standard deviation in Fig. 4. The optimal regularization strength α are found to be $\alpha = 30$ for MNIST and $\alpha = 10$ for CIFAR-10. As it can be seen, a high learning rate leads to performance degradation and model instability, and the base model can't reach the benchmark (0.9 and 0.4, respectively). However, if we use the stability constraint within the training, a stable behavior over multiple runs is maintained.

Crowd Counting. Next we evaluate our regularizer in the natural regression problem of crowd counting. After predicting the density map for an image, the pixel-wise difference between the density map and group-truth map is measured with ℓ_2 loss. We follow MCNN [41] in setting up the original training hyper-parameters, including the optimizer (Adam) and the learning rate ($\eta = 1 \cdot 10^{-4}$). The results for MCNN architecture are posted in Fig. 3 (b). While the Mean Absolute Error (MAE) of the base model fluctuates around 350 and finally goes down to 320, the regularized model leads to a better counting performance, as well as reduced variance. It reaches MAE 265 on average within 100 epochs with fewer fluctuations and has a minimum MAE of around 236 over ten experiments. We also compare it with gradient clipping, whose MAE curve is quite similar to the base model, with larger variance at some epochs. Our proposed regularizer is still the best among them, as can be seen in Table 1.

The evaluation results for the more complex CSRNet [38], trained following the original protocol, can be seen in Fig. 5. Both models behave similarly in terms of average MAE, but the regularized network shows a smaller error region over multiple runs. After epoch 60, the MAE of the base model largely fluctuates, while the regularized model generally decreases with few fluctuations. A similar behavior is observed on two other metrics: Mean Squared Error (MSE) and mean Normalized Absolute Error (NAE) in Fig. 5. This may indicate that our approach is indeed helpful for the stability of the training process in the sense of stochastic dataset sampling and random initialization.

Table 1. Crowd Counting using MCNN. Accuracy (in MAE) after 10 independent runs, for 100 epochs each.

Method	Mean MAE	Best MAE
Base	328.386	284.518
Grad-clipping	301.421	281.763
Ours	**265.477**	**236.446**

Generative Adversarial Network. We now investigate our regularizer applied to the naturally unstable GAN training. We choose SNGAN and BigGAN [45] for our experiments. We use hinge loss in both models which is the default loss function in BigGAN. We test SNGAN with SGD optimizer and BigGAN with Adam. At every epoch, we reset the history matrix, as explained in Sect. 3.3. For evaluation, we mainly compare the FID metric (less is better), which measures the diversity and quality of the generated images [46].

Motivated by the previous results, we start with a learning rate slightly higher than that of the best base model, and show the results for SNGAN in Fig. 3 (c). Around 100 epochs, there is a four-unit FID gap between them, and the blue shaded area (base model) is much larger over the whole process. We also compare to the commonly used regularization methods weight decay and gradient clipping, to further demonstrate that our proposed regularizer is not

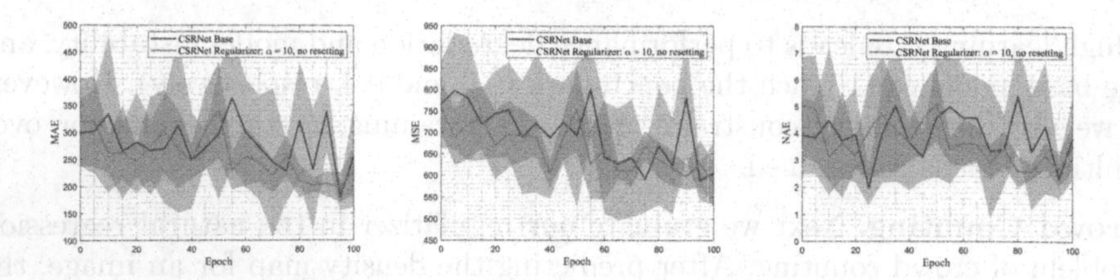

Fig. 5. Crowd Counting using CSRNet. From left to right: measured in MAE (mean absolute error), MSE (mean squared error), and NAE (MAE normalized by ground truth). Regularization significantly reduces the variance across runs.

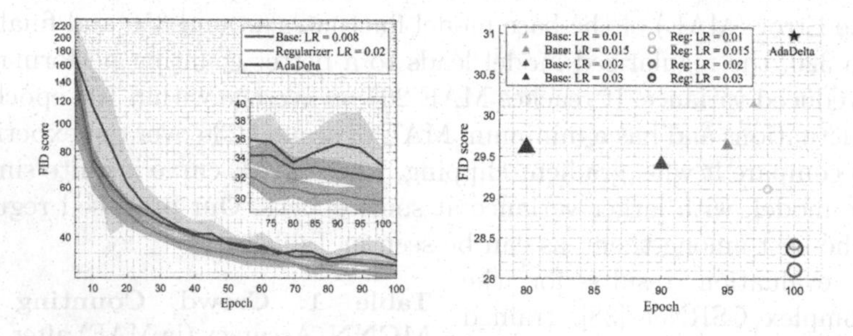

Fig. 6. Regularization Impact on FID using SNGAN. *Left:* Comparison between the best base model, our best regularized model and model trained by AdaDelta optimizer. *Right:* Lowest FID achieved within 100 epochs with and without regularizer.

equivalent to them in the GAN setup. To be similar to our stabilizing regularizer, we constrain the maximal matrix norm equal to one in gradient clipping. For the coefficient of weight decay, we alter among 0.1, 0.01, and 0.001. Since there is no significant difference in the results, we finally choose 0.01 for the comparison. As can be seen in the same figure, gradient clipping fails to improve the performance of SNGAN together with an even larger standard deviation. Weight decay is better than the other two models but still not as good as our regularized one. Note that even though we observe a large error region at epoch 10 or 20, it doesn't mean we can actually reach FID 30 here. This is merely due to the distortion of high-FID outliers on the symmetric standard deviation measure.

We also test our regularizer at different learning rates, and the quantitative results are reported in Fig. 3 (d). Similar to the conclusion from the regression task, there is a negligible difference if we set the learning rate as the best for the base model. As the learning rate increases, the FID increases with a significantly larger variance, while the regularizer can maintain stable results, and even improve FID. Again, we see that the regularizer does not have the same effect on training as learning rate tuning, and offers new and better operating points.

Figure 6 left compares the best-regularized model with a learning-rate-setting-free optimizer, AdaDelta [27]. While the regularized model has an average FID below 30, the other two models converge around or over 32. We can conclude from here that a moderately higher learning rate with a regularizer will benefit the training for SNGAN, and it is necessary to find such an appropriate learning rate. To summarize the results for SNGAN, we provide the lowest FID every model can reach after 100 epochs in Fig. 6 right. The best FID appears earlier if we increase the learning rate for the base model, indicating that the model is usually under-fitting. Adding the regularizer allows for using even higher learning rates, and the resulting FID exceeds the base model.

Table 2. BigGAN Training. FID for CIFAR-100 when the learning rate is the default setting (top), and increased by a factor of 3 (bottom). With the higher learning rate, the regularized model can reach the lowest FID at 50k iterations.

$\eta_0 = 2 \cdot 10^{-4}$

Iters	55k	60k	65k	70k	75k	80k	85k
Base	9.641	9.363	9.078	**8.850**	8.900	8.972	8.865
Reg	10.097	9.896	9.710	9.553	9.479	9.5470	9.508

$\eta_1 = 6 \cdot 10^{-4}$

Iters	40k	45k	50k	55k	60k	65k	70k
Base	10.280	9.516	9.629	10.708	14.114	53.463	91.141
Reg	9.396	9.062	**8.605**	9.048	10.802	27.608	82.379

We now test with the BigGAN architecture on the CIFAR-100 dataset. Table 2 shows the FID for two learning rates: default $\eta_0 = 2 \cdot 10^{-4}$, and three times higher at $\eta_1 = 6 \cdot 10^{-4}$. The behavior for default η_0 similar to previous tasks, which reveals that the regularizer malfunctions in this case, yielding a slightly higher FID than the base model. Increasing to η_1, the training process becomes fragile, and both models collapse after 65k iterations. Despite this, our regularized model can still reach an FID of 8.605, while the base model reaches only 9.516. Compared to the default setting, the lowest FID is achieved already at 50k iterations with the regularizer at η_1, as opposed to 70k iterations with the base model at η_0. This observation indicates that the proposed regularizer can help with the stability of the network in some cases.

5 Discussion

While our formulation in Sect. 2 makes no assumption on the loss functions, we do not observe any significant impact of our regularizer on classification tasks such as CIFAR-100. We conjecture that regression, GAN training, and classification (using cross-entropy) losses introduce different types of dynamics, and only a subset can be stabilized with the proposed method.

One possible explanation for this phenomenon could lie in the zero crossing of the loss gradient around the optimum: for ℓ_1 and ℓ_2 loss functions, too big an update step close to the correct answer causes the gradient to invert and symmetrically regain its magnitude, as shown in Fig. 7. This behavior may be a source of oscillation, that our method can compensate well. Such symmetry is not present in the one-sided cross-entropy loss, where the gradient (magnitude) monotonously decreases until the predicted score saturates. Even a big update step never causes a gradient to invert. In the case of GANs, even though technically a real/fake classification task, this symmetry might get introduced by the gradient reversal of the adversarial training.

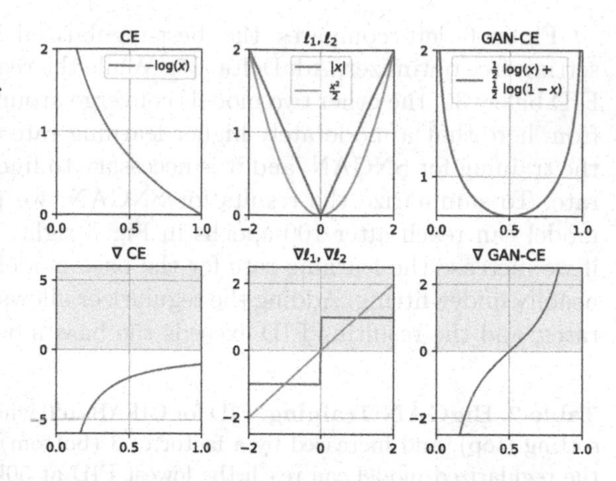

Fig. 7. Loss profiles. From left to right: Cross-entropy (CE), ℓ_1, ℓ_2, and GAN discriminator training (CE with balanced adversarial samples) and their corresponding gradients. Updates with big steps cause gradient inversion and/or an increase in gradient magnitude for ℓ_1, ℓ_2 and GAN losses. This is not the case for CE, where the gradient decreases without zero crossing.

6 Conclusion

In this work, we formulated the training of neural networks as a dynamic system and analyzed the stability behavior from the control theory point of view. Based on this theory, we develop a regularizer to stabilize the training process. The regularization strategy is based on the accumulated gradient information of the task-specific loss. Experimental results on different tasks, network architectures, and datasets show that our method stabilizes the training process across multiple independent runs and improves the task performance in conjunction with an appropriate slightly higher learning rate. That being said, we further found that the effect of the stabilization is orthogonal to a reduction of the learning rate. In many cases, we observe higher consistency and better results can be achieved with the appropriate parameters. For future work, it would be interesting to investigate why the stabilization has little effect on classification problems. Furthermore, we aim to analyse the effect of stabilization on other layers of the network, and see if it is beneficial to stabilize more than one layer.

References

1. Cauchy, A.: Methode generale pour la resolution des systemes d'equations simultanees. C.R. Acad. Sci. Paris **25**, 536–538 (1847)
2. Nesterov, Y.: A method for unconstrained convex minimization problem with the rate of convergence $o(1/k^2)$ (1983)
3. Sutton, R.S.: Two problems with backpropagation and other steepest-descent learning procedures for networks. In: Proceedings of the Eighth Annual Conference of the Cognitive Science Society. Erlbaum, Hillsdale (1986)
4. Duchi, J.C., Hazan, E., Singer, Y.: Adaptive subgradient methods for online learning and stochastic optimization. In: J. Mach. Learn. Res. (2011)
5. McMahan, H.B., Streeter, M.J.: Delay-tolerant algorithms for asynchronous distributed online learning. In: NIPS (2014)
6. Recht, B., Ré, C., Wright, S.J., Niu, F.: Hogwild: a lock-free approach to parallelizing stochastic gradient descent. In: NIPS (2011)
7. Zhang, S., Choromańska, A., LeCun, Y.: Deep learning with elastic averaging SGD. In: NIPS (2015)
8. Kingma, D.P., Ba, J.: Adam: a method for stochastic optimization. CoRR abs/1412.6980 (2015)
9. Dozat, T.: Incorporating nesterov momentum into Adam (2016)
10. Linnainmaa, S.: Taylor expansion of the accumulated rounding error. BIT Numer. Math. **16**, 146–160 (1976)
11. Griewank, A.: Who invented the reverse mode of differentiation (2012)
12. LeCun, Y., Bottou, L., Orr, G., Müller, K.: Efficient backprop. In: Neural Networks: Tricks of the Trade (2012)
13. Boyd, S.P., Vandenberghe, L.: Convex optimization. IEEE Trans. Autom. Control **51**, 1859–1859 (2006)
14. Helmke, U., Moore, J.: Optimization and dynamical systems. Proc. IEEE **84**, 907- (1996)
15. Kendall, A., Gal, Y., Cipolla, R.: Multi-task learning using uncertainty to weigh losses for scene geometry and semantics. In: 2018 IEEE/CVF Conference on Computer Vision and Pattern Recognition, pp. 7482–7491 (2018)
16. Ranjan, R., Patel, V., Chellappa, R.: Hyperface: a deep multi-task learning framework for face detection, landmark localization, pose estimation, and gender recognition. IEEE Trans. Pattern Anal. Mach. Intell. **41**, 121–135 (2019)
17. Popović, N., Paudel, D., Probst, T., Sun, G., Gool, L.: Compositetasking: understanding images by spatial composition of tasks. arXiv abs/2012.09030 (2020)
18. Zhong, Z., Zheng, L., Kang, G., Li, S., Yang, Y.: Random erasing data augmentation. arXiv abs/1708.04896 (2020)
19. Zhang, H., Cissé, M., Dauphin, Y., Lopez-Paz, D.: Mixup: beyond empirical risk minimization. arXiv abs/1710.09412 (2018)
20. Cubuk, E.D., Zoph, B., Shlens, J., Le, Q.V.: Randaugment: practical data augmentation with no separate search. arXiv abs/1909.13719 (2019)
21. Vahdat, A., Kautz, J.: Nvae: a deep hierarchical variational autoencoder. arXiv abs/2007.03898 (2020)
22. Kingma, D.P., Welling, M.: Auto-encoding variational Bayes. CoRR abs/1312.6114 (2014)
23. Rezende, D.J., Mohamed, S., Wierstra, D.: Stochastic backpropagation and approximate inference in deep generative models. In: ICML (2014)

24. Pascanu, R., Mikolov, T., Bengio, Y.: Understanding the exploding gradient problem. arXiv abs/1211.5063 (2012)
25. Yu, T., Kumar, S., Gupta, A., Levine, S., Hausman, K., Finn, C.: Gradient surgery for multi-task learning. arXiv abs/2001.06782 (2020)
26. Vuckovic, J.: Kalman gradient descent: adaptive variance reduction in stochastic optimization. arXiv abs/1810.12273 (2018)
27. Zeiler, M.D.: Adadelta: an adaptive learning rate method. arXiv abs/1212.5701 (2012)
28. Man, Z., Wu, H.R., Liu, S.X.F., Yu, X.: A new adaptive backpropagation algorithm based on Lyapunov stability theory for neural networks. IEEE Trans. Neural Netw. **17**, 1580–1591 (2006)
29. Kang, Q., Song, Y., Ding, Q., Tay, W.P.: Stable neural ode with Lyapunov-stable equilibrium points for defending against adversarial attacks. arXiv abs/2110.12976 (2021)
30. Ahmadi, A.A., Parrilo, P.A.: Joint spectral radius of rank one matrices and the maximum cycle mean problem. In: 2012 IEEE 51st IEEE Conference on Decision and Control (CDC), pp. 731–733. IEEE (2012)
31. Ahmadi, A.A., Jungers, R.M., Parrilo, P.A., Roozbehani, M.: Joint spectral radius and path-complete graph Lyapunov functions. SIAM J. Control. Optim. **52**, 687–717 (2014)
32. Altschuler, J.M., Parrilo, P.A.: Lyapunov exponent of rank-one matrices: ergodic formula and inapproximability of the optimal distribution. SIAM J. Control. Optim. **58**, 510–528 (2020)
33. Daubechies, I., Lagarias, J.C.: Two-scale difference equations ii. Local regularity, infinite products of matrices and fractals. SIAM J. Math. Anal. **23**, pp. 1031–1079 (1992)
34. Deaecto, G.S., Egidio, L.N.: Practical stability of discrete-time switched affine systems. In: 2016 European Control Conference (ECC), pp. 2048–2053 (2016)
35. Goodfellow, I., et al.: Generative adversarial nets. In: Advances in Neural Information Processing Systems, pp. 2672–2680 (2014)
36. Shahbazi, M., Huang, Z., Paudel, D.P., Chhatkuli, A., Van Gool, L.: Efficient conditional GAN transfer with knowledge propagation across classes. In: 2021 IEEE Conference on Computer Vision and Pattern Recognition. CVPR 2021 (2021)
37. Zhang, Y., Zhou, D., Chen, S., Gao, S., Ma, Y.: Single-image crowd counting via multi-column convolutional neural network. In: Proceedings of the IEEE Conference on Computer Vision and Pattern Recognition, pp. 589–597 (2016)
38. Li, Y., Zhang, X., Chen, D.: CSRnet: dilated convolutional neural networks for understanding the highly congested scenes. In: Proceedings of the IEEE Conference on Computer Vision and Pattern Recognition, pp. 1091–1100 (2018)
39. Farina, L., Rinaldi, S.: Positive Linear Systems: Theory and Applications (2000)
40. LeCun, Y., Bottou, L., Bengio, Y., Haffner, P.: Gradient-based learning applied to document recognition. Proc. IEEE **86**, 2278–2324 (1998)
41. Wang, Q., Gao, J., Lin, W., Li, X.: NWPU-crowd: a large-scale benchmark for crowd counting and localization. IEEE Trans. Pattern Anal. Mach. Intell. **43**, 2141–2149 (2020)
42. Sun, G., Liu, Y., Probst, T., Paudel, D., Popović, N., Gool, L.: Boosting crowd counting with transformers. arXiv abs/2105.10926 (2021)
43. Krizhevsky, A., Hinton, G., et al.: Learning multiple layers of features from tiny images (2009)
44. Miyato, T., Kataoka, T., Koyama, M., Yoshida, Y.: Spectral normalization for generative adversarial networks. arXiv preprint arXiv:1802.05957 (2018)

45. Brock, A., Donahue, J., Simonyan, K.: Large scale GAN training for high fidelity natural image synthesis. arXiv preprint arXiv:1809.11096 (2018)
46. Heusel, M., Ramsauer, H., Unterthiner, T., Nessler, B., Hochreiter, S.: GANs trained by a two time-scale update rule converge to a local Nash equilibrium. In: NIPS (2017)

Neighborhood Region Smoothing Regularization for Finding Flat Minima in Deep Neural Networks

Yang Zhao(✉) and Hao Zhang

Department of Electronic Engineering, Tsinghua University, Beijing 100081, China
zhao-yan18@mails.tsinghua.edu.cn, haozhang@tsinghua.edu.cn

Abstract. Due to diverse architectures in deep neural networks (DNNs) with severe overparameterization, regularization techniques are critical for finding optimal solutions in the huge hypothesis space. In this paper, we propose an effective regularization technique, called Neighborhood Region Smoothing (NRS). NRS leverages the finding that models would benefit from converging to flat minima, and tries to regularize the neighborhood region in weight space to yield approximate outputs. Specifically, gap between outputs of models in the neighborhood region is gauged by a defined metric based on Kullback-Leibler divergence. This metric could provide insights in accordance with the minimum description length principle on interpreting flat minima. By minimizing both this divergence and empirical loss, NRS could explicitly drive the optimizer towards converging to flat minima, and meanwhile could be compatible with other common regularizations. We confirm the effectiveness of NRS by performing image classification tasks across a wide range of model architectures on commonly-used datasets such as CIFAR and ImageNet, where generalization ability could be universally improved. Also, we empirically show that the minima found by NRS would have relatively smaller Hessian eigenvalues compared to the conventional method, which is considered as the evidence of flat minima (Code is available at https://github.com/zhaoyang-0204/nrs).

Keywords: Deep learning · Optimization · Flat minima

1 Introduction

Driven by the rapid development of computation hardwares, the scale of today's deep neural networks (DNNs) is increasing explosively, where the amount of parameters has significantly exceeded the sample size by even thousands of times [7,10,11]. These heavily overparameterized DNNs would induce huge hypothesis weight spaces. On the one hand, this provides the power to fit extremely complex or even arbitrary functions [13], on the other hand, it becomes much more challenging to seek optimal minima while resisting overfitting during training [26]. Generally, minimizing only the empirical training loss that characterizes

L. Wang et al. (Eds.): ACCV 2022, LNCS 13841, pp. 652–665, 2023.
https://doi.org/10.1007/978-3-031-26319-4_39

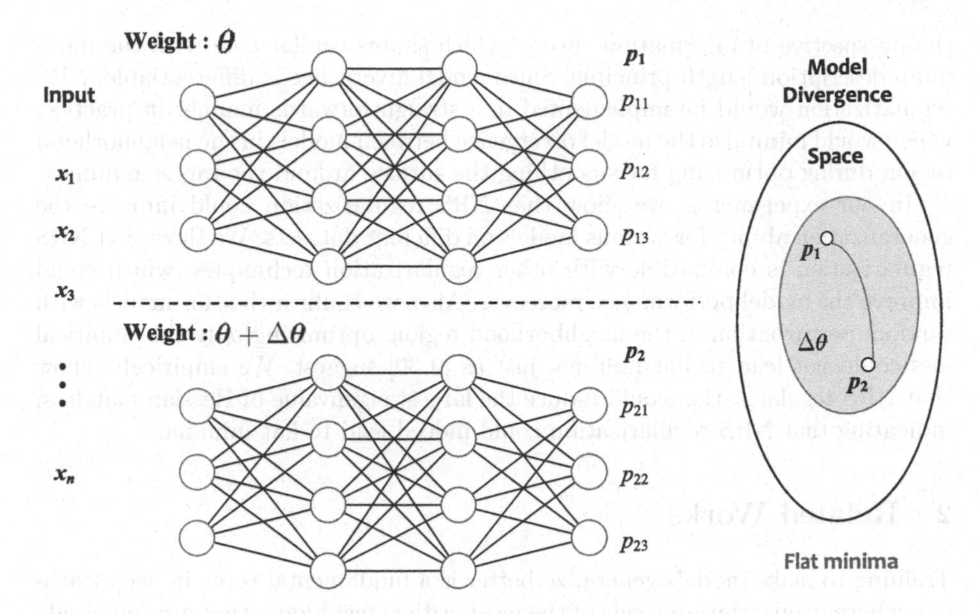

Fig. 1. Description of model divergence and flat minima. Given the same input, model divergence gauges the distance between models in the output space, and flat minima implies that models between these in the neighborhood region $\delta\boldsymbol{\theta}$ and the reference model $\boldsymbol{\theta}$ would be relatively lower compared to sharp minima. NRS would minimize the model divergence for models in this neighborhood to explicitly find flat minima during training.

the difference between the true labels and the predicted labels (such as categorical cross-entropy loss) could not provide sufficient guarantee on acquiring minima with satisfactory model generalization [20]. Regularization techniques are in higher demand than ever for guiding the optimizers towards finding minima with better model generalization.

Researchers find that minima of well-generalized models are more likely to be located at the landscape where the geometry of its neighborhood region is flat [12,16]. In other words, models would benefit from biasing the convergence towards flat minima during training. A common simple idea is to minimize the empirical loss of models with random perturbations. Unfortunately, [9,30] empirically suggest that optimization only the empirical loss of these random perturbed models would not help to flatten the loss landscape (Fig. 1).

In this paper, we show that by adding proper regularizations, such simple random perturbation could also lead to flat minima, where we propose a simple and effective regularization technique, called neighborhood region smoothing (NRS). NRS could regularize the models with random perturbation in the neighborhood region to yield the same outputs instead of their loss values. To accomplish such regularization, we firstly define the model divergence, a metric that gauges the divergence between models in the same weight space. We demonstrate that model divergence could provide interpretation of flat minima from the perspective of information theory, which shares similar core with the minimum description length principle. Since model divergence is differentiable, NRS regularization would be implemented in a straightforwards manner in practice, which would minimize the model divergence between models in the neighborhood region during optimizing for smoothing the surface around the target minima.

In our experiments, we show that NRS regularization could improve the generalization ability for various models on different datasets. We show that NRS regularization is compatible with other regularization techniques, which could improve the model performance moreover. Also, we confirm that for models with random perturbation in the neighborhood region, optimizing only the empirical loss could not lead to flat minima, just as [9,30] suggest. We empirically show that NRS regularization could reduce the largest eigenvalue of Hessian matrices, indicating that NRS regularization could indeed lead to flat minima.

2 Related Works

Training to make models generalize better is a fundamental topic in deep learning, where regularization is one of the most critical techniques that are commonly used during training.

Regularization is actually a rather broad concept involving techniques that may be beneficial to the training process in various ways. Generally, common regularization techniques would affect the model training from three aspects. Firstly, some regularization expect to reduce the searching hypothesis space, where the most conventional method is weight decay [17,19]. Secondly, some others try to complicate the target task so that the models could learn more "sufficiently". Typical methods include enlarging the input space such as cutout [5] and auto-augmentation [4], and interfering models during training like dropout [23], spatialdropout [24] regularization and stochastic depth regularization [14]. Thirdly, others implement normalizations, which includes batch normalization [15], layer normalization [1] and so on.

On the other hand, for a better understanding of generalization, researchers are also interested in studying the underlying factors that associate with the generalization ability of models. Several characteristics are demonstrated having impact on generalization including the flatness of minima, margins of classifiers [2,22], sparsity [18] and degree of feature extraction [27]. In particular, regarding the flatness of minima, despite still lack of theoretical justification and strict definition [20], empirical evidences have been found for such phenomenon. For example, [16] demonstrate that the reason large-batch training would lead to worse generalization than small-batch training is because large-batch training tends to stall at sharp minima. Moreover, by solving a minimax problem, [8,9,30] show that training could benefit from optimizing towards flat minima. Also, [28,29] explicitly regularize the gradient norm to bias training for the convergence of flat minima. All of these works have shown that model performance can be improved by adopting proper techniques that encourages flat minima.

3 Neighborhood Region Smoothing (NRS) Regularization

3.1 Basic Background

Consider inputs $\boldsymbol{x} \in \mathcal{X}$ and labels $\boldsymbol{y} \in \mathcal{Y}$ which conform the distribution $\mathscr{D}$ and a neural network model f parameterized by parameters $\boldsymbol{\theta}$ in weight space $\boldsymbol{\Theta}$ which maps the given input space $\mathcal{X}$ to the corresponding output space $\hat{\mathcal{Y}}$,

$$f(\cdot;\boldsymbol{\theta}) : \mathcal{X} \rightarrow \hat{\mathcal{Y}} \tag{1}$$

For classification tasks, $\hat{\boldsymbol{y}} \in \hat{\mathcal{Y}}$ could be the vector of the predicted probability distribution of each class.

Theoretically, given a loss function $L(\cdot)$, we expect to minimize the expected loss $L_e(\boldsymbol{\theta}) = \mathbb{E}_{\boldsymbol{x},\boldsymbol{y}\sim\mathscr{D}}[l(\hat{\boldsymbol{y}}, \boldsymbol{y}, \boldsymbol{\theta})]$. However, it is untractable in practice. We instead seek to acquire the model via minimizing the empirical loss $L_{\mathcal{S}}(\boldsymbol{\theta}) = \frac{1}{N}\sum_{i=1}^{N} l(\hat{\boldsymbol{y}}_i, \boldsymbol{y}_i, \boldsymbol{\theta})$, where the training set $\mathcal{S} = \{(\boldsymbol{x}_i, \boldsymbol{y}_i)\}_{i=0}^{N}$ is assumed to be drawn independently and identically from distribution $\mathscr{D}$. Gap between the expected loss and the empirical loss would directly lead to generalization errors of models.

For over-parametrized models, when minimizing the empirical loss on training set, the huge hypothesis weight space would be filled with numerous minima. These minima may have approximate empirical training loss but meanwhile provide diverse generalization ability. How to discriminate which minima are favorable to the model generalization is critical for getting better training performance.

In particular, minima of models with better generalization are supposed to locate at flatter surfaces, and these minima are often called flat minima. Although there may be different definitions for describing the flat minima [6,16], yet according to [12], the core interpretation behind conveys the similar idea that "a flat minimum is a large connected region in weight space where the error remains approximately constant". In other words, we expect that models parametrized by parameters in the neighborhood area of $\boldsymbol{\theta}$ could yield approximate outputs at flat minima.

3.2 Model Divergence

First, we would like to clarify a basic concept, the equivalence of two models,

Definition 1. *Two models $f(\cdot;\boldsymbol{\theta})$ and $f(\cdot;\boldsymbol{\theta}')$ are called* ***observationally equivalent*** *if $f(\boldsymbol{x};\boldsymbol{\theta}) = f(\boldsymbol{x};\boldsymbol{\theta}')$ for $\forall\ \boldsymbol{x} \in \mathcal{X}$.*

If two models are observationally equivalent, then they will definitely have the same outputs given any input in the input space, and meanwhile will lead to the same loss. But if the losses of two models are the same, it could not sufficiently ensure that they are observationally equivalent. In other words, loss is not a sufficient condition for the models to be observationally equivalent.

From previous demonstration, to find flat minima, we would expect that the reference model and the neighborhood models are as observationally equivalent as possible. So, it is natural to optimize to reduce the gap between these models, where we need to gauge such gap first. In general, this gap of outputs is usually assessed via the difference in empirical loss between these models when given the same inputs. But, this is apprarently inappropriate, because it is highly possible for models that are not observationally equivalent yield the same loss value, especially for categorical cross-entropy loss. Thus, minimize this loss gap could not fully guarantee that models will convergence to flat minima.

To this end, we would measure this gap between outputs by employing the Kullback-Leibler divergence,

Definition 2. *For model $f(\cdot;\boldsymbol{\theta})$ and model $f(\cdot;\boldsymbol{\theta}')$ in the same weight space $\boldsymbol{\Theta}$, given $\boldsymbol{x} \in \mathcal{X}$, the gap of the two models in the output space could be gauged via the Kullback-Leibler divergence,*

$$d_p(\boldsymbol{\theta},\boldsymbol{\theta}') = \mathbb{E}_{\boldsymbol{x}}[D_{\mathrm{KL}}(f(\boldsymbol{x};\boldsymbol{\theta})||f(\boldsymbol{x};\boldsymbol{\theta}'))] \tag{2}$$

where $D_{\mathrm{KL}}(\cdot||\cdot)$ denotes the KL divergence. We call $d_p(\boldsymbol{\theta},\boldsymbol{\theta}')$ the model divergence between $\boldsymbol{\theta}$ and $\boldsymbol{\theta}'$.

Note that $d_p(\boldsymbol{\theta},\boldsymbol{\theta}') \geq 0$. Clearly, for $d_p(\boldsymbol{\theta},\boldsymbol{\theta}')$, the lower the value is, the more approximate the outputs that the two models could yield. Meanwhile, it is obvious that,

Lemma 1. *The two models $f(\cdot;\boldsymbol{\theta})$ and $f(\cdot;\boldsymbol{\theta}')$ are observationally equivalent if and only if $d_p(\boldsymbol{\theta},\boldsymbol{\theta}') = 0$.*

Basically, this indicates that minimizing model divergence could in a sense serve as a sufficient condition for converging to flat minima in optimization.

Further, we would like to discuss the interpretation of flat minima from the perspective of model divergence. Based on information theory, KL divergence of the two output distributions $f(\boldsymbol{x};\boldsymbol{\theta})$ and $f(\boldsymbol{x};\boldsymbol{\theta}')$ provides insights in regards to how many additional bits are required to approximate the true distribution $f(\boldsymbol{x};\boldsymbol{\theta})$ when using $f(\boldsymbol{x};\boldsymbol{\theta}')$. Lower divergence indicates fewer information loss if using $f(\boldsymbol{x};\boldsymbol{\theta}')$ to approximate $f(\boldsymbol{x};\boldsymbol{\theta})$.

So for flat minima, since the model is expected to have approximate outputs with models in its neighborhood region, they should have low model divergence. This indicates that when using models $f(\cdot;\boldsymbol{\theta}')$ in the neighborhood to appropriate the true model $f(\cdot;\boldsymbol{\theta})$, only few extra information is required. In contrast, high model divergence would mean more extra information is required. Therefore, describing a flat minimum would require much fewer information than a sharp minimum. Remarkly, based on the minimum description length (MDL) principle, better models would benefit from simpler description. Accordingly, compared to a sharp minimum, a flat minimum will imply better model performance, since the required information description is shorter. This is actually in accordance with Occam's razor principle in deep learning.

3.3 NRS Regularization

Generally, a flat minimum suggests that for $\forall\ \delta\boldsymbol{\theta} \in B(0,\epsilon)$ where $B(0,\epsilon)$ is the Euclidean ball centered at 0 with radius ϵ, the model $f(\cdot,\boldsymbol{\theta})$ and its neighborhood

Algorithm 1. Neighborhood Region Smoothing (NRS) Regularization

Input: Training set $\mathcal{S} = \{(\boldsymbol{x}_i, \boldsymbol{y}_i)\}_{i=0}^{N}$; loss function $l(\cdot)$; batch size B; learning rate η; total steps K; neighborhood region size ϵ; model divergence penalty coefficient α.
Parameter: Model parameters $\boldsymbol{\theta}$.
Output: Model with final optimized weight $\hat{\boldsymbol{\theta}}$.
1: Parameter initialization $\boldsymbol{\theta}_0$; get the number of devices M.
2: **for** step $k = 1$ to K **do**
3: Get sample batch $\mathcal{B} = \{(\boldsymbol{x}_i, \boldsymbol{y}_i)\}_{i=0}^{B}$.
4: Shard $\mathcal{B}$ based on the number of devices $\mathcal{B} = \mathcal{B}_0 \cup \mathcal{B}_1 \cdots \cup \mathcal{B}_M$, where $|\mathcal{B}_0| = \cdots = |\mathcal{B}_M|$.
5: **Do in parallel across devices.**
6: Make a unique pseudo-random number generator κ.
7: Generate random perturbation $\delta\boldsymbol{\theta}$ within area $B(0, \epsilon)$ based on κ.
8: Create neighborhood model $f(\cdot, \boldsymbol{\theta} + \delta\boldsymbol{\theta})$.
9: Compute gradient $\nabla_{\boldsymbol{\theta}} L(\boldsymbol{\theta})$ of the final loss based on batch $\mathcal{B}_i$.
10: **Synchronize and collect the gradient $\boldsymbol{g}$.**
11: Update parameter $\boldsymbol{\theta}_{k+1} = \boldsymbol{\theta}_k - \eta \cdot \boldsymbol{g}$
12: **end for**

model $f(\cdot, \boldsymbol{\theta} + \delta\boldsymbol{\theta})$ are expected to have both low model divergence and low empirical training loss.

Therefore, we would manually add additional regularization in the loss during training for finding flat solutions,

$$\min_{\boldsymbol{\theta}} L_{\mathcal{S}}(\boldsymbol{\theta}) + \alpha \cdot d_p(\boldsymbol{\theta}, \boldsymbol{\theta} + \delta\boldsymbol{\theta}) + L_{\mathcal{S}}(\boldsymbol{\theta} + \delta\boldsymbol{\theta}) \tag{3}$$

where α denotes the penalty coefficient of model divergence $d_p(\boldsymbol{\theta}, \boldsymbol{\theta} + \delta\boldsymbol{\theta})$. In Equation 3, the final training loss $L(\boldsymbol{\theta})$ contains three items:

- The first item is the conventional empirical training loss, denoting the gap between the true labels and the predicted labels. Optimize this item would drive the predicted labels towards the true labels.
- The second item is the model divergence regularization, denoting the divergence between outputs yielded separately from the the models with parameter $\boldsymbol{\theta}$ and the model $\boldsymbol{\theta} + \delta\boldsymbol{\theta}$ in the neighborhood. Optimize this item would explicitly drive the model to yield approximate outputs in the neighborhood.
- The third item is the empirical training loss of the neighborhood model, denoting the gap between true labels and the labels predicted by this neighborhood model. Optimize this item would explicitly force the neighborhood also learn to predict the true labels.

Compared to minimize only the empirical loss of perturbed models $L(\boldsymbol{\theta} + \delta\boldsymbol{\theta})$, Equation 3 imposes much more rigorous constraint for models in the neighborhood region. This also leads to allow us to be relaxing on model selection in the neighborhood, which enables to perform a simple random perturbation.

In order to provide sufficient $\delta\boldsymbol{\theta}$ samples in neighborhood $B(0, \epsilon)$, it is best to train each input sample with a distinct $\delta\boldsymbol{\theta}$. However, this would not fit in the general mini-batch parallel training paradigm well in practice because this would require extra computation graphs for each distinct $\delta\boldsymbol{\theta}$ when deploying using common deep learning framework like Tensorflow, Jax and Pytorch. For balancing the computation accuracy and efficiency, we would generate a unique $\delta\boldsymbol{\theta}$ for each training device instead of for each input sample. In this way, each device need to generate only one neighborhood model, indicating that they will use the same computation graph for samples loaded on this device. All the mini-batch samples would fully enjoy the parallel computing in each device, which would significantly decrease the computation budget compared to training each input sample with a distinct $\delta\boldsymbol{\theta}$.

Algorithm 1 concludes the pseudo-code of the full implementation of NRS regularization. In Algorithm 1, the default optimizer is stochastic gradient descent. We will show the effectiveness of NRS algorithm in the following experimental section.

4 Experimental Results

In this section, we are going to demonstrate the effectiveness of NRS regularization by studying the image classification performances on commonly-used datasets with extensive model architectures. In our experiments, the datasets include Cifar10, Cifar100 and ImageNet, and the model architectures include VGG [21], ResNet [11], WideResNet [25], PyramidNet [10] and Vision Transformer [7]. All the models are trained from scratch to convergence, which are implemented using Jax framework on the NVIDIA DGX Station A100 with four NVIDIA A100 GPUs.

4.1 Cifar10 and Cifar100

We would start our investigation of NRS from evaluating its effect on the generalization ability of models on Cifar10 and Cifar100 dataset. Five network architectures would be trained from scratch, including both CNN models and ViT models, which are VGG16, ResNet18, WideResNet-28-10, PyramidNet-164 and Vision Transformer family[1]. For datasets, we would adopt several different augmentation strategies. One is the *Basic* strategy, which follows the conventional four-pixel extra padding, random cropping and horizontal random flipping. Meanwhile, other than the basic strategy, we would also perform more complex data augmentations to show that NRS would not conflict with other regularization techniques. Specifically, we would adopt the *Cutout* strategy when training CNN models, which would additionally perform cutout regularization [5]. But when training ViT models, we would instead adopt a *Heavy* strategy, which will additionally perform mixup regularization and train much longer compared to the *Basic* strategy.

[1] The names of all the mentioned model architectures is the same as them in their original papers.

Table 1. Testing accuracy of various models on Cifar10 and Cifar100 when using the three training strategies.

CNN Models	Cifar10		Cifar100	
VGG16	Basic	Cutout	Basic	Cutout
Baseline	$93.12_{\pm 0.08}$	$93.95_{\pm 0.11}$	$72.28_{\pm 0.17}$	$73.34_{\pm 0.22}$
RPR	$93.14_{\pm 0.21}$	$93.91_{\pm 0.17}$	$72.31_{\pm 0.21}$	$73.29_{\pm 0.19}$
NRS	$\mathbf{93.79}_{\pm 0.11}$	$\mathbf{94.77}_{\pm 0.17}$	$\mathbf{73.61}_{\pm 0.20}$	$\mathbf{75.32}_{\pm 0.19}$
ResNet18	Basic	Cutout	Basic	Cutout
Baseline	$94.88_{\pm 0.12}$	$95.45_{\pm 0.16}$	$76.19_{\pm 0.21}$	$77.03_{\pm 0.11}$
RPR	$94.90_{\pm 0.24}$	$95.41_{\pm 0.23}$	$76.25_{\pm 0.31}$	$76.98_{\pm 0.18}$
NRS	$\mathbf{95.84}_{\pm 0.15}$	$\mathbf{96.47}_{\pm 0.10}$	$\mathbf{78.67}_{\pm 0.17}$	$\mathbf{79.88}_{\pm 0.14}$
WideResNet-28-10	Basic	Cutout	Basic	Cutout
Baseline	$96.17_{\pm 0.12}$	$97.09_{\pm 0.17}$	$80.91_{\pm 0.13}$	$82.25_{\pm 0.15}$
RPR	$96.14_{\pm 0.11}$	$97.02_{\pm 0.15}$	$80.94_{\pm 0.19}$	$82.19_{\pm 0.24}$
NRS	$\mathbf{96.94}_{\pm 0.17}$	$\mathbf{97.55}_{\pm 0.13}$	$\mathbf{82.77}_{\pm 0.16}$	$\mathbf{83.94}_{\pm 0.16}$
PyramidNet-164	Basic	Cutout	Basic	Cutout
Baseline	$96.32_{\pm 0.15}$	$97.11_{\pm 0.12}$	$82.33_{\pm 0.19}$	$83.50_{\pm 0.17}$
RPR	$96.31_{\pm 0.19}$	$97.09_{\pm 0.19}$	$82.25_{\pm 0.24}$	$83.58_{\pm 0.22}$
NRS	$\mathbf{97.23}_{\pm 0.19}$	$\mathbf{97.72}_{\pm 0.11}$	$\mathbf{84.61}_{\pm 0.24}$	$\mathbf{86.29}_{\pm 0.18}$
ViT Models	Cifar10		Cifar100	
ViT-TI16	Basic	Heavy	Basic	Heavy
Baseline	$83.07_{\pm 0.08}$	$89.22_{\pm 0.18}$	$60.12_{\pm 0.15}$	$65.44_{\pm 0.20}$
RPR	$83.02_{\pm 0.13}$	$89.28_{\pm 0.14}$	$60.18_{\pm 0.21}$	$65.21_{\pm 0.17}$
NRS	$\mathbf{84.17}_{\pm 0.09}$	$\mathbf{90.47}_{\pm 0.14}$	$\mathbf{61.29}_{\pm 0.13}$	$\mathbf{66.84}_{\pm 0.22}$
ViT-S16	Basic	Heavy	Basic	Heavy
Baseline	$85.14_{\pm 0.11}$	$92.49_{\pm 0.09}$	$61.83_{\pm 0.20}$	$72.40_{\pm 0.17}$
RPR	$85.17_{\pm 0.13}$	$92.12_{\pm 0.11}$	$61.92_{\pm 0.14}$	$72.33_{\pm 0.17}$
NRS	$\mathbf{85.86}_{\pm 0.10}$	$\mathbf{93.55}_{\pm 0.16}$	$\mathbf{62.59}_{\pm 0.17}$	$\mathbf{73.17}_{\pm 0.14}$
ViT-B16	Basic	Heavy	Basic	Heavy
Baseline	$88.42_{\pm 0.12}$	$91.54_{\pm 0.11}$	$63.49_{\pm 0.17}$	$72.32_{\pm 0.21}$
RPR	$88.39_{\pm 0.14}$	$91.57_{\pm 0.12}$	$63.38_{\pm 0.15}$	$72.35_{\pm 0.17}$
NRS	$\mathbf{89.09}_{\pm 0.08}$	$\mathbf{92.23}_{\pm 0.14}$	$\mathbf{64.46}_{\pm 0.17}$	$\mathbf{73.11}_{\pm 0.13}$

We would focus our investigations on the comparisons between three training schemes. The first one would train with the standard categorical cross-entropy loss, which is $\min_{\boldsymbol{\theta}} L(\boldsymbol{\theta})$. This is our baseline. The second training scheme is to optimize the same cross-entropy loss of the models with random perturbation (RPR) in the neighborhood region instead of the true model, which is $\min_{\boldsymbol{\theta}} L(\boldsymbol{\theta}+\delta\boldsymbol{\theta})$. This one is for confirming that minimizing the loss of simple random perturbations could not be helpful to the generalization ability of models, just as papers [9,30] suggest. The last training scheme would be training with our NRS regularization. It should be noted that we would keep any other deployment the same for the three schemes during training except for the techniques mentioned in specific scheme.

For the common training hyperparameters, we would perform a grid search for acquiring the best performance for CNN models. Specifically, the base learning rate is searched over $\{0.01, 0.05, 0.1, 0.2\}$, the weight decay coefficient is searched over $\{0.0005, 0.001\}$ and the batch size is searched over $\{128, 256\}$. Also, we would adopt cosine learning rate schedule and SGD optimizer with 0.9 momentum during training. As for ViT models, we use fixed hyperparameters and use different hyperparameter deployments for the two data augmentation strategies. For *Basic* strategy, learning rate is 0.001, weight decay is 0.3, training epoch is 300, batch size is 256 and the patch size is 4×4 while learning rate is $2e - 4$, weight decay is 0.03 and training epoch is 1200 for *Heavy* strategy. Meanwhile, we would adopt the Adam optimizer during training. Additionally, for all the CNN and ViT models, we would use three different random seeds and report the average mean and variance across the testing accuracies of the three seeds.

For RPR scheme, it involves one extra hyperparameter, which represents the radius of the neighborhood region ϵ. So similarly, we would perform a grid search over $\{0.05, 0.1, 0.5\}$ as well. As for NRS strategy, it involves two extra hyperparameters, the radius of the neighborhood region ϵ and the penalty coefficient of model divergence α. We would adopt the same search for ϵ, and α is searched over $\{0.5, 1.0, 2.0\}$. Notably, grid search of ϵ and α in both RPR and NRS would be performed on the basis of hyperparameters of the best model acquired by the first training strategy.

Table 1 shows the corresponding testing accuracies of Cifar10 and Cifar100, where all the reported results are the best results during the grid search of hyperparameters. We could see that in the table, all the testing accuracies have been improved by NRS regularization to some extent compared to the baseline, which confirms its benefit for model training. We also try NRS regularization on the recent Vision Transformer model. We could find that when performing *Basic* augmentation, the testing accuracy of ViT models would be much lower than that of CNN models. This is because training ViT models generally requires plenty of input samples. In this case, NRS could also improve the generalization ability of the such models. And when performing *Heavy* augmentation, training performance would be improved significantly, where again, performance could be improved further when training with the NRS scheme.

From the results, we could find that NRS regularization would not conflict with optimizers and current regularizations like dropout (in VGG16) and batch normalization, which is important for practical implementation. Additionally, we also confirm that simply optimizing the models with neighborhood random perturbation like RPR could indeed have no effect on improving the generalization ability of models.

Table 2. Testing accuracy of various models on ImageNet dataset when using standard training strategy (Baseline) and NRS regularization strategy.

	ImageNet	
VGG16	Top-1 Accuray	Top-5 Accuray
Baseline	$73.11_{\pm 0.08}$	$91.12_{\pm 0.08}$
NRS	$\mathbf{73.52}_{\pm 0.09}$	$\mathbf{91.39}_{\pm 0.09}$
ResNet50	Top-1 Accuray	Top-5 Accuray
Baseline	$75.45_{\pm 0.15}$	$93.04_{\pm 0.07}$
NRS	$\mathbf{76.29}_{\pm 0.13}$	$\mathbf{93.53}_{\pm 0.10}$
ResNet101	Top-1 Accuray	Top-5 Accuray
Baseline	$77.15_{\pm 0.11}$	$93.91_{\pm 0.07}$
NRS	$\mathbf{78.02}_{\pm 0.10}$	$\mathbf{94.44}_{\pm 0.08}$

4.2 ImageNet

Next, we would check the effectiveness of NRS regularization on a large-scale dataset, ImageNet. Here, we would take VGG16, ResNet50 and ResNet101 model architectures as our experimental targets. For datasets, all the images would be resized and cropped to 224×224, and then randomly flipped in the horizontal direction. For common hyperparameters, instead of performing a grid search, we would fix the batch size to 512, the base learning rate to 0.2, the weight decay coefficient to 0.0001. During training, we would smooth the label with 0.1. All models would be trained for a total of 100 epochs with three different seeds.

Our investigation would focus on the comparisons between two training strategies. One is the standard training with categorical cross-entropy loss, which is our baseline. The other one is trained with NRS regularization. For hyperparameters in NRS regularization, we would fix the ϵ to 0.1 and α to 1.0 according to previous tuning experience. Table 2 shows the corresponding results.

As we could see in Table 2, the testing accuracy could be improved again to some extent when using NRS regularization. This further confirms that NRS regularization could be beneficial to the generalization ability of models.

5 Further Studies of NRS

5.1 Parameter Selection in NRS

In this section, we would investigate the influence of the two parameters ϵ and α in NRS regularization on the results. The investigation is conducted on Cifar10 and Cifar100 using WideResNet-28-10. We train the models from scratch using NSR with the same common hyperparameters of the best models acquired by the baseline strategy. And then, we would perform the grid search for the two parameters using the same scheme as in previous section.

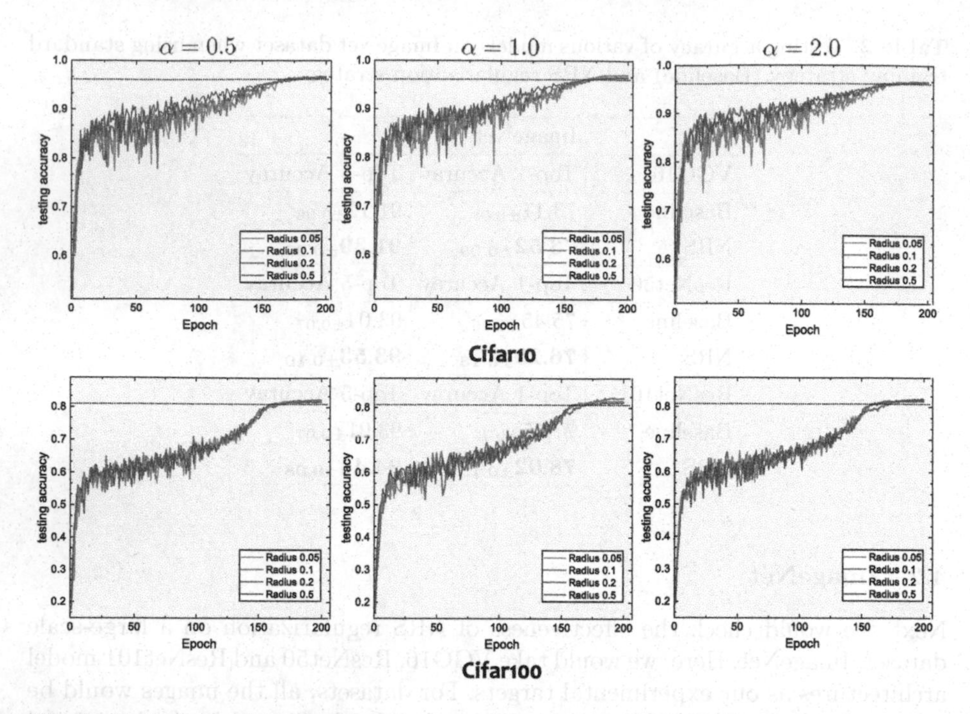

Fig. 2. Evolutions of testing accuracy on Cifar10 and Cifar100 during training when trained with different parameters in NRS. The black dash lines refer to the reference lines which are the best testing accuracy trained using standard strategy.

Figure 2 shows the evolution of testing accuracy during training when using the corresponding different parameters. We could see that actually for all the deployed ϵ and α, the generalization ability could be somewhat improved on both Cifar10 and Cifar100 datasets compared to the baseline (the black reference line in the figure). This again demonstrates the effectiveness of NRS regularization. From the figure, we could find that the model could achieve the best performance when $\epsilon = 0.1$ and $\alpha = 1.0$ for Cifar10 dataset and $\epsilon = 0.2$ and $\alpha = 1.0$ for Cifar100 dataset. Also, we could find that the generalization improvement on Cifar100 would generally be larger than that on Cifar10.

5.2 Eigenvalues of Hessian Matrix

In this section, we would investigate the intrinsic change of models when using the NSR regularization. In general, the eigenvalues of Hessian matrix are considered to have connections with the flatness of minima. Specifically, the largest eigenvalue of Hessian matrix of flat minima could be larger than that of sharp minima.

Since the dimension of the weight space is so huge, solving the Hessian matrix in a direct manner is nearly impossible. Therefore, we would employ the approximate method introduced in [3]. It calculates the diagonal block of Hessian matrix recursively from the deep layers to the shallower layers.

We would still use WideResNet-28-10 model and Cifar10 as our investigation target. We would use the best models acquired by the three training strategies in the previous section. Also, we would calculate the eigenvalues of Hessian matrix for the last layer in each model. Table 3 reports the results.

Table 3. The largest eigenvalue of Hessian matrix of models trained with three different strategies.

WideResNet-28-10	λ_{max}
Baseline	49.15
RPR	51.79
NRS	**12.42**

As can be seen from the table, using NRS regularization can significantly reduce the largest eigenvalue of the Hessian matrix compared to using the other two strategies. Also, using RPR strategy indeed could not lead to flat minima. This again verify that using NRS regularization would find flat minima during training.

6 Conclusion

In this paper, we propose a simple yet effective regularization technique, called Neighborhood Region Smoothing, for finding flat minima during training. The key idea of NRS is regularizing the neighborhood region of models to yield approximate outputs. Using outputs in NRS could give stronger regularization than using loss values, so simple random perturbation in the neighborhood region would be effective. We define model divergence to gauge the gap between outputs of models in the neighborhood region. In this way, NRS regularization is achieved by explicitly minimizing both the empirical loss and the model divergence. In our experiments, we show that using NRS regularization could improve the generalization ability of a wide range of models on diverse datasets compared to two other training strategies. We also investigate to give the best hyperparameters in NRS on Cifar10 and Cifar100 dataset. Finally, smaller eigenvalue of Hessian matrix confirms that NRS regularization could indeed to flat minima.

Acknowledgements. We would like to thank all the reviewers and the meta-reviewer for their helpful comments and kindly advices. We would like to thank Jean Kaddour from University College London for the kindly discussions.

References

1. Ba, L.J., Kiros, J.R., Hinton, G.E.: Layer normalization. arXivPreprint abs/1607.06450 (2016)
2. Bartlett, P.L., Foster, D.J., Telgarsky, M.: Spectrally-normalized margin bounds for neural networks. In: Advances in Neural Information Processing Systems, pp. 6240–6249 (2017)
3. Botev, A., Ritter, H., Barber, D.: Practical gauss-newton optimisation for deep learning. In: Proceedings of the 34th International Conference on Machine Learning. ICML 2017, vol. 70, pp. 557–565 (2017)
4. Cubuk, E.D., Zoph, B., Mané, D., Vasudevan, V., Le, Q.V.: Autoaugment: learning augmentation policies from data. arXivPreprint abs/1805.09501 (2018)
5. Devries, T., Taylor, G.W.: Improved regularization of convolutional neural networks with cutout. arXivPreprint abs/1708.04552 (2017)
6. Dinh, L., Pascanu, R., Bengio, S., Bengio, Y.: Sharp minima can generalize for deep nets. In: Proceedings of the 34th International Conference on Machine Learning. ICML 2017, vol. 70, pp. 1019–1028 (2017)
7. Dosovitskiy, A., et al.: An image is worth 16x16 words: transformers for image recognition at scale. In: 9th International Conference on Learning Representations. ICLR 2021 (2021)
8. Du, J., et al.: Efficient sharpness-aware minimization for improved training of neural networks. arXivPreprint abs/2110.03141 (2021)
9. Foret, P., Kleiner, A., Mobahi, H., Neyshabur, B.: Sharpness-aware minimization for efficiently improving generalization. In: 9th International Conference on Learning Representations. ICLR 2021 (2021)
10. Han, D., Kim, J., Kim, J.: Deep pyramidal residual networks. In: 2017 IEEE Conference on Computer Vision and Pattern Recognition. CVPR 2017, pp. 6307–6315 (2017)
11. He, K., Zhang, X., Ren, S., Sun, J.: Deep residual learning for image recognition. In: 2016 IEEE Conference on Computer Vision and Pattern Recognition. CVPR 2016, pp. 770–778 (2016)
12. Hochreiter, S., Schmidhuber, J.: Flat minima. Neural Comput. **9**(1), 1–42 (1997)
13. Hornik, K., Stinchcombe, M.B., White, H.: Multilayer feedforward networks are universal approximators. Neural Netw. **2**(5), 359–366 (1989)
14. Huang, G., Sun, Yu., Liu, Z., Sedra, D., Weinberger, K.Q.: Deep networks with stochastic depth. In: Leibe, B., Matas, J., Sebe, N., Welling, M. (eds.) ECCV 2016. LNCS, vol. 9908, pp. 646–661. Springer, Cham (2016). https://doi.org/10.1007/978-3-319-46493-0_39
15. Ioffe, S., Szegedy, C.: Batch normalization: accelerating deep network training by reducing internal covariate shift. In: Proceedings of the 32nd International Conference on Machine Learning. ICML 2015, vol. 37, pp. 448–456 (2015)
16. Keskar, N.S., Mudigere, D., Nocedal, J., Smelyanskiy, M., Tang, P.T.P.: On large-batch training for deep learning: generalization gap and sharp minima. In: 5th International Conference on Learning Representations. ICLR 2017 (2017)
17. Krogh, A., Hertz, J.A.: A simple weight decay can improve generalization. In: Advances in Neural Information Processing Systems, pp. 950–957 (1991)
18. Liu, S.: Learning sparse neural networks for better generalization. In: Proceedings of the Twenty-Ninth International Joint Conference on Artificial Intelligence. IJCAI 2020, pp. 5190–5191. ijcai.org (2020)

19. Loshchilov, I., Hutter, F.: Decoupled weight decay regularization. In: 7th International Conference on Learning Representations. ICLR 2019 (2019)
20. Neyshabur, B., Bhojanapalli, S., McAllester, D., Srebro, N.: Exploring generalization in deep learning. In: Advances in Neural Information Processing Systems, pp. 5947–5956 (2017)
21. Simonyan, K., Zisserman, A.: Very deep convolutional networks for large-scale image recognition. In: 3rd International Conference on Learning Representations. ICLR 2015 (2015)
22. Sokolic, J., Giryes, R., Sapiro, G., Rodrigues, M.R.D.: Robust large margin deep neural networks. IEEE Trans. Sig. Process. **65**(16), 4265–4280 (2017)
23. Srivastava, N., Hinton, G.E., Krizhevsky, A., Sutskever, I., Salakhutdinov, R.: Dropout: a simple way to prevent neural networks from overfitting. J. Mach. Learn. Res. **15**(1), 1929–1958 (2014)
24. Tompson, J., Goroshin, R., Jain, A., LeCun, Y., Bregler, C.: Efficient object localization using convolutional networks. In: IEEE Conference on Computer Vision and Pattern Recognition. CVPR 2015, pp. 648–656 (2015)
25. Zagoruyko, S., Komodakis, N.: Wide residual networks. In: Proceedings of the British Machine Vision Conference 2016. BMVC 2016 (2016)
26. Zhang, C., Bengio, S., Hardt, M., Recht, B., Vinyals, O.: Understanding deep learning requires rethinking generalization. In: 5th International Conference on Learning Representations. ICLR 2017 (2017)
27. Zhao, Y., Zhang, H.: Quantitative performance assessment of CNN units via topological entropy calculation. In: The Tenth International Conference on Learning Representations. ICLR 2022, Virtual Event, 25–29 April 2022 (2022)
28. Zhao, Y., Zhang, H., Hu, X.: Penalizing gradient norm for efficiently improving generalization in deep learning. In: International Conference on Machine Learning. ICML 2022, 17–23 July 2022, Baltimore, Maryland, USA (2022)
29. Zhao, Y., Zhang, H., Hu, X.: Randomized sharpness-aware training for boosting computational efficiency in deep learning. arXivPreprint abs/2203.09962 (2022)
30. Zheng, Y., Zhang, R., Mao, Y.: Regularizing neural networks via adversarial model perturbation. In: IEEE Conference on Computer Vision and Pattern Recognition. CVPR 2021, pp. 8156–8165 (2021)

19. Loshchilov, I., Hutter, F.: Decoupled weight decay regularization. In: 7th International Conference on Learning Representations, ICLR 2019 (2019)
20. Neyshabur, B., Bhojanapalli, S., McAllester, D., Srebro, N.: Exploring generalization in deep learning. In: Advances in Neural Information Processing Systems, pp. 5947–5956 (2017)
21. Simonyan, K., Zisserman, A.: Very deep convolutional networks for large-scale image recognition. In: 3rd International Conference on Learning Representations, ICLR 2015 (2015)
22. Sokolic, J., Giryes, R., Sapiro, G., Rodrigues, M.R.D.: Robust large margin deep neural networks. IEEE Trans. Sig. Process. 65(16), 4265–4280 (2017)
23. Srivastava, N., Hinton, G.E., Krizhevsky, A., Sutskever, I., Salakhutdinov, R.: Dropout: a simple way to prevent neural networks from overfitting. J. Mach. Learn. Res. 15(1), 1929–1958 (2014)
24. Tompson, J., Goroshin, R., Jain, A., LeCun, Y., Bregler, C.: Efficient object localization using convolutional networks. In: IEEE Conference on Computer Vision and Pattern Recognition, CVPR 2015, pp. 648–656 (2015)
25. Zagoruyko, S., Komodakis, N.: Wide residual networks. In: Proceedings of the British Machine Vision Conference 2016, BMVC 2016 (2016)
26. Zhang, C., Bengio, S., Hardt, M., Recht, B., Vinyals, O.: Understanding deep learning requires rethinking generalization. In: 5th International Conference on Learning Representations, ICLR 2017 (2017)
27. Zhao, Y., Zhang, H.: Quantitative performance assessment of CNN units via topological entropy calculation. In: The Tenth International Conference on Learning Representations, ICLR 2022, Virtual Event, 25–29 April 2022 (2022)
28. Zhao, Y., Zhang, H., Hu, X.: Penalizing gradient norm for efficiently improving generalization in deep learning. In: International Conference on Machine Learning, ICML 2022, 17–23 July 2022, Baltimore, Maryland, USA (2022)
29. Zhao, Y., Zhang, H., Hu, X.: Randomized sharpness-aware training for boosting computational efficiency in deep learning. arXiv preprint arXiv:2203.09962 (2022)
30. Zheng, Y., Zhang, R., Mao, Y.: Regularizing neural networks via adversarial model perturbation. In: IEEE Conference on Computer Vision and Pattern Recognition, CVPR 2021, pp. 8156–8165 (2021)

Author Index

L. Wang et al. (Eds.): ACCV 2022, LNCS 13841, pp. 667–669, 2023.
https://doi.org/10.1007/978-3-031-26319-4

Table 2. Testing accuracy of various models on ImageNet dataset when using standard training strategy (Baseline) and NRS regularization strategy.

	ImageNet	
VGG16	Top-1 Accuray	Top-5 Accuray
Baseline	$73.11_{\pm 0.08}$	$91.12_{\pm 0.08}$
NRS	$\mathbf{73.52}_{\pm 0.09}$	$\mathbf{91.39}_{\pm 0.09}$
ResNet50	Top-1 Accuray	Top-5 Accuray
Baseline	$75.45_{\pm 0.15}$	$93.04_{\pm 0.07}$
NRS	$\mathbf{76.29}_{\pm 0.13}$	$\mathbf{93.53}_{\pm 0.10}$
ResNet101	Top-1 Accuray	Top-5 Accuray
Baseline	$77.15_{\pm 0.11}$	$93.91_{\pm 0.07}$
NRS	$\mathbf{78.02}_{\pm 0.10}$	$\mathbf{94.44}_{\pm 0.08}$

4.2 ImageNet

Next, we would check the effectiveness of NRS regularization on a large-scale dataset, ImageNet. Here, we would take VGG16, ResNet50 and ResNet101 model architectures as our experimental targets. For datasets, all the images would be resized and cropped to 224×224, and then randomly flipped in the horizontal direction. For common hyperparameters, instead of performing a grid search, we would fix the batch size to 512, the base learning rate to 0.2, the weight decay coefficient to 0.0001. During training, we would smooth the label with 0.1. All models would be trained for a total of 100 epochs with three different seeds.

Our investigation would focus on the comparisons between two training strategies. One is the standard training with categorical cross-entropy loss, which is our baseline. The other one is trained with NRS regularization. For hyperparameters in NRS regularization, we would fix the ϵ to 0.1 and α to 1.0 according to previous tuning experience. Table 2 shows the corresponding results.

As we could see in Table 2, the testing accuracy could be improved again to some extent when using NRS regularization. This further confirms that NRS regularization could be beneficial to the generalization ability of models.

5 Further Studies of NRS

5.1 Parameter Selection in NRS

In this section, we would investigate the influence of the two parameters ϵ and α in NRS regularization on the results. The investigation is conducted on Cifar10 and Cifar100 using WideResNet-28-10. We train the models from scratch using NSR with the same common hyperparameters of the best models acquired by the baseline strategy. And then, we would perform the grid search for the two parameters using the same scheme as in previous section.

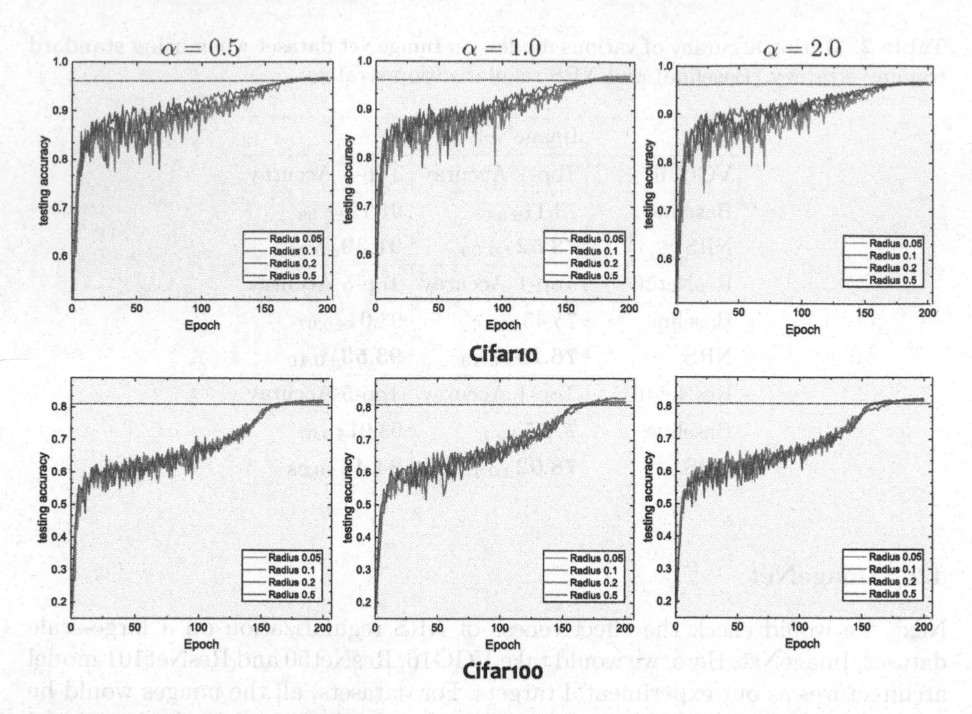

Fig. 2. Evolutions of testing accuracy on Cifar10 and Cifar100 during training when trained with different parameters in NRS. The black dash lines refer to the reference lines which are the best testing accuracy trained using standard strategy.

Figure 2 shows the evolution of testing accuracy during training when using the corresponding different parameters. We could see that actually for all the deployed ϵ and α, the generalization ability could be somewhat improved on both Cifar10 and Cifar100 datasets compared to the baseline (the black reference line in the figure). This again demonstrates the effectiveness of NRS regularization. From the figure, we could find that the model could achieve the best performance when $\epsilon = 0.1$ and $\alpha = 1.0$ for Cifar10 dataset and $\epsilon = 0.2$ and $\alpha = 1.0$ for Cifar100 dataset. Also, we could find that the generalization improvement on Cifar100 would generally be larger than that on Cifar10.

5.2 Eigenvalues of Hessian Matrix

In this section, we would investigate the intrinsic change of models when using the NSR regularization. In general, the eigenvalues of Hessian matrix are considered to have connections with the flatness of minima. Specifically, the largest eigenvalue of Hessian matrix of flat minima could be larger than that of sharp minima.

Since the dimension of the weight space is so huge, solving the Hessian matrix in a direct manner is nearly impossible. Therefore, we would employ the approxi-

Printed in the United States
by Baker & Taylor Publisher Services

Printed in the United States
by Baker & Taylor Publisher Services